Lecture Notes in Computer Science 12664

More information about this subseries at http://www.springer.com/series/7412

Alberto Del Bimbo · Rita Cucchiara ·
Stan Sclaroff · Giovanni Maria Farinella ·
Tao Mei · Marco Bertini ·
Hugo Jair Escalante · Roberto Vezzani (Eds.)

Pattern Recognition

ICPR International Workshops and Challenges

Virtual Event, January 10–15, 2021
Proceedings, Part IV

 Springer

Editors
Alberto Del Bimbo (iD)
Dipartimento di Ingegneria
dell'Informazione
University of Firenze
Firenze, Italy

Stan Sclaroff (iD)
Department of Computer Science
Boston University
Boston, MA, USA

Tao Mei
Cloud & AI, JD.COM
Beijing, China

Hugo Jair Escalante (iD)
Computational Sciences Department
National Institute of Astrophysics,
Optics and Electronics (INAOE)
Tonantzintla, Puebla, Mexico

Rita Cucchiara (iD)
Dipartimento di Ingegneria "Enzo Ferrari"
Università di Modena e Reggio Emilia
Modena, Italy

Giovanni Maria Farinella (iD)
Dipartimento di Matematica e Informatica
University of Catania
Catania, Italy

Marco Bertini (iD)
Dipartimento di Ingegneria
dell'Informazione
University of Firenze
Firenze, Italy

Roberto Vezzani (iD)
Dipartimento di Ingegneria "Enzo Ferrari"
Università di Modena e Reggio Emilia
Modena, Italy

ISSN 0302-9743   ISSN 1611-3349 (electronic)
Lecture Notes in Computer Science
ISBN 978-3-030-68798-4   ISBN 978-3-030-68799-1 (eBook)
https://doi.org/10.1007/978-3-030-68799-1

LNCS Sublibrary: SL6 – Image Processing, Computer Vision, Pattern Recognition, and Graphics

This Springer imprint is published by the registered company Springer Nature Switzerland AG
The registered company address is: Gewerbestrasse 11, 6330 Cham, Switzerland

Foreword by General Chairs

It is with great pleasure that we welcome you to the post-proceedings of the 25th International Conference on Pattern Recognition, ICPR2020 Virtual-Milano. ICPR2020 stands on the shoulders of generations of pioneering pattern recognition researchers. The first ICPR (then called IJCPR) convened in 1973 in Washington, DC, USA, under the leadership of Dr. King-Sun Fu as the General Chair. Since that time, the global community of pattern recognition researchers has continued to expand and thrive, growing evermore vibrant and vital. The motto of this year's conference was *Putting Artificial Intelligence to work on patterns*. Indeed, the deep learning revolution has its origins in the pattern recognition community – and the next generations of revolutionary insights and ideas continue with those presented at this 25th ICPR. Thus, it was our honor to help perpetuate this longstanding ICPR tradition to provide a lively meeting place and open exchange for the latest pathbreaking work in pattern recognition.

For the first time, the ICPR main conference employed a two-round review process similar to journal submissions, with new papers allowed to be submitted in either the first or the second round and papers submitted in the first round and not accepted allowed to be revised and re-submitted for second round review. In the first round, 1554 new submissions were received, out of which 554 (35.6%) were accepted and 579 (37.2%) were encouraged to be revised and resubmitted. In the second round, 1696 submissions were received (496 revised and 1200 new), out of which 305 (61.4%) of the revised submissions and 552 (46%) of the new submissions were accepted. Overall, there were 3250 submissions in total, and 1411 were accepted, out of which 144 (4.4%) were included in the main conference program as orals and 1263 (38.8%) as posters (4 papers were withdrawn after acceptance). We had the largest ICPR conference ever, with the most submitted papers and the most selective acceptance rates ever for ICPR, attesting both the increased interest in presenting research results at ICPR and the high scientific quality of work accepted for presentation at the conference.

We were honored to feature seven exceptional Keynotes in the program of the ICPR2020 main conference: David Doermann (Professor at the University at Buffalo), Pietro Perona (Professor at the California Institute of Technology and Amazon Fellow at Amazon Web Services), Mihaela van der Schaar (Professor at the University of Cambridge and a Turing Fellow at The Alan Turing Institute in London), Max Welling (Professor at the University of Amsterdam and VP of Technologies at Qualcomm), Ching Yee Suen (Professor at Concordia University) who was presented with the IAPR 2020 King-Sun Fu Prize, Maja Pantic (Professor at Imperial College UK and AI Scientific Research Lead at Facebook Research) who was presented with the IAPR 2020 Maria Petrou Prize, and Abhinav Gupta (Professor at Carnegie Mellon University and Research Manager at Facebook AI Research) who was presented with the IAPR 2020 J.K. Aggarwal Prize. Several best paper prizes were also announced and awarded, including the Piero Zamperoni Award for the best paper authored by a student, the BIRPA Best Industry Related Paper Award, and Best Paper Awards for each of the five tracks of the ICPR2020 main conference.

The five tracks of the ICPR2020 main conference were: (1) Artificial Intelligence, Machine Learning for Pattern Analysis, (2) Biometrics, Human Analysis and Behavior Understanding, (3) Computer Vision, Robotics and Intelligent Systems, (4) Document and Media Analysis, and (5) Image and Signal Processing. The best papers presented at the main conference had the opportunity for publication in expanded format in journal special issues of *IET Biometrics* (tracks 2 and 3), *Computer Vision and Image Understanding* (tracks 1 and 2), *Machine Vision and Applications* (tracks 2 and 3), *Multimedia Tools and Applications* (tracks 4 and 5), *Pattern Recognition Letters* (tracks 1, 2, 3 and 4), or *IEEE Trans. on Biometrics, Behavior, and Identity Science* (tracks 2 and 3).

In addition to the main conference, the ICPR2020 program offered workshops and tutorials, along with a broad range of cutting-edge industrial demos, challenge sessions, and panels. The virtual ICPR2020 conference was interactive, with real-time live-streamed sessions, including live talks, poster presentations, exhibitions, demos, Q&A, panels, meetups, and discussions – all hosted on the Underline virtual conference platform.

The ICPR2020 conference was originally scheduled to convene in Milano, which is one of the most beautiful cities of Italy for art, culture, lifestyle – and more. The city has so much to offer! With the need to go virtual, ICPR2020 included interactive **virtual tours** of Milano during the conference coffee breaks, which we hoped would introduce attendees to this wonderful city, and perhaps even entice them to visit Milano once international travel becomes possible again.

The success of such a large conference would not have been possible without the help of many people. We deeply appreciate the vision, commitment, and leadership of the ICPR2020 Program Chairs: Kim Boyer, Brian C. Lovell, Marcello Pelillo, Nicu Sebe, René Vidal, and Jingyi Yu. Our heartfelt gratitude also goes to the rest of the main conference organizing team, including the Track and Area Chairs, who all generously devoted their precious time in conducting the review process and in preparing the program, and the reviewers, who carefully evaluated the submitted papers and provided invaluable feedback to the authors. This time their effort was considerably higher given that many of them reviewed for both reviewing rounds. We also want to acknowledge the efforts of the conference committee, including the Challenge Chairs, Demo and Exhibit Chairs, Local Chairs, Financial Chairs, Publication Chair, Tutorial Chairs, Web Chairs, Women in ICPR Chairs, and Workshop Chairs. Many thanks, also, for the efforts of the dedicated staff who performed the crucially important work behind the scenes, including the members of the ICPR2020 Organizing Secretariat. Finally, we are grateful to the conference sponsors for their generous support of the ICPR2020 conference.

We hope everyone had an enjoyable and productive ICPR2020 conference.

Rita Cucchiara
Alberto Del Bimbo
Stan Sclaroff

Preface

The 25th International Conference on Pattern Recognition Workshops (ICPRW 2020) were held virtually in Milan, Italy and rescheduled to January 10 and January 11 of 2021 due to the Covid-19 pandemic. ICPRW 2020 included timely topics and applications of Computer Vision, Image and Sound Analysis, Pattern Recognition and Artificial Intelligence. We received 49 workshop proposals and 46 of them have been accepted, which is three times more than at ICPRW 2018. The workshop proceedings cover a wide range of areas including Machine Learning (8), Pattern Analysis (5), Healthcare (6), Human Behavior (5), Environment (5), Surveillance, Forensics and Biometrics (6), Robotics and Egovision (4), Cultural Heritage and Document Analysis (4), Retrieval (2), and Women at ICPR 2020 (1). Among them, 33 workshops are new to ICPRW. Specifically, the ICPRW 2020 volumes contain the following workshops (please refer to the corresponding workshop proceeding for details):

- CADL2020 – Workshop on Computational Aspects of Deep Learning.
- DLPR – Deep Learning for Pattern Recognition.
- EDL/AI – Explainable Deep Learning/AI.
- (Merged) IADS – Integrated Artificial Intelligence in Data Science, IWCR – IAPR workshop on Cognitive Robotics.
- ManifLearn – Manifold Learning in Machine Learning, From Euclid to Riemann.
- MOI2QDN – Metrification & Optimization of Input Image Quality in Deep Networks.
- IML – International Workshop on Industrial Machine Learning.
- MMDLCA – Multi-Modal Deep Learning: Challenges and Applications.
- IUC 2020 – Human and Vehicle Analysis for Intelligent Urban Computing.
- PATCAST – International Workshop on Pattern Forecasting.
- RRPR – Reproducible Research in Pattern Recognition.
- VAIB 2020 – Visual Observation and Analysis of Vertebrate and Insect Behavior.
- IMTA VII – Image Mining Theory & Applications.
- AIHA 2020 – Artificial Intelligence for Healthcare Applications.
- AIDP – Artificial Intelligence for Digital Pathology.
- (Merged) GOOD – Designing AI in support of Good Mental Health, CAIHA – Computational and Affective Intelligence in Healthcare Applications for Vulnerable Populations.
- CARE2020 – pattern recognition for positive teChnology And eldeRly wEllbeing.
- MADiMa 2020 – Multimedia Assisted Dietary Management.
- 3DHU 2020 – 3D Human Understanding.
- FBE2020 – Facial and Body Expressions, micro-expressions and behavior recognition.
- HCAU 2020 – Deep Learning for Human-Centric Activity Understanding.
- MPRSS - 6th IAPR Workshop on Multimodal Pattern Recognition for Social Signal Processing in Human Computer Interaction.

- CVAUI 2020 – Computer Vision for Analysis of Underwater Imagery.
- MAES – Machine Learning Advances Environmental Science.
- PRAConBE - Pattern Recognition and Automation in Construction & the Built Environment.
- PRRS 2020 – Pattern Recognition in Remote Sensing.
- WAAMI - Workshop on Analysis of Aerial Motion Imagery.
- DEEPRETAIL 2020 - Workshop on Deep Understanding Shopper Behaviours and Interactions in Intelligent Retail Environments 2020.
- MMForWild2020 – MultiMedia FORensics in the WILD 2020.
- FGVRID – Fine-Grained Visual Recognition and re-Identification.
- IWBDAF – Biometric Data Analysis and Forensics.
- RISS – Research & Innovation for Secure Societies.
- WMWB – TC4 Workshop on Mobile and Wearable Biometrics.
- EgoApp – Applications of Egocentric Vision.
- ETTAC 2020 – Eye Tracking Techniques, Applications and Challenges.
- PaMMO – Perception and Modelling for Manipulation of Objects.
- FAPER – Fine Art Pattern Extraction and Recognition.
- MANPU – coMics ANalysis, Processing and Understanding.
- PATRECH2020 – Pattern Recognition for Cultural Heritage.
- (Merged) CBIR – Content-Based Image Retrieval: where have we been, and where are we going, TAILOR – Texture AnalysIs, cLassificatiOn and Retrieval, VIQA – Video and Image Question Answering: building a bridge between visual content analysis and reasoning on textual data.
- W4PR - Women at ICPR.

We would like to thank all members of the workshops' Organizing Committee, the reviewers, and the authors for making this event successful. We also appreciate the support from all the invited speakers and participants. We wish to offer thanks in particular to the ICPR main conference general chairs: Rita Cucchiara, Alberto Del Bimbo, and Stan Sclaroff, and program chairs: Kim Boyer, Brian C. Lovell, Marcello Pelillo, Nicu Sebe, Rene Vidal, and Jingyi Yu. Finally, we are grateful to the publisher, Springer, for their cooperation in publishing the workshop proceedings in the series of Lecture Notes in Computer Science.

December 2020   Giovanni Maria Farinella
                                                                        Tao Mei

Challenges

Competitions are effective means for rapidly solving problems and advancing the state of the art. Organizers identify a problem of practical or scientific relevance and release it to the community. In this way the whole community can contribute to the solution of high-impact problems while having fun. This part of the proceedings compiles the best of the competitions track of the *25th International Conference on Pattern Recognition (ICPR)*.

Eight challenges were part of the track, covering a wide variety of fields and applications, all of this within the scope of ICPR. In every challenge organizers released data, and provided a platform for evaluation. The top-ranked participants were invited to submit papers for this volume. Likewise, organizers themselves wrote articles summarizing the design, organization and results of competitions. Submissions were subject to a standard review process carried out by the organizers of each competition. Papers associated with seven out the eight competitions are included in this volume, thus making it a representative compilation of what happened in the ICPR challenges.

We are immensely grateful to the organizers and participants of the ICPR 2020 challenges for their efforts and dedication to make the competition track a success. We hope the readers of this volume enjoy it as much as we have.

November 2020

Marco Bertini
Hugo Jair Escalante

ICPR Organization

General Chairs

Rita Cucchiara   Univ. of Modena and Reggio Emilia, Italy
Alberto Del Bimbo   Univ. of Florence, Italy
Stan Sclaroff   Boston Univ., USA

Program Chairs

Kim Boyer   Univ. at Albany, USA
Brian C. Lovell   Univ. of Queensland, Australia
Marcello Pelillo   Univ. Ca' Foscari Venezia, Italy
Nicu Sebe   Univ. of Trento, Italy
René Vidal   Johns Hopkins Univ., USA
Jingyi Yu   ShanghaiTech Univ., China

Workshop Chairs

Giovanni Maria Farinella   Univ. of Catania, Italy
Tao Mei   JD.COM, China

Challenge Chairs

Marco Bertini   Univ. of Florence, Italy
Hugo Jair Escalante   INAOE and CINVESTAV National Polytechnic Institute of Mexico, Mexico

Publication Chair

Roberto Vezzani   Univ. of Modena and Reggio Emilia, Italy

Tutorial Chairs

Vittorio Murino   Univ. of Verona, Italy
Sudeep Sarkar   Univ. of South Florida, USA

Women in ICPR Chairs

Alexandra Branzan Albu   Univ. of Victoria, Canada
Maria De Marsico   Univ. Roma La Sapienza, Italy

Demo and Exhibit Chairs

Lorenzo Baraldi   Univ. Modena Reggio Emilia, Italy
Bruce A. Maxwell   Colby College, USA
Lorenzo Seidenari   Univ. of Florence, Italy

Special Issue Initiative Chair

Michele Nappi   Univ. of Salerno, Italy

Web Chair

Andrea Ferracani   Univ. of Florence, Italy

Corporate Relations Chairs

Fabio Galasso   Univ. Roma La Sapienza, Italy
Matt Leotta   Kitware, Inc., USA
Zhongchao Shi   Lenovo Group Ltd., China

Local Chairs

Matteo Matteucci   Politecnico di Milano, Italy
Paolo Napoletano   Univ. of Milano-Bicocca, Italy

Financial Chairs

Cristiana Fiandra   The Office srl, Italy
Vittorio Murino   Univ. of Verona, Italy

Contents – Part IV

HCAU 2020 - The First International Workshop on Deep Learning for Human-Centric Activity Understanding

IADS - Integrated Artificial Intelligence In Data Science

IML - International Workshop on Industrial Machine Learning

FGVRID - Fine-Grained Visual Recognition and re-Identification

Workshop on Fine-Grained Visual Recognition and Re-identification

Workshop Description

The ubiquitous surveillance cameras are generating huge amount of videos. Automatic video content analysis and recognition are thus desirable for effective utilization of those data. Fine-Grained Visual Recognition and Re-Identification (FGVRID) aims to accurately identify visual objects and match re-appearing targets, e.g., persons and vehicles from a large set of images and videos. It has the potential to offer an unprecedented possibility for intelligent video processing and analysis, as well as to explore the promising applications on public security.

The FGVRID workshop wishes to bring together researchers from fine-grained visual categorization, as well as person/ vehicle ReID communities, and to foster discussions and exchange of ideas between them. FGVRID is not a traditional search or classification task due to its goal of accurately identifying visual objects. First, proper detection algorithms should be designed to locate objects and their parts in videos before proceeding to the identification step. Second, the visual appearance of an object is easily affected by many factors like viewpoint changes and camera parameter differences, etc. Third, annotating the fine-grained identity or category cues is expensive and time consuming. Finally, to cope with the large-scale data, scalable indexing or feature coding algorithms should be designed to ensure the online recognition efficiency. Aiming to seek novel solutions and possibilities in FGVRID, this workshop will have in-depth discussions on those issues and aims to go beyond toy datasets and small-scale algorithms. A total of 12 submissions were received and after a single-blind reviewing process including 2-3 reviewers per paper, 7 papers were accepted. The acceptance rate was 58%.

The topics of FGVRID were equitably represented with 7 presentations focusing on semi-supervised learning, effective video representations, fine-grained classification, new datasets for fine-grained visual recognition. Four speakers were invited to give talks in the workshop. Each speaker gave 40mins talk on fine-grained visual recognition and person/vehicle ReID or related topics.

Organization

General Chairs

Shiliang Zhang   Peking University
Guorong Li   University of Chinese Academy of Sciences
Weigang Zhang   Harbin Institute of Technology, Weihai
Qingming Huang   University of Chinese Academy of Sciences
Nicu Sebe   University of Trento

Program Committee

Yuankai Qi   University of Adelaide
Zhe Xue   Beijing University of Posts and Telecommunications
Hantao Yao   Institute of Automation, Chinese Academy of Sciences
Yifan Yang   University of Chinese Academy of Sciences
Jianing Li   Peking University
Dechao Meng   Institute of Computing Technology

Densely Annotated Photorealistic Virtual Dataset Generation for Abnormal Event Detection

Rico Montulet and Alexia Briassouli[✉][iD]

Department of Data Science and Knowledge Engineering, Maastricht University,
Maastricht, The Netherlands
rico@montulet.nl, alexia.briassouli@maastrichtuniversity.nl

Abstract. Many timely computer vision problems, such as crowd event detection, individual or crowd activity recognition, person detection and re-identification, tracking, pose estimation, segmentation, require pixel-level annotations. This involves significant manual effort, and is likely to face challenges related to the privacy of individuals, due to the intrinsic nature of these problems, requiring in-depth identifying information. To cover the gap in the field and address these issues, we introduce and make publicly available a photorealistic, synthetically generated dataset, with detailed dense annotations. We also publish the tool we developed to generate it, that will allow users to not only use our dataset, but expand upon it by building their own densely annotated videos for many other computer vision problems. We demonstrate the usefulness of the dataset with experiments on unsupervised crowd anomaly detection in various scenarios, environments, lighting, weather conditions. Our dataset and the annotations provided with it allow its use in numerous other computer vision problems, such as pose estimation, person detection, segmentation, re-identification and tracking, individual and crowd activity recognition, and abnormal event detection. We present the dataset as is, along with the source code and tool to generate it, so any modification can be made and new data can be created. To our knowledge, there is currently no other photorealistic, densely annotated, realistic, synthetically generated dataset for abnormal crowd event detection, nor one that allows for flexibility of use by allowing the creation of new data with annotations for many other computer vision problems. **Dataset and source code available:** https://github.com/RicoMontulet/GTA5Event.

1 Introduction

The State of the Art (SoA) deep learning methods in computer vision achieve high accuracy by leveraging large, diverse and correctly annotated datasets, with the most detailed annotations desired being at a pixel level. For problems like crowd event detection, pose estimation, person detection, recognition, segmentation, re-identification, tracking, activity recognition, the production of detailed

---

Funded under the H2020 project MindSpaces, Grant number # 825079.

A. Del Bimbo et al. (Eds.): ICPR 2020 Workshops, LNCS 12664, pp. 5–19, 2021.
https://doi.org/10.1007/978-3-030-68799-1_1

ground truth requires great manual effort, is very time-consuming, error-prone and labor intensive. This is even more so the case in tasks requiring pixel-level accuracy, such as fine-grained activity recognition, pose estimation [9], person re-identification and tracking, as well as activity/event recognition. Moreover, the advent of privacy regulations such as GDPR (https://gdpr-info.eu/) has led to the removal of datasets of individuals that have not given explicit consent, and makes the creation of new annotated datasets challenging. The current Covid-19 related restrictions on large gatherings and crowds of people are posing additional obstacles to the creation of benchmark datasets. However, the need for data with high quality annotations is continuously increasing, for training data, or augmentation of existing training data. A solution to this problem that is gaining increasing attention is the creation of realistic synthetic datasets using commercial video game engines, for the creation of highly realistic data with dense, high quality annotations in varying lighting and environmental conditions.

In this work we create photorealistic videos using the Rockstar Advanced Game Engine (RAGE) in the video game GTA V [38], as it allows for the creation of densely annotated and very realistic datasets. Its license allows for this, and specifically states: *"The publisher of Grand Theft Auto V allows non-commercial use of footage from the game as long as certain conditions are met, such as non-commercial use and not distributing spoilers"* [2]. The engine provides great flexibility, allowing for the generation of videos with wide ranging, detailed and realistic activities of individuals and groups of people in different indoors and outdoors environments, lighting and weather conditions. Our dataset comprises of 54 videos with resolution of 2560 × 1440, from 54 unique locations. Each video has 450 frames recorded from a static camera at varying heights. Detailed ground-truth data is provided for every frame, comprising of weather conditions, time of day, person segmentation, bone coordinates, depth maps and the type of group of people. The videos are rendered at different frames per second to simulate different frame rates on common security cameras. In this work, we choose to apply the generated datasets to the problem of unsupervised, abnormal crowd event detection. The motivation for this is the long-standing lack of high quality, densely annotated data for this problem, as explained in Sect. 2. We demonstrate the usefulness of our dataset in experiments on unsupervised event detection, with annotations that can also be used for a number of other computer vision problems. To our knowledge, there is no other dataset providing annotations for such a wide range of vision problems.

This paper is structured as follows: Sect. 2 describes the related work on synthetic dataset generation, and the datasets available. Section 3 describes the process for generating our synthetic dataset, and the resulting annotations and Sect. 4 shows how it can be used for the successful, unsupervised detection of abnormal events in a variety of environments, while conclusions are drawn in Sect. 5.

2   Related Work

The role of synthetic datasets as supplements to existing training data, or as data in and of themselves, is receiving increased attention [36, 37, 43, 44], as deep learning requires extensive high quality annotated data to perform well, which is not always easy to obtain in the real world. To this end, various synthetic datasets have been recently generated to solve different computer vision problems, with the works in [6, 25, 37, 38, 40, 49] all using synthetic data. Several of them, namely [25, 37, 38, 49] use the GTA V Rockstar Advanced Game Engine (RAGE), similarly to our work, but focusing on different computer vision problems. The reason for choosing RAGE is the high quality of the resulting graphics, as well as the policy of the game engine, which allows for non-commercial use of its footage [2]. Data generation tools and datasets that use virtual worlds to generate annotated image datasets are described below.

2.1   Related Tools for Synthetic Data Generation

CARLA. CARLA is an open-source platform from Intel for developing and testing autonomous driving systems [16] with various environments, sensors, and full control over data. Scene segmentation has been achieved using CARLA in [41], and LiDAR object detection uses CARLA in [17]. A challenge was also setup in 2019 with realistic driving scenarios, for autonomous driving benchmarks (https://carlachallenge.org/).

Unity ML. The Unity game engine allows the development of realistic virtual game environments for applications like Deep Learning, with Google's DeepMind recently having used it to train its deep learning models [1]. Unity ML [3, 26] use machine learning agents to create realistic and varied environments.

Unreal. Unreal [34] is a game engine for virtual environments that also generates realistic images to train deep learning methods. It has been used for AirSim [42], a simulator for drones and autonomous vehicles, and other annotated datasets [33], to train deep learning methods for autonomous vehicles. It has also been used to generate a synthetic dataset for 3D object detection and pose estimation [45].

Blender. Blender, a tool generating 3D scenes for video games, has also been used for training data for computer vision. In [13], an open-source modular pipeline, presented for photorealistic 3D scenes and images, is tested on object segmentation. Medical imaging, and specifically robotic surgery has recently benefited from data created with Blender [10], with the paper receiving a best paper award in 2020.

Europilot. Europilot, an environment based on Euro Truck Simulator 2, simulates all aspects of a driving vehicle: acceleration, breaking, steering and collision detection etc. It also offers visual rendering of the scene for computer vision purposes and is used for training autonomous vehicles [20].

Grand Theft Auto 5. Grand Theft Auto 5 (GTA V) poses a different paradigm, as it generates very photorealistic video game data. One of the main reasons it is selected specifically for computer vision tasks is the excellent quality of its graphics. Another work that uses GTA V for generating photorealistic data is Richter et al. [37,38], who injected their own software inbetween the game and the graphics card, so as to collect information about geometry and textures. In contrast to [37,38], our tool uses native RAGE functions, which allows us to get the annotations directly from the game environment, to change the scene, set weather conditions, and customize the behavior of individuals and groups.

Table 1. Abnormal event detection datasets.

Dataset	# of frames	Resolution	Events
UCSD Ped1 [27], 2014	14000	238 × 158	Abnormal object in one frame
UCSD Ped2 [27], 2014	4560	360 × 240	Abnormal object in one frame
UMN [46], 2014	3855	320 × 240	Staged crowd events
Subway entrance [4], 2008	86535	512 × 382	Few abnormalities
Subway exit [4], 2008	38940	512 × 382	Few abnormalities
CUHK avenue [29], 2013	30652	640 × 360	Few abnormalities
Street scene [35], 2020	203257	1280 × 720	Few abnormalities in street behavior
Ours: GTAV event, 2020	24000	2560 × 1440	Limitless crowd events, abnormal crowd behaviors

2.2   Related Synthetic Data

A wide range of synthetically generated data has been produced for computer vision problems https://github.com/unrealcv/synthetic-computer-vision, but not for the application that we are examining, namely crowd event detection. Moreover, unlike our dataset, existing ones do not offer the flexibility to be used for several other applications, from recognition of individual or group activities/interactions in a variety of scenarios and environments, to person tracking, segmentation re-identification and others, as detailed below.

Synthetic Datasets for Optical Flow. Synthetic datasets have been used to develop accurate optical flow algorithms since 1987 [21], with Yosemite [7] (1994) being one of the most widely used. In [30], what makes a good synthetic dataset is described, with an extensive overview of existing synthetic optical flow benchmarking datasets, including recent ones like Flying Chairs [15], and SYNTHIA [39]. It should be noted that the SoA optical flow Flownet2 [24] used in this work has also been developed using synthetically generated data, which allowed it to outperform the SoA.

Synthetic Human, Crowd Datasets. Previous efforts on synthetic data generation for crowd simulation [5,31] focus on crowd group dynamics, but not on the quality of the graphics, making them less appropriate for deep learning, which requires large amounts of high quality annotated data. Recently, a dataset and tool for crowd counting was published [48], which only generates crowd images, and not crowd videos, nor crowd event scenarios.

In [12], a dataset for human activity recognition has been generated, but with Unity, contains one person per activity, and no abnormal crowd events. Motion tracking and activity recognition can take place using the synthetic data in [19], however it features only one person per frame, and has a blank background.

Human segmentation and depth estimation datasets have also been synthetically generated recently [47], based on motion capture data. However, they use Blender [11], and comprise of single person images rather than continuous video. A large scale synthetic image dataset of images of street scenes with dense semantic segmentation maps, generated by the Unity game engine, is SYNTHIA. It has been used for semantic image segmentation, image-to-image translation [23,28] and adversarial domain adaptation [22], among others.

Our tool generates densely annotated crowds and events, but can also be used for the generation of individual or small group activities, tracking, pose, segmentation, providing solutions for a wider range of computer vision problems than existing datasets. At the same time, it provides densely annotated benchmarking data for abnormal crowd event detection, for which existing real-world datasets have been limited in size, quality and amount of events (see Sect. 2.3).

2.3   Real-World Datasets for Abnormal Crowd Event Detection

In this work we use our dataset for the problem of abnormal crowd event detection, although it can also be used for other problems, like person detection, segmentation, re-identification, tracking, individual or crowd activity recognition. We focus on crowd event detection due to the long-standing and well-documented lack of datasets for this problem. Existing datasets are small, of poor resolution, with few abnormal events and often with inconsistent annotations [35].

In Table 1 we present real-world datasets on crowd event detection, most often used for benchmark comparisons. The frequently encountered, UCSD pedestrian [27], shows pedestrians walking outdoors, with a few anomalies like a bike passing through them etc. It is small in size, contains a few abnormalities, and events are based on changes in a frame, rather than changes in behavior and motion. The University of Minnessotta (UMN) dataset [46] is even smaller, with 11 videos and 3 scenes, with simple, staged events. Only two long real-world videos, Subway and Mall, are presented in [4], with few, specific events, making them inadequate for robust testing. CUHK Avenue [29] contains data from a surveillance camera, with few anomalies, caused mostly by individual actions rather than crowd behaviors. Recently, StreetScene [35] was made available, containing a far larger number of frames at higher resolution, with the corresponding annotations. However, this dataset does not contain crowd events or abnormal individual/crowd behaviors, as it focuses on abnormalities related to street scenes and rules, such as pedestrians crossing illegally or bicycles on sidewalks.

Our dataset features significantly more frames than most of the above datasets, with the exception of StreetScene, however the number of frames in our dataset can be directly increased by using our tool. Our dataset also features a wide range of weather conditions and environments, as well as events related to abnormalities in crowd behavior and motion, rather than anomalies related to appearance changes in one frame (as in UCSD, UMN). Thus it is more appropriate for detecting abnormalities in videos, in the behavior of crowds but also individuals. It does not contain annotations only for events, but also for person segmentation, tracking, re-identification, pose classification and more, detailed in Sect. 3.2. Its graphics are of high quality, as they have been generated with the GTA V engine [14], which can produce a wide range of high quality, photo-realistic scenes.

3   Dataset Creation and Description

3.1   Dataset Creation: Interacting with the Virtual World

Our method uses the plugin Scripthook, developed by Alexander Blade, that allows developers to interact with the Rockstar Advanced Game Engine (RAGE). The scene generation and data collection is done by a plugin written in C# that controls the environment using Scripthook. Virtual scenes, weather conditions, character models, lighting conditions, movement and behaviours can all be changed using config file. In order to create a diverse dataset, different locations and camera placements needed to be explored. Conveniently, GTA V has a massive 252 square kilometer map with a vast amount of different locations. Every location is unique with a high degree of similarity to real world locations. From beaches to shopping malls to mountain ranges, the possibilities are vast.

In this work, a subset of 54 locations were used, where stationary cameras have been placed at various heights and angles, and a region of interest was selected in them. There are 704 unique person models available in the game, with different skin color, body shape (height, weight) and hair styles, with varied types of clothing for each person model. The people can move in several ways, such as walking, standing, crouching, standing having a conversation etc.

Groups of people of various sizes are spawned in the designated Region of Interest (ROI) for each location. These groups range in size from 5 to 25 people and there are 1 to 5 groups. These parameters can be changed in a configuration file to suit any needs. Every person within a group gets a specific task, whether to talk to one of their group members, wander around the scene, make a phone-call or just stand there. Once groups have been spawned the weather is randomly changed, the camera is set to render at a random FPS, and finally, the time of day is randomized. All of this happens while the game is paused. This is achieved by setting the timescale parameter in the game engine to zero. The locations along with group sizes, group locations, person locations, weather, time and FPS information is available. Figure 1 provides an overview of the distribution of the generated actions, weather conditions, fps for each video, and times of day, showing that our solution can cover numerous scenarios. There are 24,301 frames in total, with 1,825,493 annotated bounding boxes and a total of 177,372,250 bones. There are 5719 people in total having an average trajectory length of 319 frames. All locations cover a ROI of 27,635 square units, with an average ROI of 512 square units per location, where one 3D unit was found to be approximately 0.85 m [14].

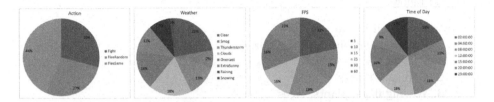

Fig. 1. The distribution of group behaviours, weathers, frames per second used to render the videos and of the times during which the videos were recorded.

Fig. 2. Our frame annotations: from left to right, the bounding boxes of people along with their ID and distance to the camera, the bone coordinates of the people in the scene, the trajectories people have walked, the pixel-wise segmentation of people, and lastly pixel-wise depth information relative to the camera plane.

Now that the scene is set up, the timescale is set to 1.0, and the game starts to render the scene. Meanwhile, the script saves images and annotations, which include RGB, depth, segmentation information, person pose information (bone locations), bounding boxes, and person IDs. Thus, the post-processing script can find trajectories, tighten bounding boxes, segmentations, and depth estimates. Figure 2 shows a sample synthetically generated frame using our method, with annotations corresponding to person detection (bounding boxes), bones, trajectories, person segmentation and depth maps.

3.2   Dataset Description

The resulting dataset contains detailed scene information and annotations, described below:

1. a bitmap stencil image
   - 0: Environment objects like floor, stairs, buildings
   - 1: Persons
   - 2: Cars and trucks
   - 3: Waving flags, plants, trees
   - 4: Beach sand, grass
   - 7: sky
2. a depth map, as a single channel image with float values in range [0, 1], where 1 is close to the camera and 0 is far (http://www.adriancourreges.com/blog/2015/11/02/gta-v-graphics-study/)
3. location information in a json file, containing:
   - location name
   - camera position
   - camera rotation
   - player position
   - player rotation
   - ROI of the location
   - PedGroups: contains the initial positions of all the people that belong to this group
   - PedCenters: contains the original centers around which the people were spawned
   - PedIdGroup: contains the handles of all the spawned peds, and the cluster center they belong to
   - fps: frames per second the video was rendered at
   - Action: the action that happens at a specific frame
   - Current time: the time the video was recorded in game time
   - Currentweather: the weather of that scene [0 = ExtraSunny, 1 = Clear, 2 = Clouds, 3 = Smog, 5 = Overcast, 6 = Raining, 7 = Thunderstorm, 13=Snow]
   - CamFOV: the field of view of the recording camera (always 50)
4. Per frame annotations that contain:
   - handle which is unique ID for every person
   - their distance from the camera in meters
   - normalized onscreen [0,1] bone coordinates

4   Unsupervised Abnormal Event Detection Using GTA V

The datasets created using our method can be very useful in the problem of abnormal crowd event detection, where there is a lack of densely annotated high resolution benchmark data, as explained in Sect. 2.3. We consider realistic crowd event scenarios in various indoors/outdoors environments and weather conditions, and specifically events such as crowd dispersion or scattering, crowd fleeing and a fight breaking out in a crowd, with some examples shown in Fig. 3.

Fig. 3. Examples from our photorealistic crowd event datasets. Left to right: School yard before students merge towards the exit and leave, villa garden before a crowd leaves, beach before people run away, mall after a crowd disperses.

4.1   Cumulative Sum Method for Abnormal Event Detection

The event taking place, as well as the time it takes place, is unknown beforehand, and is characterized mostly by a change in the crowd motion. For this reason, we choose to detect possible events by analyzing the statistics of the crowd optical flow over time. The basic assumption is that current crowd behavior is "normal", and a significant deviation from it would be abnormal, which is often the case in several realistic scenarios like the ones considered in our work.

We use sequential statistical change detection, namely the Cumulative Sum (CUSUM) approach [8,18], to detect a change between normal and abnormal crowd motion, as it is can effectively and quickly detect changes between statistical distributions. The first w_0 frames ($w_0 = 15$ here) are considered to characterize baseline (normal) crowd behavior/motion and the most recent $w_0$characterize "current" crowd behavior/motion. Their motion is found by estimating dense optical flow using SoA FlowNet2 [24] and its empirical distribution through the histogram of the optical flow. It should be noted that FlowNet2 also uses synthetically generated data to improve its accuracy, and has surpassed previous SoA optical flow estimation methods because of this.

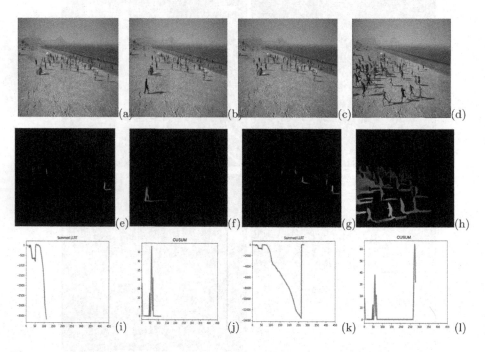

Fig. 4. Beach crowd events. Frames: (a) before 1st event (b) after 1st event (person entering, walking in a specific direction) (c) before 2nd event, (d) after 2nd event (crowd running, dispersing). Optical flow: (e) before 1st event, (f) after 1st event (g) before 2nd event, (h) after 2nd event. Change in the values of: (i) Summed LLRT for 1st event, (j) CUSUM for 1st event, (k) Summed LLRT for 2nd event, (l) CUSUM for 2nd event.

The histogram of the optical flow approximates its statistical distribution, and is denoted at frame k with $f_k(\bar{r})$ at each pixel $\bar{r} = (x, y)$, while $f_0(\bar{r})$ represents the distribution of the baseline motion. These two distributions are used to estimate the log-likelihood ratio L_k, a commonly used test statistic for detecting changes between statistical distributions, given by:

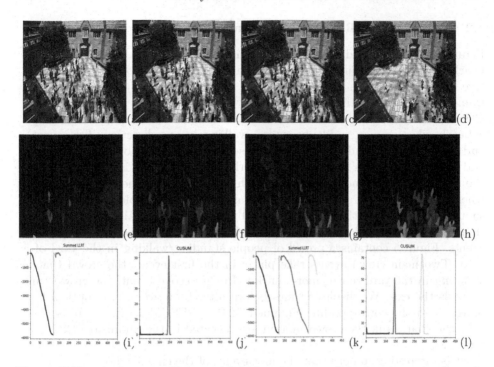

Fig. 5. Villa crowd events. Frames: (a) before 1st event (b) after 1st event (crowd moving more in a stochastic way) (c) before 2nd event, (d) after 2nd event (crowd running towards exit). Optical flow: (e) before 1st event, (f) after 1st event (g) before 2nd event, (h) after 2nd event. Change in the values of: (i) Summed LLRT for 1st event, (j) CUSUM for 1st event, (k) Summed LLRT for 2nd event, (l) CUSUM for 2nd event.

$$L_k = \ln\left(\frac{f_k(\bar{r})}{f_0(\bar{r})}\right) \tag{1}$$

The CUSUM test uses the log-likelihood ratio (LLRT) as test statistic L_k for the test at frame k, expressed in the computationally efficient iterative form [32]:

$$T_k = \max(0, T_{k-1} + L_k), \tag{2}$$

where we set $S_k = T_{k-1} + L_k$ as the Summed LLRT, and initialize $T_0 = 0$. When the data deviates from f_0, T_k increases significantly, and a change can be detected at that point. There is no theoretically founded method for setting the T_k threshold, so in this work, a value equal to 100 was empirically found to provide accurate results. In the case of multiple events, when an event is detected, T_k is re-set to 0 and the entire process restarts. This can be seen in the Summed LLRT and CUSUM plots in Figs. 4–5 where several events take place. These statistics, the videos and their optical flow, can be seen evolving in real time, for the videos examined here, as well as other videos generated by our tool, on our video demos on our GitHub.

4.2 Experimental Results: Abnormal Event Detection

Figures 4–5 show two different crowd event scenarios, on a sunny day at the beach, with the sea water moving, and in a villa yard surrounded by moving tree leaves. As detailed in the captions of Figs. 4–5, they display characteristic frames before events, the optical flow between them, the Summed LLRT and CUSUM values until those frames. In Fig. 4 we see the results of a crowd on a beach where a man suddenly enters, walking in one direction (first event), and the crowd later suddenly runs away (second event). The optical flow before and after the events is shown, and is clearly different, which is also reflected in the Summed LLRT and CUSUM for each event, whose values change sharply when the person enters the beach area, and when the people start to run. It should be noted that the motion caused by people slowly walking around, and small background motions from the sea, did not affect the detection of the actual events. Figure 5 contains a crowd of people standing/walking in the yard of a villa. Two main crowd events take place: in the first event, the crowd that is standing in the yard moves more quickly. In the second event, the crowd runs towards the exit. We display frames before/after both events, the optical flow images and the corresponding Summed LLRT, CUSUM in Fig. 5. It is clear that the changes in the crowd motion are reflected in the Summed LLRT and CUSUM.

It is interesting to note that the first event, of the crowd's motion becoming more stochastic/random, is not easy to detect visually, however our method detects it. We provide videos of the sequential detection of events in these videos, and additional examples with different scenarios and environments generated by our tool, in our GitHub. These results show CUSUM can robustly detect abnormal crowd behaviors, originating from changes in behavior that correspond to changes in motion, in a variety of environments and scenarios.

5   Conclusions

In this work, we have presented a method for generating high-resolution pho-torealistic synthetic datasets using the game engine from GTA V. The method used allows for the control of the scenes in great detail, and results in detailed annotations for a wide variety of scenes, events, activities, and crowds. Apart from data, we provide the tool itself for generating additional datasets on our GitHub[1], which can be used to solve a wide range of computer vision problems such as activity recognition, person detection and tracking, event detection and others. We demonstrate the usefulness of our dataset for abnormal crowd event detection, as there is a significant lack of annotated datasets for this problem. A sequential statistical change detection method for finding changes in the statistical distribution of datasets is applied to the optical flow of our synthetic scenes. The optical flow is estimated using SoA deep learning Flownet2, which provides dense accurate flow estimates. The results show accurate and quick detection of

---

[1] https://github.com/RicoMontulet/GTA5Event.

changes in different scenarios, indoors and outdoors, at different times of day and in different environmental conditions. Future work will expand upon the detection of abnormal events on more extensive synthetic and real datasets, but also on the use of our data for problems like person tracking, re-identification, activity and interaction recognition.

References

1. How Google's DeepMind will train its AI inside Unity's video game worlds (2018). https://web.archive.org/web/20180927024638/www.fastcompany.com/90240010/deepminds-ai-will-learn-inside-unitys-video-game-worlds
2. Policy on posting copyrighted rockstar games material (Oct 2020). https://tinyurl.com/RockstarPrivacyPolicy. Accessed 15 Sept 2020
3. Unity Machine Learning Agents (2020). https://unity.com/products/machine-learning-agents
4. Adam, A., Rivlin, E., Shimshoni, I., Reinitz, D.: Robust real-time unusual event detection using multiple fixed-location monitors. IEEE Trans. Pattern Anal. Mach. Intell. **30**(3), 555–560 (2008)
5. Andrade, E., Fisher, B.: Simulation of crowd problems for computer vision. In: 1st International Workshop on Crowd Simulation (V-CROWDS '05), pp. 71–80 (2005)
6. Bąk, S., Carr, P., Lalonde, J.-F.: Domain adaptation through synthesis for unsupervised person re-identification. In: Ferrari, V., Hebert, M., Sminchisescu, C., Weiss, Y. (eds.) ECCV 2018, Part XIII. LNCS, vol. 11217, pp. 193–209. Springer, Cham (2018). https://doi.org/10.1007/978-3-030-01261-8_12
7. Barron, J.L., Fleet, D.J., Beauchemin, S.S.: Performance of optical flow techniques. Int. J. Comput. Vis. **12**, 43–77 (1994)
8. Basseville, M., Nikiforov, I.: Detection of Abrupt Changes: Theory and Application. Prentice-Hall Inc., Englewood Cliffs (1993)
9. Cao, Z., Hidalgo Martinez, G., Simon, T., Wei, S., Sheikh, Y.A.: OpenPose: real-time multi-person 2D pose estimation using part affinity fields. IEEE Trans. Pattern Anal. Mach. Intell. **43**(1), 172–186 (2019)
10. Cartucho, J., Tukra, S., Li, Y., Elson, D.S., Giannarou, S.: VisionBlender: a tool to efficiently generate computer vision datasets for robotic surgery. In: Computer Methods in Biomechanics and Biomedical Engineering: Imaging and Visualization (2020)
11. Community, B.O.: Blender - a 3D modelling and rendering package. Blender Foundation, Stichting Blender Foundation, Amsterdam (2018). http://www.blender.org
12. De Souza, C.R., Gaidon, A., Cabon, Y., Lpez, A.M.: Procedural generation of videos to train deep action recognition networks. In: 2017 IEEE Conference on Computer Vision and Pattern Recognition (CVPR), pp. 2594–2604 (2017)
13. Denninger, M., et al.: BlenderProc: reducing the reality gap with photorealistic rendering. In: Robotics: Science and Systems (RSS) Workshops (2020)
14. Doan, A.D., Jawaid, A.M., Do, T.T., Chin, T.J.: G2D: from GTA to Data (2018)
15. Dosovitskiy, A., et al.: FlowNet: learning optical flow with convolutional networks. In: IEEE International Conference on Computer Vision (ICCV), pp. 2758–2766 (2015)
16. Dosovitskiy, A., Ros, G., Codevilla, F., Lopez, A., Koltun, V.: CARLA: an open urban driving simulator. In: 1st Annual Conference on Robot Learning, pp. 1–16 (2017)

17. Dworak, D., Ciepiela, F., Derbisz, J., Izzat, I., Komorkiewicz, M., Wjcik, M.: Performance of LiDAR object detection deep learning architectures based on artificially generated point cloud data from CARLA simulator. In: 2019 24th International Conference on Methods and Models in Automation and Robotics (MMAR), pp. 600–605 (2019)

18. Einmahl, J., McKeague, I.: Empirical likelihood based hypothesis testing. Bernoulli **9**, 267–290 (2003)

19. Elanattil, S., Moghadam, P.: Synthetic human model dataset for skeleton driven non-rigid motion tracking and 3D reconstruction (2019)

20. Gyuri, I.: Europilot: A toolkit for controlling Euro Truck Simulator 2 with Python to develop self-driving algorithms (2017). https://github.com/marshq/europilot

21. Heeger, D.J.: Model for the extraction of image flow. J. Opt. Soc. Am. A **4**(8), 1455–1471 (1987)

22. Hoffman, J., et al.: CyCADA: cycle-consistent adversarial domain adaptation. In: International Conference on Machine Learning, pp. 1989–1998 (Jul 2018)

23. Huang, X., Liu, M.-Y., Belongie, S., Kautz, J.: Multimodal unsupervised image-to-image translation. In: Ferrari, V., Hebert, M., Sminchisescu, C., Weiss, Y. (eds.) ECCV 2018, Part III. LNCS, vol. 11207, pp. 179–196. Springer, Cham (2018). https://doi.org/10.1007/978-3-030-01219-9_11

24. Ilg, E., Mayer, N., Saikia, T., Keuper, M., Dosovitskiy, A., Brox, T.: FlowNet 2.0: evolution of optical flow estimation with deep networks. In: 2017 IEEE Conference on Computer Vision and Pattern Recognition (CVPR), pp. 1647–1655 (2017)

25. Johnson-Roberson, M., Barto, C., Mehta, R., Sridhar, S.N., Rosaen, K., Vasudevan, R.: Driving in the matrix: Can virtual worlds replace human-generated annotations for real world tasks? arXiv preprint arXiv:1610.01983 (2016)

26. Juliani, A., et al.: Unity: a general platform for intelligent agents. arXiv preprint arXiv:1809.02627 (2020). https://github.com/Unity-Technologies/ml-agents

27. Li, W., Mahadevan, V., Vasconcelos, N.: Anomaly detection and localization in crowded scenes. IEEE Trans. Pattern Anal. Mach. Intell. **36**(1), 18–32 (2014)

28. Liu, M., Breuel, T., Kautz, J.: Unsupervised image-to-image translation networks. In: Proceedings of the 31st International Conference on Neural Information Processing Systems, Red Hook, NY, USA, pp. 700–708 (2017)

29. Lu, C., Shi, J., Jia, J.: Abnormal event detection at 150 fps in matlab. In: Proceedings of the 2013 IEEE International Conference on Computer Vision, ICCV '13, IEEE Computer Society, USA, pp. 2720–2727 (2013)

30. Mayer, N., et al.: What makes good synthetic training data for learning disparity and optical flow estimation? Int. J. Comput. Vis. **126**, 942–960 (2018)

31. Oghaz, M.M., Argyriou, V., Remagnino, P.: Learning how to analyse crowd behaviour using synthetic data. In: Proceedings of the 32nd International Conference on Computer Animation and Social Agents, pp. 11–14 (2019)

32. Page, E.S.: Continuous inspection scheme. Biometrika **41**, 100–115 (1954)

33. Pollok, T., Junglas, L., Ruf, B., Schumann, A.: UnrealGT: using unreal engine to generate ground truth datasets. In: Bebis, G., et al. (eds.) ISVC 2019, Part I. LNCS, vol. 11844, pp. 670–682. Springer, Cham (2019). https://doi.org/10.1007/978-3-030-33720-9_52

34. Qiu, W., et al.: UnrealCV: virtual worlds for computer vision. In: ACM Multimedia Open Source Software Competition (2017)

35. Ramachandra, B., Jones, M.J.: Street scene: a new dataset and evaluation protocol for video anomaly detection. In: IEEE Winter Conference on Applications of Computer Vision, WACV 2020, Snowmass Village, CO, USA, 1–5 March 2020 (2020)

36. Ranjan, A., Hoffmann, D.T., Tzionas, D., Tang, S., Romero, J., Black, M.J.: Learning multi-human optical flow. Int. J. Comput. Vis. (IJCV) **128**, 873–890 (2020). http://humanflow.is.tue.mpg.de

37. Richter, S.R., Hayder, Z., Koltun, V.: Playing for benchmarks. In: Proceedings of the IEEE International Conference on Computer Vision, pp. 2213–2222 (2017)

38. Richter, S.R., Vineet, V., Roth, S., Koltun, V.: Playing for data: ground truth from computer games. In: Leibe, B., Matas, J., Sebe, N., Welling, M. (eds.) ECCV 2016, Part II. LNCS, vol. 9906, pp. 102–118. Springer, Cham (2016). https://doi.org/10.1007/978-3-319-46475-6_7

39. Ros, G., Sellart, L., Materzynska, J., Vazquez, D., Lopez, A.M.: The SYNTHIA dataset: a large collection of synthetic images for semantic segmentation of urban scenes. In: 2016 IEEE Conference on Computer Vision and Pattern Recognition (CVPR), pp. 3234–3243 (2016)

40. Ros, G., Sellart, L., Materzynska, J., Vazquez, D., Lopez, A.M.: The SYNTHIA dataset: a large collection of synthetic images for semantic segmentation of urban scenes. In: Proceedings of the IEEE Conference on Computer Vision and Pattern Recognition, pp. 3234–3243 (2016)

41. Saleh, F.S., Aliakbarian, M.S., Salzmann, M., Petersson, L., Alvarez, J.M.: Effective use of synthetic data for urban scene semantic segmentation. In: Ferrari, V., Hebert, M., Sminchisescu, C., Weiss, Y. (eds.) ECCV 2018, Part II. LNCS, vol. 11206, pp. 86–103. Springer, Cham (2018). https://doi.org/10.1007/978-3-030-01216-8_6

42. Shah, S., Dey, D., Lovett, C., Kapoor, A.: Airsim: high-fidelity visual and physical simulation for autonomous vehicles. In: Field and Service Robotics (2017). https://arxiv.org/abs/1705.05065

43. Song, S., Yu, F., Zeng, A., Chang, A.X., Savva, M., Funkhouser, T.: Semantic scene completion from a single depth image. In: Proceedings of 30th IEEE Conference on Computer Vision and Pattern Recognition (2017)

44. Tremblay, J., et al.: Training deep networks with synthetic data: bridging the reality gap by domain randomization. In: Proceedings of the IEEE Conference on Computer Vision and Pattern Recognition (CVPR) Workshops (June 2018)

45. Tremblay, J., To, T., Birchfield, S.: Falling things: a synthetic dataset for 3D object detection and pose estimation. In: 2018 IEEE/CVF Conference on Computer Vision and Pattern Recognition Workshops (CVPRW), pp. 2119–21193 (2018)

46. UMN: University of Minnesota dataset. http://mha.cs.umn.edu/proj_events.shtml

47. Varol, G., et al.: Learning from synthetic humans. In: Proceedings of the IEEE Conference on Computer Vision and Pattern Recognition (CVPR) (July 2017)

48. Wang, Q., Gao, J., Lin, W., Yuan, Y.: Learning from synthetic data for crowd counting in the wild. In: Proceedings of IEEE Conference on Computer Vision and Pattern Recognition (CVPR), pp. 8198–8207 (2019)

49. Xiang, S., Fu, Y., You, G., Liu, T.: Attribute analysis with synthetic dataset for person re-identification. arXiv preprint:2006.07139 (2020)

Unsupervised Domain Adaptive Re-Identification with Feature Adversarial Learning and Self-similarity Clustering

Tianyi Yan[1,2], Haiyun Guo[1(✉)], Songyan Liu[1], Chaoyang Zhao[1], Ming Tang[1], and Jinqiao Wang[1]

[1] National Laboratory of Pattern Recognition, Institute of Automation, Chinese Academy of Sciences, Beijing 100190, China
{tianyi.yan,haiyun.guo,songyan.liu,chaoyang.zhao,tangm, jqwang}@nlpr.ia.ac.cn
[2] School of Artificial Intelligence, University of Chinese Academy of Sciences, Beijing 100190, China

Abstract. In this paper, we propose a novel unsupervised domain adaptation re-ID framework by fusing feature adversarial learning and self-similarity clustering. Different from most of the existing works which only regard the source domain data as network pretraining data, we use the source domain data both in network pretraining and finetuing stage. Concretely, we construct an feature adversarial learning module to learn domain invariant feature representations. The feature extractor network is optimized in an adversarial training manner through minimizing the discrepancy of feature representations between source and target domains. To further enhance the discriminability of the feature extractor network, we design the self-similarity clustering module to mine the implicit similarity relationships among the unlabeled samples of the target domain. By unsupervised clustering, we can generate pseudo-identity labels for the target domain data, which are then combined with the labeled source data together to train the feature extractor network. Additionally, we present a relabeling algorithm to construct correspondence between two groups of pseudo-identity labels generated by two iterative clusterings. Experimental results validate the effectiveness of our method.

Keywords: Unsupervised domain adaptive re-identification · Feature adversarial learning · Self-similarity clustering

1 Introduction

Object re-identification (re-ID), which aims to spot an object of interest from multiple non-overlapping cameras, is an important but difficult task in video surveillance applications. Recently, vehicle re-ID and person re-ID have become

A. Del Bimbo et al. (Eds.): ICPR 2020 Workshops, LNCS 12664, pp. 20–35, 2021.
https://doi.org/10.1007/978-3-030-68799-1_2

hot research topics both in industry and academia and great progress has been made by deep learning based methods, which usually depend on lots of labeled training data. However, due to the domain bias, models trained on the source domain may have a severe performance drop on the target one. The poor generalization ability severely hinders the wide application of re-ID models. Since it is often costly and unfeasible to obtain image annotation for the target domain, a better solution for this issue is unsupervised domain adaptation.

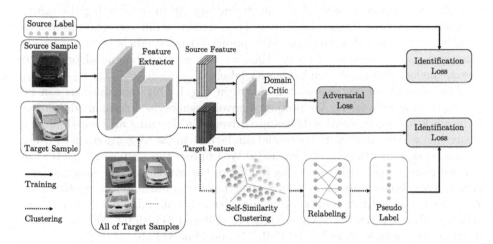

Fig. 1. Overview of our proposed approach. The feature extractor is pre-trained on the source dataset. At the beginning of each epoch, we extract the target domain feature maps, obtain the pseudo-identity labels by self-similarity clustering, and relabeling the pseudo-identity labels.

Existing unsupervised domain adaptation methods for object re-ID can be roughly categorized into two kinds: some methods focus on utilizing Generative Adversarial Networks (GAN) [4] to translate the style of person images from the source domain to the target one and train the model with translated images [5,8]; others try to mine the discriminative cues in the target domain with models pre-trained on source domain and fine-tune the models with mined relationship constraints between samples [10] or estimated pseudo labels [3,12]. Current leading methods mostly adopt a pseudo label estimation scheme via a clustering algorithm.

In this paper, we propose a novel unsupervised domain adaptation method for object re-ID by fusing feature adversarial learning and self-similarity clustering. First, we introduce the feature adversarial learning to the re-ID task. Specifically, we adopt a domain critic network to maximize the discrepancy of feature representations between the source and target domains. Meanwhile, the feature extractor network is trained to minimize the discrepancy. Through this adversarial training, the samples from the source and target domains can be mapped to the domain-invariant feature representations, which is beneficial to transfer the knowledge from the source domain to the target one. In this way, we can make full use of the source domain data to develop domain generalizable re-ID models. Besides, to further improve the discriminability of the feature extractor network, we propose the self-similarity clustering module, to fully explore the implicit semantic similarity relationships between the unlabeled samples of the target domain. By grouping the unlabeled data into several clusters via k-means clustering algorithm [15], we can generate pseudo-identity labels for them and then combine them with the labeled samples of the source domain for training. Considering the randomness of the clustering algorithm in selecting the initial clustering center, the same sample may be assigned to different clusters at two iterative clustering, thus the generated pseudo-identity labels of two iterative clustering may mismatch. To solve this, we design a relabelling algorithm to match the two groups of pseudo-identity labels.

The overview of our approach is shown in Fig. 1. Our contributions can be summarized as follows: (1) We propose a novel unsupervised domain adaptation framework for object re-ID with feature adversarial learning and self-similarity clustering, which can mine the potential similarities in the target domain by using the knowledge from the source domain. (2) We improve the clustering-based cross-domain learning methods by making full use of the source domain data and propose the novel relabeling algorithm to settle the cluster mismatching problem. (3) We achieve competitive results for unsupervised domain adaptation on popular vehicle re-ID benchmarks, such as VeRi-776 [19] and VehicleID [20], and person re-ID benchmarks, such as Market1501 [21] and DukeMTMC-ReID [22].

2   Related Work

2.1   Person and Vehicle Re-ID

A majority of research efforts on person re-ID are devoted to the fully supervised learning in one domain. Some researchers regard re-ID as deep metric learning problems [17]. Besides, many researchers pay attention to develop alignment based person re-ID methods [23,24]. Recently, many works employ extra pose estimation or human parsing module to locate body parts [25,26] and try to achieve the pixel-level alignment [27,28].

Vehicle re-ID has attracted increasing attention in recent years. Guo *et al.* [29] made use of the hierarchical structure of vehicle labels to construct a novel coarse-to-fine loss function for vehicle re-ID. Later on, Guo *et al.* [30] proposed a two-level attention network to learn the discriminative local features. Although the above methods have achieved great progress in supervised setting, their performances degrade significantly on cross-domain setting.

2.2   Unsupervised Domain Adaptation

The common method for unsupervised domain adaptation was to reduce the discrepancy of the features from the source and target domain [2,32]. Recently, several works solved the problem by mapping the data of source and target to the domain-invariant feature space [1,33]. Shen *et al.* [1] proposed an adversarial learning method to reduce the Wasserstein distance between the feature distributions of the source and target domains.

2.3 Unsupervised Domain Adaptation Methods for Re-ID

Recently, several works attempted to address the unsupervised domain adaptation methods for re-ID on the basis of image translation [34,36]. For example, Deng *et al.* [5] proposed Similarity Preserving cycle-consistent Generative Adversarial Network (SPGAN) to translate the images from different domains; Wei *et al.* [6] proposed Person Transfer Generative Adversarial Network (PTGAN) to bridge the domain gap with the identity preserved; Peng *et al.* [8] proposed a Vehicle Transfer Generative Adversarial Network (VTGAN) to translate the vehicle images from different domains. Moreover, Fu *et al.* [3] proposed a person re-ID framework based on self-similarity grouping.

In this paper, we firstly introduce the feature adversarial learning [1] to unsupervised domain adaptation for re-ID problem and validate its effectiveness on this task through extensive experiments. Besides, we improve the cluster-based methods by using both the source domain data and target domain data in network finetuning stage and the experimental results show the effectiveness of our approach.

3 Methodology

3.1 Problem Definition

In the unsupervised domain adaptation setting for re-ID problem, we have a labeled source dataset $X^s = \{(x_i^s, y_i^s)\}_{i=1}^{n^s}$ of n^s samples from the source domain $\mathcal{D}_s$, and an unlabeled target dataset $X^t = \{x_j^t\}_{j=1}^{n^t}$ of n^t samples from the target domain $\mathcal{D}_t$. We refer to [1] and assume that the two domains share the same feature space but follow different marginal data distributions, $\mathbb{P}_{x^s}$ and $\mathbb{P}_{x^t}$, respectively. The goal is to learn a discriminative feature embedding space for the target domain.

3.2 Feature Extractor Pre-training

We first pre-train the feature extractor network on the source domain. Given an instance $x \in \mathbb{R}^m$ from the source domain, the feature extractor network learns a function $f_g : \mathbb{R}^m \to \mathbb{R}^d$ that maps the instance to a d-dimensional representation with the corresponding network parameter θ_g. The overall identification loss is the weighted sum of the cross-entropy loss, triplet loss [18], and center loss [14]:

$$\mathcal{L}_{ide}(x^s, y^s) = \frac{1}{n^s} \sum_{x^s \in X^s} (\mathcal{L}_{CE}(x^s, y^s) + \lambda_{tri}\mathcal{L}_{tri}(x^s, y^s) + \lambda_{cen}\mathcal{L}_{cen}(x^s, y^s)) \quad (1)$$

3.3   Domain-Invariant Feature Adversarial Learning

We use WDGRL [1] as the feature adversarial learning framework. The goal is to reduce the discrepancy between the source and target domains. We use the domain critic network, whose target is to estimate the Wasserstein distance between the source and target representation distributions. Given the feature representation $h = f_g(x)$ computed by the feature extractor network, the domain critic network learns a function $f_w : \mathbb{R}^d \to \mathbb{R}$ that maps the feature representation to a real number with parameter θ_w. If all of the parameters of domain critic network f_w are 1-Lipschitz, then the Wasserstein distance can be approximated by maximizing the domain critic loss $\mathcal{L}_{wd}$ for θ_w:

$$\mathcal{L}_{wd}(x^s, x^t) = \frac{1}{n^s} \sum_{x^s \in X^s} f_w(f_g(x^s)) - \frac{1}{n^t} \sum_{x^t \in X^t} f_w(f_g(x^t)) \tag{2}$$

In order to enforce the Lipschitz constraint, gradient penalty $\mathcal{L}_{grad}$ is applied:

$$\mathcal{L}_{grad}(\hat{h}) = (\|\nabla_{\hat{h}} f_w(\hat{h})\|_2 - 1)^2 \tag{3}$$

where $\hat{h}$ is either the source and the target domain representations $h^s = f_g(x^s)$, $h^t = f_g(x^t)$, or the random points along the straight line between h^s and h^t.

So the feature adversarial learning scheme is to solve the following minimax problem:

$$\min_{\theta_g} \max_{\theta_w} (\mathcal{L}_{wd}(x^s, x^t) - \gamma \mathcal{L}_{grad}(\hat{h})) \tag{4}$$

where γ is the balance weight.

3.4   Self-similarity Clustering

We can generate the pseudo-identity labels for target domain samples by gathering them into several clusters and combine them with the labeled samples of the source domain for supervised training. As shown in Fig. 1, for all of the training samples of the target domain, we extract their features after the global average pooling, then apply PCA [37] on them to reduce the dimensions, and use k-means [15] algorithm to cluster the features. Then for each sample from the target domain x_i^t, we can obtain the pseudo-identity label $\hat{y}_i^t$ by clustering. The self-similarity clustering is not time-consuming because of the low-dimensional features, so it can be iteratively done at the beginning of each epoch. Then the pseudo-identity labels are updated.

3.5   Relabeling Algorithm

Due to the randomness of k-means clustering algorithm in selecting initial cluster center, the same sample may be assigned with different pseudo-identity labels according to two iterative clustering results, which will mislead the computation

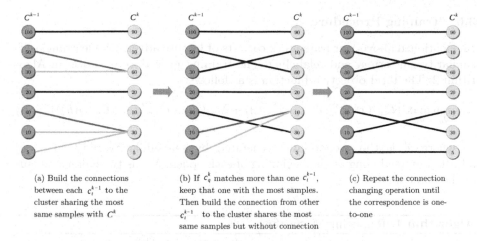

(a) Build the connections between each c_t^{k-1} to the cluster sharing the most same samples with C^k

(b) If c_q^k matches more than one c_t^{k-1}, keep that one with the most samples. Then build the connection from other c_t^{k-1} to the cluster shares the most same samples but without connection

(c) Repeat the connection changing operation until the correspondence is one-to-one

Fig. 2. Relabeling algorithm. The circles represent the clusters and the numbers represent the quantity of the samples in the clusters. The blue lines represent the one-to-one connections, the green lines represent the one-to-many connections but with the most samples at left, and the orange lines represent other connections. Note that the green and blue lines will be kept and the orange lines will be discarded (Color figure online)

of cross-entropy loss and center loss. To solve this, we must construct correspondence between two iterative clustering results, so we propose the following relabeling algorithm.

Given the unlabeled target dataset $X^t = \{x_j^t\}_{j=1}^{n_t}$, self-similarity clustering can give the pseudo-identity labels at the beginning of each epoch. We define the results of kth clustering operation as $C^k = \{c_q^k\}_{q=1}^N$, where c_q^k is a list of the samples in the qth cluster and N is the number of clusters. The results of the last clustering operation is $C^{k-1} = \{c_t^{k-1}\}_{t=1}^N$ and the target of relabeling is to obtain a one-to-one cluster matching map $M^k = \{m_t^k\}_{t=1}^N$ from C^{k-1} to C^k to make the label-unchanged samples as many as possible, where m_t^k is the index of corresponding cluster of c_t^{k-1} in C^k.

The example of our proposed relabeling algorithm is shown in Fig. 2. First, we build the connections from each cluster in C^{k-1} to the cluster which shares the most same samples in C^k, as shown in Fig. 2(a). Next, if there are more than one clusters in C^{k-1} connecting to one cluster in C^k, only keep the connection from the cluster with the most samples, and build the connections from other clusters in C^{k-1} to the cluster sharing the most same samples in C^k but without the connection, as shown in Fig. 2(b). Finally, repeat the connection change operation until the correspondence is one-to-one, as shown in Fig. 2(c). The details of the relabeling algorithm are shown in Algorithm 1.

3.6   Training Procedure

As mentioned above, our framework consists of feature adversarial learning, self-similarity clustering, and relabeling. The training procedure is shown in Algorithm 2. The total objective function is as follows:

$$\min_{\theta_g} \max_{\theta_w} ((\mathcal{L}_{ide}(x^s, y^s) + \mathcal{L}_{ide}(x^t, \hat{y}^t)) + \lambda_{cri}(\mathcal{L}_{wd}(x^s, x^t) - \gamma \mathcal{L}_{grad}(\hat{h}))) \qquad (5)$$

where $\hat{y}^t$ is the pseudo-label generated by self-similarity clustering and adjusted by relabeling at the beginning of each epoch, λ_{cri} is the balance weight of the domain critic loss.

Algorithm 1. Relabeling Algorithm

Require: $(k-1)^{th}$ clustering operation result $C^{k-1} = \{c_t^{k-1}\}_{t=1}^N$, where t is the index of the $(k-1)^{th}$ clustering results.
Require: k^{th} clustering operation result $C^k = \{c_q^k\}_{q=1}^N$, where q is the index of the kth clustering results.

1: $S \leftarrow N \times N$ matrix, where $S[t, q]$ represents the quantity of same samples of c_t^{k-1} and c_q^k.
2: $L^{k-1} = \{l_t^{k-1}\}_{t=1}^N$.
3: **for** $t = 1, ..., N$ **do**
4:     $l_t^{k-1} \leftarrow$ The list of indices that would sort $S[t, :]$ in descending order.
5: **end for**
6: $M^k = \{m_t^k\}_{t=1}^N \leftarrow \{l_t^{k-1}[1]\}_{t=1}^N$.
7: $R \leftarrow$ The indices not in M^k.
8: **while** R is not empty **do**
9:     $\widetilde{M} = \{\tilde{m}_q\}_{q=1}^N$.
10:     **for** $q = 1, ..., N$ **do**
11:         $m_q' \leftarrow$ The list of indices t of $m_t^{k-1} == q$.
12:     **end for**
13:     **for** $\tilde{m}_q$ in $\widetilde{M}$ **do**
14:         **if** $length(\tilde{m}_q) > 1$ **then**
15:             Sort $\tilde{m}_q$ by the quantity of samples of $c_{\tilde{m}_q}^{k-1}$.
16:             **for** m_t^{k-1} in m_q' **do**
17:                 $m_t^{k-1} \leftarrow \min_{r \in R} l_t^{k-1}[r]$.
18:             **end for**
19:         **end if**
20:     **end for**
21:     $R \leftarrow$ The indices not in M^k.
22: **end while**
23: **return** $M^k = \{m_t^k\}_{t=1}^N$.

---

4 Experiment

4.1 Datasets

We conduct the experiments on the vehicle re-ID benchmarks, VehicleID [20] and VeRi-776 [19], and person re-ID benchmarks, Market-1501 [21] and DukeMTMC-reID [22]. In the ablation studies, we use VehicleID as the source domain and VeRi-776 as the target domain. While comparing with state of the art, all of the four datasets are used.

4.2 Implementation Details

Supervised Pre-training. As described in Sect. 3, we first train a model on the source domain by following the training strategy described in [16,17]. Specifically, we use ResNet-50 [38] as the backbone model. We resize the input images to 256×256 for vehicle images and 256×128 for person images. For each mini-batch, we randomly sample $P = 16$ identities and $K = 8$ images for each identity from the training set, so that the mini-batch size is 128. During training, we use the Adam [39] with weight decay 0.0005 to optimize the parameters for 120 epoches. The initial learning rate is set to 3×10^{-6} and rise to 3×10^{-4} after

---

Algorithm 2. Cross Domain Re-ID Training Procedure

Require: Source data $X^s = \{(x_i^s, y_i^s)\}_{i=1}^{n^s}$; Target data $X^t = \{x_j^t\}_{j=1}^{n^t}$; Minibatch size m; Critic training step n; Balance weight $\lambda_{tri}, \lambda_{cen}, \lambda_{cri}, \gamma$; Learning rate for domain critic α_1 and feature extractor α_2.

  1: Initialize feature extractor with the pre-trained model on source domain θ_g.
  2: Initialize domain critic with random weights θ_w.
  3: **while** θ_g and θ_w not converged **do**
  4:    **if** at the beginning of the epoch **then**
  5:       $H^t = \{h_j^t\}_{j=1}^{n^t}$.
  6:       **for** $j = 1, ..., n^t$ **do**
  7:          $h_j^t = f_g(x_j^t)$.
  8:       **end for**
  9:       Cluster, H^t generate the pseudo-identity labels $\hat{Y}^t = \{\hat{y}_j^t\}_{j=1}^{n^t}$.
10:       Relabel, generate the new target dataset $\hat{X}^t = \{(x_j^t, \hat{y}_j^t)\}_{j=1}^{n^t}$.
11:    **end if**
12:    Sample minibatch $\{(x_i^s, y_i^s)\}_{i=1}^{m}$ from X^s, $\{(x_j^t, \hat{y}_j^t)\}_{j=1}^{m}$ from $\hat{X}^t$.
13:    **for** $t = 1, ..., n$ **do**
14:       $h^s \leftarrow f_g(x^s), h^t \leftarrow f_g(x^t)$.
15:       Sample h as the random points along straight lines between h^s and h^t.
16:       $\hat{h} \leftarrow \{h, h^s, h^t\}$.
17:       $\theta_w \leftarrow \theta_w + \alpha_1 \nabla_{\theta_w}[\mathcal{L}_{wd}(x^s, x^t) + \gamma \mathcal{L}_{grad}(\hat{h})]$.
18:    **end for**
19:    $\theta_g \leftarrow \theta_g - \alpha_2 \nabla_{\theta_g}[\mathcal{L}_{ide}(x^s, y^s) + \mathcal{L}_{ide}(x^t, \hat{y}^t) + \lambda_{cri}\mathcal{L}_{wd}(x^s, x^t)]$.
20: **end while**
21: **return** θ_g.

---

10 epochs, and it also decays by 0.1 after the 40th and 70th epoch. The balance weight of triplet loss and center loss is set as $\lambda_{tri} = 1$ and $\lambda_{cen} = 5 \times 10^{-4}$.

Cross-Domain Training. Most of the settings of cross-domain training are the same as the supervised pre-training except the follows. For each iteration, we sample the source dataset mini-batch and target dataset mini-batch separately. Both of the mini-batches are sampled with $P = 16$ identities and $K = 4$ images for each identity from the training set, so that their mini-batch sizes are all 64. We feed the feature before the global average pooling of ResNet-50 into the domain critic network, which consists of 4 convolutional layers. While training the domain critic network, we set the critic training step $n = 5$, the weight of the gradient penalty $\gamma = 10$. Moreover, the weight of the feature adversarial loss is set as $\lambda_{cri} = 1$. For self-similarity clustering, we extract the feature after the

Table 1. Results of different ablation studies. All of the results are obtained by using VehicleID as the source domain and VeRi-776 as the target domain. "Supervised Training" represents training the supervised model by using the label of the target domain. "Direct Transfer" means directly applying the source-trained model on the target domain. "Adv. Learn" and "Sim. Clu." means feature adversarial learning and self-similarity clustering, respectively

Method	Train. Set for Ide. Loss	Num Clus.	Relabel	mAP	R-1	R-5
Supervised train (Upper Bound)	Target (Real Label)	-	-	68.1	93.4	96.8
Direct transfer	Source	-	-	24.0	66.2	73.8
Adv. Learn	Source	-	-	30.1	68.9	76.9
Sim. Clu.	Target	576		30.3	72.5	77.9
		576	✓	31.4	76.6	83.4
	Target + Source	576		35.0	74.2	79.0
		576	✓	35.6	72.9	77.4
		100		26.7	69.4	75.4
		100	✓	27.1	69.7	75.9
		500		35.5	73.2	77.8
		500	✓	36.2	73.5	78.0
		1000		37.7	75.9	81.1
		1000	✓	39.8	75.8	81.3
		2000		37.8	76.7	82.3
		2000	✓	41.7	79.0	83.7
		3000		36.9	77.3	82.8
		3000	✓	40.7	79.0	83.3
Adv. Learn + Sim. Clu.	Target + Source	2000	✓	**44.9**	**82.5**	**87.4**

global average pooling and apply PCA [37] to reduce the dimension from 2048 to 256. So the time of clustering is much less than the model training. Our model is implemented on PyTorch [40] platform and trained with four NVIDIA RTX 2080Ti GPUs. All our experiments on different datasets follow the same settings as above.

4.3   Ablation Study

Usefulness of Feature Adversarial Learning. As shown in Table 1, feature adversarial learning can obtain 7% mAP gain compared with "Direct Transfer". Besides, when combining with self-similarity clustering, feature adversarial learning can still obtain 3.2% mAP gain.

Effectiveness of Using Source Data. Next, we compare the effect of using source data to compute the identification loss. As shown in Table 1, "Train. Set for Ide. Loss" is "Target" means that we only use the target data and its pseudo-identity labels to compute the identification loss, while "Target + Source" means that we add the identification loss computed by the source data together. The experimental results show that using source data can obtain 5% mAP gain approximately . The experimental results show that even the model has been pre-trained by the source domain data, it is still needed in the cross-domain training procedure.

Table 2. Comparison of state-of-the-art unsupervised domain adaptation methods for vehicle re-ID on VeRi-776 dataset. **Bold** indicates the best

Method	mAP	Rank-1	Rank-5
CycleGAN [35]	25.1	65.3	72.6
SPGAN [5]	34.8	71.3	77.8
DAVR [8]	26.4	62.2	73.7
Ours	**44.9**	**82.5**	**87.4**

Choice of the Number of Clusters. At first, we set the number of clusters as the real number of the identities of the target domain dataset, which is 576 for VeRi-776 dataset. Then we test the performance while changing the number of clusters to 100, 500, 1000, 2000, and 3000. We find that, whatever cluster number we set, we can always improve the domain adaptation performance. And the results are best on 2000. The results show that the number of clusters is an important factor in the framework, but the best choice is different from the real number of identities of the target domain. The same trend is also observed in [12,41]. Since the intra-class variation of re-ID problem can be quite large due to factors like viewpoint, illumination, it's explainable that there are several clusters within one class, which is also the motivation of many metric learning methods like [42]. Specifically, we set 2000 for VeRi-776, 20000 for VehicleID, 2000 for Market1501, and 4000 for DukeMTMC-ReID.

Effectiveness of Relabeling Algorithm. We compare the performance between the models with and without the relabeling algorithm. The experimental results show that the relabeling algorithm can provide consistent improvement on different numbers of clusters. The gain also rises with the number of clusters growing because the label mismatch will be more serious with the number of clusters growing. The relabeling algorithm can bring 3.9% mAP gain when the number of clusters is 2000, which is the best setting.

4.4   Comparison with State of the Art

Vehicle Re-ID. We compare our method with previous state-of-the-art unsupervised domain adaptation methods for vehicle re-ID on VeRi-776 and VehicleID datasets. First, we compare our method with some image-translation methods, including CycleGAN [35] and SPGAN [5]. Besides, we also compare our method with DAVR [8], which is a customized framework for cross-domain vehicle re-ID. The experimental results are shown in Table 2 and Table 3, and the CMC curves are shown in Fig. 3, which show the effectiveness of our method.

Table 3. Comparison of state-of-the-art unsupervised domain adaptation methods for vehicle re-ID on VehicleID dataset. **Bold** indicates the best

Method	Test Set = 800			Test Set = 1600		
	mAP	Rank-1	Rank-5	mAP	Rank-1	Rank-5
CycleGAN [35]	39.3	32.7	46.9	37.9	31.3	45.2
SPGAN [5]	61.8	54.8	69.7	59.4	52.6	66.8
DAVR [8]	54.0	49.5	68.7	49.7	45.2	64.0
Ours	**64.4**	**59.0**	**69.8**	**62.9**	**56.7**	**69.0**
Method	Test Set = 2400			Test Set = 3200		
	mAP	Rank-1	Rank-5	mAP	Rank-1	Rank-5
CycleGAN [35]	33.4	26.9	40.0	31.9	26.0	38.1
SPGAN [5]	56.7	49.5	64.4	54.7	47.7	61.9
DAVR [8]	45.2	40.7	59.0	42.9	38.7	55.9
Ours	**59.6**	**53.1**	**67.2**	**56.9**	**50.0**	**64.4**

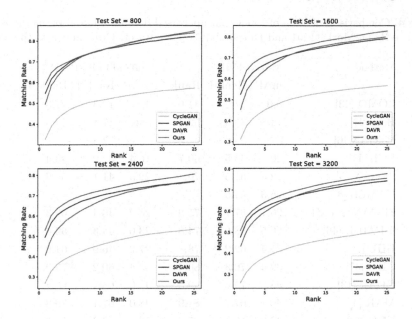

Fig. 3. CMC curves of comparison results on VehicleID dataset

Person Re-ID. Next, we train our model for person re-ID, and test the performance on Market1501 [21] and DukeMTMC-ReID [22] datasets. Similar to SSG, we divide the feature into the upper part and bottom part. Then we cluster the features of the whole bodies, upper parts, and bottom parts to obtain three pseudo-identity labels per sample. We compute the losses on the three pseudo-identity labels separately and add them up to compute the total identification loss on the target domain. The comparisons are shown in Table 4. Compared to the leading performance brought by MMCL, our method brings about 1.7% of mAP improvement on Market1501 and 4.9% of mAP improvement on DukeMTMC-ReID. Compared to the improvement on vehicle re-ID, this improvement may seem marginal. However the improvement brought by other methods is about the same. For example, PCB-R-PAST improved the performance by 0.6% on Market-1501 (54.6% *v.s.* 53.7%) and 5.3% on DukeMTMC (54.3% *v.s.* 49.0%). SSG improved the performance by 4.6% on Market-1501 (58.3% *v.s.* 53.7%) and 4.7% on DukeMTMC (53.4% *v.s.* 49.0%). Thus the improvement shown in Table 4 is sufficient to validate the effectiveness of our proposed method.

Table 4. Comparison of state-of-the-art unsupervised domain adaptation methods for person re-ID on Market1501 and DukeMTMC-reID dataset. **Bold** indicates the best

Method	Market1501			DukeMTMC-ReID		
	mAP	Rank-1	Rank-5	mAP	Rank-1	Rank-5
LOMO [43]	8.0	27.2	41.6	4.8	12.3	21.3
Bow [21]	14.8	35.8	52.4	8.3	17.1	28.8
PTGAN [6]	-	38.6	-	-	27.4	-
PUL [11]	20.5	45.5	60.7	16.4	30.0	43.4
SPGAN [5]	22.8	51.5	70.1	22.3	41.1	56.6
CAMEL [45]	26.3	54.5	-	-	-	-
SPGAN + LMP [5]	26.7	57.7	75.8	26.2	46.4	62.3
TJ-AIDL [46]	26.5	58.2	74.8	23.0	44.3	59.6
HHL [7]	31.4	62.2	78.8	27.2	46.9	61.0
ARN [47]	39.4	70.3	80.4	33.4	60.2	73.9
UDAP [13]	53.7	75.8	89.5	49.0	68.4	80.1
MAR [9]	40.0	67.7	81.9	48.0	67.1	79.8
ENC [10]	43.0	75.1	87.6	40.4	63.3	75.8
UCDA-CCE [48]	34.5	64.3	-	36.7	55.4	-
PCB-R-PAST [49]	54.6	78.4	-	54.3	72.4	-
SSG [3]	58.3	80.0	90.0	53.4	73.0	80.6
MMCL [50]	60.4	**84.4**	**92.8**	51.4	72.4	**82.9**
Sim. Clu. (Ours)	59.6	81.3	90.4	54.7	75.3	82.1
Adv. Learn + Sim. Clu. (Ours)	**61.7**	83.4	90.8	**56.3**	**77.5**	82.7

5   Conclusion

In this paper, we propose a novel unsupervised domain adaptation framework for re-ID problem with feature adversarial learning and self-similarity clustering. The feature adversarial learning and self-similarity clustering module are jointly optimized and are complementary for learning an efficient feature space for unsupervised cross domain re-ID task. The experimental results show the effectiveness of our method on both vehicle re-ID and person re-ID problems.

References

1. Shen, J., Qu, Y., Zhang, W., et al.: Wasserstein distance guided representation learning for domain adaptation. In: Proceedings of the AAAI Conference on Artificial Intelligence, New Orleans, pp. 4058–4065 (2018)
2. Sun, B., Feng, J., Saenko, K.: Return of frustratingly easy domain adaptation. In: Proceedings of the AAAI Conference on Artificial Intelligence, Phoenix, pp. 2058–2065 (2016)

3. Fu, Y., Wei, Y., Wang, G., et al.: Self-similarity grouping: a simple unsupervised cross domain adaptation approach for person re-identification. In: Proceedings of the IEEE Conference on Computer Vision and Pattern Recognition, Seoul, pp. 6111–6120 (2019)
4. Goodfellow, I.J., Pouget-Abadie, J., Mirza, M., et al.: Generative adversarial nets. In: Proceedings of Advances in Neural Information Processing Systems, Montreal, p. 27 (2014)
5. Deng, W., Zheng, L., Kang, G., et al.: Image-image domain adaptation with preserved self-similarity and domain-dissimilarity for person re-identification. In: Proceedings of the IEEE Conference on Computer Vision and Pattern Recognition, Salt Lake City, pp. 994–1003 (2018)
6. Wei, L., Zhang, S., Gao, W., et al.: Person transfer GAN to bridge domain gap for person re-identification. In: Proceedings of the IEEE Conference on Computer Vision and Pattern Recognition, Salt Lake City, pp. 79–88 (2018)
7. Zhong, Z., Zheng, L., Li, S., Yang, Y.: Generalizing a person retrieval model hetero- and homogeneously. In: Ferrari, V., Hebert, M., Sminchisescu, C., Weiss, Y. (eds.) ECCV 2018, Part XIII. LNCS, vol. 11217, pp. 176–192. Springer, Cham (2018). https://doi.org/10.1007/978-3-030-01261-8_11
8. Peng, J., Wang, H., Fu, X.: Cross Domain Knowledge Learning with Dual-branch Adversarial Network for Vehicle Re-identification. arXiv preprint arXiv:1905.00006 (2019)
9. Yu, H., Zheng, W., Wu, A., et al.: Unsupervised person re-identification by soft multilabel learning. In: Proceedings of the IEEE Conference on Computer Vision and Pattern Recognition, Long Beach, pp. 2143–2152 (2019)
10. Zhong, Z., Zheng, L., Luo, Z., et al.: Invariance matters: exemplar memory for domain adaptive person re-identification. In: Proceedings of the IEEE Conference on Computer Vision and Pattern Recognition, Long Beach, pp. 598–607 (2019)
11. Fan, H., Zheng, L., Yan, C., et al.: Unsupervised person re-identification: clustering and fine-tuning. ACM Trans. Multimed. Comput. Commun. Appl. **14**, 83 (2018)
12. Ge, Y., Chen, D., Li, H.: Mutual mean-teaching: pseudo label refinery for unsupervised domain adaptation on person re-identification. In: Proceedings of the International Conference on Learning Representations (2020, in press)
13. Song, L., Wang, C., Zhang, L., et al.: Unsupervised domain adaptive re-identification: Theory and practice. arXiv preprint arXiv:1807.11334 (2018)
14. Wen, Y., Zhang, K., Li, Z., Qiao, Yu.: A discriminative feature learning approach for deep face recognition. In: Leibe, B., Matas, J., Sebe, N., Welling, M. (eds.) ECCV 2016, Part VII. LNCS, vol. 9911, pp. 499–515. Springer, Cham (2016). https://doi.org/10.1007/978-3-319-46478-7_31
15. Arthur, D., Sergei, V.: k-means++: the advantages of careful seeding. In: Proceedings of the Eighteenth Annual ACM-SIAM Symposium on Discrete Algorithms, Society for Industrial and Applied Mathematics (2007)
16. Luo, H., Gu, Y., Liao, X., et al.: Bag of tricks and a strong baseline for deep person re-identification. In: Proceedings of The IEEE Conference on Computer Vision and Pattern Recognition Workshops (2019)
17. Luo, H., Jiang, W., Gu, Y., et al.: A strong baseline and batch normalization neck for deep person re-identification. IEEE Trans. Multimed. **22**(10), 2597–2609 (2019)
18. Schroff, F., Kalenichenko, D., Philbin, J.: FaceNet: a unified embedding for face recognition and clustering. In: Proceedings of The IEEE Conference on Computer Vision and Pattern Recognition (2015)

19. Liu, X., Liu, W., Ma, H., et al.: Large-scale vehicle re-identification in urban surveillance videos. In: Proceedings of the International Conference on Mulimedia & Expo, pp. 1–6 (2016)
20. Liu, H., Tian, Y., Wang, Y., et al.: Deep relative distance learning: tell the difference between similar vehicles. In: Proceedings of the IEEE Conference on Computer Vision and Pattern Recognition, pp. 2167–2175 (2016)
21. Zheng, L., Shen, L., Tian, L., et al.: Scalable person re-identification: a benchmark. In: Proceedings of International Conference on Computer Vision (2015)
22. Zheng, Z., Zheng, L., Yang, Y.: Unlabeled samples generated by GAN improve the person re-identification baseline in vitro. In: Proceedings of The IEEE International Conference on Computer Vision (2017)
23. Sun, Y., Zheng, L., Yang, Y., Tian, Q., Wang, S.: Beyond part models: person retrieval with refined part pooling (and a strong convolutional baseline). In: Ferrari, V., Hebert, M., Sminchisescu, C., Weiss, Y. (eds.) ECCV 2018, Part IV. LNCS, vol. 11208, pp. 501–518. Springer, Cham (2018). https://doi.org/10.1007/978-3-030-01225-0_30
24. Wang, G., Yuan, Y., Chen, X., et al.: Learning discriminative features with multiple granularities for person re-identification. In: Proceedings of ACM Multimedia Conference on Multimedia Conference, pp. 274–282 (2018)
25. Zheng, L., Huang, Y., Lu, H., et al.: Pose invariant embedding for deep person re-identification. IEEE Trans. Image Process. **28**(9), 4500–4509 (2019)
26. Zhu, K., Guo, H., Liu, Z., Tang, M., Wang, J.: Identity-guided human semantic parsing for person re-identification. In: Vedaldi, A., Bischof, H., Brox, T., Frahm, J.-M. (eds.) ECCV 2020, Part III. LNCS, vol. 12348, pp. 346–363. Springer, Cham (2020). https://doi.org/10.1007/978-3-030-58580-8_21
27. Qian, X., Qian, X., et al.: Pose-normalized image generation for person re-identification. In: Ferrari, V., Hebert, M., Sminchisescu, C., Weiss, Y. (eds.) ECCV 2018, Part IX. LNCS, vol. 11213, pp. 661–678. Springer, Cham (2018). https://doi.org/10.1007/978-3-030-01240-3_40
28. Zheng, Z., Yang, X., Yu, Z., et al.: Joint discriminative and generative learning for person re-identification. In: Proceedings of The IEEE Conference on Computer Vision and Pattern Recognition (2019)
29. Guo, H., Zhao, C., Liu, Z., et al.: Learning coarse-to-fine structured feature embedding for vehicle re-identification. In: Proceedings of AAAI Conference on Artificial Intelligence, pp. 6853–6860 (2018)
30. Guo, H., Zhu, K., Tang, M., et al.: Two-level attention network with multi-grain ranking loss for vehicle re-identification. IEEE Trans. Image Process. **28**, 4328–4338 (2019)
31. Yao, Y., Zheng, L., Yang, X., Naphade, M., Gedeon, T.: Simulating content consistent vehicle datasets with attribute descent. In: Vedaldi, A., Bischof, H., Brox, T., Frahm, J.-M. (eds.) ECCV 2020, Part VI. LNCS, vol. 12351, pp. 775–791. Springer, Cham (2020). https://doi.org/10.1007/978-3-030-58539-6_46
32. Sun, B., Saenko, K.: Deep CORAL: correlation alignment for deep domain adaptation. In: Hua, G., Jégou, H. (eds.) ECCV 2016, Part III. LNCS, vol. 9915, pp. 443–450. Springer, Cham (2016). https://doi.org/10.1007/978-3-319-49409-8_35
33. Ganin, Y., Lempitsky, V.: Unsupervised domain adaptation by backpropagation. In: Proceedings of the International Conference on Machine Learning (2017)
34. Isola, P., Zhu, J., Zhou, T., et al.: Image-to-image translation with conditional adversarial networks. In: Proceedings of the IEEE Conference on Computer Vision and Pattern Recognition (2017)

35. Zhu, J., Park, T., Isola, P., et al.: Unpaired image-to-image translation using cycle-consistent adversarial networks. In: Proceedings of International Conference on Computer Vision (2017)
36. Choi, Y., Choi, M., Kim, M., et al.: StarGAN: unified generative adversarial networks for multi-domain image-to-image translation. In: Proceedings of the IEEE Conference on Computer Vision and Pattern Recognition (2018)
37. Karl, P.: LIII. on lines and planes of closest fit to systems of points in space. London, Edinb. Dublin Philos. Mag. J. Sci. **2**, 559–572 (1901)
38. He, K., Zhang, X., Ren, S., et al.: Deep residual learning for image recognition. In: Proceedings of The IEEE International Conference on Computer Vision (2016)
39. Kingma, D., Ba, J.: Adam: a method for stochastic optimization. In: Proceedings of International Conference for Learning Representations (2015)
40. Paszke, A., Gross, S., Massa, F., et al.: PyTorch: an imperative style, high-performance deep learning library. In: Proceedings of Advances in Neural Information Processing Systems (2019)
41. Yan, X., Misra, I., Gupta, A., et al.: ClusterFit: Improving Generalization of Visual Representations. arXiv preprint arXiv:1912.03330 (2019)
42. Huang, C., Li, Y., Change Loy, C., et al.: Learning deep representation for imbalanced classification. In Proceedings of The IEEE Conference on Computer Vision and Pattern Recognition (2016)
43. Liao, S., Hu, Y., Zhu, X., et al.: Person re-identification by local maximal occurrence representation and metric learning. In: Proceedings of the IEEE Conference on Computer Vision and Pattern Recognition (2015)
44. Peng, P., Xiang, T., Wang, Y., et al.: Unsupervised cross-dataset transfer learning for person re-identification. In: Proceedings of the IEEE Conference on Computer Vision and Pattern Recognition (2016)
45. Yu, H., Wu, A., Zheng, W.: Crossview asymmetric metric learning for unsupervised person re-identification. In: Proceedings of International Conference on Computer Vision (2017)
46. Wang, J., Zhu, X., Gong, S., et al.: Transferable joint attribute-identity deep learning for unsupervised person re-identification. In: Proceedings of the IEEE Conference on Computer Vision and Pattern Recognition (2018)
47. Li, Y., Yang, F., Liu, Y., et al.: Adaptation and re-identification network: an unsupervised deep transfer learning approach to person re-identification. In: Proceedings of the IEEE Conference on Computer Vision and Pattern Recognition (2018)
48. Qi, L., Wang, L., Huo, J., et al.: A novel unsupervised camera-aware domain adaptation framework for person re-identification. In: Proceedings of The IEEE International Conference on Computer Vision (2019)
49. Zhang, X., Cao, J., Shen, C., et al.: Self-training with progressive augmentation for unsupervised cross-domain person re-identification. In: Proceedings of The IEEE International Conference on Computer Vision (2019)
50. Wang D., Zhang S.: Unsupervised person re-identification via multi-label classification. In: Proceedings of the IEEE/CVF Conference on Computer Vision and Pattern Recognition (2020)

A Framework for Jointly Training GAN with Person Re-Identification Model

Zhongwei Zhao[1], Ran Song[1(✉)] (iD), Qian Zhang[1], Peng Duan[2], and Youmei Zhang[3]

[1] School of Control Science and Engineering, Shandong University, Jinan, China
ransong@sdu.edu.cn
[2] School of Computer Science, Liaocheng University, Liaocheng, China
[3] School of Mathematics and Statistics, Qilu University of Technology, Jinan, China

Abstract. To cope with the problem caused by inadequate training data, many person re-identification (re-id) methods exploited generative adversarial networks (GAN) for data augmentation, where the training of GAN is typically independent of that of the re-id model. The coupling relation between them which probably brings in a performance gain of re-id is thus ignored. In this work, we propose a general framework to jointly train GAN and the re-id model. It can simultaneously achieve the optima of both the generator and the re-id model, where the training is guided by each other through a discriminator. The re-id model is boosted for two reasons: 1) The adversarial training that encourages it to fool the discriminator; 2) The generated samples that augment the training data. Extensive results on benchmark datasets show that for the re-id model trained with the identification loss as well as the triplet loss, the proposed joint training framework outperforms existing methods with separated training and achieves state-of-the-art re-id performance.

Keywords: Person re-identification · Generative adversarial networks · Joint training

1 Introduction

Person re-identification (re-id) is of great importance in security and surveillance. It concentrates on searching for the same person among non-overlapping cameras given a specific query person [13,37]. This problem remains unresolved due to the challenges of viewpoint variations, illumination conditions and background differences.

In recent years, person re-id has witnessed considerable improvement [3,28,34,35,38] due in part to the development of convolutional neural networks (CNN) [18,36]. Despite the remarkable performance of these methods,

Supported by the National Natural Science Foundation of China under Grant U1913204, Shandong Major Scientific and Technological Innovation Project 2018CXGC1503, and Qilu Young Scholars Program of Shandong University No. 31400082063101.

© Springer Nature Switzerland AG 2021
A. Del Bimbo et al. (Eds.): ICPR 2020 Workshops, LNCS 12664, pp. 36–51, 2021.
https://doi.org/10.1007/978-3-030-68799-1_3

the demand of large-scale annotated training data restrain their practical applications [31] as the collection and annotation of data are costly. In order to escape the laborious manual effort, generative models, especially GAN [7], have recently been introduced into re-id for the purpose of data augmentation [32,44,46]. For instance, Zheng et al. [44] adopted DCGAN to generate unlabeled images and trained the re-id networks in a semi-supervised manner, which mitigated the reliance on annotated samples. Due to the lack of annotated data, the generated samples are usually obscure and of poor visual quality, which probably misleads the re-id network. To address this problem, Qian et al. [23] enriched the diversity of the poses in training samples with a pose transfer model [33] based on conditional-GAN [11]. Domain adaption methods [5,32,46] exploited Cycle-GAN [47] to deal with the image style variations caused by the utility of different cameras, which serves as a data augmentation approach to smoothing the style disparities within the captured images or introduce new training samples from other datasets. The aforementioned methods with GANs applied as a data augmentation approach share the same two-stage pipeline as shown in Fig. 1 (a). First, a generator is trained and then the generated samples is fed into the re-id network for data augmentation. The generation process and the training of re-id model are completely separated with each other, and the improvement of the re-id model comes directly and only from the extra training images.

A natural question is that can we train the generator and the re-id model jointly to make them benefit each other, rather than simply using the generated samples to boost re-id. The method proposed by Liu et al. [19] seems like a model for jointly training, as it uses a predefined, imperfect re-id network to guide the training of a generator. However, it is noted that the guider re-id model is not the one that trained in the second stage, and the generated samples are used to train a new, desired re-ID network. Therefore, it still follows the two-stage pipeline, where the generator and re-id network are trained independently.

An intuitive way of joint training is to integrate the discriminator with the re-id model by forcing the discriminator to output class labels [26] as illustrated in Fig. 1 (b). However, such naive model will incur three problems: First, the generated images have conflict identity as the generated images should be identified as fake samples and categorized to the correct class. Thus requires specific design of the re-id network (e.g. Feature Matching GAN [4]); Second, the discriminator is employed as a classifier which will make some state-of-the-art re-id models (e.g. triplet-loss models [9]) inapplicable. Third and more importantly, the study in [4] theoretically proved that a discriminator cannot converge to the optimum when it has to play incompatible roles (e.g. a classifier and a discriminator) simultaneously.

This work aims to bridge the gap between generator and person re-id model, and make the joint training of them possible. An end-to-end framework is presented to incorporate the generator and the re-id model together for joint training and mutual promotion. In the proposed architecture, both the generator and the re-id model are subject to additional losses from each other through a discriminator. As shown in Fig. 1 (c), this forms a GAN with three components rather than a conventional one with two components. Such framework could establish

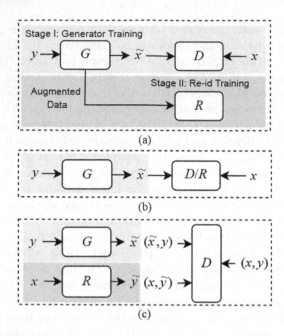

Fig. 1. Illustration of the three types of frameworks for re-id training with GAN. G, D and R denote generator, discriminator and re-id model, respectively. (a) The two-stage separate training framework; (b) An intuitive joint training framework; (c) The proposed joint training framework.

connection between the generator and the re-id model, which enables the two models to not only improve themselves by fooling the discriminator, but benefit from each other's updating.

Specifically, as shown in Fig. 1 (c), we use the re-id network to output a predicted label $\tilde{y}$ when taking an image x as input. Meanwhile, the generator behaves oppositely, which produces a generated image $\tilde{x}$ based on a true label y. The image-label pairs produced by these two branches, namely $(x, \tilde{y})$ and $(\tilde{x}, y)$, will both be regarded as fake by the discriminator. In this way, the generator and the re-id model will try to optimize themselves to fool the discriminator. And the above incompatible issue can be addressed as the discriminator plays a role of distinguishing whether an image-label pair is from the real labeled dataset or not. Therefore, benefiting from such a three-player adversarial training, the generator and the re-id model can mutually benefit each other within one framework. Besides, the well-generated samples can also be used to train the re-id model as augmented data and further yield a performance gain in terms of identification accuracy as demonstrated by our experiments.

In summary, the major contributions of the work are twofold:

- We investigate the incompatibility issue of sample generation and re-id accuracy in a GAN. An adversarial training framework is presented to make the generator and the re-id model mutually benefit each other.
- Extensive results on benchmark datasets show that the proposed approach outperforms existing GAN-based re-id methods. The re-id model can be benefited in two manners: the adversarial loss and the samples generated by the gradually improved generator.

2   Related Work

2.1   Generative Adversarial Networks

A generative adversarial networks (GAN) [7] consists of two sub-networks, a generator G and a discriminator D. The generator G outputs a sample $G(z)$ after taking in a random noise $z \sim p_z(z)$. The discriminator D decides whether a sample comes from $G(z)$ or a true data distribution. The basic mechanism of GAN is to use the generator G to fool the discriminator D.

To control the identity of the generated samples, conditional GAN [22] takes in additional conditions, such as labels or images, into the generator and discriminator to get a pre-defined output. Conditional GAN is widely used for pose transfer [21,33] or image-to-image translation [11]. CycleGAN [47] has two generator-discriminator pairs, which can transfer images between two domains.

Salimans et al. [26] proposed the discriminator D of GAN is used as a classifier by modifying the output layer. However, such framework can hardly improve the classification accuracy and the quality of generated samples at the same time. This problem was investigated in [4], which proved that the goals of classification and discrimination indeed cannot achieve optimal simultaneously, because D plays two incompatible roles, a discriminator and a classifier, at the same time. The two conflict goals prevent the convergence of the whole network. To address this problem, Li et al. [12] proposed Triple-GAN, which introduces a classifier into GAN, instead of modifying the discriminator to be a classifier as in [26]. This framework allows the generator and classifier to converge to their optimal state at the same time.

2.2   Person Re-Identification

Most recent state-of-the-art re-id models were presented based on CNN, which can be classified into two models: identification loss model and triplet loss model [43]. The identification loss model [15,41] treats re-id as a classification problem, which uses the trained model to extract the features of testing sets, and adopts a specific distance metric, such as Euclidean distance, to measure the distances between queries and galleries. The triplet loss model [2,9] aims to develop a distance metric with a deep learning architecture. In this work, we employ the identification loss model and triplet loss model as baselines, and provide an adversarial training strategy to enable the generator and re-id model to benefit each other's training.

2.3   GANs for Person Re-Identification

In the existing GAN-based re-id methods, GAN is mainly used to generate extra samples to boost the training of re-id models. These methods can be categorized into two types: 1) Generate new samples from the original datasets. For instance, Zheng *et al.* [44] employed DCGAN [24] to generate fake images, which are fed into the re-id networks together with labeled real images. To improve the discriminative capacity of re-id networks, Liu *et al.* [19] generated images for each identity using conditional GAN. 2) Transfer images from one domain to another. This is mainly achieved by CycleGAN [47]. In [46], CycleGAN was trained to transfer images into different camera styles, and thus prevented the overfitting of re-id network. With CycleGAN, Deng *et al.* [5] developed some similarity and dissimilarity restrictions for unsupervised learning. Wei *et al.* [32] separated the foreground from background to avoid the influence of background noise and thus achieved a better domain transformation performance. All aforementioned methods follow the conventional two-stage pipeline, *i.e.*, first train a generator and then feed the generated samples into the re-id network for the purpose of data augmentation. The training processes of generator and re-id model are completely separated and independent with each other.

Besides, there existed some methods recently which try to combine a generator with an identification model by sharing the feature extraction network for the purpose of feature distilling. Then, such network will be split and serve as the enhanced feature extractor in a re-id framework. Ge*et al.* [6] proposed a Feature Distilling Generative Adversarial Network (FD-GAN) to learn identity-related and pose-unrelated representations based on a Siamese structure. Li *et al.* [16] focused on open-world re-id problem, which has the different problem settings with ours. Open world re-id is to re-identify or track only a handful of target people on a watch list (gallery set). The feature extractor in this work learns to tolerate the attack by imposters by sharing weights with a binary classifier, which distinguishes whether the training images is an imposters or not.

3   Proposed Joint Training Framework

We introduce the joint training framework based on a typical conditional GAN. Given a ground-truth person image-label pair (x, y), the generator takes the label y and outputs a generated image $\tilde{x}$. The discriminator is supposed to distinguish the generated example $(\tilde{x}, y)$ from the true image-label pair (x, y). In the proposed joint training framework, a re-id model is introduced and transforms such a two-player game into a three-player game. Re-id models, which generally predict a label $\tilde{y}$ for an input person image, also can form a image-label pair $(x, \tilde{y})$. Such that the objective of discriminator evolves from distinguishing single image to distinguishing image-label pairs.

When the example x is misclassified by the re-id model (*i.e.* $y \neq \tilde{y}$), the output image-label pair $(x, \tilde{y})$ produced by the re-id model should be recognized as fake by the discriminator. For better understanding, we can regard the misclassification of re-id model as its attempt to fool the discriminator with $(x, \tilde{y})$.

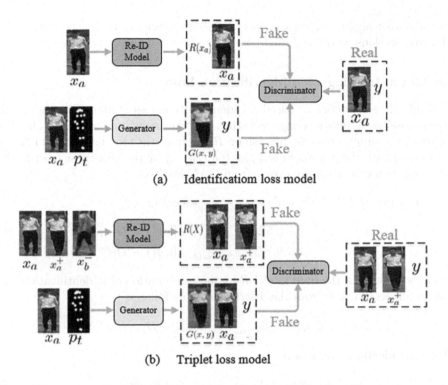

(a) Identificatiom loss model

(b) Triplet loss model

Fig. 2. Illustration of proposed joint training frameworks for identification loss model and triplet loss model.

The discriminator guides the training of re-id model by correcting its misclassification on the most confusing person identity. Such a guidance is possible because the discriminator learns jointly from the true and fake image-label pairs (*i.e.*, (x, y), $(\tilde{x}, y)$ and $(x, \tilde{y})$), and gradually increase its capability to distinguish the misclassification by re-id model. Inversely, from the perspective of generator, it will be refined to generate images matching the label information. Thus we provide three inputs for the discriminator (*i.e.*, the ground truth pair (x, y), the fake pair $(\tilde{x}, y)$ from the generator and the fake pair $(x, \tilde{y}$ from re-id model), as illustrated in Fig. 1(c). When the framework achieves optima, the three image-label pairs: (x, y), $(\tilde{x}, y)$ and $(x, \tilde{y})$, are supposed to be identical from the perspective of re-id, indicating that the re-id model and generator learn the right mappings form *image to label* and from *label to image*, respectively.

In the following, we present the construction details of the joint training framework by taking two different typical re-id models as example, identification loss model and triplet loss model. The generator is borrowed from the state-of-the-art one in [19,23], as our work focuses on proving the benefits of the joint training instead of proposing a new generator architecture. In detail, the generator G used in our work transfers the input person image x to a new image $\tilde{x} = G(x, y, p_t)$ by assigning another pose p_t while keeping the identity of the person y unchanged. For convenience, we adopt a brief form $\tilde{x} = G(x, y)$ in the following sections as the assigned pose p_t is unrelated to the training process.

3.1 Joint Training with Identification Loss

A standard re-id model R based on identification loss is shown in Fig. 2 (a), which classifies a person image x with a specific identity $\tilde{y} = R(x)$. The discriminator is supposed to distinguish $(x, R(x))$ from the true labeled data (x, y). Combining with the generated samples $(G(x, y), y)$ by the generator branch, the value function can naturally evolve from GAN loss to

$$\min_{R,G} \max_{D} \; U(R, G, D) = \mathbb{E}_{(x,y) \sim p_{data}(x,y)}[\log D(x, y)]$$
$$+ 0.5\mathbb{E}_{(x,y) \sim p_{data}(x,y)}[\log(1 - D(x, R(x)))] \qquad (1)$$
$$+ 0.5\mathbb{E}_{(x,y) \sim p_{data}(x,y)}[\log(1 - D(G(x, y), y))].$$

Hence, we can train the re-id model R by the combination of a identification loss and an adversarial loss from the discriminator as follows:

$$\mathcal{L}(R) = \mathcal{L}_{id}(R) + \mathbb{E}_{x \sim p_{data}(x)}[\log(1 - D(x, R(x)))], \qquad (2)$$

where the identification loss is

$$\mathcal{L}_{id}(R) = -\mathbb{E}_{x,y \sim p_{data}(x,y)} \log R(y|x). \qquad (3)$$

As proved in [12], Eq. (1) will enforce R and G to converge to their optima at the same time. When such optimal states are achieved, the three inputs for discriminator, namely, (x, y), $(G(x, p_t), y)$ and $(x, R(x))$ will be indistinguishable from each other, and the re-id model could be assumed to learn the correct mapping from image to label. The results will be given in the experimental section to show that the adversarial term in Eq. (2) improves re-id effectively. More details of the training procedure will be provided in the supplementary.

3.2 Joint Training with Triplet Loss

In this section, we discuss the case where the re-id model is trained with a triplet loss [9], which generally yields better re-id performance in the literature. As shown in Fig. 2 (b), unlike the identification loss, the triplet loss requires that the re-id model takes three images as input, which are a person image x_a (treated as an anchor with label a), a positive example x_a^+ (having the same label a as the anchor x_a) and a negative example x_b^- (having a different label b with the anchor x_a). Rather than predicting a specific class in the identification loss case, the only output of the triplet loss model is the distance between the inputs, including an intra-class distance $d(x_a, x_a^+)$ and an inter-class distance $d(x_a, x_b^-)$.

To construct a formally identical input for the discriminator in joint training and fit with the methodology of triplet loss, we form the image pair $X \in \{(x_a, x_a^+), (x_a, x_b^-)\}$ and use a new ground truth y for X:

$$y = \begin{cases} 1, & \text{if } X = (x_a, x_a^+) \\ -1, & \text{if } X = (x_a, x_b^-). \end{cases} \tag{4}$$

The image-label pair from the re-id model $(x, \tilde{y}) = (x, R(x))$ in the above can be formulated in $(X, \tilde{y}) = (X, R(X))$:

$$\tilde{y} = R(X) = tanh[10 * (d(x_a, x_b^-) - d(x_a, x_a^+) - m)]y, \tag{5}$$

where sgn denotes the sign function. Such that $\tilde{y}$ can be treated as an evaluation of the re-id model upon the triplet input. Therefore, the image-label pair from the generator $(\tilde{x}, y) = (G(x, y), y)$ is rewritten as $(\tilde{X}, y) = ((x_a, G(x_a, y)), y)$. For convenience, we only show one possible training status in Fig. 2 (b). More details of the training procedure will be provided in the supplementary.

Similar to the case in identification loss, the adversarial loss is defined by

$$\min_{R,G} \max_{D} U(R, G, D) = \mathbb{E}_{(X,y) \sim p_{data}(X,y)}[\log D(X, y)]$$
$$+ 0.5\mathbb{E}_{(X \sim p_{data}(X))}[\log(1 - D(X, R(X)))] \tag{6}$$
$$+ 0.5\mathbb{E}_{(x_a,y) \sim p_{data}(x_a,y)}[\log(1 - D((x_a, G(x_a, y)), y))].$$

Therefore, the loss function of the triplet loss re-id model is the combination of a triplet loss and an adversarial loss:

$$\mathcal{L}(R) = \mathcal{L}_{tri}(R) + \mathbb{E}_{X \sim p_{data}(X)}[\log(1 - D(X, R(X)))], \tag{7}$$

where

$$\mathcal{L}_{tri}(R) = \sum_{\substack{x_a, x_a^+, x_b^- \\ a \neq b}} max(0, m + d_{x_a, x_a^+} - d_{x_a, x_b^-}). \tag{8}$$

4   Experimental Results

In this section, we conduct experiments to test the effectiveness of the proposed method and made comparison with state-of-the-art approaches. The re-id performance is evaluated with mean Average Precision (mAP) and rank-1 accuracy on two large benchmarks, including Market-1501 [40] and DukeMTMC-reID [25]. Besides, MARS [39] is employed as auxiliary data to enhance the training of generator. The three datasets are detailed as follows.

Market-1501. This dataset contains 32,668 images, and each identity has several images from six disjoint cameras. The training and testing sets contain 750 and 751 identities, respectively. There are average 17.2 images per identity in the training set. In testing, 3,368 images from 750 identities are used as queries.

Table 1. The re-id performance of Adv. *vs.* Adv. + Aug.

Methods	Market-1501		DukeMTMC	
	mAP	rank-1	mAP	rank-1
Basel.(R)	47.78	73.90	44.99	65.22
Basel.(R) + Adv.	54.31	79.90	47.40	67.70
Basel.(R) + Adv. + Aug.	61.02	83.32	51.14	72.17
Basel.(R, Tri)	59.62	79.47	52.13	74.66
Basel.(R, Tri) + Adv.	61.17	82.10	54,22	76.89
Basel.(R, Tri) + Adv. + Aug.	62.41	84.15	55.03	77.57

DukeMTMC-reID. This dataset is a large multi-camera dataset. In total, there are 36,411 labeled images, and more than 2,700 people were labeled with unique identities in 8 cameras. The training set consists of 16,522 images from 702 identities, and testing set contains 17,661 images from 702 identities. 2,228 of them are used as query images.

MARS. This dataset is an extension of Market-1501. It is the first large scale video based person re-id dataset. All the bounding boxes and tracklets are generated automatically. In our setting, we use MARS as an auxiliary dataset, which provides us abundant paired pose-changed pictures.

4.1 Implementation Details

Identification Loss Models. To make fair comparison with [19,23,44], which are typical two-stage training frameworks, two types of identification loss baselines are tested in the experiments. The network is realized by the convolutions of ResNet-50 [8] (denoted as **Basel.(R)** with fine-tuned parameters on re-id. A fully connected layer, whose size is corresponding to the number of identities in each dataset, is added after convolution. Images are randomly horizontal flapped and resized to 256×256, then randomly cropped to 224×224 before being fed into the network. We use stochastic gradient descent (SGD) to train re-id models with a batch size of 128. We train 300 epochs in total. The initial learning rate is set to 0.001 and reduced by a factor of 0.1 after 25 epochs. For testing, we extract 2048-dim features before the fully connected layer for each person image.

Triplet Loss Models. We adopt triplet loss baselines using ResNet-50 (denoted as **Basel.(R, Tri)**) for ablation studies. The architectures of the network are unchanged in the triplet loss case. Besides, we use a more strong baseline [20] whose backbone is also the ResNet-50 (denoted as **Basel.(SR, Tri)**) for comparing to these state-of-the-art re-id methods. Images are randomly horizontal flapped and resized to 256×128. The model is optimized by Adam optimizer, with a base learning rate of 2e−4. We train 300 epochs in total. The margin m is set to 0.3.

Table 2. The re-id performance of separate training *vs.* joint training.

Methods	Market-1501		Duke-MTMC	
	mAP	rank-1	mAP	rank-1
Basel.(R)	47.78	73.90	44.99	65.22
Separate training	58.02	79.85	48.54	68.71
Joint training	61.02	83.32	51.14	72.17
Basel.(R, Tri)	59.62	79.47	52.13	74.66
Separate training	61.04	82.95	54.04	76.22
Joint training	62.41	84.15	55.03	77.57

GAN Architecture. We adopt the pose transfer generator architecture proposed by [21] and set the number of convolution blocks to 5. For discriminator D, we adopt the same architecture as DCGAN [24]. Besides, in D, we concatenate labels y after every block. For generator and discriminator, we use Adam optimizer with $\beta_1 = 0.5$ and $\beta_2 = 0.999$. The initial learning rate is set to $1e - 4$, then reduced by a factor of 0.5 when achieving steady state. During training, the input images of generator and discriminator are resized to 128×64. Besides, we enhance generator by introducing an auxiliary discriminator. Following the training strategy for generator of our main competitor [19], we use MARS dataset to enhance the generator. and thus introduce an auxiliary discriminator. Here we assume that MARS has same data distribution as the target test dataset, such that the auxiliary discriminator would not influence the training of the proposed framework.

4.2 Ablation Study

We conduct experiments to analyze the behavior of the joint training framework with different settings in re-id model, backbone network and training strategy.

Adversarial Loss. Our first attempt to train the re-id model in the joint framework using the adversarial loss only and without using the augmented data from the generator, such that the additional information only comes from the adversarial training with GAN. As shown in Table 1 (*Adv.* means that the re-id model is trained with the adversarial loss without data augmentation. *Aug.* means that the re-id model is trained with trained augmented data from the generator.), this leads to a promising improvement above the baseline, indicating that the joint training indeed refines the mapping from *image to label*. Moreover, the improvement from the adversarial loss only (without data augmentation) is even competitive to or superior over the results from the conventional data augmentation methods like [19, 44] (see Table 3), indicating the effectiveness of the proposed joint framework.

Table 3. Comparison to state-of-the-art methods. Group 1: re-id models without GAN. Group 2: conventional two-stage separate training frameworks based on GAN. Group 3: the proposed method with different backbone networks and re-id losses. The best, second best and third best results are highlighted in red, blue and green, respectively.

Methods	Market-1501		DukeMTMC	
	mAP	rank-1	mAP	rank-1
Methods without GAN				
Basel.(R) [41]	47.78	73.90	44.99	65.22
Basel.(R,Tri)	59.62	79.47	52.13	74.66
Basel.(SR, Tri) [20]	85.90	94.50	76.40	86.40
IDE [46]	65.87	85.66	51.83	72.31
R-50-A [23]	69.32	87.26	52.48	72.80
LOMO + XQDA [17]	–	–	17.04	30.75
Re-ranking [45]	63.63	77.11	–	–
Multiloss [14]	64.40	83.90	–	–
TriNet [9]	69.14	84.92	–	–
SVDNet [29]	62.10	82.30	–	–
PCB [30]	77.40	92.30	66.10	81.70
HA-CNN [15]	75.70	91.20	63.80	80.50
MLFN [1]	74.30	90.00	62.80	81.00
DuATM [27]	76.62	91.42	64.58	81.82
Separate training methods				
Basel.(R) + LSRO [44]	56.23	78.06	47.13	67.68
Basel.(R) + dMpRL [10]	58.59	80.37	48.58	68.24
Basel.(R) + PT [19]	57.68	79.75	48.06	68.64
IDE + ST [46]	71.55	89.49	57.61	78.32
R-50-A + PN-GAN [23]	72.58	89.43	53.20	73.58
DG-Net [42]	86.00	94.80	74.80	86.60
Our joint training methods				
Basel.(R) + ours	61.02	83.32	51.14	72.17
Basel.(R,Tri) + ours	62.41	84.15	55.03	77.57
Basel.(SR, Tri) + ours	87.60	95.10	77.00	88.00

Data Augmentation. The next attempt is to utilize the new samples generated in the joint training to improve re-id performance. This is because the generator could produce extra training samples, known as fake images, that can be used to train re-id networks further. As shown in Table 1, by introducing such data augmentation into training the re-id model, the performance is further improved compared to that without data augmentation, indicating that the adversarial loss and data augmentation could improve re-id in different ways.

Condition Image

Generated Samples

Fig. 3. The sample images in the red and blue boxes are generated by joint training and separate training, respectively. (Color figure online)

Joint Training vs. *Separated Training.* We compare the performance of the re-id model with joint training and separated training. For the separated training, we use our framework to generate augmented samples, then feed these samples as well as the original training data to another untrained re-id model. The comparison results are shown in Table 2. Hence, in the process of separated training, the re-id model is boosted only by the well-generated samples. By contrast, in addition to the augmented data that enriches the training samples, the joint training makes the re-id model learn a better mapping from image to label through the adversarial loss. Taking the results of Basel.(R) on Market-1501 as example, our re-id gains can be obtained in two ways: 1) The joint training increases 3.00% mAP and 3.47% rank-1 accuracy from an additional adversarial loss imposed to re-id network; 2) 11.24% mAP and 5.95% rank-1 gains from data augmentation by the generated fake samples. Combining 1) and 2), our framework increases 14.24% mAP and 9.42% rank-1 accuracy compared to the baseline. The observations clearly demonstrate the superiority of the proposed joint learning framework over conventional separate training one. Also, as shown in Fig. 3 the images generated by joint training have a better performance compared with separate training.

4.3 Comparison to State of the Art

We compare our results on Market-1501 and DukeMTMC to those of the recent state-of-the-art re-id methods which use GANs for data augmentation. These methods include a domain adaptation method [46], two pose-transfer method [19,23] and two semi-supervised methods [10,44]. As aforementioned, they share the same two-stage pipeline, and propose to train the generator and re-id network separately. The results in Table 3 show that our joint training framework generally perform better than those two-stage work, especially when using the strong ResNet-50 (denoted as **Basel.(SR, Tri)**) as backbone. It is noted IDE*+ST [46] and PN-GAN [23] perform better than ours sometimes (e.g., using ResNet-50 or DenseNet-169 ais the backbone), because of using very different designs of network model and loss function.

Moreover, we also make comparison with the existing re-id methods without GANs. The results are shown on the upper part of Table 3. It can be seen that our method outperforms most of them, except the three elaborate ones in [1,15,27]. Recall that our goal is to provide a joint training framework to make the generator and re-id model compatible and mutually beneficial, so the performance can be further improved by combining with more advanced models.

5   Conclusion

In this paper, we proposed to train the re-id and the GAN models jointly. Compared to the scheme where they are trained separately, the joint training improves the performance of both the re-id model and the GAN by considering the feedback through the discriminator. Extensive results on benchmark datasets proved the superiority of the proposed method over existing GAN-based re-id methods.

References

1. Chang, X., Hospedales, T.M., Xiang, T.: Multi-level factorisation net for person re-identification. In: The IEEE Conference on Computer Vision and Pattern Recognition (CVPR) (June 2018)
2. Chen, W., Chen, X., Zhang, J., Huang, K.: Beyond triplet loss: a deep quadruplet network for person re-identification. In: The IEEE Conference on Computer Vision and Pattern Recognition (CVPR), vol. 2 (2017)
3. Chen, Y.C., Zhu, X., Zheng, W.S., Lai, J.H.: Person re-identification by camera correlation aware feature augmentation. IEEE Trans. Pattern Anal. Mach. Intell. **40**(2), 392–408 (2018)
4. Dai, Z., Yang, Z., Yang, F., Cohen, W.W., Salakhutdinov, R.R.: Good semi-supervised learning that requires a bad GAN. In: Advances in Neural Information Processing Systems, pp. 6510–6520 (2017)
5. Deng, W., Zheng, L., Kang, G., Yang, Y., Ye, Q., Jiao, J.: Image-image domain adaptation with preserved self-similarity and domain-dissimilarity for person reidentification. In: Proceedings of the IEEE Conference on Computer Vision and Pattern Recognition (CVPR), vol. 1, p. 6 (2018)
6. Ge, Y., et al.: FD-GAN: pose-guided feature distilling GAN for robust person re-identification. In: Advances in Neural Information Processing Systems, pp. 1229–1240 (2018)
7. Goodfellow, I., et al.: Generative adversarial nets. In: Advances in Neural Information Processing Systems, pp. 2672–2680 (2014)
8. He, K., Zhang, X., Ren, S., Sun, J.: Deep residual learning for image recognition. arXiv preprint arXiv:1512.03385 (2015)
9. Hermans, A., Beyer, L., Leibe, B.: In defense of the triplet loss for person re-identification. arXiv preprint arXiv:1703.07737 (2017)

10. Huang, Y., Xu, J., Wu, Q., Zheng, Z., Zhang, Z., Zhang, J.: Multi-pseudo regularized label for generated data in person re-identification. IEEE Trans. Image Process. **28**(3), 1391–1403 (2018)
11. Isola, P., Zhu, J.Y., Zhou, T., Efros, A.A.: Image-to-image translation with conditional adversarial networks. arXiv preprint (2017)
12. Li, C., Xu, T., Zhu, J., Zhang, B.: Triple generative adversarial nets. In: Advances in Neural Information Processing Systems, pp. 4088–4098 (2017)
13. Li, J., Zhang, S., Tian, Q., Wang, M., Gao, W.: Pose-guided representation learning for person re-identification. IEEE Trans. Pattern Anal. Mach. Intell. 1 (2019)
14. Li, W., Zhu, X., Gong, S.: Person re-identification by deep joint learning of multi-loss classification. In: Proceedings of the 26th International Joint Conference on Artificial Intelligence, pp. 2194–2200. AAAI Press (2017)
15. Li, W., Zhu, X., Gong, S.: Harmonious attention network for person re-identification. In: CVPR, vol. 1, p. 2 (2018)
16. Li, X., Wu, A., Zheng, W.-S.: Adversarial open-world person re-identification. In: Ferrari, V., Hebert, M., Sminchisescu, C., Weiss, Y. (eds.) ECCV 2018, Part II. LNCS, vol. 11206, pp. 287–303. Springer, Cham (2018). https://doi.org/10.1007/978-3-030-01216-8_18
17. Liao, S., Hu, Y., Zhu, X., Li, S.Z.: Person re-identification by local maximal occurrence representation and metric learning. In: Proceedings of the IEEE Conference on Computer Vision and Pattern Recognition, pp. 2197–2206 (2015)
18. Liu, H., Feng, J., Qi, M., Jiang, J., Yan, S.: End-to-end comparative attention networks for person re-identification. IEEE Trans. Image Process. **26**(7), 3492–3506 (2017)
19. Liu, J., Ni, B., Yan, Y., Zhou, P., Cheng, S., Hu, J.: Pose transferrable person re-identification. In: Proceedings of the IEEE Conference on Computer Vision and Pattern Recognition, pp. 4099–4108 (2018)
20. Luo, H., Gu, Y., Liao, X., Lai, S., Jiang, W.: Bag of tricks and a strong baseline for deep person re-identification. In: Proceedings of the IEEE Conference on Computer Vision and Pattern Recognition Workshops (2019)
21. Ma, L., Jia, X., Sun, Q., Schiele, B., Tuytelaars, T., Van Gool, L.: Pose guided person image generation. In: Advances in Neural Information Processing Systems, pp. 406–416 (2017)
22. Mirza, M., Osindero, S.: Conditional generative adversarial nets. arXiv:1411.1784 (2014)
23. Qian, X., et al.: Pose-normalized image generation for person re-identification. In: Ferrari, V., Hebert, M., Sminchisescu, C., Weiss, Y. (eds.) ECCV 2018, Part IX. LNCS, vol. 11213, pp. 661–678. Springer, Cham (2018). https://doi.org/10.1007/978-3-030-01240-3_40
24. Radford, A., Metz, L., Chintala, S.: Unsupervised representation learning with deep convolutional generative adversarial networks. arXiv preprint arXiv:1511.06434 (2015)
25. Ristani, E., Solera, F., Zou, R., Cucchiara, R., Tomasi, C.: Performance measures and a data set for multi-target, multi-camera tracking. In: Hua, G., Jégou, H. (eds.) ECCV 2016, Part II. LNCS, vol. 9914, pp. 17–35. Springer, Cham (2016). https://doi.org/10.1007/978-3-319-48881-3_2
26. Salimans, T., Goodfellow, I., Zaremba, W., Cheung, V., Radford, A., Chen, X.: Improved techniques for training GANs. In: Advances in Neural Information Processing Systems, pp. 2234–2242 (2016)

27. Si, J., et al.: Dual attention matching network for context-aware feature sequence based person re-identification. In: The IEEE Conference on Computer Vision and Pattern Recognition (CVPR) (June 2018)

28. Su, C., Yang, F., Zhang, S., Tian, Q., Davis, L.S., Gao, W.: Multi-task learning with low rank attribute embedding for multi-camera person re-identification. IEEE Trans. Pattern Anal. Mach. Intell. **40**(5), 1167–1181 (2018)

29. Sun, Y., Zheng, L., Deng, W., Wang, S.: SVDNet for pedestrian retrieval. arXiv preprint **1**, 6 (2017)

30. Sun, Y., Zheng, L., Yang, Y., Tian, Q., Wang, S.: Beyond part models: person retrieval with refined part pooling (and a strong convolutional baseline). In: Ferrari, V., Hebert, M., Sminchisescu, C., Weiss, Y. (eds.) ECCV 2018, Part IV. LNCS, vol. 11208, pp. 501–518. Springer, Cham (2018). https://doi.org/10.1007/978-3-030-01225-0_30

31. Wang, X., Zheng, W.S., Li, X., Zhang, J.: Cross-scenario transfer person reidentification. IEEE Trans. Circuits Syst. Video Technol. **26**(8), 1447–1460 (2016)

32. Wei, L., Zhang, S., Gao, W., Tian, Q.: Person transfer GAN to bridge domain gap for person re-identification. In: Proceedings of the IEEE Conference on Computer Vision and Pattern Recognition, pp. 79–88 (2018)

33. Yan, Y., Xu, J., Ni, B., Zhang, W., Yang, X.: Skeleton-aided articulated motion generation. In: Proceedings of the 2017 ACM on Multimedia Conference, pp. 199–207. ACM (2017)

34. Zhang, W., He, X., Lu, W., Qiao, H., Li, Y.: Feature aggregation with reinforcement learning for video-based person re-identification. IEEE Trans. Neural Netw. Learn. Syst. **30**(12), 3847–3852 (2019)

35. Zhang, W., He, X., Yu, X., Lu, W., Zha, Z., Tian, Q.: A multi-scale spatial-temporal attention model for person re-identification in videos. IEEE Trans. Image Process. **29**, 3365–3373 (2020)

36. Zhang, W., Hu, S., Liu, K., Zha, Z.: Learning compact appearance representation for video-based person re-identification. IEEE Trans. Circuits Syst. Video Technol. **29**(8), 2442–2452 (2019)

37. Zhang, W., Ma, B., Liu, K., Huang, R.: Video-based pedestrian re-identification by adaptive spatio-temporal appearance model. IEEE Trans. Image Process. **26**(4), 2042–2054 (2017)

38. Zhao, R., Oyang, W., Wang, X.: Person re-identification by saliency learning. IEEE Trans. Pattern Anal. Mach. Intell. **39**(2), 356–370 (2017)

39. Zheng, L., et al.: MARS: a video benchmark for large-scale person re-identification. In: Leibe, B., Matas, J., Sebe, N., Welling, M. (eds.) ECCV 2016, Part VI. LNCS, vol. 9910, pp. 868–884. Springer, Cham (2016). https://doi.org/10.1007/978-3-319-46466-4_52

40. Zheng, L., Shen, L., Tian, L., Wang, S., Wang, J., Tian, Q.: Scalable person re-identification: a benchmark. In: Proceedings of the IEEE International Conference on Computer Vision, pp. 1116–1124 (2015)

41. Zheng, L., Yang, Y., Hauptmann, A.G.: Person re-identification: Past, present and future. arXiv preprint arXiv:1610.02984 (2016)

42. Zheng, Z., Yang, X., Yu, Z., Zheng, L., Yang, Y., Kautz, J.: Joint discriminative and generative learning for person re-identification. In: Proceedings of the IEEE Conference on Computer Vision and Pattern Recognition, pp. 2138–2147 (2019)

43. Zheng, Z., Zheng, L., Yang, Y.: A discriminatively learned CNN embedding for person reidentification. ACM Trans. Multimed. Comput. Commun. Appl. (TOMM) **14**(1), 13 (2017)

44. Zheng, Z., Zheng, L., Yang, Y.: Unlabeled samples generated by GAN improve the person re-identification baseline in vitro (Oct 2017)
45. Zhong, Z., Zheng, L., Cao, D., Li, S.: Re-ranking person re-identification with k-reciprocal encoding. In: CVPR (2017)
46. Zhong, Z., Zheng, L., Zheng, Z., Li, S., Yang, Y.: Camera style adaptation for person re-identification. In: Proceedings of the IEEE Conference on Computer Vision and Pattern Recognition, pp. 5157–5166 (2018)
47. Zhu, J.Y., Park, T., Isola, P., Efros, A.A.: Unpaired image-to-image translation using cycle-consistent adversarial networks (Oct 2017)

Interpretable Attention Guided Network for Fine-Grained Visual Classification

Zhenhuan Huang[1], Xiaoyue Duan[1], Bo Zhao[1], Jinhu Lü[1],
and Baochang Zhang[1,2(✉)]

[1] Beihang University, Beijing, China
[2] Shenzhen Academy of Aerospace Technology, Shenzhen, China
{16231192,17375262,zhaobo0706,bczhang}@buaa.edu.cn, jhlu@iss.ac.cn

Abstract. Fine-grained visual classification (FGVC) is challenging but more critical than traditional classification tasks. It requires distinguishing different subcategories with the inherently subtle intra-class object variations. Previous works focus on enhancing the feature representation ability using multiple granularities and discriminative regions based on the attention strategy or bounding boxes. However, these methods highly rely on deep neural networks which lack interpretability. We propose an Interpretable Attention Guided Network (IAGN) for fine-grained visual classification. The contributions of our method include: i) an attention guided framework which can guide the network to extract discriminitive regions in an interpretable way; ii) a progressive training mechanism obtained to distill knowledge stage by stage to fuse features of various granularities; iii) the first interpretable FGVC method with a competitive performance on several standard FGVC benchmark datasets.

Keywords: FGVC · Interpretable attention · Knowledge distillation · Progressive training mechanism

1 Introduction

Recently, a steady progress has been achieved in generic object recognition with the help of both large-scale annotated datasets and sophisticated model design. However, it is still a challenging task to recognize fine-grained object categories (e.g., bird species [22], car models [12] and aircraft [14]) which attract extensive research attention. Fine-grained objects are visually similar in global structure by a rough glimpse, while they can be classified into different categories when looking into details, so learning discriminative feature representations from pivotal parts matters in fine-grained image recognition. Existing fine-grained recognition

The work was supported in part by National Natural Science Foundation of China under Grants 62076016 and 61672079. This work is supported by Shenzhen Science and Technology Program KQTD2016112515134654. Baochang Zhang is the correspondence author who is also with Shenzhen Academy of Aerospace Technology, Shenzhen, China.

A. Del Bimbo et al. (Eds.): ICPR 2020 Workshops, LNCS 12664, pp. 52–63, 2021.
https://doi.org/10.1007/978-3-030-68799-1_4

methods can be divided into two groups. One group firstly locates the discriminative parts of the object and then classifies based on the discriminative regions. Additional bounding box annotations on objects or parts which cost a fortune to collect are commonly required in these two-stage methods [2,11,16]. The other group manages to automatically lead model to focus on discriminative regions via an attention mechanism in an unsupervised manner, which neglects extra annotations. However, these methods [8,17,27,28] usually need additional network structure (e.g. attention mechanism), and thus generate attention without any interpretability.

In this paper, a novel fine-grained image recognition framework named Interpretable Attention Guided Network (IAGN) is introduced together with a progressive training mechanism. In addition to the standard classification backbone network, we introduce an interpretable attention generation method to automatically learn discriminative regions, as shown in Fig. 1. An input image is first carefully augmented by shuffling patches to emphasize discriminative local details. Interpretation attention generation method automatically localizes discriminative regions using in-place back propagation. On the other hand, our data augmentation method and progressive training mechanism further lead our model to recognize from global structure to local details. Moreover, we also introduce a knowledge distilling mechanism to teach the lower network layers with soft targets generated by a higher network layers, which have a broader receptive field and encode higher-level semantics. Main contributions of this paper can be summarized as follows:

- An Interpretable Attention Guided Network (IAGN) is introduced for fine-grained visual recognition. It generates attention to lead our model to localize discriminative regions in an interpretable manner;
- A progressive training mechanism is obtained to distill knowledge stage by stage to fuse features of various granularities;
- Our IAGN achieves new state-of-the-art or competitive performances on all three standard FGVC benchmark datasets.

2 Related Work

2.1 Fine-Grained Classification

Fine-grained image classification methods have been largely improved thanks to the latest development and research findings of convolutional neural networks (CNNs). While some methods attempt to obtain a better visual representation directly from the original image, other techniques try to locate the discriminative regions or parts and learn their features based on the attention generated by the network. Compared with the earlier part/attention based methods, recent research focus has shifted from strongly-supervised learning with annotations of key areas [1,11,26], to weakly-supervised learning with only the supervision of category labels [8,23,28].

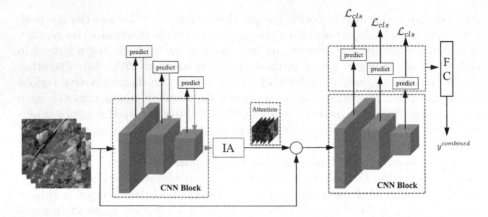

Fig. 1. Overview of the interpretable attention guided network. Green and blue blocks denote CNN blocks which share same weights. IA block denotes the interpretable attention generation block. The data that flows along orange arrow represents the process of interpretable attention generation which is not needed in back propagation. Back propagation only flows along black arrows. (Color figure online)

Recent studies based on weakly-supervised learning mainly address attention to finding the most discriminative parts, more complementary parts and parts of multiple granularities. In order to integrate and fuse information from these discriminative parts better, some fusion methods are put forward. Fu et al. [8] find that region detection and fine-grained feature learning can promote each other, and thus build a reinforced attention proposal network to obtain discriminative attention regions and multi-scale feature representation based on these regions. Zheng et al. [28] apply a channel grouping network to jointly learn part proposals and feature representations on each part, and classify these features to predict the categories of the input image. Sun et al. [17] propose an attention based network, which first apply a one-squeeze multi-excitation module and then put forward a multi-attention multi-class constraint to help to extract multiple region features. Yang et al. [23] introduce a novel self-supervision mechanism which locates informative regions effectively without bounding boxes and part annotations.

Inspired by these previous studies, we propose a progressive training mechanism which can distill knowledge stage by stage to fuse features from different granularities and enhance the classification performance. Besides, inspired by the jigsaw puzzle solution, which has been utilized in previous works [5,19,21] and can split the images into pieces to help the network exploit local regions, we adopt a data augmentation method so that our network would focus more on the discriminative local parts.

2.2   Interpretable Neural Networks

Neural Network has achieved huge success in many fields including computer vision, natural language processing and so on these years. However, Neural Network has always been regarded as a "black box" lacking interpretability - we give the network an input, and then get a decision-making result as a feedback, but nobody knows clearly about the decision-making process. Owing to this, it is difficult to convince users of the reliability of Neural Network, resulting in many constraints in its application, especially in security sensitive fields.

The interpretability of neural networks can be divided mainly into two categories: ante-hoc interpretability and post-hoc interpretability. Many recent studies have focused on the latter, which promotes our understanding of neural networks by attempting to interpret trained network models. Zeiler et al. [24] use deconvolutional networks to visualize what patterns activate each unit. Zhou et al. [30] utilize global average pooling in CNN to generate Class Activation Maps (CAM), visualizing discriminative regions which CNN draws attention to when classifying the images. Later they further propose a framework called "Network Dissection" [29], which quantifies the interpretability of CNN by evaluating the corresponding relationship between a single hidden unit and a series of semantic concepts.

Our method is based on the Gradient-weighted Class Activation Mapping (Grad-CAM) [18] method. This technique produces visual explanations for discriminative region decisions of the network, thus making it more interpretable.

3   Method

In this section, the proposed Interpretable Attention Guided Network (IAGN) is described. As shown in Fig. 1, the whole framework of our IAGN includes four parts, which are detailedly described as below.

3.1   Data Augmentation Method

In natural language processing [4], shuffling the order of sequence would help the neural network find discriminative words while neglecting irrelevant ones. Similarly, in the FGVC task where local features (more details) instead of global features determine the classification result, shuffling regions of image would promote neural networks to learn from discriminative region details. As shown in Fig. 2, our data augmentation method is proposed to disrupt the spatial layout of local image regions. Given an input image I, we first uniformly partition the image into $N \times N$ patches denoted by matrix R. $R_{i,j}$ denotes an image patch where i and j are the horizontal and vertical indices respectively ($1 \leq i, j \leq N$). In order to destruct global structure but avoid destroying semantics to some extent, patches would be shuffled in their 2D neighbourhood. For the ith row of R, a new position vector p_i of size N is generated, where the ith element $q_{i,j} = i + d$, where $d \sim U(-k, k)$ is a random variable following a uniform

Fig. 2. Example images for fine-grained recognition and the corresponding shuffled images by data augmentation method.

distribution in the range of $[-k, k]$. Here, k is a hyperparameter $1 \leq k < N$ defining the neighbourhood range. Then we sort the position vector and get a new permutation σ_i^{row} of patches in i^{th} row subjected to:

$$\forall j \in 1, 2, \ldots, N, |\sigma_i^{row}(j) - j| \leq 2k, \tag{1}$$

where $\sigma_i^{row}(j)$ denotes new vertical index of original image patch $R_{i,j}$.

Similarly, for column j, we can get a permutation σ_j^{col} of patches in j^{th} column subjected to:

$$\forall i \in 1, 2, \ldots, N, |\sigma_j^{col}(i) - i| \leq 2k, \tag{2}$$

where $\sigma_j^{col}(i)$ denotes new horizontal index of original image patch $R_{i,j}$.

Therefore, the original image patch at location (i, j) will be placed at location $(\sigma_i^{row}(j), \sigma_j^{col}(i))$. Till now, our data augmentation method has destructed the global structure and ensured that the local region jitters inside its neighbourhood with a tunable hyperparameter. Since the global structure has been destructed, to recognize these randomly shuffled images, the classification network has to find the discriminative regions and learn the tiny differences among categories.

3.2   Interpretable Attention

To endow the network with the ability to extract discriminative regional features, [7,8,28] crop image or generate attention map via subordinate network for the part localization. To interpret the localization process, we utilize the Grad-CAM [18] technique in our network to generate an interpretable attention.

Convolutional layers naturally retain the spatial information which is lost in fully-connected layers. The neurons in these layers extract semantic class-specific

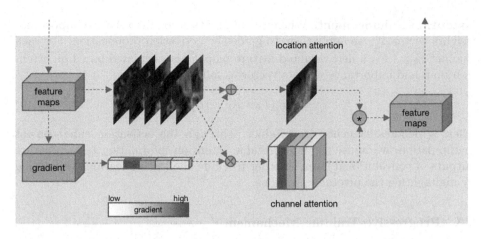

Fig. 3. The framework of interpretable attention generation block is composed of location attention and channel attention. $\otimes$ denotes channel wise product operation. $\oplus$ denotes channel wise product and add operation. $*$ denotes element wise product.

information in the image (object parts). We use the Grad-CAM technique to obtain the 'importance values' to each neuron in certain layers through back-propogation of the gradient information for a particular decision of interest. Then we will construct our attention map via these 'importance values' which can be interpreted as contribution of each neuron for the final decision. As shown in Fig. 3, in order to obtain the class-discriminative attention map $A_k^c \in R^{u \times v}$ of width u and height v for any class C, we first compute the gradient of the score for class C, y^c (before the softmax), with respect to feature map activations F_k of the convolutional layer at stage k, i.e. $\frac{\partial y^c}{\partial F_k}$. These gradients flowing back are global-average-pooled over the width and height dimensions (indexed by i and j respectively) to obtain the channel wise importance weights vector α_k^c:

$$\alpha_k^c = \frac{1}{Z} \sum_i \sum_j \frac{\partial y^c}{\partial F_k}, \tag{3}$$

where i^{th} element of α_k^c denotes importance value of ith channel in feature map activations F_k.

After generating channel-wise importance weights vector α_k^c, we apply it to feature maps and sum in channel dimension to obtain attention map A_k^c:

$$A_k^c = \sum_{ch=1}^{N_k} softmax(\alpha_k^c) \odot F_k, \tag{4}$$

where $\odot$ denotes channel wise product, and N_k denotes the number of channels in feature map activations F_k.

Attention Enhancement. When we obtain attention map A_k^c and importance weights vector α_k^c, we apply them to guide the activation propogation. For each channel $F_{k,i} \in F_k$, a new weighted feature map $\hat{F}_k$ is calculated based on attention map and importance weights vector as follows:

$$\hat{F}_k = (F_k * A_k^c) \odot \alpha_k^c, \tag{5}$$

where $*$ denotes element wise product. Through the activation enhancement manipulation, we infuse the interpretable attention information to the feature outputs of convolutional layer, in order to guide the feature learning processing by highlighting the pivotal activations.

3.3 Progressive Training Mechanism

As an analogy to recognition process of human, we adopt a progressive training mechanism where we train our network from the higher stage to the lower stage progressively. At the higher stage, the larger receptive field and stronger representation ability enable the network to represent high level semantics (say global structure). While at lower stage, receptive field and representation ability are limited, the network would be forced to exploit discriminative information from local details (i.e., object textures). Compared to training the whole network directly, this progressive training mechanism allows the model to take a glance at global structure first and then locate discriminative information from local regions for further prediction instead of learning all the granularities simultaneously.

For the training of the outputs from each stages and the output from the concatenated features, we adopt cross entropy (CE) $\mathcal{L}_{CE}$ between ground truth label y and prediction probability distribution for loss computation as follows:

$$\mathcal{L}_{CE}(y^l, y) = -\sum_c y_c * log(y_c^l), \tag{6}$$

and

$$\mathcal{L}_{CE}(y^{concat}, y) = -\sum_c y_c * log(y_c^{concat}). \tag{7}$$

As depicted in Algorithm 1, at each iteration, a batch of data D will be used for S steps, and we only train the output of a certain stage at each step in series except for the first step when we concatenate all outputs. As steps go on, the scale of data augmentation increases.

Inference. During the inference, only the original images will be input into our model and the data augmentation method is unnecessary. In this case, the model combines outputs of the last three stages to obtain y^{concat} for final prediction as shown in Fig. 1. For the ensemble purpose, the prediction from various stages contains the information of different granularity, which leads to a better performance when we combine all outputs together with equal weights.

Algorithm 1. Progressive Training

Require:
    The set of training images for current batch, D_n;
    The set of labels for current batch, y_n;
    Neural network on former batches, E_n;
    Number of progressive training steps, N
Ensure:
    Neural network on current batch, E_{n+1}
1: **for** each $i \in [1, N]$ **do**
2:     generate shuffled images D'_n using training batch D_n;
3:     compute classification scores y_i for current batch at stage s_i;
4:     compute cross entropy loss $\mathcal{L}^i_{CE} = -\sum_c y_c * log(y^i_c)$
5:     compute gradients and update parameters;
6: **end for**
7: $E_n \rightarrow E_{n+1}$
8: **return** E_{n+1}

Knowledge Distillation. Considering our progressive training mechanism, it is natural to embed knowledge distillation [10] in our training process. The intention of proposed progressive training strategy is to learn from global structure to local feature for finer recognition. With the help of knowledge distillation, transferring knowledge from stage to stage can be facilitated by a wild margin.

4 Experiments

We evaluate the performance of our proposed IAGN on three standard fine-grained object recognition datasets: CUB-200–2011 (CUB) [22], Stanford Cars (CAR) [12] and FGVC-Aircraft (AIR) [14]. We do not use any bounding box/part annotations in all our experiments.

4.1 Implementation Details

We perform all experiments using PyTorch [15] with version higher than 1.3 over a cluster of GTX 2080 GPUs. We evaluate our proposed method on the widely used backbone network for classification, ResNet-50 [9]. This network is pre-trained on ImageNet dataset. The category label of the image is the only annotation used for training. The input images are resized to a fixed size of 512 × 512 and are randomly cropped into 448 × 448. Random rotation and horizontal flip are applied for data augmentation. All above settings are standard in the literature. For all the experiments in this paper, prediction head is plugged into the last three convolutional stages in ResNet-50 backbone, and outputs of these three stages will be concatenated. For training, shuffled scale for data augmentation is set to [1, 2, 4, 8] while concatenated outputs and outputs from stage 5, 4, 3 are used for back propagation respectively at each step. We train our model for up to 150 epochs with batch size as 10, weight decay as 0.0005 and a momentum as 0.9. We use stochastic gradient descent (SGD) optimizer and batch normalization as the regularizer reduced by following the cosine annealing schedule. When testing, image shuffling is disabled, and combined output is the only feature used for location attention generation and final prediction. The input images are center cropped and then fed into the backbone classification network for final predictions.

4.2 Performance Comparison

The results on CUB-200-2011, Stanford Cars, and FGVC-Aircraft are presented in Table 1. Both the accuracy of y^{concat} and the combined accuracy of all four outputs are listed.

We achieve competitive results on this dataset in a much easier experimental procedure, since only one network is needed during testing. The end-to-end feature encoding methods achieve good performance on birds, while their advantages diminish when dealing with rigid objects. The localization and classification subnets achieve competitive performance on various datasets, usually with a large number of network parameters. For instance, the RA-CNN [8] consists of three independent VGGNets and two localization sub-networks. By comparison, without extra annotations, our end-to-end approach achieves state-of-the-art and performs consistently well on both rigid and non-rigid objects. Our method outperforms RA-CNN [8] and MGE-CNN [25] by 3.8% and 0.6%, even though they build several different networks to learn information of various granularities, and train the classification of each network separately and then combine their information for testing. This result proves the advantage and validity of our method which exploit multi-granularity information gradually in one network.

Table 1. Comparison results on three different standard datasets.

Method	Backbone	Accuracy (%)		
		CUB-200-2011	*Stanford Cars*	*FGVC-aircraft*
FT ResNet [20]	ResNet50	84.1	91.7	88.5
B-CNN [13]	ResNet50	84.1	91.3	84.1
KP [6]	VGG16	86.2	92.4	86.9
RA-CNN [8]	VGG19	85.3	92.5	–
MA-CNN [28]	VGG19	86.5	92.8	89.9
MC-Loss [3]	ResNet50	87.3	93.7	92.6
DCL [10]	ResNet50	87.8	94.5	**93.0**
MGE-CNN [25]	ResNet50	88.5	93.9	–
S3N [7]	ResNet50	88.5	94.7	92.8
IAGN	ResNet50	88.7	94.0	91.8
IAGN (combined)	ResNet50	**89.1**	**94.8**	92.5

4.3 Visualization

We visualize the feature maps of the last three convolution layers in Fig. 4, and we can find that the feature map responses of IAGN are concentrated in discriminative regions. At different stages, the discriminative parts can be consistently highlighted by IAGN model, which demonstrates that our IAGN method is robust. Furthermore, Fig. 4 obviously shows that our model cares about global structure at higher stages and focuses on local details (say discriminative parts) at lower stages. This exactly meets our expectations of the critical advantage of our proposed mechanism.

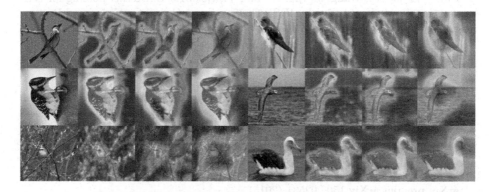

Fig. 4. Visualization of interpretable attention generated by the last three stages of our proposed IAGN. The second, third and forth columns corresponds to the attention of the third, the forth and the last stage respectively.

5 Conclusion

In this paper, we propose a novel network named IAGN for fine-grained visual classification. The attention guided framework automatically localizes discriminative regions in an interpretable way. Besides, our data augmentation method and progressive training mechanism further lead the network to implement classification in a global-to-local pattern. Furthermore, knowledge distillation is introduced stage by stage to improve the performance of feature fusion of various granularities. Our method does not require extra regions supervision information and can be trained end-to-end. Extensive experiments against state-of-the-art methods exhibit the superior performances of our method on various fine-grained recognition tasks while maintaining an excellent interpretability.

References

1. Berg, T., Belhumeur, P.N.: POOF: part-based one-vs.-one features for fine-grained categorization, face verification, and attribute estimation. In: IEEE Conference on Computer Vision and Pattern Recognition, pp. 955–962 (2013)

2. Berg, T., Liu, J., Woo Lee, S., Alexander, M.L., Jacobs, D.W., Belhumeur, P.N.: Birdsnap: large-scale fine-grained visual categorization of birds. In: IEEE Conference on Computer Vision and Pattern Recognition, pp. 2011–2018 (2014)
3. Chang, D., et al.: The devil is in the channels: mutual-channel loss for fine-grained image classification. IEEE Trans. Image Process. **29**, 4683–4695 (2020)
4. Chen, Y., Bai, Y., Zhang, W., Mei, T.: Destruction and construction learning for fine-grained image recognition. In: IEEE Conference on Computer Vision and Pattern Recognition, pp. 5157–5166 (2019)
5. Cho, T.S., Avidan, S., Freeman, W.T.: A probabilistic image jigsaw puzzle solver. In: IEEE Conference on Computer Vision and Pattern Recognition, pp. 183–190. IEEE (2010)
6. Cui, Y., Zhou, F., Wang, J., Liu, X., Lin, Y., Belongie, S.: Kernel pooling for convolutional neural networks. In: IEEE Conference on Computer Vision and Pattern Recognition, pp. 2921–2930 (2017)
7. Ding, Y., Zhou, Y., Zhu, Y., Ye, Q., Jiao, J.: Selective sparse sampling for fine-grained image recognition. In: IEEE International Conference on Computer Vision (2020)
8. Fu, J., Zheng, H., Mei, T.: Look closer to see better: recurrent attention convolutional neural network for fine-grained image recognition. In: IEEE Conference on Computer Vision and Pattern Recognition, pp. 4438–4446 (2017)
9. He, K., Zhang, X., Ren, S., Sun, J.: Deep residual learning for image recognition. In: IEEE Conference on Computer Vision and Pattern Recognition, pp. 770–778 (2016)
10. Hinton, G., Vinyals, O., Dean, J.: Distilling the knowledge in a neural network. arXiv preprint arXiv:1503.02531 (2015)
11. Huang, S., Xu, Z., Tao, D., Zhang, Y.: Part-stacked CNN for fine-grained visual categorization. In: IEEE Conference on Computer Vision and Pattern Recognition, pp. 1173–1182 (2016)
12. Krause, J., Stark, M., Deng, J., Fei-Fei, L.: 3D object representations for fine-grained categorization. In: 4th International IEEE Workshop on 3D Representation and Recognition. Sydney, Australia (2013)
13. Lin, T.Y., RoyChowdhury, A., Maji, S.: Bilinear CNNs for fine-grained visual recognition. In: IEEE Transactions on Pattern Analysis and Machine Intelligence (2017)
14. Maji, S., Kannala, J., Rahtu, E., Blaschko, M., Vedaldi, A.: Fine-grained visual classification of aircraft. Technical report (2013)
15. Paszke, A., et al.: PyTorch: an imperative style, high-performance deep learning library. In: Advances in Neural Information Processing Systems, pp. 8026–8037 (2019)
16. Peng, Y., He, X., Zhao, J.: Object-part attention model for fine-grained image classification. IEEE Trans. Image Process. **27**(3), 1487–1500 (2017)
17. Rodríguez, P., Gonfaus, J.M., Cucurull, G., Roca, F.X., Gonzàlez, J.: Attend and rectify: a gated attention mechanism for fine-grained recovery. In: Ferrari, V., Hebert, M., Sminchisescu, C., Weiss, Y. (eds.) ECCV 2018, Part VIII. LNCS, vol. 11212, pp. 357–372. Springer, Cham (2018). https://doi.org/10.1007/978-3-030-01237-3_22
18. Selvaraju, R.R., Cogswell, M., Das, A., Vedantam, R., Parikh, D., Batra, D.: Grad-CAM: visual explanations from deep networks via gradient-based localization. In: IEEE International Conference on Computer Vision, pp. 618–626 (2017)

19. Son, K., Hays, J., Cooper, D.B.: Solving square jigsaw puzzles with loop constraints. In: Fleet, D., Pajdla, T., Schiele, B., Tuytelaars, T. (eds.) ECCV 2014, Part VI. LNCS, vol. 8694, pp. 32–46. Springer, Cham (2014). https://doi.org/10. 1007/978-3-319-10599-4_3

20. Wang, Y., Morariu, V.I., Davis, L.S.: Learning a discriminative filter bank within a CNN for fine-grained recognition. In: IEEE Conference on Computer Vision and Pattern Recognition, pp. 4148–4157 (2018)

21. Wei, C., et al.: Iterative reorganization with weak spatial constraints: solving arbitrary jigsaw puzzles for unsupervised representation learning. In: IEEE Conference on Computer Vision and Pattern Recognition, pp. 1910–1919 (2019)

22. Welinder, P., et al.: Caltech-UCSD Birds 200. Technical Report CNS-TR-2010-001, California Institute of Technology (2010)

23. Yang, Z., Luo, T., Wang, D., Hu, Z., Gao, J., Wang, L.: Learning to navigate for fine-grained classification. In: Ferrari, V., Hebert, M., Sminchiscscu, C., Weiss, Y. (eds.) ECCV 2018, Part XIV. LNCS, vol. 11218, pp. 438–454. Springer, Cham (2018). https://doi.org/10.1007/978-3-030-01264-9_26

24. Zeiler, M.D., Fergus, R.: Visualizing and understanding convolutional networks. In: Fleet, D., Pajdla, T., Schiele, B., Tuytelaars, T. (eds.) ECCV 2014, Part I. LNCS, vol. 8689, pp. 818–833. Springer, Cham (2014). https://doi.org/10.1007/ 978-3-319-10590-1_53

25. Zhang, L., Huang, S., Liu, W., Tao, D.: Learning a mixture of granularity-specific experts for fine-grained categorization. In: IEEE International Conference on Computer Vision, pp. 8331–8340 (2019)

26. Zhang, N., Donahue, J., Girshick, R., Darrell, T.: Part-based R-CNNs for fine-grained category detection. In: Fleet, D., Pajdla, T., Schiele, B., Tuytelaars, T. (eds.) ECCV 2014, Part I. LNCS, vol. 8689, pp. 834–849. Springer, Cham (2014). https://doi.org/10.1007/978-3-319-10590-1_54

27. Zhao, B., Wu, X., Feng, J., Peng, Q., Yan, S.: Diversified visual attention networks for fine-grained object classification. IEEE Trans. Multimed. 19(6), 1245–1256 (2017)

28. Zheng, H., Fu, J., Mei, T., Luo, J.: Learning multi-attention convolutional neural network for fine-grained image recognition. In: IEEE International Conference on Computer Vision, pp. 5209–5217 (2017)

29. Zhou, B., Khosla, A., Lapedriza, A., Oliva, A., Torralba, A.: Object detectors emerge in deep scene CNNs. arXiv preprint arXiv:1412.6856 (2014)

30. Zhou, B., Khosla, A., Lapedriza, A., Oliva, A., Torralba, A.: Learning deep features for discriminative localization. In: IEEE Conference on Computer Vision and Pattern Recognition, pp. 2921–2929 (2016)

Use of Frequency Domain for Complexity Reduction of Convolutional Neural Networks

Kamran Chitsaz[1], Mohsen Hajabdollahi[1], Pejman Khadivi[2(✉)], Shadrokh Samavi[1,3],
Nader Karimi[1], and Shahram Shirani[3]

[1] ECE Department, Isfahan University of Technology, Isfahan, Iran
[2] CS Department, Seattle University, Seattle, USA
khadivip@seattleu.edu
[3] ECE Department, McMaster University, Hamilton, Canada

Abstract. The implementation of convolutional neural networks (CNNs) is not easy because of the high number of parameters that these networks have. Researchers have applied numerous approaches to reduce the complexity of convolutional networks. Quantization of the weights and pruning are two complexity reduction methods. A new paradigm for accelerating CNNs operations and simplification of the network is to perform all the computations in the Fourier domain. Using a fast Fourier transform (FFT) can simplify the operations by converting the convolution operation into multiplication. Different approaches can be taken for the simplification of computations in FFT. Our approach in this paper is to let the CNN operate in the FFT domain by splitting the input. There are problems in the computation of FFT using small kernels. Splitting is an effective solution for small kernels. The splitting reduces the redundancy that is caused by the overlap-and-add, and hence, the network's efficiency is increased. Hardware implementation of the proposed FFT method and complexity analysis of the hardware demonstrate the proper performance of the proposed approach.

Keywords: Hardware implementation · CNN acceleration · Spectrum domain computation · Fast Fourier Transform (FFT) · Splitting

1 Introduction

Convolutional neural networks (CNN) have been shown to be effective deep learning models for machine learning problems, especially in the field of computer vision. CNNs are used in a wide range of machine learning problems such as image classification, semantic segmentation, scene understanding, and medical image analysis [1–3]. Using CNNs, a nonlinear model is trained to map an input space to a corresponding output space. This model has a large number of parameters that can cause problems in the implementation of this model. This problem is exacerbated in situations where there is a lack of available hardware resources [4].

Among a pool of parameters in CNNs, a lot of them are redundant [5]. A minor part of the network conducts a significant portion of the computations of a CNN model. During the past decade, researchers have been trying to simplify the CNNs from the

© Springer Nature Switzerland AG 2021
A. Del Bimbo et al. (Eds.): ICPR 2020 Workshops, LNCS 12664, pp. 64–74, 2021.
https://doi.org/10.1007/978-3-030-68799-1_5

computation and memory requirement perspectives and using a variety of techniques [5, 6].

Performing the convolutional computations of a CNN in the Fourier domain, i.e., using fast Fourier transform (FFT), is a promising technique that accelerates the network operation by a significant reduction in the computations [14, 15]. Using FFT has improved the resource utilization and computational time of the CNN. However, in FFT based computations, the problem of handling sizeable intermediate feature maps is less investigated.

In this paper, a new FFT-based method for improving the CNN computations during inference is proposed in which CNN process patches of an input image. As a result, FFT is computed on image parts, and better memory management can be possible, and the number of computations is reduced. By modifying the splitting size, the number of redundant operations can be reduced, and the network can be processed according to the available resources.

Numerical results demonstrate that the proposed solution significantly reduces the number of multiplications and additions in a CNN network. Furthermore, the proposed method is implemented in hardware. Hardware implementation of the proposed FFT method and complexity analysis of the hardware demonstrate the proper performance of the proposed approach.

The rest of this paper is organized as follows. Section 2 is dedicated to the literature review. In Sect. 3, a brief review of FFT based CNN processing is described. The proposed method is introduced in Sect. 4. Experimental results are shown in Sect. 5, and finally, in Sect. 6, concluding remarks are offered.

2   Literature Review

Since the emerging of different CNN models, researchers have been working on simplifying their structure and reduce their redundancy. CNNs can be simplified from different perspectives. Quantization techniques are aiming to reduce the bit width that is used in the representation of the network parameters [6, 7]. Pruning is another technique that is used to remove elements of the network that are not useful. Designers apply the pruning procedure at different levels, such as connections, nodes, channels, and filters [5, 8].

Various studies explore the design of networks with simple structures. Neural architecture search (NAS) procedures are methods that search for efficient structures manually or automatically [9, 10]. In some recent research works, the problem of design a simple structure is studied in the context of hardware implementation. In these methods, different techniques to have an efficient hardware implementation is under consideration [11]. Parallelization, data and resource sharing, and pipelining are examples of hardware-based techniques to implement a suitable structure [12, 13].

Recently, the implementation of CNNs in the spectrum domain has been substantially explored by the research community [14, 15]. Performing the computations of a CNN network in the Fourier domain can be very beneficial. Some techniques have been proposed for the efficient implementation of FFT on hardware [16, 17]. When CNN convolutional processing is performed in the FFT domain, in both training and inference phases, the number of multiply and add operations is reduced significantly. While during the training phase, FFT-based design results in faster operation of the network, in inference time, FFT can be used to design an efficient hardware implementation of the system [17, 18].

Most of the previously published research in the field of FFT-based CNN design is focused on replacing the convolution process by an element-wise production in the Fourier domain [14, 19]. In [20], with respect to the advantages of the Fourier transform's linearity, the computational complexities for 3D filters are reduced. Furthermore, in [21], a tile-based FFT is proposed to reduce the number of computations. However, in [20] and [21], FFT implementation on dedicated hardware is not addressed directly. In [17] and [18], an FFT based approach using overlap and addition method is proposed to reduce the memory access and computational complexity and implement dedicated hardware.

3 CNN Computing Based on FFT

Convolution is one of the major parts of a CNN, and a significant portion of the operations conducted by this deep learning model is dedicated to convolutions. Suppose that we have a sample input $d \in \mathbb{R}^{N \times N}$ and a sample filter weight $w \in \mathbb{R}^{K \times K}$ with $K < N$. A single convolution is described by the following equation:

$$y = d \otimes w \tag{1}$$

In which $\otimes$ is the convolution operation, d is the image, and w is the kernel. In the Fourier domain, Eq. (1) will be changed to the following equation:

$$Y = FFT(d) \odot FFT(Padding(w)) = D \odot W \tag{2}$$

where $\odot$ represents a Hadamard product, and FFT represents the fast Fourier transform:

$$D_{k,l} = \sum_{n=0}^{N-1} \sum_{m=0}^{N-1} d_{m,n} e^{-\frac{2\pi i}{N}(nl+mk)} \tag{3}$$

Padding is required to pad w with zeros to make the size of w equal to the size of d. In CNNs, $K \ll N$, and hence, substantial padding is required. FFT can be calculated based on the Cooley-Tukey algorithm with the cost of $O(N^2 \log(N))$ and the cost of the Hadamard product is N^2. Therefore, the total cost of operations in Eq. (2) is $O(N^2 \log(N))$.

When the kernel is small, Eq. (2) results in some redundancy that is explored in [16, 17], and [18]. In order to eliminate this redundancy, the Overlap-and-Add convolution (OaAconv) is proposed in the literature [16–18], which is illustrated in Fig. 1. In this method, the input image is split into patches with the same size as the CNN kernel. FFT operations are conducted on these small patches, and after entry wise multiplications, they are transformed inversely to the spatial domain using an overlap addition. In Fig. 1, overlapped regions in the output, y, are shaded in gray. As illustrated in Fig. 1, input image in this method should be split into patches with the same size of the filters, which are very small in the typical CNN structures. Also, redundancy is observed during overlapped addition, which conducted with a number comparable to the image patches. Therefore, a large number of operations are required for overlapped additions.

4 FFT Based Computation Using Splitting

In this section, we describe the proposed method and we analyze its computational complexity and memory requirements.

4.1 Proposed Method

The block diagram of the proposed method is illustrated in Fig. 2. In the first step, the input image is split into some patches. These patches can be of any size and can be selected based on the hardware constraints and other requirements. Let us assume that each split of the image is represented by d_i and the total number of splits is η. Then, the set of all the splits is represented by d, as follows:

$$\mathcal{S}_d = \{d_1, d_2, \dots, d_\eta\} \tag{4}$$

Note that the input image can be reconstructed using the concatenation operation, $\mathcal{C}$, applied to all the splits $d_i \in d$. In other words,

$$d = \mathcal{C}(\mathcal{S}_d) = \mathcal{C}(\{d_1, d_2, \dots, d_\eta\}) \tag{5}$$

Let us assume that the size of each sample patch is $S \times S$, and a filter kernel W has the size of $K \times K$. Since convolutional operations require some padding p_i, d_i should be padded by p_i and a patch set $\widetilde{d}$ with sample patch $\widetilde{d}_i$ is created as follows.

$$\widetilde{\mathcal{S}}_d = \{(d_1 \& p_1), (d_2 \& p_2), \dots, (d_\eta \& p_\eta)\} \tag{6}$$

where, $\&$ is the padding operator. Equivalently, the padded patch d_i is represented by $\widetilde{d}_i$ and we have:

$$\widetilde{\mathcal{S}}_d = \left\{ \widetilde{d}_1, \widetilde{d}_2, \dots, \widetilde{d}_\eta \right\} \tag{7}$$

Therefore, the final patch $\widetilde{d}_i$ has the size of $(S + K - 1) \times (S + K - 1)$. In other words, as illustrated in Fig. 2, patches are extracted from the input image by overlapping with the size of $\lfloor \frac{K}{2} \rfloor$. Then, the convolutional operations of Eq. (1), can be rewritten as the convolution of patches:

$$y = \widetilde{\mathcal{S}}_d \otimes w = \left\{ \widetilde{d}_1 \otimes w, \widetilde{d}_2 \otimes w, \dots, \widetilde{d}_\eta \otimes w \right\} \tag{8}$$

We still have some redundancy due to overlapping operations. We choose small size kernels in CNNs, and we can select relatively large patches, which will make the redundancy negligible. According to Fig. 2, in each step, a patch with its padding is extracted from the input image and then is transformed into the FFT domain. Also, a convolutional filter w should be transformed into the FFT domain. Due to the different size of the patches and filters, a zero-padding p_0 is also applied to the filter before the FFT operation:

$$\widetilde{w} = w \& p_0 \tag{9}$$

Input (d)   Filter (W)

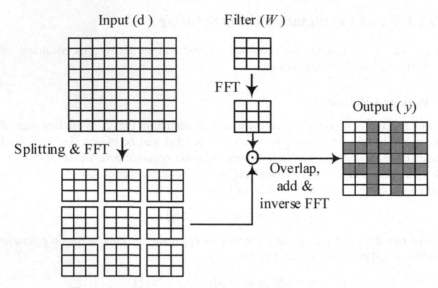

Fig. 1. Overlap and add method for an FFT-based convolution.

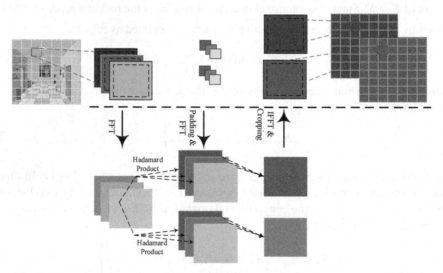

Fig. 2. The proposed procedure for the FFT-based processing of CNN using splitting.

In the Fourier domain, as discussed in Section III, convolution is substituted by Hadamard product. In the proposed method, Hadamard product is applied to each split as follows:

$$\tilde{Y}_i = FFT\left(\tilde{d}_i\right) \odot FFT\left(\tilde{w}\right) = \tilde{D}_i \odot \tilde{W} \tag{10}$$

The Hadamard production has lower complexity than a convolution procedure because Hadamard is an entry-wise process. All of the patches are handled based on Eq. (10), and inverse FFT is applied to the resulted $\tilde{Y}_i$'s.

$$\tilde{y}_i = IFFT\left(\tilde{Y}_i\right) = IFFT\left(\tilde{D}_i \odot \tilde{W}\right) \tag{11}$$

After the IFFT transform, the resulted patch is larger than its actual size. Hence, we need to crop the patch to make it suitable for concatenation. The result of this process is the *convolved* image y:

$$y = \mathcal{C}\left(\left\{Crop\left(\tilde{y}_1\right), \ldots, Crop\left(\tilde{y}_\eta\right)\right\}\right) \tag{12}$$

where, $\mathcal{C}$ is the concatenation operator and $Crop(\cdot)$ is the crop function.

The major problem in the implementation of CNN on hardware is handling feature maps. There are lots of feature maps in the CNNs, and this problem is aggravated when the input is relatively large. The issue of handling the large feature maps is not appropriately addressed in FFT-based methods. This problem is addressed in the proposed approach by using a splitting method and simultaneously reducing the number of redundant operations. In the proposed splitting approach, different splitting sizes can be used, and hence, better handling of the large feature maps is possible. Also, the splitting method can be used in case of other filters, which reduce the number of memory access to read input data.

4.2 Computational Complexity and Memory Access Analysis

If we consider Eq. (1), convolutional operations in the spatial domain need $O\left((N - K + 1)^2 K^2\right)$ multiplications and about $O\left((N - K + 1)^2 K^2\right)$ additions. Also, an element of input data (with stride 1) is accessed K times, and the total memory accesses is $O\left(N^2 + K^2\right)$ times. In FFT-based CNN networks, computation of FFT needs $O\left(N^2 \log N\right)$ number of multiplications and additions, and a cost of multiplication is N^2 for Hadamard product calculations. Hence, the total time complexity of $O\left(N^2 \log N\right)$ can be considered for the convolutional part of the CNN, in FFT-based designs. Furthermore, in this method, the memory storage requirement is $2 N^2$. In [16–18], overlap and add approach is proposed and it is stated that the total complexity of their method is $O\left(N^2 \log K\right)$, while the memory storage requirement is $O\left(N^2 + K^2\right)$.

In the proposed method, an image patch with the size of $S \times S$ is used. Therefore, the FFT procedure has the cost of $O(S^2 \log S^2)$ for every patch and there are $\frac{N^2}{S^2}$ patches in an image. Hence, the total cost is $O\left(N^2 \log S^2\right)$. The size of the memory requirement is $O(N^2 + S^2)$ and each location of memory containing data is accessed only once. If patches have the same size as the kernels, the time complexity of the proposed method is the same as the OaAConv method time complexity. However, as mentioned before, if S is larger than K some portion of the calculations performed in the proposed method will be negligible. Let us assume that $|MUL_{OaAConv}|$ be the number of multiplications that are required in the OaAConv method. Then, we have

$$|MUL_{OaAConv}| = \frac{N^2\left(2(2K - 1)^2 \log(2K - 1)^2 + (2K - 1)^2\right)}{K^2} \tag{13}$$

The above equation shows the total number of multiplications in the FFT procedure using the Cooley-Tukey algorithm, Hadamard product, and inverse FFT. In addition to the multiplications, FFT and inverse FFT procedures need some additions. Let us assume that $|ADD_{OaAConv}|$ be the total number of additions. Then, it can be shown that

$$|ADD_{OaAConv}| = \frac{N^2 \left(2(2K-1)^2 \log(2K-1)^2 + K^2 - K\right)}{K^2} \tag{14}$$

We can perform similarly for the proposed method to count the number of multiplications and additions, which we name them as $|MUL_{SplitConv}|$ and $|ADD_{SplitConv}|$. Number of multiplications can be calculated using the following equation:

$$|MUL_{SplitConv}| = \frac{N^2 \left(2(S+K-1)^2 \log(S+K-1)^2\right)}{S^2} \tag{15}$$

and number of additions can be calculated using Eq. (16):

$$|ADD_{SplitConv}| = \frac{N^2 \left(2(S+K-1)^2 \log(S+K-1)^2\right)}{S^2} \tag{16}$$

Based on Eqs. (13) to (16), it is possible to visualize and compare the total number of operations based on different parameters. Numerical results are illustrated in Figs. 3, 4, and 5.

The total number of multiplications and additions in OaAConv and the proposed method are compared in Fig. 3 when the value of K is changing from 3 to 13. In this Fig. 3, N is equal to 640 and S is equal to 16. Same quantities are illustrated in Fig. 4 for different N values. In Fig. 4, K is 9 and S is 16.

It can be observed that the number of total operations in the proposed method is less than the OaAConv method. Furthermore, by increasing the kernel size, the difference between the two approaches is increased in terms of the number of multiply and add operations. Moreover, larger values of S results in better improvement from this perspective.

From the memory analysis perspective, it is worth mentioning that by using relatively small size patches, small size feature maps are generated, and memory management problems can be handled more efficiently. In the conventional FFT method, all the input data elements should be accessed simultaneously, which causes a challenge, especially for large input images. Thanks to using the patches of input data, convolutional operations can be conducted with more independence. Also, a patch can be fetched only once, and better efficiency is achieved.

5   Experimental Results

To evaluate the performance of the proposed method, we implement our approach on a hardware platform. We first implemented the proposed FFT based convolution using Verilog hardware description language. A Xilinx FPGA XC6VLX240T is then used as a target device using Xilinx Vivado tools. The implemented codes on the FPGA are evaluated by its software counterpart using Python programming language.

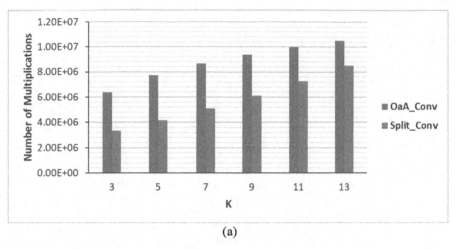

(a)

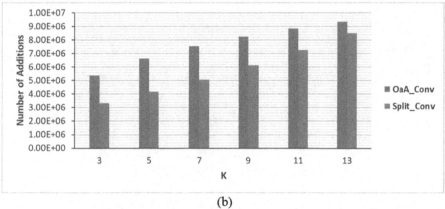

(b)

Fig. 3. (a) Number of multiplications and (b) number of additions versus K.

Table 1. Complexity of the spatial and proposed convolution with a single channel as input

	# of LUT	# of Registers	# of DSP	Latency (cycle)
Spatial Conv.	63531	74079	612	104
SplitConv.	49200	58708	256	395

For testing, 8 × 8 patches are considered for our proposed method. At first, an FFT based convolution for 8 × 8 blocks with a single channel as the input is designed. The synthesis report from our implementation is illustrated in Table 1. A convolution method in the spatial domain is also implemented, and detailed synthesis results are reported

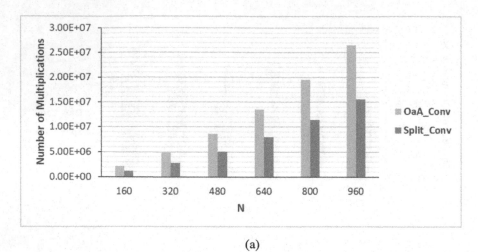

(a)

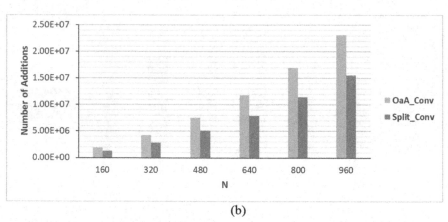

(b)

Fig. 4. (a) Number of multiplications and (b) number of additions versus N

in Table 1. In single-channel input, the proposed method has not any improvement. For a better comparison of the performance of the above two approaches, the spatial convolution is designed so that both methods need the same resources.

Based on Table 1, convolutional layers on the VGG16 network are implemented, and their run times are analyzed and compared in Fig. 5. It can be observed that the run time of our approach is shorter than the spatial domain convolution. Although in the single-channel input, as illustrated in Table 1, the proposed method has not any improvement, but by increasing the number of filter channels, a significant improvement can be observed for the proposed method.

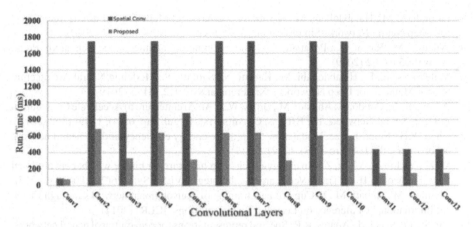

Fig. 5. Run time of implemented methods for computing convolution in VGG layers.

6 Conclusion

In this paper we proposed an efficient method for FFT based CNN processing. The proposed method splits the input image into patches, and all of the FFT related operations are conducted on them. We showed that by using the splitting method, the number of multiplications and additions was reduced compared to state of the art OaAConv. Hardware simulations demonstrated that a suitable implementation resulted. For the convolutional layers in the VGG16 network, a noteworthy improvement was observed compared to the spatial convolution. Also, the splitting approach is very beneficial for reducing the cycles and operations that convolutional networks require and prevents multiple access to the same data.

References

1. Lecun, Y., Bengio, Y., Hinton, G.: Deep learning. Nature 521)7553(, 436–444 (2015)
2. Shelhamer, E., Long, J., Darrell, T.: Fully convolutional networks for semantic segmentation. In: IEEE Conference on Computer Vision and Pattern Recognition (CVPR), pp. 3431–3440 (2015)
3. Hajabdollahi, M., Esfandiarpoor, R., Sabeti, E., Karimi, N., Soroushmehr, S.M.R., Samavi, S.: Multiple abnormality detection for automatic medical image diagnosis using bifurcated convolutional neural network. Biomed. Signal Process. Control **57**, 101792 (2020)
4. Sze, V., Chen, Y.-H., Yang, T.-J., Emer, J.S.: Efficient processing of deep neural networks: a tutorial and survey. Proc. IEEE **105**(12), 2295–2329 (2017)
5. Guo, Y., Yao, A., Chen, Y.: Dynamic network surgery for efficient DNNs. In: Advances in Neural Information Processing Systems, pp. 1379–1387 (2016)
6. Lin, D.D., Talathi, S.S., Annapureddy, V.S.: Fixed point quantization of deep convolutional networks. In: International Conference on Machine Learning, pp. 2849–2858 (2016)
7. Wu, J., Leng, C., Wang, Y., Hu, Q., Cheng, J.: Quantized convolutional neural networks for mobile devices. In: IEEE Conference on Computer Vision and Pattern Recognition (CVPR), pp. 4820–4828 (2016)

8. Pasandi, M.M., Hajabdollahi, M., Karimi, N., Samavi, S.: Modeling of Pruning Techniques for Deep Neural Networks Simplification, arXiv Prepr. arXiv2001.04062 (2020).

9. Wistuba, M., Rawat, A.., Pedapati, T.: A survey on neural architecture search, arXiv Prepr. arXiv1905.01392 (2019)

10. Malekhosseini, E., Hajabdollahi, M., Karimi, N., Samavi, S.: Modeling Neural Architecture Search Methods for Deep Networks arXiv Prepr. arXiv1912.13183 (2019)

11. Ardakani, A., Condo, C., Ahmadi, M., Gross, W.J.: An Architecture to Accelerate Convolution in Deep Neural Networks. IEEE Trans. Circuits Syst. Regul. Pap. **65**(4), 1349–1362 (2018)

12. Chen, Y.H., Emer, J., Sze, V.: dataflow to optimize energy efficiency of deep neural network accelerators. IEEE Micro **37**(3), 12–21 (2017)

13. Wang, Y., Lin, J., Wang, Z.: An energy-efficient architecture for binary weight convolutional neural networks. IEEE Trans. Very Large Scale Integr. (VLSI) Syst. 26(2), 280–293 (2017)

14. Mathieu, M., Henaff, M., LeCun, Y.: Fast training of convolutional networks through FFTS. In: International Conference on Learning Representations (ICLR) (2014)

15. Rippel, O., Snoek, J., Adams, R.P.: Spectral representations for convolutional neural networks. In: Advances in Neural Information Processing Systems, pp. 2449–2457 (2015)

16. Abtahi, T., Kulkarni, A., Mohsenin, T.: Accelerating convolutional neural network with FFT on tiny cores. IEEE International Symposium on Circuits and Systems (ISCAS), pp. 1–4 (2017)

17. Abtahi, T., Shea, C., Kulkarni, A., Mohsenin, T.: Accelerating Convolutional Neural Network with FFT on Embedded Hardware.. IEEE Trans. Very Large Scale Integr. Syst. 26(9), 1737–1749 (2018)

18. Highlander, T., Rodriguez, A.: Very Efficient Training of Convolutional Neural Networks using Fast Fourier Transform and Overlap-and-Add, arXiv preprint arXiv:1601.06815, 2015.

19. Lavin, A., Gray, S.: Fast algorithms for convolutional neural networks. In: Proceedings of the IEEE Conference on Computer Vision and Pattern Recognition, pp. 4013–4021 (2016)

20. Nguyen-Thanh, N., Le-Duc, H., Ta, D.-T., Nguyen, V.-T.: Energy-efficient techniques using FFT for deep convolutional neural networks. In: International Conference on Advanced Technologies for Communications (ATC), pp. 231–236 (2016)

21. Lin, J., Yao, Y.: A fast algorithm for convolutional neural networks using tile-based fast fourier transforms. Neural Process. Lett. **50**(2), 1951–1967 (2019). https://doi.org/10.1007/s11063-019-09981-z

From Coarse to Fine: Hierarchical Structure-Aware Video Summarization

Wenxu Li[1,2], Gang Pan[1(✉)], Chen Wang[1], Zhen Xing[1], Xiaozhou Zhou[3], Xiaoxuan Dong[1], and Jiawan Zhang[1]

[1] Tianjin University, Tianjin, China
{pangang,tjuwangchen,xiaoxuandong,jwzhang}@tju.edu.cn
[2] Imperial College London, London, UK
wl1520@ic.ac.uk
[3] Dartmouth College, Hanover, USA
Xiaozhou.Zhou@Dartmouth.edu

Abstract. Hierarchical structure is a common characteristic of some kinds of videos (e.g., sports videos, game videos): the videos are composed of several actions hierarchically and there exists temporal dependencies among segments of different scales, where action labels can be enumerated. Our ideas are based on two intuition: First, the actions are the fundamental units for people to understand these videos. Second, the process of summarization is naturally one of observation and refinement, i.e., observing segments in video and hierarchically refining the boundaries of an important action according to video hierarchical structure. Based on above insights, we generate action proposals to exploit the structure and formulate the summarization process as a hierarchical refining process. We also train a hierarchical summarization network with deep Q-learning (HQSN) to achieve the refining process and explore temporal dependency. Besides, we collect a new dataset that consists of structured game videos with fine-grain actions and importance annotations. The experimental results demonstrate the effectiveness of our framework.

Keywords: Video summarization · Reinforcement learning · Video understanding

1 Introduction

With the rapid growth of video materials on the Internet, how to effectively deal with these materials to obtain users' interest is imminent. Video summarization is an approach to generate short video clips that meet the needs of users from raw videos.

At present, there are two main categories of summarization methods: domain-specific approaches [1,17,24,34] and domain-agnostic approaches [8–10,13,22,32, 33,35,36]. Domain-specific approaches utilize some domain-prior knowledge to process videos in a specific domain (e.g., editing conventions of broadcast videos

© Springer Nature Switzerland AG 2021
A. Del Bimbo et al. (Eds.): ICPR 2020 Workshops, LNCS 12664, pp. 75–87, 2021.
https://doi.org/10.1007/978-3-030-68799-1_6

or detecting key-event), which means that transferring from one to another is difficult. Domain-agnostic approaches aim to be as agnostic as possible to the domain and are evaluated on the open-domain datasets [5,9,28] composed of user videos.

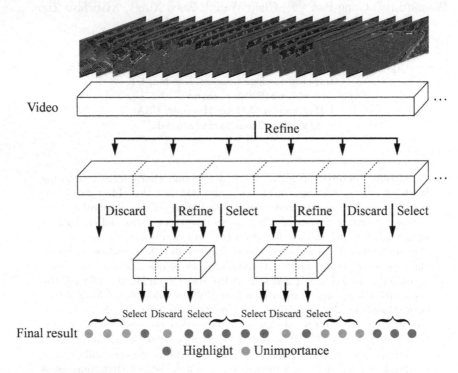

Fig. 1. Video summarization is a process of observation and refinement. Splitting the segment and refining the segment can allow us find precise boundary; Observing the current state and timely stopping the refining process (selecting and discarding in the figure) can avoid wasting unnecessary time.

The common characteristics of some types of videos are hierarchy and dependency (e.g., sports videos, game videos or tutorial videos). As we know, these videos often consist of several shots, each of which is composed of several events or actions. And strong temporal dependencies exist in structure units of different scales (e.g., excellent goals and playback shots in soccer videos). The structure information is essential to understand these videos. However, there are few approaches to achieve the influence of video hierarchical structure on the summarization results. To better bring about the hierarchical structure, our key ideas are based on two intuitions: (1) The actions are the basic units to understand. (2) The process of video summarization is a continuous and iterative procedure of observation and refinement. When humans summarize a video, browsing the whole video to find out the approximate locations of highlights is usually the first step. Then we refine the video segment hierarchically to get precise boundaries. And when it is no need to refine the current segment, humans can selectively discard or select the whole segment to stop refining.

Based on the above discussion, a hierarchical structure-aware summarization framework is proposed, which consists of two parts: **action proposal generation** and **hierarchical decision-making process** (observation and refinement). The first part utilizes the structure information directly. The second part does not use the external structure information but uses the hierarchical model that is designed to fit in the structured video. These two parts are independent of each other but both follow the structure-aware principle.

Action Proposal Generation. Former methods usually employ uniform segmentation or KTS [22] which is based on frame-to-frame similarity to segment a video into a sequence of fundamental units. Then a subset of fundamental units is selected by the summarization methods. In the summarization process, fundamental units are crucial for summarization methods. For example, improper segmentation may split a complete important event into two segments, which might lead to both segments predicted to be negative. The natural choice for video with clear hierarchical structure is to detect actions as fundamental units. Because action contains complete semantic information, the destruction of semantic information is avoided. Specifically, the Boundary Sensitive Network (BSN) [15] is applied to segment videos by grouping frames with the same semantic information into a single segment (i.e., action).

Hierarchical Decision-Making. Our model is formulated as a recurrent neural network-based agent that interacts with a video over time. And the agent is required to learn a policy for sequentially observing and refining hypotheses about highlight. Applying this into a recurrent neural network-based architecture, we train a hierarchical summarization network with deep Q-learning (HQSN). The network based on reinforcement learning (RL) can capture the interdependence as different combinations of structure units are explored. Our RL-based HQSN works very differently from most existing methods. Most existing methods take the approach of building frame-level or segment-level classifiers, running them exhaustively over the video at a single temporal scale, and applying post-processing such as duration priors and knapsack algorithm. In the absence of post-processing, our RL-based hierarchical method learns policies for which scale to observe and when to stop refining. It is able to do so while observing only a fraction of the multi-scale feature, thus improving the time efficiency. In addition, the multi-scale design in the refining process can provide more structural semantic information. We show that our HQSN is able to reason effectively on temporal bounds of important action and achieve the best performance on our dataset.

In summary, the main contributions of this paper can be concluded as:

- Propose segmenting the video by action instance to avoid destroying semantic information.
- Improve former reinforcement learning based on summarization methods to reach higher performance in time complexity and accuracy.

– Collect a new structured dataset that consists of game videos with fine-grain
actions and importance annotations.

2   Related Work

Video Summarization. The domain-specific approaches aim at videos in a
specific domain by utilizing some domain prior knowledge (e.g., detecting audience cheer in soccer videos [17]) or adapting to specific video characteristics
at the model design level (e.g., exploring individual local motion regions for
surveillance video [34]). The domain-specific approaches can process some types
of videos efficiently and accurately but can not be transferred to another domain.

The domain-agnostic approaches focus more on open-domain user videos.
And they always follow a similar paradigm that consists of three steps: video
segmentation based on temporal similarity such as Kernel Temporal Segmentation (KTS) [22], importance score prediction, and segment selection. One
group of works [9,10,13,33,35] aims at using supervised approaches to predict
importance scores directly. And another group of works [25,38] aims at using
unsupervised/weakly approaches to satisfy some prior hypotheses like representativeness, coverage or diversity. Zhao et al. [35] improve prior approaches by
utilizing the shot boundaries message and propose a structure-adaptive video
summarization method HSA-RNN. But they utilize the video structure on shot-
level and do not explicitly consider the hierarchy during classification (summarization) process. However, Otani et al. [19] have revealed a severe problem
about the former domain-agnostic paradigm that randomly generated summaries
achieve comparable performance to the state-of-the-art on the former benchmark
datasets [9,28]. In CoVieW 2019 challenge [1], multitask models [11,14,20,26] have
been proposed to deal with the video summarization task with model fusion.
These models can utilize relationships among different aspects of video understanding tasks (action recognition, scene recognition, audio).

Temporal Action Proposal. Traditionally, many previous methods [29,30]
generate proposals in a sliding window fashion, which are typically inefficient in
terms of computation cost. Most recent methods [2,6,7,27] adopt top-down fashion to generate proposals, which are more dependent on the predefined duration
and interval, such as dictionary learning [2] and recurrent neural network [6].
These methods avoid computation cost brought by the sliding window but lose
the boundary precision and duration flexibility. Instead, some methods [15,37]
adopt bottom-up fashion. Zhao et al. [37] generate proposals post-processed by
a watershed algorithm, but lack of confidence scores for retrieving. Recently,
BSN [15] generates proposals via evaluating the confidence of whether a proposal contains an action within its region, and has shown great performance.

Reinforcement Learning (RL). RL-based algorithms have been applied in
many computer vision tasks such as image captioning [23], object tracking [31].
Our model mainly draws inspiration from RL-based works that have formulated

---

[1] http://cvlab.hanyang.ac.kr/coview2019.

the computer vision tasks as a searching process [16]. And our HQSN has same DQN paradigm as DQSN [38] but we further proposed a hierarchical refining process and history buffer that stores the history decision. Besides value-based DQSN [38], policy-based DSN [38] has also been proposed for RL-based video summarization. Both of them formulate the video summarization as a sequential decision-making problem but use different RL algorithms. In addition, their reward signal is also different. While DQSN [39] gets reward from an external classification network predicting the video-level category, DSN [38] design an unsupervised reward function that jointly accounts for diversity and representativeness of generated summaries. However, these reward functions introduce some prior hypotheses about what is important (diversity-representativeness or class-driven). Therefore, their ability to adapt to other domains is limited.

3   Proposed Approach

The proposed structure-aware framework contains two key steps. First, we employ BSN [15] to produce a set of temporal action proposals. The temporal action proposals are considered as fundamental units (FUs) of the video structure. Second, a multi-scale feature tree is extracted from the sequence of FU features as the input to HQSN. Deep reinforcement learning is employed to train HQSN to complete the hierarchical refining process.

3.1   Hierarchical Summarization Network

Given a sequence of FUs, a multi-scale feature tree is built with temporal convolution on the top of the FUs sequence. The parameters of temporal convolution are trained with RL paradigm, which means agent need to learn multi-scale representation (observation) of the video. Leaf nodes of multi-scale feature tree are the FUs and non-leaf nodes are convoluted from the next-level nodes. The temporal convolution's stride (3 in this paper) is set to equal the kernel size of the temporal convolution so that each node can correspond to one or more FUs separately and has no intersection with each other. Nodes on each level capture the information with different temporal receptive field. Therefore the video summarization can be formulated as a hierarchical refining process by top-down traversal of the tree, as shown in Fig. 1.

Refining Process. The summarization process can be defined as a Markov Decision Process (MDP), which is formulated as 5-tuple (S, A, P_a, R_a, γ). S is a finite set of state s_t. And s_t at time t is the union of sel_t and $unsel_t$, where sel_t is a sequence of FUs that the agent has selected by time-step t and $unsel_t$ is a partial sequence of nodes that the agent has not selected by time-step t. Initially, $unsel_0$ is composed of the top nodes of the tree. The decision object is always the first node in the $unsel_t$. At each time-step t, the agent chooses an action $a_t \in A$ from policy $\pi(s_t)$ to decide to select, discard or refine the object node. More specifically, A is a finite set of actions composed of three actions: 1 for selecting all the FUs corresponding to the object node, 0 for discarding the FUs and 2 for refining the object node. If $a_t = 2$, all the children of the object node will be added to the $unsel_t$, which means the child nodes need to be considered at next time-step. After transition from s_t to s_{t+1} by updating sel_t and $unsel_t$, the agent receives a reward r_t due to the action a_t and the state s_t. The agent needs to learn a policy $\pi(s_{t+1})$ that balances time complexity and accuracy.

Training Process. HQSN's architecture is shown in Fig. 2. The input of the GRU [4] has three parts corresponding to three RNN networks: sel_t, the sequence of leaf nodes (FUs) in $unsel_t$ and the sequence of non-leaf nodes in $unsel_t$. These three RNNs provide information from different aspects. The selected RNN can provide information about agent's previous decisions. It is necessary because the selection of an FU would have implication on the selection of others. And the inputs of the other two subnets are leaf nodes in $unsel_t$ and non-leaf nodes in $unsel_t$, respectively. We explicitly distinguish leaf nodes from non-leaf nodes by different RNNs because the agent needs to learn a different policy for selecting leaf nodes and non-leaf nodes. Intuitively, when the agent makes a decision for a high-level node, the policy of the agent should be more conservative. And the input features are first mapped to an embedding space via a fully connected (FC) layer.

HQSN predicts the action-value function $Q_\theta(s_t, a)$, where θ is parameters of HQSN. The goal of our HQSN is to minimise mean-squared error between approximate action-value $Q_\theta(s_t, a)$ and true action-value $Q_\pi(s_t, a)$. We perform ϵ-greedy policy to explore the structure, i.e., with probability ϵ select a random action a_t otherwise choose an optimal action $a_t = \arg\max_a Q_\theta(s_t, a)$. With choosing an optimal action, agent takes actions to maximize discounted reward $R_t = \sum_{t'=t}^{T} \gamma^{t'-1} r_{t'}$. We employ experience replay by storing the agent's experience $e_t = (s_t, a_t, r_t, s_{t+1})$ into reply buffer $D_t = \{e_1, ..., e_2\}$ with fixed capacity. During learning, we sample minibatches (s, a, r, s') from replay buffer D. According to [18], HQSN updates at iteration i with the following loss function:

$$L(\theta) = \mathbb{E}_{(s,a,r,s')\sim D}[(R - Q_\theta(s, a))^2] \tag{1}$$
$$\text{s.t. } R = r + \gamma \max_{a'} Q_{\theta^-}(s', a'),$$

where γ is the discount factor and θ^- are the parameters of target network.

3.2   Reward Function

We use a supervised reward function to deal with the class imbalance environment (i.e., the number of key frames is much fewer than those non-key frames) in video summarization task. FU is the fundamental element of the reward function. For each FU, the label g_{t_n} is set to 1 if the proportion of important frames contained in the FU is greater than 0.5, otherwise it is set to 0. After the agent choosing an action a_t, several FUs might be selected or discarded (if $a_t = 2$, no FU is affected). Obviously, the reward r_t depends on the labels of each affected

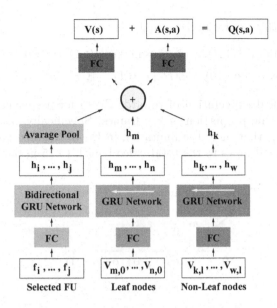

Fig. 2. A schematic of our HQSN architecture. The network has three sub-networks corresponding to three subsets of the state, respectively. f_t represents fundamental unit (FU) features; h_t represents hidden states of RNN; plus represents feature concatenating; $v_{t,l}$ represents the t-th non-leaf node at l-th level in the multi-scale feature tree.

FU and the action a_t. Thus, we first define the reward as the sum of the fundamental reward function $r^f(g_{t_n}, a_t)$. Moreover, to reduce time costs, we add a time penalty item to penalize the agent at every time step. The overall reward function consists of two parts:

$$r_t = \sum_{n=1}^{N^f} r^f(g_{t_n}, a_t) - 0.1, \qquad (2)$$

where -0.1 is time penalty item, N^f is the number of affected FUs and $r^f(g_{t_n}, a_t)$ represents a mapping function from (g_{t_n}, a_t) to values.

Intuitively, if the label $g_{t_n} = a_t$ we reward the action a_t (i.e., $r^f(g_{t_n}, a_t) > 0$), otherwise we penalize the action a_t (i.e., $r^f(g_{t_n}, a_t) < 0$). The inappropriate reward setting (i.e., a specific action is prone to get higher reward for all states) would lead to instable and inaccurate result in our experiment. It is because the inappropriate reward and class imbalance environment would mislead the agent to study a better random strategy instead of a summarization strategy. Thus, we introduce some limits for r^f to make the mathematical expectation of rewards the same regardless of the random strategy that the agent follows.

Mathematically, this limitation is formulated as

$$\text{s.t. } \min(r^f(1,1), r^f(0,0) > \max(r^f(0,1), r^f(0,0))) \tag{3}$$
$$\mathbb{E}(r^f \mid a = 0) = \mathbb{E}(r^f \mid a = 1) = \mathbb{E}(r^f \mid a = 2) = 0,$$

where mathematical expectation of rewards $\mathbb{E}(r^f)$ for a specific action can be estimated by the the proportion of key frames. Specifically, one of the reward $r^f(g_{t_n}, a_t)$ settings that meet the limitation in Eq. 3 is shown in Table 1, where α is the ratio of keyframes to the total video length (0.15 in our dataset).

Table 1. The fundamental reward setting.

g_{t_n} a_t	0	1
0	$\alpha/(1-\alpha)$	$\alpha/(\alpha-1)$
1	-1	1

4 Experiments

4.1 Datasets

Figure 3 presents some examples of our dataset and annotations. The current key characteristics of different datasets are shown in Table 2. As discussed in Sect. 1, we only focus on structured videos None of existing datasets satisfies our requests for the following reasons because all the current datasets contain a large number of unstructured user videos.

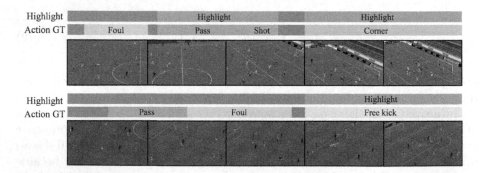

Fig. 3. Two exemplar annotations and video clips from our dataset. The game video dataset consists of long and trimmed videos with fine-grain action and importance annotations.

We create a new dataset from game *Football Manager* 2016 that can automatically generate and label a soccer game according to game settings.

This dataset contains 26 matches of different backgrounds, teams and camera positions with covering 7 types of actions: pass/interception, tackle, shoot/save, foul, free kick, corner ball, throw-in. There are several reasons why we choose videos of the soccer game as a test dataset. First, soccer game videos have clear structures: a game video is composed of several shots and each shot is composed of several actions and there is a strong temporal dependency among important actions (e.g. good pass and good goal in the game). Second, the game system could automatically and precisely output importance and action annotations at frame-level. Finally, all actions in the game are well defined and unambiguous. The dataset is divided into training, validation and testing sets by ratio of 8:1:1.

Table 2. The comparison between datasets.

Dataset	Video quantity	Length	Annotation granularity
CoVieW	1500	2–10	5 s
CoSum [5]	51	1–12 min	Shot ($\approx$6 s)
TVsum [28]	50	2–10 min	2 s
SumMe [9]	25	1–6 min	Frame (0.04 s)
Our dataset	52	30–35 min	Frame (0.05 s)

Table 3. Summarization results (precision, recall and f-score) of various methods on our dataset.

FUs	Uniform segments			Action proposal		
Metrics	P	R	F	P	R	F
Random [19]	0.110	0.155	0.129	0.091	0.124	0.105
Human	–	–	–	0.652	0.720	0.684
DppLSTM [33]	0.424	0.596	0.495	0.464	0.605	0.525
DR-DSN [38]	0.132	0.186	0.154	0.118	0.153	0.133
DR-DSN$_{sup}$	0.434	**0.610**	**0.507**	0.454	0.588	0.512
DQSN$_{sup}$ [39]	**0.461**	0.526	0.492	**0.477**	0.571	0.519
HQSN	0.460	0.527	0.491	0.465	**0.605**	**0.526**

4.2   Video Summarization

Experiment Setting. To verify the influence of the action proposals, as shown in Table 3, we set up two groups of comparison experiments: the fundamental units are produced by action proposal generation and uniform segmentation (112-frame segments). A 112-frame segment and an action proposal are on the same scale. We compare our method with RL-based methods (DR-DSN [38], DQSN [39]) with different reward function and deep learning based methods (dppLSTM [33]). For evaluation, we calculate the F-score as the metric to assess the similarity between automatic summaries and ground truth summaries. Compare to the DQSN that employ DQN directly, HQSN changed to hierarchical decision-making and added a historical buffer. So the "DQSN$_{sup}$" is equivalent to our HQSN without hierarchy design and history buffer, which means the ablation study has been contained on the above experiments.

Implement Details. We implement our model using PyTorch [21]. We adopt two-stream network I3D [3] to extract features and employ BSN [15] to generate action proposal. The dimension of embedding space and hidden units of GRU is 256. The number of layers of GRU is 2. The discount factor γ is 0.9 and ϵ decreases exponentially from 1 and stops at 0.1 at 100000 step. The capacity of M and minibatch of transitions is set to 100000 and 16, respectively. We also use Adam [12] optimizer and clip the norm of gradients to 10 to avoid exploding gradients. The number of layers of the feature tree is 2.

For all RL-based methods include HQSN, we set different fixed random seeds and repeat the same step for 5 times. The final F-score result is the average of several experiment results.

Result Discussion. The results of our method against other methods are shown as Table 3. Our findings are summarised as follows: (1) Utilizing video structures to create action proposals as fundamental units is effective for both deep learning based methods and RL-based methods. For RL-based methods, the improvement is more obvious, which means reinforcement learning is more sensitive to structural units where semantic information is corrupted. In addition, action proposal generation is not required but can improve the performance of current approaches. (2) DR-DSN [38] with diversity-representativeness reward does not perform well. Zhou et al. [38] made a prior assumption about what is important: diversity and representativeness. However, this prior assumption might not fit in soccer videos with very similar frames. (3) Comparing with other RL-based methods, MQSN performs better. It is because our multi-scale decision-making design can fit in the hierarchical structure better. Furthermore, the set of selected FUs can provide information about the agent's previous decisions, which improves the performance. (4) RL-based methods get higher performance than deep learning based methods. The superiority of RL can thus be explained by the fact that RL-based framework can better capture the interdependencies among FUs. (5) The random experiment proves that our dataset is unambiguous, which means our dataset does not have the same issue as observed in [19].

5   Conclusion

We presented a novel structure-aware framework that utilized the video structure to optimize the summarization process. Our proposed method is not limited to a specific domain that we introduced and can be transferred to any domain with clear structures. Experimental results showed that our method is not only outperforming former RL-based alternatives but also highly competitive with deep learning based approaches. The current unsupervised/weakly reward function still have a lot of restrictions because it introduces some prior assumptions about what is important. The advantage of RL without gradient supervision was not fully utilized. In the future, we will focus on how to train an agent with weakly domain-adaptive reward function getting.

References

1. Bettadapura, V., Pantofaru, C., Essa, I.: Leveraging contextual cues for generating basketball highlights. In: Proceedings of the 24th ACM international conference on Multimedia, pp. 908–917. ACM (2016)
2. Caba Heilbron, F., Carlos Niebles, J., Ghanem, B.: Fast temporal activity proposals for efficient detection of human actions in untrimmed videos. In: Proceedings of the IEEE Conference on Computer Vision and Pattern Recognition, pp. 1914–1923 (2016)
3. Carreira, J., Zisserman, A.: Quo vadis, action recognition? a new model and the kinetics dataset. In: Proceedings of the IEEE Conference on Computer Vision and Pattern Recognition, pp. 6299–6308 (2017)
4. Cho, K., et al.: Learning phrase representations using rnn encoder-decoder for statistical machine translation. In: Proceedings of the Empirical Methods in Natural Language Processing, pp. 1724–1734 (2014)
5. Chu, W.S., Song, Y., Jaimes, A.: Video co-summarization: video summarization by visual co-occurrence. In: Proceedings of the IEEE Conference on Computer Vision and Pattern Recognition, pp. 3584–3592 (2015)
6. Escorcia, V., Caba Heilbron, F., Niebles, J.C., Ghanem, B.: DAPs: deep action proposals for action understanding. In: Leibe, B., Matas, J., Sebe, N., Welling, M. (eds.) ECCV 2016. LNCS, vol. 9907, pp. 768–784. Springer, Cham (2016). https://doi.org/10.1007/978-3-319-46487-9_47
7. Gao, J., Yang, Z., Chen, K., Sun, C., Nevatia, R.: Turn tap: temporal unit regression network for temporal action proposals. In: Proceedings of the IEEE International Conference on Computer Vision, pp. 3628–3636 (2017)
8. Gong, B., Chao, W.L., Grauman, K., Sha, F.: Diverse sequential subset selection for supervised video summarization. In: Proceedings of Advances in Neural Information Processing Systems, pp. 2069–2077 (2014)
9. Gygli, M., Grabner, H., Riemenschneider, H., Van Gool, L.: Creating summaries from user videos. In: Fleet, D., Pajdla, T., Schiele, B., Tuytelaars, T. (eds.) ECCV 2014. LNCS, vol. 8695, pp. 505–520. Springer, Cham (2014). https://doi.org/10.1007/978-3-319-10584-0_33
10. Gygli, M., Grabner, H., Van Gool, L.: Video summarization by learning submodular mixtures of objectives. In: Proceedings of the IEEE Conference on Computer Vision and Pattern Recognition, pp. 3090–3098 (2015)
11. Jiang, Y., Cui, K., Peng, B., Xu, C.: Comprehensive video understanding: video summarization with content-based video recommender design. In: Proceedings of the IEEE International Conference on Computer Vision Workshops, pp. 1–8 (2019)
12. Kingma, D.P., Ba, J.: Adam: a method for stochastic optimization. In: Proceedings of the International Conference on Learning Representations, pp. 1–15 (2014)
13. Kulesza, A., Taskar, B., et al.: Determinantal point processes for machine learning. Found. Trends in Mach. Learn. 5(2–3), 123–286 (2012)

14. Kwon, H., Shim, W., Cho, M.: Temporal u-nets for video summarization with scene and action recognition. In: Proceedings of the IEEE International Conference on Computer Vision Workshops, pp. 1–4 (2019)
15. Lin, T., Zhao, X., Su, H., Wang, C., Yang, M.: Bsn: boundary sensitive network for temporal action proposal generation. In: Proceedings of the European Conference on Computer Vision, pp. 3–19 (2018)
16. Mathe, S., Pirinen, A., Sminchisescu, C.: Reinforcement learning for visual object detection. In: Proceedings of the IEEE Conference on Computer Vision and Pattern Recognition, pp. 2894–2902 (2016)
17. Merler, M., et al.: Automatic curation of sports highlights using multimodal excitement features. IEEE Trans. Multimedia **21**(5), 1147–1160 (2018)
18. Mnih, V., et al.: Playing atari with deep reinforcement learning. In: Neural Information Processing Systems Deep Learning Workshop, pp. 1–9 (2013)
19. Otani, M., Nakashima, Y., Rahtu, E., Heikkila, J.: Rethinking the evaluation of video summaries. In: Proceedings of the IEEE Conference on Computer Vision and Pattern Recognition, pp. 7596–7604 (2019)
20. Park, J., Lee, J., Jeon, S., Sohn, K.: Video summarization by learning relationships between action and scene. In: Proceedings of the IEEE International Conference on Computer Vision Workshops, pp. 1–8 (2019)
21. Paszke, A., et al.: Pytorch: an imperative style, high-performance deep learning library. In: Advances in neural information processing systems, pp. 8026–8037 (2019)
22. Potapov, D., Douze, M., Harchaoui, Z., Schmid, C.: Category-specific video summarization. In: Fleet, D., Pajdla, T., Schiele, B., Tuytelaars, T. (eds.) ECCV 2014. LNCS, vol. 8694, pp. 540–555. Springer, Cham (2014). https://doi.org/10.1007/978-3-319-10599-4_35
23. Ren, Z., Wang, X., Zhang, N., Lv, X., Li, L.J.: Deep reinforcement learning-based image captioning with embedding reward. In: Proceedings of the IEEE Conference on Computer Vision and Pattern Recognition, pp. 290–298 (2017)
24. Ringer, C., Nicolaou, M.A.: Deep unsupervised multi-view detection of video game stream highlights. In: Proceedings of the 13th International Conference on the Foundations of Digital Games, pp. 1–6. ACM (2018)
25. Rochan, M., Wang, Y.: Video summarization by learning from unpaired data. In: Proceedings of the IEEE Conference on Computer Vision and Pattern Recognition, pp. 7902–7911 (2019)
26. Seong, H., Hyun, J., Kim, E.: Video multitask transformer network. In: Proceedings of the IEEE International Conference on Computer Vision Workshops, pp. 1–9 (2019)
27. Shou, Z., Wang, D., Chang, S.F.: Temporal action localization in untrimmed videos via multi-stage cnns. In: Proceedings of the IEEE Conference on Computer Vision and Pattern Recognition, pp. 1049–1058 (2016)
28. Song, Y., Vallmitjana, J., Stent, A., Jaimes, A.: Tvsum: summarizing web videos using titles. In: Proceedings of the IEEE Conference on Computer Vision and Pattern Recognition, pp. 5179–5187 (2015)
29. Wang, L., Qiao, Y., Tang, X.: Action recognition and detection by combining motion and appearance features. THUMOS14 Action Recogn. Challenge **1**(2), 2 (2014)
30. Yuan, J., Ni, B., Yang, X., Kassim, A.A.: Temporal action localization with pyramid of score distribution features. In: Proceedings of the IEEE Conference on Computer Vision and Pattern Recognition, pp. 3093–3102 (2016)

31. Yun, S., Choi, J., Yoo, Y., Yun, K., Young Choi, J.: Action-decision networks for visual tracking with deep reinforcement learning. In: Proceedings of the IEEE Conference on Computer Vision and Pattern Recognition, pp. 2711–2720 (2017)
32. Zhang, K., Chao, W.L., Sha, F., Grauman, K.: Summary transfer: exemplar-based subset selection for video summarization. In: Proceedings of the IEEE Conference on Computer Vision and Pattern Recognition, pp. 1059–1067 (2016)
33. Zhang, K., Chao, W.-L., Sha, F., Grauman, K.: Video summarization with long short-term memory. In: Leibe, B., Matas, J., Sebe, N., Welling, M. (eds.) ECCV 2016. LNCS, vol. 9911, pp. 766–782. Springer, Cham (2016). https://doi.org/10.1007/978-3-319-46478-7_47
34. Zhang, S., Zhu, Y., Roy-Chowdhury, A.K.: Context-aware surveillance video summarization. IEEE Trans. Image Proces. 25(11), 5469–5478 (2016)
35. Zhao, B., Li, X., Lu, X.: Hsa-rnn: hierarchical structure-adaptive rnn for video summarization. In: Proceedings of the IEEE Conference on Computer Vision and Pattern Recognition, pp. 7405–7414 (2018)
36. Zhao, B., Xing, E.P.: Quasi real-time summarization for consumer videos. In: Proceedings of the IEEE Conference on Computer Vision and Pattern Recognition, pp. 2513–2520 (2014)
37. Zhao, Y., Xiong, Y., Wang, L., Wu, Z., Tang, X., Lin, D.: Temporal action detection with structured segment networks. In: Proceedings of the IEEE International Conference on Computer Vision, pp. 2914–2923 (2017)
38. Zhou, K., Qiao, Y., Xiang, T.: Deep reinforcement learning for unsupervised video summarization with diversity-representativeness reward. In: Thirty-Second AAAI Conference on Artificial Intelligence, pp. 7582–7589 (2018)
39. Zhou, K., Xiang, T., Cavallaro, A.: Video summarisation by classification with deep reinforcement learning. In: Proceedings of the British Machine Vision Conference, pp. 1–13 (2018)

ADNet: Temporal Anomaly Detection in Surveillance Videos

Halil İbrahim Öztürk[1,2]([✉]) and Ahmet Burak Can[1]

[1] Department of Computer Engineering, Hacettepe University, Ankara, Turkey
{halil_ozturk,abc}@hacettepe.edu.tr
[2] Havelsan, Ankara, Turkey
hozturk@havelsan.com.tr

Abstract. Anomaly detection in surveillance videos is an important research problem in computer vision. In this paper, we propose ADNet, an anomaly detection network, which utilizes temporal convolutions to localize anomalies in videos. The model works online by accepting consecutive windows of video clips. Features extracted from video clips in a window are fed to ADNet, which allows to localize anomalies in videos effectively. We propose the AD Loss function to improve abnormal segment detection performance of ADNet. Additionally, we propose to use F1@k metric for temporal anomaly detection. F1@k is a better evaluation metric than AUC in terms of not penalizing minor shifts in temporal segments and punishing short false positive temporal segment predictions. Furthermore, we extend UCF Crime [29] dataset by adding two more anomaly classes and providing temporal anomaly annotations for all classes. Finally, we thoroughly evaluate our model on the extended UCF Crime dataset. ADNet produces promising results with respect to F1@k metric. Code and dataset extensions are publicly at https://github.com/hibrahimozturk/temporal_anomaly_detection.

Keywords: Temporal anomaly detection · Temporal anomaly localization · Surveillance videos

1 Introduction

In today's physical world, surveillance systems are placed everywhere to provide security, such as shopping malls, offices, homes, rail stations. Detection of abnormal situations in surveillance systems is crucial to make early intervention possible e.g., helping people harmed by a traffic accident, explosion, shooting. Sometimes preventing growth of undesired events like panic, fight, or fire can be possible by detecting anomalies timely in surveillance video streams. However, analyzing video streams in real-time and detecting abnormal cases require excessive human resources and prone to errors due to lose of human attention with time. Since human observation is not an effective solution, automatic anomaly detection approaches that leverage artificial intelligence mechanisms are needed in surveillance systems.

© Springer Nature Switzerland AG 2021
A. Del Bimbo et al. (Eds.): ICPR 2020 Workshops, LNCS 12664, pp. 88–101, 2021.
https://doi.org/10.1007/978-3-030-68799-1_7

Detecting abnormal cases in real world videos is a complex problem, because it is hard to define abnormal cases objectively. Abnormality definition changes according to context or environment, such as seeing a bicycling man in a building can be abnormal while it is not on a road. However, some events can usually be considered as abnormal in most environments like panic in a crowd, fighting persons, vandalism, explosion, fire, etc. Therefore, most anomaly detection studies [15,17,19,29,36] usually focus on detecting these kind of events. General approach in these studies is to learn normal behavior in the environment and define everything out of the normal as abnormal. This approach might produce some false positives but has potential to recognize anomaly cases that are not included in the training data set. Other possible approach is to teach anomaly classes in a data set to the machine learning model and try to recognize them. In this way the approach becomes an action recognition/localization solution and loses ability to recognize unseen anomaly classes.

Fig. 1. Ground truth (GT) and prediction (pred) timelines (a) from Explosion category (b) from Robbery category. Red presents abnormal classes.

In this paper, we propose ADNet, an anomaly detection network to localize anomaly events in temporal space of videos. ADNet uses temporal convolutions to localize abnormal cases in temporal space. While other temporal convolutional networks [6,16] require to get a whole video to produce a result, ADNet utilizes a sliding window approach and gets clips of a video in sequential order. Thus, varying sized videos can be handled effectively and loss of information due to fitting of the whole video to a fixed-size input can be prevented. Window based anomaly localization provides better performance when determining anomalies in videos and also enables real-time processing of surveillance videos and provides an indirect data augmentation mechanism to increase training data. After extracting features from each video clip by using a spatio-temporal deep network, ADNet takes features of all clips in a window and produce a separate anomaly score for each clip. We further define AD Loss function to increase anomaly detection accuracy. When ADNet is trained with the proposed AD Loss, performance of detecting abnormal segments increases.

We also propose to use segment based F1@k metric to measure effectiveness of an anomaly detection model instead of clip based AUC metric. AUC is not a good metric for this purpose since it does not take into account temporal order of clips. It penalizes minor shifts in temporal segments and can not effectively punish short false positive segments in temporal space. F1@k is calculated on IoU with k percentage of temporal segments obtained by thresholding anomaly scores of clips with 0.5 value. It is better in terms of measuring how correctly predicted temporal segment matches with ground truth temporal segment.

As a last contribution we add two more anomaly classes to UCF Crime [29] dataset, which consists of 13 anomaly classes. Figure 1 shows ground truth and prediction timelines for two different anomaly classes. UCF Crime [29] data set provides temporal anomaly annotations in the test set, but only provides video level annotations in the training set. We provide temporal annotations for all anomaly classes of the data set. Finally, we thoroughly evaluate our model on UCF Crime dataset and extended dataset version, UCF Crime V2. ADNet produces promising abnormal segment detection score with **28.32 F1@10** while the baseline model's [29] score is **4.13 F1@10**. For the normal and abnormal segments, ADNet has 58.16 F1@10 score while baseline model's score is 45.20 F1@10.

2 Related Works

This work focuses on anomaly detection but it also utilizes action localization and recognition approaches. Thus, this section outlines the related works on anomaly detection, action recognition, and temporal action localization fields.

Anomaly Detection aims to detect events which are highly deviates from normal events. Kratz and Nishino [15] propose a statistical approach to detect anomalies in extremely crowd scenes. Mahadevan et al. [19] propose a framework and a dataset to detect spatial and temporal anomalies in crowded scenes by using Gaussian Mixture Model on temporal space and discriminant saliency on spatial space. Li et al. [17] studies localization of anomaly on the same dataset. Yuan et al. [36] propose structural context descriptor (SCD) and 3D discrete cosine transform (DCT) multi-object tracker to detect and localize anomalies in crowded scenes. Sultani et al. [29] propose model which is our baseline and UCF Crime dataset that has videos annotated with normal and abnormal labels. Baseline model has trained with multiple instance learning (MIL) and special loss on weakly supervised UCF Crime dataset to predict clip wise label. Videos are split to 32 segments. Each segment has extracted features, which are average of extracted clip features in the segment. Positive and negative bags consist of positive and negative segments, respectively. The model is trained with proposed loss. Shah et al. [24] propose CADP dataset and a method for traffic anomaly detection. [10,35] studies anomaly detection of vehicles from traffic videos or streams.

Action Recognition tries to classify actions on trimmed videos. Tran et al. [31] proposes Convolutional 3D (C3D) network which has 3D convolutions to capture spatio-temporal features. Carreira and Zisserman [3] propose Two-Stream Inflated 3D (I3D) network to learn better temporal features with different architecture by using 3D convolutions. I3D has two sub networks as [26], where they accept RGB and Optic Flow inputs. Classification probabilities are averaged before final prediction. ADNet uses only RGB branch of I3D to work more efficiently. Temporal Shift Module (TSM) Lin et al. [18] propose to predict action classes with pseudo-3D convolutions efficiently. Shifting some channels in one ore two direction of ResNet [9], Temporal Segment Networks (TSN) [34], provides temporal information to classify actions on Kinetics [13] and Something-Something[8] datasets. In order to decrease FLOPs, [23, 30, 32] use pseudo-3D convolution that is combination of 2D convolution in spatial space and 1D convolution in temporal space. Tran et al. [33] propose Channel Separated Network (CSN), which decreases FLOPs by using 3D group convolutions instead of 3D convolutions.

Temporal Action Localization is that detecting and segmenting actions of untrimmed videos in temporal space. Farha and Gall [6] proposes Temporal Convolutional Networks (TCN) by inspiration of a speech synthesis method, WaveNet [21]. 1D convolutions are applied on temporal space of extracted features of consecutive frames. Encoder-Decoder TCN (ED-TCN) achieves 68 F1@10 score, while Dilated TCN achieves 52.2 F1@10 score on 50 Salads [28] dataset. ED-TCN and Dilated TCN accept features from VGG-style [27] spatial networks. Iqbal et al. [12] proposes improving temporal action localization via transfer learning from I3D action recognition network [3] with BoW-Network. BoW-Network creates visual words, which has 4000 components with k-means algorithm on Thumos [11] dataset. Farha and Gall [6] proposes Multi-Stage TCN (MS-TCN) network and a loss function, which is combination of classification and smoothing loss. MS-TCN achieves 76.3 F1@10 score on 50 Salads [28] dataset. Ding and Xu [4] propose Temporal Convolutional Feature Pyramid Network (TCFPN) and Iterative Soft Boundary Assignment (ISBA). ISBA generates temporal segments of given actions without temporal boundaries by using transcripts of video in Hollwood Extended [1] dataset. TCFPN is trained with weakly-supervised data as in [20, 25]. Gao et al. [7] uses sliding windows with different sizes to predict action scores at CTAP, collected windows are fed to temporal convolutional network. Buch et al. [2] proposes SST that uses GRU instead of sliding window approch, the GRU accepts C3D clip embeddings as input.

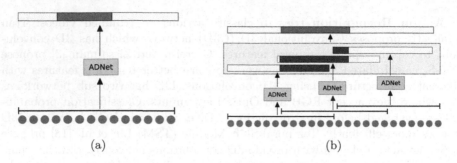

(a)   (b)

Fig. 2. Temporal convolutional networks *(a)* accepts clip features x_i which are presented with blue circles. Instead of convolving over clip features of a video as in *(a)*, clip features are divided to equal windows to pass ADNet *(b)*. Windows intersect with previous windows, output scores in overlapped parts are averaged to get final output scores. Abnormal results are presented with red colour.

3   Method

In this section we introduce Anomaly Detection Network (ADNet) for anomaly detection and segmentation in temporal space. Input to ADNet is a set of features $x_{1:C} = (x_1, x_2, ..., x_C)$ extracted using a spatio-temporal CNN for each clip of given video, where C is number of clips for a video. A clip consists of a number of consecutive frames of video. ADNet outputs anomaly probabilities $y_{1:C} = (y_1, y_2, ..., y_C)$ for each clip, where $y_t \in [0, 1]$. Target values are $a_{1:C} = (a_1, a_2, ..., a_C)$ where $a \in \{0, 1\}$. 0 and 1 indicates normal and abnormal classes. Labels are generated from output probabilities by applying 0.5 threshold value, which can be changed in inference time to adjust the model for different anomaly conditions. We describe ADNet in Sect. 3.1. Then we describe Temporal Sliding Window in Sect. 3.2. At the end of method section, we discuss the proposed loss in Sect. 3.3.

3.1   Anomaly Detection Network

To predict anomalies in a timeline by using temporal information, we adapt and improve MS-TCN [6] model which makes classification and temporal segmentation. We input all clip features of a video without splitting to small parts.

We define a sequence of stages, which has a series of blocks that consists of convolutional layers. Each stage outputs anomaly scores, y_t, for each clip features, x_t. Input of next stage is the output of previous stage, except the first stage. Input features are shrink with D^H 1×1 convolutions on D^0 channels, where D^0 is the dimension of an input feature and D^H is the number of channels of features outputted by hidden layers of stages. Let V^{HxT} be output of each layer except the last layer in a stage. Dilated convolutions in layers are convolved on temporal dimension of V^{HxT}. Dilation rate of convolution is calculated with 2^l formula, where l is order of layer in stage (i.e. $0, 1, 2, .., L$). Dilated convolutions increase receptive field of network with small kernel size. The activation function applied after convolutions is ReLU. Thus, 1×1 convolution follows ReLU. Residual connections between before and after convolutions in a block avoids gradient vanishing problem. Output of a stage is defined as follows:

$$Y^s = Sigmoid(W * V + b) \tag{1}$$

where $*$ is the convolution operation, W is filter kernel and b is bias value of kernel, $Y^s \in \mathbb{R}^{1xT}$ is the vector of anomaly scores outputted by s_{th} stage.

We input the output previous stage, Y^{s-1}, to the next stage, Y^s. At the start of the stage, 1×1 convolution is applied to increase channel size from 1 to D^H.

3.2   Temporal Sliding Window

In the previous section, we input the whole clip features $x_{1,...,T}$ to the network. In order to provide generalization by data augmentation we split consecutive clip features to windows, where each window contains W consecutive clip features, $(x_1, x_2, ..., x_W)$. If video has less clip features than window width, $T < W$, we pad window with empty clip features x_0, which filled with 0 values. Let $x = (x_1, x_2, x_0, ..., x_0)$ be an input and $m = (1, 1, 0, ...0)$ be the mask for x, where $x \in \mathbb{R}^{D_0 xW}$ and $m \in \mathbb{R}^{1xW}$. Information flow from padded empty clip features is blocked by masking outputs of convolutions as follows:

$$V^l = (W_2 * ReLU(W_1 * V^{l-1} + b_1) + b_2) \cdot m \tag{2}$$

where V^l is the output of lth layer, (W_1, W_2, b_1, b_2) are parameters of convolution, $\cdot$ is dot product.

This makes also online anomaly detection and segmentation possible, while they are impossible in the network accept whole clip features as input. Window stride size is set to half of the window size to obtain smoother results in inference, while increasing number of training clip windows. In other words, we augment data with both of splitting videos to windows and half stride windows during training. Start position of window i is calculated with $w_i^{start} = W/2 * (i - 1)$, end position formula is $w_i^{end} = W + W/2 * (i - 1)$. We use same window size for a network in both train and test stages. In the inference, we average anomaly scores of overlapping windows, as in Fig. 2.

Each layer in a stage except the last layer outputs V^{TxD_H}, where dimension of output is same for the convolutions. To match input dimension with output dimension, we pad input P, where $P = \lfloor K/2 \rfloor * 2^L$, $P < W$, K is kernel size. If padding P is equal to input size of window W, more than half of the inputs of convolution would be padding elements. Window width can not be more than the receptive field, where $ReceptiveField = 2^{l+1} - 1$. Therefore, to avoid information loss, we determine maximum number of layers for a window width W as follows:

$$L = \left\lceil log_2 \frac{W}{\lfloor K/2 \rfloor} \right\rceil \tag{3}$$

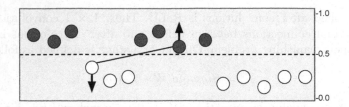

Fig. 3. AD Loss increases distance between nearest opposite pair in manner of anomaly score for each clip. Red circles represent abnormal clips, white circles represent normal clips

3.3 Loss

Intuition behind the loss is similar to VSE++ [5], increasing distance between hard pairs which are closest wrong embeddings in terms of cosine distance. We define hard pair as clips from opponent classes which have closest anomaly score outputs as in Fig. 3. Hard pairs change from in each of step of training. Combination of L_{MSE} mean squared error loss and L_{AD} anomaly detection loss produces the final loss value as in Formula 4. Contribution of L_{AD} is controlled with $0 < \lambda < 1$ parameter in Formula 4. L_{AD} is calculated as in Formula 5, where y_A is score of abnormal input, y_{HN} hard normal of the abnormal input, y_N is score of normal input, and y_{HA} hard abnormal of the normal input. α parameter controls ideal distance between hard pairs.

$$L_s = L_{MSE} + \lambda * L_{AD} \tag{4}$$

$$L_{AD} = max(-(y_A - y_{HN} - \alpha) - (y_{HA} - y_N - \alpha), 0) \tag{5}$$

As we mentioned in Sect. 3.1, each stage outputs anomaly probability for each input clip feature of given video. We calculate loss for each stage output as in Formula 6. The summation of losses are minimized in training.

$$L = \sum_s L_s \tag{6}$$

3.4 Implementation Details

We extract clip features from I3D [3] by applying average pooling to activations before classification layer. Video clips for I3D are generated with 16 frame temporal slide. We chose 16 temporal slide instead of 1 to decrease inference time. TSM [18] is second feature extractor used in our study. We use Adam [14] optimizer with 5e−4 learning rate in all experiments . We trained and tested our models on PyTorch [22] framework. Kernel size and channel size of ADNet are 3 and 64 respectively in all settings. Labels of clips are determined by distribution of normal and abnormal classes.

4   Experiments

4.1   Evaluation Metrics

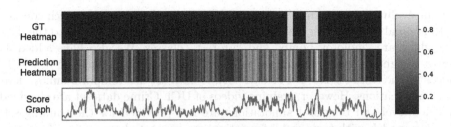

Fig. 4. First two rows are heatmaps of ground truth, predicted anomaly scores for *Assault 10* video from the test set. Last row is graph of anomaly scores which is between 0 and 1. AUC score is 74.83, F1@25 score is 24.99

The baseline method [29] and most other studies use AUC metric to measure anomaly detection performance. Abnormal cases happen in a segment of timeline. AUC metric evaluates performance of each clip independently, in other words, avoiding the temporal orders of the clips. For this reason AUC cannot do the necessary punishment when short false positives arises. An appropriate metric should not penalize minor shifts between predicted segments and ground truth segments while penalizing over-segmentation errors. However, clip wise AUC metric penalizes minor shifts. Therefore, we approach to the problem as temporal action localization. We adapt F1@k metric proposed in [16] for evaluating anomaly detection performance, which overcomes the weaknesses of AUC metric.

Abnormal segments form small part of ground truth in a test sample. For this reason, either wrong predictions in abnormal segments or small false positives in normal segments do not sufficiently affect AUC score. Figure 4 shows timelines of ground truth and prediction segmentation for a test video. While AUC score of the test video is **74.83**, F1@25 score is **24.99**. This example shows the robustness of F1@k metric comparing to AUC metric for this problem.

F1@k is calculated by k percentage intersection over union (IOU) between predicted temporal segments and ground truth temporal segments. However, [16] does not include background segments to F1@k metric evaluation. In order to increase penalization for over-segmentation of abnormal segments, we do not consider segments with normal events as background segments, we include them to evaluation.

As mentioned before, we extract features for each clip from spatio-temporal networks. Let $C_i = (f_{n*i}, ..., f_{n*(i+1)})$ be a clip where f_j is j_{th} frame of given video. Since accepted number of frames from feature extractors can be different, number of clips can be different for given video. To evaluate fairly, we produce frame level labels or scores from clip level results by copying clip scores y_i to frame scores $\hat{y}_j$ of the clip, $\hat{y}_{n*i,\ldots,n*(i+1)} = y_i$. We make evaluation in frame-level instead of clip level.

4.2 Dataset

For the evaluation of the model, we use UCF Crime data set [29], which consists of 13 anomaly classes. We have added two different anomaly classes to the data set, which are "molotov bomb" and "protest" classes. We also have added 33 videos to fighting class. In total, we have added 216 videos to the training set, 17 videos to the test set. Test set of UCF Crime data set has temporal annotations and classifications. However, training videos of UCF Crime data set are classified in video-level and temporal annotations are not provided for the training set. To train models with temporal information, we annotated anomalies of training videos in temporal domain. In order to annotate efficiently, annotators have used seconds as basis and assumed that the frames in a second all belong to the same class. Since we extend the dataset with new anomaly classes and temporal annotations for training videos, we name new version of the dataset as UCF Crime V2.

Since baseline model has been trained without new anomaly classes, we do not evaluate the baseline on UCF Crime v2 but we evaluate on UCF Crime v1. We investigate effects of window size, number of layer, feature extractor and loss functions on UCF Crime v2 dataset.

4.3 Results

In this section we present and discuss results of evaluation experiments.

Table 1. Comparison of number of layers. I3D is feature extractor in the experiments.

Methods	UCF Crime V2		
	F1@10	F1@25	F1@50
ADNet W64-S5-L8	53.03	47.03	33.95
ADNet W64-S5-L7	53.82	48.24	34.02
ADNet W64-S5-L6	**58.73**	**52.38**	**40.98**
ADNet W64-S5-L5	48.12	39.95	29.10
ADNet W64-S5-L4	51.35	42.19	28.17

Table 2. Comparison of different window sizes (W: window, S: number of stages, L: number of layers in a stage). I3D is feature extractor in the experiments.

Methods	UCF Crime V2		
	F1@10	F1@25	F1@50
ADNet W32-S3-L5	50.78	43.22	32.05
ADNet W64-S5-L6	**58.73**	**52.38**	**40.98**
ADNet W128-S8-L7	55.61	50.04	36.60
ADNet w/o Window	51.62	43.86	32.30

Effect of Temporal Sliding Window. We start evaluation by showing the effect of temporal sliding window method. We compare different window widths and without window of ADNet in Table 2. Experiments in the table have been made on UCF Crime V2 dataset. Number of layers in this experiment is set according to window width and Formula 3. Window size and number of stages might have different values, which are experimented in our ablation study. Table 2 presents three window sizes: 32, 64 and 128 clips in a window. Number of stages for each window width are selected by taking into account the best result of that window. Generally, number of stages increases in parallel to window width. As we mentioned before, temporal sliding window augments data during training. Decreasing window size provides better augmentation, but temporal information is collapsed. There is a trade-off between data augmentation and temporal knowledge. Table 2 shows that the best result is achieved with 64 temporal window width. All of the temporal sliding window results are better than straight forward ADNet (w/o Window) in terms of F1@k scores. This means that temporal sliding window improves performance of ADNet in UCF Crime v2 dataset.

Effect of Number of Layers. We have compared window widths where number of layers is calculated with Formula 3. In this part we discuss effect of number of layers to performance. Table 1 shows results of experiments with fixed window width and number of stages. ADNet with more layers achieves better results. However, a layer number more than the maximum layer number calculated by Formula 3 cause information loss as mentioned in Sect. 3.

Table 3. Comparison of loss functions. I3D is feature extractor in the experiments.

Methods	UCF Crime V2			
	Abnormal Segments		Normal Segments	
	F1@10	F1@25	F1@10	F1@25
ADNet (MSE)	29.00	19.33	**71.23**	**66.44**
ADNet (MSE+AD)	**32.16**	**20.70**	56.34	50.54

Loss Function. We propose the AD loss function, which tries to maximize distance between hard pairs from opponent classes. Table 3 shows results for MSE loss and MSE and anomaly detection (AD) loss. In these experiments, window width is 64, number of stages is 5, and number of layers is 6. The parameters are selected based on the previous experiments. λ parameter in Formula 5 controls contribution of AD loss to total loss, λ is set to 0.5 in these experiment. The results show that MSE+AD loss is more successful than MSE loss at abnormal segments, but MSE+AD loss is worse than MSE loss at normal segments. For this reason we use MSE loss only in other experiments.

Feature Extractor. We have experimented on two different spatio-temporal feature extractors. In the previous experiments, we used I3D [3] as the feature extractor. Temporal Shift Module (TSM) [18] is more a efficient action recognition network than I3D network while performance on Kinetics [13] dataset is similar. To extract features from video clips, we have used TSM as an alternative

Table 4. Comparison of different spatio-temporal feature extractors

Methods	UCF Crime V2			
	Abnormal Segments		Normal Segments	
	F1@10	F1@25	F1@10	F1@25
ADNet [I3D]	29.0	19.33	**71.23**	**66.44**
ADNet [TSM]	**33.50**	**22.78**	61.71	57.52

to I3D. Table 4 shows the results for I3D and TSM networks, where parameters of ADNet are as follows, window width is 64, number of stages is 5 and number of layers is 6. Although I3D and TSM get different sized input clips and produce outputs in different formats, this experiment shows us that ADNet can utilize different feature extractor networks. As a result of this experiment, we observed that I3D features is more useful in ADNet network for segmenting normal events while TSM features is better for segmenting abnormal events.

Table 5. Performance comparison of state-of-the art methods

Methods	UCF Crime v1								
	Abnormal Segments			Normal Segments			All Segments		
	F1@10	F1@25	F1@50	F1@10	F1@25	F1@50	F1@10	F1@25	F1@50
Baseline Network [C3D]	4.13	1.65	0	63.27	56.36	46.54	45.20	39.64	32.32
MLP [C3D]	7.34	1.86	0.65	65.96	63.15	54.75	49.40	44.83	38.15
ED-TCN [I3D]	21.18	12.63	4.88	61.60	53.59	36.71	47.81	39.61	25.85
ADNet [I3D] (Ours)	**28.32**	**18.71**	**9.44**	**71.23**	**66.44**	**55.48**	**58.16**	**51.85**	**41.29**

Comparison with State-of-the-Arts. Temporal annotations of training set of UCF Crime data set has not been available until our study. Baseline model [29] has not been trained on temporarily annotated train set. In order to compare ADNet with models trained on temporarily annotated training set, we have trained two models which accept features extracted from I3D network. First model is Multi Layer Perceptron (MLP) with 3 layers as in baseline network [29] which generates predictions clip-wise. Second model is Encoder Decoder Temporal Convolutional Network (ED-TCN) proposed in [16] with temporal sliding window method as in Sect. 3.2. According to the results presented in Table 5, our proposed model achieves better scores than baseline network and other models in all categories on UCF Crime v1 test set. While baseline model achieves **4.13 F1@10** score at abnormal segments, our model achieves **28.32 F1@10** score. Window width, number of stages and number of layers of ADNet in Table 5 are 64, 5, and 6, respectively. Ground truths and predictions in timeline of two test videos from Explosion and Robbery categories are in Fig. 1.

5   Conclusion

We proposed a temporal anomaly detection network, which enables to localize anomalies in videos with temporal convolutions. In our knowledge, this is
the first approach to formulate anomaly detection problem in a similar way to
action localization problem. We also introduced AD loss function, which enabled
to have better detection performance in abnormal classes. We evaluated and
discussed the effects of model parameters, which are window width, number of
layers, number of stages, feature extractor, and loss functions. We also extended
the UCF Crime anomaly dataset with two additional anomaly classes and temporal annotations of training videos. Extensive evaluations of the model shows
that the model has promising results on real world anomaly videos. Window
based operation of the model allows processing of online video streams. We also
investigated the evaluation metrics in terms of measuring anomaly detection
performance. Since F1@k does not penalize minor shifts and does punish short
false positive temporal segment predictions, we concluded that F1@k metric is
better than AUC metric for measuring anomaly detection performance. In the
future works, scene context information and relations between objects can be
utilized to improve anomaly detection performance.

Acknowledgements. This work was supported in part by the Scientific and Technological Research Council of Turkey (TUBITAK) under Grant No. 114G028 and
1198E098.

References

1. Bojanowski, P., et al.: Weakly supervised action labeling in videos under ordering
   constraints. In: Fleet, D., Pajdla, T., Schiele, B., Tuytelaars, T. (eds.) ECCV 2014.
   LNCS, vol. 8693, pp. 628–643. Springer, Cham (2014). https://doi.org/10.1007/
   978-3-319-10602-1_41
2. Buch, S., Escorcia, V., Shen, C., Ghanem, B., Carlos Niebles, J.: Sst: single-stream
   temporal action proposals. In: Proceedings of the IEEE conference on Computer
   Vision and Pattern Recognition, pp. 2911–2920 (2017)
3. Carreira, J., Zisserman, A.: Quo vadis, action recognition? a new model and the
   kinetics dataset. In: proceedings of the IEEE Conference on Computer Vision and
   Pattern Recognition, pp. 6299–6308 (2017)
4. Ding, L., Xu, C.: Weakly-supervised action segmentation with iterative soft boundary assignment. In: Proceedings of the IEEE Conference on Computer Vision and
   Pattern Recognition, pp. 6508–6516 (2018)
5. Faghri, F., Fleet, D.J., Kiros, J.R., Fidler, S.: Vse++: Improving visual-semantic
   embeddings with hard negatives. arXiv preprint arXiv:1707.05612 (2017)
6. Farha, Y.A., Gall, J.: Ms-tcn: Multi-stage temporal convolutional network for
   action segmentation. In: Proceedings of the IEEE Conference on Computer Vision
   and Pattern Recognition, pp. 3575–3584 (2019)
7. Gao, J., Chen, K., Nevatia, R.: Ctap: complementary temporal action proposal generation. In: Proceedings of the European conference on computer vision (ECCV),
   pp. 68–83 (2018)

8. Goyal, R. et al.: The something something video database for learning and evaluating visual common sense. In: ICCV, vol. 1, p. 5 (2017)

9. He, K., Zhang, X., Ren, S., Sun, J.: Deep residual learning for image recognition. In: Proceedings of the IEEE conference on computer vision and pattern recognition, pp. 770–778 (2016)

10. Huang, X., He, P., Rangarajan, A., Ranka, S.: Intelligent intersection: Two-stream convolutional networks for real-time near-accident detection in traffic video. ACM Trans. Spat. Algorithms Syst. (TSAS) **6**(2), 1–28 (2020)

11. Idrees, H.: The thumos challenge on action recognition for videos in the wild. Comput. Vis. Image Underst. **155**, 1–23 (2017)

12. Iqbal, A., Richard, A., Gall, J.: Enhancing temporal action localization with transfer learning from action recognition. In: Proceedings of the IEEE International Conference on Computer Vision Workshops (2019)

13. Kay, W., et al.: The kinetics human action video dataset. arXiv preprint arXiv:1705.06950 (2017)

14. Kingma, D.P., Ba, J.: Adam: a method for stochastic optimization. arXiv preprint arXiv:1412.6980, 2014

15. Kratz, L., Nishino, K.: Anomaly detection in extremely crowded scenes using spatio-temporal motion pattern models. In: 2009 IEEE Conference on Computer Vision and Pattern Recognition, pp. 1446–1453. IEEE (2009)

16. Lea, C., Flynn, M.D., Vidal, R., Reiter, A., Hager, G.D.: Temporal convolutional networks for action segmentation and detection. In: proceedings of the IEEE Conference on Computer Vision and Pattern Recognition, pp. 156–165 (2017)

17. Li, W., Mahadevan, V., Vasconcelos, N.: Anomaly detection and localization in crowded scenes. IEEE Trans. Pattern Anal. Mach. Intell. **36**(1), 18–32 (2013)

18. Lin, J., Gan, C., Han, S.: Tsm: temporal shift module for efficient video understanding. In: Proceedings of the IEEE International Conference on Computer Vision, pp. 7083–7093 (2019)

19. Mahadevan, V., Li, W., Bhalodia, V., Vasconcelos, N.: Anomaly detection in crowded scenes. In: 2010 IEEE Computer Society Conference on Computer Vision and Pattern Recognition, pp. 1975–1981. IEEE (2010)

20. Nguyen, P., Liu, T., Prasad, G., Han, B.: Weakly supervised action localization by sparse temporal pooling network. In: Proceedings of the IEEE Conference on Computer Vision and Pattern Recognition, pp. 6752–6761 (2018)

21. Oord, A.V.D.: Wavenet: a generative model for raw audio. arXiv preprint arXiv:1609.03499 (2016)

22. Paszke, A., et al.: Pytorch: an imperative style, high-performance deep learning library. In: Advances in Neural Information Processing Systems, pp. 8024–8035 (2019)

23. Qiu, Z., Yao, T., Mei, T.: Learning spatio-temporal representation with pseudo-3d residual networks. In: Proceedings of the IEEE International Conference on Computer Vision, pp. 5533–5541 (2017)

24. Shah, A.P., Lamare, J.B., Nguyen-Anh, T., Hauptmann, A.: Cadp: a novel dataset for cctv traffic camera based accident analysis. In: 2018 15th IEEE International Conference on Advanced Video and Signal Based Surveillance (AVSS), pp. 1–9. IEEE (2018)

25. Shou, Z., Gao, H., Zhang, L., Miyazawa, K., Chang, S.F.: Autoloc: weakly-supervised temporal action localization in untrimmed videos. In: Proceedings of the European Conference on Computer Vision (ECCV), pp. 154–171 (2018)

26. Simonyan, K., Zisserman, A.: Two-stream convolutional networks for action recognition in videos. In: Advances in Neural Information Processing systems, pp. 568–576 (2014)
27. Simonyan, K., Zisserman, A.: Very deep convolutional networks for large-scale image recognition. arXiv preprint arXiv:1409.1556 (2014)
28. Stein, S., McKenna, S.J.: Combining embedded accelerometers with computer vision for recognizing food preparation activities. In: Proceedings of the 2013 ACM International Joint Conference on Pervasive and Ubiquitous Computing, pp. 729–738 (2013)
29. Sultani, W., Chen, C., Shah, M.: Real-world anomaly detection in surveillance videos. In: Proceedings of the IEEE Conference on Computer Vision and Pattern Recognition, pp. 6479–6488 (2018)
30. Sun, L., Jia, K., Yeung, D.Y., Shi, B.E.: Human action recognition using factorized spatio-temporal convolutional networks. In: Proceedings of the IEEE International Conference on Computer Vision, pp. 4597–4605 (2015)
31. Tran, D., Bourdev, L., Fergus, R., Torresani, L., Paluri, M.: Learning spatiotemporal features with 3d convolutional networks. In: Proceedings of the IEEE International Conference on Computer Vision, pp. 4489–4497 (2015)
32. Tran, D., Wang, H., Torresani, L., Ray, J., LeCun, Y.: Paluri, M.: A closer look at spatiotemporal convolutions for action recognition. In: Proceedings of the IEEE conference on Computer Vision and Pattern Recognition, pp. 6450–6459 (2018)
33. Tran, D., Wang, H., Torresani, L., Feiszli, M.: Video classification with channel-separated convolutional networks. In: Proceedings of the IEEE International Conference on Computer Vision, pp. 5552–5561 (2019)
34. Wang, L., et al.: Temporal segment networks: towards good practices for deep action recognition. In: Leibe, B., Matas, J., Sebe, N., Welling, M. (eds.) ECCV 2016. LNCS, vol. 9912, pp. 20–36. Springer, Cham (2016). https://doi.org/10.1007/978-3-319-46484-8_2
35. Yao, Y., Xu, M., Wang, Y., Crandall, D.J., Atkins, E.M.: Unsupervised traffic accident detection in first-person videos. arXiv preprint arXiv:1903.00618 (2019)
36. Yuan, Y., Fang, J., Wang, Q.: Online anomaly detection in crowd scenes via structure analysis. IEEE Trans. Cybern. 45(3), 548–561 (2014)

Soft Pseudo-labeling Semi-Supervised Learning Applied to Fine-Grained Visual Classification

Daniele Mugnai[(⊠)], Federico Pernici, Francesco Turchini,
and Alberto Del Bimbo

University of Florence, Florence, Italy
{daniele.mugnai,federico.pernici,francesco.turchini,
alberto.delbimbo}@unifi.it

Abstract. Pseudo-labeling is a simple and well known strategy in Semi-Supervised Learning with neural networks. The method is equivalent to entropy minimization as the overlap of class probability distribution can be reduced minimizing the entropy for unlabeled data. In this paper we review the relationship between the two methods and evaluate their performance on Fine-Grained Visual Classification datasets. We include also the recent released iNaturalist-Aves that is specifically designed for Semi-Supervised Learning. Experimental results show that although in some cases supervised learning may still have better performance than the semi-supervised methods, Semi Supervised Learning shows effective results. Specifically, we observed that entropy-minimization slightly outperforms a recent proposed method based on pseudo-labeling.

Keywords: Fine-grained visual classification · Semi-supervised learning · Entropy minimization

1 Introduction

Fine-Grained Visual Categorization (FGVC) aims to distinguish between image classes such as species of birds, dogs, flowers or even models of cars. This is much harder than general-purpose classification as only few subtle key features matter. A further issue of FGVC is that data annotation is very expensive and it requires domain experts. The data annotation problem can be partially alleviated using Semi-Supervised Learning (SSL) by leveraging large set of unlabeled data and few labeled ones [1]. Except for a very recent paper [2], SSL has not been investigated in FGVC. However, this topic is getting increasing support and attention to such an extent that a dataset for this specific problem has been released[1]. SSL has shown to be a suitable learning paradigm for leveraging unlabeled data to reduce the cost of large labeled datasets [3].

---

[1] The Semi-Supervised iNaturalist-Aves Dataset: https://github.com/cvl-umass/semi-inat-2020.

© Springer Nature Switzerland AG 2021
A. Del Bimbo et al. (Eds.): ICPR 2020 Workshops, LNCS 12664, pp. 102–110, 2021.
https://doi.org/10.1007/978-3-030-68799-1_8

A common assumption in SSL is that the decision boundaries of the classifier should not pass through high-density regions of the marginal data distribution [1]. One way to impose this constraint is to force the output of the classifier to have low-entropy predictions on the unlabeled data [4]. This strategy is known as *entropy minimization* and is particularly interesting because *pseudo-label* SSL [5], the simplest algorithms in SSL, does entropy minimization implicitly by constructing hard labels from the most confident class predictions on unlabeled data. Class predictions are subsequently used as training targets in a standard supervised learning paradigm and optimized according to the cross-entropy loss. Entropy minimization can be considered a soft version of the pseudo-labeling method.

In this paper we review the theoretical relationship between the two methods and evaluate SSL *entropy minimization*, on several FGVC datasets including the recent Semi-Supervised iNaturalist-Aves and compare the results with [2] in which a pseudo-label based learning method is used.

Experimental results show that although in some cases supervised learning may still have better performance than the semi-supervised methods, Semi Supervised Learning shows effective results. Specifically, we observed that entropy-minimization slightly outperforms a recent proposed method based on pseudo-labeling.

2   Related Work

2.1   Fine-Grained Visual Classification

Fine-grained visual classification is an important and well studied problem. There are two main paradigms of FGVC approaches: (1) with localization-classification sub-networks and (2) with end-to-end feature encoding [6].

The first family of methods trains a localization sub-network to locate key part regions. Then the classification sub-network uses the information of fine-grained regions captured by the localization sub-network to further enhance its classification capability. These methods aim at learning distinct features present in different parts of the object, e.g. the differences between the beak and tail of bird species. [7–9].

The second family of methods tends to achieve fine-grained recognition by developing powerful deep models for discriminative feature representation. Global methods analyze the whole input image and look for important parts exploiting different strategies for pre-training [10], augmentation [11,12], or pooling [13–15].

Other techniques do not explicitly enforce the objects parts assumption and rely on general purpose-classification. The work [16] exploits the large external information in private JFT dataset that includes 300M labeled images [17] and trains very large architectures of about 550M parameters [18]. They show that fine-grained labels of JFT achieve better transfer learning for FGVC datasets. The work [19] evaluates the transformer model [20] in FGVC benchmarks achieving significant performance improvements. All these methods make extensive use of labeled dataset and/or take advantage of high computational power or external datasets.

2.2 Semi-Supervised Learning

Semi-Supervised Learning (SSL) combines labeled data with unlabeled data during training [1]. Evaluation of SSL algorithms [21–23] is mainly conducted on small-scale datasets such as CIFAR-10 [24] and SVHN [25]. Evaluation of SSL approaches on more challenging datasets is not very common. Mean Teacher [3] represents the state-of-the-art on ILSVRC-2012 [26] when using only 10% of the labels and [2] achieved the state state-of-the-art result on FGVC datasets.

Pseudo-labeling [5] is a simple and well known strategy in SSL with neural networks. For unlabeled data, a pseudo-label is the class which has the maximum predicted probability, and is considered as if it were the true class. This method is equivalent to entropy minimization [4,5]. The conditional entropy of the class probabilities can be exploited to measure class overlapping. The overlap of class probability distribution can be reduced minimizing the entropy for unlabeled data. However, pseudo-label method is unable to correct its own errors [3,27]. In this paper we evaluate entropy minimization [4] with Convolutional Neural Networks models and compare it with pseudo-label SSL [2]. The comparison is evaluated according to six FGVC datasets.

SSL shares similarities with the related topic of Semi-Supervised Domain adaptation (SSDA), in which the test and train samples come from two different distributions [28,29]. However, standard evaluation of SSL algorithms typically do not consider this case. Finally, SSDA is a key strategy for learning incrementally from unlabeled and non-stationary video streams [30–33].

2.3 Semi-Supervised FGVC

To the best of our knowledge, [2] is the only approach evaluating FGVC datasets in a SSL learning context and it proposes a pseudo-label based technique to leverage unlabeled data. The method, after each training, generates pseudo-labels on the unlabeled set to be added to the labeled training samples; it select the top-k most-confident label greater than a threshold value.

The Semi-Supervised iNaturalist-Aves dataset (FGVC7) has been recently released. It presents some of the challenges encountered in a realistic setting, such as fine-grained similarity between classes, significant class imbalance, and domain mismatch between the labeled and unlabeled data. As reported by the panel of the competition all participating teams applied the pseudo-label method [5] and the state-of-the-art method [27] provides similar performance but is computationally more expensive. Other recent state-of-the-art methods [3,34,35] are also exploited but do not improve the performance.

3   Problem Formulation

In this section we briefly review the formulation of SSL and the relationship between entropy minimization and pseudo-labeling. In Semi-Supervised

Learning [1] we are provided with a dataset of K classes containing both labeled and unlabeled examples. The dataset $\mathcal{D}$ is divided in two parts: a labeled subset $\mathcal{D}_l = \{(\mathbf{x}_i, y_i)\}_{i=1}^{|D_l|}$ and an unlabeled subset $\mathcal{D}_u = \{(\mathbf{x}_j)\}_{j=1}^{|D_u|}$, where $|D_l|$ and $|D_u|$ are respectively the number of examples of the labeled and unlabeled datasets. Semi-Supervised Learning (SSL) aims to improve model performance by incorporating a large amount of unlabeled data during training. Formally, the goal of SSL is to leverage the unlabeled data $\mathcal{D}_u$ to produce a prediction function f^θ, with trainable parameters θ, that is more accurate than using the labeled data D_l only.

3.1   Pseudo-label

Unlabeled samples are treated as labeled samples, and training proceeds with the standard supervised loss function:

$$\widehat{y}_i^k = \begin{cases} 1 & \text{if } k = \text{argmax} f_k^\theta(x_j) \\ 0 & \text{otherwise.} \end{cases}$$

In this way pseudo-labels of unlabeled samples are considered as if they were true labels. The Cross-Entropy loss calculated on pseudo-labeled samples is:

$$\mathcal{L}_{pl} = -\frac{1}{|D_u|} \sum_{j=1}^{|D_u|} \sum_{i=1}^{K} \widehat{y}_i^j \log f_i^\theta(x_j).$$

So the overall objective function is:

$$\hat{\theta} = \underset{\theta}{\text{argmin}} \left(-\frac{1}{|D_l|} \sum_{j=1}^{|D_l|} \sum_{i=1}^{K} y_i^j \log(f_i^\theta(x_j)) - \lambda \frac{1}{|D_u|} \sum_{j=1}^{|D_u|} \sum_{i=1}^{K} \widehat{y}_i^j \log f_i^\theta(x_j) \right), \quad (1)$$

where the first term is the standard cross-entropy loss in which y_i^j is the label of the j-th sample of the class i and λ weights the contribution of the second term.

3.2   Entropy Minimization

One common assumption in many SSL methods is that decision function boundary should not pass through high-density regions of the marginal data distribution. One way to enforce this, is requiring the classifier to output low-entropy predictions on unlabeled data [4]. This encourages the network to make confident (i.e., low-entropy) predictions on unlabeled data regardless of the predicted class, discouraging the decision boundary from passing near data points where it would otherwise be forced to produce low-confidence predictions. This effect can be achieved by adding a simple loss term which minimizes the entropy of the prediction function $f^\theta(x)$:

$$\hat{\theta} = \underset{\theta}{\text{argmin}}(\mathcal{L}_{ce} + \lambda H).$$

In which the entropy H calculated on unlabeled data is:

$$H = -\frac{1}{|D_u|} \sum_{j=1}^{|D_u|} \sum_{i=1}^{K} f_i^\theta(x_j) \log f_i^\theta(x_j).$$

$$\hat{\theta} = \underset{\theta}{\operatorname{argmin}} \left(-\frac{1}{|D_l|} \sum_{j=1}^{|D_l|} \sum_{i=1}^{K} y_i^j \log(f_i^\theta(x_j)) - \lambda \frac{1}{|D_u|} \sum_{j=1}^{|D_u|} \sum_{i=1}^{K} f_i^\theta(x_j) \log f_i^\theta(x_j) \right).$$

(2)

As can be noticed Eqs. 1 and 2, are equivalent: the hard pseudo-label $\hat{y}_i^k$ is replaced by soft one in term of the network output $f_i^\theta(x_j)$. According to this, pseudo-labeling is closely related with entropy minimization.

4   Experimental Results

4.1   Datasets

Table 1. A summary information for the evaluated FGVC datasets. Semi-Supervised iNaturalist-Aves explicitly includes unlabeled data for classes outside the known ones (out-of-class).

Dataset	Classes #	Train #	Test #	Unl. in Class #	Unl. out of Class #
FGVC–Aircraft	100	6667	3333	–	–
CUB-200-2011	200	5994	5794	–	–
Stanford Cars	196	8144	8041	–	–
Oxford flowers	102	2040	6149	–	–
Stanford Dogs	120	12000	8580	–	–
iNaturalist-Aves	200	4000	4000	26640	122208

We compare supervised, pseudo-label and entropy minimization approaches on a variety of fine-grained datasets. As we describe in the following, these datasets have many challenging peculiarities. The six datasets we employed are:

- *FGVC Aircraft* [36], which includes 100 aircraft classes and a total of 10,000 images with small background noise but higher inter-class similarity.
- *Stanford Cars* [37], which consists of 16,185 images from 196 different models of cars.
- *Cub-200-2011* [38], which contains 11,788 images from 200 bird species, with large intra-class variation but small inter-class variation.
- *Oxford flower 102* [39], that is a dataset consisting of 102 flower categories, commonly occurring in the United Kingdom. Each class consists of 40 to 258 images. The images have large scale, pose and light variations.

- *Stanford Dogs Dataset* [40], which contains images of 120 breeds of dogs from around the world with 150–252 images for each breed.
- *SSL iNaturalist Aves dataset* which is composed of images of 200 different species of birds, split as in Table 1. We used the 3,959 labeled and the 26,640 unlabeled in-class images to train our model. The dataset also includes 122,208 unlabeled, out-of-class images we did not use.

A summary of the evaluated datasets with their split is shown in Table 1.

Table 2. Evaluation based on the datasets of Table 1. We reported the standard classification accuracy.

Method	Aircraft	Birds	Cars	Flowers	Dogs	iNaturalist	#Parameters
Supervised	*87.38*	**85.50**	88.35	**94.05**	75.49	*47.65*	25M (ResNet50)
SSLFGC [2]	–	78.06	88.52	*93.83*	**88.43**	–	56M (Inception)
EntMin	84.45	78.63	90.23	88.95	71.30	46.4	25M (ResNet50)
EntMin	**89.78**	*80.50*	**90.85**	88.75	*82.28*	**49.7**	44M (ResNet101)

4.2 Comparison with SSL Methods

We compare two SSL baselines. The results are reported in Table 2. For the pseudo-labeling method, we follow the approach in [2], which employs a Inception-ResNetV2 network [41]. For entropy minimization, we use two different architectures: Resnet50 and ResNet101 [42]. The number of parameters for each model is reported in Table 2. According to [2], we split the training set of each dataset into 75% for training and 25% for validation, while test set is used as unlabeled data. Data augmentation is performed with basic image transformations as random crop, center crop and random horizontal flip. We also report the results of the supervised baseline not using unlabeled data. As expected, We find that the supervised baseline outperforms the semi-supervised ones in *Aircraft*, *Birds*, *Flowers*, *Dogs* and *iNaturalist*. In some cases even SSLFGC, which accounts for more than double the parameters number of ResNet50 architecture, shows lower performances.

Regarding SSL, entropy minimization has slightly overall better performances, with the exception of the *Flowers* and the *Dogs* datasets in which SSLFGC achieves better results. Better performances of entropy minimization can be ascribed to the ability of the method to correct its own mistakes; with pseudo label method, wrong classifications can be quickly amplified resulting in confident but erroneous labels on the unlabeled data points.

References

1. van Engelen, J.E., Hoos, H.H.: A survey on semi-supervised learning. Mach. Learn. **109**(2), 373–440 (2019). https://doi.org/10.1007/s10994-019-05855-6

2. Nartey, O.T., Yang, G., Wu, J., Asare, S.K.: Semi-supervised learning for fine-grained classification with self-training. IEEE Access **8**, 2109–2121 (2019)
3. Tarvainen, A., Valpola, H.: Mean teachers are better role models: weight-averaged consistency targets improve semi-supervised deep learning results. In: Advances in Neural Information Processing Systems, pp. 1195–1204 (2017)
4. Grandvalet, Y., Bengio, Y.: Semi-supervised learning by entropy minimization. In: Advances in Neural Information Processing Systems, pp. 529–536 (2005)
5. Lee, D.-H.: Pseudo-label: the simple and efficient semi-supervised learning method for deep neural networks. In: ICML: Workshop: Challenges in Representation Learning (WREPL), p. 2013. Atlanta, Georgia, USA (2013)
6. Wei, X.-S., Wu, J., Cui, Q.: Deep learning for fine-grained image analysis: a survey. arXiv preprint arXiv:1907.03069 (2019)
7. Ge, W., Lin, X., Yu, Y.: Weakly supervised complementary parts models for fine-grained image classification from the bottom up. In: Proceedings of the IEEE Conference on Computer Vision and Pattern Recognition, pp. 3034–3043 (2019)
8. Korsch, D., Bodesheim, P., Denzler, J.: Classification-specific parts for improving fine-grained visual categorization. In: Fink, G.A., Frintrop, S., Jiang, X. (eds.) DAGM GCPR 2019. LNCS, vol. 11824, pp. 62–75. Springer, Cham (2019). https://doi.org/10.1007/978-3-030-33676-9_5
9. Zhang, L., Huang, S., Liu, W., Tao, D.: Learning a mixture of granularity-specific experts for fine-grained categorization. In Proceedings of the IEEE International Conference on Computer Vision, pp. 8331–8340 (2019)
10. Cui, Y., Song, Y., Sun, C., Howard, A., Belongie, S.: Large scale fine-grained categorization and domain-specific transfer learning. In: Proceedings of the IEEE Conference on Computer Vision and Pattern Recognition, pp. 4109–4118 (2018)
11. Touvron, H., Vedaldi, A., Douze, M., Jégou, H.: Fixing the train-test resolution discrepancy. In: Advances in Neural Information Processing Systems, pp. 8252–8262 (2019)
12. Krause, J., et al.: The unreasonable effectiveness of noisy data for fine-grained recognition. In: Leibe, B., Matas, J., Sebe, N., Welling, M. (eds.) ECCV 2016. LNCS, vol. 9907, pp. 301–320. Springer, Cham (2016). https://doi.org/10.1007/978-3-319-46487-9_19
13. Lin, T.-Y., RoyChowdhury, A., Maji, S.: Bilinear CNN models for fine-grained visual recognition. In Proceedings of the IEEE International Conference on Computer Vision, pp. 1449–1457 (2015)
14. Simon, M., Rodner, E., Darrell, T., Denzler, J.: The whole is more than its parts? from explicit to implicit pose normalization. IEEE Trans. Pattern Anal. Mach. Intell. **42**, 749–763 (2018)
15. Zheng, H., Fu, J., Zha, Z.-J., Luo, J.: Learning deep bilinear transformation for fine-grained image representation. In: Advances in Neural Information Processing Systems, pp. 4277–4286 (2019)
16. Ngiam, J., Peng, D., Vasudevan, V., Kornblith, S., Le, Q.V., Pang, R.: Domain adaptive transfer learning with specialist models. arXiv preprint arXiv:1811.07056 (2018)
17. Sun, C., Shrivastava, A., Singh, S., Gupta, A.: Revisiting unreasonable effectiveness of data in deep learning era. In: Proceedings of the IEEE International Conference on Computer Vision, pp. 843–852 (2017)
18. Real, E., Aggarwal, A., Huang, Y., Le, Q.V.: Regularized evolution for image classifier architecture search. In: Proceedings of the AAAI Conference on Artificial Intelligence, vol. 33, pp. 4780–4789 (2019)

19. Dosovitskiy, A., et al.: An image is worth 16x16 words: transformers for image recognition at scale (2020)
20. Vaswani, A.: Attention is all you need. In: Advances in Neural Information Processing Systems, pp. 5998–6008 (2017)
21. Oliver, A., Odena, A., Raffel, C.A., Cubuk, E.D., Goodfellow, I.: Realistic evaluation of deep semi-supervised learning algorithms. In: Advances in Neural Information Processing Systems, pp. 3235–3246 (2018)
22. Miyato, T., Maeda, S., Koyama, M., Ishii, S.: Virtual adversarial training: a regularization method for supervised and semi-supervised learning. IEEE Trans. Pattern Anal. Mach. Intell. **41**(8), 1979–1993 (2018)
23. Athiwaratkun, B., Finzi, M., Izmailov, P., Wilson, A.G.: There are many consistent explanations of unlabeled data: Why you should average. In: International Conference on Learning Representations (2018)
24. Krizhevsky, A., et al.: Learning multiple layers of features from tiny images (2009)
25. Netzer, Y., Wang, T., Coates, A., Bissacco, A., Wu, B., Ng, A.Y.: Reading digits in natural images with unsupervised feature learning (2011)
26. Russakovsky, O., et al.: Imagenet large scale visual recognition challenge. Int. J. Comput. Vision **115**(3), 211–252 (2015)
27. Zhai, X., Oliver, A., Kolesnikov, A., Beyer, L.: S4l: self-supervised semi-supervised learning. In: Proceedings of the IEEE International Conference on Computer Vision, pp. 1476–1485 (2019)
28. Yao, T., Pan, Y., Ngo, C.-W., Li, H., Mei, T.: Semi-supervised domain adaptation with subspace learning for visual recognition. In: Proceedings of the IEEE Conference on Computer Vision and Pattern Recognition, pp. 2142–2150 (2015)
29. Saito, K., Kim, D., Sclaroff, S., Darrell, T., Saenko, K.: Semi-supervised domain adaptation via minimax entropy. In Proceedings of the IEEE International Conference on Computer Vision, pp. 8050–8058 (2019)
30. Pernici, F., Bruni, M., Del Bimbo, A.: Self-supervised on-line cumulative learning from video streams. Computer Vision and Image Understanding, pp. 102983 (2020)
31. Pernici, F., Del Bimbo, A.: Unsupervised incremental learning of deep descriptors from video streams. In: 2017 IEEE International Conference on Multimedia & Expo Workshops (ICMEW), pp. 477–482. IEEE (2017)
32. Lisanti, G., Masi, I., Pernici, F., Del Bimbo, A.: Continuous localization and mapping of a pan-tilt-zoom camera for wide area tracking. Mach. Vis. Appl. **27**(7), 1071–1085 (2016)
33. Salvagnini, P., et al.: Information theoretic sensor management for multi-target tracking with a single pan-tilt-zoom camera. In: IEEE Winter Conference on Applications of Computer Vision, pp. 893–900. IEEE (2014)
34. Berthelot,D., et al.: Mixmatch: a holistic approach to semi-supervised learning. In: Advances in Neural Information Processing Systems, pp. 5049–5059 (2019)
35. Sohn, K., et al.: Fixmatch: simplifying semi-supervised learning with consistency and confidence. arXiv preprint arXiv:2001.07685 (2020)
36. Maji, S., Rahtu, E., Kannala, J., Blaschko, M., Vedaldi, A.: Fine-grained visual classification of aircraft. arXiv preprint arXiv:1306.5151 (2013)
37. Krause, J., Stark, M., Deng, J., Fei-Fei, L.: 3D object representations for fine-grained categorization. In: Proceedings of the IEEE International Conference on Computer Vision Workshops, pp. 554–561 (2013)
38. Wah, C., Branson, S., Welinder, P., Perona, P., Belongie, S.: The caltech-ucsd birds-200-2011 dataset (2011)

39. Nilsback, M.-E., Zisserman, A.: Automated flower classification over a large number of classes. In: Indian Conference on Computer Vision, Graphics and Image Processing, December 2008
40. Khosla, A., Jayadevaprakash, N., Yao, B., Fei-Fei, L.: Novel dataset for fine-grained image categorization. In: First Workshop on Fine-Grained Visual Categorization, IEEE Conference on Computer Vision and Pattern Recognition, Colorado Springs, CO, June 2011
41. Szegedy, C., Ioffe, S., Vanhoucke, V., Alemi, A.: Inception-v4, inception-resnet and the impact of residual connections on learning (2016)
42. He, K., Zhang, X., Ren, S., Sun, J.: Deep residual learning for image recognition (2015)

HCAU 2020 - The First International Workshop on Deep Learning for Human-Centric Activity Understanding

Workshop on Deep Learning for Human-Centric Activity Understanding

Workshop Description

Understanding human activity and thus effectively collaborating with humans is critical for some artificial intelligence systems. It is a very challenging problem which involves multiple tasks such as human action detection, person tracking, pose estimation, human-object interaction, and so on. Each of them has been independently developed into a research sub-area. Among them, however, there may exist some connections, which can be leveraged to boost the recognition. The purpose of this workshop is to bring together the research on human activity understanding, which hopefully can trigger more discussions on cross-task recognition and inspire new research ideas for human-centric activity understanding. This workshop encourages multi-task pattern recognition research, such as joint action detection and person tracking, joint event segmentation and recognition, joint pose tracking and estimation, and so on.

This workshop received 17 submissions in total. It is a single-blind review process, and each paper was reviewed by two reviewers. Eventually, there are 10 papers were accepted and included in this edition. They covered various aspects of human activity understanding, including human action recognition, human pose estimation and recognition, human activity anticipation, anomaly event detection, and son on.

November 2020

Lamberto Ballan
Jingen Liu
Ting Yao
Tianzhu Zhang

Organization

Program Chairs

Lamberto Ballan	University of Padova, Italy
Jingen Liu	JD.com AI Research (Silicon Valley), USA
Ting Yao	JD.com AI Research (Beijing), China
Tianzhu Zhang	University of Science and Technology of China, China

Program Committee

Juan Carlos Niebles	Stanford University, USA
Chong-Wah Ngo	City University of Hong Kong, China
Ling Shao	IIAI, Abu Dhabi
Zheng Sou	Facebook, USA
Wu Liu	JD.com AI Research (Beijing), China
Waqas Sultani	ITU, Pakistan
Huijuan Xu	UC Berkely, USA
Xinchao Wang	Stevens Institute of Technology, USA
Yu Kong	Rochester Institute of Technology, USA
Yi Yao	SRI International, USA
Mohamed Elhoseiny	KAUST
Mohammad Hosseini	Comcast Research, USA

Spot What Matters: Learning Context Using Graph Convolutional Networks for Weakly-Supervised Action Detection

Michail Tsiaousis[1]([✉])[iD], Gertjan Burghouts[2][iD], Fieke Hillerström[2][iD], and Peter van der Putten[1][iD]

[1] Leiden University, Niels Bohrweg 1, 2333 Leiden, CA, The Netherlands
`tsiaousis.michail@gmail.com, p.w.h.van.der.putten@liacs.leidenuniv.nl`
[2] TNO, Oude Waalsdorperweg 63, 2597 The Hague, AK, The Netherlands
`{gertjan.burghouts,fieke.hillerstrom}@tno.nl`

Abstract. The dominant paradigm in spatiotemporal action detection is to classify actions using spatiotemporal features learned by 2D or 3D Convolutional Networks. We argue that several actions are characterized by their context, such as relevant objects and actors present in the video. To this end, we introduce an architecture based on self-attention and Graph Convolutional Networks in order to model contextual cues, such as actor-actor and actor-object interactions, to improve human action detection in video. We are interested in achieving this in a weakly-supervised setting, i.e. using as less annotations as possible in terms of action bounding boxes. Our model aids explainability by visualizing the learned context as an attention map, even for actions and objects unseen during training. We evaluate how well our model highlights the relevant context by introducing a quantitative metric based on recall of objects retrieved by attention maps. Our model relies on a 3D convolutional RGB stream, and does not require expensive optical flow computation. We evaluate our models on the DALY dataset, which consists of human-object interaction actions. Experimental results show that our contextualized approach outperforms a baseline action detection approach by more than 2 points in Video-mAP. Code is available at https://github.com/micts/acgcn.

Keywords: Weakly-supervised action detection · Graph convolutional networks · Relational reasoning · Actor-context relations

1 Introduction

Human action recognition is an important part of video understanding, with potential applications in robotics, autonomous driving, surveillance, video retrieval and healthcare. Given a video, spatiotemporal action detection aims

---

M. Tsiaousis—This work was carried out during an internship at TNO.

A. Del Bimbo et al. (Eds.): ICPR 2020 Workshops, LNCS 12664, pp. 115–130, 2021.
https://doi.org/10.1007/978-3-030-68799-1_9

to localize all human actions in space and time, and classify the actions being performed. The dominant paradigm in action detection is to extend CNN-based object detectors [13,18] to learn appearance and motion representations in order to jointly localize and classify actions in video. The desired output are action tubes [9]: sequences of action bounding boxes connected in time throughout the video.

In contrast to object detection, action detection requires learning of both appearance and motion features. Although spatiotemporal features are essential for action recognition and detection, they might prove insufficient for actions that share similar characteristics in terms of appearance and motion. For example, spatiotemporal features might not be sufficient to differentiate the action "Taking Photos" in Fig. 1 from a similar one, such as "Phoning", since both share similar characteristics in space and time (i.e. similar posture, motion around the head). As humans, we make use of context to put actions and objects in perspective, which can be an important cue to improve action recognition. Such contextual cues can refer to actor-object and actor-actor interactions. For instance, a person holding a camera is more likely to perform the action "Taking Photos" than "Phoning", and vice versa. CNNs are able to capture such abstract or distant visual interactions only implicitly by stacking several convolutional layers, which increases the overall complexity and number of parameters. Hence, an approach to explicitly model contextual cues would be beneficial.

We introduce an approach to explicitly learn contextual cues, such as actor-actor and actor-object interactions, to aid action classification for the task of action detection. Our model, inspired by recent work on graph neural networks [12,27,32], learns context by performing relational reasoning on a graph structure using Graph Convolutional Networks (GCN) [12]. A high-level overview is illustrated in Fig. 1. Given a detected actor in a short video clip, we construct a graph with an actor node encoding actor features, and context nodes encoding context features, such as objects and other actors in the scene. The graph's adjacency matrix consists of relation values encoding the importance of context nodes to the actor node, and is learned during training via gradient descent. Graph convolutions accumulate the learned context to the actor to obtain contextualized/updated actor features for action classification. Our model aids explainability by visualizing the learned adjacency matrix as an attention map that highlight the relevant context for recognizing the action.

We are interested in an approach to learn these contextual cues using as few annotated data as possible. Recent works [7,23,25,33] that model contextual cues for action detection rely on full supervision in terms of actor bounding box annotations. However, extensive video annotation is time consuming and expensive [15]. In this work, we are interested in learning context for the task of weakly-supervised action detection, i.e. action detection when only a handful of annotated frames are available throughout the action instance. Following the setting of sparse spatial supervision [31], we train our contextual model by using up to five actor bounding box annotations throughout the action instance.

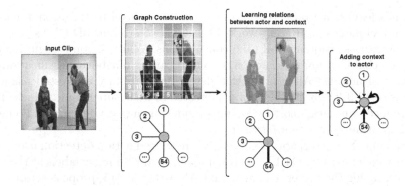

Fig. 1. Given spatiotemporal features extracted from an input clip, we construct a graph with an actor (grey) node and context (numbered) nodes in order to model relations, such as actor-actor and actor-object interactions. Graph convolutions accumulate the learned context to the actor to obtain updated actor features for classification.

We evaluate our models on the challenging Daily Action Localization in YouTube (DALY) dataset [31], which consists of 10 action classes of human-object interactions (e.g. Drinking, Phoning, Brushing Teeth), and is annotated based on sparse spatial supervision. Therefore, DALY is a suitable test bed to model context for the task of weakly-supervised action detection.

Our contributions are as follows: 1) We introduce an architecture employing Graph Convolutional Networks [12] in order to model contextual cues to improve classification of human actions in videos; 2) Our model aids explainability by visualizing the graph's adjacency matrix in the form of attention maps that highlight the learned context, even in a zero-shot setting, i.e. for actions and objects unseen during training; 3) We achieve 1) and 2) in a weakly-supervised setting, i.e. when annotated data are sparse throughout the action instance; 4) We introduce an intuitive metric based on recall of retrieved objects in attention maps, in order to quantitatively evaluate how well the model highlights the important context. As attention maps are often used for qualitative inspection only, this metric may be of general use beyond our use case.

In Sect. 3, we present the baseline model and our approach on learning context by performing reasoning on a graph structure using GCN [12]. Additionally, we discuss training of these models using sparse spatial supervision [31]. We conduct experiments and report results in Sect. 4. In Sect. 5, we perform a qualitative and quantitative analysis of attention maps.

2   Related Work

2.1   Action Recognition and Detection

Action recognition aims to classify the action taking place in a video. Early approaches relied on two-stream 2D CNNs [21] operating on RGB and optical flow inputs, or on detectors to classify bounding boxes of components in keyframes leaving action classification at the video level to traditional machine learning techniques [3]. Recent works focus on (two-stream) 3D CNNs [4,17,24], which perform spatiotemporal (3D) convolutions. While action recognition considers the classification task, action detection carries out both classification and detection of actions. Although action detection is usually addressed using full supervision [7,9,23,25,30,33], we are interested in weak supervision, which allows us to reduce annotation cost by training models using very few action bounding box annotations per action instance.

Sivan and Xiang [22] approach weakly-supervised action detection using Multiple Instance Learning (MIL). Their approach requires binary labels at the video level, indicating the presence of an action. Mettes et al. [15] propose action annotation using points, instead of boxes. Chéron et al. [5] present a unified framework for action detection by incorporating varying levels of supervision in the form of labels at the video level, a few bounding boxes, etc. Weinzaepfel et al. [31] introduce DALY and the setting of sparse spatial supervision, in which, up to five bounding boxes are available per action instance. Chesneau et al. [6] produce full-body actor tubes inferred from detected body parts, even when the actor is occluded or part of the actor is not included in the frame.

2.2 Visual Relational Reasoning

There has been recent research on augmenting deep learning models with the ability to perform visual relational reasoning.

Santoro et al. [20] propose the relation network, which models relations between pairs of feature map pixels for visual question answering. This idea has been extended for action detection [23,25] to model actor-context relations. Similar to [23,25], we treat every $1 \times 1 \times 1$ location of the feature map as context. In these works, learned actor-context relations are used either directly [23] or to highlight features of the feature map [25] for action classification. In contrast, we encode actor-context relations as edges in a graph, and graph convolutions output updated actor features for action classification.

With regard to visual attention, non-local neural networks [28] compute the output of a feature map pixel as a weighted sum of all input pixels. For action detection, Girdhar et al. [7] extend the Transformer architecture [26] for action detection. Although a Transformer can be represented as a graph neural network and vice versa, we argue that a graph representation is simpler and more intuitive compared to the Transformer representation of Queries, Keys and Values. Whilst [7] use two transformations of a feature map to represent context as Keys and Values, this representation does not have a direct interpretation in a graph. In contrast, we use a single transformation to obtain context features, which correspond to context nodes in the graph. Furthermore, our model does not require residual connections [11] nor Layer Normalization [2], two essential components of the Transformer.

In this work, we apply Graph Convolutional Networks (GCN) [12], which provide a structured and intuitive way to model relations between nodes in a graph. Recently, GCN have been used for visual relational reasoning for the tasks of action recognition [29] and group activity recognition [32]. Zhang et al. [33] employ GCN [12] for action detection, where nodes represent detected actors and objects. As we do not require an external object detector, our approach is suitable to reason with respect to arbitrary context and objects which cannot be detected, e.g. because the object detector has not been trained to do so.

Whilst aforementioned approaches [7,23,25,33] rely on full actor supervision during training, our work focuses on learning contextual cues using weak supervision in terms of actor bounding box annotations. Although similar attention maps are also presented in [7,23,25], our work is the first to generate them in a zero-shot setting, and introduce a metric to quantitatively evaluate them.

3   Learning Context with GCN

We propose an approach to learn contextual cues, such as actor-actor and actor-object interactions, by performing relational reasoning on a graph structure using Graph Convolutional Networks (GCN) [12]. We expect such contextual cues to improve action classification, as they can be discriminative for the actions performed, and provide more insight into what the model has learnt, which benefits interpretability. Moreover, we are interested in learning context using weak supervision in terms of actor bounding box annotations. An overview of our proposed GCN model is illustrated in the second branch of Fig. 2. The input is a short video clip with at least one actor performing an action. A 3D convolutional network extracts spatiotemporal features for the input clip, up to a convolutional layer. We treat every $1 \times 1 \times 1$ spatiotemporal location of the output feature map as context, while actor features are extracted by RoI Pooling [8] on the detected actor's bounding box and subsequent 3D convolutional layers. We construct a graph consisting of context nodes and an actor node, with connections drawn from every context node to the actor node. Relation values between nodes are encoded in the adjacency matrix, and learned using a dot-product self-attention mechanism [26,27]. Graph convolutions accumulate the learned context to the actor in order to obtain updated/contextualized actor features for action classification. We compare the GCN model to a baseline model that uses no context, and classifies the action using the feature representation corresponding to the actor bounding box.

In this section, we first present the 3D convolutional backbone network to learn spatiotemporal features using weak supervision. Next, we present our approach on learning contextual cues for action detection using GCN and we provide implementation details.

3.1   Feature Extraction with Weak Supervision

Backbone Network. Spatiotemporal features for the whole input clip are extracted using a 3D convolutional backbone network. There are several 3D architectures in literature [4,17,24]. We opt for I3D [4], which is widely used and has demonstrated very positive results in action recognition. The input is

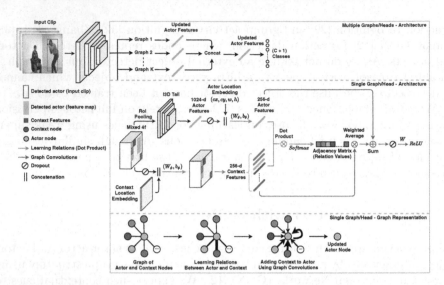

Fig. 2. The lowest part shows the graph representation of the GCN model for a single graph, while its implementation using matrix operations is shown in the middle part. The top part illustrates the construction of multiple graphs (multi-head attention) in order to learn different types of actor-context relations.

a sequence of frames of size $C \times T \times H \times W$, where C denotes the number of channels, T is the number of input frames, and H and W represent the height and width of the input sequence. Features are extracted up to `Mixed_4f` layer, which has an output feature map of size $D' \times T' \times H' \times W'$, where D' denotes the number of feature channels, $T' = \frac{T}{8}$, $H' = \frac{H}{16}$, and $W' = \frac{W}{16}$.

Actor Feature Extraction with Weak Supervision. We are interested in learning spatiotemporal features using only a handful of annotated frames per action instance. For an annotated frame, also called keyframe, annotation is in the form of an action bounding box and corresponding class label. Due to the limited number of available annotated frames, training a Region Proposal Network (RPN) [18] to produce actor box proposals would be sub-optimal. To this end, we train our models using sparse spatial supervision as introduced in [31]. In detail, a Faster R-CNN [18] detects all actors in each frame, and detections are tracked throughout the action instance using a tracking-by-detection approach [30], which produces class-agnostic action tubes. In practice, we use tubes provided by [31]. Tubes are labeled based on spatiotemporal Intersection over Union (IoU) with sparse annotations, i.e. ground truth tubes comprised of up to 5 bounding boxes throughout the action instance. Tubes with spatiotemporal IoU greater than 0.5 are assigned to the action class of the ground truth tube with the highest IoU. If no such ground truth tube exists, the action tube is labeled as background. The backbone is augmented with a RoI pooling layer [8] to extract features for each actor for action classification. Boxes of each tube are appropriately scaled and mapped to the output feature map of Mixed_4f layer, with a temporal stride of four frames. For each action tube, RoI pooling extracts actor features of size $D' \times T' \times 7 \times 7$. Actor features are then passed through I3D tail consisted of 3D convolutional layers Mixed_5b and Mixed_5c. Finally, a spatiotemporal (3D) average pooling layer reduces the size to $D'' \times 1 \times 1 \times 1$.

3.2   Graph Convolutional Networks

Learning Relations. Our graph consists of two types of nodes: context nodes and actor nodes. Context node features, $f'_j \in \mathbb{R}^{D' \times 1 \times 1 \times 1}$, $j = 1, 2, \ldots, M$, $M = T'H'W'$, correspond to every $1 \times 1 \times 1$ spatiotemporal location of the output feature map of Mixed_4f layer. Actor node features, $a'_i \in \mathbb{R}^{D'' \times 1}$, $i = 1, 2, \ldots, N$, where N is the number of detected actors in the input clip, are extracted as described in Sect. 3.1. Relations between actor features and context features, shown in orange arrows in Fig. 2, are learned using a dot-product self-attention operation [26,27], after projecting the features in a lower dimensional space using a linear transformation. Formally,

$$e_{ij} = \theta(a'_i)^T \cdot \phi(f'_j) \tag{1}$$

where

$$a_i = \theta(a'_i) = \mathbf{W}_\theta a'_i + \mathbf{b}_\theta \tag{2}$$
$$f_j = \phi(f'_j) = \mathbf{W}_\phi f'_j + \mathbf{b}_\phi \tag{3}$$

Equations 2–3 are transformations for actor features and context features, respectively, with $\mathbf{W}_\theta \in \mathbb{R}^{D'' \times D}$, $\mathbf{W}_\phi \in \mathbb{R}^{D' \times D}$; $\mathbf{b}_\theta, \mathbf{b}_\phi \in \mathbb{R}^{D \times 1}$; $D < D', D''$. In matrix form, $\mathbf{A} \in \mathbb{R}^{N \times D}$ for transformed actor features and $\mathbf{F} \in \mathbb{R}^{M \times D}$ for transformed context features. The graph is represented by an adjacency matrix, $\mathbf{G} \in \mathbb{R}^{N \times M}$, where $g_{ij} \in \mathbf{G}$ denotes the relation or attention value, indicating the importance of context feature, f_j, to actor feature, a_i. Consequently, $\mathbf{G}$ is a directed graph connecting every context node to every actor node. Relation or attention values, g_{ij}, are obtained by applying softmax normalization on e_{ij} (output of dot-product) across context features

$$g_{ij} = \frac{\exp(e_{ij})}{\sum_k \exp(e_{ik})} \tag{4}$$

Graph Convolutions. Having defined the graph and a mechanism for learning actor-context relations, we perform reasoning on the graph in order to obtain updated actor features. This is achieved by accumulating information from context nodes to the actor node using graph convolutions. Updated actor features, $\mathbf{Z} \in \mathbb{R}^{N \times D}$, are obtained by

$$\mathbf{Z} = \sigma\left(\left(\mathbf{GF} + \mathbf{A}\right)\mathbf{W}\right) \tag{5}$$

The operation is shown in blue arrows in Fig. 2. The weighted average of $\mathbf{F}$ with the relation values $\mathbf{G}$ produces weighted context features. Adding actor features $\mathbf{A}$ to the resulting representation imposes identity links for all actor nodes in the graph. The output is passed through a learnable linear transformation $\mathbf{W} \in \mathbb{R}^{D \times D}$ and a non-linear activation function $\sigma(\cdot)$ implemented as ReLU [10].

In order to capture multiple types of relations between the actor and the context, we perform multi-head attention [26] by constructing multiple graphs at a given layer and merging their outputs using concatenation or summation. Weight matrices $\mathbf{W}_\theta, \mathbf{W}_\phi, \mathbf{W}$ are independent across graphs. Finally, in order to encode updated actor features on a higher level, we stack multiple GCN layers by providing the output of multiple graphs as input to the next GCN layer.

Location Embedding. Location information, such as the position of an actor with respect to other actors and objects, is important for modeling contextual cues. However, such information, encoded indirectly by regular convolutions, is lost when applying convolutions on a graph structure.

We incorporate location information in both context features and actor features. For context features, we concatenate coordinates (x, y) along the channel dimension before applying $\mathbf{W}_\phi$, indicating the location of the feature on the output feature map. For actor features, we concatenate coordinates (cx, cy, w, h) before applying $\mathbf{W}_\theta$, corresponding to the average center, width and height of the actor tube across the input clip. Coordinates are normalized in $[-1, 1]$.

3.3 Implementation Details

We implement our models in PyTorch [16]. I3D is pre-trained on ImageNet [19] and then on the Kinetics [4] action recognition dataset, while the external detector is pre-trained on the MPII Human Pose dataset [1]. The input is a clip of 32 RGB frames with spatial resolution of 224×224. The output feature map of Mixed_4f layer has $D' = 832$ channels, while actor features have $D'' = 1024$. Transformations $\mathbf{W}_\theta, \mathbf{W}$ are implemented as fully connected layers and $\mathbf{W}_\phi$ as a 3D convolutional layer with kernel size $1 \times 1 \times 1$. We set $D = 256$. We apply 3-dimensional dropout to context features before $\mathbf{W}_\phi$. Additionally, 1-dimensional dropout is applied to actor features before $\mathbf{W}_\theta$ in the first GCN layer, before $\mathbf{W}$ in all GCN layers and prior to the final classification layer (in both GCN and baseline model). Dropout probability is 0.5 in all cases. All fully connected layers are initialized using a Normal distribution according to [10]. We set the gain parameter to 1 for $\mathbf{W}_\theta$ and to $\sqrt{2}$ for the rest of the fully connected layers. $\mathbf{W}_\phi$ is initialized using a Uniform distribution according to [10] in the range $(-b + 0.01, b - 0.01)$ for the first GCN layer, and in the range $(-b, b)$ for subsequent layers, using a gain of $\frac{1}{\sqrt{3}}$. Biases of all layers are initialized to zero.

Models are optimized using SGD and cosine learning rate annealing, with learning rate $2.5 \cdot 10^{-4}$ over 150 epochs, and $4.7 \cdot 10^{-5}$ over 450 epochs, for the baseline and GCN model, respectively. We use a batch size of 3 clips, where each clip is randomly sampled from a video in the training set. Tubes of each clip are scored using the softmax scores produced by the model. During inference, we sample 10 32-frame clips from each video, and tubes are scored by averaging the softmax scores across the clips. The same clips are sampled in order to facilitate fair comparison between different models. Training time is approximately one day for GCN and less than half a day for the baseline on a GTX 1080 Ti GPU.

4 Experiments

In this section, we first describe the DALY dataset and the evaluation metric used throughout the experiments. Next, we conduct experiments to evaluate the performance of the GCN model, and we compare it with the baseline and the state of the art on DALY. Finally, we evaluate the GCN model using minimal spatial supervision i.e. one bounding box per action instance.

4.1 Dataset and Evaluation Metric

We develop and evaluate our models on the Daily Action Localization in Youtube (DALY) [31] dataset. It consists of 510 videos of 10 human actions, such as "Drinking", "Phoning" and "Brushing Teeth". In this paper, we do not perform temporal localization, and we assume that the temporal boundaries of each action instance within a video are known. An action instance has an average duration of 8 s and may contain more than one person performing an action. Each of the 10 classes contains an interaction between a person and an object that define the action taking place. There are 31 training videos and 20 test videos per class. We fine-tune our models by holding out a subset of the training set as a validation set, consisted of 10 videos from each class. We evaluate models using Video-mAP at 0.5 IoU threshold (Video-mAP@0.5) [9].

4.2 Evaluation of Architecture Choices

In this section, we experimentally evaluate the GCN model with respect to several architecture choices. Specifically, we experiment with up to two GCN layers and up to three graphs per layer. Additionally, we compare concatenation and summation as merging functions to combine the output of multiple graphs. Finally, we measure the impact of including the location embedding and the I3D tail (convolutional layers Mixed_5b and Mixed_5c) to extract actor features.

Results with respect to different number of layers and graphs are shown in Table 1, along with the number of parameters for every configuration (I3D parameters are not included). Note that for two GCN layers, the first layer always employs concatenation as a merging function. Building multiple graphs is beneficial for model performance, for both functions. It is interesting that mAP increases for a 2-layer GCN model with concatenation, but not with summation. Concatenation outperforms summation in nearly all configurations. For the rest of the experiments, we choose a 2-layer, 2-graph GCN model with concatenation, which provides a good trade-off between performance and number of parameters.

Table 1. Validation mAP with respect to different number of layers, number of graphs per layer and merging functions to combine the output of multiple graphs. The number of model parameters are provided for every configuration.

# Layers	# Graphs	Merging function	# Parameters	Val. mAP
1	1	–	543K	49.39
1	2	Sum	1.084M	51.39
		Concat	1.086M	51.3
1	3	Sum	1.624M	50.7
		Concat	1.630M	50.98
2	1	–	887K	47.28
2	2	Sum	1.903M	50.1
		Concat	1.906M	51.82
2	3	Sum	3.050M	49.98
		Concat	3.055M	52.09

In order to measure the impact on model performance obtained by the location embedding and I3D tail, we remove them from the architecture and examine the difference in model performance. By removing the location embedding, the model has no information of the actor's location relatively to other actors and objects, and relations are calculated based solely on visual features. This results in a decrease of 1.1 points in mAP (50.7), which indicates that modeling spatial actor-context relations improves performance. By removing the I3D tail, actor features are extracted from the output feature map of Mixed_4f layer. This results in a significant decrease of more than four points in mAP (47.42), highlighting the importance of using the I3D tail to encode actor features.

4.3   Comparison with Baseline and State of the Art

We compare the GCN model with the baseline model and the state-of-the-art [6,31] on the DALY test set in Table 2.

The baseline model classifies actor features obtained from I3D (see Sect. 3.1) using a linear layer that outputs classification scores for C action classes and a background class. Results are shown in Table 2. Across five repetitions, the GCN model outperforms the baseline model by 2.24 (3.7%) points in mean mAP, and by 2.94 (4.9%) points in maximum mAP. The left-hand side of Fig. 3 illustrates per-class average precision for the baseline and GCN model. GCN performs comparably or better than the baseline model in all classes except "TakingPhotosOrVideos". On the right-hand side of Fig. 3, we visualize t-sne [14] actor feature embeddings, colored by the respective action, for the GCN (top) and baseline model (bottom). The GCN model produces tighter and more distinct clusters compared to the baseline model.

Table 2. Comparison of GCN model with the baseline model and state-of-the-art on the test set. We report model architecture and input modalities (RGB, Optical Flow).

Model	Architecture	Input	Test mAP
Weinzaepfel et al. [31]	Fast R-CNN (VGG-16)	RGB, OF	61.12
Chesneau et al. [6]	Fast R-CNN (VGG-16)	RGB, OF	63.51
Baseline (Ours)	I3D	RGB	59.58 ($\pm$ 0.22) (59.79)
GCN (Ours)	I3D	RGB	61.82 ($\pm$ 0.51) (62.73)

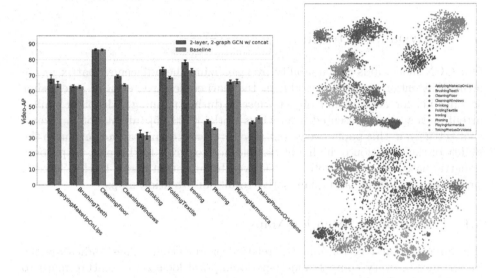

Fig. 3. Per-class Video-AP on the test set across five repetitions of GCN and baseline model, and t-SNE actor feature embeddings of GCN (top) and baseline model (bottom).

Comparing the GCN model with the state-of-the-art [6, 31], we obtain slightly improved performance in comparison to [31], while Chesneau et al. [6] achieve better performance by 1.69 points in mean mAP and 0.78 points in maximum mAP. We argue that this is due to the following reasons. Firstly, our models are trained using fewer videos, since we hold out a part of the training set as a validation set for fine-tuning. Secondly, [6, 31] train their models on the region proposals produced by the detector (see Sect. 3.1), while we train our models on the data provided by [31], which are only the final detections of the detector. Consequently, we train our models using fewer videos and boxes compared to [6, 31]. It is worth noting that, in contrast to [6, 31], our model does not employ expensive optical flow computation.

4.4   Reducing Annotation to One Bounding Box

We examine model performance when minimal spatial supervision is used, i.e. one bounding box per action instance. We label tubes based on spatial IoU with the ground truth box of a randomly selected keyframe for each action instance. A GCN model is then trained using the newly labeled tubes. Using only a single keyframe to label tubes, we obtain a small decrease in mAP, from 61.82 to 61.07. On the other hand, when one keyframe is used during evaluation too, performance decreases by 3.15 points in mAP. The reason for such a decrease is that mAP is not adequately estimated using only one evaluation keyframe.

5   Analysis of Attention

Our GCN model aids explainability by visualizing the adjacency matrix in the form of attention maps that highlight the learned context, even in a zero-shot setting, i.e. for actions and objects unseen during training. Although similar attention maps are presented in previous works [7,23,25] (albeit not in a zero-shot setting), in this paper, we go one step further to quantitatively evaluate the ability of the attention to highlight the relevant context. To this end, we propose a metric based on recall of objects retrieved by attention maps. In this Section, we present qualitative and quantitative results of attention maps.

5.1   Evaluation of Attention Maps

The adjacency matrix contains the relation or attention values, indicating the importance of every context node (spatiotemporal location of feature map) to the actor node. By visualizing the adjacency matrix, we obtain an attention map that highlights, for a given actor, the important context regions the model pays most attention to. The map is interpolated to the original input size and overlaid on the input clip. Our model is even able to generalize its attention in a zero-shot setting i.e. for actions that the model has not been trained to recognize. To achieve this, we train a GCN model by excluding two action classes, and then visualize the attention maps for the excluded classes.

Qualitative Evaluation. Figure 4 illustrates examples of attention maps, where each one is the combination of four adjacency matrices (2 GCN layers; 2 graphs per layer) by summing their values along the spatial dimensions. Each example contains four attention maps, representing time progression along the input clip. The last row of Fig. 4 contains zero-shot attention maps for classes "Ironing" and "TakingPhotosOrVideos". The attention maps show that our GCN model highlights relevant context, such as objects, hands and faces, and is also able to track objects along time. Finally, our model is able to highlight relevant objects (e.g. Iron, Camera) for actions unseen during training (last row of Fig. 4).

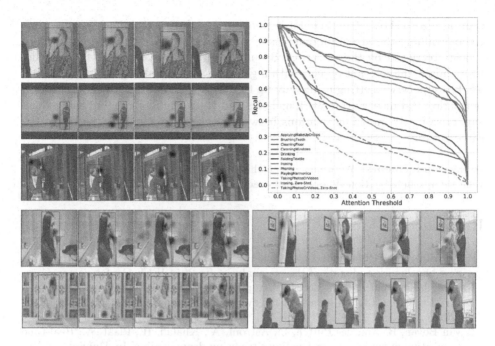

Fig. 4. Per-class object recall curves, along with visualizations of attention maps (last row illustrates zero-shot cases) for actions "Drinking", "CleaningFloor", "Cleaning-Windows", "BrushingTeeth", "FoldingTextile", "Ironing", "TakingPhotosOrVideos".

Quantitative Evaluation. We evaluate how well the attention maps highlight relevant objects by introducing a metric based on recall of objects retrieved by the attention. DALY provides object bounding box annotations on annotated frames. Given the attention map produced for a detected actor, we sum the attention values inside the object's bounding box. An instance is a true positive if the sum of values is larger than a threshold, and a false negative, otherwise.

A per-class quantitative evaluation of attention maps is shown in Fig. 4, with recall on the y-axis and the attention threshold on the x-axis. Dashed curves correspond to zero-shot cases. Our metric suggests that objects are retrieved by the attention with relatively high recall, even for large attention thresholds, which shows the effectiveness of our model to highlight the relevant context.

6   Conclusion

We propose an approach using Graph Convolutional Networks [12] to model contextual cues, such as actor-actor and actor-object interactions, to improve action detection in video. On the challenging DALY dataset [31], our model outperforms a baseline, which uses no context, by more than 2 points in Video-mAP, performing on par or better in all action classes but one. The learned adjacency matrix, visualized as an attention map, aids explainability by highlighting the learned context, such as objects relevant for recognizing the action, even in a zero-shot setting, i.e. for actions unseen during training. We quantitatively evaluate the attention maps using our proposed metric based on recall of objects retrieved by the attention. Results show the effectiveness of our model to highlight the relevant objects with high recall. All the above are achieved in a weakly-supervised setting using only up to five or even one actor box annotation per action instance. Future work includes end-to-end model training using weak supervision and modeling relations between consecutive clips and videos of the same action.

Acknowledgements. We would like to thank Philippe Weinzaepfel for providing us with the predicted action tubes of their tracking-by-detection model.

References

1. Andriluka, M., Pishchulin, L., Gehler, P., Schiele, B.: 2D human pose estimation: new benchmark and state of the art analysis. In: 2014 IEEE Conference on Computer Vision and Pattern Recognition, pp. 3686–3693 (2014)
2. Ba, J., Kiros, J.R., Hinton, G.E.: Layer normalization. ArXiv abs/1607.06450 (2016)
3. van Boven, B., van der Putten, P., Åström, A., Khalafi, H., Plaat, A.: Real-time excavation detection at construction sites using deep learning. In: Duivesteijn, W., Siebes, A., Ukkonen, A. (eds.) IDA 2018. LNCS, vol. 11191, pp. 340–352. Springer, Cham (2018). https://doi.org/10.1007/978-3-030-01768-2_28
4. Carreira, J., Zisserman, A.: Quo vadis, action recognition? a new model and the kinetics dataset. In: 2017 IEEE Conference on Computer Vision and Pattern Recognition, pp. 4724–4733 (2017)
5. Chéron, G., Alayrac, J.B., Laptev, I., Schmid, C.: A flexible model for training action localization with varying levels of supervision. In: Advances in Neural Information Processing Systems 31, pp. 942–953. Curran Associates, Inc. (2018)
6. Chesneau, N., Rogez, G., Alahari, K., Schmid, C.: Detecting parts for action localization. ArXiv abs/1707.06005 (2017)
7. Girdhar, R., João Carreira, J., Doersch, C., Zisserman, A.: Video action transformer network. In: 2019 IEEE/CVF Conference on Computer Vision and Pattern Recognition, pp. 244–253 (2019)
8. Girshick, R.: Fast R-CNN. In: 2015 IEEE International Conference on Computer Vision, pp. 1440–1448 (2015)
9. Gkioxari, G., Malik, J.: Finding action tubes. In: 2015 IEEE Conference on Computer Vision and Pattern Recognition, pp. 759–768 (2015)
10. He, K., Zhang, X., Ren, S., Sun, J.: Delving deep into rectifiers: surpassing human-level performance on imagenet classification. In: 2015 IEEE International Conference on Computer Vision, pp. 1026–1034 (2015)
11. He, K., Zhang, X., Ren, S., Sun, J.: Deep residual learning for image recognition. In: 2016 IEEE Conference on Computer Vision and Pattern Recognition, pp. 770–778 (2016)
12. Kipf, T.N., Welling, M.: Semi-supervised classification with graph convolutional networks. In: 5th International Conference on Learning Representations, OpenReview.net (2017)

13. Liu, W., Anguelov, D., Erhan, D., Szegedy, C., Reed, S., Fu, C.-Y., Berg, A.C.: SSD: single shot MultiBox detector. In: Leibe, B., Matas, J., Sebe, N., Welling, M. (eds.) ECCV 2016. LNCS, vol. 9905, pp. 21–37. Springer, Cham (2016). https://doi.org/10.1007/978-3-319-46448-0_2
14. van der Maaten, L., Hinton, G.: Visualizing data using t-SNE. J. Mach. Learn. Res. **9**, 2579–2605 (2008)
15. Mettes, P., Snoek, C.G.: Pointly-supervised action localization. Int. J. Comput. Vision **127**(3), 263–281 (2019)
16. Paszke, A., et al.: Pytorch: an imperative style, high-performance deep learning library. In: Advances in Neural Information Processing Systems 32, pp. 8024–8035. Curran Associates, Inc. (2019)
17. Qiu, Z., Yao, T., Mei, T.: Learning spatio-temporal representation with pseudo-3D residual networks. In: 2017 IEEE International Conference on Computer Vision (ICCV), pp. 5534–5542 (2017)
18. Ren, S., He, K., Girshick, R., Sun, J.: Faster R-CNN: towards real-time object detection with region proposal networks. IEEE Trans. Pattern Anal. Mach. Intell. **39**(6), 1137–1149 (2017)
19. Russakovsky, O., et al.: Imagenet large scale visual recognition challenge. Int. J. Comput. Vision **115**(3), 211–252 (2015)
20. Santoro, A., et al.: A simple neural network module for relational reasoning. In: Advances in Neural Information Processing Systems 30, pp. 4967–4976. Curran Associates, Inc. (2017)
21. Simonyan, K., Zisserman, A.: Two-stream convolutional networks for action recognition in videos. In: Advances in Neural Information Processing Systems 27, pp. 568–576. Curran Associates, Inc. (2014)
22. Siva, P., Xiang, T.: Weakly supervised action detection. In: Proceedings of the British Machine Vision Conference. BMVA Press (2011)
23. Sun, C., Shrivastava, A., Vondrick, C., Murphy, K., Sukthankar, R., Schmid, C.: Actor-centric relation network. In: Ferrari, V., Hebert, M., Sminchisescu, C., Weiss, Y. (eds.) ECCV 2018. LNCS, vol. 11215, pp. 335–351. Springer, Cham (2018). https://doi.org/10.1007/978-3-030-01252-6_20
24. Tran, D., Wang, H., Torresani, L., Ray, J., LeCun, Y., Paluri, M.: A closer look at spatiotemporal convolutions for action recognition. In: 2018 IEEE/CVF Conference on Computer Vision and Pattern Recognition, pp. 6450–6459 (2018)
25. Ulutan, O., Rallapalli, S., Srivatsa, M., Manjunath, B.S.: Actor conditioned attention maps for video action detection. In: 2020 IEEE Winter Conference on Applications of Computer Vision (WACV), pp. 516–525 (2020)
26. Vaswani, A., et al.: Attention is all you need. In: Advances in Neural Information Processing Systems 30, pp. 5998–6008. Curran Associates, Inc. (2017)
27. Velickovic, P., Cucurull, G., Casanova, A., Romero, A., Liò, P., Bengio, Y.: Graph attention networks. ArXiv abs/1710.10903 (2018)
28. Wang, X., Girshick, R., Gupta, A., He, K.: Non-local neural networks. In: 2018 IEEE/CVF Conference on Computer Vision and Pattern Recognition, pp. 7794–7803 (2018)
29. Wang, X., Gupta, A.: Videos as space-time region graphs. In: Ferrari, V., Hebert, M., Sminchisescu, C., Weiss, Y. (eds.) ECCV 2018. LNCS, vol. 11209, pp. 413–431. Springer, Cham (2018). https://doi.org/10.1007/978-3-030-01228-1_25
30. Weinzaepfel, P., Harchaoui, Z., Schmid, C.: Learning to track for spatio-temporal action localization. In: 2015 IEEE International Conference on Computer Vision, pp. 3164–3172 (2015)

31. Weinzaepfel, P., Martin, X., Schmid, C.: Towards weakly-supervised action localization. ArXiv abs/1605.05197 (2016)
32. Wu, J., Wang, L., Wang, L., Guo, J., Wu, G.: Learning actor relation graphs for group activity recognition. In: 2019 IEEE/CVF Conference on Computer Vision and Pattern Recognition, pp. 9956–9966 (2019)
33. Zhang, Y., Tokmakov, P., Hebert, M., Schmid, C.: A structured model for action detection. In: 2019 IEEE/CVF Conference on Computer Vision and Pattern Recognition, pp. 9967–9976 (2019)

Social Modeling Meets Virtual Reality: An Immersive Implication

Habib Ullah[1], Sultan Daud Khan[2], Mohib Ullah[3(✉)], and Faouzi Alaya Cheikh[3]

[1] College of Computer Science and Engineering, University of Ha'il, Hail, Saudi Arabia
[2] National University of Technology, Islamabad, Pakistan
[3] Department of Computer Science, Norwegian University of Science and Technology, Gjøvik, Norway
mohib.ullah@ntnu.no

Abstract. The development of novel techniques for social modeling in the context of surveillance applications has significantly reduced manual processing of large and continuous video data. These techniques for social modeling widely cover crowd motion analysis since the impact of social modeling on crowd is significant. However, existing crowd motion analysis methods face a number of problems including limited availability of crowd data representing a specific behavior and weaknesses of proposed models to explore the underlying patterns of crowd behavior. To cope with these problems, we propose a novel method based on energy modeling and social interaction of individual particles in crowd to detect unusual behavior. Our method describes collective dissipative interactions among particles in a crowd scene. We reveal the changing patterns about the crowd behavior states, to support the conversion between different social behaviors during evolution. To further improve the performance of our method, virtual reality can be considered to consolidate the acquisition of data associated with a particular behavior. Therefore, we provide theoretical background of immersive implication considering virtual reality that can expose individuals to virtual crowds and acquire useful data on human motion and behaviors in crowds. The experimental evaluation of our energy and social interaction driven method shows convincing results.

Keywords: Virtual reality · Crowd analysis · Social modeling · Anomaly detection

1 Introduction

Video analysis of crowded scenes in different places such as public parks, universities, stadiums and stations has received significant interest in surveillance applications that aim to detect different unusual behavior [25,28]. An unusual

© Springer Nature Switzerland AG 2021
A. Del Bimbo et al. (Eds.): ICPR 2020 Workshops, LNCS 12664, pp. 131–140, 2021.
https://doi.org/10.1007/978-3-030-68799-1_10

behavior occurs less frequently and is aberrant from those usual ones that frequently take place in crowd scenes. Manual analysis of surveillance videos to identify such behaviors is a highly time consuming process. In situations where a stampede or panic is reported, often many days of recordings need to be checked. It is, therefore, highly possible that due to manual error such abnormalities might cost a lot, and so, automatic systems are particularly beneficial in such situations. The description of unusual activities in complex crowd scenes like pedestrian walkways and stadiums is an arduous task, as an unusual behavior can be anything ranging from a panic situation to a stampede. In fact, it is difficult to develop an exhaustive list of all unusual behaviors. Therefore, a technique developed for any unusual behavior detection must be significantly driven by generalization capabilities to handle any such situation.

In the literature, different methods [4,24,26] are proposed for unusual crowd behavior detection. For instance, Kasudiya et al. [15] proposed the use of a wireless sensor network for monitoring the crowd and their behavior. Singh et al. [23] proposed aggregation of ensembles for detecting an anomaly in video data showing crowded scenes, which leverage the existing capability of pre-trained ConvNets and a pool of classifiers. Their method uses an ensemble of different fine-tuned convolutional neural networks based on the hypothesis that different network architectures learn different levels of semantic representation from crowd videos and thus an ensemble will enable enriched feature sets to be extracted. Behera et al. [3] investigated that the behavior of a dense crowd can be approximated using well-known physics-based models. They proposed a computer vision guided expert system with the help of a Langevin equation-based force model to represent the linear flow of the crowd, particularly in situations when the density is high. Hatirnaz et al. [12] introduced optical flow based features for abnormal crowd behaviour detection. They annotated surveillance videos using a new semantic metadata model based on multimedia standards using semantic web technologies. In this way, globally inter-operable metadata about abnormal crowd behaviours are generated. Then based on crowd behaviours a novel concept-based semantic search interface is proposed. Ullah et al. [27] used a modified variant of the social force model to highlight potential particles of interest for crowd behavior analysis. Hu et al. [13] proposed a parallel spatial-temporal convolution neural networks model to detect and localize the abnormal behavior in video surveillance. For crowd behavior analysis, Cheung et al. [5] proposed a method composed of two components: a procedural simulation framework to generate crowd movements and behaviors, and a procedural rendering framework to generate different videos or images.

Unusual crowd behavior detection is a challenging problem due to complex social relationship among individuals in the form of groups. The methods we discussed in the literature do not fully exploit social modeling to develop robust model for unusual crowd behavior detection due to the associated issues. For instance, interaction modeling is a problem when huge number of objects are present in a crowded scene. Also, within the same scene, the behavior of individuals may vary greatly, due to which, distinguishing between anomalies and a normal event is a difficult task. To address these issues, we propose a novel method considering the energy and social interaction based modeling. Our method is characterized by unique social interaction representation since we take into account the characteristics of neighboring particles. Our method can be extended to any type of unusual crowd behavior detection. Additionally, we present theoretical background related to virtual reality to improve the performance of our proposed method. In fact, virtual reality is a powerful platform to understand the hidden patterns in crowd scenes.

The rest of the paper is organized as follows: Sect. 2 illustrates the proposed methodology and describes the working of each of the components. In Sect. 3, we present the theoretical background about the use of virtual reality to improve the performance of our proposed method. In Sect. 4, we describe the experiments performed for the validation of the proposed model and compare its performance with other reference methods. Section 5 concludes our proposed method.

2   Crowd Analysis

We present an energy model inspired by [18] to calculate the energy of the interacting particles. Our interaction model takes into account the collisions that occur in unusual crowd behavior. This model is based on the concept to describe collective dissipative interactions among particles in a crowd scene. Our modeling of particle interactions in the regime of crowd analysis redefined the entropy through information theory. This broadened the picture and allows the concept of entropy to permeate the area of crowd motion analysis.

We initialize a set of particles over the video frame. The particle initialization represents the localization of pixels positions uniformly spread over the video. We then track the initialized particles using the Lucas-Kanade optical flow technique. Since we are interested only in the particles associated with the moving entities instead of static regions, we propose an energy model in Eq. 1 for retaining the moving particles.

$$\eta = \frac{1}{|p-1|}(|1 - \sum_{i=1}^{N}(p)^{q_i}|) + \sum_{i=1}^{N}\frac{sin(q_i)}{2sin|q_i - p|} \tag{1}$$

In the equation, p represents the velocity of central particle and q_i represents the velocity of a neighboring particle in a localized region. In fact, our model moves away from the traditional way of handling interaction and energy dissipation related to unusual crowd behavior. Our model is well-defined for particle interactions and it gives physical meaning to the associated parameters. We assume that the crowd scene is associated with N particles. An individual particle moves with a velocity p. According to Eq. 1 we filter out static particles and retain particles associated with motion in crowd scene. Our model resembles with the model of kinetic theory of gases; the collisions of particles with each other are elastic. If we consider crowd as a system, then this crowd system with unusual behavior would be totally out of equilibrium system. The initial state and all subsequently intermediate states of the crowd system are a non-equilibrium thermodynamic system. The particles interact with each other and we will simplify that interaction by assuming that they only do it by collisions since we are targeting unusual crowd behavior. In general, in this system the particles lose and gain energy just by the collisions. Considering the forced state to the equilibrium state, these particles will collide with each other multiple times in such a way that they can lose energy. Then, those particles dissipate energy to the surroundings due to the shocks, but we will assume that the interaction is made in a peculiar way; when they collide with different energies there is no loss of energy, this is when the difference exceeds a certain amount.

The benefits of interaction can drive the evolution of a crowd structure that represents different behaviors. The social interaction of individuals in a group is an emerging methodology to research the theory of crowd behavior. In light of this, we mimic group theory, and then develop our method to solve the complex problems associated with unusual crowd behaviors. Group theory is based on the studies of collective phenomena such as flocks of birds, colonies of ants, and swarms of bees to model biological swarms. These models have been applied as nature-inspired algorithms for solving complex problems in the real world. In fact, the influence of biological analogies is attested by subfields of computer science, such as artificial neural networks, genetic algorithms, and evolutionary computation. For example, the particle swarm optimization (PSO) model [16] is inspired by the social behavior of bird flocking, the ant colony model (ACO) [8] is based on the division and cooperation of ants foraging, the artificial bee colony (ABC) model [14] is constructed by mimicking the cooperative behavior of bee colonies, and the social spider optimization (SSO) model [7] is based on the simulation of cooperative behavior of social-spiders. Each of these algorithms has its own characteristics. Their modeling processes and effective mechanisms inspire us when developing our own method for crowd behavior detection.

We consider the retained particles according to Eq. 1 to find out particles representing unusual crowd behavior. For this purpose, our proposed method is based on social behavior modeling [10] to detect unusual crowd behavior. We model the mechanism between Leaders and Followers particles to infer local dynamics. In this way, our method reveals the changing patterns about the crowd behavior states, to support the conversion between different social behaviors during evolution, to demonstrate crowd behaviors from social grouping perspective, and to avoid the integration of one crowd behavior with another. In fact, our model analyzes the behavior of social groups, the changing patterns of different behaviors, and to set the changing patterns as behavior's criterion of a crowd state. In addition, the mathematical model of our method is deduced from the group theory, crowd dynamics, and the crowd motion pattern theory.

According to the social model [10], the position and velocity of a central particle p_i are $xp_i(t)$ and $vp_i(t)$, respectively. The position and velocity of a surrounding particle q_j is $xq_j(t)$ and $vq_j(t)$. In the $(t+1)th$ iteration, the position and velocity of particle i will be updated as,

$$vp_i(t+1) = w \times vp_i(t) + (\frac{1}{N}\sum_{j=1}^{N} vq_j(t) - vp_i(t)) \tag{2}$$

$$xp_i(t+1) = xp_i(t) + (vp_i(t+1) \times \delta(t)) \tag{3}$$

In the equation, w is a parameter which is set equal to a fixed value during the experiments. The above equations do not take into account the orientation information of particles. The particle status process belongs to the exploration behavior of the central and surrounding particles. This process reveals the change rule of social population in term of localized crowd, and supports the conversion between different social behaviors during evolution. Therefore, such equations should be adjusted to include velocity, position, and angle. To consider all the information into a unified equation ($M_{p,q}$), the modeling is performed as presented in Eq. 4,

$$M_{p,q} = \prod_{j=1}^{N} \frac{\int_{p,q_j}^{p,q_j} 2wvp_i + (\varphi + sin\theta)d\theta}{4d^2} \tag{4}$$

where N is the total number of surrounding particles q_j and vp_i is the velocity of the central particle p_i. φ is the prior average orientation information of all the surrounding particles q_j. θ is the angle between the central particle p_i and a surrounding particle q_j. d is Eucleadien distance between particle p_i and q_j. Equation 4 classifies each particle p_i as belonging to unusual crowd behavior or normal crowd behavior. This equation renders the classification characteristics of each particle. Therefore, it could be treated as features to identify and detect crowd behavior as a whole.

3   Crowd and Virtual Reality

Mathematical modeling of social behaviors of crowd present significant success in the field. However, some limitations still exist to fully characterize crowd representing different behaviors. These limitations include limited availability of crowd data representing a specific behavior, weaknesses of proposed models to explore the underlying patterns of crowd, the perception and understanding of an unusual crowd behavior from first person point of view to ensure people safety, and the impact of varying densities of crowd on individuals. To improve the performance of proposed models and to cope with the key challenges, virtual reality platforms can be exploited since virtual reality has been successfully used for wide range of applications including medical education [2], rehabilitation [11], and data visualization [9,17] to name a few. In fact virtual reality is a powerful platform to acquire useful data on human motion and behaviors in crowds. It allows exposing individuals to virtual crowds: only one individual is required to observe holistic behavior in crowded situations. In the virtual reality, stimuli can be properly controlled and repeated over several individuals. These characteristics have made virtual reality a significant tool to perform experiments in socio-psychology, spatial-cognition, and motion control. In fact, this platform is an effective tool to study how we navigate in crowds. To effectively use virtual reality platform, we have to ensure the validity of obtained data in various crowd situations, perform proper modeling so that locomotion trajectories in virtual environment are similar to reality trajectories, minimize the affect of interaction loop in virtual reality on user behavior, and ensure that the visual feedback in virtual reality enables participants to make realistic navigation decisions.

To address the aforementioned limitations considering virtual reality, two different approaches can be considered to deal with unusual crowd behaviors. The macroscopic approach considers the crowd as a single entity and the microscopic approach considers that global crowd pattern emerge from local interactions between individuals. The use of virtual reality in analyzing crowd motion have a wide range of applications, from training personnel to ensure people safety in crowd, to architecture in building analysis and emergency evacuation studies. The significant impact of virtual reality can be used to extract more and more realistic crowd patterns to enlarge the available patterns associated with a particular crowd behavior. In this context, we define realism as a match between extracted crowd patterns from virtual environment and real patterns extracted from real crowd data. To provide realism to crowd motion analysis for unusual behavior identification, there is a need to understand and model how humans move and behave during local interactions with their neighborhood in virtual conditions. To extract patterns from virtual environment, we have to consider the underlying modeling of human motion and interactions in various situations (several kinds of motion or interactions) and take into account multiple factors such as sociological or psychological ones.

Data can be extracted from virtual environment if we consider uncertainties on people states and motivations and other uncontrolled factors. The virtual environment [21] is based on local interactions between individuals. Many interactions occur between walkers, with many factors of influence. There is a need for observations of individuals facing interactions in crowds to better understand them and improve the level of realism in virtual environment. Considering the presented information about virtual reality in the context of crowd behavior analysis, virtual reality can be used as an experimental tool to perform such observations with an accurate control of experimental conditions.

4   Experiemental Evaluation

We assess the performance of our social interaction based method for unusual crowd behavior detection on a standard dataset available publicly that is UMN [1]. The UMN dataset (as shown in Fig. 1) from the university of Minnesota consists of videos representing unusual crowd behavior. There are three different indoor and outdoor scenes showing 11 different scenarios of unusual crowd behavior. In total, there are 7739 frames and the resolution of each frame is 320×240 pixels. The initial part of each video consists of normal behaviors of pedestrian walking and standing.

We compare the results with four closely related reference methods: spatio-temporal anomaly model (STAM) [20], abnormal behavior model (ABM) [6] and crowd influence model (CIM) [22]. For quantitative evaluation of unusual crowd behavior detection, the equal error rate (EER) for frame-level and the detection rate (DR) for pixel-level analysis are calculated to measure the overall performance. In the literature, the frame-level criterion is mostly used by researchers. The frame-level criterion only measures temporal localization accuracy. It could cause errors due to lucky detection of unusual crowd behavior. Therefore, it assigns a perfect score to a model that detects unusual behavior at a random location of a frame. By considering this fact, it seems that the pixel-level criterion is much better evaluation criterion. Therefore, we use both the temporal and spatial accuracies to rule out random detection. Both criteria are based on true-positive rates (TPR) and false positive rates (FPR). The presence and absence of unusual behaviors are represented by a positive and a negative, respectively. This is compared to the frame level ground-truth, to determine the number of true and false-positive frames. Similarly, pixels related to the unusual crowd behavior are compared to the pixel-level ground-truth to determine the number of true-positive and false-positive.

Fig. 1. UMN dataset. Four different scenes are shown representing unusual crowd behavior where people are running in different directions randomly.

Table 1. UMN dataset. Equal error rate (EER) and detection rate (DR) for the reference methods and our proposed method are presented in the first row and the second row, respectively. Lower EER and higher DR represent better performance.

Dataset	STAM	ABM	CIM	Prop.
UMN	21	34	19	17
	55	61	68	71

For quantitative performance, we calculated the average EER and average DR for UMN dataset reported in Table. For better performance, the EER rate should be lower and the DR rate should be higher. As can be seen in the table, our proposed method outperforms all the reference methods: STAM [20], ABM [6], and CIM [22]. These results show that there is a significant advantage of our proposed method that uses energy and interaction based modeling bringing forth strong generalization capabilities. In fact, the reference methods based on shallow features cannot cope with the adaptively changing sparse movement of the people flows where dynamic motion and occlusions exist. Also the reference methods finds raw features without taking into account the appearance information. Furthermore, STAM [20], ABM [6], and CIM [22] fail to encode unique motion patterns because informative movements only occur in specific regions of the videos. Our proposed method represents high quality description of unusual crowd behavior with the energy and social interaction components. Therefore, we outperform the reference methods in both frame-level and pixel-level analysis. Presenting results based on both criteria reveal the effectiveness of our method.

To use virtual reality for crowd behavior analysis, Unity [19] cross-platform engine can be considered as a powerful tool for crowd behavior research. It can be explored for the ability to create three-dimensional visual scenes and to measure responses to the visual stimuli for testing the hypotheses in a manner and scale that were previously unfeasible (Table 1).

5   Conclusion

We proposed a novel method for unusual crowd behavior detection. Our proposed method represents high quality description of unusual behavior in term of energy and interaction modeling. We also provided the background of using virtual reality platform to improve the performance by extracting of crowd patters from virtual environment. The performance of our proposed method is tested on a standard dataset and compared to three closely related reference methods. The performance metrics EER and DR show that our method outperforms all the reference methods in both frame-level and pixel-level analysis.

As a future work, we would also extend our proposed method to take into account the benefits of virtual reality platforms.

References

1. Unusual crowd activity dataset of university of minnesota. http://mha.cs.umn. edu/movies/crowd-activity-all.avi
2. Ammanuel, S., Brown, I., Uribe, J., Rehani, B.: Creating 3D models from radiologic images for virtual reality medical education modules. J. Med. Syst. **43**(6), 166 (2019)
3. Behera, S., Dogra, D.P., Bandyopadhyay, M.K., Roy, P.P.: Estimation of linear motion in dense crowd videos using langevin model. Expert Systems with Applications, p. 113333 (2020)
4. Cao, H., Sankaranarayanan, J., Feng, J., Li, Y., Samet, H.: Understanding metropolitan crowd mobility via mobile cellular accessing data. ACM Trans. Spatial Algorithms Syst. (TSAS) **5**(2), 1–18 (2019)
5. Cheung, E., Wong, A., Bera, A., Wang, X., Manocha, D.: LCrowdV: generating labeled videos for pedestrian detectors training and crowd behavior learning. Neurocomputing **337**, 1–14 (2019)
6. Chibloun, A., El Fkihi, S., Mliki, H., Hammami, M., Thami, R.O.H.: Abnormal crowd behavior detection using speed and direction models. In: 2018 9th International Symposium on Signal, Image, Video and Communications (ISIVC), pp. 197–202. IEEE (2018)
7. Cuevas, E., Cienfuegos, M., ZaldíVar, D., Pérez-Cisneros, M.: A swarm optimization algorithm inspired in the behavior of the social-spider. Expert Syst. Appl. **40**(16), 6374–6384 (2013)
8. Dorigo, M., Maniezzo, V., Colorni, A.: Ant system: optimization by a colony of cooperating agents. IEEE Trans. Syst. Man Cybern. Part B (Cybernetics) **26**(1), 29–41 (1996)
9. El Beheiry, M., Doutreligne, S., Caporal, C., Ostertag, C., Dahan, M., Masson, J.B.: Virtual reality: beyond visualization. J. Mol. Biol. **431**(7), 1315–1321 (2019)
10. Feng, X., Wang, Y., Yu, H., Luo, F.: A novel intelligence algorithm based on the social group optimization behaviors. IEEE Trans. Syst. Man Cybern. Syst. **48**(1), 65–76 (2016)
11. Hartney, J.H., Rosenthal, S.N., Kirkpatrick, A.M., Skinner, J.M., Hughes, J., Orlosky, J.: Revisiting virtual reality for practical use in therapy: patient satisfaction in outpatient rehabilitation. In: 2019 IEEE Conference on Virtual Reality and 3D User Interfaces (VR), pp. 960–961. IEEE (2019)
12. Hatirnaz, E., Sah, M., Direkoglu, C.: A novel framework and concept-based semantic search interface for abnormal crowd behaviour analysis in surveillance videos. Multimedia Tools and Applications, pp. 1–39 (2020)
13. Hu, Z.P., Zhang, L., Li, S.F., Sun, D.G.: Parallel spatial-temporal convolutional neural networks for anomaly detection and location in crowded scenes. J. Visual Commun. Image Representation **67**, 102765 (2020)
14. Karaboga, D.: An idea based on honey bee swarm for numerical optimization. Technical report, Technical report-tr06, Erciyes university, engineering faculty, computer (2005)
15. Kasudiya, J., Bhavsar, A., Arolkar, H.: Wireless sensor network: a possible solution for crowd management. In: Somani, A.K., Shekhawat, R.S., Mundra, A., Srivastava, S., Verma, V.K. (eds.) Smart Systems and IoT: Innovations in Computing. SIST, vol. 141, pp. 23–31. Springer, Singapore (2020). https://doi.org/10.1007/ 978-981-13-8406-6_3

16. Kennedy, J., Eberhart, R.: Particle swarm optimization. In: Proceedings of ICNN'95-International Conference on Neural Networks, vol. 4, pp. 1942–1948. IEEE (1995)
17. Kvinge, H., Kirby, M., Peterson, C., Eitel, C., Clapp, T.: A walk through spectral bands: using virtual reality to better visualize hyperspectral data. In: Vellido, A., Gibert, K., Angulo, C., Martín Guerrero, J.D. (eds.) WSOM 2019. AISC, vol. 976, pp. 160–165. Springer, Cham (2020). https://doi.org/10.1007/978-3-030-19642-4_16
18. López-Carrera, B., Yáñez-Márquez, C.: A simple model for the entropy of a system with interacting particles. IEEE Access **7**, 108969–108979 (2019)
19. Messaoudi, F., Simon, G., Ksentini, A.: Dissecting games engines: the case of unity3d. In: 2015 International Workshop on Network and Systems Support for Games (NetGames), pp. 1–6. IEEE (2015)
20. Ojha, N., Vaish, A.: Spatio-temporal anomaly detection in crowd movement using sift. In: 2018 2nd International Conference on Inventive Systems and Control (ICISC), pp. 646–654. IEEE (2018)
21. Olivier, A.H., Bruneau, J., Cirio, G., Pettré, J.: A virtual reality platform to study crowd behaviors. Transp. Res. Procedia **2**, 114–122 (2014)
22. Pan, L., Zhou, H., Liu, Y., Wang, M.: Global event influence model: integrating crowd motion and social psychology for global anomaly detection in dense crowds. J. Electron. Imaging **28**(2), 023033 (2019)
23. Singh, K., Rajora, S., Vishwakarma, D.K., Tripathi, G., Kumar, S., Walia, G.S.: Crowd anomaly detection using aggregation of ensembles of fine-tuned convnets. Neurocomputing **371**, 188–198 (2020)
24. Ullah, H.: Crowd Motion Analysis: Segmentation, Anomaly Detection, and Behavior Classification. Ph.D. thesis, University of Trento (2015)
25. Ullah, H., Altamimi, A.B., Uzair, M., Ullah, M.: Anomalous entities detection and localization in pedestrian flows. Neurocomputing **290**, 74–86 (2018)
26. Ullah, H., Ullah, M., Conci, N.: Dominant motion analysis in regular and irregular crowd scenes. In: Park, H.S., Salah, A.A., Lee, Y.J., Morency, L.-P., Sheikh, Y., Cucchiara, R. (eds.) HBU 2014. LNCS, vol. 8749, pp. 62–72. Springer, Cham (2014). https://doi.org/10.1007/978-3-319-11839-0_6
27. Ullah, M., Ullah, H., Conci, N., De Natale, F.G.: Crowd behavior identification. In: 2016 IEEE International Conference on Image Processing (ICIP), pp. 1195–1199. IEEE (2016)
28. Zitouni, M.S., Sluzek, A., Bhaskar, H.: Visual analysis of socio-cognitive crowd behaviors for surveillance: a survey and categorization of trends and methods. Eng. Appl. Artif. Intell. **82**, 294–312 (2019)

Pickpocketing Recognition in Still Images

Prisa Damrongsiri$^{(\boxtimes)}$ and Hossein Malekmohamadi

Institute of Artificial Intelligence, De Montfort University, Leicester LE1 9BH, UK
p2508775@my365.dmu.ac.uk, hossein.malekmohamadi@dmu.ac.uk

Abstract. Human activity recognition (HAR) is a challenging topic in the computer vision field. Pickpocketing is a type of human criminal actions. It needs extensive research and development for detection. This paper researches how it's possible of pickpocketing recognition in still images. This paper takes consideration both of classification and detection. We develop our models from state-of-art pre-trained models: VGG16, ResNet50, ResNet101, and ResNet152. Moreover, we also include a convolutional block attention module (CBAM [27]) in the model. The attention mechanism enhances model performances by focusing on informative features. For classification, the highest accuracy (89%) is ResNet152 with CBAM [27] (ResNet152+CBAM). We also examine pickpocketing detection on RetinaNet [14] and YOLOv.3 [34]. The mean average precision (mAP) of pickpocketing detection is consistent with Redmon et al. [34]. RetinaNet's precision (80 mAP) is higher than YOLOv.3 (78 mAP), but YOLOv.3 is much faster detection. ResNet152+CBAM model detection on RetinaNet approach provides the highest mAP. However, it is much slower detection than YOLOv.3 (only 10 ms). This paper proves that It is possible to implement pickpocketing on still images in a reliable time and with outstanding accuracy. This proposed model possibly apply to the other HAR tasks.

1 Introduction

Pickpocketing is one of the human actions in the form of crime. Criminals pick the victim's wallet or valuable items in a crowded public. In some cases, pickpocketing is smooth and easy to pick things without unaware of the victim. Pickpocketing is a global crisis. Four hundred thousand pickpocketing incidents are happening per day in the world [7]. It is very challenging to detect and track criminals. Moreover, getting the victim's items back is rarely possible. Currently, the transfer learning technique is widely used in the computer vision field. Several papers have studied HAR based on videos or still images. The development in the HAR field is successful. HAR consists of various techniques such as pose analysis, human-object detection, and using Bag-of-visual-words (BOVW). Pickpocketing has not been addressed in prior research works. Pickpocketing is an action which behaves in the same pattern as other usual activities such as moving objects, bending. Therefore, there is a research gap in pickpocketing recognition. This paper studies on pickpocketing recognition based on still images. The related works and development are described in the next section of this paper.

© Springer Nature Switzerland AG 2021
A. Del Bimbo et al. (Eds.): ICPR 2020 Workshops, LNCS 12664, pp. 141–152, 2021.
https://doi.org/10.1007/978-3-030-68799-1_11

2   Background

Recent works on HAR are based on either videos or still images. Using videos has the benefit of utilising spatiotemporal features compared to still images. Videos recognition deliver more accuracy than image-based models. However, it requires a higher processing power. The still images are prevalent in social media. Furthermore, the apparent reason is still frame take much less memory. Image-based HAR divided into three categories: human body information, human-object interaction, and scene context information. Several state-of-art CNN models trained on a large scale and comprehensive context dataset (ImageNet, Coco, and CIFAR10). Some papers applied these pre-trained models to HAR. Sreela et al. [23] used a transfer learning technique (ResNet) for feature extraction. Then, classify 14 human actions by SVM. Furthermore, Mohammadi et al. [17] improved performance action recognition by employing the attention module and ensemble technique to combine the state-of-art CNN models. Yan et al. [28] added soft attention module into the state-of-art CNN model (VGG16). The soft attention enhances action recognition by learning scene information of an image. Additionally, a few articles are researching on crime detection area. Koshti et al. [2] detected anomalies behaviour (Normal and Abnormal classification) via video-based by Two Stream Inflated 3D ConvNets (I3D). UCF-crime [29] video surveillance is the primary dataset of Koshti et al. [2]. UCF-crime [29] consists of 13 anomalies such as robbery, assault, shooting, burglary, stealing, arrest, fighting, shoplifting, arson, explosion, vandalism, abuse, and road accident. I3D-ResNet-50 [2] is the principal model of Koshti et al. [2], which are consisted of ResNet50 as the backbone and following with bottleneck blocks to reduce channels. Moreover, I3D-ResNet-50 [2] also apply batch normalisation in each convolutional layer. Evaluation of I3D-ResNet-50 [2] provided the highest accuracy on CNN classifier (84.28% accuracy). In the meanwhile, some paper researched on suspicious shoplifting in [4]. UCF101 [22] and UCF-crime [24] are investigated unusual shoplifting behaviour. The video dataset was extracted in 4 stages: (1) strict crime moment detection, (2) comprehensive crime moment detection, (3) pre-crime behaviour, and (4) suspicious behaviour. They [4] used 3D-CNN for feature extraction and classification. The result of the model provided high accuracy (75%) in suspicious behaviour detection. Further, Japanese Telecom developed the AI Guardsman to detect shoplifters based on pose estimation (OpenPose [8]) which identify a person's body for suspicious behaviour and also make an alarm to the shopkeeper in case shoplifting has been committed [25].

Generally, state-of-art deep neural network architectures are widely used in HAR such as VGG [21] and ResNet [6]. VGG [21] proposed in the paper 'Very Deep Convolutional Networks for Large Scale Image Recognition.' VGG [21] used small filters to reduce the number of parameters. VGG [21] consists of 5 main convolution blocks. There are two main types which are different weight layers: VGG16 (16 layers) and VGG19 (19 layers). VGG16 achieved 90.1% top-5 test accuracy in ImageNet [30]. On the other hand, ResNet [6] developed in 'Deep Residual Learning for Image Recognition.' A deeper neural network is more challenging for training. The vanishing gradient [9] problem is one of the problems in the deeper neural network. This problem is solved in ResNet [6] by adding the skipping connection, which is called a residual module. ResNet [6] also provides different layers: ResNet18, ResNet34, ResNet50, ResNet101, and ResNet152. Evaluation ImageNet [30] of ResNet provides accuracy 92% in ResNet50 and ResNet101. Moreover, ResNet152 achieved in 93% top-5 test accuracy. Currently, several papers employed transfer learning technique with these state-of-art models. Hence, this paper also employed VGG16, ResNet50, ResNet101, and ResNet152 as the pre-trained models for training the pickpocketing dataset.

Frequently, HAR also considered action detection. Action detection used the same approaches as object detection. There are two types of detector: one-stage and two-stage detector. Two-stage detector proposes a set of regions of interest (ROI) in the first part of model. Next, the model classified them in the region candidates. However, One-stage detector skips the region proposal stage. It runs detection and classification at the same. One stage is faster and simpler, but it is lower precision than the two-stage detector. The popular one stage detectors are YOLO [34], SSD [20], and RetinaNet [14]. Sometime one-stage detector has a problem of imbalance data between the foreground and background. The issue is solved by Focal loss [14]. The focal loss was applied in RetinaNet [14] model, which is introduced by Facebook AI Research (FAIR). RetinaNet model consists of 2 powerful concepts: focal loss and feature pyramid network [13], which are the feature extraction part of RetinaNet [14] architecture.

Currently, the attention mechanism is prevalent in the human action recognition field. The attention determines the informative feature. The useful key point is provided in a specific position of the image. The attention mechanism is developed in several techniques. Structured spatial attention by Khandelwal et al. [12] represented that previous attention values of image feature provided valuable information. Bastidas et al. [1] applied the Channel attention by using soft attention on each channels. Moreover, there are some papers combined both of Spatial attention [12] and Channel attention [1] into the network [18,26,27]. Woo et al. [27] proposed Convolutional Block Attention Module (CBAM). Woo et al. [27] sequentially added channel [1] and spatial [12] module. CBAM [27] provided high accuracy compare with other attention techniques. The channel attention [1] focuses on the 'what' import feature in the input image. It aggregates spatial information of a feature map by using both average pooling and maximum pooling features. On the other hand, spatial attention module [1] utilises the inter-spatial relationship of features. It focuses on 'where' is an informative part. These related works supported that how possible to research on pickpocketing recognition. The development of pickpocketing recognition is described in the next section.

3 Methodology

3.1 Dataset Preparation

The dataset in this paper divides into two classes: pickpocketing and non-pickpocketing. Each label has 500 images (total 1000 images). Source of pickpocketing images are from a social network. Main informative feature of pickpocketing consist of 2 persons, thief and victim. Frequently, pickpocketing action occur in crowd. So, it represents crowd as background of these input images. In mean while, Non-pickpocketing images are associated with people in the crowd and human activity. This dataset is broken into 3 parts: (1) 60% training (600 images), (2) 20% validation (200 images), (3) 20% testing (200 images). Example of images show in Fig. 1 and 2.

Fig. 1. Example of pickpocketing images

Fig. 2. Example of non-pickpocketing images

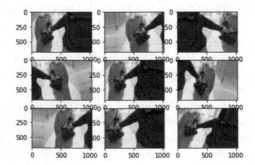

Fig. 3. Data augmentation is the technique to increase data (zoom, flip, rotate)

Data augmentation is a technique to generate more images from raw data by zoom, horizontal flip, shift. The augmentation technique helps to increase the size of the dataset and also solve the over-fitting issue. Therefore, This paper also applies this data augmentation technique on training. The example of augmentation shows in Fig. 3. According to this paper also researches on pickpocketing detection, pickpocketing annotation is required. The annotation are in the form of Xmin, Xmax, Ymin, Ymax, and label. The annotation preparation of this paper is manually to annotate by Labeling [36]. The criminal's hand and the victim's body is an annotation on this paper which is shown in Fig. 4.

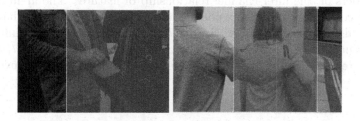

Fig. 4. Bounding box of pickpocketing

3.2   Design

The development of pickpocketing recognition consists of 2 components: classification and detection. Considering training CNN models from scratch is complicated and high time-consuming for training an efficient model. Therefore, this paper employs the state-of-art CNN of VGG16, ResNet50, ResNet101, and ResNet152. These models are retrained with the pickpockcting dataset by transferring learning techniques.

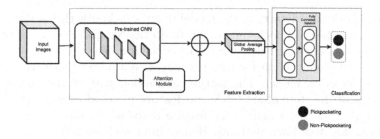

Fig. 5. Pickpocketing classification architecture consists of a pre-trained model and CBAM [27] attention module. Both of them represent in feature extraction part. After that, it connects with fully connected layers

Pickpocketing classification is a binary classifier. The model consists of 2 main components which is shown in Fig. 5. Pre-trained CNN model, CBAM attention, and global average pooling layer are feature extraction. Five fully connected layers connect to this feature extraction section. The first FC layer has 2048 nodes and halve nodes in next layer so on and so forth (2048, 1024, 512, 256, 128). Each layer drop out nodes (=0.5) to avoid over-fitting issue. Finally, these models are trained and validated by cross-entropy loss. Softmax function classifies classes of pickpocketing and non-pickpocketing in terms of probability. The notable point of this paper is including CBAM [27] (Convolutional Block Attention Mechanism) at the end of CNN. CBAM [27] is sequentially combination of Channel Attention Module (CAM [27]) and the Spatial Attention Module (SAM [27]). CAM [27] is looking for 'what' is significant parts of the input feature. CAM [27] considers on the inter-channel relationship of features. The architecture of CAM [27] combines both of max pooling and average pooling. And share to the same neural network. Meanwhile, SAM [27] focuses on 'where' is an informative part. It is utilising the inter-spatial relationship of features. SAM [27] architecture concatenate max pooling and average pooling. Then, pass it convolutional block.

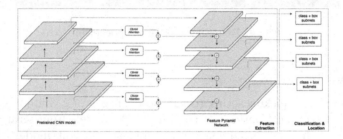

Fig. 6. Pickpocketing detection on RetinaNet consists of a pre-trained model and attention module. Then, it connects with feature pyramid network which is a feature extraction part

For pickpocketing detection, the model detects pickpocketing action via still images. It identifies pickpocketing action by RetinaNet [14] and YOLOv.3 [34]. Both of them is one-stage detectors. Pickpocketing detection's input requires input image and annotation for training model. RetinaNet [14] is the state-of-art object detection model. The feature extraction part of RetinaNet [14] is combination of the state-of-art CNN and feature pyramid network (FPN) [13]. RetinaNet [14] generates multi-scale feature maps in the form of the bottom-up pathway and top-down pathway. Hence, improved resolution and semantic information are addressed in this approach. In this case, we reuse the pre-trained model from the pickpocketing classification section. Those pre-trained models are the backbone of model detection on RetinaNet approach [14]. Then, it connects to 2 sub-networks for classification and regression boxes. RetinaNet Architecture of pickpocketing detection shows in Fig. 6. In the meanwhile, YOLOv.3 [34] approach is also a fast single-stage detector which is fast detection. YOLOv.3 [34] divides the image into the grid cell. Each grid cell predicts bounding boxes and class probabilities. YOLOv.3 [34] makes detection at three different scales same as RetinaNet [14].

4   Experiments

As mentioned earlier, pickpocketing recognition develops both classification and detection. Then, they are analysed in the form of experiments. The first experiment due to classification and the second experiment analyses on detection. Table 1 represents pickpocketing classification models on the first experiment. We train four different CNN pre-trained models: VGG16, ResNet50, ResNet101, and ResNet152. Moreover, each of them also examines including a CBAM [27] module in the model. These backbones use the same classification architecture (5 layers of Fully Connected Layers). Evaluation of pickpocketing classification is measured with testing data (200 images). The evaluation of this classification is in the form of accuracy.

Table 1. VGG16, ResNet50, ResNet101, and ResNet152 Backbones represent different parameters for classification (first experiment)

Backbone	Parameters
VGG16 without attention	18,552,770
ResNet50 without attention	30,544,962
ResNet101 without attention	49,589,314
ResNet152 without attention	65,279,042
VGG16+CBAM	18,618,980
ResNet50+CBAM	31,595,940
ResNet101+CBAM	50,640,292
ResNet152+CBAM	66,330,020

In Table 2 represents the model detections on RetinaNet [14] and YOLOv.3 [34]. The backbone on the RetinaNet [14] approach is reused from the first experiment (Classification part). They consist of 4 different backbones: VGG16, ResNet50, ResNet101, and ResNet152. Moreover, it also includes CBAM [27] attention module. The performance of the pickpocketing detection is evaluated same data from first experiment. The measurement is in the form of mAP (Mean Average Precision) at the IOU threshold = 50.

Evaluation of the first experiment (pickpocketing classification) found on Table 3. It is in the form of accuracy. The result represents a comparison of the original model and including CBAM attention in the model. Adding CBAM enhances the higher accuracy in all models. The evaluation represents that more complexity of model it provides higher accuracy. The highest accuracy found in ResNet152+CBAM (89%).

Table 2. Difference of 9 backbones are used for pickpocketing detection (second experiment)

Detector approaches	Backbone	Parameters
RetinaNet	VGG16 without attention	23,407,725
RetinaNet	ResNet50 without attention	36,582,957
RetinaNet	ResNet101 without attention	55,427,309
RetinaNet	ResNet152 without attention	71,117,037
RetinaNet	VGG16+CBAM	23,556,915
RetinaNet	ResNet50+CBAM	37,763,539
RetinaNet	ResNet101+CBAM	56,807,891
RetinaNet	ResNet152+CBAM	72,497,619
YOLOv.3	DarkNet53	65,252,682

Table 3. Accuracy of pickpocketing classification (first experiment)

Pretrained network	No attention module	CBAM
VGG16	83%	85%
Resnet50	85%	86%
Resnet101	85%	88%
Resnet152	86%	89%

Pickpocketing detection's performance is shown in Table 4. Experiment evaluates in 9 different backbones on single-stage detectors (RetinaNet [14] and YOLOv.3 [34]). The performance of these models are in the mean average precision (mAP) ans detection time (ms). Including CBAM in the architecture can improve performance of all models same as classification part. The highest mAP is ResNet152+CBAM (80.2%) on RetinaNet [14] approach. However, the ResNet152+CBAM [27] network (242.3 ms) take more inference time than YOLOv.3 [34]. Although, YOLOv.3 [34] is lower mAP than ResNet152+CBAM [27], it [34] provides the good mAP (78%). They are comparable. Additionally, the strength point of YOLOv.3 is the fastest detection which is 10ms. Pickpocketing performance should take consider both precision and time. Therefore, YOLOv.3 [34] is the proposed model for pickpocketing detection on this paper.

Table 5 is accuracy model classification of related works [2,31]. Currently, there is no paper researched on pickpocketing recognition. However, there are some related papers on anomaly classification. Koshti et al. [2] developed I3D-ResNet-50 [2] to classify 2 labels: normal and abnormal situation. The I3D-ResNet-50 [2] is evaluated with UCF-Crime [29]. It achieved high accuracy(84.28%).

Table 4. Mean average precision (mAP) of pickpocketing detection

Pretrained network	mAP	Inference time (ms)
VGG16	68	444.6
ResNet50	75	190.6
ResNet101	77.15	212.3
ResNet152	78.07	236.4
VGG16+CBAM	68	451.3
ResNet50+CBAM	76	195.6
ResNet101+CBAM	78.14	219.6
ResNet152+CBAM	80.2	242.3
YOLOv.3	78	10

Table 5. Comparison classification of crime classification model of related works (second experiment)

Model	Dataset	Accuracy
I3D-ResNet-50 [2]	UCF-Crime	84.28%
InceptionV3-VGG16 [31]	UCF-Crime	88.74%

On the other hand, InceptionV3-VGG16 [31] used two-stream CNN. It is proposed in paper: 'two-stream CNN Architecture for Anomalous Event Detection in Real-World Scenarios' by Majhi et al. [31]. The InceptionV3-VGG16 [31] consists of two-stream architecture: Spatial stream and Temporal Stream. InceptionV3 is used in Spatial stream, and VGG16 is used in Temporal stream. The InceptionV3-VGG16 [31] two-stream CNN developed with UCF-Crime [29] same as [2]. Furthermore, The InceptionV3-VGG16 [31] produced higher accuracy (88.74%).

According to our model classification using different input dataset with these related papers [2,31], it cannot directly compare the accuracy of our model with these related papers. 89% accuracy of our model (ResNet152+CBAM from Table 3) does not imply that our model provides higher performance than I3D-ResNet-50 [2] and InceptionV3-VGG16 [31]. However, It imply that our model classification is comparable with the associated paper on crime recognition area.

Moreover, Table 6 shows the model detection performance of related works. Anomaly detection and localisation of criminal behaviour are quite challenging. However, few papers studied in this area. M2Det [32] is performed for gun detection in a paper: 'Gun Detection in Surveillance Videos using Deep Neural Networks' by Lim et al. [32]. M2Det [32] is state-of-art architecture on object detection which is a single-stage detector based on a multi-level feature pyramid network. Lim et al. trained M2Det [32] with UCF-crime [29] and Granada [35] in various environments. M2Det [32] of Lim et al. [32] got 44.2% mAP. Currently, there is no paper research on pickpocketing. Therefore, it cannot directly

Table 6. Crime model detection of related work

Model	Dataset	mAP
M2Det [32]	UCF-Crime, Granada	44.2

compare the benchmark of our model with M2Det [32]. However, the mAP of our model on YOLOv.3 [34] (Table 4) is comparable with related paper [32]. It means that it is possible to research pickpocketing detection.

5   Conclusion

This paper investigates the possibility of developing pickpocketing recognition in still images. Eight models are implemented on classification and nine models for detection. The model development uses state-of-art pre-trained models: VGG16, ResNet50, ResNet101, and ResNet152. Moreover, we included the Convolutional Block Attention Module (CBAM [27]) at the end of the feature extraction section to improve accuracy. The highest accuracy is found on ResNet152+CBAM (89%). Next step, this paper also researches on pickpocketing detection in the image. On the RetinaNet approaches, we reused the backbones of pickpocketing classification. Then it is combined with the Feature pyramid network (FPN). ResNet152+CBAM provides the highest precision (80 mAP). However, RetinaNet takes a lot of inference time. Time is another important factor in crime recognition. YOLOv.3 is another state-of-the-art single-stage detector which claims fast detection. Therefore, pickpocketing detection is also developed YOLOv.3 approach. The inference time of YOLOv.3 is much improved from RetinaNet approach. Although the mean average precision of YOLOv.3 (78 mAP) is lower than ResNet152+CBAM [27], it is comparable. Consequently, YOLOv.3 is the proposed model detection of this paper. It also proves that applying of CBAM into the model enhances the model performance. Therefore, including CBAM with YOLOv.3 is next step of development, which is potential to enhance higher performance. Moreover, our proposed model classification (ResNet152+CBAM) reasonably applies to other HAR such as robbery, fighting.

References

1. Bastidas, A., Tang, H.: Channel attention networks (2019)
2. Koshti, D., Kamoji, S., Kalnad, N., Sreekumar, S., Bhujbal, S.: Video anomaly detection using inflated 3D convolution network. In: 2020 International Conference on Inventive Computation Technologies (ICICT), Coimbatore, India, pp. 729–733 (2020). https://doi.org/10.1109/ICICT48043.2020.9112552
3. Elgendy, M.: Deep Learning For Vision Systems, 1st edn. (n.d.)
4. Guillermo, A., José, R., José, C.: Suspicious behavior detection on shoplifting cases for crime prevention by using 3D convolutional neural networks, pp. 1–7 (2020). https://arxiv.org/abs/2005.02142. Accessed 27 July 2020

5. Guo, G., Lai, A.: A survey on still image based human action recognition. Pattern Recogn. **47**(10), 3343–3361 (2014)
6. He, K., Zhang, X., Ren, S., Sun, J.: Deep residual learning for image recognition (2015)
7. Henrys, S.: Pickpocketing Statistics (2019). Safes International. https://www. safesinternational.com/items/Safes-International-News/pickpocketing-statistics
8. Hidalgo, G., et al.: OpenPose: whole-body pose estimation (2019)
9. Hochreiter, S.: The vanishing gradient problem during learning recurrent neural nets and problem solutions. Int. J. Uncertain. Fuzziness Knowl. Based Syst. **6**(2), 107–116 (1998)
10. Hu, J., Shen, L., Albanie, S., Sun, G., Wu, E.: Squeeze-and-excitation networks (2019)
11. Image-net.org: 2010 ImageNet Large Scale Visual Recognition Competition, ILSVRC 2010 (2010). http://www.image-net.org/challenges/LSVRC/2010
12. Khandelwal, S., Sigal, L.: AttentionRNN: a structured spatial attention mechanism (2019)
13. Lin, T., Dollár, P., Girshick, R., He, K., Hariharan, B., Belongie, S.: Feature pyramid networks for object detection (2016)
14. Lin, T., Goyal, P., Girshick, R., He, K., Dollar, P.: Focal loss for dense object detection. IEEE Trans. Pattern Anal. Mach. Intell. **42**(2), 318–327 (2020)
15. Liu, W., et al.: SSD: single shot MultiBox detector (2015)
16. Marcelino, P.: Transfer learning from pre-trained models (2020). https://towardsdatascience.com/transfer-learning-from-pre-trained-models-f2393f124751. Accessed 2 Jul 2020
17. Mohammadi, S., Ghofrani Majelan, S., Shokouhi, S.B.: Ensembles of deep neural networks for action recognition in still images, pp. 1–4 (2020)
18. Park, J., Woo, S., Lee, J.-Y., Kweon, I.S.: BAM: bottleneck attention module (2018)
19. Peng, X., Wang, L., Wang, X., Qiao, Y.: Bag of visual words and fusion methods for action recognition: comprehensive study and good practice. Comput. Vis. Image Underst. **150**, 109–125 (2016)
20. Redmon, J., Divvala, S., Girshick, R., Farhadi, A.: You only look once: unified, real-time object detection (2015)
21. Simonyan, K., Zisserman, A.: Very deep convolutional networks for large-scale image recognition (2014)
22. Soomro, K., Zamir, A., Shah, M.: UCF101: a dataset of 101 human actions classes from videos in the wild (2012)
23. Sreela, S., Idicula, S.: Action recognition in still images using residual neural network features. Proc. Comput. Sci. **143**, 563–569 (2018)
24. Sultani, W., Chen, C., Shah, M.: Real-world anomaly detection in surveillance videos (2018)
25. The Verge: This Japanese AI security camera shows the future of surveillance will be automated (2018). https://www.theverge.com/2018/6/26/17479068/ai-guardman-security-camera-shoplifter-japan-automated-surveillance. Accessed 27 Jul 2020
26. Wang, F., et al.: Residual attention network for image classification. In: CVPR (2015)
27. Woo, S., Park, J., Lee, J., Kweon, I.: CBAM: convolutional block attention module (2018)
28. Yan, S., Smith, J., Lu, W., Zhang, B.: Multibranch attention networks for action recognition in still images. IEEE Trans. Cogn. Dev. Syst. **10**(4), 1116–1125 (2018)

29. Sultani, W., Chen, C., Shah, M.: UCF-Crime (2018). https://www.crcv.ucf.edu/projects/real-world/
30. Russakovsky, O., et al.: ImageNet large scale visual recognition challenge. Int. J. Comput. Vis. **115**(3), 211–252 (2015). https://doi.org/10.1007/s11263-015-0816-y
31. Majhi, S., Dash, R., Sa, P.K.: Two-stream CNN architecture for anomalous event detection in real world scenarios. In: Nain, N., Vipparthi, S.K., Raman, B. (eds.) CVIP 2019. CCIS, vol. 1148, pp. 343–353. Springer, Singapore (2020). https://doi.org/10.1007/978-981-15-4018-9_31
32. Lim, J.: Gun detection in surveillance videos using deep neural networks. In: 2019 Asia-Pacific Signal and Information Processing Association Annual Summit and Conference (APSIPA ASC), pp. 1998–2002. IEEE, Lanzhou (2019)
33. Kanehisa, R., Neto, A.: Firearm detection using convolutional neural networks. In: Proceedings of the 11th International Conference on Agents and Artificial Intelligence, pp. 707–714. SCITEPRESS - Science and Technology Publications, Prague (2019)
34. Redmon, J., Farhadi, A.: YOLOv3: an incremental improvement, p. 6 (2018)
35. Olmos, R., Tabik, S., Herrera, F.: Automatic handgun detection alarm in videos using deep learning. arXiv:1702.05147 (2017)
36. tzutalin/labelImg (2018). https://github.com/tzutalin/labelImg

t-EVA: Time-Efficient t-SNE Video Annotation

Soroosh Poorgholi[✉], Osman Semih Kayhan, and Jan C. van Gemert

Computer Vision Lab, Delft University of Technology, Delft, The Netherlands
s.poorgholi74@gmail.com, {O.S.Kayhan,J.C.vanGemert}@tudelft.nl

Abstract. Video understanding has received more attention in the past few years due to the availability of several large-scale video datasets. However, annotating large-scale video datasets are cost-intensive. In this work, we propose a time-efficient video annotation method using spatio-temporal feature similarity and t-SNE dimensionality reduction to speed up the annotation process massively. Placing the same actions from different videos near each other in the two-dimensional space based on feature similarity helps the annotator to group-label video clips. We evaluate our method on two subsets of the ActivityNet (v1.3) and a subset of the Sports-1M dataset. We show that t-EVA (https://github.com/spoorgholi74/t-EVA) can outperform other video annotation tools while maintaining test accuracy on video classification.

Keywords: Video annotation · t-SNE · Action recognition

1 Introduction

The availability of large-scale video datasets [17,19,20], has made video understanding in various tasks such as action recognition [25,37,41], object tracking [42,47,48] an attractive topic of research. Various supervised methods, [8,37,41], have improved video classification and temporal localization accuracy on large-scale video datasets such as ActivityNet (v1.3) [17]; however, labeling videos on such a large-scale dataset, requires a great deal of human effort. Therefore, other methods aim to train the networks for tasks such as video action recognition in a semi-supervised [1,46] manner without having the full labels. To decrease the dependency on the quality and amount of annotated data, [12,15] investigated pre-training features with internet videos with noisy labels in a weakly supervised manner. However, these methods cannot achieve higher accuracy on video classification tasks than supervised models on large-scale video datasets such as Kinetics [20]. Instead of using such techniques, we focus on reducing the annotation effort for adding more training data.

Fully-supervised models require much annotated data that is unavailable as videos are unlabeled by nature, and annotating them is labor-intensive. Large scale datasets [6,17,20] use strategies like *Amazon Mechanical Turk* (AMT) to annotate the videos. [20] uses majority voting between multiple AMT workers

© Springer Nature Switzerland AG 2021
A. Del Bimbo et al. (Eds.): ICPR 2020 Workshops, LNCS 12664, pp. 153–169, 2021.
https://doi.org/10.1007/978-3-030-68799-1_12

to accept annotation of a single video. Using such methods is not efficient for video annotation on a large scale as it costs a lot in terms of time and money. MuViLab [2], an open-source software, enables the oracle to annotate multiple parts of a video simultaneously. However, these methods do not exploit the structure of the video data.

We introduce an annotation tool that helps the annotator group-label videos based on their latent space feature similarity in a 2-dimensional space. Transferring the high-dimensional features obtained from 3D ConvNet to two dimensions using t-SNE gives the annotator an easy view to group label the videos both, temporal labels and classification labels. The annotation speed depends on the quality of the extracted features and how well they are placed together in the t-SNE plot. If the classes are well-separated in the t-SNE plot, group labeling becomes faster for the oracle.

We evaluated our method on two subsets of ActivityNet (v1.3 datasets) [17] and a subset of

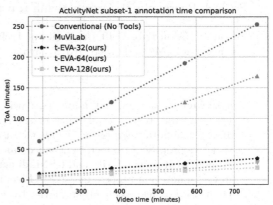

Fig. 1. Comparison of annotation time using different tools versus video time for the ActivityNet [17] subset-1. Our annotation method (t-EVA) outperforms the conventional (no specific tools) annotation and MuViLab [2] in annotation time. With a window size of 128 time-steps (128-TS), our method can annotate 769 min of video in 21 min. The MuViLab and conventional annotation numbers are extrapolated.

Sports-1M dataset [19] with 15 random classes. *Conventional annotation* refers to humans watching the videos and annotating the temporal boundaries of the human actions in videos without any specific tool. *MuViLab* is a more advanced open-source tool that extracts short clips from each video and plays them simultaneously in a grid-like figure beside each other. Oracle can annotate the video by selecting multiple short clips at the same time and assigning the specific class. We show that t-EVA outperforms conventional annotation techniques (with no specific tools) and MuViLab [2] in time of annotation (ToA) by a large margin on the ActivityNet dataset while still being able to keep the test accuracy on video classification task within a close range of using the original ground truth annotations (Fig. 1).

2   Related Work

Video Understanding. In the past the focus was on the use of specific hand-designed features such as HOG3D [21] SIFT-3D [33], optical flow [34] and iDT [40]. Among these methods, iDT and Optical flow is being used in combination with CNNs in different architectures such as two-stream networks [36]. Later some attempts used 2D CNNs and extract features from video frames and combine them with different temporal integration functions [14,45]. The introduction of 3D convolution [35,37] in CNNs which extends the 2D CNNs in temporal dimension showed promising results in the task of action recognition in large-scale video datasets. 3D CNNs in different variations such as single stream and multiple-stream are among state of the art in the task of video understanding [4,10,13,18,28,32,38].

Dimensionality Reduction. Dimensionality reduction (DR) is an essential tool for high-dimensional data analysis. In linear DR methods such as PCA, the lower-dimension representation is a linear combination of the high-dimensional axes. Non-linear methods, on the other hand, are more useful to capture a more complex high-dimensional pattern [22]. In general, non-linear DR tries to maintain the local structure of the data in the transition from high-dimension to low-dimension and tends to ignore larger distances between the features [5]. t-Distributed Stochastic Neighbor Embedding (t-SNE) introduced by [39] is a non-linear DR technique which is used more for visualization. [24] shows that t-SNE is able to distinct well-separable clusters in low-dimensional space. Moreover, some works have been proposed for more effective use of t-SNE. [5] proposes a tool to support interactive exploration and visualization of high-dimensional data. An alternative to t-SNE is using UMAP [29] for dimensionality reduction. However, t-SNE is better studied, shows good results, and has the benefit of high-speed optimization [31]. Therefore, t-EVA uses t-SNE to reduce the dimensionality of the feature representations.

  Data Annotation is essential for supervised models. Different tools have been proposed for making an easy annotation tool for videos and images. However, they usually do not exploit the structure of the data, which is especially useful in videos [2,3,7]. Some works [11,23,26,44] have been done to make the process of image annotation easier. [23] offers a real-time framework for annotating internet images, and [11] uses multi-instances learning to learn the classes and image attributes together; however, none of these methods use a deep representation of data. In more recent works [44] uses *Deep Multiple Instance Learning* to automatically annotate images and [26] uses semi-supervised t-SNE and feature space visualization in lower dimension to provide an interactive annotation environment for images. [9] proposed a general framework for annotating images and videos. However, to the best of our knowledge, our method is the first video annotation platform that can *exploit the structure* of video using latent space feature similarity to increase the annotation speed.

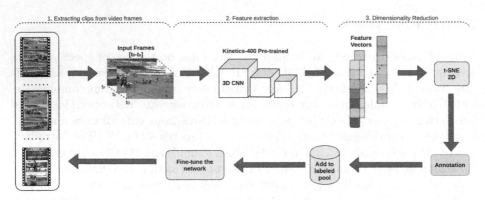

Fig. 2. t-EVA pipeline: 1) Video clips are extracted from n consecutive frames $[t_0\text{-}t_n]$ (time-steps). 2) Spatio-temporal features are extracted from the last layer of a 3D ConvNet before the classifier layer. 3) High dimensional features are projected to two dimensions using t-SNE and are plotted on a scatter plot. 4) Oracle annotates the clips represented in the scatter plot using a lasso tool. 5) The newly annotated data is added to the labeled pool. 6) The network is fine-tuned for a certain number of epochs. This cycle is repeated until all the videos are labeled, or the annotation budget runs out.

3   t-EVA for Efficient Video Annotation

We propose incremental labeling with t-SNE based on feature similarity (Fig. 2). First, several videos are randomly selected from the unlabeled pool, and 3D ConvNet features are extracted. The feature embeddings are transferred to a two-dimensional space using t-SNE. As it can be seen in Fig. 3, the oracle has two subplots for annotation: (i) A plot in which the oracle can use a lasso tool to group label videos and (ii) Other plot with the middle frame of each clip in which the oracle can move and zoom with the cursor on the plot and observe where to annotate. After annotating the first set of videos, the video clips are moved to the labeled pool, and the 3D network is fine-tuned for a certain number of epochs with the newly labeled videos. We continue this process until all the videos are labeled, or the annotation budget finishes.

We use 3D ConvNets to extract features from the videos and split each video v into k shorter clips $v_i = [clip_1, \ldots, clip_k]$ by sampling every n non-overlapping frames $clip_i = [frame_1, \ldots, frame_n]$. Sampling in multiple time-steps enables us to capture different lengths of actions in the dataset. Afterward, each clip c_i is fed into the 3D ConvNet, for feature extraction. The features are extracted from the last convolution layer after applying global average pooling. In t-SNE, the pair-wise distances between feature vectors are used to map features to 2D. In this paper, we use the Barnes-Hut optimized t-SNE version [27], which reduces the complexity of $O(NlogN)$ where N is the number of data-points.

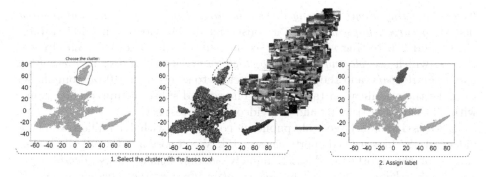

Fig. 3. A minimal representation of the annotation tool. 1) The oracle can see the scatter plot (left) and the corresponding frames from the videos (middle) in separate figures. 2) Based on the figures' inspection, the oracle can detect different clusters of an action class (kayaking) and use the lasso tool to select the cluster. 3) In the end, the oracle assigns a label and based on the assigned class name, the selected points in the scatter plot change color.

3.1 How to Annotate?

An overview of the annotation procedure can be seen in Fig. 3. First, the oracle sees the scatter plot with all points with the same color representing the unlabeled pool (Fig. 3 left) and the corresponding middle frame of each clip in the video (Fig. 3 middle). The oracle can move the cursor and zoom in the plot to inspect the frames with more details. Second, using the lasso tool, the oracle can draw a lasso around the scatter plots based on the visual similarity and inspection of the video frames. Third, oracle assigns the labels, and the network is fine-tuned for a certain number of epochs. The same process repeats until all the videos are annotated, or the annotation budget ends.

4 Experiments

In this section, we first explain the benchmark dataset and evaluation metrics. In addition, we empirically show how our t-EVA can speed up annotation for the ActivityNet dataset while keeping the video classification accuracy in a close range to the usage of the ground truth labels. We also compare our results with MuViLab [2] annotation tool. Furthermore, we qualitatively show how t-EVA can help to annotate the Sports1-M [19].

4.1 Datasets

ActivityNet (v1.3) is an untrimmed video dataset with a wide range of human activities [17]. It comprises of 203 classes with an average of 137 untrimmed videos per class in about 849 h of video. We use two subsets of the ActivityNet dataset. The first subset comprises 10 random classes, namely *preparing salad,* *kayaking, fixing bicycle, mixing drinks, bathing dog, getting a haircut, snatch, installing carpet, hopscotch, zumba* consisting of 607 videos with 407 training videos and 200 testing videos. The second subset adds another 5 handpicked classes, which are *playing water polo, high jump, discus throw, rock climbing, using parallel bars*, and they are visually close to some of the 10 random classes to make the classification task harder. The second subset comprises 950 videos with 639 videos in training and 311 videos in the test set.

Sports-1M is a large-scale public video dataset with 1.1 million YouTube videos of 487 fine-grained sports classes [19]. We choose a subset of 15 random classes of the Sports-1M dataset, namely *boxing, kyūdō, rings (gymnastics), yoga, judo, skiing, dachshund racing, snooker, drag racing, olympic weightlifting, motocross, team handball, hockey, paintball, beach soccer* with 702 videos in total. The dataset provides video level annotation for the entire untrimmed video; however, the temporal boundaries of the actions in the video are not identified. Approximately 5% of the videos contain more than one action label.

4.2   Evaluation Metrics

To evaluate our method on ActivityNet subsets, we report the *time of annotation* (ToA) as a metric to measure how fast the oracle can annotate a certain number of videos. The ToA score is an average of three times repeating each experiment by the oracle. ToA for conventional annotation and MuViLab on ActivityNet subset-1 is extrapolated since annotating 13 h of video using these methods is not feasible. We also report video classification accuracy in the form of mean average precision (mAP) for the ActivityNet subsets to measure the quality of annotation when the network is fine-tuned with our annotations versus with the ground truth annotations. mAP is used instead of a confusion matrix since some videos of ActivityNet contain more than one action [17].

For the Sports-1M [19] dataset, we perform a qualitative analysis of the t-SNE projections. To motivate our design choices beyond qualitative results, we introduce a realistic annotation emulation metric to estimate the quality of t-SNE projections on a global and local level. To report how well the t-SNE projection can separate the classes at a global level, we use a measure of cluster homogeneity, and completeness. Homogeneity measures if the points in a cluster only belong to one class and completeness measures if all points from one class are grouped in the same cluster. In an ideal t-SNE projection, all the points in each cluster belong to one class (homogeneity $= 1.0$), and all the points from a class are in the same cluster (completeness $= 1.0$), which makes the annotation process much faster. For clustering, K-Means clustering with K being the number of classes is used. We use the K-Means clustering algorithm because it is fast and has less hyperparameters to choose.

Since ToA can be a subjective metric, to evaluate the generalization of t-EVA and to emulate the oracle's annotation speed better, we also use a measure of local homogeneity using K-nearest neighbors (KNN) with $K = 4$ as in [26].

KNN can be used to estimate the local homogeneity between the features in lower dimensions. Higher KNN accuracy results in higher local homogeneity and better grouping; meaning, the oracle can annotate the videos faster.

4.3 Implementation Details

Feature Extraction. We use the 3D ResNet-34 architecture [16], pre-trained on Kinetics-400, as a feature extractor for all the experiments owing to their good performance and usage of RGB frames only. As in [16], each frame is resized spatially to 112×112 pixels from the original resolution. Each video is transferred to clips by sampling every 32 consecutive frames. The feature extractor in every forward pass takes a clip in the form of a 5D tensor as an input. Each dimension of the input tensor represents the batch size, input color channels, number of frames, spatial height, and width, respectively. Namely, an input tensor for a clip sampled at 32 frames can be shown as (1, 3, 32, 112, 112). The features are extracted after the final 3D average pooling with an $8 \times 4 \times 4$ kernel before the classifier layer. The dimensions of the feature vectors are $k \times 512$ with k being the total number of clips and later reduced to $k \times 2$ using t-SNE.

t-SNE. For dimensionality reduction, a Barnes-Hut implementation of t-SNE with two components are used from the scikit-learn library [30]. The perplexity is set to 30, and the early exaggeration parameter is 12, with a learning rate of 200. The cost function is optimized for 2500 iterations.

Training. After annotating each set of videos, the network is fine-tuned for a certain number of epochs. For training, the same 3D ResNet-34 [16] architecture is used. The sample duration is chosen as 32 frames for each clip, and the input batch size is 32. Stochastic gradient descends (SGD) is used as the optimizer with a learning rate of 0.1, weight decay of $1e-3$, and momentum of 0.9.

4.4 Results on ActivityNet

ActivityNet Subset-1. First, we put all the 407 videos in the unlabeled pool. Then, we divide the videos randomly into four different sets of unlabeled videos. The clips are generated with 32 consecutive frames, and the features are extracted using the 3D Resnet-34. After annotating each set of unlabeled videos, the network is fine-tuned for 20 epochs with the labeled videos. To note that, previously labeled videos are also used in the later epochs. The process continues until the network reaches 100 epochs. Between epoch 60 and 100, the network is fine-tuned using all 407 videos. Meanwhile, we refine the labels of the videos.

The videos are annotated incrementally, each time one set is labeled. Table 1 shows that the annotation time drops after every iteration of annotation and fine-tuning. Before fine-tuning the network, the labeling of the first set takes 600 s. ToA reduces 150 s at epoch 60 when the network is fine-tuned with previously labeled videos. Because of the incremental labeling and fine-tuning, the network learns to extract better features from the videos, which can be better grouped in the t-SNE plot. It is also expected that the oracle spends more time annotating the first few unlabeled set as the network is not yet fine-tuned. The quality of annotation at the early stage significantly impacts the next iterations of extracted features.

Table 1. Oracle's time of annotation (ToA) is shown on subset 1 of the ActivityNet (v1.3) dataset with 10 classes containing 407 videos ($\sim$13 h). At every 20 iterations from 0 to 60, 102 new videos are annotated, and the network is fine-tuned for 20 epochs. From epoch 60 to 100, no new video is added. The previous video labels are refined by the oracle as the network can extract better features. The network is fine-tuned on the existing labeled videos until epoch 100. It can be seen with incremental annotation and fine-tuning the annotation time in the later epochs drops.

Epoch	0	20	40	60	80	100
ToA (s)	600	552	516	450	240	180

Annotation Speed. To evaluate the annotation speed, we choose three methods: conventional, MuViLab [2], and t-EVA.

One way to increase the annotation speed of t-EVA is by putting more videos on the screen for the oracle to annotate. However, it does not make the labelling process easier. Since ActivityNet videos on average have 30 frames per second (FPS), every 32 time-steps that we sample represent almost 1 s ($\sim\frac{32}{30}$) of video. Putting all of the 407 videos (13 h) overflows the screen with the frames and makes the annotation harder for the oracle. One way to prevent overflowing the figures with thousands of frames is to increase the time-steps for sampling frames from each clip to the point that the network can still preserve the clips' temporal coherency. This way, we can show all of the videos on the 2D plot with fewer points. Consequently, we design three different t-EVA in terms of the number of time steps as t-EVA-32, 64, and 128.

Table 2. Comparison of time gain when annotating with different methods on a subset-1 of ActivityNet containing 769 min of video. Our method (t-EVA) with 128 time-steps outperforms conventional, and MuViLab [2] methods with labeling 769 min of video in 21 min. Using more consecutive frames increases annotation speed.

	Conventional	MuViLab	t-EVA-32	t-EVA-64	t-EVA-128
Time Gain	3×	4.5×	18×	24×	36×

First, we choose ActivityNet subset-1 with a total duration of 769 min. We annotated 30 min of videos using MuViLab and Conventional methods and extrapolated the result to match the total duration of ActivityNet subset-1. Additionally, the entire subset-1 is annotated using different variants of t-EVA, and we compare the annotation speed of all these methods (Table 2). The results

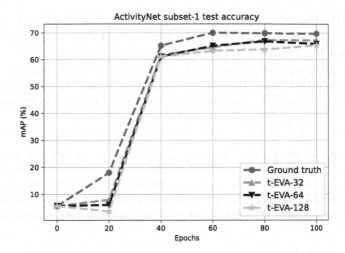

Fig. 4. Comparison of video classification performance in the form of mAP (%) between fine-tuning the 3D ConvNet on ground truth label versus fine-tuning with our annotation acquired using different time-steps (TS). Fine-tuning the 3D ConvNet on the annotation generated by our method can achieve comparable video classification accuracy to the ground truth.

show that labeling 769 min of video takes approximately 21 min with the t-EVA-32 method. t-EVA-32 outperforms both conventional and MuViLab methods on ActivityNet subset-1 in annotation speed by a large margin by respectively 4 to 6 times faster. With t-EVA-64 and 128, time gain can reach respectively 24 and 36 times more. Conventional annotation and MuViLab do not take advantage of the temporal dimension of videos for annotation. Nevertheless, our method exploits the spatio-temporal features and places similar actions near each other in the t-SNE plot for the oracle to annotate the actions.

We also evaluate the performance of the network on the test set of ActivityNet subset-1. In Fig. 4, we compare the classification performance of the networks: (i) fine-tuned with original ground truth labels and (ii) fine-tuned by using newly annotated videos by 32, 64, and 128 time-steps. Annotating the videos with t-EVA method can achieve a classification performance of 67.2% with 32-TS, 65.9% with 64-TS, and 65.4% with 128-TS, which is comparable to the training with ground truth labels (blue) by 69.7% mAP.

Table 3 shows the speed-accuracy trade off between t-EVA and ground-truth annotation. When the original ground truth labels are used for fine-tuning the network, we obtain 69.7% of mAP. 407 videos can be labeled in 42 min with t-EVA-32 by losing only 2.5% of performance in comparison to using ground truth labels. When the time-steps are increased as 64 and 128, the annotation speed decreases respectively to 31 and 21 min, yet the classification performance also reduces by 3.8% and 4.3%. Using 128 time-steps (t-EVA-128) reduces test accuracy while increasing the annotation speed. The decrease in accuracy compared to the 32-TS version is expected since the annotation is more prone to noise when the time-step is increased to 128 frames. With 128-TS for each clip, every point in the scatter plot represents 4 s of the video while it represents 1 s in the 32-TS version. Namely, labeling points wrongly in the 128 version (t-EVA-128) brings more significant consequences in the fine-tuning process. However, Table 3 indicates that using 128-TS (t-EVA-128) compared to the 32-TS (t-EVA-32) increases the annotation speed twice while the mAP score decreases less than 2%.

Table 3. Comparison of video classification performance (mAP) and ToA (time of annotation) on ActivityNet subset-1. This subset contains 407 videos in about 13 h of video. Our method in 32 time-steps (t-EVA-32) and 128 time-steps (t-EVA-128) achieves comparable test accuracy to the ground truth accuracy and requires a much shorter time to annotate. There is a trade-off between annotation speed and performance.

Method	GT	t-EVA-32	t-EVA-64	t-EVA-128
mAP	69.7%	67.2%	65.9%	65.4%
ToA (min)	–	42	31	21

4.5   Generalization

To further demonstrate the generalization of our method, we conduct the same annotation experiment on a more challenging subset of ActivityNet (v1.3) with 15 classes and a subset of Sports-1M [19] with 15 random classes.

ActivityNet (v1.3) Subset-2. Subset 2 of ActivityNet (v1.3) contains 637 training videos and 311 test videos. The first iteration of features is extracted from the 637 training videos and is annotated in 15 min by the oracle using t-EVA. After 20 epochs of fine-tuning, the new features are extracted, and the labels are fine-tuned again by the oracle. After this stage, the network is fine-tuned for 80 epochs. After fine-tuning for 100 epochs, our method reaches a test accuracy of 66.4%, while the training with ground-truth labels achieves an accuracy of 68.3% on the video classification task.

The 4-NN accuracy of the final features is 92.4%, which shows the quality of the extracted features is sufficient for the oracle to annotate. t-EVA can also perform well on the ActivityNet subset-2. The fact validates that our method can also generalize on a more challenging subset of ActivityNet.

Sports-1M. We further validate our method on a subset of Sports-1M [19] dataset with 15 random classes. We randomly sample 200 videos (∼860 min) from the total 702 videos available in the 15 classes. The features are extracted from 200 videos, and ground truth labels of the two-dimensional features can be seen in Fig. 5. Using 4-NN, we obtain an accuracy of 92.3%, which shows the features can be annotated based on similarity. Using our method, we were able to annotate 860 min of video in 28 min, giving us a time gain of 30.7. t-EVA indicates an extensive time gain on the Sports-1M dataset.

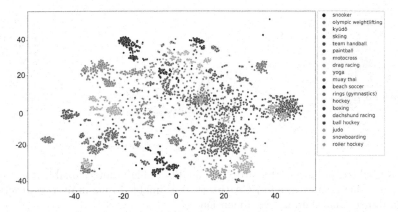

Fig. 5. t-SNE projection of extracted features from 200 videos from the Sports-1M [19] dataset with ground truth labels as colors. 200 videos are from 15 random classes; however, some videos contain more than one activity class. The 4-NN accuracy, which emulates the quality of the projection through measuring local homogeneity, is 92.3%, indicating such a figure is annotate-able by the oracle.

5   Ablation Study

In this section, we conduct an ablation study to motivate our design choices in the following aspects: (i) dimensionality reduction method, (ii) t-SNE parameter selection, and (iii) 2D versus 3D backbone for feature extraction.

5.1   Dimensionality Reduction

We investigate using PCA as a linear dimensionality method and t-SNE as a non-linear dimensionality method for visualizing the high-dimensional features in two dimensions. We use the extracted feature from the ActivityNet subset-1 with 407 videos. Figure 6-b shows qualitatively that PCA is not able to group similar features and separate unalike features from the videos in the transition to a lower dimension, making the annotation more difficult. However, Fig. 6-a, shows that t-SNE projection can maintain the local structure of each class while separating the features from different classes. To report the quality of projection in quantitative measures, we use KNN with $K = 4$. The 4-NN classification accuracy in Fig. 6 for the t-SNE projection is 80.6%, and for the PCA projection is 58.2%. Therefore, PCA, a linear dimensionality method, cannot reduce the feature dimension while placing similar classes near each other.

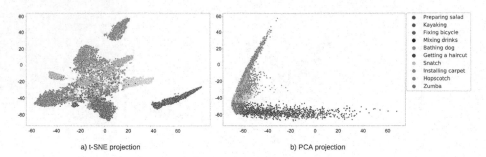

Fig. 6. Visual comparison of the projection quality of high-dimensional features to two dimensions using t-SNE (a) and PCA (b). PCA is unable to maintain the structure of the high-dimensional data in two dimensions.

5.2   t-SNE Parameters

We investigate using different perplexity parameters for the t-SNE projection. [39] recommend using perplexity parameter between [5–50], however larger and denser datasets requires relatively higher perplexity. With low perplexity, the local structure of data in each video dominates the action grouping from multiple video [43], but our goal is to group multiple actions from different videos. To emulate the t-SNE projection quality for the annotation, we report homogeneity and completeness scores with different perplexities in Table 4. Perplexity 30 shows the highest homogeneity and completeness scores, meaning that t-SNE projection with perplexity 30 can separate the classes better than projecting with the other perplexity parameters. Therefore, using t-SNE with perplexity 30 makes the group labeling process easier for the oracle.

5.3   2D-3D Comparison

We investigate replacing the 3D ConvNet with a 2D CNN to compare the quality of the feature embedding. For 3D ConvNet, 3D ResNet-34 pre-trained on Kinetics [20] and for the 2D CNN ResNet-50 pre-trained on Kinetics [20] are used. We chose Resnet-50 instead of Resnet-34 for the 2D CNN because the Kinetics pre-trained weights were only available for ResNet-50. To experiment, we sample every 32 consecutive frames (time-steps) as a clip in the 3D ConvNet, and for the 2D CNN, we choose one frame for every 32 frames to represent that specific window. The experiment is done on the subset-1 of the Activity-Net dataset with 10 classes. It can be seen in Fig. 7 that we start the experiments with 32 time-steps. With 32 time-steps, we can see the 2D CNN can capture the same action in different videos but can not place them together as well as the 3D ConvNet. Therefore, the colors representing the classes are better gathered

Table 4. Comparison of homogeneity and completeness scores as a measure to emulate the quality of t-SNE projection on a global-level. Higher homogeneity means all the points in a cluster belong to the same class. Higher completeness means all the points belonging to a class are in the same cluster. t-SNE perplexity parameter as 30 gives the highest homogeneity and completeness score.

	px-5	px-15	px-30	px-50	px-100	px-120
Homogeneity	44.7%	58.7%	62.5%	61.3%	61.7%	61.5%
Completeness	42.5%	56.1%	60%	58.5%	59%	58.8%

nearby in the 3D ConvNet, making the annotation process faster than the 2D CNN projection. Moreover, by increasing the time-steps for frame sampling, the 2D CNN, even with deeper architecture, starts losing the temporal coherency between the data-points because 2D CNN only focuses on the spatial information between the frames. Focusing only on spatial information can still work in lower time-steps (32-TS) because the frames from the same action contain similar spatial information. However, using spatial information alone becomes problematic in higher time-steps as increasing the time-steps reduces the spatial similarity between the frames.

To evaluate our findings quantitatively, we use K-NN accuracy as a quantitative emulation for the quality of features for annotation. Table 5 shows that increasing the number of frames in the clips degrades the 4-NN accuracy of 2D CNN dramatically from 93% to 75%. However, 3D CNN only loses around 5% from 32 time steps to 128. The local homogeneity decreases more drastically in 2D CNNs compared to 3D CNNs, which makes annotation more difficult for the oracle. In other words, the 2D CNN alone can not maintain the temporal structure of the data in higher time-steps. Thus, in the t-EVA method, 3D features are extracted to use for group labeling.

Table 5. Comparison of 4-NN accuracy of extracted features from a 2D CNN (ResNet-50) and a 3D ConvNet (3D ResNet-34) on subset-1 of ActivityNet [17]. Increasing time-steps cause the 2D CNN to lose the spatial similarity between the frames and fail to group them in the t-SNE plot, while the 3D ConvNet can still group similar actions even in higher time-steps.

	32-TS	64-TS	128-TS
2D CNN	93.1%	89.3%	74.6%
3D CNN	100%	97.6%	95.2%

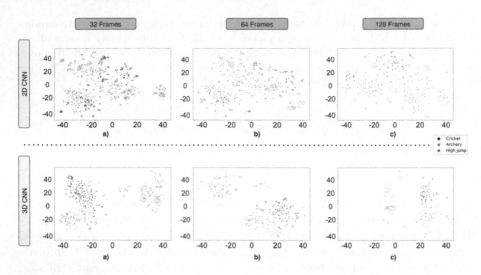

Fig. 7. Comparison of t-SNE projection of extracted features from a 2D CNN versus a 3D ConvNet for videos from 3 action classes of ActivityNet dataset [17]. Increasing the time-steps for sampling clips from the videos causes the 2D CNN to lose the clips' spatial information. However, the features from the 3D ConvNet can maintain the coherency between the clips.

6   Conclusion

This paper introduced a smart annotation tool, t-EVA, for helping the oracle to group label videos based on their latent space feature similarity in two-dimensional space. Our experiments on subsets of large-scale datasets shows that t-EVA can be useful in annotating large-scale video datasets, especially if the annotation budget and time are limited. Our method can outperform the conventional annotation method, and MuViLab [2] time-wise in the order of magnitude with a minor drop in the video classification accuracy. Besides, t-EVA is a modular tool, and its components can be easily replaced by other methods. To illustrate, 3D ResNet can be changed to another feature extractor.

t-EVA method has a trade-off between annotation speed and network performance. Increasing time steps can reduce the annotation time; however, the network's accuracy may also decrease.

t-EVA can be sensitive to the initial state of the feature extractor. If the feature extractor can not separate classes well, it can take a longer time to annotate the videos initially. After fine-tuning the network with new labels for a few epochs, the labeling time can reduce again. Besides, putting more video frames in the t-SNE plot can overflow the screen and make the annotation process harder for the oracle.

References

1. Ahsan, U., Sun, C., Essa, I.A.: DiscrimNet: semi-supervised action recognition from videos using generative adversarial networks. CoRR abs/1801.07230 (2018). http://arxiv.org/abs/1801.07230
2. Alessandro Masullo, L.D.: Muvilab (2019). https://github.com/ale152/muvilab
3. Gupta, A.K.: ImgLab (2017). https://github.com/NaturalIntelligence/imglab
4. Carreira, J., Zisserman, A.: Quo vadis, action recognition? A new model and the kinetics dataset. CoRR abs/1705.07750 (2017). http://arxiv.org/abs/1705.07750
5. Chatzimparmpas, A., Martins, R.M., Kerren, A.: t-viSNE: interactive assessment and interpretation of t-SNE projections. IEEE Trans. Vis. Comput. Graph. **26**(8), 2696–2714 (2020). https://doi.org/10.1109/tvcg.2020.2986996
6. Damen, D., et al.: Scaling egocentric vision: the EPIC-KITCHENS dataset. CoRR abs/1804.02748 (2018). http://arxiv.org/abs/1804.02748
7. darrenl, o.c.: labelimg (2017). https://github.com/tzutalin/labelImg
8. Diba, A., et al.: Temporal 3d ConvNets: new architecture and transfer learning for video classification. CoRR abs/1711.08200 (2017). http://arxiv.org/abs/1711.08200
9. Dutta, A., Zisserman, A.: The via annotation software for images, audio and video. In: Proceedings of the 27th ACM International Conference on Multimedia, MM 2019, pp. 2276–2279. Association for Computing Machinery, New York (2019). https://doi.org/10.1145/3343031.3350535
10. Feichtenhofer, C., Fan, H., Malik, J., He, K.: SlowFast networks for video recognition. In: Proceedings of the IEEE/CVF International Conference on Computer Vision (ICCV) (October 2019)
11. Wang, G., Forsyth, D.: Joint learning of visual attributes, object classes and visual saliency. In: 2009 IEEE 12th International Conference on Computer Vision, pp. 537–544 (2009)
12. Ghadiyaram, D., Feiszli, M., Tran, D., Yan, X., Wang, H., Mahajan, D.: Large-scale weakly-supervised pre-training for video action recognition. CoRR abs/1905.00561 (2019). http://arxiv.org/abs/1905.00561
13. Girdhar, R., Carreira, J., Doersch, C., Zisserman, A.: Video action transformer network. CoRR abs/1812.02707 (2018). http://arxiv.org/abs/1812.02707
14. Girdhar, R., Ramanan, D., Gupta, A., Sivic, J., Russell, B.: ActionVLAD: learning spatio-temporal aggregation for action classification. In: Proceedings of (CVPR) Computer Vision and Pattern Recognition, pp. 3165–3174 (July 2017)
15. Girdhar, R., Tran, D., Torresani, L., Ramanan, D.: Distinit: Learning video representations without a single labeled video. CoRR abs/1901.09244 (2019). http://arxiv.org/abs/1901.09244
16. Hara, K., Kataoka, H., Satoh, Y.: Learning spatio-temporal features with 3d residual networks for action recognition. CoRR abs/1708.07632 (2017). http://arxiv.org/abs/1708.07632
17. Heilbron, F.C., Escorcia, V., Ghanem, B., Niebles, J.C.: ActivityNet: a large-scale video benchmark for human activity understanding. In: 2015 IEEE Conference on Computer Vision and Pattern Recognition (CVPR), pp. 961–970 (2015)
18. Hussein, N., Gavves, E., Smeulders, A.W.M.: Timeception for complex action recognition. CoRR abs/1812.01289 (2018). http://arxiv.org/abs/1812.01289
19. Karpathy, A., Toderici, G., Shetty, S., Leung, T., Sukthankar, R., Fei-Fei, L.: Large-scale video classification with convolutional neural networks. In: CVPR (2014)

20. Kay, W., et al.: The kinetics human action video dataset. CoRR abs/1705.06950 (2017). http://arxiv.org/abs/1705.06950

21. Kläser, A., Marszalek, M., Schmid, C.: A spatio-temporal descriptor based on 3d-gradients. In: BMVC (2008)

22. Lee, J.A., Verleysen, M.: Nonlinear Dimensionality Reduction. Springer, New York (2007). https://doi.org/10.1007/978-0-387-39351-3

23. Li, J., Wang, J.Z.: Real-time computerized annotation of pictures. IEEE Trans. Pattern Anal. Mach. Intell. **30**(6), 985–1002 (2008)

24. Linderman, G.C., Steinerberger, S.: Clustering with t-SNE, provably. CoRR abs/1706.02582 (2017). http://arxiv.org/abs/1706.02582

25. Liu, K., Liu, W., Gan, C., Tan, M., Ma, H.: T-C3D: temporal convolutional 3d network for real-time action recognition. In: AAAI (2018)

26. Luus, F.P.S., Khan, N., Akhalwaya, I.: Active learning with TensorBoard Projector. CoRR abs/1901.00675 (2019). http://arxiv.org/abs/1901.00675

27. van der Maaten, L.: Barnes-Hut-SNE (2013)

28. Martinez, B., Modolo, D., Xiong, Y., Tighe, J.: Action recognition with spatial-temporal discriminative filter banks (2019)

29. McInnes, L., Healy, J., Melville, J.: UMAP: uniform manifold approximation and projection for dimension reduction (2018)

30. Pedregosa, F., et al.: Scikit-learn: machine learning in Python. J. Mach. Learn. Res. **12**, 2825–2830 (2011)

31. Pezzotti, N., et al.: GPGPU linear complexity t-SNE optimization. IEEE Trans. Visual. Comput. Graph. (Proc. VAST 2019) **26**(1), 1172–1181 (2020). http://graphics.tudelft.nl/Publications-new/2020/PTMHLLEV20

32. Qiu, Z., Yao, T., Mei, T.: Learning spatio-temporal representation with pseudo-3d residual networks. CoRR abs/1711.10305 (2017). http://arxiv.org/abs/1711.10305

33. Scovanner, P., Ali, S., Shah, M.: A 3-dimensional sift descriptor and its application to action recognition. In: Proceedings of the 15th ACM International Conference on Multimedia, MM 2007, pp. 357–360. Association for Computing Machinery, New York (2007). https://doi.org/10.1145/1291233.1291311

34. Sevilla-Lara, L., Liao, Y., Güney, F., Jampani, V., Geiger, A., Black, M.J.: On the integration of optical flow and action recognition. CoRR abs/1712.08416 (2017). http://arxiv.org/abs/1712.08416

35. Shou, Z., Wang, D., Chang, S.: Action temporal localization in untrimmed videos via multi-stage CNNs. CoRR abs/1601.02129 (2016). http://arxiv.org/abs/1601.02129

36. Simonyan, K., Zisserman, A.: Two-stream convolutional networks for action recognition in videos. CoRR abs/1406.2199 (2014). http://arxiv.org/abs/1406.2199

37. Tran, D., Bourdev, L.D., Fergus, R., Torresani, L., Paluri, M.: C3D: generic features for video analysis. CoRR abs/1412.0767 (2014). http://arxiv.org/abs/1412.0767

38. Tran, D., Wang, H., Torresani, L., Feiszli, M.: Video classification with channel-separated convolutional networks. CoRR abs/1904.02811 (2019). http://arxiv.org/abs/1904.02811

39. van der Maaten, L., Hinton, G.: Visualizing high-dimensional data using t-SNE. J. Mach. Learn. Res. **9**, 2579–2605 (2008)

40. Wang, H., Schmid, C.: Action recognition with improved trajectories. In: 2013 IEEE International Conference on Computer Vision, pp. 3551–3558 (2013)

41. Wang, L., et al.: Temporal segment networks: towards good practices for deep action recognition. CoRR abs/1608.00859 (2016). http://arxiv.org/abs/1608.00859

42. Wang, Q., Zhang, L., Bertinetto, L., Hu, W., Torr, P.H.S.: Fast online object tracking and segmentation: a unifying approach. CoRR abs/1812.05050 (2018). http://arxiv.org/abs/1812.05050
43. Wattenberg, M., Viégas, F., Johnson, I.: How to use t-SNE effectively (2016)
44. Wu, J., Yu, Y., Huang, C., Yu, K.: Deep multiple instance learning for image classification and auto-annotation (June 2015)
45. Xu, Z., Yang, Y., Hauptmann, A.G.: A discriminative CNN video representation for event detection. CoRR abs/1411.4006 (2014). http://arxiv.org/abs/1411.4006
46. Zeng, M., Yu, T., Wang, X., Nguyen, L.T., Mengshoel, O.J., Lane, I.: Semi-supervised convolutional neural networks for human activity recognition. CoRR abs/1801.07827 (2018). http://arxiv.org/abs/1801.07827
47. Zhang, Y., Wang, C., Wang, X., Zeng, W., Liu, W.: A simple baseline for multi-object tracking (2020)
48. Zhu, Z., Wang, Q., Li, B., Wu, W., Yan, J., Hu, W.: Distractor-aware Siamese networks for visual object tracking. CoRR abs/1808.06048 (2018). http://arxiv.org/abs/1808.06048

Vision-Based Fall Detection Using Body Geometry

Beddiar Djamila Romaissa[1,2](✉)(iD), Oussalah Mourad[1](✉)(iD), Nini Brahim[2](✉),
and Bounab Yazid[1](✉)

[1] Center for Machine Vision and Signal Analysis, University of Oulu, Oulu, Finland
ad_beddiar@esi.dz, {Djamila.Beddiar,Mourad.Oussalah,Yazid.Bounab}@oulu.fi
[2] Research Laboratory on Computer Science's Complex Systems,
University Laarbi Ben M'hidi, Oum El Bouaghi, Algeria
brahim_nini@yahoo.fr

Abstract. Falling is a major health problem that causes thousands of deaths every year, according to the World Health Organization. Fall detection and fall prediction are both important tasks that should be performed efficiently to enable accurate medical assistance to vulnerable population whenever required. This allows local authorities to predict daily health care resources and reduce fall damages accordingly. We present in this paper a fall detection approach that explores human body geometry available at different frames of the video sequence. Especially, the angular information and the distance between the vector formed by the head -centroid of the identified facial image- and the center hip of the body, and the vector aligned with the horizontal axis of the center hip, are then used to construct distinctive image features. A two-class SVM classifier is trained on the newly constructed feature images, while a Long Short-Term Memory (LSTM) network is trained on the calculated angle and distance sequences to classify falls and non-falls activities. We perform experiments on the Le2i fall detection dataset and the UR FD dataset. The results demonstrate the effectiveness and efficiency of the developed approach.

Keywords: Fall detection · Elderly assistance · SVM classification · Deep Learning · LSTM · Pretrained models

1 Introduction

Performing regular daily life activities by elderly population can cause fall. This is due to inherent factors such as age-related biological changes, neurological

This work is partly supported by the Algerian Residential Training Program Abroad Outstanding National Program (PNE) that supported the first author stay at University of Oulu and the European YougRes project (# 823701), which are gratefully acknowledged.

© Springer Nature Switzerland AG 2021
A. Del Bimbo et al. (Eds.): ICPR 2020 Workshops, LNCS 12664, pp. 170–185, 2021.
https://doi.org/10.1007/978-3-030-68799-1_13

disorders, physiological health profile and environmental conditions. In general, fall can result because of sudden loss of balance, stability, dizziness or vertigo during daily life movements. It can also be caused by chronic diseases, cognitive impairment, the use of walking aid or multiple medications, gait and visual deficit [21]. Falling is an abnormal human activity that occurs infrequently and unpredictably. It is defined in [1] as an event which results in a person coming to rest inadvertently on the ground or the floor or any other lower level.

It is acknowledged that fall is one of the major public health problems in the world that should be carefully addressed and appears to be the second leading cause of accidental or unintentional injury deaths [1]. Therefore, Fall detection, Fall classification and Fall prediction are recognized as important research directions in the study of falls and are among the hottest topics in health-care. Indeed, the availability of efficient methods to identify and, possibly, predict fall occurrence can have a huge public impact since it may significantly minimize damages, enable efficient medical assistance and provide daily health care for vulnerable population. Moreover, missing to identify falls can expose the individual to serious health and safety risks. It has an obvious effect on individual autonomy, independence and life quality. It is to note that experiencing fall many times, may lead to Basophobia, also called fear of falling [12]. This syndrome can cause many other disorders such as lack of mobility and independence, and/or social isolation. On the other hand, reducing the time interval between falling and rescuing is essential in order to minimize the negative consequences of falls, which raises the importance of fall prediction task. Motivated by the importance of detecting falls and the observation that the vector formed by the head and the center hip of the body is aligned horizontally and in parallel to the ground during a falling posture, while it is perpendicular to the ground axis in a sitting or standing posture, we present in this paper a novel machine learning like approach for Fall Detection. Our approach relies on the calculation of the angle and the distance between the vector formed by the head and the center hip of the body to the vector formed on the horizontal axis of the center hip of the body. For each video sequence, we calculate, the above mentioned angle and distance among all the frames. The computed angles and distances form the new feature sets that characterize the video sequences. We train an LSTM network on these features to recognize fall and non-fall activities. Furthermore, we construct new images using these angle and distance sequences so that each video sequence is represented by one image of its corresponding angles and distances. Then, a two-class SVM is trained on these images to detect fall and non-fall activities. We use the Le2i dataset [5] and the UR FD dataset [13] to evaluate the performance of our method, different metrics have been employed to quantify the quality of the designed SVM classifier in detecting falls. The experimental results indicate that our approach is practical and achieves good accuracy in detecting falls. The main contributions of our proposal can be summarized into:

– A new method for downsampling the videos using optical flow in order to keep only frames with significant motion is put forward. This allows us to reduce the number of frames of the video to be executed in subsequent reasoning.

- A new research dataset related to our manual annotation task containing the 2D coordinates of both center hip of the body and the head centroid (of facial representation) available from each frame of the video data is made available to research community.
- Calculating the angle and the distance between the head centroid and the center hip of the body.
- Tracking the variation of the above angular estimation across all frames of the re-sampled video.
- A new SVM-binary classification that distinguishes fall from non-fall scenarios using the sequence of angles and distances pertaining to each video has been devised.
- Contribution of other potential feature sets are explored and exploited for representing video sequences.
- A comparison between LSTM and SVM classification results has been carried out.

The rest of this paper is organized as follows. First, we briefly provide background and previous research related to vision-based Fall detection (FD) in Sect. 2. Section 3 outlines our approach. Then, we describe and discuss in Sect. 4 the experimental results of our proposal on the publicly available datasets. Finally, we conclude our paper and set future directions for fall detection in Sect. 5.

2 Related Work

Fall detection techniques can be categorized into three major classes: ambient-based, wearable-based and vision-based systems [21]. Ambient-based systems use light, proximity, motion, and vibration sensors to collect daily life activities data and detect falls. Wearable-based systems rely on the sensors embedded in particular devices that the subject should wear in order to track his/her motion [12]. Additionally, vision-based systems use RGB or depth cameras to record the subject's activities, in indoor or outdoor environments [12]. The recorded images or videos are analyzed later to detect falls. Motivated by robustness, efficiency, ease of use and installation of the last methods, the approach that we present in this paper relates to vision-based FD. Thereby, we briefly report here some of the existing vision-based FD methods.

Roughly speaking, Vision-based FD approaches focus on meaningful fall related features extracted from the video frames such as silhouettes, body shape and skeleton information. These features are then used as input to some machine learning classifier such as SVM, KNN, Hidden Markov Models (HMM), among others, to train and later automatically detect fall and non-fall cases. For instance, [15] extracts distinctive features of human silhouettes to construct new action representations. The authors model the actions using a bag-of-words and conduct the classification using an extreme learning machine (ELM). Authors in [9] suggest robust features called History Triple Features using a generalization of the Radon Transform. Furthermore, SVM based approaches have proven their efficiency for fall detection tasks in many alternative works see, for instance, [7,8,11]. In [7], five distinct features are employed (aspect ratio, change in aspect ratio, fall angle, center speed and head speed). Authors in [8] use a normalized motion energy image to model the silhouette shape deformation features, while [11] proposes a novel descriptor, called Trajectory Snippet Histograms, to model the rapid motions change. They used Bag of Words to describe each video clip and train an SVM for unusual videos classification. In addition, shape and motion features are tracked to detect falls using a single camera based system in [18]. Likewise, [22] proposes to analyze dynamic appearance, shape and motion features of the target person and then characterize the human falls with simple velocity statistics of moving features. Authors in [4] suggest a vision-based fall detection system for elderly living alone. The system relies on the optical flow estimation to estimate the speed of motion and to deduce the fall activity accordingly, while comparing the last positions of the target.

On the other hand, many vision-based research is devoted to fall detection using Kinect sensors. This is because depth cameras can overcome some privacy issues related to traditional camera systems. For instance, [16] proposes a real-time fall detection system based on 3D Kinect depth maps. These depth maps are used to extract 3D silhouettes features. Similarly, [20] employs Kinect sensor to acquire point cloud images and extract energy fall features. Other researchers demonstrate that using Kinect sensor alone does not provide sufficient coverage and, therefore, cannot yield robust and efficient fall detection capabilities.

With the advance in Deep Learning (DL) approaches, many researchers put forward DL based approach for fall detection tasks. For instance, [6] proposes a real-time fall detection approach that allows the capture of RGB video streams, individual's position estimation and, thereby, fall detection likelihood, which then generates potential alert messages to caregivers with registered audio and video. In [19], the authors present a novel FD method based on Convolutional Neural Networks (CNN) using optical flow images. Moreover, transfer learning is widely used to take advantage of pre-trained models by reusing their network weights or fine-tuning the classification layers. For instance, [3] was able to efficiently detect falls using a CNN Alexnet architecture. In [10], the authors present a two-stream approach based on MobileVGG network. Similarly, the authors of [10] combine an improved lightweight VGG network and the motion characteristics of the human body. Likewise, a 3D CNN-based method combined with long short-term memory (LSTM) is also presented in [14]. The 3D CNN is used to extract motion and spatial features while the LSTM-based spatial visual attention scheme is incorporated to locate the fall in each frame. Authors in [2] present a fall detection system based on LSTM, using location features from the group of available joints in the human body. Inspired by the aforementioned work, we focus in this paper on vision-based fall detection using LSTM for classification of angle and distance features, that are extracted from video sequences. Transfer learning is performed to take advantage of the strong ability of the Resnet50 model in extracting significant features that were later fed to our two-class SVM classifier.

3   Proposed Method

The starting point in our developed methodology consists in identifying relevant features that can genuinely distinguish fall from non-fall activities. In this respect, we noticed that when a person is sitting or standing, the head and the center hip form a vector which is perpendicular to the horizontal axis passing through the center hip, as illustrated in "Fig. 1(a)" and "Fig. 1(b)". The horizontal axis is defined as a straight line parallel to the X_axis and passing through the center hip. In contrast, when a person is in a lying or falling posture, this vector is approximately aligned and in parallel to the horizontal axis of the center hip of the body. In addition, sitting slumped to one side leads to forming an angle of around 45° or 120° between the mentioned vector and the horizontal axis as shown in "Fig. 2". The angle value depends on the degree of slump sitting. However, the posture is considered lying or falling when this value is close to 0° or 180° as shown in "Fig. 3".

(a)   (b)

Fig. 1. Samples from the Le2i fall detection dataset representing the angle α in (a) sitting and (b) standing postures. The value of α is around 90° in both postures.

(a)   (b)

Fig. 2. Samples from the Le2i fall detection dataset representing the angle α in (a) bending to the left posture and (b) bending to the right posture. The value of α is around 120° and 45° respectively.

To illustrate our approach mathematically, we refer to the head centroid by the point $H(x_h, y_h)$ and to the center hip of the body by the point $B(x_b, y_b)$. Let $\vec{U}$ be the vector from H to B. Similarly, let $\vec{V}$ be the vector joining the point B to the point $C(x_c, y_c)$. The point C is defined such that $x_c > x_b$ and $y_c = y_b$.

Fig. 3. Samples from the Le2i fall detection dataset representing the angle α in a falling posture. The value of α is around $180°$.

Relying on this observation, we calculate for each video, the angle α formed between $\vec{U}$ and $\vec{V}$ and the distance γ between the head and the center hip of the body (i.e: the magnitude of the vector $\vec{U}$) for all its frames. These notations are used along the paper. Each video is therefore characterized by a feature vector containing the sequence of the computed angles and distances.

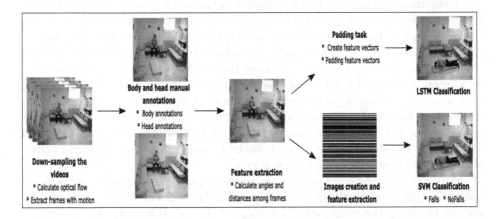

Fig. 4. The pipeline of our proposed fall detection approach

The first step of our approach consists in down-sampling the videos to reduce the number of their frames by keeping only frames which contain motion. This helps us to reduce both the computational and man-power burden effort in the next steps. Then, we use these frames to manually annotate the head and the body. We then calculate their centroid and center hip respectively. Next, we calculate the angle α and the distance γ and represent each video with a sequence of angles α_i and a sequence of distances γ_i, where i represents the frame's index. Once these sequences are created for each video, we distinguish two scenarios. In the first scenario we train an LSTM network on these features (angles, distances) and classify videos accordingly into falls and non-falls. For this purpose, we also devised a data augmentation strategy that handles the potential mismatch of input size to the LSTM network. For this purpose, we performed a simple padding task where the feature vector (angle and distance value) of the first frame is duplicated and concatenated in order to yield the same dimension as the largest sequence. In the second scenario, we create two sets of images that are fed later to a pretrained network in order to extract distinctive features. The extracted features are then trained with a two-class SVM classifier to detect falls from daily life activities. "Figure 4" outlines the pipeline of our proposed fall detection approach. It is summarized in 4 steps as follows.

3.1   Down-Sampling the Videos

The length of the Le2i dataset video sequences vary from 30 s to 4 min with a frame rate of 25 frames per second. Indeed, the extraction of video frames can result into more than 1000 frames, which makes the process of manual head annotations very exhausting. Therefore, instead of using all the frames, we down-sample each video frames in order to reduce the intensive computational processing. Inspired by [17] where a video summarization technique based on motion analysis using optical flow computations to identify global extrema and local minimum between two maximums in the motion was presented, we also use the optical flow (OF) to down-sample our videos. For that, we exploit the Horn-Schunck method to estimate the motion in the video sequences. We observe that falls, in general, occur fast and can be characterized by a significant motion change among frames. Therefore, we keep those frames that represent meaningful motion change and construct a re-sampled video sequence using these frames, with a rate of 25 frames per second. It is to note that the new re-sampled videos contain less frames than the original videos. For example, a video of 1607 frames is reduced to 578 frames and another video of 1283 frames is downsampled to 583 frames. The downsampling rate is variable and depends on the mean values of the optical flow components of each video.

More specifically, to automate this downsampling process, we first estimate the optical flow among all the frames using the Horn-Schunck method that consists in resolving the constraint: $I_x.u + I_y.v + I_t = 0$. Where I_x, I_y, I_t are the spatiotemporal image brightness derivatives, while u and v correspond to the horizontal and the vertical optical flow components, respectively. Then, we calculate the mean of both horizontal *(Vx)* and vertical *(Vy)* components of the OF, which we call *meanVx* and *meanVy* respectively. Subsequently, the mean squared normalized error performance (MSE) is computed to estimate the similarity between the horizontal\vertical components of the OF of each frame and the mean value of horizontal and vertical components of the OF, respectively (*similarityVx* and *similarityVy*). "Equation (1)" demonstrates how to calculate such similarity using the MSE values.

$$similarityVz_i = \frac{1}{P} . \sum_{p-1}^{P} (Vz_i(p) - meanVz(p))^2 \qquad (1)$$

Where z refers to either x or y component, P refers to the pixels of the frame, i refers to its index and p to a particular pixel of the frame i.

Frames that have a similarity $similarityVx_i$ (resp. $similarityVy_i$) above or equal their mean similarity $meanSimVx$ (resp. $meanSimVy$) are preserved while others are removed to construct the re-sampled video. Besides, the maintained frames should respect the conditions given by "(2)".

$$\begin{cases} similarityVx_i >= meanSimVx \\ similarityVy_i >= meanSimVy \end{cases} \tag{2}$$

Our videos are resampled and the number of frames of each video is reduced to almost the half. Our aim behind this is to facilitate the next step of our proposal which is done manually.

3.2 Body and Head Manual Annotations

Once, the videos are re-sampled, and as our work focuses on the features extracted from the body geometry, we manually annotate the individual's head position in video frame and calculate its centroid to avoid subsequent problems originated in the tracking algorithm. More specifically, the annotation of each frame contains the frame's index, the localization of the head presented in terms of bounding box and the coordinates of the head centroid. In addition, we manually annotate the human body in video frames and estimate the center hip by calculating the centroid of the shape that surrounds the individual's body. Figure 5 illustrates the manual annotation of the body and the head, where the blue point corresponds to an estimation of the center hip of body in (a) and the centroid of the head in (b). The samples were taken from the Le2i dataset.

(a)   (b)

Fig. 5. Samples from the Le2i fall detection dataset representing the manual annotation of a) the center hip of the body and b) the head.

The head centroid and the center hip of the body are used later to calculate their associated distance γ and the angle α between the vector $\vec{U}$ and the vector formed by the horizontal axis corresponding to the x coordinate of the center hip called $\vec{V}$.

For the angle calculus, we can calculate its cosine value and deduce the corresponding angle. The cosine is calculated using the law of cosines and the Euclidean norm is used to calculate the magnitude of vectors. "Equation (3)" illustrates how we calculate the cosine of the angle α. $\overrightarrow{HC}$ refers to the vector between the head centroid and the axis point C and $\left\|\overrightarrow{X}\right\|$ is the Euclidean norm of the vector $\overrightarrow{X}$.

$$cos(\alpha) = \frac{-\left\|\overrightarrow{HC}\right\|^2 + \left\|\overrightarrow{U}\right\|^2 + \left\|\overrightarrow{V}\right\|^2}{2.\left\|\overrightarrow{U}\right\|.\left\|\overrightarrow{V}\right\|} \qquad (3)$$

We therefore calculate the distance between the head and the center hip of the body among all the video frames using the Euclidean norm.

3.3 Feature Extraction

As mentioned above, we discern two scenarios. In the first one, we construct our feature vectors using angles and distances. The angles and the distances are calculated between the vectors $\overrightarrow{U}$ and $\overrightarrow{V}$ among all frames of the re-sampled videos. Therefore, each video is characterized either by the feature vector $V = \{\alpha_1, \alpha_2, \alpha_3 \dots \alpha_i\}$ or $V = \{[1, \alpha_1, \gamma_1], [2, \alpha_2, \gamma_2], [3, \alpha_3, \gamma_3] \dots [i, \alpha_i, \gamma_i]\}$ where i is the index of the video frame. Since the video sequences do not contain the same number of frames, these feature vectors are of different lengths and could not be fed directly to the classifier which require the input size to be fixed. For that, we perform a padding strategy that allows to keep all the vectors of the same dataset with the same size. The new length is calculated to be the maximum value of the vectors' lengths. So, each vector is extended to the new length by adding the new values to its beginning. In order to not influence the feature vector with random new values, we fill them out using the first value of the vector: angle and distance of the first frame. Feature vectors are then fed to a classifier: SVM or LSTM for fall detection. For example, for a video V_1 characterized by $V_1 = \{[1, \alpha_1, \gamma_1], [2, \alpha_2, \gamma_2], [3, \alpha_3, \gamma_3] \dots [k, \alpha_k, \gamma_k]\}$ where K represents the number of its frames. Let us refer to the maximum value of all video lengths with Max, where $K <= Max$. We add $(Max\text{-}K)$ elements of value $[\alpha_1, \gamma_1]$ at the beginning of V_1, so V_1 becomes $V_1 = \{[1, \alpha_1, \gamma_1], \{[2, \alpha_1, \gamma_1], \dots , \{[Max\text{-}K, \alpha_1, \gamma_1], \{[1+Max\text{-}K, \alpha_1, \gamma_1] [2+Max\text{-}K, \alpha_2, \gamma_2], [3+Max\text{-}K, \alpha_3, \gamma_3] \dots [k, \alpha_k, \gamma_k]\}$.

In the second scenario, we use only the angles calculated above to construct the first set of images (gray level images). Hence, the feature vector $V = \{\alpha_1, \alpha_2, \alpha_3 \dots \alpha_i\}$ is employed to construct the newly created gray level image for video V. However, we concatenate angles and distances to construct the second set of images (RGB images). The newly created RGB image for the video V from the second set is made using the feature vector $V = \{[1, \alpha_1, \gamma_1], [2, \alpha_2, \gamma_2], [3, \alpha_3, \gamma_3] \dots [i, \alpha_i, \gamma_i]\}$ where values of i build the first channel, values α_i build the second channel and values γ_i build the third channel respectively. Each video is characterized by an image from the first set and an image from the second set. This way, these images encode the angle sequences and the distance sequences taking into account the temporal aspect of the video illustrated by the first channel (the video frames). We give examples of created gray-level and RGB images of falls in "Fig. 6(a)", "Fig. 6(b)" and non-fall activities in "Fig. 6(c)" and "Fig. 6(d)".

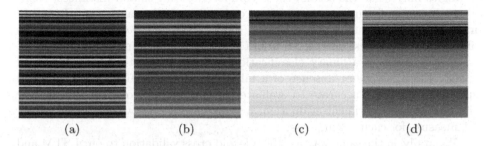

(a)   (b)   (c)   (d)

Fig. 6. Samples of created images from only angles (a and c) and created images from angles and distances (b and d). (a) and (b) represent falls while (c) and (d) represent non-falls.

3.4   Classification

We train a Long short-term memory (LSTM) network using the sequence of both angles and distances in the first scenario to detect falls and non-falls. Besides, to make the learning faster, we construct our LSTM model using a biLSTM layer that allows us to access data in both forward and reverse directions.

To detect falls in the second scenario, distinctive features are extracted from our two sets of feature images using a pretrained model. The first set consists in images constructed from angles only while the second set consists in images constructed using frames, angles and distance sequences. In our approach, we use activations of the Resnet50 and the AlexNet networks as our features. Then, we feed them to a two-class SVM classifier to distinguish between falls and daily life activities.

4   Experimental Results and Discussion

Fall is a kind of unpredictable action that occurs infrequently. Due to rarity of occurrence of falls, most existing fall detection datasets are set up by simulated fall data. The lack of such benchmark datasets and real fall data make the evaluation process of fall detection systems hard and less convincing. We evaluate our approach on the publicly available Le2i FD [5] and UR Fall Detection [13] datasets. In the subsequent section, we present these two datasets as well as the evaluation metrics employed to evaluate the performance of our proposal and, finally, the experimental results we obtained.

4.1 Experimental Setup

We evaluate firstly the results obtained from our SVM and LSTM models that are trained on two configurations of features. The first feature set is composed of angles only while the second set is composed of angles and distances. Next, we evaluate the features extracted from our constructed images using Resnet50 model to those extracted using AlexNet model. We compare as well the results obtained from images constructed from (1) angles and; (2) angles and distances.

Fall detection is a binary classification problem where the classifier should specify the existence or absence of a fall in a video sequence. Sensitivity and specificity are most accurate to evaluate the performance of such system. For the purpose of experiment evaluation, we calculate Accuracy, Precision, Recall, F1_measure for each testing case.

We apply, in the same way as [19], a k-fold cross validation to our LSTM and SVM models with k = 5 where the UR fall detection and the Le2i datasets were randomly split into k equal size subsets. At each iteration of the 5 iterations, we compose the training and validation sets with 4 subsets and one single set, respectively. We use a data augmentation process to transform training and test sets with an optional pre-processing stage such as resizing which helped us to resize the images of the data-store in order to make them compatible with the input size of the pretrained model. Therefore, at each epoch, the training set is modified slightly to get better results and to avoid overfitting. The results are computed across the combination of all the iterations.

4.2 Datasets

The Le2i fall detection dataset contains 221 videos of 131 falls and 90 daily life activities (ADL). The different activities are recorded by a single fixed camera with a frame rate of 25 frames/s and a resolution of 320×240 pixels. All the activities are simulated by several actors and are gathered at four different locations: Home, Office, Coffee room and Lecture room. The dataset illustrates many difficulties of realistic video sequences of an elderly home or office such as variable illumination and occlusion. The manual annotations of 191 videos was given, with extra information representing the ground-truth of the fall position and the localization of the body in the image sequence.

The UR fall detection dataset contains 70 (30 falls + 40 activities of daily living) sequences [13]. Two Microsoft Kinect cameras were used to record fall events from two different perspectives where ADL were recorded with only one camera. This results into 60 fall sequences and 40 non-fall activities.

4.3 Experiment Results

For the Le2i dataset, we achieve a sensitivity of 100% for the first set composed of angles only and the second set of features composed of angles and distances, both sets trained on the LSTM network. We achieve a best sensitivity of 100%

Table 1. Performance results for our FD approach (feature images) on the Le2i subset using an AlexNet and a Resnet50 models for feature extraction

Features	Accuracy	Precision	Recall	F_score
Features+SVM				
Angle	0.692	0.875	0.700	0.778
Angle+Distance	0.731	0.842	0.800	0.821
Features+LSTM				
Angle	0.731	0.731	**1.000**	0.845
Angle+Distance	0.769	0.769	**1.000**	0.870
Feature images+AlexNet+SVM				
Angle	0.769	0.818	0.900	0.857
Angle+Distance	0.885	0.952	0.909	0.931
Feature images+Resnet50+SVM				
Angle	**0.962**	0.952	**1.000**	**0.975**
Angle+Distance	**0.962**	**1.000**	0.950	0.974

Table 2. Performance results for our FD approach (feature images) on the UR fall detection dataset using an AlexNet and a Resnet50 models for feature extraction

Features	Accuracy	Precision	Recall	F_score
Features+SVM				
Angle	0.700	0.727	0.727	0.727
Angle+Distance	0.850	0.818	0.900	0.857
Features+LSTM				
Angle	0.600	0.579	**1.000**	0.734
Angle+Distance	0.850	0.800	**1.000**	0.889
Feature images+AlexNet+SVM				
Angle	0.950	**1.000**	0.917	**0.957**
Angle+Distance	0.920	0.923	0.923	0.923
Feature images+Resnet50+SVM				
Angle	**0.960**	**1.000**	0.833	0.907
Angle+Distance	**0.960**	**1.000**	0.900	0.947

for SVM trained on features extracted with Resnet50 from images built using angles. It is clear from Table 1 that the results obtained from LSTM trained on angle and distance features are higher than the results obtained from angle features only in terms of accuracy, precision and F_score.

Table 1 illustrates the results obtained for both sets of images (constructed from angles versus angles+distances) using the activations of the AlexNet and the Resnet50 models as well as results of training an LSTM straight on the

Table 3. Comparison between performance results of our FD approach with other existing approaches on the Le2i dataset

Approaches	Accuracy	Precision	Recall	F_score
Combined curvlets+HMM [23]	**97.02%**	–	98.00%	–
OF+CNN [19]	97.00%	–	93.60%	–
Ours: **Angle+Distance+Resnet50+SVM**	96.20%	**100%**	95.00%	**97.40%**
Ours: **Angle+AlexNet+SVM**	76.90%	81.80%	90.00%	85.70%
Ours: **Angle+Distance+LSTM**	76.90%	76.90%	**100%**	87.00%

Table 4. Comparison between performance results of our FD approach with other existing approaches on the UR Fall detection dataset

Approaches	Accuracy	Precision	Recall	F_score
Combined curvlets+HMM [23]	**96.88%**	–	–	–
OF+CNN [19]	95.00%	–	100%	–
Ours: **Angle+Distance+Resnet50+SVM**	96.00%	100%	90.00%	94.70%
Ours: **Angle+AlexNet+SVM**	95.00%	100%	91.70%	**95.70%**
Ours: **Angle+Distance+LSTM**	85.00%	80.00%	**100%**	88.90%

features for the Le2i dataset. We can see from this table that the results obtained from the images constructed from the angles and the distances (RGB images) are higher than the results obtained from images constructed from angles only when using Resnet50. This can lead to conclude that Resnet50 performs well on RGB images and gives more significant features. However, we notice the opposite in the results obtained for both sets of images using the activations of the AlexNet model.

Similarly for the UR fall detection dataset, a sensitivity of 100% for training the LSTM on the first set composed of angles only and the second set of features composed of angles and distances. The sensitivity obtained for SVM classification of images extracted from our newly created images using the Alexnet activations as features is better than the one obtained from the Resnet50 activations for both sets of features (angles, angles+distances). However, the accuracy and the precision values are higher while using the Resnet50 activations.

We can observe from Table 1 and Table 2 as well that using the new images gave us better results than directly feeding the features vectors: angles and distances to our LSTM and SVM models. The results were by far improved by creating these images and extracting significant features from them using pre-trained models.

We compare our results with [19,23] since we used the same protocol evaluation and same metrics, although we acknowledge the difficulty in performing reliable comparison with other state-of-art works. Table 3 and Table 4 illustrate our results versus the results obtained by [19,23] on the Le2i dataset and UR FD dataset respectively, where we present different variants of our approach. We can see from these tables that we outperformed the aforementioned state-of-the-art results. This can be justified by the fact that our approach is not dependent on the background and illumination changes unlike [19] who are using optical flow and [23] who are combining SVM with hidden markov models. Both models depends on the RGB videos and can be influenced by illumination or occlusion.

5 Conclusion and Future Directions

We present in this paper an effective vision-based approach for fall detection based on angles calculation. Our approach allows us to construct gray-scale images of calculated angles between the head, the center hip of the target subjects and the horizontal axis passing through the center hip. Another set of images is constructed using angles and distances between the head and the center hip of the body as well. These constructed sets of images constitute our distinctive features for fall detection task. Next, an SVM classifier is used along with a pretrained model to classify the constructed images into falls and daily life activities. We compare in this paper the features extracted using both the Resnet50 and the Alexnet models. We use the Le2i dataset and the UR fall detection dataset to evaluate the performance of our approach using the accuracy, precision, recall and F_score evaluation metrics. Experimental results show that the performances of our proposed approach are comparable to that of the state-of-the-art fall detection methods and outperform [19,23]. However, some limitations are also noticed. For instance, it will be desirable to improve the approach to distinguish between lying and falling postures. Besides, in the future, we would also like to automatically annotate the head and the body center hip positioning of individuals from video sequences. On the other hand, there is a room for improvement in the training pipeline through a better selection of training samples inputted to our SVM and LSTM classifiers, better optimization of LSTM parameters and through pursuing a cross-dataset based approach. We have, for instance, noticed the prospect of performing a cross-view evaluation to investigate the performance of the approach when different perspectives are studied.

References

1. World health organization, who global report on falls prevention in older age. Technical report (2007)
2. Adhikari, K., Bouchachia, H., Nait-Charif, H.: Long short-term memory networks based fall detection using unified pose estimation. In: 12th International Conference on Machine Vision, ICMV 2019, vol. 11433, p. 114330H. International Society for Optics and Photonics (2020)
3. Anishchenko, L.: Machine learning in video surveillance for fall detection. In: 2018 Ural Symposium on Biomedical Engineering, Radioelectronics and Information Technology (USBEREIT), pp. 99–102. IEEE (2018)

4. Bhandari, S., Babar, N., Gupta, P., Shah, N., Pujari, S.: A novel approach for fall detection in home environment. In: 2017 IEEE 6th Global Conference on Consumer Electronics (GCCE), pp. 1–5. IEEE (2017)
5. Charfi, I., Miteran, J., Dubois, J., Atri, M., Tourki, R.: Optimized spatio-temporal descriptors for real-time fall detection: comparison of support vector machine and Adaboost–based classification. J. Electron. Imaging 22(4), 041106 (2013)
6. Ciabattoni, L., Foresi, G., Monteriù, A., Pagnotta, D.P., Tomaiuolo, L.: Fall detection system by using ambient intelligence and mobile robots. In: 2018 Zooming Innovation in Consumer Technologies Conference (ZINC), pp. 130–131. IEEE (2018)
7. Debard, G., et al.: Camera-based fall detection using real-world versus simulated data: how far are we from the solution? J. Ambient Intell. Smart Environ. 8(2), 149–168 (2016)
8. Feng, W., Liu, R., Zhu, M.: Fall detection for elderly person care in a vision-based home surveillance environment using a monocular camera. Sig. Image Video Process. 8(6), 1129–1138 (2014). https://doi.org/10.1007/s11760-014-0645-4
9. Goudelis, G., Tsatiris, G., Karpouzis, K., Kollias, S.: Fall detection using history triple features. In: Proceedings of the 8th ACM International Conference on PErvasive Technologies Related to Assistive Environments, pp. 1–7 (2015)
10. Han, Q., et al.: A two-stream approach to fall detection with MobileVGG. IEEE Access 8, 17556–17566 (2020)
11. Iscen, A., Armagan, A., Duygulu, P.: What is usual in unusual videos? Trajectory snippet histograms for discovering unusualness. In: Proceedings of the IEEE Conference on Computer Vision and Pattern Recognition Workshops, pp. 794–799 (2014)
12. Khan, S.S., Hoey, J.: Review of fall detection techniques: a data availability perspective. Med. Eng. Phys. 39, 12–22 (2017)
13. Kwolek, B., Kepski, M.: Human fall detection on embedded platform using depth maps and wireless accelerometer. Comput. Meth. Prog. Biomed. 117(3), 489–501 (2014)
14. Lu, N., Wu, Y., Feng, L., Song, J.: Deep learning for fall detection: three-dimensional CNN combined with LSTM on video kinematic data. IEEE J. Biomed. Health Inform. 23(1), 314–323 (2018)
15. Ma, X., Wang, H., Xue, B., Zhou, M., Ji, B., Li, Y.: Depth-based human fall detection via shape features and improved extreme learning machine. IEEE J. Biomed. Health Inform. 18(6), 1915–1922 (2014)
16. Mastorakis, G., Makris, D.: Fall detection system using kinect as infrared sensor. J. Real Time Image Proc. 9(4), 635–646 (2014)
17. Mendi, E., Clemente, H.B., Bayrak, C.: Sports video summarization based on motion analysis. Comput. Electr. Eng. 39(3), 790–796 (2013)
18. Nguyen, V.A., Le, T.H., Nguyen, T.T.: Single camera based fall detection using motion and human shape features. In: Proceedings of the 7th Symposium on Information and Communication Technology, pp. 339–344 (2016)
19. Nunez-Marcos, A., Azkune, G., Arganda-Carreras, I.: Vision-based fall detection with convolutional neural networks. Wirel. Commun. Mob. Comput. 2017, (2017)
20. Peng, Y., Peng, J., Li, J., Yan, P., Hu, B.: Design and development of the fall detection system based on point cloud. Procedia Comput. Sci. 147, 271–275 (2019)
21. Ramachandran, A., Karuppiah, A.: A survey on recent advances in wearable fall detection systems. BioMed Res. Int. 2020 (2020)

22. Yun, Y., Gu, I.Y.H.: Human fall detection in videos via boosting and fusing statistical features of appearance, shape and motion dynamics on Riemannian manifolds with applications to assisted living. Comput. Vis. Image Underst. **148**, 111–122 (2016)
23. Zerrouki, N., Houacine, A.: Combined curvelets and hidden Markov models for human fall detection. Multimed. Tools Appl. **77**(5), 6405–6424 (2018)

Comparative Analysis of CNN-Based Spatiotemporal Reasoning in Videos

Okan Köpüklü[✉], Fabian Herzog, and Gerhard Rigoll

Institute for Human-Machine Communication, Technical University Munich,
Munich, Germany
okan.kopuklu@tum.de

Abstract. Understanding actions and gestures in video streams requires temporal reasoning of the spatial content from different time instants, i.e., spatiotemporal (ST) modeling. In this survey paper, we have made a comparative analysis of different ST modeling techniques for action and gesture recognition tasks. Since Convolutional Neural Networks (CNNs) are proved to be an effective tool as a feature extractor for static images, we apply ST modeling techniques on the features of static images from different time instants extracted by CNNs. All techniques are trained end-to-end together with a CNN feature extraction part and evaluated on two publicly available benchmarks: The Jester and the Something-Something datasets. The Jester dataset contains various dynamic and static hand gestures, whereas the Something-Something dataset contains actions of human-object interactions. The common characteristic of these two benchmarks is that the designed architectures need to capture the full temporal content of videos in order to correctly classify actions/gestures. Contrary to expectations, experimental results show that Recurrent Neural Network (RNN) based ST modeling techniques yield inferior results compared to other techniques such as fully convolutional architectures. Codes and pretrained models of this work are publicly available (https://github.com/fubel/stmodeling).

Keywords: Spatiotemporal modeling · CNNs · RNNs · Activity understanding · Action/gesture recognition

1 Introduction

Deep learning has been successfully applied in the area of image processing, providing state of the art solutions for many of its problems such as super-resolution [20], image denoising [22], and classification [4]. Due to the outstanding performance of two-dimensional (2D) Convolutional Neural Networks (CNNs) on processing static images, many attempts have been made to generalize 2D CNN architectures to capture the spatiotemporal (ST) structure of videos [29,37]. Until recently, 2D CNNs were the only options for video analysis tasks since lack of large scale video datasets made it impossible to train 3D CNNs properly.

© Springer Nature Switzerland AG 2021
A. Del Bimbo et al. (Eds.): ICPR 2020 Workshops, LNCS 12664, pp. 186–202, 2021.
https://doi.org/10.1007/978-3-030-68799-1_14

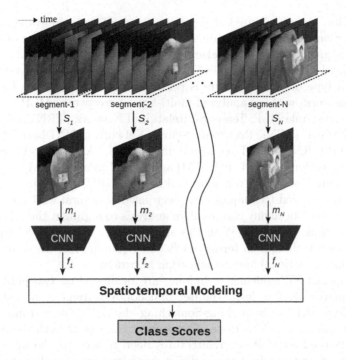

Fig. 1. Spatio-Temporal Modeling Architecture: One input video containing an action/gesture is divided into N segments. Afterwards, equidistant frames (m_1, m_2, ... m_N) are selected from the segments and fed to a 2D CNN for feature extraction. Extracted features are fed to a ST modeling block, which produces the final class score of the input video. In this example, action of "taking something from somewhere" is depicted (a sequence from the Something-Something dataset).

With the availability of large scale video datasets such as Kinetics [2], deeper and wider 3D CNN architectures can be successfully trained to achieve better performance compared to 2D CNNs [11]. More importantly, 3D CNNs can capture the ST patterns in videos inherently without requiring additional mechanisms. However, their drawback is that the input size should always remain the same for 3D CNNs such as 16 or 32 frames, which makes them not suitable for capturing temporally varying actions. This is not a problem for activity recognition tasks for Kinetics [2] or UCF-101 [31] datasets, as videos can be successfully classified using even very small snippets of the complete video. However, there are tasks where the designed architectures need to observe the complete video at once in order to make successful decisions. For these tasks, 2D CNN based architectures are still useful as the complete videos can be sparsely sampled with the desired number of segments and features of the selected frames can be extracted. Still, these architectures need an extra mechanism to provide ST modeling of the extracted features.

This work aims to analyze and compare various techniques for ST modeling of the features extracted by a 2D CNN from sparsely sampled frames of action/gesture videos. Figure 1 depicts the used ST modeling architecture. A complete action/gesture video is divided into a predefined number of segments. From each segment, a frame is selected (randomly in training and equidistant in testing) and fed into the 2D CNN to extract its features. In order to understand which type of action/gesture is performed, an ST modeling technique is used. In this work, we have analyzed multi-layer perceptron (MLP) based techniques such as simple MLP, Temporal Relational Network (TRN) and Temporal Segment Network (TSN), Recurrent Neural Network (RNN) based techniques such as vanilla RNN, gated recurrent unit (GRU), long short-term memory (LSTM), bidirectional LSTM (B-LSTM) and convolutional LSTM (ConvLSTM) techniques, and finally fully convolutional network (FCN) techniques.

Although analyzed techniques have been used at several works in the literature, there has not been any comparative analysis to highlight the advantages of each ST modeling technique. With this survey paper, we try to fill this gap by comparing each technique in terms of efficiency (i.e. number of parameters and floating-point operations) and classification accuracy.

The proposed ST modeling techniques are evaluated on two publicly available benchmarks: (i) The Jester dataset that contains dynamic and static hand gesture videos, (ii) the Something-Something dataset that contains videos of various human-object interactions. The common aspect of both these videos is that the proposed recognition architectures need to analyze the full content of the video in order to make a successful recognition, which makes them perfect benchmarks for analyzing ST modeling techniques.

2   Related Work

Deep learning architectures for ST modeling have been extensively studied in recent years, particularly in the context of action and gesture recognition [15,29,37,40]. Karpathy et al. [15] suggest several CNN architectures that fuse information across the temporal domain and applied the resulting models to the Spots-1M classification and UCF Action Recognition data sets. To speed up the training, they proposed a CNN-based multi-resolution architecture that could slightly improve the final results. Two stream CNNs [7,29] fuse a spatial network processing the video frames with a temporal network using optical flow to obtain a common class score. These methods rely on separately processing the spatial and temporal components of the video, which can be a disadvantage. 3D convolutional neural networks, on the other hand, can be used to inherently learn the spatiotemporal structure of videos [11,16]. Tran et al. [34] apply a 3D CNN architecture to obtain spatiotemporal feature volumes of input videos. To reduce training complexity, Sun et al. [32] propose a factorization of 3D spatiotemporal kernels into sequential 2D spatial kernels and separately handle sequence alignment. Although a sparse sampling strategy can be applied to the input value to span a larger time duration [18], all 3D architectures have the disadvantage that the input size needs to be fixed, which limits their capability of handling data sets with varying video lengths.

Recurrent neural networks are a natural choice for processing dynamic length video sequences, and several modern architectures have been proposed for action recognition in videos. LSTM [12] has been used in various video understanding tasks. Donahue *et al.* [5] employ an LSTM after CNN-based feature extraction on the individual frames to learn spatiotemporal components and apply the architecture on the UCF Action Recognition data set. Similarly, Baccouche *et al.* [1] use 3D convolutional neural networks together with an LSTM network. Liu *et al.* [21] suggest to modify the Vanilla LSTM architecture to learn spatiotemporal domains. GRU [3] is a popular variant of LSTM architecture which is actively used in video recognition tasks such as [6]. There have been many other variants of LSTM architecture, which are summarized in [10]. Another recurrent method is the Differentiable RNN [36] generated by salient motion patterns in consecutive video frames.

Although LSTM structure is proven to be stable and powerful in modeling long range temporal relations in various studies [9,33], it handles spatiotemporal data using only full connections where no spatial information is encoded. ConvLSTM [38] addresses this problem by using convolutional structures in both the input-to-state and state-to-state transitions. ConvLSTM is first introduced for precipitation nowcasting task [38], and later used for many other applications such as video saliency prediction [30], medical segmentation [39].

Fully Convolutional Networks (FCN) is first proposed for image segmentation task [23] and currently majority of segmentation architectures are based on FCNs. Later FCN architectures have been used at many other tasks such as object detection [25,26]. The idea of using convolution layers can also be applied to the task of ST modeling.

Methods like Temporal Segment Networks [37] enable processing longer videos by segmenting the input video into a certain number of segments, selecting short-length snippets randomly from each segment and finally fusing individual prediction scores. These prediction scores are the result of a spatial convolutional network operating on the samples frames and a temporal convolutional network operating on optical flow components. However, it must be noted that averaging is applied for the fusion of extracted features in TSN, which causes to lose the temporal order of the features. Similarly, Temporal Relation Networks [40] extract a number of ordered frames from the input video, which are then passed through a convolutional neural network for feature extraction. Different from TSN, TRN keeps the order of the extracted features and tries to discover possible temporal relations at multiple time scales.

3   Methodology

In this section, we first describe the complete ST modeling architecture, which is based on a 2D CNN feature extraction part and one ST modeling block. Afterward, we investigate different ST modeling techniques in detail that can be used within this architecture. Finally, we will give the training details used in the experiments.

3.1   ST modeling Architecture

As illustrated in Fig. 1, a video clip V that contains a complete action/gesture is divided into N segments. Each segment is represented as $S_n \in \mathbb{R}^{w \times h \times c \times m}$ of $m \geq 1$ sequential frames with 224×224 spatial resolution and $c = 3$ channels. RGB modality is used in all of the trainings. Afterward, within segments, equidistant frames are selected and passed to a 2D CNN model for feature extraction. Extracted features are first pooled and transformed to a fixed size of 256 (except for TSN where features are transformed to *number-of-classes*) via a one-layer Multi-layer Perceptron (MLP) except for ConvLSTM and 3D-FCN techniques. For these two techniques, no pooling is applied at the feature extraction and number of channels is transformed to 256 by using a 1×1 2D convolution layer.

For feature extraction, two different CNN models are used: (i) SqueezeNet [13] with simple bypass and (ii) Inception with Batch Normalization (BNInception) [14]. The reason to choose these models is that the performance of the investigated ST modeling techniques can be evaluated with a lightweight CNN feature extractor (SqueezeNet) and relatively more complex and heavyweight CNN feature extractor (BNInception). In this way, *CNN-model-agnostic* performance of evaluated techniques can be observed.

Extracted features are finally fed to an ST modeling block, which produces the final class scores of the input video clip. Next, we are going to investigate different ST modeling techniques in detail that are used in this block.

3.2   Multi-layer Perceptron (MLP) Based Techniques

MLP-based ST modeling techniques are simple but effective to incorporate temporal information. These techniques make use of MLPs once or multiple times. Extracted features are then fed to these MLP-based ST modeling blocks keeping their order intact. The intuition is that MLPs can capture the temporal information of the sequence inherently without knowing that it is a sequence at all.

Simple MLP. As illustrated in Fig. 2, extracted features are concatenated preserving their order. Then, the concatenated single $N \times 256$ dimensional vector is fed to a 2-layer MLP with 512 and *Number-of-classes* neurons. Finally, the output is fed to a softmax layer to get class conditional scores. This is a simple but effective approach. Combined with other modalities such as optical flow, infrared and depth, competitive results can be achieved [17].

Temporal Segment Network (TSN). TSN aims to achieve long-range temporal structure modeling using sparse sampling strategy [37]. When the original paper was written, TSN achieved state-of-the-art performance on two activity recognition datasets, namely the UCF-101 [31] dataset and the HMDB [19] dataset.

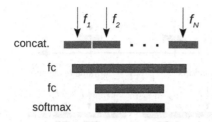

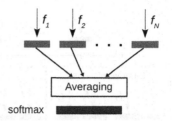

Fig. 2. Simple MLP technique. Extracted features are concatenated keeping their order same to form $N \times 256$ dimensional vector. This vector is fed to a 2-layer MLP to get final class scores.

Fig. 3. Temporal Segment Network (TSN) architecture. Extracted frame features are transformed to *Number-of-classes* dimension and averaged to get class conditional scores.

The original TSN architecture uses optical flow and RGB modalities, as well as different consensus methods such as evenly averaging, maximum, and weighted averaging. Among them, evenly averaging achieved the best results in the original experiments. Therefore, we have also experimented with evenly averaging for RGB modality only.

The corresponding TSN approach is depicted in Fig. 3. Unlike other ST modeling techniques, the extracted frame features are transformed into a fixed size of *number-of-classes* instead of 256. Afterward, all extracted features are averaged and fed to a softmax layer to get class conditional scores.

Although TSN achieved state-of-the-art performance on UCF-101 and HMDB benchmarks at the time, it achieves inferior performance in the Jester and Something-Something benchmarks. The reason is that averaging causes loss of temporal information. This does not create a huge problem for the UCF-101 and HMDB benchmarks as temporal order is not critical for these. Correct classification can even be achieved using only one frame of the complete video. However, the Jester and Something-Something datasets require the incorporation of the complete video in order to infer correct class scores.

Temporal Relation Network (TRN). TRNs [40] aim to discover possible temporal relations between observations at multiple time scales. The main inspiration for this work comes from the relational reasoning module for visual question answering [27]. The pairwise temporal relations (2-frame relations) on the observations of the video V are defined as

$$T_2(V) = h_\phi \left(\sum_{i<j} g_\theta(f_i, f_j) \right),$$

where the input is the features of the n selected frames of the video $V = \{f_1, f_1, \ldots, f_n\}$, in which f_i represents the feature of the ith frame segment extracted by a 2D CNN. Here, h_ϕ and g_θ represent the feature fusing functions, which are MLPs with parameters ϕ and θ, respectively. For these functions, the exact

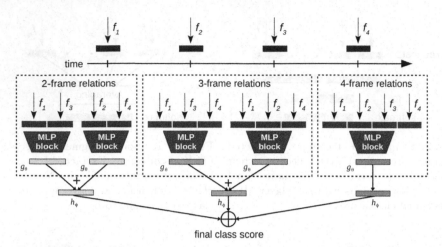

Fig. 4. Illustration of Temporal Relation Networks. Features extracted from different segments of a video by a 2D CNN are fed into different frame relation modules. Only a subset of the 2-frame, 3-frame, and 4-frame relations are shown in this example (4 segments), as there are higher frame relations included according to the segment size.

same MLP block as depicted in Fig. 2 is used. These two-frame temporal relations functions are further extended to higher frame relations, but the order of the segments should always be kept same in order to learn temporal relations inherently. Finally, all frame relations can be incorporated in order to get a single final output $MT_N(V) = T_2(V) + T_3(V) + \ldots + T_N(V)$, which is referred as multiscale TRN, where each T_d captures temporal relationships between features of d ordered frames. Figure 4 depicts the overall TRN architecture.

3.3 Recurrent Neural Networks (RNN) Based Techniques

Recurrent neural networks (RNNs) are a special type of artificial neural networks and consist of recurrently connected hidden layers which are capable of capturing temporal information. Furthermore, they allow the input and output sequences to vary in size. It is important to note that the hidden layer parameters do not depend on the time step but are shared across all RNN slices. The ability to keep information from previous time steps makes the hidden layer work like a memory. General M-layered RNN architecture is depicted in Fig. 5.

In our experiments, we use two different vanilla RNNs, based on the hyperbolic tangent activation function, and the rectified linear unit (ReLU) activation function, respectively. Vanilla RNN with hyperbolic tangent activation function can be described by following equations

$$h_t = \tanh\left(W_{hh}h_{t-1} + W_{xh}x_t\right)$$
$$y_t = W_{hy}h_t$$

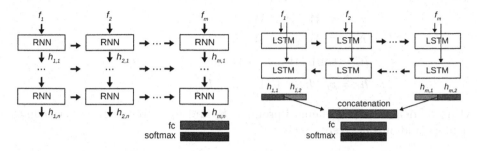

Fig. 5. M-layered architecture of Recurrent Neural Networks.

Fig. 6. The data flow for Bidirectional LSTM architecture.

Generally, we feed the output of the last node to a fully connected layer to obtain a vector size of the number of classes in the dataset. We also proceed in the same manner for all other RNN types except for the *Bidirectional LSTM*.

Long Short-Term Memory (LSTM). LSTMs [12] are recurrent neural networks consisting of a cell, an input gate, a forget gate, and an output gate. The input gate i_t decides how much the current x_t contributes to the overall output. The cell c_t is responsible for remembering the previous state information, and also uses the results of the forget gate f_t, which decides how much of the previous cell c_{t-1} flows into the current cell. As the name suggests, the forget gate can completely erase the previous state if necessary. Finally, the output gate determines the contribution of the current cell c_t. All in all, the standard LSTM can be described by the following equations

$$i_t = \sigma \left(W_{xi} x_t + W_{hi} h_{t-1} + W_{ci} \circ c_{t-1} + b_i \right)$$
$$f_t = \sigma \left(W_{xf} x_t + W_{hf} h_{t-1} + W_{cf} \circ c_{t-1} + b_f \right)$$
$$c_t = f_t \circ c_{t-1} + i_t \circ \tanh \left(W_{xc} x_t + W_{hc} h_{t-1} + b_c \right)$$
$$o_t = \sigma \left(W_{xo} x_t + W_{ho} h_{t-1} + W_{co} \circ c_t + b_o \right)$$
$$h_t = o_t \circ \tanh \left(c_t \right)$$

where '∘' denotes the Hadamard product.

Gated Recurrent Units (GRU). GRUs [3] are very similar to LSTMs and consist of two gates - an update gate z_t and a reset gate r_t. However, unlike LSTMs, GRUs do not have their own memory control mechanism. Instead, the entire hidden layer information is directed to the next time step. The advantage of GRUs compared to LSTMs is their simplicity in structure, which significantly reduces the number of parameters to be learned. GRU can be described by the following equations

$$z_t = \sigma \left(W_{xz} x_t + W_{hz} h_{t-1} + b_z \right)$$
$$r_t = \sigma \left(W_{xr} x_t + W_{hr} h_{t-1} + b_r \right)$$
$$\tilde{h}_t = \tanh \left(W_{xh} x_t + W_{hh} (r_t \circ h_{t-1}) + b_h \right)$$
$$h_t = z_t \circ h_{t-1} + (1 - z_t) \circ \tilde{h}_t$$

where 'o' denotes the Hadamard product; $\tilde{h}_t$ and h_t represent the intermediate memory and output, respectively.

Bidirectional LSTM (BLSTM). BLSTMs [28] are a special form of LSTMs, but are trained in both directions. The fully connected layer is obtained by concatenating two halved outputs $h_{1,1}$ and $h_{m,2}$, namely the first output of the positive time direction and the last output of the negative time direction. The data flow for BLSTM architecture is depicted in Fig. 6. We also investigate the effect of the hidden size by reducing it to half of the hidden size value we used for the other RNN-structures. This allows us to make meaningful comparisons with the latter. The reduction of the hidden layer size means that the vector size remains unchanged before the last fully connected layer. Consequently, the same number of output neurons is used for the classification.

Convolutional LSTM (ConvLSTM). The main drawback of conventional LSTM (also GRU and vanilla RNN) in handling spatiotemporal data is that input-to-state and state-to-state transitions are made by full connections, where no spatial information is encoded. To overcome this drawback, convolutional LSTM proposes to use convolutional structures for the mentioned transitions. The main equations of ConvLSTM are given as

$$i_t = \sigma \left(W_{xi} * x_t + W_{hi} * h_{t-1} + W_{ci} \circ c_{t-1} + b_i \right)$$
$$f_t = \sigma \left(W_{xf} * x_t + W_{hf} * h_{t-1} + W_{cf} \circ c_{t-1} + b_f \right)$$
$$c_t = f_t \circ c_{t-1} + i_t \circ \tanh \left(W_{xc} * x_t + W_{hc} * h_{t-1} + b_c \right)$$
$$o_t = \sigma \left(W_{xo} * x_t + W_{ho} * h_{t-1} + W_{co} \circ c_t + b_o \right)$$
$$h_t = o_t \circ \tanh \left(c_t \right)$$

where '∗' and 'o' denote the convolution operator and Hadamard product, respectively. In order to make use of ConvLSTM technique for ST modeling, we have made some modifications. First, to keep spatial resolution, we have removed the final pooling layer of our feature extractor 2D CNN. Then, the output features are concatenated in time dimension forming a $D \times W \times H$ tensor. The last output of ConvLSTM is average pooled and fed to a final fully connected layer and a softmax layer to get class conditional scores.

3.4 Fully Convolutional Network (FCN) Based Techniques

As the name implies, all of the layers of a fully convolutional network are convolutional layers. FCNs do not contain any linear (fully connected) layers at the

Table 1. Details of 2D fully convolutional ST modeling architecture.

Layer/stride	Filter size	Output size
Input		$1 \times N \times 256$
Conv1/s(1,2)	3×3	$64 \times N \times 128$
Conv2/s(1,2)	3×3	$64 \times N \times 64$
Conv3/s(1,2)	3×3	$128 \times N \times 32$
Conv4/s(1,2)	3×3	$128 \times N \times 16$
Conv5/s(1,2)	3×3	$256 \times N \times 8$
Conv6/s(1,1)	1×1	$NumCls \times N \times 8$
AvgPool/s(1,1)	$N \times 8$	$NumCls$

Table 2. Details of 3D fully convolutional ST modeling architecture.

Layer/stride	Filter size	Output size
Input		$256 \times D \times W \times H$
Conv1/s(2,1,1)	$3 \times 3 \times 3$	$64 \times D/2 \times W \times H$
Conv2/s(2,1,1)	$3 \times 3 \times 3$	$128 \times D/4 \times W \times H$
Conv3/s(2,1,1)	$3 \times 3 \times 3$	$256 \times D/8 \times W \times H$
Conv4/s(1,1,1)	$1 \times 1 \times 1$	$NumCls \times D/8 \times W \times H$
AvgPool/s(1,1,1)	$D/8 \times W \times H$	$NumCls$

end, which is the typical use for classification task. In order to utilize FCNs as ST modeling technique, output features coming from the 2D CNNs can be concatenated over a dimension such that convolution operation can be performed over the concatenated tensor. If the features are pooled, concatenated features form a 2D tensor with over which 2D convolutional layers can operate. If pooling is not applied at feature extraction stage, concatenated features form a 2D tensor over which 3D convolutional layers can operate.

2D-FCN. The inputs to 2D-FCN are the concatenated feature vectors of each segment resulting $N \times 256$ such that each row represents features from a segment. The input volume enters a series of 2D convolutions with stride $(1, 2)$, which keeps the temporal dimension (i.e. the number of segments) intact throughout convolution operations. The kernel size is set to 3×3 with the same padding for all convolutions. After applying five convolutions, 2D convolution with 1×1 kernel is applied where the number of channels equals the number of classes. Finally, average pooling with $N \times 8$ is applied to get class conditional scores. After each convolution, batch normalization and ReLU is applied. The details of the used 2D-FCN are given in Table 1.

3D-FCN. In order to make use of spatial information, similar to ConvLSTM, we have not used a pooling layer at the end of feature extractor 2D CNN, and output features are concatenated in depth dimension to create $D \times W \times H$ tensor. This tensor enters a series of 3D convolutions with stride $(2, 1, 1)$ in order to keep the spatial resolution same. The kernel size is set to $3 \times 3 \times 3$ with the same padding for all convolutions. After applying three convolutions, 3D convolution with $1 \times 1 \times 1$ kernel is applied to reduce the number of channels to number of classes, which is pooled later to get class scores. After each convolution, batch normalization and ReLU is applied. The details of the used 3D-FCN are given in Table 2.

The proposed approach in fact very similar to rMCx models in [35] except for that 3D convolutions are applied at the very end. It is again similar to ECO architecture [41], but 2D features in ECO architecture are extracted again at an early stage of the 2D CNN feature extractor. Although, 3D CNN architectures can be used for varying video lengths, we can use 3D convolution layers as ST modeling technique since we are using a fixed segment size for all the input videos.

3.5   Training Details

Given the ST modeling architecture in Fig. 1, the CNN architecture used to extract frame features plays a critical role in the performance of the overall architecture. In order to get *CNN-model-agnostic* performance of the applied ST modeling techniques, the SqueezeNet and BNInception models are used. For both models, features are transformed to 256-dimensional vectors (*Number-of-classes*-dimensional vectors for only TSN) via an MLP after global pooling layer except for ConvLSTM and FCN3D. For these two ST modeling approach, spatial resolution (13×13 and 7×7 for SqueezeNet and BNInception) is preserved by removing the final pooling layer, and 1×1 convolution layer is used to transform the number of channels to 256. For all experiments, CNN models pretrained on ImageNet dataset are used.

Learning: Stochastic gradient descent (SGD) with standard categorical cross-entropy loss is applied. For momentum and weight decay, 9×10^{-1} and 5×10^{-4} are used, respectively. The learning rate is initialized with 1×10^{-3} and reduced twice with a factor of 10^{-1} after validation loss converges.

Regularization: Several regularization techniques are applied in order to reduce over-fitting and achieve a better generalization. Weight decay of $\gamma = 5 \times 10^{-4}$ is applied to all parameters of the architecture. A dropout layer is applied after the global pooling layer of 2D CNN architectures with a ratio of 0.3. Moreover, data augmentation of multiscale random cropping is applied for both datasets and random horizontal flip applied for Something-Something dataset. Random horizontal flipping is not performed for the trainings of Jester dataset since this changes the annotations of some classes.

Implementation: The complete ST modeling architecture is implemented and trained (end-to-end) in PyTorch.

4   Experiments

4.1   Datasets

The Jester-V1 dataset is currently the largest hand gesture dataset that is publicly available [24]. It is an extensive collection of segmented video clips that contain humans performing pre-defined hand gestures in front of a laptop camera or webcam. The dataset consists of 148092 video clips under 27 classes, which is split into training, validation and test sets containing 118562, 14787 and 14743 videos, respectively. For the experiments, the validation set is used as the labels of the test set are not made available by dataset providers.

The Something-Something-V2 dataset is a collection of segmented video clips that show humans performing pre-defined basic actions with everyday objects [8]. It allows researchers to develop machine learning models capturing a fine-grained understanding of basic actions. The dataset consists of 220847 video clips under 174 classes, which is split into training, validation and test sets containing 168913, 24777 and 27157 videos, respectively. For the experiments, the validation set is used as the labels of the test set are not made available by dataset providers.

The duration of gesture clips in Jester dataset is concentrated between 30–40 frames. However, the Something-Something dataset has videos with relatively varying temporal dimension between 20 and 70 frames, which is the reason why 3D CNN architectures accepting fixed-size inputs are not suitable for this benchmark. In order to recognize video clips correctly, the used architectures should incorporate information coming from all parts of the videos.

4.2   Resource Efficiency Analysis

For real-time systems, the resource efficiency of the applied ST modeling techniques is as essential as the achieved classification accuracy. Therefore, we have investigated the number of parameters and floating-point operations (FLOPs) of each technique, which can be found in Table 3. For all calculations, we have used 8-segment case for Jester dataset.

Out of all ST modeling techniques, TSN comes for free since it requires no parameters and there is only averaging operation. However, temporal information is lost due to averaging, which results in inferior performance compared to simple-MLP or TRN techniques.

ConvLSTM and 3D-FCN does not employ pooling at the end of feature extractor, which requires highest number of FLOPs compared to others. The number of FLOPs for SqueezeNet is $13^2/7^2 = 3.48$ times higher than BNInception due to the output resolution of feature maps. In terms of number of parameters, ConvLSTM, 3D-FCN and TRN-multiscale requires the highest number with 4.73M, 1.56M and 2.34M parameters, respectively.

On the other hand, the resource efficiency of the feature extractors (i.e. 2D CNNs) are also important. The BNInception architecture contains 11.30M parameters and requires 1894 MFLOPs to extract features of a 224×224 frame.

Table 3. Comparison of different ST modeling techniques over classification accuracy, number of parameters and computation complexity (i.e., number of Floating Point Operations - FLOPs). Methods are evaluated using 8 and 16 segments on validation sets of Jester-V1 and Something-Something-V2 datasets. The number of parameters and FLOPs are calculated for only ST modeling blocks excluding CNN feature extractors for Jester dataset using 8 segments. FLOPs values of ConvLSTM and 3D-FCN are reported separately for BNInception (left) and SqueezeNet (right) since their spatial resolution is 7×7 and 13×13, respectively.

Model	MFLOPs	Params	Accuracy (%)					
			Jester (8 seg.)		Something (8 seg.)		Something (16 seg.)	
			Squeez.	BNIncep.	Squeez.	BNIncep.	Squeez.	BNIncep.
Simple-MLP	2.13	1.06M	87.28	92.80	31.89	46.35	33.96	47.01
TSN	0.001	0.00M	72.84	82.74	20.91	37.28	22.15	36.22
TRN-multiscale	11.95	2.34M	88.39	93.20	33.73	**46.91**	34.38	**47.73**
RNN_tanh	3.16	0.14M	70.51	79.53	16.12	25.17	14.48	21.64
RNN_ReLU	3.16	0.14M	78.33	88.15	21.40	36.01	15.84	24.88
LSTM	8.42	0.53M	84.28	90.80	25.24	39.04	28.25	42.83
GRU	6.32	0.40M	83.10	90.86	25.40	40.69	30.24	43.31
B-LSTM	6.33	0.40M	84.87	91.12	25.04	39.35	27.88	42.41
ConvLSTM	1849.70/6379.54	4.73M	89.57	93.38	31.31	46.40	32.86	46.64
2D-FCN	39.07	0.56M	88.11	93.64	27.72	39.17	29.95	40.56
3D-FCN	152.07/524.49	1.56M	**90.19**	**94.07**	**37.10**	46.66	**37.59**	47.37

On the other hand, the SqueezeNet architecture contains only 1.24M parameters and requires 338 MFLOPs to extract the features of a same-sized frame.

4.3   Results Using Jester Dataset

For the Jester dataset, the spatial content for all classes are the same: A hand in front of a camera performing a gesture. Therefore, a designed architecture should capture the form, position, and the motion of the hand in order to recognize the correct class.

Comparative results of different ST-modeling techniques for Jester dataset can be found in Table 3. Inspired from [17], we have used eight segments for this benchmark as it achieves the best performance for MFF architecture. Compared to BNInception, architectures with SqueezeNet have 5–10% inferior classification accuracy for the same ST modeling technique. However, the technique-wise comparison remains similar within the same 2D CNN backbone.

Out of all ST modeling techniques, TRN-multiscale, 2D-FCN, 3D-FCN and ConvLSTM stand out for classification accuracy. Considering the resource efficiency, the simple-MLP model can also be preferred over TRN-multiscale. Surprisingly, RNN-based methods except ConvLSTM, which first come to mind for modeling sequences, perform worse than these techniques.

The superiority of ConvLSTM over other RNN based techniques and superiority of 3D-FCN over 2D-FCN validates the importance of the spatial content. 3D-FCN is the best performing technique in terms of accuracy for Jester dataset.

However, preserving the spatial resolution brings the burden of increased computation and number of parameters. As expected, TSN yields the lowest classification accuracy as the averaging operation causes a loss of temporal information.

4.4 Results Using Something-Something Dataset

Compared to the Jester dataset, the Something-Something dataset contains much more classes with more complex spatial content. In order to identify the correct class label, the designed architectures need to extract the spatial content and temporally link this content successfully. Therefore, the frame feature extractors (i.e., 2D CNNs) are critical for the overall performance.

Comparative results of different ST-modeling techniques for the Something-Something dataset can be found in Table 3. Beside 8-segment architectures, we have also made experiments for 16-segment architectures as the spatial complexity of the dataset is higher compared to Jester. Due to this complexity, architectures with SqueezeNet have 10% to 15% inferior classification accuracy compared to architectures with BNInception. However, similar to Jester dataset, the technique-wise comparison remains similar within the same 2D CNN backbone.

Compared to 8-segments, 16-segment architectures perform better. However, performance improvement is not as drastic as the effect of feature extractors. This shows that the main complexity of this task comes from the complexity of scenes, not the complexity of finer temporal details. In order to get better performance on Something-Something dataset, more complex architectures with deeper and wider structure can be preferred.

Out of all ST modeling techniques, 3D-FCN, ConvLSTM and TRN-multiscale again stand out for classification accuracy. Specifically, 3D-FCN with SqueezeNet performs best outperforming the second best model TRN-multiscale by 3.37%. Similar performance difference cannot be observed when BNInception used as feature extractor. We conjecture that this is due to the higher feature resolution of the SqueezeNet architecture.

Moreover, ConvLSTM also outperforms all other RNN based techniques by 3–6% showing the importance of spatial information again for this task. On the other hand, 2D-FCN cannot achieve the performance it reached in the Jester dataset and performs inferior to GRU and LSTM. Similar to the Jester dataset, Vanilla RNN and TSN yield lowest accuracies.

5   Conclusion

In this work, we have analyzed various techniques for CNN-based spatiotemporal modeling and compared them based on a consistent 2D CNN feature extraction of sparsely sampled frames. The individual methods were then evaluated on the Jester and Something-Something datasets. It has been shown that the CNN models used for feature extraction and the number of frames sampled affect the results. For the Jester dataset, the 3D-FCN technique achieves the best results using both SqueezeNet and BNInception. On the Something-Something dataset, again 3D-FCN and TRN-multiscale techniques outperform all other models while ConvLSTM performs similar results. It has also been shown that simple vanilla RNNs are unable to understand the complex spatiotemporal relationships of the data. All the more complex RNNs tested perform very similarly.

Interestingly, the TSN model, which showed state-of-the-art performance on the UCF-101 and HMDB benchmarks, performs rather poorly in our experiments, which shows the importance of maintaining the temporal information. Among all techniques, ConvLSTM and 3D-FCN requires the highest number of FLOPs, since they do not employ pooling at the feature extractor and preserve spatial resolution. For number of parameters, again ConvLSTM, 3D-FCN and TRN-multiscale techniques are the most expensive ones. While some models like TRN, LSTM, GRU, and B-LSTM can benefit from an increase in the number of segments, Vanilla RNNs and the TSN model can suffer from overfitting. One possibility for future research would be to develop resource efficient ST modeling techniques that preserves the spatial resolution of the extracted features.

Acknowledgements. We gratefully acknowledge the support of NVIDIA Corporation with the donation of the Titan Xp GPU used for this research.

References

1. Baccouche, M., Mamalet, F., Wolf, C., Garcia, C., Baskurt, A.: Sequential deep learning for human action recognition. In: Salah, A.A., Lepri, B. (eds.) HBU 2011. LNCS, vol. 7065, pp. 29–39. Springer, Heidelberg (2011). https://doi.org/10.1007/978-3-642-25446-8_4
2. Carreira, J., Zisserman, A.: Quo Vadis, action recognition? A new model and the kinetics dataset. In: Proceedings of the IEEE Conference on Computer Vision and Pattern Recognition, pp. 6299–6308 (2017)
3. Cho, K., van Merrienboer, B., Bahdanau, D., Bengio, Y.: On the properties of neural machine translation: encoder-decoder approaches, pp. 103–111 (2014)
4. Deng, J., Dong, W., Socher, R., Li, L.J., Li, K., Fei-Fei, L.: ImageNet: a large-scale hierarchical image database. In: 2009 IEEE Conference on Computer Vision and Pattern Recognition, pp. 248–255. IEEE (2009)
5. Donahue, J., et al.: Long-term recurrent convolutional networks for visual recognition and description, pp. 2625–2634 (2015)
6. Dwibedi, D., Sermanet, P., Tompson, J.: Temporal reasoning in videos using convolutional gated recurrent units. In: Proceedings of the IEEE Conference on Computer Vision and Pattern Recognition Workshops, pp. 1111–1116 (2018)
7. Feichtenhofer, C., Pinz, A., Zisserman, A.: Convolutional two-stream network fusion for video action recognition. In: Proceedings of the IEEE Conference on Computer Vision and Pattern Recognition, pp. 1933–1941 (2016)
8. Goyal, R., et al.: The "something something" video database for learning and evaluating visual common sense, vol. 1, no. 2, p. 3 (2017)
9. Graves, A.: Generating sequences with recurrent neural networks. arXiv preprint arXiv:1308.0850 (2013)

10. Greff, K., Srivastava, R.K., Koutník, J., Steunebrink, B.R., Schmidhuber, J.: LSTM: a search space odyssey. IEEE Trans. Neural Netw. Learn. Syst. **28**(10), 2222–2232 (2016)
11. Hara, K., Kataoka, H., Satoh, Y.: Can spatiotemporal 3D CNNs retrace the history of 2D CNNs and ImageNet? In: Proceedings of the IEEE Conference on Computer Vision and Pattern Recognition, pp. 6546–6555 (2018)
12. Hochreiter, S., Schmidhuber, J.: Long short-term memory. Neural Comput. **9**(8), 1735–1780 (1997)
13. Iandola, F.N., Han, S., Moskewicz, M.W., Ashraf, K., Dally, W.J., Keutzer, K.: SqueezeNet: AlexNet-level accuracy with 50x fewer parameters and <0.5 mb model size. arXiv preprint arXiv:1602.07360 (2016)
14. Ioffe, S., Szegedy, C.: Batch normalization: accelerating deep network training by reducing internal covariate shift. arXiv preprint arXiv:1502.03167 (2015)
15. Karpathy, A., Toderici, G., Shetty, S., Leung, T., Sukthankar, R., Fei-Fei, L.: Large-scale video classification with convolutional neural networks. In: Proceedings of the IEEE Conference on Computer Vision and Pattern Recognition, pp. 1725–1732 (2014)
16. Köpüklü, O., Kose, N., Gunduz, A., Rigoll, G.: Resource efficient 3D convolutional neural networks. arXiv preprint arXiv:1904.02422 (2019)
17. Köpüklü, O., Kose, N., Rigoll, G.: Motion fused frames: data level fusion strategy for hand gesture recognition. In: Proceedings of the IEEE Conference on Computer Vision and Pattern Recognition Workshops, pp. 2103–2111 (2018)
18. Köpüklü, O., Rigoll, G.: Analysis on temporal dimension of inputs for 3D convolutional neural networks. In: 2018 IEEE International Conference on Image Processing, Applications and Systems (IPAS), pp. 79–84. IEEE (2018)
19. Kuehne, H., Jhuang, H., Garrote, E., Poggio, T., Serre, T.: HMDB: a large video database for human motion recognition. In: 2011 International Conference on Computer Vision, pp. 2556–2563. IEEE (2011)
20. Ledig, C., et al.: Photo-realistic single image super-resolution using a generative adversarial network. In: Proceedings of the IEEE Conference on Computer Vision and Pattern Recognition, pp. 4681–4690 (2017)
21. Liu, J., Shahroudy, A., Xu, D., Wang, G.: Spatio-temporal LSTM with trust gates for 3D human action recognition. In: Leibe, B., Matas, J., Sebe, N., Welling, M. (eds.) ECCV 2016. LNCS, vol. 9907, pp. 816–833. Springer, Cham (2016). https://doi.org/10.1007/978-3-319-46487-9_50
22. Liu, P., Zhang, H., Zhang, K., Lin, L., Zuo, W.: Multi-level wavelet-CNN for image restoration. In: The IEEE Conference on Computer Vision and Pattern Recognition (CVPR) Workshops (June 2018)
23. Long, J., Shelhamer, E., Darrell, T.: Fully convolutional networks for semantic segmentation. In: Proceedings of the IEEE Conference on Computer Vision and Pattern Recognition, pp. 3431–3440 (2015)
24. Materzynska, J., Berger, G., Bax, I., Memisevic, R.: The Jester dataset: a large-scale video dataset of human gestures. In: Proceedings of the IEEE/CVF International Conference on Computer Vision (ICCV) Workshops (October 2019)
25. Redmon, J., Divvala, S., Girshick, R., Farhadi, A.: You only look once: unified, real-time object detection. In: Proceedings of the IEEE Conference on Computer Vision and Pattern Recognition, pp. 779–788 (2016)
26. Ren, S., He, K., Girshick, R., Sun, J.: Faster R-CNN: towards real-time object detection with region proposal networks. In: Advances in Neural Information Processing Systems, pp. 91–99 (2015)

27. Santoro, A., et al.: A simple neural network module for relational reasoning. In: Advances in Neural Information Processing Systems, pp. 4967–4976 (2017)
28. Schuster, M., Paliwal, K.K.: Bidirectional recurrent neural networks. IEEE Trans. Sig. Process. **45**(11), 2673–2681 (1997)
29. Simonyan, K., Zisserman, A.: Two-stream convolutional networks for action recognition in videos. In: Ghahramani, Z., Welling, M., Cortes, C., Lawrence, N.D., Weinberger, K.Q. (eds.) Advances in Neural Information Processing Systems, vol. 27, pp. 568–576. Curran Associates, Inc. (2014)
30. Song, H., Wang, W., Zhao, S., Shen, J., Lam, K.-M.: Pyramid dilated deeper ConvLSTM for video salient object detection. In: Ferrari, V., Hebert, M., Sminchisescu, C., Weiss, Y. (eds.) ECCV 2018. LNCS, vol. 11215, pp. 744–760. Springer, Cham (2018). https://doi.org/10.1007/978-3-030-01252-6_44
31. Soomro, K., Zamir, A.R., Shah, M.: UCF101: a dataset of 101 human actions classes from videos in the wild. arXiv preprint arXiv:1212.0402 (2012)
32. Sun, L., Jia, K., Yeung, D.Y., Shi, B.E.: Human action recognition using factorized spatio-temporal convolutional networks. In: Proceedings of the IEEE International Conference on Computer Vision, pp. 4597–4605 (2015)
33. Sutskever, I., Vinyals, O., Le, Q.V.: Sequence to sequence learning with neural networks. In: Advances in Neural Information Processing Systems, pp. 3104–3112 (2014)
34. Tran, D., Bourdev, L., Fergus, R., Torresani, L., Paluri, M.: Learning spatiotemporal features with 3D convolutional networks. In: Proceedings of the IEEE International Conference on Computer Vision, pp. 4489–4497 (2015)
35. Tran, D., Wang, H., Torresani, L., Ray, J., LeCun, Y., Paluri, M.: A closer look at spatiotemporal convolutions for action recognition. In: Proceedings of the IEEE Conference on Computer Vision and Pattern Recognition, pp. 6450–6459 (2018)
36. Veeriah, V., Zhuang, N., Qi, G.J.: Differential recurrent neural networks for action recognition. In: Proceedings of the IEEE International Conference on Computer Vision, pp. 4041–4049 (2015)
37. Wang, L., et al.: Temporal segment networks: towards good practices for deep action recognition. In: Leibe, B., Matas, J., Sebe, N., Welling, M. (eds.) ECCV 2016. LNCS, vol. 9912, pp. 20–36. Springer, Cham (2016). https://doi.org/10.1007/978-3-319-46484-8_2
38. Xingjian, S., Chen, Z., Wang, H., Yeung, D.Y., Wong, W.K., Woo, W.C.: Convolutional LSTM network: a machine learning approach for precipitation nowcasting. In: Advances in Neural Information Processing Systems, pp. 802–810 (2015)
39. Zhang, L., et al.: Spatio-temporal convolutional LSTMs for tumor growth prediction by learning 4D longitudinal patient data. IEEE Trans. Med. Imaging **39**(4), 1114–1126 (2019)
40. Zhou, B., Andonian, A., Oliva, A., Torralba, A.: Temporal relational reasoning in videos. In: Ferrari, V., Hebert, M., Sminchisescu, C., Weiss, Y. (eds.) ECCV 2018. LNCS, vol. 11205, pp. 831–846. Springer, Cham (2018). https://doi.org/10.1007/978-3-030-01246-5_49
41. Zolfaghari, M., Singh, K., Brox, T.: ECO: efficient convolutional network for online video understanding. In: Ferrari, V., Hebert, M., Sminchisescu, C., Weiss, Y. (eds.) ECCV 2018. LNCS, vol. 11206, pp. 713–730. Springer, Cham (2018). https://doi.org/10.1007/978-3-030-01216-8_43

Generalization of Fitness Exercise Recognition from Doppler Measurements by Domain-Adaption and Few-Shot Learning

Biying Fu[1,2(✉)] 🆔, Naser Damer[1,2] 🆔, Florian Kirchbuchner[1,2] 🆔, and Arjan Kuijper[1,2] 🆔

[1] Fraunhofer Institute for Computer Graphics Research IGD, Fraunhoferstr. 5, 64283 Darmstadt, Germany
biying.fu@igd.fraunhofer.de
[2] Mathematical and Applied Visual Computing, TU Darmstadt, Darmstadt, Germany

Abstract. In previous works, a mobile application was developed using an unmodified commercial smartphone to recognize whole-body exercises. The working principle was based on the ultrasound Doppler sensing with the device built-in hardware. Applying such a lab environment trained model on realistic application variations causes a significant drop in performance, and thus decimate its applicability. The reason of the reduced performance can be manifold. It could be induced by the user, environment, and device variations in realistic scenarios. Such scenarios are often more complex and diverse, which can be challenging to anticipate in the initial training data. To study and overcome this issue, this paper presents a database with controlled and uncontrolled subsets of fitness exercises. We propose two concepts to utilize small adaption data to successfully improve model generalization in an uncontrolled environment, increasing the recognition accuracy by two to six folds compared to the baseline for different users.

Keywords: Human activity recognition · Mobile sensing · Ultrasound sensing

1 Introduction

Human activity recognition (HAR) covers a wide range of application areas. Understanding human actions in daily living enables application designers to build assisting smart home applications for elderly care [2], security applications with video surveillance [14], applications for Quantified-self [6,16], or associate physiological signals with emotions [9] to build interactive applications. HAR is the key to enable human-centered application designs and natural interaction in a smart environment [5].

© Springer Nature Switzerland AG 2021
A. Del Bimbo et al. (Eds.): ICPR 2020 Workshops, LNCS 12664, pp. 203–218, 2021.
https://doi.org/10.1007/978-3-030-68799-1_15

However, human motion is highly complex and possesses a high degree of freedom. Learning a generalized model for all possible variations of a human motion is very challenging. Thus, training a model on limited amount of individuals under constrained environment often leads to a large performance drop when applying the model on individuals/environments disjoint from the training data. This reduction in performance originates from the large variations between the controlled training data and the real-world application scenarios. It is caused by the diversity and complexity of the users actions, the device hardware or other environmental variations. However, all possible reasons lead to a degradation in the usability of the proposed application. One possible solution is to reduce the inherent difference between the development dataset and the real-world dataset by making the development data resemble the real-world data. However, due to the diversity in the real-world applications, there is no generalized model that is applicable in all possible situations.

This work addresses sport exercise recognition from a stationary smartphone using the Doppler measurement. We propose a set of methods to improve the generalization of a pre-trained model (trained on controlled data) to scenarios containing a combination of unseen environments, individuals, and devices. To achieve this, we propose and investigate two concepts, along with a clear baseline that demonstrate the generalization problem. We have developed a mobile application that aims at collecting data that is used to deal with this challenge. Our application is based on the built-in hardware of a commercial smartphone to measure whole-body exercise activities. The main contribution of this work is grouped as follows:

- A novel database for investigating micro-Doppler motion in relation to whole-body exercise data with built-in smartphone hardware. The database contains sessions in controlled environment, as well as a disjoint subset containing variations of environments, individuals, and devices.
- Propose and adapt two concepts (with variations) to improve the recognition generalization. These concepts are based on domain adaptions, as well as few-shot learning. Both concepts proved to enhance the generalizability on data variations in comparison to a clear baseline.

The structure of the paper is organized as follows: in Sect. 2 we provide some state-of-the-art researches on model generalization to fit new data and categorized these approaches under two main categories (retrain required and not). In Sect. 3 we introduce one of the main contributions of this paper by presenting our collected database and motivate the need for such a database. We then introduce the sensing principle and describe the details about the database and what it enables us to study. In Sect. 4, we propose the baseline model and the new approaches targeting our problem, under two main concepts. Section 5 introduces evaluation results of our proposed individual approaches, along the baseline. We further introduce a discussion on certain methods and provide some guidelines in design choice for such an application. Finally, we conclude our work in Sect. 6 and provide relevant future research directions.

2   Related Works

Andrew Ng once stated [13], that the most frequent fail of inference model in reality is the inherent difference between your development set and test set. To overcome this problem, effort can be put into developing highly-representative development/training databases. Though, it is impossible to make the development set identical to the test set (application scenario), due to the large variety and complexity in human activities. Common methods to adapt the pre-trained model on individual new data samples without enforcing much restrictions on the development set is thus desirable. We distinguish between two main categories: with and without retraining the base model to adapt to new data.

Retraining of the Base Model. Domain adaptation builds on transferring knowledge from similar domains to cope with unknown target domains. This method is especially useful, when you do not have enough annotated datasets for the particular problem at hand. The goal is to extract knowledge from related, known datasets, and use this knowledge to learn the new task at hand.

Wang [18] benefited from using similar labeled source domain data to annotate the target domain that has only a few or even no labels. They evaluated their approach on acceleration dataset from different body positions as different domains. To alleviate the problem of negative knowledge transfer, they proposed an unsupervised similarity measure to choose the *right* source domain with respect to the target domain. Khan and Roy et al. [8] proposed a CNN based transductive transfer learning model to adapt action recognition classifiers trained in one context to be applicable to a different contextual domain. The limitation is that the set of activities being monitored is the same in both context domains, as they are transferring knowledge from individual convolutional layers. Evaluated on their acquired smartphone and smartwatch acceleration dataset on 15 users and 8 activities, they demonstrated the ability of their proposed methods on transferring knowledge from smartphone to smartwatch domain and vice versa.

These methods pose less constraints on the target domain, however are difficult to train, as the knowledge transfer is solely based on the source domains. Thus the choice of the appropriate source domain is critical for the performance of the inference model on the unknown target domain. A retraining is required to relate the source domain to the new target domain due to knowledge transfer.

No Retraining of the Base Model. Few-shot learning is currently an active research field in machine learning. The ability of deep neural networks to learn complex correlations and patterns from a vast dataset is proven. However, current deep learning approaches suffer from the problem of poor sample efficiency. To make a model learn on a new class, sufficient amount of labeled samples from this class is required to avoid overfitting.

Few-shot learning methods increase the model generalizability with limited data. They have been mostly used in image classification tasks [10,15]. Feng [4] recently applied few-shot learning-based classification on human activity recognition tasks. They applied a deep learning framework to automatically extract features and perform classification. However, instead of transferring knowledge in a common feature space, their proposed method tends to perform knowledge transfer in terms of model parameter transfer. Based on two benchmark datasets, they evaluated their proposed technique. A metric to measure the cross-domain class-wise relevance is introduced to mitigate the challenging issue of negative transfer. These datasets consist of only sparse sensory input, mostly acceleration sensor data attached to different body parts.

We consider this category to be the most realistic in designing applications for human activity recognition tasks with sensory data. Since we can not adopt to the complexity and diversity of all persons actions during the training phase, we need the network to have the ability to adapt to individual users by introducing only a small amount of this users data to optimize the trained network. Few-shot classification methods are methods that can be leveraged to unseen classes, even when less labels from these classes are available without retraining the base model.

Exercise detection on personal devices is often applied to track daily ambulation activities [11]. For tracking of more stationary activities involving whole-body interaction, a remote system is better than wearing a smartphone on the body, as the detection is more unobtrusive. However, the complexity and diversity in human action makes it difficult to develop one single system to fits all situations. To overcome this issue, it requires more advanced machine learning methods to improve the model generalization.

3   Ultrasonic Smartphone Exercises: Sensing and Database

In this paper, we contribute a database collected with built-in hardware of commercial smartphones. This database deals with exercise data using Doppler sensing by utilising the smartphone as a sonar device. Using smartphone to collect activity data is not novel per se. Most existing databases are, however, focused on acceleration data using the smartphone as a wearable device. Popular databases are for example provided in the works [1,11]. Though tracking and recognizing for various aerobic training exercises are popular, there exists limited research on recognizing more stationary exercises, such as strength-based training without the use of wearable. These exercises are essential for rehabilitation purposes [17]. Typically they are even harder to track, as they rely on coordinated movement of specific body parts. Instead of building customized hardware designs such as in [6,16] to target these exercises, we leveraged the existent infrastructure of a commercial mobile device to build an ubiquitous application. The set of stationary and strength-based exercises are depicted in Fig. 1.

3.1   Sensing Principle

By emitting a 20 kHz continuous audio signal from the built-in speaker, we turned the commercial smartphone into an active sonar. The echo encoded with

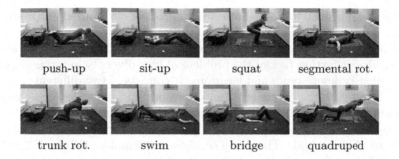

push-up   sit-up   squat   segmental rot.

trunk rot.   swim   bridge   quadruped

Fig. 1. Figure shows the eight workout exercises we present in our database. The smartphone is placed roughly 50 cm away from the body of the performing user.

the Doppler modulation is received from the device internal microphone. Doppler measurement allows us to catch relative movement in close range to the sensing device. A positive Doppler is received with a relative speed towards the device and a negative Doppler vice versa. The device speaker can typically emit tones up to 22 kHz on a commodity audio system without performance degradation. Thus, we can detect a one-sided Doppler speed up to $17.4 \frac{m}{s}$ (2.05 kHz). To extract the Doppler features, a Short Time Fourier Transformation (STFT) [3] is used to convert the time signal to the spectrum domain. The parameter of STFT determines the resolution in time and frequency domain. The selected frequency resolution corresponds to a relative speed of $3 \frac{cm}{s}$ (3.6 Hz) with a time resolution of 46.5 ms. A sliding window approach of 6 s windows and an overlap of 50% are chosen to prepare data samples used to train the classification task. To further reduce the computational effort, we restrict the spectrum amplitudes within frequency band between 19.5 kHz to 20.5 kHz, as other signals beyond the motion information are irrelevant. A typical *push-up* exercise takes around 2–3 s each [19], depending on the individual fitness. We conclude that with the current system setting, we are able to resolve both the slowest and fastest movement of the targeted exercise set.

3.2 Database

This database allows us to study the body motion in relation to Doppler profiles from built-in hardware of various commercial smartphones. The effect of fine-grained movements from both limbs and arms cause micro-Doppler patterns in addition to the main Doppler reflection. Studying these micro-Doppler events enhances the ability of recognizing more complex and naturalistic human activities including whole-body interactions. To the best of our knowledge, there does not exist such a database so far. Due to the similarity of ultrasound sensing and electromagnetic waves in physical characteristics, this database can also be leveraged for machine learning practitioners to design radar-based applications and gain useful insight without the additional cost of customized hardware. The database is available in GitHub[1].

The presented database not only helps to design applications using a specific mobile device, it further tackles the problem of user diversity in terms of various aspects, such as device types and locations. It consists of two different setups, in order to investigate the effects of various methods on improving model generalization for different data distributions. Data of the first setup is called Lab-Data. The Lab-Data consists of data collected in a laboratory setup as depicted in Fig. 1 from 14 individuals. The group consists of 4 females (157 cm–172 cm) and 10 males (172 cm–193 cm). The affinity towards sport exercises varies across the test participants. The built-in microphone of the sensing device is placed 50 cm apart, facing the exercising individuals on the floor, aligned with the hip. For each individual, two separate sessions were collected with 10 repetitions of each exercise class. Left and right variations for exercises such as *segmented rotation, trunk rotation*, and *quadruped* are counted as one repetition. *Swim* is performed in average for 30 s in each session to reach similar time duration comparable to the other exercise types. In order to collect the micro-Doppler motions from the arms, the device is aligned with the shoulder for *swim* and *trunk rotation*. The duration of each session is approximately 7–9 min in average. The smartphone used for data acquisition has the brand Samsung Galaxy A6 (2018) and the placement of the sensing device to the exercising body is constrained to the same position for all participants.

The goal of the second setup is to be leveraged on testing various finetuning approaches, as these data are collected under individual, different hardware, and uncontrolled environments independent of the Lab setup. This part of data is called the Uncontrolled-Data. It consists of data collected from five different individuals. Due to logistic and privacy constrains related to the experimental setup, the second setup contains a smaller number of participants compared to the Lab-Data. The hardware device is not limited to the smartphone used in the Lab-Data. Each individual was asked to collect eight individual sessions distributed over several days in their familiar surroundings and without any supervision. Each session has a comparable length to the collected data from the Lab-Data. Some general statistics about the participants and the hardware devices used, are listed in Table 1.

Table 1. Description of software and hardware setups for the Uncontrolled-Data.

Participant	Sex	Height	Exercise frequency	Device	Location
P1	Male	180 cm	Frequent	SONY Xperia XZ2 Compact	Env. 1
P2	Male	181 cm	Frequent	Samsung Galaxy A5 (2017)	Env. 2
P3	Male	181 cm	Frequent	SONY Xperia Z5 Dual	Env. 3
P4	Male	182 cm	Frequent	Samsung Galaxy A6 (2018)	Env. 4
P5	Female	168 cm	Less frequent	SONY Xperia XZ2 Compact	Env. 1

[1] https://github.com/fbiying87/Ultrasound_mobile_database.git.

The data acquisition app is installed on the individual mobile device. The participants from the uncontrolled setup were asked to collected data from their home environment to simulate the real-world scenarios. Figure 2 illustrates the different data acquisition environments that affect the signal strength of the underlying hardware device. In contrast to the Lab setup, the other apartments all have wooden floors which makes the back reflected signal strength stronger compared to the Lab setup.

Environment 1 Environment 2 Environment 3 Environment 4

Fig. 2. Illustration depicts the four different data acquisition setups from the individual test participants.

This paper aims at investigating methods to improve the generalizability of pre-trained models on new individuals under more realistic conditions. Based on the underlying method, we split the data as follows:

- Basic training data contain all individuals and sessions of the Lab-Data.
- Subject development data contains 4 sessions (out of 8) of each of the users in the uncontrolled setup, the Uncontrolled-Data.
- Testing data contains the 4 sessions (out of 8) of each of the users in the uncontrolled setup, the Uncontrolled-Data, that were not used in the Subject development data.

4   Methods

This paper is motivated by our observations from our previous experiences in realistic use of mobile device to detect workout exercises [7]. However, previous work faced a major usability challenge as it did not adapt well on individuals and environments unseen in the training phase. This issue is the main target of the methods presented in this paper. To state the problem, we observed the input signal of the same participant performing under two different environments, as depicted in Fig. 3. Despite the overall speed and appearance remain similar, the strength and noise embedded reveal a strong difference in both settings. We noticed a strong decay in signal power due to the material-dependent attenuation of the transmit power.

Realistically, we can not train a classifier adapting to every possible sensing environment, unless our training data unrealistically contain unlimited variations. The quality of hardware devices integrated in the smartphone may also

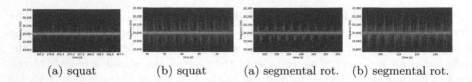

(a) squat   (b) squat   (a) segmental rot.   (b) segmental rot.

Fig. 3. (a) Visualization of the Doppler profiles collected in a constrained laboratory environment (Lab-Data) for the Participant P4. (b) Visualization of the Doppler profiles collected in Env. 4 for the same Participant from the Uncontrolled-Data.

introduce strong variations in the signal power. But the basic physical characteristics remain. To adapt to new, real-world circumstances, we need to individually finetune the trained model. In this paper, we investigate several approaches to improve the model generalizability given limited individual data.

4.1   The Baseline Method

The base inference model is built using a stacked bidirectional LSTM network. To baseline our proposed solution, we need to demonstrate the exercises detection performance when the uncontrolled environment is not considered. The model contains 2 stacked bidirectional LSTM layers with 128 hidden nodes in each LSTM cell. For each input node, a slice of spectrogram with the frequency bands (ranging from 19.5 kHz to 20.5 kHz) from a time step resolution (46.5 ms) is provided to the network. The bidirectional structure permits the network to look forward and backward in time to extract fine-grained sequence information using micro-Doppler from the spectrum domain.

An 2-D instance normalization layer is applied on the sample spectrogram prior to the network input in order to reduce the hardware specific power dependencies. Output includes the eight activity classes plus the *none* class describing all the transitions and noisy samples between two successive action classes. The learning rate is set to 0.001 and the Adam Optimizer is used to optimize the network parameters. Cross entropy loss is used as the cost function to minimize the loss of the misclassification error from the training samples. Batch-wise training is used to construct similar training procedures compared to other network structures.

The training contains the Lab-Data only. Subject development data with 4 sessions from the Uncontrolled-Data is used in the validation stage, while the 4 sessions of the remaining Uncontrolled-Data is used in the test phase. Batch-wise approach is applied, while each batch contains 15 samples randomly selected from each class.

4.2   Our Proposed Method 1: Domain Adaptation

To improve the model generalization ability, the first method we propose is from the domain adaptation (DA) network family. DA is commonly used to transfer knowledge from labeled source domain to target domain where data is unlabeled or only partially labeled. In our case, the source domain refers to the Lab-Data, while the target domain refers to the Uncontrolled-Data collected by the different individuals under changed conditions. By applying this architecture, we aim at adopting the base feature extractors to be sensitive to deterministic features from the Uncontrolled-Data domain with the knowledge draw from the source domain. In such a way, the classification performance does improve for the uncontrolled setup.

The metric to measure the similarity of both distributions is based on the maximum mean discrepancy (MMD) on the final feature level. The model architecture of the adaptation network is depicted in Fig. 4. Common approaches using DA do not require to include target labels. However, without any label information from the target domain, the knowledge transfer does not work well on the targeted use-case, as both domains are quite dissimilar. In order to improve the knowledge transfer characteristic, we included partial labels (50%) from the subject development set of the Uncontrolled-Data to increase the performance on feature level adaptation. An instance normalization layer is used prior to the ConvNet to mitigate the hardware dependent effects of the transmit power from different smartphone models.

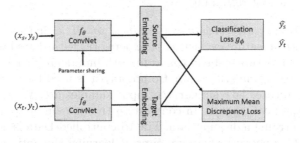

Fig. 4. Depicts the domain adaptation model. Source and target data are projected into the same embedding space using the common base ConvNet. Adaptation is realized by minimizing the classification losses for source and target and reducing differences in both feature distributions.

A base ConvNet is used to extract common features from source and target domain. The ConvNet structure consists of 4 successive convolutional layers, each followed by a batch normalization layer, leaky rectified linear unit (ReLU) activation layer with the negative slope coefficient of 0.2 and a max pooling layer to reduce the input dimensions. The successive layers increased the number of filters from 32-32-64-64. The same ConvNet structure is used for all networks as the feature extraction component in this work. The objective is to minimize the combination of three different losses as given in Eq. (1): both cross entropy losses from the source and target domain classification, and the MMD loss of the feature embeddings originated from the source and target domain in the same embedding space.

$$\mathcal{L} = \mathcal{L}(g_\phi(f_\theta(\mathbf{x}_s)), y_s) + \mathcal{L}((g_\phi(f_\theta(\mathbf{x}_t)), y_t) + \mathcal{L}_{MMD} \qquad (1)$$

The network is trained with 100 epochs. Each batch composites of 15 samples from each of the 9 classes. Adam optimizer with a learning rate of 0.001 is set to learn the network hyperparameters. Since the base feature extraction network needs to be optimized on both data domains, a retraining of the base model is required. We applied the Lab-Data as the source domain and the uncontrolled subject development set as the target domain. The adapted model is evaluated on the testing data of the uncontrolled setup.

This method is good to adopt the feature extraction layers to work for features from different domains. A negative aspect is that if both domains differ too much, it could lead to a negative adaptation and causing the performance on the source data to decrease. To address this issue, we present the following methods which do not need to retrain the pre-trained inference model to finetune to new individuals. We benefit from a few labels of the new datasets to label this unknown dataset.

4.3   Our Proposed Method 2: Few-Shot Classification

This section deals with three methods from the research domain of the few-shot classification learning. The network is trying to learn common features within a subset of tasks without retraining.

F1: Siamese Network with Few-Shot Classification. The Siamese network consists of two identical feature extraction base networks with shared weight parameters. The learned feature embeddings from both inputs are then compared with each other to form a similarity score. Commonly it is used for verification tasks and the score indicates how similar two input samples are. We extended it for the few-shot classification task. In contrast to the domain adaptation method, this method can train on disjoint data. The Uncontrolled-Data is not used in the training stage. Our network structure aims at learning the optimum separation between all multiple classes at once.

During training time, each training batch consists of 15 query samples from each class. The support set consists of 5 support samples from each class. They are fed through the ConvNet f_θ to get the feature embeddings. A mean embedding is calculated for each class to reduce the computational complexity. A similarity score is determined for the query sample embedding $f_\theta(\mathbf{x}_q)$ and the individual support class embeddings $f_\theta(\mathbf{x}_i)$, where $i = 1..C$ and C represents the number of classes. The euclidean distance measure is further fed through a dense layer to learn the final similarity score $\delta_i = \sigma(\phi(|f_\theta(\mathbf{x}_q) - f_\theta(\mathbf{x}_i)|))$. This parametric dense layer is optimized by a pair of input samples every time step. Based on the similarity measure between the query sample and all other support classes $(\delta_1, \delta_2, \delta_3, ..., \delta_C)$, the objective of the Siamese network is to maximize the maximum likelihood estimation of data pairs.

Adam optimizer is used to train the network parameter with a learning rate of 0.0005. 500 Epochs were performed during the training phase with 9-ways-5-shots. In the evaluation phase, the number of supports each class can be tuned between 5 and 10 samples each. We perform 100 iterations to obtain the average accuracy on the test set.

The similarity measure of the Siamese network here is learnt by a parametric dense layer, in the next two paragraphs we introduce two other alternatives from the few-shot classification task where a non-parametric distance metric-based learning is used in the classification stage. These methods are more robust against small difference in source and target domains, and thus more generalized.

F2: Prototypical Few-Shot Classification Network. Inspired by [15], we adopted the prototypical network as the second method (F2) in the few-shot classification problem. In comparison to the first few-shot method (F1) with Siamese network, this network is non-parametric in the classification stage. This approach is a feature metric-based learning approach. Instead of transferring network parameters, this learning approach is based on learning the similarity distance between the feature embeddings from new target data to the prototypes of the samples from the same setup. This method is similar to a clustering-based method to find the k-nearest neighbours used for the classification.

The negative log-likelihood (NLL) function is applied to the negative euclidean distance of the embedding vectors to each class center embeddings (prototype). The objective is to reduce the NLL loss of the query output to the true classification label. Adam optimizer is used to optimize the weights of the ConvNet hyperparameters. A learning rate of 0.001 is used to learn the hyperparameters. 500 Epochs were performed to train 5-ways-10-shots classification tasks. 15 query images each class are used to build the query set. In the evaluation phase, the number of class recognition task has increased to the total 9 classes. The number of supports from each class is varied between 5 and 10 samples each. We perform 100 iterations to obtain the average accuracy on the test set.

The disadvantage of using euclidean measure to express the similarity is that this measure is not bounded. In contrast to Siamese network, there is non parametric learning after the feature extraction ConvNet structure. The last method uses the cosine similarity measure which is a bounded metric.

F3: Local Descriptor Correlated Few-Shot Classification Network. this method of the few-shot classification (F3) is inspired by Li [12]. The author introduced a new distance metric to label the query sample. Instead of using image-level feature based measure, they introduced a local descriptor based image-to-class measure. We think this architecture will work on the 2D time-frequency spectrum, as this network is sensitive to local features within a global context. In contrast to object images, the Doppler spectrum contains less fine-grained contextual information, but local features caused by repetitive movements and micro-movements from limbs and arms are clearly observable. In this paper, we examine the local descriptor correlation from the structural information extracted from the micro-Doppler range caused by different whole-body activities.

The local features from the ConvNet output is correlated with all other local descriptors of the support embeddings each class using a cosine similarity.

This provides a locally feature-based image-to-class mapping. The advantage of the cosine metric is because the cosine distance measures the pattern similarity without being largely effected by the magnitude. The objective is to reduce the cross entropy loss for this multiclass classification problem. Adam optimizer with a learning rate of 0.001 is chosen. The parameters for the few-shot learning and the batch-wise composition are the same as in the method F2 for the prototypical network.

In few-shot classification learning task, the training data can be disjoint of the test data. We used the Lab-Data to build the training tasks and used the subject development data as support set and the testing data from the uncontrolled setup to evaluate the model.

In this section, we introduced three methods from few-shot learning: (F1) Siamese, (F2) ProtoNet, and (F3) LocalNet. These methods are theoretically suitable to enhance the generalization for our use-case, since we want to mitigate the retraining phase for similar tasks.

5   Evaluation and Discussion

The database details are introduced in Sect. 3. In this section, we present and discuss the results of the proposed approaches in comparison to the baseline. We aim to optimize the trained model with Lab-Data under controlled condition to be generalizable to individuals under the uncontrolled setups. The accuracy of correctly classified activities of the baseline model and models which require adaptation with the Uncontrolled-Data during the training phase are displayed in Table 2.

Table 2. The accuracy of correctly classified activities for baseline method and methods with model retraining is depicted. The term P_i indicates the ID of the participants. Enabling the knowledge from the Uncontrolled-Data to modify the base feature extraction increases the performance on individual finetuning.

Method	Ratio of labels used from Uncontrolled-Data	P1	P2	P3	P4	P5
Baseline	–	16.25%	46.35%	11.17%	25.73%	15.5%
DA	0%	35,20%	37,80%	20.14%	57.67%	38.04%
DA	50%	77.61%	85.39%	58.61%	78.86%	75.87%
DA	100%	**87.13%**	**98.14%**	**84.10%**	**76.92%**	**96.85%**

5.1   Baseline Results

Given the Lab-Data in the base training, the trained inference model does not generalize well on test data collected under uncontrolled conditions, if not provided in the base training. The stacked bidirectional LSTM network failed to cope with data from real-world environments, as both data distributions differ too much. According to the baseline result provided in Table 2, in most of the cases the accuracy equals to a uniform distribution. The model performs no better than random guessing on the Uncontrolled-Data. In our application, the total number of classes is 9, random guessing corresponds to an average accuracy of $\frac{1}{9} = 11.1\%$.

5.2   Our Proposed Approaches

The Results of Using the Domain Adaptation (DA) Method. are depicted in Table 2. Here we successively increased the amount of target labels from the subject development set to be included into the training to improve the performance of the domain adaptation. Without including any target labels, the network only minimizes the distribution of the source and target embeddings in an unsupervised way. The performance is only 10–20% points better compared to the baseline. By incorporating more label information from the target domain, the knowledge transfer improves on the target domain classification. With only 50% of the target labels, the results increased about 40% points and with 100% of the labels, the results increased about two to six folds in average.

Results of Few-Shot Classification Networks. Here, we evaluated three alternatives of the few-shot classification approach. These models typically do not need to retrain the pre-trained inference model, as the tasks are disjoint. According to Table 3, a general tendency is that the classification accuracy increases at least 5% points with increasing number of support samples per class used in the evaluation. This is intuitive, due to an increased reliability and an improved decision boundary related to more support samples. The Siamese network (F1) provides similar results compared to the prototypical network (F2), as both network architectures work with euclidean similarity scores on the sample-base.

Table 3. The accuracy of correctly classified activities for the three alternatives of few-shot classification task is shown. The number @(5 or 10) indicates the number of support samples each class used in the evaluation. The term P_i indicates the ID of the participants. The performance increases in general with the increasing number of supports. The LocalNet with the cosine similarity measure outperforms the other methods, as it includes the image-to-class feature correlation.

No	Method	P1	P2	P3	P4	P5
–	Baseline	16.25%	46.35%	11.17%	25.73%	15.5%
F1	Siamese@5	65.53%	83.77%	72.81%	65.61%	68.33%
F2	ProtoNet@5	67.06%	79.9%	67.85%	77.54%	65.92%
F3	LocalNet@5	**85.28%**	**79.9%**	**64.85%**	**94.33%**	**86.98%**
F1	Siamese@10	69.69%	90.07%	75.65%	69.08%	72.56%
F2	ProtoNet@10	74.18%	88.59%	69.31%	87.88%	70.80%
F3	LocalNet@10	**89.55%**	**97.84%**	**67.51%**	**98.00%**	**91.63%**

Their performances are increased around 50% points compared to the baseline model. The image-to-class measure in the LocalNet (F3) performs the best as depicted in Table 3 with an increase of two to six folds in average compared to the baseline.

5.3 Discussion

The performance of activity recognition based on ultrasound sensing using a mobile device is subject to many variables. To deploy a fixed application to real-world scenarios is therefore not easy and often has to overcome some difficulties. In many cases, the performance drops due to the dissimilarity in both domains.

To overcome this issue, we investigated several methods in this paper. We provide a database collected under various conditions to allow researchers perform experiments on it to solve the problem of lack of generalization on new individuals. The base data consists of Lab-Data under controlled environment and same sensing device. Uncontrolled setups from five different individuals are used to evaluate the methods for finetuning on individual dataset.

Finetuning a base model on new domains requires sufficient amount of labeled samples from the Uncontrolled-Data. Otherwise, the model would overfit adopting on this small data amount. However, labels are most difficult to acquire and the individual labeling process might be error prone. In such cases, domain adaption method can be leveraged, where no label information of the target domain is required. Though, such network could benefit from including a small amount of the target labels in case both domains differ as investigated in our use-cases.

Few-shot classification is suitable for adopting finetuning on limited data without retraining. This method can cope with individual hardware characteristics without modifying the base training. By comparing knowledge extracted from support samples of different categories, an unknown sample is able to assign to the correct category under the assumption that samples of similar categories are close in the embedding space. As no feature adaptation from the target domain is applied, this model requires both domains behave similarly.

To leverage few-shot classification, the user has to pre-label a small amount of individual sessions before the model is adopted to this user. These labels are used to classify the new samples based on certain distance metrics. The developer does not need to modify the feature extraction network to individually adapt to each new user. In case of the domain adaptation, the developer needs to modify the base feature extraction according to the user data. It further assumes the similarity of both domains in order to avoid negative knowledge transfer.

6 Conclusion

In this paper, we investigated different approaches to improve the generalizability of pre-trained classification models under controlled condition in uncontrolled real-world scenarios based on a mobile application for workout exercise recognition. We first presented a database to enable us analysing this problem and building novel solutions. This database allows us to study the body motion in relation to Doppler profiles from built-in hardware of different commercial smartphones. The gap between the development setup and the real-world scenarios, as we prove, often lead to performance drop and bad usability. The reason can be manifold, as it could rely on individual difference, hardware specifics or environmental changes. Our database is built to overcome this gap. We proposed two methods: domain adaptation and few-shot classification, to resolve the issue of lack of model generalizability. Our evaluations showed that the baseline method fails when faced with realistic data. Our proposed concept of using domain adaption without including the target labels improved the baseline only by 10–20% points in most cases. However, this method benefits from including target labels, as with increasing amount of target labels in the training phase, the recognition performance increases by two to six folds compared to the baseline. Our proposed solution that is based on few-shot classification improved the accuracy to the same range, however, without the effort of retraining.

Acknowledgment. This research work has been funded by the German Federal Ministry of Education and Research and the Hessian Ministry of Higher Education, Research, Science and the Arts within their joint support of the National Research Center for Applied Cybersecurity ATHENE.

References

1. Anguita, D., Ghio, A., Oneto, L., Parra, X., Reyes-Ortiz, J.L.: A public domain dataset for human activity recognition using smartphones. In: ESANN (2013)
2. Chen, L., Nugent, C.D., Wang, H.: A knowledge-driven approach to activity recognition in smart homes. IEEE Trans. Knowl. Data Eng. **24**(6), 961–974 (2011)
3. Durak, L., Arikan, O.: Short-time Fourier transform: two fundamental properties and an optimal implementation. IEEE Trans. Signal Process. **51**(5), 1231–1242 (2003)
4. Feng, S., Duarte, M.F.: Few-shot learning-based human activity recognition. Expert Syst. Appl. **138**, 112782 (2019)
5. Fu, B., Damer, N., Kirchbuchner, F., Kuijper, A.: Sensing technology for human activity recognition: a comprehensive survey. IEEE Access **8**, 83791–83820 (2020)
6. Fu, B., Jarms, L., Kirchbuchner, F., Kuijper, A.: ExerTrack-towards smart surfaces to track exercises. Technologies **8**, 17 (2020)
7. Fu, B., Kirchbuchner, F., Kuijper, A.: Unconstrained workout activity recognition on unmodified commercial off-the-shelf smartphones. In: The 13th PErvasive Technologies Related to Assistive Environments Conference, p. 10 (2020)
8. Khan, M.A.A.H., Roy, N., Misra, A.: Scaling human activity recognition via deep learning-based domain adaptation. In: 2018 IEEE International Conference on Pervasive Computing and Communications (PerCom), pp. 1–9. IEEE (2018)
9. Kim, K.H., Bang, S.W., Kim, S.R.: Emotion recognition system using short-term monitoring of physiological signals. Med. Biol. Eng. Comput. **42**(3), 419–427 (2004). https://doi.org/10.1007/BF02344719
10. Koch, G., Zemel, R., Salakhutdinov, R.: Siamese neural networks for one-shot image recognition. In: ICML Deep Learning Workshop, Lille, vol. 2 (2015)

11. Kwapisz, J.R., Weiss, G.M., Moore, S.A.: Activity recognition using cell phone accelerometers. ACM SigKDD Explor. Newslett. **12**(2), 74–82 (2011)
12. Li, W., Wang, L., Xu, J., Huo, J., Gao, Y., Luo, J.: Revisiting local descriptor based image-to-class measure for few-shot learning. In: Proceedings of the IEEE Conference on Computer Vision and Pattern Recognition, pp. 7260–7268 (2019)
13. Ng, A.: Machine learning yearning (96) (2017). http://www.mlyearning.org/
14. Saxena, S., Brémond, F., Thonnat, M., Ma, R.: Crowd behavior recognition for video surveillance. In: Blanc-Talon, J., Bourennane, S., Philips, W., Popescu, D., Scheunders, P. (eds.) ACIVS 2008. LNCS, vol. 5259, pp. 970–981. Springer, Heidelberg (2008). https://doi.org/10.1007/978-3-540-88458-3_88
15. Snell, J., Swersky, K., Zemel, R.: Prototypical networks for few-shot learning. In: Advances in Neural Information Processing Systems, pp. 4077–4087 (2017)
16. Sundholm, M., Cheng, J., Zhou, B., Sethi, A., Lukowicz, P.: Smart-Mat: recognizing and counting gym exercises with low-cost resistive pressure sensing matrix. In: UbiComp 2014, pp. 373–382. ACM Press, New York (2014)
17. U.S. Department of Health and Human Services: Physical Activity Guidelines for Americans (2008). https://health.gov/paguidelines/pdf/paguide.pdf
18. Wang, J., Zheng, V.W., Chen, Y., Huang, M.: Deep transfer learning for cross-domain activity recognition. In: Proceedings of the 3rd International Conference on Crowd Science and Engineering, pp. 1–8 (2018)
19. Yoo, W.g.: Effect of exercise speed and isokinetic feedback on the middle and lower serratus anterior muscles during push-up exercises. J. Phys. Ther. Sci. **26**(5), 645–646 (2014)

Local Anomaly Detection in Videos Using Object-Centric Adversarial Learning

Pankaj Raj Roy[1(✉)], Guillaume-Alexandre Bilodeau[1], and Lama Seoud[2]

[1] LITIV, Department of Computer and Software Engineering,
Polytechnique Montréal, Montréal, Canada
{pankaj-raj.roy,gabilodeau}@polymtl.ca
[2] Institute of Biomedical Engineering, Polytechnique Montréal, Montréal, Canada
lama.seoud@polymtl.ca

Abstract. We propose a novel unsupervised approach based on a two-stage object-centric adversarial framework that only needs object regions for detecting frame-level local anomalies in videos. The first stage consists in learning the correspondence between the current appearance and past gradient images of objects in scenes deemed normal, allowing us to either generate the past gradient from current appearance or the reverse. The second stage extracts the partial reconstruction errors between real and generated images (appearance and past gradient) with normal object behaviour, and trains a discriminator in an adversarial fashion. In inference mode, we employ the trained image generators with the adversarially learned binary classifier for outputting region-level anomaly detection scores. We tested our method on four public benchmarks, UMN, UCSD, Avenue and ShanghaiTech and our proposed object-centric adversarial approach yields competitive or even superior results compared to state-of-the-art methods.

Keywords: Video anomaly detection · Object-based · Adversarial learning

1 Introduction

Detecting anomalies in surveillance videos allows designing safer living environments by identifying potential risks, unsafe interactions between users or confusing urban signage. Similarly to previous work [11,18,25], we define abnormal event detection as the identification of spatio-temporal image regions in a video that deviate from the learned normal ones. We focus on detecting abnormal events on a per individual/object basis, also known as local anomalies. They are independent from other surrounding spatio-temporal events. Thus, we only need to consider the image regions corresponding to objects possessing the abnormal behaviour. We want to detect those events with just a small number of frames.

Supported by grants from IVADO and NSERC funding programs.

A. Del Bimbo et al. (Eds.): ICPR 2020 Workshops, LNCS 12664, pp. 219–234, 2021.
https://doi.org/10.1007/978-3-030-68799-1_16

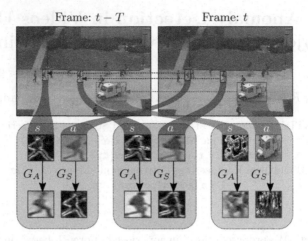

Fig. 1. Generated images from our proposed method. G_S generates the past spatial gradient image s from the appearance a of a region in a frame t and G_A does the reverse. For an abnormal region (in red), the images are not generated correctly compared to the normal regions (in blue). (Color figure online)

Recently, a solution based on a Convolutional Auto-Encoder (CAE) was proposed in [4] for detecting local anomalies, which takes less memory for building the networks and for which the training is significantly faster compared to holistic methods that consider the whole image, not just objects' bounding boxes. This is seen as an object-centric approach, due to the fact that it ignores background information and learns to classify local anomalies solely based on local information of the objects. However, it relies on K-means clustering combined with a one-versus-rest SVM classification scheme, that requires a predefined knowledge on the number of clusters which might vary depending on the scenario. Moreover, the CAE models are trained separately and are, therefore, unable to learn the relation between different local information like appearance and gradient.

To tackle these issues, we propose a new method for detecting local anomalies in videos that uses a novel unsupervised two-staged object-centric adversarial framework. The first stage of our method learns the normal local gradient-appearance correspondences and the second stage learns to classify events in an unsupervised manner. The local gradient-appearance correspondence is learned by relating the gradients of a previous frame with the visual appearance of an object in the current frame.

Our method first uses a pretrained object detector to extract all the regions of interest in a frame, and then, extracts the spatial gradients in the previous frame at the location of the detected objects in the current frame. After that, we train the components of our generative framework: 1) two cross-domain generators, where one learns to predict the past gradients by taking the appearance and the other one learns the reverse, and 2) two discriminators that discriminate between the real and generated appearance and real and generated gradients, respectively.

This first stage results in the training of two cross-domain transformers. For the second stage, we apply the cross-domain transformers for generating gradient and appearance images from the normal real appearance and gradient images. Then, we compute the partial mean-squared reconstruction errors (PMSRE) between real and the generated images and train a generative adversarial network (GAN) with a generator, which generates fake PMSRE, and a discriminator, a binary classifier, that determines whether they are real or not.

During inference, when an anomaly occurs as illustrated in the Fig. 1, one or both of the transformers (G_A and G_S) will not be able to correctly predict the past spatial gradients or/and the current appearance, thus, indicating the likelihood of that region of being abnormal. We tested our proposed approach on four public datasets with local anomalies: UMN [15], UCSD [14], Avenue [11] and ShanghaiTech [12]. The results show that our method is better or competitive with the state-of-the-art on all datasets.

Our contributions are the following: 1) We propose a novel two-stage object-centric adversarial approaches for local anomaly detection in videos, 2) we employ an unsupervised cross-domain GAN trained using pixel-level regions of the objects having normal behaviour in videos and 3) we propose an adversarially-learned binary classifier that classifies normal from abnormal PMSRE.

2   Related Work

Before the success of deep learning methods, most of the authors were relying on manually predefined feature extraction. For example, in [16], the authors extract histograms of oriented tracklets from a few consecutive frames. Features derived from optical flow are also used for detecting abnormal events through the use of a covariance matrix [25]. The success of these methods to detect abnormal events depends on the quality of the extracted hand-crafted features, and thus, the quality of the detection is heavily influenced by them. Besides, considering engineered features instead of pixel data to learn normal/abnormal classification implies loss of valuable spatial and temporal information.

Instead of analyzing image regions, another approach is to classify as normal or abnormal foreground object trajectories. For instance, in [7], the authors compute trajectories for normal events, apply sparse reconstruction analysis on them to learn the normal patterns, and detect any abnormal trajectories as outliers. For abnormal trajectory detection of road users, an unsupervised approach via a deep Auto-Encoder (DAE) was proposed in [20] for learning the normal trajectories and detecting the abnormal ones as outliers. In a following work [19], the use of a GAN in a discriminative manner was applied for classifying abnormal trajectory reconstruction errors produced by a pretrained DAE. This inspired us for applying an adversarial approach for detecting abnormal events in videos. Despite the fact that the problem becomes simpler when converting events into trajectories, this approach suffers from the fundamental issue of losing appearance information and relying on an external mechanism for obtaining trajectory data. Normal trajectories are also scene specific.

The authors of [18] proposed to use a CAE to learn the normal appearance and motion features extracted using a Canny edge detector and optical flow. Once the CAE is fully trained, it is then used on every frame for constructing the regularization of reconstruction errors (RRE), which will later be used for detecting anomalies. Nevertheless, this approach trained solely on the normal samples might start to generalize over the abnormal ones, thus affecting the classification performance. Recently, an object-centric approach using CAE models was proposed in [4] in which the generated latent appearance and motion features (motion features are actually computed from past and future gradient images) of objects are used for classifying local anomalies. However, the CAE models, one for appearance and two for motion, are trained separately and several SVM classifiers are required for doing the anomaly classification. We took inspiration from this work by applying the object-centric input images to the framework of a cross-domain GAN proposed in [6], which allows us to better learn the gradient and appearance correspondence by jointly training two image generators in an end-to-end manner. We also improve classification using adversarial learning of reconstruction errors following [19]. Moreover, we only relies on the previous frame for gradient, thus making it applicable in real-time scenarios.

To solve the aforementioned issues with CAEs, some authors proposed to use GANs for training a discriminator in an unsupervised fashion by using the generator to generate abnormal data during the learning process [22]. The trained discriminator can then later be used as a binary classifier for detecting anomalies. However, these methods cannot handle the spatio-temporal aspect in the video, and thus, perform poorly when the appearance change over time. To tackle this problem, authors in [10,17] proposed to use a GAN that can learn to produce the future frame of the scene with normal events, and, when an abnormal event occurs in a scene, the generator will not produce the correct subsequent frame, thus allowing the detection of abnormal events. The downside of these methods is that they rely on the optical flow methods and cannot be generalized across different scenes.

3   Proposed Method

In order to alleviate the shortcomings of existing methods, our proposed approach incorporates an object-centric mechanism into an adversarially-learned prediction-based method, to learn to distinguish local anomalies based on the appearance/gradient of regions of object of interest, thus ignoring background information.

Our method first detects all the objects in a video frame by using a multiclass object detector, and extracts the spatial gradient image of the corresponding regions of interest. Then, we train a GAN, called GARDiN (Gradient-Appearance Relation Discovery Network) inspired from DiscoGAN [6] to discover the object-centric cross-domain relations between past spatial gradients (that capture shapes and patterns) and the current visual appearance using the extracted data. This allows GARDiN to learn how the shape/appearance of a region evolves over time. This jointly trains two cross-domain object-centric generators that each transforms an image from one domain (appearance/gradient) into another domain (gradient/appearance), and two discriminators that each learns to discriminate a specific real domain against the generated fake one. After that, we construct normal partial mean squared reconstruction errors (PMSRE) produced by comparing the real appearance and gradient images of the regions and the generated ones. Inspired by [19], we then use these PMSRE for training a discriminator that acts as a binary reconstruction error classifier through a typical GAN-based approach in which the generator learns to generate realistic PMSRE, while the discriminator discriminates them from the real ones.

Once fully trained, we apply GARDiN to obtain PMSRE and directly use our adversarially-learned PMSRE classifier to discover whether the object-centric region is normal or not. We named our anomaly detection system GARDiN video anomaly detector (GARDiN-VAD).

3.1 Object Detection and Gradient Extraction

In order to detect multiple objects in a video frame, we use the pretrained multiclass object detector, CenterNet [28], which is currently one of the best and readily available in machine learning frameworks. This detector is both reasonably accurate and fast enough for an anomaly detection system. Also, since it does not rely on implicit anchors, it can detect well small objects, which is crucial for detecting anomalies in a crowded scene. Note that, to obtain the appearance images, we transform all the detected objects into grayscale and resize them into $64 \times 64 \times 1$.

In addition to detecting the spatial locations of objects frame by frame in a given video, we also compute an object-centric past spatial gradient image for each object as in [4], which is defined as the 2D spatial gradient produced by the Sobel operator on the region in the previous frame using the bounding box coordinates of the object in the current frame. This past spatial gradient image enables our adversarial framework to implicitly learn the change of object shape and position as the object moves. Moreover, compared to an optical flow image, it is significantly less expensive to compute and generalizes well, as it ignores the specific motion direction, thus facilitating the unsupervised learning of normal motion patterns. Note that, because the change caused by motion is small in two consecutive frames in a 25 fps rate video, we use a temporal spacing of T frames when computing the past spatial gradient image ($T = 3$ in our experiments).

3.2 Gradient-Appearance Relation Discovery Network (GARDiN)

Inspired by cross-domain GAN [6] that discover the relationship between images across different domains, we propose to apply this idea for learning the correspondence between the appearance and the gradient of an object moving in a video. Thus, we define appearance and gradient as two distinct domains in which the goal is to discover the relationship between images belonging to each of these domains.

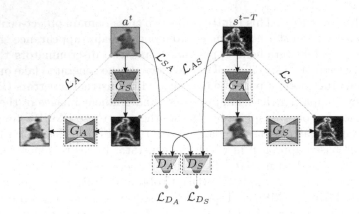

Fig. 2. Adversarial framework of GARDiN. During the training process, for a frame t, G_S and G_A learn to generate images s^{t-T} and a^t respectively from spatial gradient domain S and from appearance domain A by using the reconstruction losses $\mathcal{L}_A$, $\mathcal{L}_S$, $\mathcal{L}_{AS}$ and $\mathcal{L}_{SA}$, while the appearance discriminator D_A and spatial gradient discriminator D_S classify the real against the generated ones with $\mathcal{L}_{D_A}$ and $\mathcal{L}_{D_S}$.

Formulation. As illustrated in Fig. 2, our GAN is composed of two generators that either transform an appearance image into a gradient image or a gradient image into an appearance image, and two discriminators that each discriminates the real appearance/gradient image against the transformed one. More specifically, considering the two domains, appearance A and spatial gradient S, the generator G_S maps the images from A to S and generator G_A from S to A. For instance, in a video frame t, given an input appearance image a^t from domain A of an object, $G_S(a^t)$ should produce an image resembling the real corresponding spatial gradient image s^{t-T} from domain S. In addition to that, $G_A(G_S(a^t))$ should reproduce the original input image a^t. The same applies for an input spatial gradient image s^{t-T} from domain S. Therefore, for a given video frame t, we can formulate four reconstruction losses:

$$
\begin{aligned}
\mathcal{L}_{AS} &= d\left(G_S\left(a^t\right), s^{t-T}\right) \\
\mathcal{L}_{SA} &= d\left(G_A\left(s^{t-T}\right), a^t\right) \\
\mathcal{L}_A &= d\left(G_A\left(G_S\left(a^t\right)\right), a^t\right) \\
\mathcal{L}_S &= d\left(G_S\left(G_A\left(s^{t-T}\right)\right), s^{t-T}\right),
\end{aligned}
\tag{1}
$$

where $d()$ is a custom distance function, $\mathcal{L}_{AS}$ and $\mathcal{L}_{SA}$ deal with the transformation of an image from $A \rightarrow S$ and from $S \rightarrow A$ respectively, $\mathcal{L}_A$ and $\mathcal{L}_S$ apply to the reconstruction of the given input image using the two generators in sequence. Empirically, we found that combining different distance measures such as L1, L2 and SSIM [26] yields the best performance, as demonstrated in our ablation study. Thus, the distance $d(I_1, I_2)$ between two images I_1 and I_2 is given by the following equation:

$$d(I_1, I_2) = d_{L1}(I_1, I_2) + d_{L2}(I_1, I_2) + d_{ss}(I_1, I_2) \tag{2}$$

where

$$d_{L1}(I_1, I_2) = \frac{1}{n} \sum_{x,y} |I_1(x, y) - I_2(x, y)|$$

$$d_{L2}(I_1, I_2) = \frac{1}{n} \sqrt{\sum_{x,y} (I_1(x, y) - I_2(x, y))^2} \tag{3}$$

$$d_{ss}(I_1, I_2) = \frac{1}{2} (1 - \text{SSIM}(I_1, I_2)),$$

where n is the number of pixels in the images.

The total loss, the Gradient-Appearance Consistency loss $\mathcal{L}_{\text{GAC}}$, is given by

$$\mathcal{L}_{\text{GAC}} = \mathcal{L}_{AS} + \mathcal{L}_{SA} + \mathcal{L}_A + \mathcal{L}_S. \tag{4}$$

This loss ensures the transformation consistency of images between the two domains A and S by the generators.

Now, for making our framework adversarial, we consider two discriminators D_A and D_S that distinguish between the transformed images and the real images from domains A and S respectively. We use the following adversarial losses:

$$\begin{aligned}
\mathcal{L}_{D_S} &= \mathbb{E}_{s^{t-T} \sim p_s} \left[\log D_S\left(s^{t-T}\right)\right] + \mathbb{E}_{a^t \sim p_a} \left[\log\left(1 - D_S\left(G_S\left(a^t\right)\right)\right)\right] \\
\mathcal{L}_{D_A} &= \mathbb{E}_{a^t \sim p_a} \left[\log D_A\left(a^t\right)\right] + \mathbb{E}_{s^{t-T} \sim p_s} \left[\log\left(1 - D_A\left(G_A\left(s^{t-T}\right)\right)\right)\right],
\end{aligned} \tag{5}$$

in which p_a and p_s describe the distributions of the input images a^t and s^{t-T} respectively. In theory, G_S and G_A try to generate realistic spatial gradient and appearance images by minimizing $\mathcal{L}_{D_S}$ and $\mathcal{L}_{D_A}$ accordingly. Conversely, D_S and D_A try to discriminate the real images against the generated ones by maximizing $\mathcal{L}_{D_S}$ and $\mathcal{L}_{D_A}$.

Consequently, by incorporating the reconstruction and adversarial losses, we obtain the full objective of GARDiN as follows:

$$\mathcal{L}_{\text{GARDiN}} = \mathcal{L}_{\text{GAC}} + \mathcal{L}_{D_S} + \mathcal{L}_{D_A}. \tag{6}$$

This enables the model to predict the past spatial gradient of an object by looking at the current appearance and predict the current appearance by looking at the past gradient. Therefore, we hypothesize that, when an anomaly occurs, one or both of the generators will make an incorrect prediction, thus allowing its detection.

Architecture. The network architecture of our GAN is based on DiscoGAN [6]. Both generators (G_S and G_A) share the same U-net like architecture using skip-connections. Considering the input image size ($64 \times 64 \times 1$) for each generator of GARDiN, the encoder part is composed of 6 2-strided convolutional layers followed by the decoder made out of 6 2-strided transpose convolutional layers, outputting an image of size (64 × 64 × 1) using Sigmoid activation. The numbers of filters in the encoder are $\{32, 64, 128, 256, 256, 256\}$ and in the decoder $\{256, 256, 128, 64, 32, 1\}$. Similarly, both discriminators (D_S and D_A) follow the PatchGAN architecture [5] with 4 2-strided convolutional layers using $\{32, 64, 128, 256\}$ filters and a 1-strided convolutional output layer, which produces an output of size (4 × 4 × 1) allowing to classify by overlapping patches. During the elaboration of our method, we noticed that using 4 × 4 filters for all convolutional layers gives the best results. To tackle the issue of vanishing gradients, we incorporate leaky ReLU after each convolutional layer, except the last one, followed by instance normalization [24] between each convolutional layer.

3.3   Partial Mean Squared Reconstruction Errors (PMSRE)

Once we have trained our generators G_S and G_A for predicting object-centric spatial gradient and appearance images, the next step is to compute the reconstruction errors that will be used for classification in order to predict anomalies. As previously illustrated in Fig. 1, we noticed that, most of the time, an anomaly occurs locally in one or more spatial locations in the reconstructed gradient and/or appearance images. Thus, we found it more appropriate to perform the classification on partial reconstruction errors instead of on a global one. Empirically, in both the gradient and appearance domains, we observed that dividing the pixel-level reconstruction errors into 4 blocks rendered the best performance when training our adversarial binary classifier. With an input appearance image a of an object and the predicted appearance image a^*, we get the following partial mean squared error for an image block B_k:

$$e_k(a, a^*) = \frac{1}{h \cdot w} \sum_{i=1}^{h} \sum_{j=1}^{w} \left(a_{ij} - a_{ij}^*\right)^2 \tag{7}$$

where $h = w = 32$ is the size of a block. By following the same logic for the input spatial gradient image s and the predicted one s^*, we obtain the one dimensional partial reconstruction errors vector e:

$$e = [e_1(a, a^*), e_1(s, s^*), e_2(a, a^*), e_2(s, s^*), e_3(a, a^*), e_3(s, s^*), e_4(a, a^*), e_4(s, s^*)]. \tag{8}$$

3.4   Adversarial Classification of the PMSRE

Inspired by ALREC [19], we incorporate the idea of adversarially training a binary discriminator that learns to discriminate real e against fake ones generated by a generator which, in turn, learns to generate realistic e. In the inference mode, only the discriminator is used for predicting whether e of an object is normal or not.

Formulation. The main idea behind the use of a GAN for detecting abnormal e is to learn the distribution of normal e. The generator G, using an input Gaussian noise z, should be able to generate realistic normal e at the end of the training. In the learning process, e will become more and more realistic, and we assume that this should allow the discriminator D to learn the boundary between normal and abnormal e. Therefore, the discriminator D should learn to classify the close to be real generated samples as fake while ensuring the detection of the real ones as real. We use the Focal Loss, FL, for imposing more weight on harder samples than on the easier ones [8]:

$$\text{FL}(p) = -\alpha \, (1 - p)^\gamma \log(p), \tag{9}$$

where p is the prediction probability depending on the ground-truth label, α affects the offset for class imbalance and γ helps to adjust the level of focus on hard samples. Using this, the full objective function for training our adversarial binary classifier using normal e examples is the following:

$$\mathcal{L}_C = \mathbb{E}_{e \sim p_e} \left[\text{FL}\left(D\left(e\right)\right)\right] + \mathbb{E}_{z \sim p_z} \left[\text{FL}\left(1 - D\left(G\left(z\right)\right)\right)\right]. \tag{10}$$

We use label 0 for identifying fake/generated e and 1 for the real/normal e. In inference mode, we directly apply the trained D to produce region-level abnormality score $s_e = 1 - D(e)$, which is a prediction probability that varies between 0 (normal) and 1 (abnormal). Following the same experimental protocol as [10], we normalize s_e scores between 0 and 1 for each sequence independently.

Architecture. We are using an architecture based on a fully-connected neural network, which enables the learning of complex pattern of e. The generator G, taking an input noise of size 16, is composed of 5 dense hidden layers with $\{64, 128, 128, 256, 256\}$ units respectively. The discriminator D is made out of 5 dense hidden layers with respective $\{256, 256, 128, 128, 64\}$ units. In addition, to avoid overfitting, leaky ReLU and Dropout are used between each hidden layer, and the output layer follows a Sigmoid activation.

3.5 Abnormal Events Detection

The last step of our proposed method is to convert the region-level anomaly detection to the frame-level, to find the subsequences of a video, if any, that contain an anomaly. To obtain the frame-level anomaly score s_f, we simply take the region-level anomaly score s_e which produces the maximum value. As in [4], we also use a Gaussian filtering technique with a standard deviation of 10 for smoothing the frame-level scores temporally throughout the sequence. A threshold can then be used to determine whether the frames are normal or not.

4   Experiments

4.1   Datasets and Evaluation Procedure

We conducted experiments on four publicly available datasets with varying definition and complexity of anomalies: UMN [15], UCSD Pedestrian [14], CUHK

Avenue [11] and ShanghaiTech [12]. For all datasets, the training videos are assumed to be normal. UMN features 11 videos with 3 different scenes where anormal events are people running. We used the normal portion of 6 videos, 2 videos per scene, as training set and all the videos for the testing set as done by previous works. UCSD Pedestrin comprises two datasets: Ped1 and Ped2. Ped1 is composed of 34 training and 36 testing videos, with 40 abnormal events. Ped2 is made of 16 training and 12 testing videos, with 12 anomalies. Anomalies are the presence of skateboarders, cyclist, wheelchairs and vehicles in the pedestrian walkway areas. CUHK Avenue is composed of 16 training and 21 testing videos, where the test set contains 47 anomalies involving person running, loitering and leaving/throwing objects. Finally, ShanghaiTech is a highly challenging anomaly dataset with 13 different scenes involving diverse viewpoints and illuminations, resulting in a total of 330 training and 107 testing videos. Globally, there are 130 abnormal events in the test set with numerous types of anomalies, like people fighting, a person jumping, robbing, cyclists, etc.

To evaluate our method, we adopted the frame-level Area Under Curve (AUC) metric. To do so, we apply the Receiver Operation Characteristic (ROC) on the frame-level anomaly ground-truth labels with respect to our predicted frame-level anomaly scores s_f by progressively modifying the classification threshold. As in [10], to compute the global dataset AUC score, we first got the frame-level scores per video by applying our trained method, then we combined all the scores temporally and, lastly, we computed the AUC on the concatenated scores.

4.2   Experimental Setup

Our method is implemented using Python 3 and Keras. For detecting multiple objects in the frames using CenterNet [28], we used the model provided for the Hourglass-104 backbone with the pretrained weights from the MS-COCO dataset [9], providing 81 different object classes. To reduce missing detections, we allow all classes with a confidence of at least 0.3.

We train GARDiN for 200 epochs of randomly shuffled mini-batches of size 64 using a learning rate starting from 10^{-2} and following a polynomial decay of power 2 every 25 epochs. To train the anomaly classifier, we follow the same training mechanism, but with a starting learning rate of 10^{-4} decaying every 10 epochs for a maximum of 50 epochs of randomly shuffled mini-batches of 256 samples. For the classifier's Focal Loss, we have empirically chosen $\alpha = 0.1$ and $\gamma = 10$. To train both frameworks, we use Adam optimizer with $\beta_1 = 0.5$ and $\beta_2 = 0.999$. To stabilize the adversarial training process for both GANs, we slightly smoothed the labels when training the discriminators with real samples.

4.3   Results

Table 1 presents our frame-level AUC anomaly detection results on the four datasets. We also included in the table some recent state-of-the-art methods

Table 1. Frame-level abnormal event detection AUC results (in %) on four dataset. * Means that results were recalculated to follow the procedure in [10].

Method	UMN	Ped1	Ped2	Avenue	ST
Conv-AE [3]	–	81.0	90.0	70.2	60.9
Discriminative [1]	91.0	–	–	78.3	–
ConvLSTM-AE [13]	–	75.5	88.1	77.0	–
Deep-Cascade [21]	99.6	–	–	–	–
STAE-optflow [27]	–	**87.1**	88.6	80.9	–
Deep Conv-AEs [18]	–	56.9	84.7	77.2	–
Future frame pred [10]	–	83.1	95.4	84.9	72.8
OC Conv-AEs* [2,4]	99.6	–	–	86.6	78.6
M-A Correspond [17]	–	–	96.2	86.9	–
GARDiN-VAD (ours)	**99.7**	85.2	**97.5**	**87.3**	**81.1**

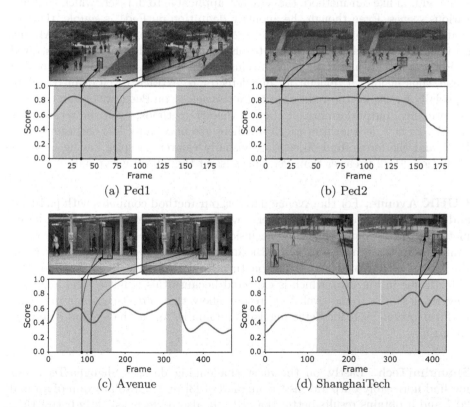

(a) Ped1   (b) Ped2

(c) Avenue   (d) ShanghaiTech

Fig. 3. Qualitative anomaly detection results on test videos from each dataset. Red curves show the frame-level anomaly score and the areas in cyan represent the ground-truth abnormal frames. Black and red arrows point to the ground-truth bounding boxes and the detected regions by GARDiN-VAD respectively. (Color figure online)

evaluated on at least one of the considered datasets. Figure 3 illustrates anomaly detections on Ped1, Ped2, Avenue and ShanghaiTech using our proposed method.

UMN. For the UMN dataset, our approach significantly outperforms [1] and is on par with [4,21] by achieving a near perfect result. More specifically, our proposed GARDiN-VAD can accurately detect the people escape instances on all three different scenarios, while discarding the background information. This illustrates the applicability of our object-centric adversarial approach for detecting real-world crowd panic events.

UCSD Pedestrian. On Ped1, we note a notable improvement in the AUC score using GARDiN-VAD compared to some anomaly detection counterparts [13,18] and almost on par results with [3,10,14]. However, we notice that the spatio-temporal-based method of [27] largely surpasses GARDiN-VAD. In fact, the spatio-temporal auto-encoder in [27] was only evaluated on videos having a single scene and, unlike our method, they are not applicable to datasets which contains various scenes. Even though the anomaly definition on Ped1 is simple, the fact that the image resolution is only 158×238 makes the input data for GARDiN significantly more noisy compared to other datasets. Moreover, as illustrated in Fig. 3(a), there might be heavy occlusions in some portions of the crowded scene, thus making it difficult for our detector to extract a well defined region, especially for the far top-right objects. Nevertheless, on Ped2, we obtain the best performance, outperforming all state-of-the-art methods, even the recent ones [10,17]. In fact, as shown in Fig. 3(b), there are noticeably less occlusions than Ped1 and the foreground objects are visually clearer. Despite having a better resolution, Ped2 is easier than Ped1 because of the lateral viewpoint.

CUHK Avenue. For the Avenue dataset, our method competes with [4,10,17] and significantly outperforms others, which shows that object-centric-based methods can be more robust to occlusions and camera jittering. However, it cannot detect anomalies involving the interaction between multiple objects in the scene, as illustrated by a person throwing a bag which goes outside the video frame in Fig. 3(c), which is expected because this is not a local anomaly. Nevertheless, the frame-level AUC results show that object-centric approaches perform overall well for detecting local anomalies on a challenging side-view scenario.

ShanghaiTech. Lastly, on the most challenging dataset ShanghaiTech, our method noticeably outperforms the method of [3] by an absolute gain of around 20% and it obtains results better than [10]. It also performs slightly better than the other object-centric method [2,4] while relying only on the past and present observations. Although sometimes, depending on the camera angle, cyclists can have lower anomaly scores than pedestrians as shown in Fig. 3(d), methods relying only on the regions of the objects will be able to detect local anomalies across different scenes, mainly due to the fact that they exclude background information and are less context-dependent.

Table 2. Ablation study AUC results (in %) on Ped2.

(a) GARDiN losses using One-Class SVM with L2 distance metric.

$\mathcal{L}_{AM}$	✓	✓	✓	✓
$\mathcal{L}_{MA}$	—	✓	✓	✓
$\mathcal{L}_{A}$	—	—	✓	✓
$\mathcal{L}_{M}$	—	—	—	✓
AUC	79.3	81.2	82.7	**85.2**

(b) Distance metrics in GARDiN losses using One-Class SVM.

d_{L_1}	✓	✓	—	—	✓	✓	✓
d_{L_2}	—	✓	—	—	✓	✓	✓
d_{ss}	—	—	✓	—	✓	—	✓
d_{nr}	—	—	—	✓	—	✓	✓
AUC	85.2	87.1	84.5	83.6	**91.3**	88.7	89.8

(c) Classification methods of reconstruction errors.

Method	L1	L2	SSIM	PMSRE
OC-SVM [23]	88.9	91.3	92.1	93.4
One-vs-rest-SVMs [4]	89.4	92.1	92.8	94.3
DAE [20]	—	—	—	95.2
ALREC [19]	—	—	—	95.9
ALREC-FL (ours)	—	—	—	**97.5**

4.4 Ablation Study

We chose Ped2 for conducting our ablation study, presented in Table 2, since we can train our models faster on it and the anomaly definition generalizes well across other datasets. First, as summarized in Table 2(a), to validate the loss function $\mathcal{L}_{MAC}$ in Eq. (4) which ensures object-centric gradient and appearance consistency when training GARDiN, we used the L2 distance measure between real and generated images in the loss functions $\mathcal{L}_{AM}$, $\mathcal{L}_{MA}$, $\mathcal{L}_A$ and $\mathcal{L}_M$, and used a simple outlier detection technique based on a One-Class SVM for detecting anomalies. By testing various combinations, we observe that the fusion of all the losses produce the best result. Secondly, to find the most appropriate distance metric that will be used in $\mathcal{L}_{MAC}$, we also tested several combinations of distances based on L1 (d_{L_1}), L2 (d_{L_2}), SSIM (d_{ss}) and PSNR inspired by [10] (d_{nr}). Table 2(b) shows the importance of combining image quality distance metrics for better assessment of the correspondence between gradient and appearance of objects. Lastly, we tested several unsupervised classification approaches

Table 3. GARDiN AUC results (in %) on UCSD (Ped1 and Ped2), Avenue and Shang-haiTech using one of two different detectors.

Detector	Ped1	Ped2	Avenue	ShanghaiTech
RetinaNet [8]	83.6	97.4	83.1	80.3
CenterNet [28]	**85.2**	**97.5**	**87.3**	**81.1**

[4,19,20,23] for distinguishing normal and abnormal reconstruction errors of generated images from GARDiN. The AUC results in Table 2(c) confirm that the adversarial method outperforms others by a large margin. We observe that incorporating Focal Loss [8] when training ALREC [19] (ALREC-FL) notably improves the performance for abnormal PMSRE detection.

We also evaluated the impact of the object detector on the overall AUC results. We compared CenterNet [28] and RetinaNet [8] for this task. Table 3 shows that using a higher performance detector (CenterNet) noticeably improves the anomaly detection performance, mostly on Ped1 and Avenue datasets which contain the highest amount of occlusions and noise among the studied datasets.

4.5   Inference Running Time

On a Intel i5-9400F machine with 16 GB RAM using Nvidia RTX 2070 GPU with 8 GB VRAM and considering an average number of objects in a video frame of 5, the preprocessing step for the detection of objects and the extraction of their gradient images takes about 75 ms per frame. In inference mode, the running time of the combined GARDiN and ALREC-FL frameworks is approximately 5 ms per frame. Thus, the overall pipeline of our proposed method consumes roughly 80 ms for a single frame, leading to a running speed of 12.5 FPS.

5   Conclusion

In this paper, we propose GARDiN-VAD: a novel unsupervised approach for local anomaly detection in videos based on object-centric adversarial learning trained using normal training samples only. First, we extract the appearance and the gradient of all the objects in the scenes by using the pretrained CenterNet object detector. Then, we train GARDiN, composed of two generators and two discriminators to learn the relationship between appearance and gradient. After that, we train ALREC-FL with PMSRE to classify abnormal PMSRE caused by abnormal appearance-gradient relationships. On four public benchmarks, our method yields competitive results, superior to state-of-the-art approaches.

References

1. Del Giorno, A., Bagnell, J.A., Hebert, M.: A discriminative framework for anomaly detection in large videos. In: Leibe, B., Matas, J., Sebe, N., Welling, M. (eds.) ECCV 2016. LNCS, vol. 9909, pp. 334–349. Springer, Cham (2016). https://doi.org/10.1007/978-3-319-46454-1_21
2. Feng, J.: A implementation of Object-Centric VAD using Tensorflow (2019). https://github.com/fjchange/object_centric_VAD
3. Hasan, M., Choi, J., Neumann, J., Roy-Chowdhury, A.K., Davis, L.S.: Learning temporal regularity in video sequences. In: CVPR. IEEE (2016)
4. Ionescu, R.T., Khan, F.S., Georgescu, M.I., Shao, L.: Object-centric auto-encoders and dummy anomalies for abnormal event detection in video. In: CVPR. IEEE (2019)
5. Isola, P., Zhu, J.Y., Zhou, T., Efros, A.A.: Pix2Pix. In: CVPR. IEEE (2017)
6. Kim, T., Cha, M., Kim, H., Lee, J.K., Kim, J.: Learning to discover cross-domain relations with generative adversarial networks. In: ICML. JMLR.org, March 2017
7. Li, C., Han, Z., Ye, Q., Jiao, J.: Visual abnormal behavior detection based on trajectory sparse reconstruction analysis. Neurocomputing **119**, 94–100 (2013)
8. Lin, T.Y., Goyal, P., Girshick, R., He, K., Dollár, P.: Focal loss for dense object detection. In: ICCV. IEEE (2017)
9. Lin, T.-Y., et al.: Microsoft COCO: common objects in context. In: Fleet, D., Pajdla, T., Schiele, B., Tuytelaars, T. (eds.) ECCV 2014. LNCS, vol. 8693, pp. 740–755. Springer, Cham (2014). https://doi.org/10.1007/978-3-319-10602-1_48
10. Liu, W., Luo, W., Lian, D., Gao, S.: Future frame prediction for anomaly detection - a new baseline. In: CVPR. IEEE (2018)
11. Lu, C., Shi, J., Jia, J.: Abnormal event detection at 150 FPS in MATLAB. In: ICCV. IEEE (2013)
12. Luo, W., Liu, W., Gao, S.: A revisit of sparse coding based anomaly detection in stacked RNN framework. In: ICCV. IEEE (2017)
13. Luo, W., Liu, W., Gao, S.: Remembering history with convolutional LSTM for anomaly detection. In: ICME. IEEE (2017)
14. Mahadevan, V., Li, W., Bhalodia, V., Vasconcelos, N.: Anomaly detection in crowded scenes. In: CVPR. IEEE (2010)
15. Mehran, R., Oyama, A., Shah, M.: Abnormal crowd behavior detection using social force model. In: CVPR Workshops. IEEE (2009)
16. Mousavi, H., Mohammadi, S., Perina, A., Chellali, R., Murino, V.: Analyzing tracklets for the detection of abnormal crowd behavior. In: WACV. IEEE (2015)
17. Nguyen, T.N., Meunier, J.: Anomaly detection in video sequence with appearance-motion correspondence. In: ICCV. IEEE (2019)
18. Ribeiro, M., Lazzaretti, A.E., Lopes, H.S.: A study of deep convolutional auto-encoders for anomaly detection in videos. Pattern Recogn. Lett. **105**, 13–22 (2018)
19. Roy, P., Bilodeau, G.A.: Adversarially learned abnormal trajectory classifier. In: CRV. IEEE (2019)
20. Roy, P.R., Bilodeau, G.-A.: Road user abnormal trajectory detection using a deep autoencoder. In: Bebis, G., et al. (eds.) ISVC 2018. LNCS, vol. 11241, pp. 748–757. Springer, Cham (2018). https://doi.org/10.1007/978-3-030-03801-4_65
21. Sabokrou, M., Fayyaz, M., Fathy, M., Klette, R.: Deep-cascade: cascading 3D deep neural networks for fast anomaly detection and localization in crowded scenes. IEEE Trans. Image Process. **26**, 1992–2004 (2017)

22. Sabokrou, M., et al.: AVID: adversarial visual irregularity detection. In: Jawahar, C.V., Li, H., Mori, G., Schindler, K. (eds.) ACCV 2018. LNCS, vol. 11366, pp. 488–505. Springer, Cham (2019). https://doi.org/10.1007/978-3-030-20876-9_31
23. Schölkopf, B., Platt, J.C., Shawe-Taylor, J., Smola, A.J., Williamson, R.C.: Estimating the support of a high-dimensional distribution. Neural Comput. **13**, 1443–1471 (2001)
24. Ulyanov, D., Vedaldi, A., Lempitsky, V.: Instance normalization: the missing ingredient for fast stylization. CoRR (2016)
25. Wang, Z., Hou, C., Li, B., Chen, T., Yao, L., Song, M.: Global abnormal event detection in video via motion information entropy. In: AT-RASC (2018)
26. Wang, Z., Bovik, A.C., Sheikh, H.R., Simoncelli, E.P.: Image quality assessment: from error visibility to structural similarity. IEEE Trans. Image Process. **13**, 600–612 (2004)
27. Zhao, Y., Deng, B., Shen, C., Liu, Y., Lu, H., Hua, X.S.: Spatio-temporal AutoEncoder for video anomaly detection. In: ACM on Multimedia Conference. ACM Press (2017)
28. Zhou, X., Wang, D., Krähenbühl, P.: Objects as points. CoRR (2019)

A Hierarchical Framework for Motion Trajectory Forecasting Based on Modality Sampling

Yifan Ma[1], Bo Zhang[1(✉)] (iD), Nicola Conci[2], and Hongbo Liu[1]

[1] College of Information Science and Technology, Dalian Maritime University,
Dalian, China
rupertsdrop.myf@gmail.com, {bzhang,lhb}@dlmu.edu.cn
[2] University of Trento, Trento, Italy
nicola.conci@unitn.it

Abstract. In this paper, we present a hierarchical framework for multi-modal trajectory forecasting, which can provide for each pedestrian in the scene the distributions for the next moves at every time step. The overall architecture adopts a standard encoder-decoder paradigm, where the encoder is based on a self-attention mechanism to extract the temporal features of motion histories, while the decoder is built upon a stack of LSTMs to generate the future path sequentially. The model is learned in a discriminative manner, with the purpose of differentiating among varied motion modalities. To this end, we propose a clustering strategy to construct the so-called transformation set. The transformation set collaborates with the hierarchical LSTMs in the decoder, in order to approximate the real distributions in the training data. Experimental results demonstrate that the proposed framework can not only predict the future trajectory accurately, but also provide multi-modal trajectory distributions explicitly.

Keywords: Trajectory prediction · Multi-modality modeling · Hierarchical LSTM · Discriminative learning

1 Introduction

Trajectory prediction can be applied to a wide range of scenarios, covering visual surveillance and intelligent monitoring in public places, with application to early warning, abnormal event detection, and collision avoidance. However, predicting the future locations of a moving pedestrian is a challenging problem, which implies understanding pedestrian dynamics in order to infer long-term time series of the pedestrians' paths. In fact, it is likely that pedestrians' behaviors are likely to exhibit multi-modal characteristics. As an example, when a pedestrian approaches a traffic junction, he has multiple options, such as going straight, turning left/right, or coming back. This situation can be described by a

A. Del Bimbo et al. (Eds.): ICPR 2020 Workshops, LNCS 12664, pp. 235–249, 2021.
https://doi.org/10.1007/978-3-030-68799-1_17

multi-modal distribution that depends on personal preferences, intentions, and environmental factors.

With the rapid development of deep networks, generative models such as Variational Auto Encoders (VAEs) and Generative Adversarial Networks (GANs) have shown outstanding capabilities in multi-modal trajectory forecasting. Although these approaches are able to generate multiple trajectories, they can not approximate the data distribution explicitly. To address this problem, we propose a novel architecture that exploits hierarchical LSTMs, which can model the probability of the next move, discriminating the diversity of motion modes.

The overall framework consists of a common encoder-decoder paradigm. The encoder adopts the self-attention mechanism to extract the features of motion history along the temporal dimension. The decoder is implemented using multiple LSTMs organized hierarchically. The bottom LSTM serves for computing the explicit distribution of the next position at each time instant, while the central LSTM is used to fine-tune the predicted trajectory point. At the top layer, a standard LSTM is adopted to store the motion histories, initializing the bottom LSTM in the next time step, forming a recurrent structure.

Specifically, providing the exact distribution for trajectory generation is the distinguishing novelty in our work. To achieve this, we first propose a clustering strategy to construct the so-called transformation set, which is a more representative format of the motion modalities. The trajectory points in the transformation set have the chances to be taken into account for trajectory generation. The model is trained in a discriminative manner, which can better differentiate varied motion modes. At every time instant, we expect that the model can promote the sampling probabilities at the points, which can better represent motion modalities in the transformation set, while suppressing the probabilities at other positions.

To sum up, the main contributions of this work are:

- we propose a novel hierarchical LSTM-based framework for trajectory forecasting, which can not only obtain accurate prediction performances, but also present the multi-modal characteristics of the future plausible paths;
- we present a clustering strategy which is used to construct the representative form of the motion modalities in the training set. The obtained transformation set will be utilized in the discriminative training procedure, for the purpose of better perceiving the differences among varied trajectory modalities;
- our framework is able to explicitly describe the multi-modal distributions of the future locations at every time step, which can directly guide the sequence generation procedure.

The rest of the paper is organized as follows: in Sect. 2, we review the recent literature in trajectory forecasting. In Sect. 3, we first present how to construct the transformation set, and then illustrate the detailed architectures of the encoder and decoder. Experimental results are recorded in Sect. 4, to evaluate the performances in terms of multi-modal trajectory generation and prediction accuracy. Conclusions are drawn in Sect. 5.

2 Related Work

A detailed overview on the recent literature in the domain of trajectory forecasting can be found in a recent survey paper [5]. In the following, we will present the most recent advancement in the area, with particular reference to trajectory clustering, sequence modeling, and multi-modality modeling, which is the focus of the current work.

Trajectory Clustering: clustering is an efficient way to arrange large quantity of trajectory data, which can provide useful hints about the description of the different motion modes observed in a dataset. Trajectory clustering has been investigated for many years. Regarding the unsupervised approaches, Zhong et al. [15] applied the standard k-means algorithm on real trajectories, for the purpose of data-driven crowd modeling. Zhou et al. [16] proposed a coherent filtering approach to detect coherent motion patterns in crowded environments. As for supervised approaches, Lawal et al. [7] exploited the support vectors to cluster pedestrian groups.

Differently from traditional clustering methods, our target is to group trajectory points at every time step, where the obtained cluster centers are further used for motion modality modeling. Thus, we do not need to consider data associations along the temporal dimension.

Sequence Modeling: early works usually adopt standard Markov-based models to learn the spatial-temporal evolution of trajectories. More recently, with the emergence of recurrent neural networks (such as RNN, LSTM, and GRU), dealing with long-term time series has become feasible. In [1], Alahi et al. proposed the Social-LSTM model, which takes into consideration also people in the neighborhood when predicting the target's future path. In [14], Vemula et al. proposed a structural RNN for trajectory prediction, which can capture the relative importance of each person in the neighborhood. In [3], Bartoli et al. proposed the context-aware recurrent neural network, which can jointly model human-human and human-space interactions for trajectory forecasting.

Compared to the existing literature, we introduce a novel hierarchical framework to model the prediction portion of trajectories, which is clearly different from the standard single-layer structure (such as the Markov chains, RNN, LSTM, and GRU). This helps to introduce discriminative learning in the training process, which enables the decoder to better distinguish varied motion modalities.

Multi-modality Modeling: Most approaches presented above provide a single prediction of the target's next move. In order to model the multi-modal characteristics of human behaviors, a variety of probabilistic models have been proposed to handle the behavior uncertainties. In particular, generative models are widely adopted. In [8], Lee et al. proposed the so-called DESIRE framework for multi-modality modeling, where the core sample generation module is implemented using a conditional VAE (CVAE), which can output multiple plausible trajectories. The obtained trajectories are further ranked and refined using an inverse optimal control scheme. In [11], Rhinehart et al. proposed a re-parameterized push-forward policy to guide vehicles in an autonomous driving scenario, where the LIDAR map was exploited. This model can find a proper balance between what is defined 'diverse' and 'precise', achieving promising performances on the KITTI benchmark dataset. Another popular branch is the adoption of GAN-based models. In [4], Gupta et al. adopted a GAN for multi-modal trajectory generation. In [12], the so-called Sophie model was proposed, which uses a similar GAN-based mechanism, and improves the state of the art by including an attention model.

Differently from the generative models mentioned above, which are not able to provide the exact trajectory distributions, our work aims at modeling trajectory distributions along the temporal dimension explicitly, thus providing an alternative way for multi-modal trajectory forecasting.

3   Methodology

In this section, we will present the details of the proposed framework. First, we introduce the construction of the so-called transformation set, which resembles the motion modalities. Then, we present the encoder structure, used to exact the temporal features for the observation sequences. Finally, we illustrate the architecture of the decoder, and its corresponding organization into stacked LSTMs.

3.1   Transformation Set Construction

At any given time instant, we use $traj$ to represent the trajectory points in the training set, which contains N elements. Specifically, only the prediction portion are participated in the clustering procedure. Let $target$ be the selected trajectory points, and $cluster$ be the clustering results, where $cluster[i]$ represents the i-th cluster, and $cluster[i][j]$ represents the j-th element in the i-th cluster. The centers of all the clusters will be considered as the representative points, stored in rep. The details are presented in Algorithm 1. The function $push(A,B)$ represents the addition of a trajectory point B into a set A. Line 8 first computes the Euclidean distances between $traj[num]$ and all the elements in the $target$ set, and then returns the corresponding minimum distance and the element index.

It should be noticed that the threshold θ plays an essential role in the clustering performances. The higher θ implies more elements in a cluster, which will lead to modality collapse, while the lower θ indicates more clusters, which will result viceversa in modality redundancy. Algorithm 1 will be executed sequentially when rolling out trajectories along the temporal dimension. The obtained results are gathered and annotated as the transformation set, which contains representative points better describing the motion modalities. In Fig. 1, we present an example of the transformation set, where the toy dataset [2] is used for demonstration. There are six groups of synthetic trajectories, all starting from one specific point located along a circle. When approaching the circle center, they split

---

Algorithm 1. transformation set construction.

1: **Input:** $traj, target, cluster, threshold\ \theta$
2: **Output:** rep
3: **Initialization:** $target[1] \Leftarrow traj[1]$;
4: push($cluster[1]$,$traj[1]$);
5: **begin:**
6: $i \Leftarrow 1$;
7: **for** $num \Leftarrow 2$ to N
8:   $(dist, index) \Leftarrow \min\{\|traj[num], target\|\}$;
9:   **if** $(dist > \theta)$
10:     push($target, traj[num]$);
11:     push($cluster[++i], traj[num]$);
12:   **else**
13:     push($cluster[index], traj[num]$);
14:   **endif**
15: **endfor**
16:
17: **for** $i \Leftarrow 1$ to length($cluster$)
18:   $rep[i] \Leftarrow$ average($cluster[i]$);
19: **return** rep;
20: **end**

---

into three different subgroups (Fig. 1(a)). The blue and the orange dots represent elements at different time steps in the transformation set. It can be seen from the figures that there are totally 18 varied motion modalities in the *toy* dataset. Performances vary by changing the value of the threshold. Setting $\theta = $ 5e–3 (Fig. 1 (b)) achieves the best clustering performances, where points in the transformation set at any time instant perfectly match the motion modalities. For example, when increasing θ to 5e–2 (see Fig. 1(c)), some motion modalities disappear (the number of orange points decreases from 18 to 11). If we further increase $\theta = $ 8e–2 (Fig. 1(d)) we notice it can not even describe the distributions of the original data properly.

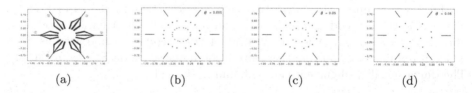

     (a)   (b)   (c)   (d)

Fig. 1. The transformation set. We show the clustering performances using different values of θ. (a) Multi-modal trajectory distributions in the *toy* dataset; (b) $\theta = $ 5e–3; (c) $\theta = $ 5e–2; (d) $\theta = $ 8e–2.

3.2 Encoder

The encoder is used to extract the features of motion history (the observation sequence) along the temporal dimension, which adopts the self-attention mechanism as presented in [13] (originally designed for natural language processing). Similarly, we consider a discrete trajectory sequence as a 'sentence', where each trajectory point can be viewed as a single 'word'. The detailed structure is shown in Fig. 2.

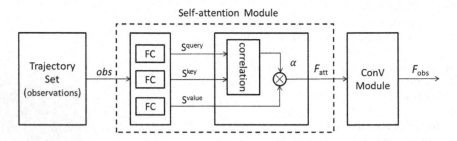

Fig. 2. The encoder architecture. The observation part in a trajectory is used as the input. The output of the encoder is annotated as F_{obs}.

For each individual, the length of the training trajectory is set to T ($T = T_{obs} + T_{pred}$, indicating the lengths of the observation and the prediction sequences, respectively). The input of the encoder is named as obs, which is a coordinate sequence with the length of T_{obs}. Each coordinate consists of a two-dimensional trajectory point (x_t, y_t). As shown in Eq. (1), the input is first processed by three separated fully-connected layers. In line with the original work presented in [13], we also use S^{query}, S^{key}, and S^{value} to represent the outputs of the FC layers.

$$
\begin{aligned}
S^{query} &= \mathrm{FC}(obs; w_q) \\
S^{value} &= \mathrm{FC}(obs; w_v) \\
S^{key} &= \mathrm{FC}(obs; w_k)
\end{aligned}
\tag{1}
$$

Then, we compute the correlation between S^{query} and S^{key} as in Eq. (2), where d is the embedding dimension. The output of the attention module is obtained using Eq. (3). The final output of the encoder is annotated as F_{obs}, which is obtained by applying the convolutional operation on F_{att} as in Eq. (4). The dimensionality alignment can be found in Table 1.

$$
\alpha = softmax(\frac{1}{\sqrt{d}}S^{query}(S^{key})^T)
\tag{2}
$$

$$
F_{att} = \alpha S^{value}
\tag{3}
$$

$$
F_{obs} = \mathrm{Conv}(F_{att})
\tag{4}
$$

Table 1. Dimensionality alignment in the encoder. c represents the number of channels.

Variable	Size
obs	$T_{obs} \times 2$
S^{query}; S^{value}; S^{key}	$T_{obs} \times c$
F_{att}	$T_{obs} \times c$
F_{obs}	$1 \times c$

3.3 Decoder

The decoder architecture is presented in Fig. 3, which mainly consists of three LSTMs organized hierarchically, together with an additional candidate module, arranged as follows:

- score LSTM: builds the multi-modal distribution of the future path;
- regression LSTM: used to fine-tune the position of the predicted trajectory point;
- memory LSTM: stores the motion history, and initialize the recurrent structure at the next time step;
- candidate module: finds suitable candidate points in the transformation set;

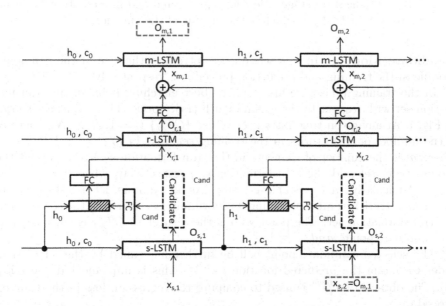

Fig. 3. The decoder architecture. The whole procedure is initiated at the s-LSTM. The output of the candidate block refers to the possible motion modality, and will be fine-tuned by the r-LSTM in order to generate the next position.

LSTM Modules. The fundamental structures of the LSTMs are presented in Eq. (5) :

$$o_{s,t} = \text{s-LSTM}(h_{s,t-1}, c_{s,t-1}, x_{s,t})$$
$$o_{r,t} = \text{r-LSTM}(h_{r,t-1}, c_{r,t-1}, x_{r,t}) \qquad (5)$$
$$o_{m,t} = \text{m-LSTM}(h_{m,t-1}, c_{m,t-1}, x_{m,t})$$

where $x_{s,t}$, $x_{r,t}$, and $x_{m,t}$ represent the inputs of the corresponding LSTM blocks, respectively, while $o_{s,t}$, $o_{r,t}$, and $o_{m,t}$ represent the corresponding outputs ($t = 1, 2, 3, \dots$). Specifically, the output of the m-LSTM at the time instant $(t-1)$ is used to initialize the input of the s-LSTM at the next time instant t, in a recurrent fashion.

We use the term *cand* to indicate the output of the candidate module, which consists of a 2-dimensional coordinate. The inputs of the LSTMs are updated using Eq. (6), (7), and (8):

$$x_{s,t} = o_{m,t-1} \qquad (6)$$

$$x_{r,t} = \text{FC}[h_{t-1} \oplus \text{FC}(cand)] \qquad (7)$$

$$x_{m,t} = \text{FC}(o_{r,t}) + cand \qquad (8)$$

where $\oplus$ indicates the concatenation operation.

For the initialization stage, the cell state c_0 and the hidden state h_0 in all the LSTMs are set to F_{obs}, which is the output of the encoder.

Candidate Module. In this section, we introduce the role of the candidate module in the training and prediction procedures, respectively.

In the training phase, we first find the best matching point in the transformation set with respect to the ground truth trajectories. The details are shown in Fig. 4. At any given time instant t, we use $\{GT_{i,t}\}$ ($i = 1, 2, \dots, N$, where N is the number of trajectories in the training set) and $\{TF_{t,j}\}$ ($j = 1, 2, \dots, m$, where m is the number of elements in the transformation set) to represent the ground truth set and the transformation set, respectively, where $t = 1, 2, \dots, T_{pred}$. Particularly, m varies at every step. As an example, let us take $GT_{1,t}$ in the ground truth set: after obtaining its coordinate, we search in the corresponding transformation set. The point who is the nearest to $GT_{1,t}$ is considered as the candidate point *cand*.

The selected candidate point will be further fine-tuned by the r-LSTM in order to obtain the predicted location loc_t^{pred}. This is implemented using Eq. (8). The obtained loc_t^{pred} is used to compute the regression loss in the training procedure.

Next, we will show how to model the sampling probability. Let us assume there are m trajectory points at the time instant t in the transformation set. We model the probability for every element in $\{TF_{t,j}\}$ as in Eq. (9), which reflects the possibility that each element can be used for trajectory generation:

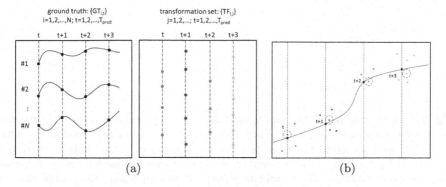

Fig. 4. Candidate matching. (a) left: ground truth; right: transformation set, where different colors represent different time instants. (b) matching procedure: at any time instant, we choose the point that is spatially nearest to the ground truth (the black dot) in the transformation set as the candidate. The dash circles indicate the selected candidates.

$$P = softmax(\frac{1}{\sqrt{d}}o_{s,t}\ f_{tf}(TF_t)); \tag{9}$$

where $o_{s,t}$ is a $1 \times L$ vector; TF_t is a $2 \times m$ matrix, storing all the coordinates, $f_{tf}()$ is implemented using the fully-connected layer, which can map TF_t into a $L \times m$ matrix. Thus, the obtained probability P is a $1 \times m$ vector corresponding to the m elements in the transformation set, which can be viewed as the sampling probability at each time step.

The objective of the training procedure is to enforce the P vector having higher values at the candidate points, while suppressing the values at other points. It implies that our model has the capability of approximating the trajectory distribution in the training set properly (see Fig. 5).

To this aim, we introduce the cross-entropy term into the loss function. The candidate point $cand$ is considered as the positive sample, while on the other hand, we randomly select another point in $\{TF_{t,j}\}$ as the negative sample. Thus, the loss at the time instant t is comprised by the regression and the cross-entropy parts. When rolling out a trajectory along the temporal dimension, the overall loss is computed according to Eq. (10):

$$loss = \frac{1}{T_{pred}} \sum_{t=1}^{T_{pred}}[(loc_t^{pred} - loc_t^{gt})^2 + C_t(P^+, P^-)] \tag{10}$$

where loc_t^{pred} and loc_t^{gt} represent the predicted location and the corresponding ground truth, respectively. The terms of P^+ and P^- represent the values in the P vector corresponding to the positive and negative samples, respectively. The function $C_t(\ ,\ )$ is used to compute the cross-entropy.

Once the training procedure is completed, the decoder can provide the probability P at every time instant. In the prediction phase, we sample according to P, and the selected point in the transformation set is then fine-tuned by the r-LSTM in order to obtain the most likely position at the next time step.

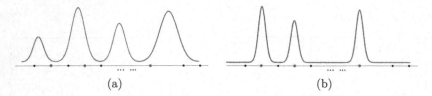

(a)   (b)

Fig. 5. A demo example of the training objective. At a given time instant t, the red dots represent the selected candidate points with respect to the motion modalities, while black dots indicate other points in the transformation set. (a) probability P during training; (b) probability P after training. We expect that the model can promote the sampling probabilities at the candidate points, while suppressing the probabilities at other positions.

4   Experimental Results

For validation, we first demonstrate that the proposed framework is able to generate multi-modal trajectories from a synthetic dataset. Next, we evaluate the trajectory prediction performances on realistic benchmark datasets, to confirm the effectiveness of our approach.

4.1   Multi-modality Evaluation on Synthetic Data

We adopt a state-of-the-art *toy* trajectory dataset [2] for evaluation, which contains multi-modal synthetic data.

There are totally six groups of synthetic trajectories, each of which is further split into three sub-directions, indicating 18 different moving modalities in total. We manually assign the label for each group, as can be seen in Fig. 1. Particularly, the red dots represent the intersection points, where trajectories start to disperse.

We randomly generate multiple sample trajectories according to a variety of prior distributions placed at each intersection point. We adopt a fix threshold $\theta = 5e{-}3$ (which has been validated in Fig. 1) when constructing the transformation set. In Fig. 6, we illustrate the evolution of P for each intersection point at different learning iterations. At the beginning, all the elements in P equal to $1/18$. With further training, the model gradually converges. At iteration 600, there still exist several unexpected modalities. While at iteration 5,000, the components in P with respect to group 2, 3, 5, and 6 have almost approximated to the real distributions in the training set. Group 1 and 4 will finally converge as well when increasing the number of iterations.

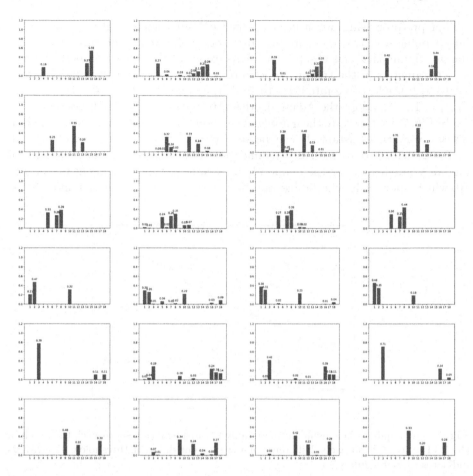

Fig. 6. Multi-modality demonstration. Row 1–6: six different groups; Column 1: the prior distributions on different directions in the training set; Column 2–4: the evolution of P in the training procedure at iteration 600, 1000, and 5,000.

4.2 Trajectory Prediction Evaluation

In this section, we carry out our experiments on the most popular benchmarks in the area of trajectory forecasting, widely used in the literature: the ETH [10] and UCY [9] datasets, for the purpose of evaluating the trajectory prediction accuracy. The ETH dataset has two subsets (namely, *eth* and *hotel*), which present multiple pedestrians moving in different backgrounds. The UCY dataset is arranged into three categories: *zara1*, *zara2*, and *univ*. These two datasets contain thousands of trajectories in total, which are captured in real scenarios. All the videos are collected from the aerial point of view, covering challenging behaviors typical of an urban scene, like people walking together, group movements, and collision avoidance.

For pre-processing, we utilize the relative displacements with respect to the initial point in a trajectory as the trajectory point representation, which can remove the impact that absolute positions exert on the prediction performance. The parameters in the training procedure are listed as follow: the batch size is set to 512; the learning rate is set to 1e–4; the epoch is set to 100; $T_{obs} = 8$ and $T_{pred} = 12$. We adopt the Adam algorithm to optimize the loss function.

In Fig. 7, we illustrate the number of elements in the obtained transformation set at every time instant. It can be seen clearly that a low value of θ (such as $\theta = 2$) will lead to modality redundancy, resulting in dramatic changes of modality numbers along the temporal dimension. In this experiment we set θ to 10, which allows achieving stable variations in the number of motion modalities.

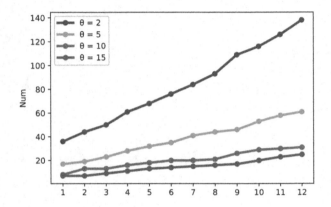

Fig. 7. The number of elements in the obtained transformation set at each time step in the ETH and UCY datasets. Different colors represent different thresholds. The X-axis indicates discrete time steps.

The average displacement error (ADE) and the final displacement error (FDE) metrics are used to evaluate the prediction accuracy. For comparison, the linear model [6], the LSTM model, the social-LSTM [1], the social-GAN [4], and the Sophie model [12] are selected as benchmarks.

For trajectory generation, we select the top-k elements in the transformation set at each time instant, which have higher values in P. These elements are then fine-tuned by the r-LSTM to forecast the next plausible positions. Among all the generated trajectories, the one who is the closest to the ground truth is used to evaluate the prediction accuracy, following the same criteria of evaluation as reported in [4].

We present the quantitative results in Table 2, where k is set to 3. As for ADE, our model achieves the best performances on *univ*, *zara1*, and *zara2*. The proposed method performs worse than the linear model on the *hotel* subset, but still much better comparing to other benchmarks. The Sophie model performs the best in the *eth* subset. On the other hand, our model obtains superior performances on the FDE metric for all the five subsets consistently. The above

Table 2. The trajectory prediction performances using the ADE and FDE metrics.

Metric	Dataset	Linear	LSTM	S-LSTM	Social-GAN	Sophie	Ours
ADE	eth	1.33	1.09	1.09	0.81	**0.70**	0.87
	Hotel	**0.39**	0.86	0.79	0.72	0.76	0.42
	Univ	0.82	0.61	0.67	0.60	0.54	**0.48**
	Zara1	0.62	0.41	0.47	0.34	**0.30**	**0.30**
	Zara2	0.77	0.52	0.56	0.42	0.38	**0.21**
Average		0.79	0.70	0.72	0.58	0.54	**0.46**
FDE	Eth	2.94	2.41	2.35	1.52	1.43	**1.36**
	Hotel	0.72	1.91	1.76	1.61	1.67	**0.67**
	Univ	1.59	1.31	1.40	1.26	1.24	**0.77**
	Zara1	1.21	0.88	1.00	0.69	0.63	**0.45**
	Zara2	1.48	1.11	1.17	0.84	0.78	**0.35**
Average		1.59	1.52	1.53	1.18	1.16	**0.72**

results illustrate that the proposed approach can predict the future trajectory accurately.

We now compare the forecasting performance using different values of k (namely, the top-k elements in the transformation set which have higher values in P), as shown in Table 3. Note that the values in the row 'baseline' are the best values obtained by the different models, as shown in Table 2. From the experimental results we can find that using $k = 2$ can reach competitive performances as well, although slightly worse than using $k = 3$. Setting k to 1 implies that our framework degenerates to a deterministic model, thus generating a single hypothesis at each time step, which corresponds to the maximum value in P. This is the main reason why using $k = 1$ performs worse as compared to other generative models (such as [4] and [12], which can handle multiple hypotheses) and deterministic models (such as [6] and [1], where the average trajectory is more likely to be used for evaluation).

Table 3. Comparisons on the ADE and FDE metrics using different values of k.

Metric/Dataset		Eth	Hotel	Univ	Zara1	Zara2	Average
ADE	$k = 1$	1.43	0.56	0.93	0.66	0.44	0.80
	$k = 2$	1.05	0.45	0.57	0.37	0.27	0.54
	$k = 3$	0.87	0.42	**0.48**	**0.30**	**0.21**	**0.46**
Baseline		**0.70**	**0.39**	0.54	0.34	0.38	0.47
FDE	$k = 1$	2.53	0.93	1.80	1.31	0.90	1.49
	$k = 2$	1.69	0.72	0.99	0.63	0.48	0.90
	$k = 3$	**1.36**	**0.67**	**0.77**	**0.45**	**0.35**	**0.72**
Baseline		1.43	0.72	1.24	0.63	0.78	0.96

5 Conclusions

In this work, we present a novel hierarchical framework for multi-modal trajectory forecasting, which is able to provide the explicit distribution of the future positions of pedestrians moving freely in a crowded scene. To achieve this, we first propose a clustering strategy to construct the so-called transformation set, which can be seen as a more representative form of the motion modalities. This transformation set works with the bottom LSTM in the decoder to describe the sampling distribution at every time instant. In order to better understand the differences among the different modalities, we train the model in a discriminative manner, aiming at promoting the sampling probabilities at the candidate points, while suppressing the probabilities at other points in the transformation set. This operation can better approximate the real trajectory distributions in the training set. We carried out our experiments on the standard benchmark datasets (*toy*, ETH, and UCY, etc.), and the experimental results demonstrate the effectiveness of the proposed framework.

Acknowledgment. This work is supported by the National Natural Science Foundation of China (Grant No. 61702073), and the China Postdoctoral Science Foundation (Grant No. 2019M661079).

References

1. Alahi, A., Goel, K., Ramanathan, V., Robicquet, A., Fei-Fei, L., Savarese, S.: Social lstm: Human trajectory prediction in crowded spaces. In the Proceedings of the IEEE International Conference on Computer Vision and Pattern Recognition, pp. 961–971. IEEE (2016)
2. Amirian, J., Hayet, J.B., Pettre, J.: Social ways: Learning multi-modal distributions of pedestrian trajectories with GANs. In the Proceedings of the IEEE International Conference on Computer Vision and Pattern Recognition Workshop, IEEE (2019)
3. Bartoli, F., Lisanti, G., Ballan, L., Bimbo, A.D.: Context-aware trajectory prediction. In the Proceedings of the IEEE International Conference on Pattern Recognition, pp. 1941–1946 (2018)
4. Gupta, A., Johnson, J., Fei-Fei, L., Savarese, S., Alahi, A.: Social GAN: socially acceptable trajectories with generative adversarial networks. In the Proceedings of the IEEE International Conference on Computer Vision and Pattern Recognition, pp. 2255–2264. IEEE (2018)
5. Hirakawa, T., Yamashita, T., Tamaki, T., Fujiyoshi, H.: Survey on vision-based path prediction. In: Streitz, N., Konomi, S. (eds.) DAPI 2018. LNCS, vol. 10922, pp. 48–64. Springer, Cham (2018). https://doi.org/10.1007/978-3-319-91131-1_4
6. Kalman, R.E.: A new approach to linear filtering and prediction problems. J. Basic Eng. **82**(1), 35–45 (1960)
7. Lawal, I., Poiesi, F., Aguita, D., Cavallaro, A.: Support vector motion clustering. IEEE Trans. Circuits Syst. Video Technol. **27**(11), 2395–2408 (2017)
8. Lee, N., et al.: Desire: Distant future prediction in dynamic scenes with interacting agents. In the Proceedings of the IEEE International Conference on Computer Vision and Pattern Recognition, pp. 336–345. IEEE (2017)

9. Lerner, A., Chrysanthou, Y., Lischinski, D.: Crowds by example. Comput. Graph. Forum **26**, 655–664 (2007)
10. Pellegrini, S., Ess, A., Schindler, K., Gool, L.V.: You'll never walk alone: Modeling social behavior for multi-target tracking. In the Proceedings of the IEEE International Conference on Computer Vision, pp. 261–268. IEEE (2009)
11. Rhinehart, N., Kitani, K., Vernaza, P.: R2p2:a reparameterized pushforward policy for diverse, precise generative path forecasting. In: the Proceedings of the European Conference on Computer Vision, pp. 772–788. (2018)
12. Sadeghian, A., Kosaraju, V., Sadeghian, A., Hirose, N., Savarese, S.: Sophie: An attentive GAN for predicting paths compliant to social and physical constraints. In: the Proceedings of the IEEE International Conference on Computer Vision and Pattern Recognition, pp. 1349–1358. IEEE (2019)
13. Vaswani, A., et al.: Attention is all you need. In: Advances in Neural Information Processing Systems, pp. 5998–6008 (2017)
14. Vemula, A., Muelling, K., Oh, J.: Social attention: modeling attention in human crowds. In the Proceedings of the IEEE International Conference on Robotics and Automation, pp. 1–7. IEEE (2018)
15. Zhong, J., Cai, W., Luo, L., Yin, H.: Learning behavior patterns from video: a data-driven framework for agent-based crowd modeling. In: the Proceedings of the International Conference on Autonomous Agents and Multi-agent Systems, pp. 801–809 (2015)
16. Zhou, B., Tang, X., Wang, X.: Coherent filtering: detecting coherent motions from crowd clutters. In: Fitzgibbon, A., Lazebnik, S., Perona, P., Sato, Y., Schmid, C. (eds.) ECCV 2012. LNCS, vol. 7573, pp. 857–871. Springer, Heidelberg (2012). https://doi.org/10.1007/978-3-642-33709-3_61

Skeleton-Based Methods for Speaker Action Classification on Lecture Videos

Fei Xu(✉), Kenny Davila, Srirangaraj Setlur, and Venu Govindaraju

Department of Computer Science and Engineering, University at Buffalo,
Buffalo, NY, USA
{fxu3,kennydav,setlur,govind}@buffalo.edu

Abstract. The volume of online lecture videos is growing at a frenetic pace. This has led to an increased focus on methods for automated lecture video analysis to make these resources more accessible. These methods consider multiple information channels including the actions of the lecture speaker. In this work, we analyze two methods that use spatio-temporal features of the speaker skeleton for action classification in lecture videos. The first method is the AM Pose model which is based on Random Forests with motion-based features. The second is a state-of-the-art action classifier based on a two-stream adaptive graph convolutional network (2S-AGCN) that uses features of both joints and bones of the speaker skeleton. Each video is divided into fixed-length temporal segments. Then, the speaker skeleton is estimated on every frame in order to build a representation for each segment for further classification. Our experiments used the AccessMath dataset and a novel extension which will be publicly released. We compared four state-of-the-art pose estimators: OpenPose, Deep High Resolution, AlphaPose and Detectron2. We found that AlphaPose is the most robust to the encoding noise found in online videos. We also observed that 2S-AGCN outperforms the AM Pose model by using the right domain adaptations.

Keywords: Action classification · Lecture video analysis · Pose estimation · Lecture video dataset

1 Introduction

Today, the number of online lecture videos is growing faster than ever. These videos are becoming a valuable educational resource for both teachers and students globally. Easy and efficient access to specific topics within the massive amount of lecture videos is enabled by applications for summarization, navigation, and retrieval of lecture video content. Methods for automated lecture video analysis facilitate these applications by extracting information from multiple channels including scene text, supplementary lecture materials, audio, transcriptions and speaker actions. In particular, many lecture videos feature handwriting of content on traditional whiteboards/chalkboards, and speaker actions such as writing or erasing can be used to facilitate the extraction of such content [1].

© Springer Nature Switzerland AG 2021
A. Del Bimbo et al. (Eds.): ICPR 2020 Workshops, LNCS 12664, pp. 250–264, 2021.
https://doi.org/10.1007/978-3-030-68799-1_18

In these scenarios, the development of accurate classifiers for speaker actions in lecture videos is important for advancing the field of lecture video analysis.

In this work, we analyze two methods that use spatio-temporal features of the speaker skeleton for action classification in lecture videos. The first method is based on our previous work, AM Pose [1], which uses Random Forests and motion-based features to classify speaker actions for further lecture video summarization on the AccessMath dataset [2]. The second method is a generic action classifier [3] which uses a two-stream adaptive graph convolutional network (2S-AGCN).

We adopt a generic framework which splits the video into small temporal segments, which we call action segments, for speaker action classification (See Fig. 1). The pose of the speaker is estimated on every frame using one of four state-of-the-art methods for pose estimation: OpenPose [4], Deep High Resolution (DHR) pose estimator [5], Detectron 2 [6], and AlphaPose [7]. Then, all poses are normalized and a feature representation is built for every action segment according to the corresponding action classifier model.

Our experiments are based on two lecture video datasets for which the speaker actions have been annotated. The first one is the AccessMath dataset [2], which has linear algebra lectures by one speaker. We significantly extended this dataset by adding videos from multiple online sources to built the second dataset. These annotations will be made publicly available.

During the course of this study, we attempt to answer the following research questions: **RQ1.** What is the performance of different state-of-the-art pose estimators when applied to online lecture videos? **RQ2.** How can we adapt generic skeleton-based action classification methods to the lecture video domain? **RQ3.** Can the 2S-AGCN model perform better than the AM Pose method and to what degree?

2   Background

Fig. 1. Summary of our lecture video speaker action classification framework. The process starts with a lecture video which is logically divided into action segments. The pose of the speaker is estimated and normalized in every video frame. A representation is built for all speaker poses in each action segment. A classifier is then used to determine the action of the speaker in every action segment.

Our work addresses the topic of skeleton-based action classification in lecture videos. In this section, we briefly describe vision-based human pose estimation methods which can be used to get the speaker skeleton in a lecture video. We then discuss skeleton-based methods for action classification. Finally, we review recent works on automated lecture video analysis.

2.1 Vision-Based Human Pose Estimator

A vision-based human pose estimator (HPE) is a method used to extract pose features from still images and videos. Chen et al. [8] describe three types of human body models that can be used in HPE: skeleton-based, contour-based, and volume-based. Here, we concentrate on skeleton-based pose representations.

Depending on how the estimation starts, Chen et al. [8] categorize pose estimators into two families: bottom-up and top-down. The bottom-up methods estimate all parts of the human body from input images and assemble them into different skeletons by fitting human body models. One example is OpenPose [4] which detects body parts and represents them using location and direction information stored in 2D vector fields. Full skeletons are produced in a greedy fashion by locally matching connected body parts (e.g. elbow candidates to hand candidates), by maximizing their confidence using the Hungarian method [9]. This is prone to failure when people are partially occluded in the image. Our previous work [1] uses OpenPose for speaker action classification on lecture videos.

The family of top-down methods, first generate bounding boxes of all people using object detectors and then extract human pose representations for each person detected. The Deep High-Resolution (DHR) learning method [5] uses multiple parallel sub-networks to produce high resolution heatmaps of human pose keypoints. The Detectron 2 method [6] uses mask R-CNN [10] to get the bounding boxes of human objects and also to detect human pose key-points within these boxes. The AlphaPose model [7] uses a regional multi-person pose estimation which deals with noisy detections of human objects. It uses a spatial transform network [11] to improve the object proposals that are fed to the single person pose estimator [12]. In general, top-down methods are likely to be slower than bottom-up methods when the image contains a large number of people [8].

Our work focuses on analyzing lecture videos where the upper body of the speaker is visible most of the time. We have considered the four pose estimators described here for speaker pose estimation in our work (see Sect. 5.1).

2.2 Action Classification

Human actions can be represented by sequential human pose changes. Some recent works address video-based action representations in different ways. Nguyen et al. [13] fuse salient maps from multiple prediction models first, and construct video representation using Spatial-Temporal Attention-aware Pooling (STAP). Yi et al. [14] combine appearance and motion saliencies to extract most salient trajectories, which are encoded by Bag of Features (BoF) or Fisher Vector (FV) as action features. In the work by Wang et al. [15], trajectories are constructed based on skeleton joints and projected as 2D images which are used for action classification by Convolutional Neural Network. The works by Yan et al. [16] and Shi et al. [3] construct skeleton-based graph representations which contain the spatial-temporal information for actions in videos. In this work, we consider action classification methods using skeleton-based action representations.

The 2s-AGCN model [3] uses adaptive graph convolutional networks with two spatio-temporal graph representations, one for joints and one for bones. For each representation, three adjacency matrices are required to decide the graph topology. The first matrix represents the physical connections from the skeleton. The second and third matrices make the graph topology adaptive because they are learned from the input data, and they represent the existence and strength of connections between any two joints/bones.

In our previous work, AM Pose [1], we used motion-based features and Random Forests for speaker action classification. These features are generated from joints and bones of the speaker skeleton, and they were designed based on observations of the speaker lecturing behavior. However, the AccessMath dataset [2] on which AM Pose was evaluated, has only one speaker. Our novel expansion includes multiple speakers. It is possible that for the same action some speakers behave differently (e.g. handedness, gestures, etc.). In this work, we take AM Pose as the baseline method, and we compare its classification accuracy against the 2S-AGCN on both the AccessMath dataset as well as our extended dataset.

2.3   Lecture Video Analysis

Lecture video analysis applications include lecture content summarization, indexing, search and navigation. Based on the type of data used for lecture video analysis, we can broadly categorize existing works into three groups.

The first group uses scene text appearing in lecture videos. Ma et al. [17] use various visual features to remove speakers/audience shots and extract slide/board frames from lecture videos for indexing. Davila and Zanibbi [18] extract keyframe-based lecture video summaries from single-shot lecture videos, and propose the Tangent-V visual search engine [19] which can use rendered LATEX from course notes as query images to search handwritten formulas in these keyframes and vice versa. Kota et al. [20] propose a framework to detect and binarize handwritten whiteboard content for lecture video summarization based on keyframes.

The second group does not rely on scene text. Instead, it uses other information channels from lecture videos such as audio or transcripts. Soares and Barrére [21] use audios of lecture videos to generate fully voiced audio chunks, and the corresponding textual and acoustic features are used with genetic algorithms for video segmentation. Shah et al. [22] use lecture video transcripts and Wikipedia texts to detect the boundaries of lecture topics for video segmentation.

The third group uses both scene text and information from other channels. Some works [23, 24] combine the speech from the instructor with the extracted lecture video slides for video retrieval. Xu et al. [1] use speaker action classification for whiteboard lecture video segmentation and keyframe selection. The handwritten content on these keyframes is then extracted for summarization. In this work, we focus on improving speaker action classification on whiteboard and chalkboard lecture videos for more effective lecture video analysis.

3   Methodology

In our current work, we use a generic framework for skeleton-based speaker action classification on lecture videos. The framework (illustrated in Fig. 1) uses small fixed-length temporal segments, called action segments, where we assume that the speaker is performing only one action. The input is a lecture video and the output is the predicted speaker action for every action segment of the video. First, we start by estimating the skeleton-based pose of the speaker on every video frame. We then normalize the pose estimations and create a representation for each action segment according to the selected method for action classification. We consider two skeleton-based speaker action classification models: AM Pose [1] and 2S-AGCN [3].

Assumptions. Each lecture video has only one instructor whose upper body is almost always visible and performs one action at any moment of the lecture. The video is recorded using a stationary camera without focus changing. The audience, if any, is never visible on the video. Finally, the lecture is based on handwritten content on a whiteboard/chalkboard.

3.1   Action Segment Extraction

The basic unit used for classification of actions in our framework is the action segment. In order to train and test our approach, we must extract the action segments of each lecture video. During normal runtime, the input can be simply divided into small sequential temporal segments of fixed length. Following the previous work AM Pose [1], we have currently fixed this length to 15 frames which roughly corresponds to half a second for all videos in our datasets.

In this work, we use 4 overlapping tracks of action segments both for training and testing videos, as illustrated in Fig. 2. The ground truth of the video contains the beginning and ending frames of each speaker action. When we sample a given action segment, we assign to it the label of the majority class of all frames contained within that segment.

3.2   Speaker Pose Estimation

Pose refers to the visual position and orientation of the human body at a given moment and can be represented in multiple ways as described in Sect. 2.1. In this work, we are concerned with skeleton-based pose estimations. Figure 3 shows different examples of speaker poses estimated using AlphaPose [7]. In Fig. 3a and 3c, the speaker face is partially or totally invisible. Figure 3b shows the joints of the upper body enumerated in the COCO-18 format [25]. We consider 4 different pose estimators in this work: OpenPose [4], DHR [5], Detectron2 [6], and AlphaPose [7]. An empirical comparison of these methods is described in Sect. 5.1.

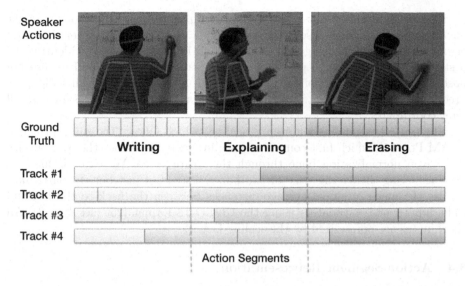

Fig. 2. Action segments used for speaker action classification. The action of the speaker, shown at the top, are labeled by frame ranges. Then, we extract one or more overlapping tracks of action segments of a fixed length (15 frames). The label of an action segment is decided using the majority class of the frames it covers.

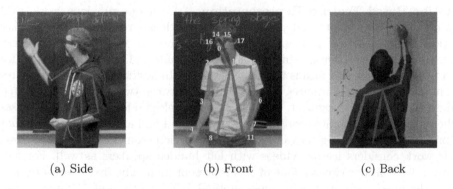

(a) Side   (b) Front   (c) Back

Fig. 3. Examples of skeleton data produced by the AlphaPose system [7] on lecture videos. We illustrate different poses such as the (a) side, (b) front and (c) the back. In (b), we have also included the original joint numbers, based on COCO-18 format [25].

3.3 Speaker Pose Normalization

Lecture videos are recorded in a variety of environments and video resolutions resulting in various ranges for the absolute speaker pose coordinates for each lecture video. In order to improve the robustness of speaker action classification, we normalize these coordinates using an affine transformation (translation and scaling) for each action segment.

First, we recognize that many frames within a given action segment might have invalid or missing pose estimations. This can be due to either detection errors or the speaker being out of the image in that frame. We identify the first valid frame where the pose estimator captures a skeleton, and we use the nose joint as the origin point to translate all coordinates from all valid pose predictions in that segment. For invalid poses, a special value is assigned to all joint coordinates.

The second normalization step is scaling, which follows the same principle as AM Pose [1], which first computes a scaling factor based on the average size of a given normalization bone through the entire video. We extend this idea by considering the average size of two bones instead of just one. In particular, we use the mean of global average distances between the neck to the left and right hip on pose estimators using the COCO-18 format [25] (see Fig. 3). All coordinates are normalized by the scaling factor.

3.4   Action Segment Representation

We consider two representations for speaker action segments in this paper. The first is based on the motion-based heuristic features from AM Pose [1]. The second uses two spatio-temporal graphs, one for joints and one for bones, of the skeletons in the action segment [3].

Motion-Based Feature Representation.   As in our previous work, AM Pose[1], we select joints and pairs of joints which are related to the speaker actions. In one action segment, joint-wise features are constructed for every pair of consecutive frames using the means, medians and covariance matrix for joint displacement and means for joint distances. In addition, pair-wise features including means and variances of displacement between two joints are computed on every frame of the segment. For both types of selected features, we omit cases where either of the joints coordinates is invalid and add a confidence value that represents the percentage of valid joints in the action segment. Unlike AM Pose, this work considers lecture videos with left handed speakers as well. For this reason, we consider two versions of this representation. The first one uses joints from the head, torso and right upper limb (RUL) for a total of 71 features and the second considers head, torso and both upper limbs (BUL) for a total of 106 features.

Graph-Based Representation.   This is the representation used by Shi et al. [3] in 2S-AGCN, which considers two types of features to represent both spatial and temporal information of action segments. The first one is joints data from pose estimation. While the original 2S-AGCN work assumes 3D poses and up to two people in the image, here we only use 2D pose estimation for a single person. The second feature is the bones in the skeleton, where bones represent physical connections between two joints. For the two joints of a bone, the one closer to the central point is the source joint and the other is the target joint. In our work, the neck is the central joint (point 1 in Fig. 3b). The displacement from source joint to target joint is used to represent the bone as a vector. For each of these two features, a spatio-temporal graph is constructed for every action segment, where the graph nodes are represented by corresponding joint or bone features, and the graph edges are the spatial and temporal connections between nodes.

Based on the observation that the lower limbs of the speakers are rarely visible in the videos, we consider 3 variations for the spatio-temporal graph of action segment. The first variation, full body, uses all joints on the speaker skeleton, while the other two use upper body joints including only the RUL or BUL.

3.5 Speaker Action Classification

In this work, we consider 8 speaker actions which are relevant to the handwritten content on whiteboards/chalkboards in lecture videos: write, pick eraser, erase, drop eraser, out, out-writing, out-erasing, and explain. Actions related to writing and erasing are key to changes in the content on the board. During the *explain* action, the speaker will mostly move around and use different gestures to emphasize important written content. The *out* action refers to the case when speaker is not visible in the image, and *out-writing/erasing* represent the hardest case to analyze where the speaker is mostly invisible but the handwritten content around the image edge is changed.

We have considered two different approaches for speaker action classification. The first is AM Pose [1] which combines the motion-based feature representation with Random Forests. The second is the 2S-AGCN model [3] which uses adaptive graph convolutional networks with spatio-temporal graph-based representations generated from both joints and bones of speaker skeleton data.

4 Datasets

We use two datasets for our experiments: AccessMath [2] and a novel extension. The speaker actions distributions on these datasets are shown in Table 1.

4.1 AccessMath

AccessMath is a set of 20 linear algebra lecture videos recorded in different classrooms with a single speaker. These videos are 1080p at 29.97 FPS and their average length is 49 min. For lecture video summarization, Davila et al. [18] used 12 videos (around 10 h) from the AccessMath dataset, with 5 videos for training (around 3.5 h) and 7 for testing. In AM Pose [1], we only annotated actions of the speaker for the 5 training videos.

4.2 Extended Dataset

We extended the original AccessMath dataset by annotating the intervals of speaker actions on lecture videos from multiple online sources. This dataset has 28 whiteboard videos and 6 chalkboard videos, for a total of 34. The resolutions of these lecture videos vary from 480p to 1080p. The video lengths vary in the range of 2.4 to 67.3 min with an average of 29.4 min. Overall, the dataset has 16.7 h of lecture videos. A total of 14 videos are from the AccessMath dataset, including the 5 training videos annotated in AM Pose.

The dataset includes 8 speakers, 6 of whom are right-handed and the other 2 are left-handed. For each speaker, half of the videos are for training and the other half are for testing. In this way, we have 17 training videos and 17 testing videos in total.

Most video frames in this dataset show the speaker lecturing. However, many videos from online sources include opening and ending credits, which are mostly text-based. We annotate such frames as *out* because speakers are not in the image.

These lecture videos are collected from YouTube channels, including the online version of the AccessMath dataset[1]. In order to ensure that all action segments (see Sect. 3.1) represent the same video length, we re-encoded all videos to 29.97 FPS. However, we notice that the original AccessMath videos and their corresponding online versions have different video length and image quality. To ensure that the annotations of the original videos are still compatible with the online versions, we use the video lengths of both versions to automatically synchronize the annotations to the online video. The annotations of the extended dataset will be publicly available[2].

Table 1. Distribution of the classes of Speaker Actions considered in this work. We consider both the AccessMath dataset and the extended dataset.

Action	AccessMath	Extended dataset	
	Training (%)	Training (%)	Testing (%)
Drop Eraser	0.8558	0.7523	0.8347
Erase	7.4659	5.1220	5.9263
Explain	25.7790	43.6089	44.9741
Out	25.6347	15.9167	13.8559
Out Erasing	0.0471	0.0537	0.0184
Out Writing	0.3163	0.2375	0.0022
Pick Eraser	0.8374	0.7371	0.9091
Write	39.0637	33.5719	33.4793

5   Experiments

In order to answer the research questions mentioned in Sect. 1, we ran three experiments: pose estimator selection (**RQ1**), adaptations of the 2S-AGCN [3]

---

[1] https://www.youtube.com/playlist?list=PLg2YxOqXd_2Ptnj2adKJRngjD1TK 7fAo5.

[2] https://kdavila.github.io/lecturemath/.

model for speaker action classification (**RQ2**), and finally a comparison between AM Pose [1] and 2S-AGCN (**RQ3**).

Table 2. Pose Estimator Selection. Cross-validation results for speaker action classification on AccessMath training videos using 4 different pose estimators as well as both the original and the noisier re-encoded videos from YouTube. Action classification is performed with AM POSE [1] using data from upper body and the right upper limb (RUL) or both upper limbs (BUL).

	Original		YouTube	
	RUL	BUL	RUL	BUL
OpenPose 1.5.1 [4]	83.59	84.01	82.72	83.15
DHR [5]	84.65	84.62	82.94	83.09
Detectron2 [6]	84.81	84.90	83.76	83.85
AlphaPose [7]	**85.23**	**85.57**	**84.23**	**84.32**

5.1 Pose Estimator Selection

To determine what is the performance of different state-of-the-art pose estimators on online lecture videos (**RQ1**), we evaluated four models: Openpose 1.5.1 [4], DHR [5], Detectron2 [6] and AlphaPose [7]. In addition, we used two versions of the AccessMath dataset [2] videos: the original and the online (YouTube). We compared the results of the original AM Pose [1] which considers features from RUL of the speaker, and the extended version which includes features from BUL (see Sect. 3.4). Following the experimental methodology from our previous work [1], we used cross-validation over the training videos.

The speaker action classification results for different pose estimators, feature sets and video sources are shown in Table 2. Except in the case when DHR is used on the original AccessMath videos, using features from BUL for action classification performs better than just considering the RUL. We observed that the additional features proposed in this work help to differentiate actions such as *out*, *explain* and *write*.

For all of the pose estimators evaluated here, the speaker action classification accuracy is lower on the YouTube version than on the original. In this case, when parts of the speaker body move too fast, they become blurred on the image. This causes errors in the pose estimation such as missing speaker skeletons. This can likely be explained by the fact that the tested pose estimators are not trained with this kind of image noise. The missing estimations can make the corresponding action segments to be incorrectly classified as *out*, which affects the classification accuracy. We observed that the average size of YouTube videos is 368 MB, and the average size of the originals is 4.28 GB.

We found that AlphaPose performs better than the other three pose estimators in terms of cross-validation action classification accuracy on both versions of the videos. In the following two experiments, we used AlphaPose to generate pose estimations.

5.2 Adaptations of the 2S-AGCN Model for Speaker Action Classification

To determine what would be required to adapt a generic skeleton-based action classifier to the lecture video domain (**RQ2**), we tested different domain adaptations using the 2S-AGCN model [3]. We evaluated the effect of pose normalization (see Sect. 3.3), using three types of pose data: raw pose, pose normalized by translation, and pose normalized by both translation and scaling. We also tested the effectiveness of each of the two streams for the adaptive graph convolutional network (AGCN), and we considered three different graph representations: full skeleton, upper body with RUL only and upper body with BUL. The second representation is used to make a fair comparison between AGCN and the original AM Pose. Same as our previous experiment, we also followed the same cross-validation protocol from AM Pose [1], but we only used the online version of the AccessMath dataset training videos.

In all conditions, the network was trained for a total of 16 epochs, with batch size of 64 and initial learning rate of 0.1 with decay factor of 0.1 every 5 epochs. Note that two streams of AGCN were trained for each cross-validation fold: One for joints, and one for bones. The average of all 5 folds are presented in Table 3. In all cases, using both streams of joints and bones (2S-AGCN) gives better classification result than using either stream alone. As shown in Table 2, using AGCN with joints representation performs better than using AM Pose with motion-based features generated from joints coordinates when the pose normalization is applied. We also observed that normalization helps more on the joint stream than bone stream. Overall, 2S-AGCN using BUL graph representations with pose translation gives the best result. In the third experiment, since we used a larger lecture video dataset with different speakers in different recording environments, we used both translation and scale for pose normalization.

Table 3. Cross-validation results for speaker action classification using the 2S-AGCN model [3]. We compare the effect of different pose normalization settings and 3 different graph representations

Streams	Graph (# Joints)	Pose normalization		
		None	Translation	Translation + Scale
Joints	RUL (12)	83.66	85.20	84.93
	BUL (14)	83.80	85.27	85.13
	Full (18)	84.25	84.84	85.15
Bones	RUL (12)	84.07	83.86	83.85
	BUL (14)	83.21	84.29	84.06
	Full (18)	83.48	83.52	83.39
Joints & Bones	RUL (12)	84.95	85.51	85.26
	BUL (14)	84.52	**85.74**	85.29
	Full (18)	85.31	85.21	85.43

5.3   Classifier: AM Pose Vs 2S-AGCN Model

To determine if the 2S-AGCN model [3] performs better than AM Pose [1] and to what degree (**RQ3**), our final experiment compared the AM Pose and the 2S-AGCN models on the extended dataset (see Sect. 4.2). Results in terms of action classification accuracy for all conditions are shown in Table 4.

The distribution of the speaker actions (see Table 1) shows that the most common actions on both datasets are *write, explain, out* and *erase*, and the extended dataset has more *explain* and less of the other three actions. We expected that the classification accuracy using the original AM Pose [1] with RUL would drop on the extended dataset as it has two left-handed speakers. However, while using features from BUL helps to classify actions better than using RUL in AM Pose, the overall difference between these two configurations is small. The per-class classification f-measures of *explain, out, write* and *erase* are 83.39%, 80.18%, 69.18%, and 4.53% for the left-handed speakers, and 83.37%, 88.32%, 87.11% and 77.61% for the right-handed ones respectively. It can be seen that the classifier performs similarly for both left and right-handed speakers on the *explain* action as it is the most common class (44.97%) in the extended dataset. It also predicts most of the *out* actions. Probably, the non-dominant upper limb provides enough information to classify these two actions. At the same time, the accuracies for the *write* and the *erase* actions are lower for left-handed speakers. In particular, the *erase* action is badly predicted, but this only represents 5.9% of the total actions (see Table 1). Overall, there are only around 2.1 h (out of 16.7 in total) of left-handed speaker videos in the extended dataset.

When the action segment contains a transition from one action to another, we label the action segment by the majority rule (see Sect. 3.1). Approximately 9.12% of the action segments in the test set match this condition. We anticipated that the action classifier might give the other action as the prediction for these action segments. Therefore, we introduced another evaluation standard (majority + secondary) which counts the prediction as correct if it is either of these two actions (see third column of Table 4). For all conditions of both methods, the difference between both evaluation standards is consistently around 3.2%. This difference corresponds to trivial action classification errors. We considered them as trivial because the corresponding action segments map to frames which are annotated with either of these two actions. For example, for AM Pose - BUL, we observed that 41.18% of the action segments containing transitions were incorrectly classified, but 83.74% of these mistakes were indeed trivial errors, and they represent 19.90% of the overall error.

Under the second evaluation standard, the incorrectly predicted action segments represent the non-trivial errors. For example, in the condition of AM Pose - BUL, 10.64% of the *write* actions are predicted as *explain* actions, which represents 22.21% of the overall classification error. On the other side, 8.55% of *explain* actions are classified as *write* actions, which is 23.96% of the total error. These confusions are probably related to the fact that sometimes speakers point at some handwritten content during *explain* and mimic *write* actions for emphasizing the content. These challenging errors represent potential areas where the action classification accuracy can be improved.

Table 4. Final Speaker Action Classification results on the LectureMath dataset. Apart from the **majority classification** result, we consider **secondary classification** for the boundary action segment. AM Pose - RUL follows work of [1], and 2S-AGCN considers both joints and bones spatio-temporal features as work of [3]

Method	Evaluation Mode	
	Majority	Majority + Secondary
AM Pose - RUL	83.57	86.77
AM Pose - BUL	83.97	87.16
Joint-AGCN (14)	84.73	87.94
Bone-AGCN (14)	84.19	87.4
2S-AGCN (14)	**85.46**	**88.68**

Overall, 2S-AGCN gives better result than AM Pose using both evaluation standards, and it also performs better than using either joint or bone alone.

6   Conclusion

In this work, we tried to answer three main research questions. First, to determine what is the performance of different state-of-the-art pose estimators on online lecture videos (RQ1), we compared four methods using cross-validation on the original and online (YouTube) versions of the AccessMath training videos [2]. We found that online lecture videos have encoding noise which caused pose estimators to miss or incorrectly predict poses. We also found that using Alpha-Pose helped to achieve higher speaker action classification accuracy on both versions of the videos.

Second, to determine what would be required to adapt a generic skeleton-based action classifier to the lecture video domain (RQ2), we tested the 2S-AGCN model [3] following the same cross-validation method with different combinations of graph representations and pose normalization methods. We found that using the graph representation of the speaker upper body normalized by translation can give the best classification accuracy on the YouTube versions of the AccessMath training videos. Consistent with the work by Shi et al. [3], we achieved higher classification accuracy by using both streams of AGCN.

Third, to determine if the 2S-AGCN model performs better than AM Pose and to what degree (RQ3), we extended the AccessMath dataset by annotating speakers actions in 29 lecture videos from different online sources. Then, we ran different configurations for both speaker action classification models on this dataset. We found that AM Pose can classify the speaker actions well, even on the larger lecture video dataset. However, the 2S-AGCN model still performs better than AM Pose by a small margin.

In the future, we intend to improve the action classification on noisy lecture videos, especially for the case where the pose estimator incorrectly predicts the speaker pose. We will continue to expand our lecture video dataset by including more speakers. We also want to use the 2S-AGCN for lecture video summarization similar to AM Pose [1] on the extended dataset. In addition, we would like to adapt this action classification framework to videos where only the hands of the speaker are visible.

Acknowledgement. This material was partially supported by the National Science Foundation under Grants No. 1640867 (OAC/DMR), and No. 1651118 (SBE).

References

1. Xu, F., Davila, K., Setlur, S., Govindaraju, V.: Content extraction from lecture video via speaker action classification based on pose information. In: 2019 International Conference on Document Analysis and Recognition (ICDAR), pp. 1047–1054. IEEE (2019)
2. Davila, K., Agarwal, A., Gaborski, R., Zanibbi, R., Ludi, S.: Accessmath: indexing and retrieving video segments containing math expressions based on visual similarity. In: 2013 Western New York Image Processing Workshop (WNYIPW), pp. 14–17. IEEE (2013)
3. Shi, L., Zhang, Y., Cheng, J., Lu, H.: Two-stream adaptive graph convolutional networks for skeleton-based action recognition. In: 2019 Conference on Computer Vision and Pattern Recognition (CVPR), pp. 12018–12027. IEEE/CVF (2019)
4. Cao, Z., Hidalgo, G., Simon, T., Wei, S., Sheikh, Y.: Openpose: realtime multi-person 2d pose estimation using part affinity fields. arXiv preprint arXiv:1812.08008 (2018)
5. Sun, K., Xiao, B., Liu, D., Wang, J.: Deep high-resolution representation learning for human pose estimation. In: 2019 Conference on Computer Vision and Pattern Recognition (CVPR), pp. 5693–5703. IEEE/CVF (2019)
6. Wu, Y., Kirillov, A., Massa, F., Lo, W.-Y., Girshick, R.: Detectron2 (2019). https://github.com/facebookresearch/detectron2
7. Fang, H.-S., Xie, S., Tai, Y.-W., Lu., C.: Rmpe: regional multi-person pose estimation. In: 2017 International Conference on Computer Vision (ICCV), pp. 2353–2362. IEEE/CVF (2017)
8. Chen, Y., Tian, Y., He, M.: Monocular human pose estimation: a survey of deep learning-based methods. Computer Vision and Image Understanding, pp. 102897 (2020)
9. Kuhn, H.W.: The hungarian method for the assignment problem. Naval Res. Logistics Q. **2**(1–2), 83–97 (1955)
10. He, K., Gkioxari, G., Dollár, P., Girshick, R.: Mask R-CNN. In: 2017 International Conference on Computer Vision (ICCV), pp. 2961–2969. IEEE/CVF (2017)
11. Max Jaderberg, Karen Simonyan, Andrew Zisserman, et al. Spatial transformer networks. In Advances in neural information processing systems, pages 2017–2025, 2015
12. Newell, A., Yang, K., Deng, J.: Stacked Hourglass Networks for Human Pose Estimation. In: Leibe, B., Matas, J., Sebe, N., Welling, M. (eds.) ECCV 2016. LNCS, vol. 9912, pp. 483–499. Springer, Cham (2016). https://doi.org/10.1007/978-3-319-46484-8_29
13. Nguyen, T.V., Song, Z., Yan, S.: Stap: spatial-temporal attention-aware pooling for action recognition. IEEE Trans. Circuits Syst. Video Technol. **25**(1), 77–86 (2014)

14. Yi, Y., Zheng, Z., Lin, M.: Realistic action recognition with salient foreground trajectories. Expert Syst. Appl. **75**, 44–55 (2017)
15. Wang, P., Li, W., Li, C., Hou, Y.: Action recognition based on joint trajectory maps with convolutional neural networks. Knowl.-Based Syst. **158**, 43–53 (2018)
16. Yan, S., Xiong, Y., Lin, D.: Spatial temporal graph convolutional networks for skeleton-based action recognition. In: Thirty-second Conference on Artificial Intelligence (AAAI) (2018)
17. Ma, D., Xie, B., Agam, G.: A machine learning based lecture video segmentation and indexing algorithm. In: Document Recognition and Retrieval XXI, vol. 9021, pp. 90210V. International Society for Optics and Photonics (2014)
18. Davila, K., Zanibbi, R.: Whiteboard video summarization via spatio-temporal conflict minimization. In: 2017 14th IAPR International Conference on Document Analysis and Recognition (ICDAR), vol. 1, pp. 355–362. IEEE (2017)
19. Davila, K., Zanibbi, R.: Visual search engine for handwritten and typeset math in lecture videos and latex notes. In: 2018 16th International Conference on Frontiers in Handwriting Recognition (ICFHR), pp. 50–55. IEEE (2018)
20. Kota, B.U., Davila, K., Stone, A., Setlur, S., Govindaraju, V.: Generalized framework for summarization of fixed-camera lecture videos by detecting and binarizing handwritten content. Int. J. Doc. Anal. Recogn. (IJDAR) **22**(3), 221–233 (2019)
21. Soares, E.R., Barrére, E.: An optimization model for temporal video lecture segmentation using word2vec and acoustic features. In: 25th Brazillian Symposium on Multimedia and the Web, pp. 513–520 (2019)
22. Shah, R.R., Yu, Y., Shaikh, A.D., Zimmermann, R.: Trace: linguistic-based approach for automatic lecture video segmentation leveraging wikipedia texts. In: 2015 IEEE International Symposium on Multimedia (ISM), pp. 217–220. IEEE (2015)
23. Yang, H., Meinel, C.: Content based lecture video retrieval using speech and video text information. IEEE Trans. Learn. Technol. **7**(2), 142–154 (2014)
24. Radha, N.: Video retrieval using speech and text in video. In: 2016 International Conference on Inventive Computation Technologies (ICICT), vol. 2, pp. 1–6. IEEE (2016)
25. Lin, T.-Y., et al.: Microsoft COCO: common objects in context. In: Fleet, D., Pajdla, T., Schiele, B., Tuytelaars, T. (eds.) ECCV 2014. LNCS, vol. 8693, pp. 740–755. Springer, Cham (2014). https://doi.org/10.1007/978-3-319-10602-1_48

IADS - Integrated Artificial Intelligence In Data Science

Workshop on Integrated Artificial Intelligence in Data Science (IADS)

Workshop Description

IADS is a forum for researchers and practitioners working on integrated models regarding artificial intelligence and data sciences for different domains and applications. As we know, artificial intelligence (AI) has now become an emerging research topic since it can be used to solve high-complexity problems and figure out the optimized solutions in many applications and domains, which also has huge potential to create a better society. The beneficial deployment of AI in science, medicine, technology, humanities, and social sciences has already been shown. Data science also referred to as pattern analytics and mining can be used to retrieve useful and meaningful information from the databases, which is helpful to make efficient decisions and strategies regarding different domains and applications. In particular, due to the exponential growth of data in recent years, the dual concept of big data and AI has given rise to many research topics, such as scale-up behavior from the former classical algorithms.

A recent challenge is also represented by the integration of multiple AI technologies, emerging from different fields (e.g. vision, security, control, bio-informatics), to develop efficient and robust systems interacting in the real world. Despite tremendous progress on core AI technologies over the last years, the integration of such competencies into larger systems that are reliable, transparent, and maintainable is still at the beginning stage. There are still numerous open issues both from theoretical and practical perspectives.

This year, we received 23 submissions for reviews, from authors belonging to 13 countries including Canada, Czech Republic, China, Estonia, Germany, India, Italy, Malaysia, Norway, Netherlands, Pakistan, Taiwan, and the U.S.A. (alphabetical order). Each paper is reviewed by at least two qualified reviewers to examine the quality of the submission. After the thoughtful peer-review process, we finally selected 16 papers for presentation at the workshop. The review process focuses on the quality of the papers, such as their scientific novelty and contributions to the existing artificial intelligence and data science problems. The acceptance of the papers was the result based on the reviewers' comments and suggestions. All the high-quality papers were accepted, and the acceptance rate was about 70%. The accepted articles represent an interesting mix of techniques to solve the current research issues regarding artificial intelligence and data sciences, such as fake review classification using supervised machine learning, defect detection of stainless steel plates using deep learning technology, deep neural networks for detecting real emotions using biofeedback and voice, data augmentation for a deep learning framework for ventricular septal defect ultrasound image classification, GIS-based framework to classify urban elements through emotional states posted on social network, A neural network model for lead optimization of MMP12 inhibitors, an empirical analysis of integrating feature extraction to automated machine learning pipeline, input-aware neural knowledge tracing machine, towards corner case detection by modeling the uncertainty of instance segmentation networks, intelligent and interactive video annotation for instance segmentation using Siamese neural networks, imputation of rainfall data using improved neural network algorithm, novelty based driver identification on RR intervals from ECG data, link prediction in social networks by variational graph autoencoder and similarity-based methods: a brief comparative analysis, a hybrid wine classification model for quality prediction, A PSO-based sanitization process with multi-thresholds model, and task-specific novel object characterization. The workshop program was completed by the invited talk titled "Advances and Challenges for the Automatic Discovery of Interesting Patterns in Data" given by Philippe Fournier-Viger from Harbin Institute of Technology (Shenzhen), China.

Last but not least, we would like to thank the IADS 2020 Program Committee, whose members made the workshop possible with their rigorous and timely review process. We would also like to thank Giovanni Maria Farinella and Tao Mei, the workshop chairs for valuable help and support.

Organization

IADS Chairs

Jerry Chun-Wei Lin   Western Norway University of Applied
                                  Sciences, Norway
Stefania Tomasiello   University of Tartu, Estonia
Gautam Srivastava   Brandon University, Canada

Program Committee

Siddhartha Bhattacharyya   Technical University of Ostrava, Czechia
Chien-Fu Cheng   Tamkang University, Taiwan
Chun-Hao Chen   National Taipei University of Science and
                                  Technology, Taiwan
Xiaochun Cheng   Middlesex University, UK
Youcef Djenouri   SINTEF Digit, Norway
Ali Dehghantanha   University of Guelph, Canada
Vicente Garcia Diaz   University of Oviedo, Spain
Liang Hu   University of Essex, UK
Tzung-Pei Hong   National University of Kaohsiung, Taiwan
Alireza Jolfaei   Macquarie University, Australia
Vincenzo Loia   University of Salerno, Italy
Huimin Lu   Kyushu Institute of Technology, Japan
Jerry M. Mendel   University of Southern California, USA
Matin Pirouz Nia   California State University, Fresno, USA
Reza Parizi   Kennesaw State University, USA
Yulei Wu   University of Exeter, UK
Xuyun Zhang   University of Auckland, New Zealand
Ji Zhang   University of Southern Queensland, Australia
Jun Wu   Shanghai Jaio Tong University, China

Fake Review Classification Using Supervised Machine Learning

Hanif Khan[1], Muhammad Usama Asghar[1], Muhammad Zubair Asghar[1],
Gautam Srivastava[2(✉)], Praveen Kumar Reddy Maddikunta[3],
and Thippa Reddy Gadekallu[3]

[1] Institute of Computing and Information Technology, Gomal University,
D.I.Khan (KP), Pakistan
h_rmarwat@yahoo.com, usama.asghar@yahoo.com, zubair@gu.edu.pk
[2] Department of Math and Computer Science, Brandon University,
Brandon, MB, Canada
srivastavag@brandonu.ca
[3] School of Information Technology and Engineering, Vellore Institute of Technology,
Vellore, India
{praveenkumarreddy,thippareddy.g}@vit.ac.in

Abstract. The revolution of social media has propelled the online community to take advantage of online reviews for not only posting feedback about the products, services, and other issues but also assists individuals to analyze user's feedback for making purchase decisions, and companies for improving the quality of manufactured goods. However, the propagation of fake reviews has become an alarming issue, as it deceives online users while purchasing and promotes or demotes the reputation of competing brands. In this work, we propose a supervised learning-based technique for the detection of fake reviews from the online textual content. The study employs machine learning classifiers for bifurcating fake and genuine reviews. Experimental results are evaluated against different evaluation measures and the performance of the proposed system is compared with baseline works.

Keywords: Social networks · Fake reviews · Machine learning

1 Introduction and Background

In the recent era of the social media revolution, individual users and other stakeholders of the online community like business companies are in continuous use of online textual content (online reviews) for selling and purchasing the products and analyzing customer feedback. There is an increasing trend of writing and posting fake reviews on social media sites to promote or demote competing products [1]. This issue not only deceives online customers while purchasing

Supported by Natural Sciences and Engineering Council of Canada (NSERC).

products but also defames the companies in terms of degraded customer satisfaction. Therefore, it is an important task to design a system that can detect and classify text into fake (spam) and real (non-spam) reviews, which will assist the online community in making purchase decisions correctly and analyzing customer feedback efficiently. Prior works on the fake review detection [2–4] have used supervised machine learning (ML) and lexicon-based techniques. Asghar *et al.* [2] have used sentiment-based scoring techniques to identify fake and real reviews posted on social media sites. However, we propose to apply the supervised ML technique, namely support vector machine (SVM), to detect fake (spam) and real (ham) text. Our work is different from the work performed by [2] in terms of applying a supervised ML approach with a labeled dataset. The proposed supervised ML technique for fake review detection is compared with the baseline works and other supervised ML classifiers.

In this study, a supervised ML technique, namely SVM, will be employed on a benchmark fake reviews dataset for the classification of text to binary classes: fake (spam) and real (ham). Additionally, for improving the efficiency of the proposed system, an optimized set of parameters will be used in the SVM classifier.

1.1 Problem Statement

The fake review (spam) detection from textual content using a supervised ML algorithm is an emerging issue due to the diverse features of spamicity in text. This study addresses the problem of fake review (spam) classification from text using the supervised ML technique. Consider a set of text reviews $= \{tr1, tr2, tr3 \dots trn\}$ taken as input, the objective is to design a predictive model that assigns a spamicity class Sc $\epsilon\{0, 1\}$ to a text review Tri, where 1 depicts spam and 0 shows ham (non-spam).

1.2 Research Questions

The following research questions are addressed in this paper:

1. RQ1. How to implement a supervised ML technique for classifying text into spam and non-spam?
2. RQ2. How to evaluate the efficiency of the supervised ML system for classifying text into non-spam (genuine) and spam (fake) classes?
3. RQ3. What is the efficiency of the proposed technique concerning similar studies for classifying spam reviews efficiently?

1.3 Aims and Objectives

Aims. The main aims of this study are as follows:

Objectives

1. To implement supervised ML technique for classifying text into spam and non-spam.
2. To evaluate the efficiency of the supervised ML method for the classification text into spam and non-spam classes.
3. To evaluate the efficiency of the proposed technique concerning similar studies for classifying spam reviews efficiently.

1.4   Research Contributions

This study makes the following contributions:

1. Classification of text into binary classes (spam and non-spam) by applying supervised ML technique.
2. Evaluating the efficiency of the proposed system for efficiently classifying text into spam and non-spam classes.
3. Comparison of the proposed system with the state-of-the-art techniques developed for spam detection.

The rest of the article is organized as follows: Sect. 2 presents a review of literature; in Sect. 3, the proposed methodology is presented; Sect. 4 focuses on results and discussion; and finally, the conclusion and future work is presented in Sect. 5.

2   Literature Review

Jain *et al.* [5] executed a profound learning method for spam classification. They utilized Long Short-Term Memory (LSTM) and Recursive Neural Network (RNN) for spam classification. SMS spam assortment and Twitter datasets are utilized for tests to improve the best results. Pre-prepared vectors are utilized by expanding the volume of information for preparing in the future.

Rating and Review Processing Method was proposed by Ghai *et al.* [6] to find the general score of reviews for spam identification, classifies a review as supportive or non-accommodating relying upon the score allotted to review. Best results depend on product survey information, which is scrapped from Amazon.com, utilizing scrapping tools in Python. Different studies are included to be implemented in the future.

Asghar *et al.* [2] work manages crossover arrangement by grouping the content as spam and non-spam in opinion spam identification system. A weighting plan is utilized for assigning priority to spam. Assessment measurements, like F-measure, Accuracy, Precision, and Recall, are utilized to assess the model's efficiency. The drawback of this work is that a constrained list of capabilities, feature selection is performed manually.

Narayan *et al.* [7] applied supervised learning techniques by using different classification algorithms to detect the spam review. Their system achieved a maximum Accuracy of 81.25%. This work can be extended by investigation of unsupervised and semi-supervised techniques.

You *et al.* [8] proposed an unsupervised method focusing on recognizing online spams, which is changed into a density-based exception detection issue. The proposed work has four stages (1) Review perspective rating count, (2) Tensor factorization calculation (3) Aspect rating-based local outlier factor model (AR-LOF), (4) Viewpoint positioning strategy to recognize spam reviews. The outcomes show that the proposed model is compelling and beats existing methodologies.

Mataoui *et al.* [9] presented a spam detection system that uses a set of selected features to process Arabic content generated on social networks. The proposed system uses a new supervised approach, which is mainly associated with the Arabic language. Interesting results with 91.73% of Precision for the unbalanced dataset are achieved. They propose to extend the work to the Algerian language.

Novel Convolutional Neural Network (CNN) was introduced by Li *et al.* [10] to detect text representation for false review identification. They experimented with a document by two classification techniques. They have analyzed with feature combinations between SVM and neural network-based methods. Experiments show that a sentence-weighted neural network is more usable than other neural network-based models. The current work can be extended by figuring the position weight of each sentence by utilization of the memory network-based model.

ML techniques and the N-gram investigation model is proposed by Ahmed *et al.* [11] for false news detection. TF-IDF and LSVM classifier are used for feature extraction. A very good result is achieved with an Accuracy of 92%. This work can be implemented on other recent datasets.

Kashti *et al.* [12] presented an active learning method for detecting false and sincere reviews. The reviews are accepted from authenticated users only. Positive, negative, and neutral reviews are classified by using NLP and text mining algorithms.

J48 classifier is used by Kokate *et al.* [13] for identifying and classifying the reviews of movies that are given by viewers. TP rate, TN rate, Accuracy, rule, condition are compared between J48 and ICRM algorithms.

Noekhah *et al.* [14] proposed an iterative calculation to distinguish counterfeit audits, survey spammers, and gather of spammers at the same time by applying a graph-based model on the Amazon dataset. Experimental results prove that the proposed algorithm outperforms all other baseline approaches in terms of Accuracy. Better results were achieved when all social and semantic highlights were collective. In the future, this work can be extended to incorporate all features by using graph structure along with a novel iterative algorithm.

Asghar *et al.* [15] addressed the issue of building a domain-dependent polarity lexicon from a set of reviews by using SWN (Senti Word Net 3.0), on account of its enough word inclusion and continuous updates. A Convolutional Neural Network model was proposed by Sun *et al.* [16] to coordinate the item-related review features through a word creation model. SVM classification methods are used for the classification task. CNN model shows better prediction performance than the SVM based model. To achieve more accurate predictions in the future, other types of reviews can be investigated.

Near Point Auto-Regressive (NPAR) algorithm is proposed by Wang *et al.* [17] to take genuine product reviews. The experimental results show that the algorithm works well in distinguishing spam opinions. Further, this research can be extended to identify time series anomalies.

Kiwanuka *et al.* [18] presented an automated, comprehensive email feature engineering framework for spam detection and classification. The first step is the development of many features from any email corpus and then extracting the features automatically using feature transformation and aggregation primitives. The framework incorporates a scalable mechanism for automated feature engineering and classification algorithms for spam classification. In the future, this work can be extended to investigate the process optimization of the developed framework to enable more efficient feature engineering and classification processes.

Mukherjee *et al.* [19] attempted to detect real-life fake reviews on a business website by analyzing yelp's filtered method. The authors have used a supervised approach to make yelp's filtered reviews for training. The results proved that yelp's filtering is reliable.

Linguistic Technique Senti wordnet, NLTK, and word count tool are proposed by Algur *et al.* [20] to find spamicity of the reviews based on the consistency and review content rating. The proposed system used relatively viable spamicity recognition.

Li *et al.* [21] proposed a more profound comprehension of the general nature of false opinions. The experiments are done within-domain tasks and extend to cross domains to investigate a progressively broad classifier. SAGE classifier achieves much better results than SVM. A ML technique was proposed by Crawford *et al.* [22] to detect fake reviews. They concentrated on supervised learning techniques. In the future, diverse component engineering strategies can be applied to various datasets to figure out which features are commonly helpful for online fake reviews.

The authors in [23] proposed a technique to help the customer in making proper purchasing choices by classifying the opinion as positive or negative. They carried out the decision by rating the behavior of the spam reviews. The authors proposed a novel and successful system which shows ordered sentiment in the form of "Chernoff's face".

The PU-learning approach was proposed by Fusilier *et al.* [24] for distinguishing deceptive opinions. Naive Bayes and SVM classifiers are used as learning algorithms in the PU-learning method. The usage of a one-class SVM classifier is suitable when there are very few examples of deceptive opinions for training. This work can be extended by integrating the PU-learning and self-training approaches.

A system was proposed by Radulescu *et al.* [25] to identify and eliminate spam comments by using learning algorithms and topic detection. The framework utilized the benchmark dataset and analyzed the presentation by three classification techniques. The decision trees classifier shows the best result for the implementation of the spam detection system.

3 Methodology

The proposed spam detection system consists of different modules like (i) dataset acquisition, (ii) noise reduction, (iii) implementing supervised learning technique, (iv) evaluating the performance of the proposed fake review (spam) detection system. Figure 1 shows the block diagram of the proposed system.

3.1 Data Acquisition

The fake review dataset is acquired from the benchmark source[1]. Table 1 gives details of the acquired dataset. The dataset is split into two parts, i.e., 20% for Testing and 80% for training. There 5573 reviews, out of which 474 are labeled as spam (fake) and 4825 as genuine (non-spam).

Table 1. Dataset statistics

Dataset	Description	Numbers of reviews	Numbers of reviews in labeled classes	
			SPAM	HAM
D1	Fake reviews (Spam)	5573	747 (13.40%)	4825 (86.57%)

Training Data. As mentioned earlier, 80% of the dataset consists of training data. In Table 2, samples of the training dataset are presented.

Table 2. Set of sample entries taken from training data

Review Id	Review	Label
1	Ok lar Joking wif u oni	HAM
2	Are you unique enough Find out from 30th August www areyouunique co uk	SPAM

Testing Data. The testing data is used for the assessment of the model, and it is given after the model is trained on the training dataset. In other words, test data is provided to the classifier to evaluate whether the classifier works satisfactorily. A 20% of the entire dataset is sliced for the testing phase. There are different methods to split the dataset into training and testing, viz (i) randomly dividing (ii) cross Endorsement, and (iii) keep in Cross Endorsement [26]. Their detail is given as follows:

(i) Randomly Dividing: In this approach, the data is segregated in the period of training and testing in a certain proportion. e.g., random samples are selected by dividing the dataset by 80:20. Randomly analysis action is much powerful than the other techniques because the dataset is more properly directed. In this study, we split the dataset randomly in the ratio of 20:80 for training and testing.

---

[1] https://www.kaggle.com/uciml/sms-spam-collection-dataset.

(ii) Cross Endorsement: In this approach, the dataset is broken down into multiple training and testing chunks by choosing a subset randomly. In the repeated partition of a dataset within testing and training, further split the data into two parts, i.e., training and endorsement.

(iii) Keep in Cross Endorsement: Data is split into three portions along with the distinct sizes, which are, validation, testing, and training. At the training stage, the model is trained, and from the endorsement to validate, the performance result of the model is evaluated, thereby ignoring the overtraining. If the result is good, then the training dataset is implemented over the validation point.

Table 3 shows sample entries of the testing dataset.

Table 3. Set of sample entries taken from testing dataset

Id	Review	Label
i	What time you coming down later	HAM
ii	FreeMsg Hey there darling its been 3 weeks now and no word back I d like some fun you up for it still? Tb ok XxX std chgs to send 150 to rcv	SPAM

First of all, we create a dataset in the CSV format file. A sample of reviews is presented in Table 2 and Table 3 as a testing and training set. We have acquired the SPAM and genuine reviews[2]. The 20:80% split of testing and training split is shown in Fig. 1 [27].

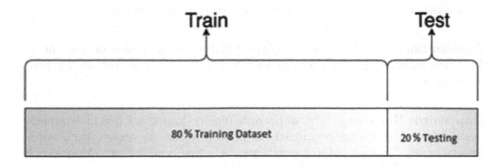

Fig. 1. Dataset division into 80; 20% for the training and testing

In Algorithm 1, pseudo-code steps are shown for splitting the dataset into training and testing.

---

[2] https://www.kaggle.com/uciml/sms-spam-collection-dataset.

Algorithm 1: Abstract of steps required to split the dataset into testing and training

---

1 Dividing Data within test/train:
2 $SetTrain\_A = [N]$
3 $SetTrain\_B = [N]$
4 $Setrain\_A = [N]$
5 $SetTest\_B = [N]$
6 $SetTotalTestSize = N * 20\%$
   $SetTotalIndex = RANDOM(0, N - 1, TotalTestSize)$
7 **for** $IND = 0$ to N **do**
8     $SetTempArray = [N]$
9   **for** $ITEMSinT\_Identifications[IND]$ **do**
10       $TempArray.Add(ITEMS)$
11   **if** $TotalIndex.contains(IND)$ **then**
12       $Test\_A.Add(TempArray)$
13       $Test\_B.Add(ReviewList[IND][1])$
14   **else**
15       $Train\_A.Add(TempArray)$
16       $Train\_B.Add(TempArray)$

3.2 Reducing Noise

Different pre-processing steps are applied to the collected dataset. These include tokenization, extra notable characters, and hashtag removal [28]. Algorithm 2 presents the pre-processing tasks using Python coding with Anaconda Framework [29].

A short description of the pre-processing steps is given below:

Tokenization: The word is transformed within small pieces of content in the tokenization step. For tokenization, NLTK tokenizer is used in Python-environment.

Stop Words Removing: The stop words play no important role in sentiment classification. Using a pre-assembled list of stop words, we remove such words systematically. Examples of stop words are: "a", "the" and "is" etc. Algorithm 2 presents a set of tasks required for pre-processing implementation.

3.3 Implementation of Supervised Machine Learning Classifier

The implementation of a spam detection system for classifying text into a fake (spam) and genuine (ham/non-spam) is accomplished by applying a supervised ML technique, namely, SVM [30].

Based on supervised ML, different studies have been conducted for fake review (spam) classification [2–4]. This work investigates state-of-the-art SVM

Algorithm 2: A set of steps required for pre-processing implementation

---

1 Pre-Processing
2 $SETTComments = [N]$
3 $SETDictionaries = [N]$
4 $SETIteratedRow = 0$
5 **for** $IteratedRow$ to $N.Count\ \text{-}1$ **do**
6 $\quad SETCWords = split(Para[IteratedRow], "")$
7 $\quad SETListOfComments = [N]$
8 $\quad SETListOfPunctuations = '.', ';', '.'$
9 $\quad SETListOfStopWords = "the", "an", "is", "are"$
10 $\quad$ **for** $each\ CHARS\ in\ CWords$ **do**
11 $\quad\quad$ **if** $CHARS\ AT\ LAST\ IS\ A\ PUNCTUATION$ **then**
12 $\quad\quad\quad$ PUNCTUATION REMOVE
13 $\quad\quad$ **else**
14 $\quad\quad\quad CHARS\ DOESNOT\ CONTAIN\ ANY\ STOP\ WORD$
15 $\quad\quad\quad ListOfComments.ADD(CHARS)$
16 $\quad\quad\quad DictionariesDOESN'TCONTAINCHARS$
17 $\quad\quad\quad Dictionaries[CHARS] = 0$

Table 4. A sample set of reviews before and after applying preprocessing steps

Review	Before clean	After clean	
		Stop words removel	Tokenizations
1)	Ok lar Joking wif u oni	lar joking u oni	Ok lar Joking wif u oni (Adjective) (Noun) (Verb)
2)	Your free ringtone is waiting to be collected Simply text the password MIX to 85069 to verify Get Usher and BritneyFML	free ringtone waiting collected simple test password MIX 85069 verify Usher Beitney FML	Your free ringtone is waiting to be collected Simply text the password MIX to 85069 to verify Get Usher and BritneyFML (Pronoun) (Adjective) (Noun) (Verb) (Adverb) (Conjuction)

with parameter setting for classifying text into fake (spam) and real (ham/non-spam). We evaluate the efficiency of the proposed system by comparing the experimental results with the baseline methods [3,4]. They used supervised ML techniques for fake review classification, whereas we implement an SVM classifier on the benchmark spam dataset with a tuned parameter setting. The SVM classifier is better enough to yield efficient results in terms of improved Accuracy while classifying textual content [2,4,30] (Table 4).

As shown in Fig. 2, firstly, the model is trained by giving input of labeled data in the form of textual reviews/tweets having classes spam and Ham (non-spam). In the next phase, the trained model is evaluated on the test module. To overcome the issue of overfitting, validation of the model is performed.

Feature Engineering. Feature engineering is an important phase while applying a supervised ML technique. Following methods of feature representation are

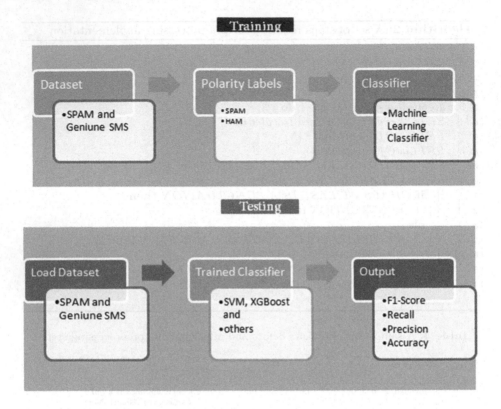

Fig. 2. Architecture of proposed system for spam detection

applied in the proposed work: i) Term Frequency (TF), ii) Term Frequency and Inverse Documents Frequency (TF and IDF), iii) Feature Vector

1. Term Frequency (TF): It calculates the numbers of incidence in a given review [31].
2. Inverse Document Frequency (IDF) and Term Frequency (TF): In the provided dataset, term frequency and inverse document frequency computations are used for showing the importance of words [32].
3. Feature Vector: It transforms the input review into the token count's matrix. [32].

Support Vector Machine (SVM). This classifier is used for nonlinear and linear problems. It's a supervised ML algorithm for classifications. The data is arranged in different classes and then the SVM works to discover the Hyper Plane, where the line divides the data into different categories. The main idea of SVM in connection with sentiment classifications is to decide about the hyperplane, which splits the corpus [33]. The mathematical representation of SVM is:

$$T = \{(y1, z1), (y2, z2), \ldots\ldots\ldots\ldots\ldots\ldots(yz, zn)\} \tag{1}$$

where T is the dataset of reviews, y, which belongs to values of z and indicates in case, the elements have a connection with that class.

3.4   Applying Other Machine Learning Classifiers

In addition to applying SVM, we also performed experiments with other ML classifiers, namely, Logistics Regression, Naive Bayesian, random forest, K-Nearest Neighbors, XGBoost, and Decision Tree classifier. These are described in the following sections.

Logistic Regression. The Logistics Regression classifier aims to classify the review into multiple categorize of polarity classes based on testing and training datasets. As it avoids overfitting, that's why Logistics Regression is the fastest predictive classifier that makes the best generalization. The performance of Logistics Regression is best on a new dataset [34]. The mathematical equation of Logistics Regression is as follows:

$$x^{\beta 0 + \beta 1 x 1 + \beta 2 x 2} \tag{2}$$

Their x is persistent, and the others represent the boundary function of equations.

Naive Bayes. The Naive Bayes classifier is used for classifications and regressions and by nature, it's a supervised ML classifier. On behalf of Bayes Theorem, Navies Bayes classifiers have a relation with the probability family of classifiers. When Naive Bayes classifiers are applied on big or small datasets, mostly good classifications results are obtained. When the input features are higher, then the Navies Bayes classifier gives a good result. Mathematically, it is formulated as:

$$P(X/Y) = \frac{P(X)}{P(Y)P(Y/X)} \tag{3}$$

Random Forest. This classifier is extra pliable than other ML classifiers on the base of hyperparameter tuning. Mostly it gives useful and efficient classification results. The Random Forest classifier is mostly used for classification and regression tasks. The result of every decision tree is the outcome of a random forest. The random forest classifier with many trees of decision as the outcome produces good generalizations [33, 39]. The numerical equation is given as follows:

$$w = \frac{1}{z} \sum_{z=1}^{x} zw\,(x') \tag{4}$$

where w is the numbers of example with replacement, zw is the tree classification of training and z is an instance of training from y, z.

K-Nearest Neighbor (KNN). K-Nearest Neighbor (KNN) is used to solve regression and classification problems. The KNN classifier is generally used on large scale classification problems in the industry. It is in agreement with specimen oriented learning. Training examples of the dataset are stored in a slot of N-Dimensional KNN, so it is called dull learner. The KNN classifies new cases by using majority votes of K-neighbor [34].

Were ith and bi are the examples and E is a consequence of prediction.

Extreme Gradient Boosting. The Extreme Gradient Boosting (XGBOOST) classifier is based on the Gradient boosting framework [35]. It provides promising results in a distributed atmosphere like Hadoop, SGE, and MPI.

Decision Tree. The decision tree classifiers are usually used for regression, classification, and other problems. It is also a supervised learning method. The nodes represent the features to be classified and the branch is the characteristic of values. The startup of classification needs position through sorting characteristics values from base nodes. Conquers as well as divide techniques are used for tree structure [34]. The decision tree equation is given as follows:

$$(y, z) = (a_1, a_2, a_3 \ldots a_{k,z}) \qquad (5)$$

the subset is denoted by z, base nodes are represented by y, and leaves of the tree are represented by a.

3.5   How the Proposed System Works?

The supervised learning technique for classifying spam and genuine reviews starts by entering input review, pre-processing the review, and finally classifying it as fake (SPAM) and genuine (HAM) using the SVM classifier. The dataset is split into a train (80%) and test (20%) modules. At the training stage, labeled data is provided to the classifier. Once the training stage of classifiers is finished, the classification of ML classifiers is validated through the assessment of the rest of the testing data. The obtained result is evaluated through multiple measures like Precision, F1-measure, Accuracy, and Recall. The working steps of the suggested method are shown in Algorithm 3.

3.6   Comparing Classifiers' Performance

To evaluate the prediction performance of the proposed system, other ML classifiers are applied, and the comparative analysis yields a qualitative evaluation of the proposed SVM classifier regarding fake and real review prediction from the textual content.

Finally, we evaluate the performance of the different classifiers in terms of different evaluation measures like Precision, Recall, Accuracy, and F1-Measures to check the classification performance result. The comparative analysis (see

Sect. 4) yields a qualitative evaluation of the proposed SVM classifier regarding fake and real review prediction from the textual content.

Based on the performance evaluation of multiple ML classifiers, applied on the dataset, we propose that (SVM) Support Vector Machine classifier gives the best classification results on the fake (spam) review dataset.

Algorithm 3: Proposed System's working steps

---

1 Result: Classified Rays with respect to Polarity
2 POLARITIES: ["SPAM" , "HAM"]
3 Classifiers: ["SVM", "KNN", "XGBoost", "DT", "RF", "NB" , "LR"]
4 Start
5 Scanned the SMS
6 $SET\ TEXT\_STREAM = Scan(Polarities)$
7 **Pre-processing**
8 **Tokenizations**
9 $SET\ TokenStream = GetTokens(TEXT\_STREAM)$
10 **Stop words Removel**
11 $SET\ PlainStream = RemoveStopWords(TokenStream)Punctuation$
12 Dividing full of Data with Train or Test.
13 SET TestLength = 20 Percent
14 $Train\_A, Train\_B, Test\_A, Test\_B = Split(PlainStream, TestLength)$
15 **Doing VectorCount (PlainStream)**
16 **TF and IDF**
17 **Applied Classifier**
18 $SET\ ModelClassfiers = GetClassifiers()$
19 $SET\ ModelClassification = ModelClassfiers : fit(Train\_A, Train\_B)$
20 **Predictions**
21 $SET\ PredictModel = ModelClassification : predicts(Text\_A)$
22 **Accuracy**
23 $SETModelAccuracy = GetAccuracy(PredictModel, Text\_A)$
24 **Confusion Matrix**
25 $SetCMatrix = GetConfusionMatrix(Test\_B, PredictModel)$
26 **Performance**
27 $GetMeasure(PreciseModel, BMeasure)$
28 $RETURN\ GetClassificationReport(Test\_B, PredictModel, POLARITIES)$

---

4   Results

In this section, the answer to each research question is provided by conducting experiments and analyzing results.

4.1   Answer to First Research Question

To find an answer for RQ1: "How to apply different ML classifiers (SVM) on a dataset of spam reviews for the prediction of spam's and genuine reviews"?, the supervised ML algorithm, which is (SVM) Support Vector Machine is applied on the dataset of fake reviews to predict fake(spam) and genuine reviews from SMS. The Support Vector Machine (SVM) algorithm is used for regression and classification problems. The dataset in training section X is used for predicting an object variable Y. The setting of parameters for the SVM classifier is presented in Table 5.

Table 5. Setting of SVM classifier parameters

Parameters	Explanation
C: float, elective (default = 1.0)	Parameter of Regularization. Regularization power is conversely proportionate to C
kernel: string, elective (default = 'rbf')	Kernel type must be Poly, Linear, Sigmoid, rbf, precomputed or callable. by default its "rbf"
degree: integer, elective (default = 3)	Degree is "poly" for function of Polynomial kernel
gamma: 'auto', 'scale' or float, elective (default = 'scale')	Coefficient of kernel for Poly, Sigmoid and rbd
coef0: float, elective (default = 0.0)	This is just important for sigmoid and Poly
shrinking heuristic: boolean, elective (default = True)	It's for shrinking heuristic
probability estimate: boolean, elective (default = False)	it's used to enable the estimate of probability
tolerance: float, elective (default = 1e−3)	Stopping criterion
size of cache: float, elective	Fix the size of the cache of Kernel
$class_w eight$: $balance$, '$dictionary$', elective	Used for the setting of Parameter Class C i to class weight
verbose: boolean, default: False	Permit output of verbose
$maximum_i ter$: integer, elective (default = −1)	−1 for no limits
$decision_f unction_f orm$: 'ovr', 'ovo', default = 'ovr'	For running of decision function in Zero vs Rest or One vs One
$break_t ies$: bool, elective (default = False)	Prediction will break ties under the values of confidence of the function of decision
$random_s tate$: Random State example, integer or None, elective (default = None)	Pseudo-random seed of number generator are used when for probability estimate rearrange the data

4.2 Answer to Second Research Question

To find an answer for RQ2: "How to evaluate the efficiency of different ML classifiers to predict spam and genuine reviews?", different ML classifiers are applied on the fake review dataset with detail are following. Experiment no 1: This experimentation is performed on the dataset having 5572 reviews, labeled as "spam and genuine". Different ML classifiers, such as RF, XGBoost, SVM, KNN, DT, NB, and LR are implemented, and performance evaluation results are reported in Table 8. We used multiple metrics like Precision, F1-score, Accuracy, and Recall. The result of the (SVM) support vector machine classifier is much finer than the other ML classifiers, i.e. Recall (99%), Accuracy (98.92%), Precision (99%), and F1-Score (99%).

Table 6. Experimental results of different ML classifiers.

Classifier	F1-score %	Recall %	Precision %	Accuracy %
K-Nearest Neighbour	0.89	0.89	0.89	83.25
Support Vector Machine	0.99	0.99	0.99	98.92
Decision Tree	0.97	0.97	0.97	97.31
XGBoost	0.97	0.97	0.97	96.77
Logistics Regression	0.97	0.97	0.97	97.13
Navies Bayes	0.98	0.98	0.98	97.67
Random Forest	0.97	0.97	0.97	96.5

Table 6 presents the classification performance of different classifiers obtained from the 5572 reviews spam dataset. It is noted that the (SVM) Support Vector Machine yields an Accuracy of 92.82%. The performance of K-Neighbor's classifier in terms of Accuracy is 83.25%. After analyzing the results of all classifiers, it is observed that SVM gives the best classification result and KNN gives the worst result on the given Dataset in the form of F1-score, Recall, Precision, and Accuracy as compared to other classifiers.

The results presented in Table 6 shows that the SVM classifier gives the best result, justified on the following literature-supported grounds.

1. Categorization problems of most text are linear separable: The data which is used in the fake review (spam) dataset, are categorized in two different labels, which are hum and spam, based on the tag of class use in the training dataset. Such types of data are linear partible, and for which, the SVM classifier gives the best classifications result in [31].
2. Input space is high dimensionality: For the learning of text classifiers, a lot of feature-space (Above 10000) is required. Such a huge feature space leads to overfitting. However, SVM has the dealing capability with such large characteristic spaces [31]. The size of our dataset is also quite large, with more than 11000-dimensional space, so based on dimensional space, we can say that SVM has yielded the best classification result on the given dataset.

Worst Performing Classifiers

Table 6 shows that the K-Nearest neighbor classifier gives the worst performance of classification as compared with others. The main reason why the KNN classifier gives the worst Accuracy is the identification of a new data class is based on a plain voting majority system, which is sometimes not acceptable because the distance of every close neighbor is mostly different against the gap from the test data. That's why KNN classifiers give low Accuracy on the given dataset. There are some other reasons for low classifications, which affects the result of different ML classifiers, for example, the size of the dataset, the sum of classes, arrangements of dataset and ratio of a dataset in training and testing and the sum of cases in the dataset [36].

Recommending Classifiers Best Performances

The results presented in Table 8 speak that the SVM classifier gives the best classification results in the form of improved F1-score, Accuracy, Precision, and Recall as compared to other ML classifiers, and therefore, it is recommended that the SVM with the parameter setting depicted in the previous section, has remained the best one for classifying reviews as fake(spam) or genuine (non-spam).

Results of Cross-validation for Different Classifiers

10-Fold Cross-Validation is performed to evaluate the performance results of different classifiers. The results available in Table 7 represent standard validation of Accuracy, standard validation of Precision macro, mean of Accuracy, mean Precision macro, standard validation of Precision macro, standard Recall macro, mean Recall macro, standard F1- macro and mean F-1 macro.

Table 7. Cross validation of different classifiers

Classifiers	Mean accuracy	Standard deviation	Mean precision macro	Standard deviation	Mean recall macro	Standard deviation	Mean F-I macro	Standard deviation
Random Forest	0.971	0.005	0.984	0.005	0.895	0.028	0.935	0.012
SVM	0.988	0.004	0.988	0.006	0.959	0.016	0.988	0.004
KNN	0.915	0.005	0.955	0.002	0.682	0.018	0.743	0.021
Logistic Regression	0.971	0.005	0.979	0.007	0.896	0.017	0.931	0.013
XB Boost	0.968	0.006	0.974	0.008	0.891	0.024	0.930	0.019
Decision Tree	0.965	0.005	0.942	0.006	0.909	0.019	0.926	0.011
NB	0.968	0.006	0.938	0.013	0.911	0.025	0.925	0.015

4.3   Answer to Third Research Question

To find an answer to RQ3: "What is the efficiency of the proposed classifier concerning the baseline method?", the efficiency of the recommended classifier is compared with the baseline studies.

Comparison with Baseline Methods

Pragna and RamBai [3] aims to detect Spam and Ham SMS through the mobile model and compared the classification of multiple ML classifiers, like SVM, SGD, DT, KNN, LR, DT, and RF. After classification, they confirmed that the SVM classifier gives the best 96.23% Accuracy, among others (see Table 8).

Renuka et al. [4] performed work to predict spam emails and compare the classification results of ML classifiers. They applied three different ML classifiers, which are NB, J48, and MLP, and reported that MLP classifiers give the best 93% Accuracy (see Table 8).

Table 8. Comparison with baselines result

Studies	Technique	Results
Pragna and RamBai [3]	SVM	96.23% (Accuray)
Renuka *et al.* [4]	MLP	93 % (Accuray)
Our work	Machine Learning Classifier: SVM	98.92% (Accuracy) 99% (Recall) 99% (Precision) 99% (F-Measure)

Proposed Work: The proposed technique fake review classification using SVM classifier achieved promising results in terms of improved Accuracy (98.93%), Recall (99%), Precision (99%), and F-measure (99%). It is evident from achieved results that the proposed technique has outperformed the baseline methods.

5   Conclusion and Future Work

This study deals with the classification of text into fake (spam) and genuine (non-spam) reviews by classifying applying supervised ML technique, namely SVM with a recommended set of parameters. Additionally, we applied other ML classifiers and evaluated their results. Before feeding the text to the ML classifier, different preprocessing techniques are applied for noise reduction. The experimental results of SVM (R: 0.99, A: 98.92, P: 0.99 and F-1 Score: 0.99) and other classifiers, namely XGBoost, K-Nearest Neighbor (KNN), Random Forest (RF), Naïve Bayes (NB), Decision Tree (DT) and Logistic Regression (LR) are evaluated and it is recommended that SVM is deemed appropriate for fake review classification than the other ML classifiers. It is further observed that the KNN has shown the worst performance (R: 0.89, A: 83.25, P: 0.89, and F-1 score: 0.89), as compared to other classifiers.

Limitations

1. In this study the dataset is imbalanced, which effects the ML classifiers to exhibit the poor performance result
2. The random splitting technique is used for splitting the dataset into testing and training.
3. Only TF-IDF feather engineering technique is used in this study.
4. Limited size (5573) of the dataset is used in this study, which affects the result of classifiers, and there is a need to increase the size of the dataset for better results [37,38].

Future Directions

1. A balanced dataset can improve the classification result of the ML classifiers.

2. Different techniques for segmentation of dataset e.g. hold-out cross-validation, cross-validation, stratified sampling, and a few more, can also be applied to evaluate the classification performance of ML classifiers.
3. Other feature engineering techniques, such as word embedding can also be investigated for better results.
4. Increasing the size of the dataset can also produce more promising results.
5. Blockchain technology can be used to identify the source of the fake news [40–42].

References

1. Asghar, M.Z., Subhan, F., Ahmad, H., et al.: Senti-eSystem: a sentiment-based eSystem-using hybridized fuzzy and deep neural network for measuring customer satisfaction. Software: Pract. Exper. **51**, 571–594 (2021). https://doi.org/10.1002/spe.2853
2. Asghar, M.Z., Ullah, A., Ahmad, S., Khan, A.: Opinion spam detection framework using hybrid classification scheme. Soft. Comput. **24**(5), 3475–3498 (2019). https://doi.org/10.1007/s00500-019-04107-y
3. Pragna, B., RamaBa, M.: Spam detection using NLP techniques. Int. J. Recent Technol. Eng. (IJRTE) **8**(2S11), 2423–2426 (2019). ISSN 2277-3878
4. Renuka, D.K., Hamsapriya, T., Chakkaravarthi, M.R., Surya, P.L.: Spam classification based on supervised learning using machine learning techniques. In: 2011 International Conference on Process Automation, Control and Computing, pp. 1–7. IEEE, July 2011
5. Jain, G., Sharma, M., Agarwal, B.: Optimizing semantic LSTM for spam detection. Int. J. Inf. Technol. **11**(2), 239–250 (2018). https://doi.org/10.1007/s41870-018-0157-5
6. Ghai, R., Kumar, S., Pandey, A.C.: Spam detection using rating and review processing method. In: Panigrahi, B.K., Trivedi, M.C., Mishra, K.K., Tiwari, S., Singh, P.K. (eds.) Smart Innovations in Communication and Computational Sciences. AISC, vol. 670, pp. 189–198. Springer, Singapore (2019). https://doi.org/10.1007/978-981-10-8971-8_18
7. Narayan, R., Rout, J.K., Jena, S.K.: Review spam detection using opinion mining. In: Sa, P.K., Sahoo, M.N., Murugappan, M., Wu, Y., Majhi, B. (eds.) Progress in Intelligent Computing Techniques: Theory, Practice, and Applications. AISC, vol. 519, pp. 273–279. Springer, Singapore (2018). https://doi.org/10.1007/978-981-10-3376-6_30
8. You, L., Peng, Q., Xiong, Z., He, D., Qiu, M., Zhang, X.: Integrating aspect analysis and local outlier factor for intelligent review spam detection. Future Gener. Comput. Syst. **102**, 163–172 (2020)
9. Mataoui, M.H., Zelmati, O., Boughaci, D., Chaouche, M., Lagoug, F.: A proposed spam detection approach for Arabic social networks content. In: 2017 International Conference on Mathematics and Information Technology (ICMIT), pp. 222–226. IEEE, December 2017
10. Li, L., Qin, B., Ren, W., Liu, T.: Document representation and feature combination for deceptive spam review detection. Neurocomputing **254**, 33–41 (2017)
11. Ahmed, H., Traore, I., Saad, S.: Detection of online fake news using N-gram analysis and machine learning techniques. In: Traore, I., Woungang, I., Awad, A. (eds.) ISDDC 2017. LNCS, vol. 10618, pp. 127–138. Springer, Cham (2017). https://doi.org/10.1007/978-3-319-69155-8_9

12. Kashti, M.R.P., Prasad, P.S.: Enhancing NLP techniques for fake review detection. Int. Res. J. Eng. Technol. (IRJET) **6**, 241–245 (2019)
13. Kokate, S., Tidke, B.: Fake review and brand spam detection using J48 classifier. IJCSIT Int. J. Comput. Sci. Inf. Technol. **6**(4), 3523–3526 (2015)
14. Noekhah, S., Fouladfar, E., Salim, N., Ghorashi, S.H., Hozhabri, A.A.: A novel approach for opinion spam detection in e-commerce. In: Proceedings of the 8th IEEE International Conference on E-Commerce with Focus on E-Trust (2014)
15. Asghar, M.Z., Khan, A., Ahmad, S., Khan, I.A., Kundi, F.M.: A unified framework for creating domain dependent polarity lexicons from user generated reviews. PLoS ONE **10**(10), e0140204 (2015)
16. Sun, C., Du, Q., Tian, G.: Exploiting product related review features for fake review detection. Math. Probl. Eng. **2016**, 1–7 (2016)
17. Wang, Y., Zuo, W., Wang, Y.: Research on opinion spam detection by time series anomaly detection. In: Sun, X., Pan, Z., Bertino, E. (eds.) ICAIS 2019. LNCS, vol. 11632, pp. 182–193. Springer, Cham (2019). https://doi.org/10.1007/978-3-030-24274-9_16
18. Kiwanuka, F.N., Alqatawna, J.F., Amin, A.H.M., Paul, S., Faris, H.: Towards automated comprehensive feature engineering for spam detection (2019)
19. Mukherjee, A., Venkataraman, V., Liu, B., Glance, N.: What yelp fake review filter might be doing? In: Seventh International AAAI Conference on Weblogs and Social Media, June 2013
20. Algur, S.P., Biradar, J.G.: Rating consistency and review content based multiple stores review spam detection. In: 2015 International Conference on Information Processing (ICIP), pp. 685–690. IEEE, December 2015
21. Li, J., Ott, M., Cardie, C., Hovy, E.: Towards a general rule for identifying deceptive opinion spam. In: Proceedings of the 52nd Annual Meeting of the Association for Computational Linguistics (Volume 1: Long Papers), pp. 1566–1576, June 2014
22. Crawford, M., Khoshgoftaar, T.M., Prusa, J.D., Richter, A.N., Al Najada, H.: Survey of review spam detection using machine learning techniques. J. Big Data **2**(1), 1–24 (2015). https://doi.org/10.1186/s40537-015-0029-9
23. Prajapati, J., Bhatt, M., Prajapati, D.J.: Detection and summarization of genuine review using visual data mining. Int. J. Comput. Appl. **975**, 8887 (2012)
24. Fusilier, D.H., Cabrera, R.G., Montes, M., Rosso, P.: Using PU-learning to detect deceptive opinion spam. In: Proceedings of the 4th Workshop on Computational Approaches to Subjectivity, Sentiment and Social Media Analysis, pp. 38–45, June 2013
25. Radulescu, C., Dinsoreanu, M., Potolea, R.: Identification of spam comments using natural language processing techniques. In: 2014 IEEE 10th International Conference on Intelligent Computer Communication and Processing (ICCP), pp. 29–35. IEEE, September 2014
26. Reitermanova, Z.: Data splitting. In: WDS, vol. 10, pp. 31–36 (2010)
27. Nabil, M., Aly, M., Atiya, A.: ASTD: Arabic sentiment tweets dataset. In: Proceedings of the 2015 Conference on Empirical Methods in Natural Language Processing, pp. 2515–2519, September 2015
28. Asghar, M.Z., Khan, A., Khan, F., Kundi, F.M.: RIFT: a rule induction framework for Twitter sentiment analysis. Arab. J. Sci. Eng. **43**(2), 857–877 (2017). https://doi.org/10.1007/s13369-017-2770-1
29. Ejaz, A., Turabee, Z., Rahim, M., Khoja, S.: Opinion mining approaches on Amazon product reviews: a comparative study. In: 2017 International Conference on Information and Communication Technologies (ICICT), pp. 173–179. IEEE, December 2017

30. Khattak, A.M., Ullah, H., Khalid, H.A., Habib, A., Asghar, M.Z., Kundi, F.M.: Stock market trend prediction using supervised learning. In: Proceedings of the Tenth International Symposium on Information and Communication Technology, pp. 85–91, December 2019

31. Joachims, T.: Text categorization with Support Vector Machines: learning with many relevant features. In: Nédellec, C., Rouveirol, C. (eds.) ECML 1998. LNCS, vol. 1398, pp. 137–142. Springer, Heidelberg (1998). https://doi.org/10.1007/BFb0026683

32. Effrosynidis, D., Peikos, G., Symeonidis, S., Arampatzis, A.: DUTH at SemEval-2018 task 2: Emoji prediction in tweets. In: Proceedings of the 12th International Workshop on Semantic Evaluation, pp. 466–469, June 2018

33. Nayak, A., Natarajan, D.: Comparative study of Naive Bayes, support vector machine and random forest classifiers in sentiment analysis of Twitter feeds. Int. J. Adv. Stud. Comput. Sci. Eng. 5, 14–17 (2016)

34. Ismail, H., Harous, S., Belkhoucshe, B.: A comparative analysis of machine learning classifiers for Twitter sentiment analysis. Res. Comput. Sci. 110, 71–83 (2016)

35. Babajide Mustapha, I., Saeed, F.: Bioactive molecule prediction using extreme gradient boosting. Molecules 21(8), 983 (2016)

36. Van der Walt, C.M., Barnard, E.: Data characteristics that determine classifier performance (2006)

37. Kwon, O., Sim, J.M.: Effects of data set features on the performances of classification algorithms. Expert Syst. Appl. 40(5), 1847–1857 (2013)

38. Reddy, G.T., et al.: Analysis of dimensionality reduction techniques on big data. IEEE Access 8, 54776–54788 (2020)

39. Maddikunta, P.K.R., Srivastava, G., Gadekallu, T.R., Deepa, N., Boopathy, P.: Predictive model for battery life in IoT networks. IET Intell. Transp. Syst. 14, 1388–1395 (2020)

40. Ch, R., Srivastava, G., Gadekallu, T.R., Maddikunta, P.K.R., Bhattacharya, S.: Security and privacy of UAV data using blockchain technology. J. Inf. Secur. Appl. 55, 102670 (2020)

41. Baza, M., Mahmoud, M., Srivastava, G., Alasmary, W., Younis, M.: A light blockchain-powered privacy-preserving organization scheme for ride sharing services. In: 2020 IEEE 91st Vehicular Technology Conference (VTC2020-Spring), pp. 1–6. IEEE, May 2020

42. MK, M., Srivastava, G., Somayaji, S.R.K., Gadekallu, T.R., Maddikunta, P.K.R., Bhattacharya, S.: An incentive based approach for COVID-19 using blockchain technology. arXiv preprint arXiv:2011.01468 (2020)

Defect Detection of Stainless Steel Plates Using Deep Learning Technology

Yu-Jen Huang[1], Ko-Wei Huang[1], and Shih-Hsiung Lee[2(✉)] iD

[1] Department of Electrical Engineering, National Kaohsiung University of Science and Technology, Kaohsiung, Taiwan
[2] Department of Intelligent Commerce, National Kaohsiung University of Science and Technology, Kaohsiung, Taiwan
shlee@nkust.edu.tw

Abstract. In the era of industry 4.0, factories around the world are developing towards automation and artificial intelligence, in which industrial detection plays an important role. After the cutting process, the surface of a stainless steel plate may produce various defects, such as scratches, chisels, and stains. Due to the characteristic of bright reflections on the surface of a stainless steel plate, the traditional manual comparison detection method is time-consuming, laborious, and prone to different detection results due to the interference of high reflection, resulting in the outflow of defective products. This paper used existing mature deep learning models for object detection, YOLOv3 (You Only Look Once) and SSD (Single Shot MultiBox Detector), which are the base network architectures for the defect detection of stainless steel plates, in order to effectively improve the accuracy of stainless steel plate detection. Through image preprocessing, the relative positions of sample defects are marked to improve data processing before training, in order that a large number of image samples can be quickly and effectively processed for training.

Keywords: Defect detection · Deep learning · Object detection

1 Introduction

Image recognition has always been one of the emphases in the development of multimedia technology. Image recognition is often applied in daily life, such as face or license plate recognition. In recent years, image recognition hardware has been gradually improved and deep learning technology has been introduced, which have resulted in great breakthroughs in the field of computer vision artificial neural networks. Deep learning [1] is a kind of multilayer perceptron representation learning method, which is constructed by non-linear modules that transform the upper level representation into higher-level and more abstract representation. A network system is composed of several simple modules or neural layers, each of which is composed of several neurons. In addition, deep learning only requires a small amount of manual intervention, which is very suitable

© Springer Nature Switzerland AG 2021
A. Del Bimbo et al. (Eds.): ICPR 2020 Workshops, LNCS 12664, pp. 289–301, 2021.
https://doi.org/10.1007/978-3-030-68799-1_20

for formulating current image visual data and unknown eigenvalues. To date, many scholars have proposed deep neural networks with different structures for object detection, and target detection based on CNN [2] has made great progress recently. In this paper, the most common defects of metal mechanical machining parts are combined with multimedia data, and the defects of stainless steel plates are simulated by deep learning technology for image recognition, which is conducive to improving production and work efficiency. In the process of metal reprocessing and cutting, there will be scratches, chisels, stains, and other defects produced on the metal surface, which must be identified and removed through detection to improve the quality of the products. In traditional detection, frozen images of high-speed moving metal machining parts are inspected manually by visual inspection combined with a stroboscope; however, as the human eye cannot conduct long-term monitoring like a camera, manual detection cannot achieve the goal of comprehensive detection. Therefore, this paper develops a defect detection system based on deep learning technology for metal cut stainless steel plates.

2   Related Works

Among the object detection algorithms with real-time performance mentioned in recent years, YOLO [3], YOLOv2 [4], YOLOv3 [5], Retinanet [6], and SSD [7] are most influential. This study uses YOLOv3 and SSD as the basic detection techniques to timely detect object types and object locations.

2.1   YOLOv3: You Only Look once V3

Both YOLOv1 and YOLOv2 have a common shortcoming, meaning that they usually fail to detect small objects. Therefore, in order to improve the ability to detect small objects, YOLOv3 was specifically improved by targeting three different scales. Moreover, the number of anchor boxes was increased, that is, the density was higher, and IOU could be promoted. In addition, the number of CNN layers was increased in the detection part. As more convolution layers were added, while the speed was relatively slower than YOLOv2, the effect was much better than YOLOv1 and YOLOv2.

2.2   SSD: Single Shot MultiBox Detector

Liu et al. proposed SSD, which set new records in terms of the performance and accuracy of object detection, scoring more than 74% (average accuracy) on standard data sets, such as Pascal VOC and COCO. Contemporary popular state-of-art detection systems generally follow the following steps: generating some hypothetical bounding boxes in advance, extracting features from these bounding boxes, and then, adding a classifier to determine whether there are objects inside them and what objects are. The feature of SSD is its deep neural network forward for object detection. The output spaces of the bounding boxes are discretized into a set of default boxes through the different feature map locations of different scales and different aspect ratios; therefore, it has a good performance in both immediacy and accuracy.

3   Stainless Steel Images and Definitions of Defects

Among all the surface images of steel products, the surface images of stainless steel plates have the highest reflection, and the material variations of each image are also very high, as shown in Fig. 1. Therefore, this paper discusses and defines the defects of stainless steel plates. As image complexity is extremely high, defects must be clearly defined, in order that the characteristics of each defect in the image can be understood. There are three kinds of defects designed in this study, which are scratches, chisels, and stains. According to their characteristics, they can be roughly divided into two categories: linear and regional.

Fig. 1. The surface image of steel product.

3.1   Linear Defects

Scratch. Scratches are characterized by linear groove marks on the surface of a stainless steel plate, as shown in Fig. 2, which may be caused by the moving drum wrapped on the line guide, side guide, or cooling bed conveying table. In appearance, scratches are characterized by long strips with slight bending, the dark lines on the scratch are obvious, and there are some bright lines surrounding the scratch.

Fig. 2. The scratch defect type.

3.2 Regional Defects

Chisel. Chisel marks are characterized by sharp, short, penetrating defects in a meteoric shape on the surface of the stainless steel plate, which are unevenly and roughly distributed, have protrusions, and are sunk into the surface of the stainless steel plate, as shown in Fig. 3. Chisel marks, which are caused by collision and mainly distributed in the head and tail of the steel coil, may result from the friction between the upper and lower layers of the steel strip when the steel coil is rotating. In addition, there are obvious bright spots in the sinking location.

Stain. Stains, which are caused by lubricating oil, are characterized by cometary shapes, speckle shapes, and broom shapes on the surface of stainless steel plates. Due to the oxidation process of oil stains, light yellow or tan marks are formed on the surface of the stainless steel plates, as shown in Fig. 4. The reason may be that there is excessive lubricating oil added for the maintenance of the mechanism, which lead to leakage and stains after rolling. In image presentation, regions with relatively dark brightness are the defective regions, which are easily identified.

Fig. 3. The chisel defect type.

Fig. 4. The stain defect type.

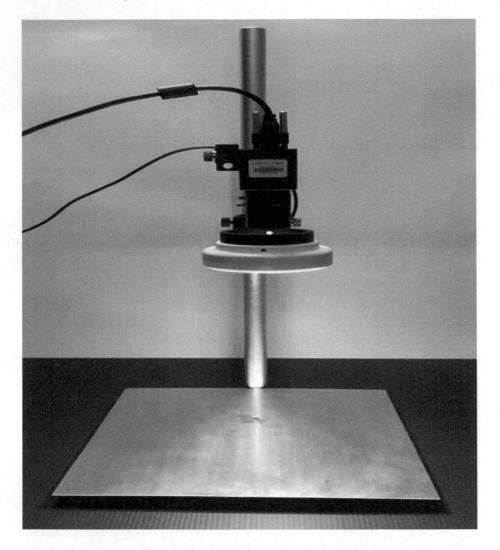

Fig. 5. The experimental environment establishment.

4   Experiment

4.1   Image Capture and Environment Establishment

Regarding image capture, this study used a single 5 megapixel CCD color camera with a 6–12 mm lens, a lighting system, and a macro frame to adjust the working distance to capture images of the stainless steel plates, as shown in Fig. 5. The width and length of the stainless steel plate surface of each cut sample were 150 mm and 150 mm, respectively. Compared with the object to be detected, the volume of the defects were smaller and more subtle. In addition, as the surface of the stainless steel plate was made of a mirror reflecting material, the image capture effect was difficult to control; therefore, a CCD camera with a high resolution of 5 megapixels was selected. In order to highlight the defects, the field of view (FOV) was reduced as far as possible in this study. Due to the relationship between the light source and the working distance (WD), a certain distance was required, thus, a 6–12 lens was adopted. The overall structure was constructed with aluminum alloy lifting brackets to obtain the appropriate working distance (20 cm) and the appropriate field of view.

4.2 Dataset

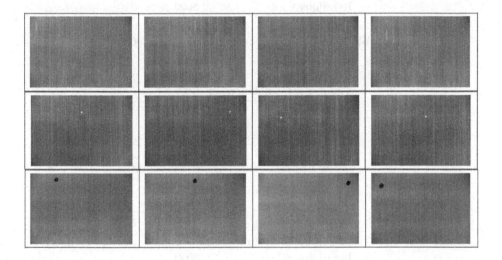

Fig. 6. Dataset.

This study purchased commercially available stainless steel plates, simulated various mechanical defects, and selected 3000 different images of stainless steel plates, including 1000 samples with scratches, 1000 samples with chisel marks, and 1000 samples with stains, for training and verification, of which 90% were training materials and 10% were verification materials, as shown in Fig. 6.

4.3 YOLOv3 Model Training and SSD 300 Model Training

This study examined three types of defects: scratch defects, chisel defects, and stain defects. The configurations of YOLOv3and SSD300 training parameters are shown in Tables 1 and 2, respectively.

Table 1. The training configuration of YOLOv3.

Parameters	Settings
Training set: validation set	9:1
Width of image	416
Height of image	416
Channels	3
Decay weight	0.005
Optimizer	Adam
Batch size	16
Learning rate	0.001
Iterations	3000

Table 2. The training configuration of SSD300.

Parameters	Settings
Training set: validation set	9:1
Width of image	300
Height of image	300
Channels	3
Decay weight	0.005
Optimizer	Adam
Batch size	16
Learning rate	0.001
Learning rate decay factor	0.94
Iterations	3000

4.4 Verification of the Classification Accuracy of the YOLOv3 Model

The experimental results are shown in Table 3. The detection rates of the various image tests were 97.8%, 96.7%, and 98.7%, respectively, among which, the over kill rate was less than 3%, and the under kill rates were 20.95%, 44.4%, and 21.3%, respectively. The average values of the under kill rates, over kill rates, and detection rates of the various surface images of stainless steel plates were 28.9%, 0.21%, and 97.7%, respectively. Figures 7, 8, and 9 show the experimental results of detecting the defects on the surfaces of the stainless steel plates, respectively. The diagrams circled in green, blue, and red are the correct judgment results of the method proposed in this paper, indicating that the model has certain stability in detecting defects.

Table 3. The result of YOLOv3 experiment.

Items	Scratch	Chisel	Stain	Average
Under kill rate	20.95%	44.4%	21.3%	28.9%
Over kill rate	0%	0.63%	0%	0.21%
Accuracy rate	97.8%	96.7%	98.7%	97.7%

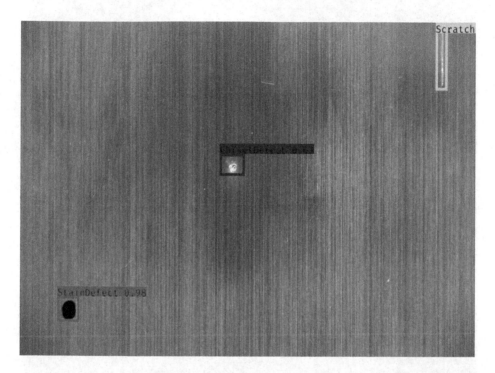

Fig. 7. The detection result of YOLOv3. (Color figure online)

4.5   Verification of the Classification Accuracy of the SSD 300 Model

The experimental results are shown in Table 4. The detection rates of the various image tests were 84.6%, 88.9%, and 90.7%, respectively, among which, the over kill rate was more than 0.3%, and the under kill rates were 70.31%, 71.9%, and 62.1%, respectively. The average values of the under kill rates, over kill rates, and detection rates of the various surface images of the stainless steel plates were 68.1%, 9.5%, and 88%, respectively. Figures 10, 11, and 12 show the experimental results of detecting the defects on the surfaces of the stainless steel plates, respectively. According to the experimental data, SSD300 is less stable than YOLOv3 in defect detection.

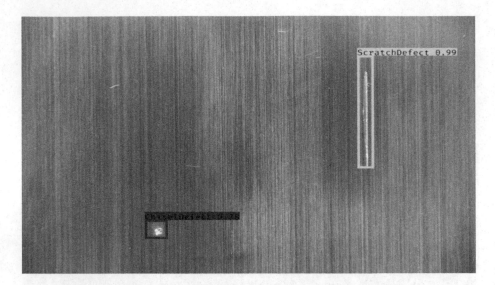

Fig. 8. The detection result of YOLOv3. (Color figure online)

Fig. 9. The detection result of YOLOv3. (Color figure online)

Table 4. The result of SSD300 experiment.

Items	Scratch	Chisel	Stain	Average
Under kill rate	70.31%	79.9%	62.1%	68.1%
Over kill rate	11.64%	9.1%	7.72%	9.5%
Accuracy rate	84.6%	88.9%	90.7%	88%

Fig. 10. The detection result of SSD300.

Fig. 11. The detection result of SSD300.

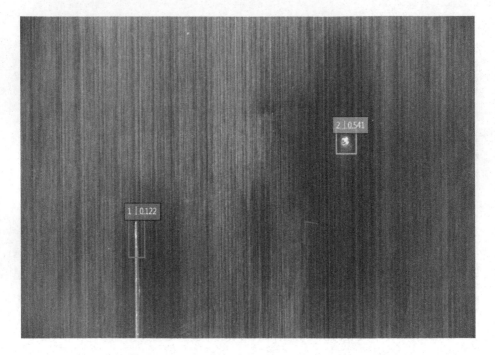

Fig. 12. The detection result of SSD300.

5   Conclusion

This paper proposed a set of stainless steel plate detection techniques to replace the current method of manual visual detection of stainless steel plate surface defects. The testing environment was established according to the sample size and material characteristics of stainless steel plates. Through deep learning technology, the stainless steel plates were judged as good or defective products, and the defective locations were marked. The detection system, as developed in this paper, can reduce the factory end visual inspection work, which reduce both labor and costs, in order to achieve the effect of real-time detection. In this study, the YOLOv3 model was used to judge the stainless steel plates to be detected, and solve the different defects of stainless steel plates. After 3000 samples of defective products were verified, the system detection rate was 97.7%. On the contrary, the detection rate of SSD 300 in the defect detection of stainless steel plates was 88%, which is slightly inferior, and mainly because the SSD 300 model is relatively poor in detecting and recognizing small objects, while YOLOv3 uses a lower-level anchor to train for detecting small objects. Through the simulation of environmental design, this paper defined the defect types of stainless steel plates; however, there are still more difficult and invisible defects, such as chisel defects. Therefore, our future study will add filters to reduce highlight interference and improve recognition accuracy.

References

1. Goodfellow, I., Bengio, Y., Courville, A.: Deep Learning. Adaption Computation and Machine Learning Series. MIT Press, Cambridge (2016)
2. Lecun, Y., Bottou, L., Bengio, Y., Haffner, P.: Gradient-based learning applied to document recognition. Proc. IEEE **86**(11), 2278–2324 (1998)
3. Redmon, J., Divvala, S., Girshick, R., Farhadi, A.: You only look once: unified, real-time object detection. In: Proceedings of the IEEE Conference on Computer Vision and Pattern Recognition, pp. 779–788 (2016)
4. Redmon, J., Farhadi, A.: YOLO9000: better, faster, stronger, arXiv preprint arXiv:1612.08242 (2016)
5. Redmon, J., Farhadi, A.: YOLOv3: an incremental improvement. arXiv arxiv:1804.02767 (2018)
6. Lin, T.Y., Goyal, P., Girshick, R., He, K., Dollar, P.: Focal loss for dense object detection. arXiv preprint arXiv:1708.02002 (2017)
7. Liu, W., Anguelov, D., Erhan, D., Szegedy, C., Reed, S.: SSD: single shot multibox detector. arXiv preprint arXiv:1512.02325 (2015)

Deep Neural Networks for Detecting Real Emotions Using Biofeedback and Voice

Mohammed Aledhari[1]([✉]), Rehma Razzak[1], Reza M. Parizi[1],
and Gautam Srivastava[2]

[1] Kennesaw State University, Marietta, GA 30060, USA
{maledhar,rparizi1}@kennesaw.edu, rrazzak@students.kennesaw.edu
[2] Department of Math and Computer Science, Brandon University,
Brandon, MB, Canada
srivastavag@brandonu.ca

Abstract. When people are in an interview, with the interview questions, people's emotions will change differently. Therefore, it is very helpful to detect people's emotions in real-time. To do so, comprehensive data collection was performed through the voice recording platform and the Empatica E4 wristband (biofeedback). Also, through using both existing feed-forward deep neural network technology and machine learning, we implemented an artificial deep neural network that aims to detect real emotions using multiple sensors: voice and biometrics. The artificial deep neural network we implemented consistently achieved an accuracy of 85% in our testing set and 79% in validation sets to determine the emotional scale. The research also assists with understanding how to detect emotional ranges and the important role that it plays in interviews and conversations.

Keywords: Emotion recognition · Machine learning · Biofeedback · Neural networks · Voice recording

1 Introduction

The brain is a complex organ that is not fully understood [4], and, with what is understood of it, there are still difficulties in getting machines to think or behave like humans, especially in regards to emotion recognition [12]. After all, emotions are states related to physiological responses and appear due to external or internal stimuli. Additionally, several different emotions can be distinguished from each other by facial expressions, and behavioral and physiological responses, with the most common being happiness, sadness, disgust, and others. Changes in physiological signals related to emotional states are involuntary, and people are often unaware of them. Therefore, physiological signal analysis can be a reliable method for emotion recognition. As such, automatic recognition of emotion has gained plenty of attention in recent years due to its high potential and numerous uses [8]. In our case, our project focuses on using emotion recognition for

© Springer Nature Switzerland AG 2021
A. Del Bimbo et al. (Eds.): ICPR 2020 Workshops, LNCS 12664, pp. 302–309, 2021.
https://doi.org/10.1007/978-3-030-68799-1_21

how people's emotions fluctuate during an interview. When people are in an interview, their emotions will change differently. Therefore, it is very beneficial to detect people's emotions in real-time. Detecting people's real-time emotions can reduce the occurrence of embarrassing situations, eliminate the ambiguity caused by participants' questions and answers, and achieve better information transmission functions and people's emotional comfort [5].

However, to achieve successful emotion recognition, we need to collect relevant data. Collecting relevant data has always been a very difficult step, and the data for this survey required researchers to obtain different types of interviews [11]. During the collection of experimental data, interviews are the most important part. Different forms of dialogue, different themes of content, and obvious changes in emotions are the most important contents for experimental investigators and participants. However, because the form of interviews is very difficult to control, many errors can make the collection of experimental data meaningless. For example, the responses and responses of participants are not obvious and very vague. To solve the aforementioned problems, help experimental investigators and participants reduce the inconsistency and ambiguity of the interview content, we chose to use the Empatica E4 wristband. We chose this approach because the Empatica E4 wristband can record people's different feedback and reactions to emotions in real-time. We also combine our usage of the Empatica E4 wristband with machine learning to convey the emotional ranges of participants from requirements elicitation interviews. In this regard, the main questions raised for this real-time monitoring sentiment experiment are: *(1) Regarding this experiment, which technology in machine learning is the most effective and can bring the greatest development to the experiment? (2) Regarding this experiment, what kind of function in machine learning can be maximized in the experiment?*

Our research uses deep ANN to convey emotional ranges of participants from requirements elicitation interviews. Our research is important because not many research uses biofeedback to detect emotion, instead, they use images or videos. Our research could also help understand what certain data looks like that portrays different emotional ranges. A model of our research solution is seen in Fig. 1.

1.1   Paper Goals and Organization

The goal of the deep neural network is to build a model that can predict emotions based on biometric inputs. We are looking to initially prove that it is possible to use a neural network to predict emotions. Following that is the improvement of the model to work in a real-time condition. The last point is that we look to build a deep CNN to determine the emotions of the recorded voice to increase the detected confidence in the future. Currently, we are meeting our initial goal of being able to prove that it is possible to use a neural network to predict emotions. Based on our dataset, we are reliably receiving greater than sixty percent accuracy on our training sets. This is a correlation between our datasets and the emotional scale which leads us to believe better accuracy may be achieved with more testing.

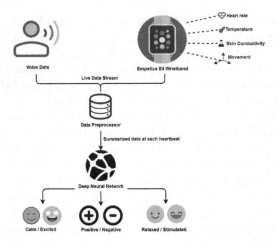

Fig. 1. The model shows the deep neural network processing data from the E4 wristband and the voice recordings in order to achieve the emotional ranges.

Following this introduction, the literature review of related works is given in Sect. 2. The methodology of the proposed work is discussed in Sect. 3. Section 4 discusses the experiment and results of the proposed work, while the conclusion is presented in Sect. 5.

2   Related Works

You learn to express your emotions long before you learn how to convey them in words. For example, as a baby, you cry when you want food instead of asking for some. Human emotion recognition is evident from the face of a human being, though certain assessments can give more confident levels of emotion. For a machine, the best way to infer the emotion is to have a combination of the most basic and bottom-level emotion detection, advanced emotion prediction, and very natural emotion pattern adjustment.

Some current methods that are used to solve emotion detection problems are classification algorithms and deep learning algorithms such as ove-vs.-rest(OVR), support vector machines (SVM), Naive Bayes (NB), K-nearest neighbor (KNN), multilayer perceptron (MLP), long-short-term-memory (LSTM), and convolutional neural network (CNN) [10]. While other methods can also be used, we are using a deep artificial neural network (ANN). ANNs are supposed to work like the human brain where the brain's synapses and neurons are represented by nodes and layers [6].

Another example comes from the work of [3], where the authors propose a real-time mobile biofeedback system to depict five basic emotions and provides the user with emotional feedback. The authors presented their empirical results for the implementation of a physiological signal-based emotion recognition system in two scenarios involving controlled and non-controlled environmental settings. Additionally, the authors selected twenty participants to take part in their experiment. The authors' proposed system is called iAware, which also uses the Empatica E4 wristband. In the authors' results, they successfully demonstrated that iAware helps increase emotional self-awareness by reducing the predictive error by 3.333% for women and 16.673% for men. These impressive results effectively demonstrate the applicability of such work and can have multiple use-cases in other fields, such as special education.

It is important to be able to recognize physiological emotion because there are important applications of it in areas such as mental illness, social communication, and the connection between humans and machines [7]. Concerning the research we are conducting, we agree with these researchers on the basis that detect and monitor signs of people's emotional changes, so you can understand what happens when participants and machine learning technologies are combined [9]. That is why creating this machine learning prototype would be beneficial to integrate with supportive hardware equipment used by demand analysts in elicitation interviews to collect better requirements. The biofeedback was used to give emotional ranges from the image surveys that the participants took. With the voice recording data, we would have no way of ensuring a certain emotion because the interview data collected had no such implications.

3   Methodology

3.1   Performing the Regression

For this work, the system data flow begins when the individual variables are measured by the wristband and written to the data files. One part of the program continually reads the IBI file and waits for an interviewee's heartbeat to be detected, while another part consolidates the individual data files as discussed in the techniques section. As soon as a heartbeat is detected, the program checks whether the indicated interval is valid or misdetected; we used a heuristic approach, declaring an interval valid if it consisted of less than 100 BVP measurements. For valid ranges, the program organized the data as discussed in the techniques section and then passed the information to the neural network.

3.2   Building the Deep ANN

We implemented our deep neural network using python and the TensorFlow [2] and Keras [1] libraries. Other libraries were used to include features such as loading .csv files, handling data, and plotting which will be discussed. Because of certain dependencies with Tensorflow, this work will only work with Python version 3.6 or older.

To begin building our deep neural network, we had to determine how our data was stored. In our case, it was stored in a CSV file with 19 features and 3 output labels. We needed to then remove any data with missing labels, which we completed by removing all data which did not have labels. This reduced our overall data to over 10,000 rows from our original dataset. We used the python library NumPy to store this data into feature and label matrices. The feature matrix had a shape of M × 19 and a y label matrix of M × 3 where M is the number of data points.

We normalized our data using MinMax scaling. This simple normalization technique takes the current value - minimum value/maximum value - minimum value over all the columns in our dataset. After all of this data setup, we were able to load everything into our neural network.

We used the Keras compile function to add in our optimization and loss functions and then began to train our data on the loaded X and y values. In neural networks and optimization function increases in accuracy based on the number of epochs, in our compilation, we set the epochs based on the constructor input, which we found best set to 45. This is also where we split our data into the training and test sets of 80–20. We then evaluate our model to see it's accuracy and loss using the Keras evaluate function.

To visualize our model we used the pyplot package from Matplotlib. The Keras evaluate function returns a history of the training and test losses and accuracies. We used this to build our visualizations to see how the optimizer and loss function worked with the given inputs. These visualizations are shown in the results section of our paper.

4   Experiment and Results

4.1   Techniques Used

We used a feed-forward deep neural network, which is a supervised network with a single input layer, multiple hidden layers, and a single output layer.

We used TensorFlow to train and fit our data to the Neural Network. Tensorflow with Keras is effective for this because it allows developers to easily train and fit their data while abstracting the mathematical intricacies involved. The model for our work is based upon the fully connected neural network above. The details of the model will be explained in our methodology section. The advantage of using this deep feed-forward neural network is that we can expect higher accuracies than in other machine learning methods. Also, because we have a large amount of data (>10k labeled entries), the neural network will be able to perform most optimally compared to others.

The data for this work was obtained from biofeedback wristbands, which measured acceleration in the x, y, and z directions, body temperature, electrodermal activity (EDA), and blood volume pulses (BVP). Using the BVP data, the wristband also extracted the timestamps of individual heartbeats(IBI) and average heart rate (HR). The wristband measured these variables at wildly different frequencies and saved them to different files, so we had to employ a data-merging technique to combine these into usable datasets. Each data file begins with an initial timestamp and the sensor rate, so the data merging program was able to calculate intermediate timestamps for each datum. Finally, the program merged the individual data, leaving a single consolidated file with all recorded data for that specific interviewee.

4.2   Results

Initially, our results were based upon a neural network designed to classify amongst 100 labels matched to the one where our value was. We found that there was low accuracy and could not find a justifiable correlation between the features and the outputs.

The results of proposed model are: *training loss* is 4.5009, *training accuracy* is 1.0000, *testing loss* is 4.5515, and *testing accuracy* is 0.8511. These numbers show how accurately the test set performs correctly on the model. In particular, the loss function is a value determined by our sparse categorical cross-entropy function, which we are trying to minimize.

We have since improved upon these results by changing to a deep neural network with a regression output instead of the classification. The results of our new neural network are improved with a 79% *accuracy* on the validation set when using mean squared error as our loss function. These results we believe do show a correlation between the features and the data. We also tested different lost functions, drop rates, and optimization functions to determine which had the best outputs for our graph.

The loss function seemed to have the most effect on our results. Since we were using a regression output, the logical choice of functions was the mean squared error, and the mean absolute error can be seen in Fig. 2. We found that mean squared error to give more consistently accurate results. The best results were achieved using the Leaky Relu function for our hidden activation layers, which consisted of 5 hidden layers with 35 hidden units each. The Adam optimizer was applied since it has a mean squared error function, making it optimal for our experiment. Our dropout rate was 0.3, which was used as our default state.

We later determined to test if the number of epochs was too low and that if we had a higher amount of epochs, the optimizer would run longer and be able to find better correlations. Figure 2 gives an example of the results from those tests. These results show that after a certain percentage, the neural network sees no further improvements. Due to Keras having many built-in optimization functions which include adam, nadam, adamax, rmsprop, and more, We tested these out to see which optimization function proved to be the best for our neural network. Figure 2 illustrates these results.

One significant issue with the heartbeat-centric system design is that the wristband does not always measure IBI events accurately. For some of the datasets, more than 80% of the data is misdetected; that is, the IBI events are not labeled. Because of this, we added a data pre-processing step that takes these misdetected ranges and attempts to isolate the individual heartbeats within them. Our first approach involved finding the average length of a heartbeat and dividing the misdetected range equally into that size, but this resulted in highly

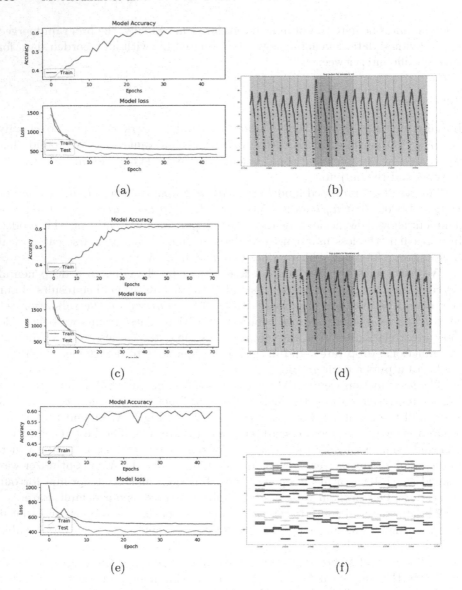

Fig. 2. Multimodal emotion recognition using variety of advanced hyperparameters part 1. (a) Model using mean squared error loss function; (b) Model using mean absolute error loss function; (c) Model using larger epochs (70 vs 45); (d) Model using adam; (e) Model using nadam; (f) Model using adamax.

unreliable results: heartbeats were not correctly divided and the regressed coefficients between heartbeats varied wildly, whereas the correctly identified ranges have smooth coefficient transitions. So, the goal became to isolate heartbeat events with smooth coefficient transitions that were successfully implemented.

5 Conclusion

In conclusion, we have been able to build a deep neural network and feed data into it. While further improvement could be made for our model, we believe this model can be highly applicable for several use-cases. Additionally, this model could benefit future researchers who are interested in this topic by providing sufficient guidelines for using emotion-recognition for interviews. We have provided detailed insights regarding our methodology, experiment, and results. Furthermore, using multiple sensor data that are biofeedback and voice for emotion-recognition would be more beneficial than images and video since biofeedback accounts for subtle changes in an individual, which can not be obtained through images or videos.

References

1. Keras api reference (2015). https://keras.io/api/
2. Abadi, M., et al.: Tensorflow: large-scale machine learning on heterogeneous distributed systems. arXiv preprint arXiv:1603.04467 (2016)
3. Albraikan, A., Hafidh, B., El Saddik, A.: iAware: a real-time emotional biofeedback system based on physiological signals. IEEE Access **6**, 78780–78789 (2018)
4. Bear, M., Connors, B., Paradiso, M.A.: Neuroscience: Exploring the Brain. Jones & Bartlett Learning, LLC, Burlington (2020)
5. Domínguez-Jiménez, J.A., Campo-Landines, K.C., Martínez-Santos, J., Delahoz, E.J., Contreras-Ortiz, S.: A machine learning model for emotion recognition from physiological signals. Biomed. Signal Process. Control **55**, 101646 (2020)
6. Kwon, S.J.: Artificial Neural Networks. Nova Science Publishers, New York (2011)
7. Marechal, C., et al.: Survey on AI-based multimodal methods for emotion detection (2019)
8. Ng, H.W., Nguyen, V.D., Vonikakis, V., Winkler, S.: Deep learning for emotion recognition on small datasets using transfer learning. In: Proceedings of the 2015 ACM on International Conference on Multimodal Interaction, pp. 443–449 (2015)
9. Picard, R.W., Vyzas, E., Healey, J.: Toward machine emotional intelligence: analysis of affective physiological state. IEEE Trans. Pattern Anal. Mach. Intell. **23**(10), 1175–1191 (2001)
10. Sharifirad, S., Jafarpour, B., Matwin, S.: How is your mood when writing sexist tweets? detecting the emotion type and intensity of emotion using natural language processing techniques. arXiv preprint arXiv:1902.03089 (2019)
11. Spoletini, P., Brock, C., Shahwar, R., Ferrari, A.: Empowering requirements elicitation interviews with vocal and biofeedback analysis. In: 2016 IEEE 24th International Requirements Engineering Conference (RE), pp. 371–376. IEEE (2016)
12. Steppan, M., Fürer, L., Schenk, N., Schmeck, K., et al.: Machine learning facial emotion recognition in psychotherapy research: a useful approach? (2020)

Data Augmentation for a Deep Learning Framework for Ventricular Septal Defect Ultrasound Image Classification

Shih-Hsin Chen[1], I-Hsin Tai[2], Yi-Hui Chen[3,4(✉)], Ken-Pen Weng[5], and Kai-Sheng Hsieh[6]

[1] Department of Information Management, Cheng Shiu University, Kaohsiung City 83347, Taiwan (R.O.C.)
shchen@csu.edu.tw
[2] Department of Pediatric Cardiology, China Medical University Chidren's Hospital, Taichung City 40447, Taiwan (R.O.C.)
ifaithgrace@icloud.com
[3] Department of Information Management, Chang Gung University, Taoyuan 33302, Taiwan (R.O.C.)
cyh@mail.cgu.edu.tw
[4] Kawasaki Disease Center, Kaohsiung Chang Gung Memorial Hospital, Kaohsiung 83301, Taiwan (R.O.C.)
[5] Congenital Structural Heart Disease Center, Department of Pediatrics, Kaohsiung Veterans General Hospital, No. 386, Dazhong 1st Road, Zuoying District, Kaohsiung City, Taiwan (R.O.C.)
kpweng@vghks.gov.tw
[6] Department of Pediatrics, Shuang-Ho Hospital-Taipei Medical University, New Taipei City 23561, Taiwan (R.O.C.)
kshsieh@hotmail.com

Abstract. Congenital heart diseases (CHD) can be detected through ultrasound imaging. Although ultrasound can be used for immediate diagnosis, doctors require considerable time to read dynamic clips; typically, physicians must continuously examine disease data from beating heart images. Most importantly, this type of diagnosis relies heavily on the expertise and experience of the diagnosing physician. This study established an ultrasound image classification with deep learning algorithms to overcome the challenges involved in CHD diagnosis. We detected the most common CHD, namely the first, second, and fourth types of ventricular septal defect (VSD). We improved the performance levels of well-known deep learning algorithms (InceptionV3, ResNet, and DenseNet). Because algorithm optimization and overfitting problems can influence the performance of deep learning algorithms, we studied some optimizer algorithms and early-stopping strategies. To enhance the solution quality, we used data augmentation methods for solving this classification problem. The selected approach was further compared with Google AutoML, which applies structure search for quality prediction. Our results revealed that the proposed deep learning algorithm was able to recognize most types of VSD. However, one type of VSD remains unconquered and warrants more advanced techniques.

© Springer Nature Switzerland AG 2021
A. Del Bimbo et al. (Eds.): ICPR 2020 Workshops, LNCS 12664, pp. 310–322, 2021.
https://doi.org/10.1007/978-3-030-68799-1_22

Keywords: Ventricular septal defect (VSD) · Echo · Deep learning · Classification · Data augmentation

1 Introduction

Due to the advance in science and technology, high-level ultrasound can detect congenital heart disease (CHD) in a fetus aged from 18 to 22 weeks. Early detection is desirable so that the corresponding medical treatment can be administered at the earliest [5]. Compared with non-Taiwanese studies on the incidence of CHD, most Taiwanese studies have been limited to statistical investigations in various hospitals. Wu et al. [19] queried Taiwan's health insurance database to obtain seven years of comprehensive data; they found that on average, 13/1000 newborn male infants and 14/1000 female infants have CHD each year in Taiwan, and further discovered that approximately 10 to 13 per 1,000 are simple heart disease. Similarly, Yeh et al. [20] analyzed the same health insurance database and calculated the 5-year survival rate of CHD to be approximately 95%, and the relative mortality rate to be 5%.

One of the mainstream methods for diagnosing heart disease is to use ultrasound examinations, which have benefits of no radioactivity, a low burden on newborns, and low cost [1,6,16]. Cardiac ultrasound scanning can be divided into black-and-white imaging and color Doppler ultrasound dynamic imaging. Black-and-white ultrasound is mainly used to observe the composition of the heart. However, color Doppler ultrasound can detect the flow and velocity of blood; in Fig. 1, the portions colored in red indicate blood flow in the direction of the ultrasonic probe, whereas blue portions depict blood flow away from the ultrasonic probe. The brighter the color is, the higher the blood flow velocity is. Detection of an abnormal condition [13] become easy with such convenient methods of diagnosis. Whereas Fig. 1a present little information, Fig. 1b represents more information. If blood flow in the heart chambers or the middle does not occur, it usually means that the atrium or ventricle has a hole (defect) or the tissue has other abnormalities. This study employed color Doppler ultrasound images.

Although ultrasound imaging is a key tool for doctors, judgement of such imaging entails manifold challenges, including the expertise, unclear images, and noisy signals [15]. Because scholars have published many breakthroughs in deep learning (DL) algorithms [9–11,18], various DL algorithms can be used in speech recognition, visual image recognition, image target analysis, new drug development and genetic testing, and other fields. A study [12] reviewed some applications of DL algorithms. In the past 3 years, some studies have used DL algorithms for detecting heart disease. In particular, scholars have published seven studies devoted to cardiac ultrasound [2–4,6,8,13,14], but only two articles have considered CHD [2,17] because the medical challenges involved in the detection of CHD are markedly complicated. Current scholarship on CHD has considerable room for improvement.

Because DL algorithm are rarely used in CHD papers [2], this study is a pioneer in CHD analysis. Wu et al. [19] reported that in Taiwan, ventricular

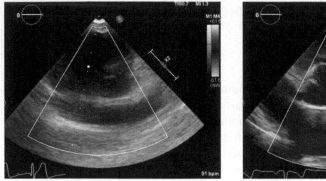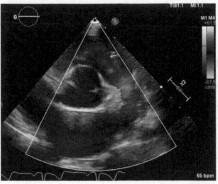

(a) No blood flow detection   (b) Doppler image in short axis

Fig. 1. Two samples of normal status

septal defect (VSD) has 401 cases per 100,000 newborns. Because VSD is the most dangerous CHD, we studied VSD and its three sub-types. We also studied whether a DL algorithm can help accurately classify the three types.

The rest of this paper is organized as follows. Section 2 depicts the studied problem and then presents methods to enhance the prediction quality, including data augmentation and tuning of the methods or parameters of each DL algorithm. In Sect. 3.1, we present the parameter configuration obtained using regression models. We conducted an experiment with favorable parameters and optimized methods. Section 5 presents a comparison of a selected DL algorithm with Google AutoML. The conclusions are presented in Sect. 5.

2 Methods

Although the main aim of this study was to demonstrate that our DL algorithm can solve the VSD classification problem, defining the three sub-types of VSD is essential. The studied types are presented in Sect. 2.1. We describe some key DL algorithms with the data augmentation approach.

2.1 Problem Definition

Three types of VSD, namely Type 1, Type 2, and Type 4, were studied. When doctors seek to observe VSD Type 1, they should apply the parasternal short-axis view. The aortic (AO) valve is visible at the center of the ultrasound image. If AO cannot be viewed clearly, doctors may not use this figure. If the hole blood flow is between 11 and 1 o'clock, it may be the first type. We present some selected images of VSD Type 1 in Fig. 2a.

Type 2 VSD can also be observed with the short axis view. The AO is centered in the ultrasound image in this type too. If the hole blood flow is between 9 and

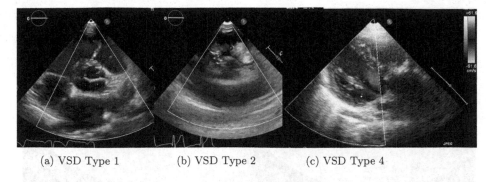

(a) VSD Type 1   (b) VSD Type 2   (c) VSD Type 4

Fig. 2. Doppler ultrasound images of VSD Type 1, Type 2, and Type 4

11 o'clock, it may be the second type. We present the diagnosis of three patients with VSD Type 2 in Fig. 2b.

Finally, when physicians seek to identify Type 4 VSD, they examine the four-chamber view, which is different from the short axis used for Type 1 and Type 2 VSD. They seek Doppler flow in the muscular portion, which means near the apex. Several positions could demonstrate the characteristic of VSD Type 4. Hence, we expect that this type is difficult to recognize. Consider Fig. 2c as an example; we can observe a spot in the 12 o'clock positions. Because various position might be possible, DL algorithms or other approaches might not be efficient in classifying this type.

2.2 Data Augmentation Approaches

We employed InceptionV3 [18], ResNet [10], and DenseNet [7] as backbone algorithms. Our code used 50 and 121 layers for ResNet and DenseNet, respectively. We validated some well-known optimizers, such as Adadelta, Adagrad, Adam, Rmsprop, and SGD. In addition, four suitable data augmentation techniques were employed. A pictorial example of the selected method is presented in Fig. 3.

3 Empirical Results of Data Augmentation

We collected 17 ultrasound videos of eight patients from Kaohsiung Chang Gung Medical Hospital. We extracted each video frame into static images in PNG images and then removed the personal information from the images. From all the images, we handpicked the figures from three categories: Normal, VSD Type 1, and VSD Type 2. VSD Type 4 was not considered in this stage. We considered 150, 59, and 41 images for Normal, VSD Type 1, and VSD Type 2, respectively. These figures were further divided into training, validation, and testing sets, and Google AutoML was employed. The data set arrangement is presented in Fig. 4.

We applied ImageAI to code the InceptionV3, ResNet, and DenseNet frameworks. Because the possible combinations of each factor are many and corresponding computation would have been burdensome, the three algorithms

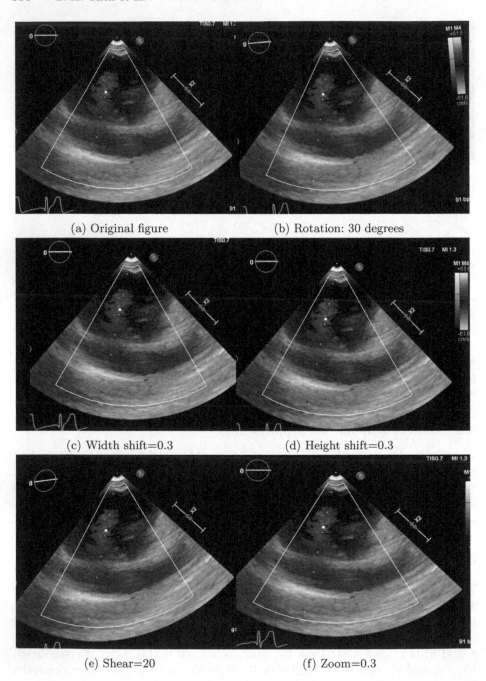

(a) Original figure   (b) Rotation: 30 degrees

(c) Width shift=0.3   (d) Height shift=0.3

(e) Shear=20   (f) Zoom=0.3

Fig. 3. Data augmentation for the ultrasound images

selected the optimizer and the parameters of the five data augmentation methods randomly for 120 sessions. Table 1 lists the possible methods and settings of the selected factors. Each algorithm ran 150 epochs together with an early-stop strategy preventing excessive computation time and overfitting. If 75 epochs do not improve the accuracy, the algorithm stopped the computation. We ran the algorithms in Tensorflow 1.11 environment with two nVidia RTX 2080 GPUs.

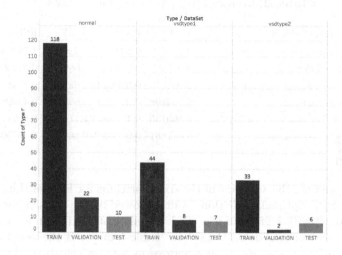

Fig. 4. Data set arrangement

Table 1. Feature and weight values in the regression model

Factor	Weight
Optimizer	Adadelta, Adagrad, Adam, Rmsprop, SGD
Height shift	0, 0.25, 0.5, 0.75, 1
Rotation	0, 30
Shear range	0, 0.15, 0.3
Width shift	0, 0.25, 0.5, 0.75, 1
Zoom	0, 0.15, 0.3, 0.45

3.1 Comparison of the Regression Models

We employed six well-known regression models for accuracy predictions, including Bayesian linear regression, decision forest regression, linear regression, and others. Among 120 records, 70% was used to train the regression models, and the remaining 30 records were used to evaluate the prediction accuracy. Four major indices were applied, namely the mean absolute error (MAE), root mean squared error (RMSE), relative absolute error (RAE), and relative squared error

(RSE). In general, the smallest possible values of the four indices are preferred. Table 2 lists the performance of the six regression models for InceptionV3. The decision forest regression exhibited the most accurate results, followed by the linear regression model. Bayesian linear regression was the least favorable model for InceptionV3.

Table 2. Regression models for the InceptionV3

Regression models	MAE	RMSE	RAE	RSE
Bayesian linear regression	0.110753	0.13145	1.228904	1.284042
Boosted decision tree regression	0.062107	0.085442	0.689128	0.542506
Decision forest regression	0.05651	0.075609	0.627033	0.424815
Linear regression	0.059498	0.077479	0.660186	0.446093
Neural network regression	0.100135	0.11018	1.111089	0.902116
Poisson regression	0.077881	0.08646	0.864158	0.555505

Table 3 presents the accuracy of the six algorithms for ResNet. Linear regression achieved the optimal prediction result, followed by the decision forest regression. Neural network regression was the least efficient prediction algorithm. The same result was obtained for DenseNet. Table 4 presents the accuracy of the six algorithms for DenseNet. Linear regression was proven to be the most efficient prediction algorithm, and neural network regression was the least efficient prediction algorithm.

Table 3. Regression models for the ResNet

Regression models	MAE	RMSE	RAE	RSE
Bayesian linear regression	0.086334	0.104971	0.703997	0.597123
Boosted decision tree regression	0.053632	0.066149	0.437331	0.237124
Decision forest regression	0.044944	0.054169	0.36649	0.159008
Linear regression	0.039802	0.049498	0.324561	0.132771
Neural network regression	0.090846	0.101825	0.740787	0.561866
Poisson regression	0.061643	0.07123	0.502662	0.274948

These comparisons revealed that the linear regression model is promising for the three benchmark algorithms. Predicting the accuracy of the DL algorithm is easy when infinite combinations of parameters are available. Moreover, the weight values could be interpreted to obtain more information. We present the interpretation of the weights in the following.

Table 4. Regression models for the DenseNet

Regression models	MAE	RMSE	RAE	RSE
Bayesian linear regression	0.0988	0.129884	0.549549	0.379376
Boosted decision tree regression	0.064489	0.107466	0.358705	0.259716
Decision forest regression	0.055332	0.096903	0.307768	0.211168
Linear regression	0.054489	0.091246	0.303079	0.187233
Neural network regression	0.103416	0.153845	0.575223	0.532262
Poisson regression	0.075248	0.113912	0.418544	0.291808

3.2 Weight Values of Linear Regression Model

In Table 5, we list the corresponding weight of the linear regression model for three well-known DL algorithms. A positive weight denotes high accuracy of the feature. Moreover, to select the most accurate optimization algorithm, we selected the highest weight value obtained from the five optimization algorithms for each DL framework. The selected method for each DL framework is indicated by symbolic marks in Table 5.

Consider InceptionV3, for example; because we only used one optimizer, the most accurate algorithm for InceptionV3 was Adam because it had the highest weight value. Adadelta yielded the least accurate result because its weight value was negative. In terms of the data augmentation results, height shift, rotation, shear, and zoom demonstrated positive effects. Because InceptionV3 presented an adverse effect, width shift was not useful. Based on the results of the data augmentation, we used a zoom value of 0.45, height shift of 0.75, shear of 0.15, and 30 rotations.

For ResNet, Adam yielded the highest accuracy, whereas Adadelta had the lowest. Data augmentation methods were not useful in enhancing the performance of ResNet. For DenseNet, we selected Adagrad as the optimizer. However, only the rotation yielded a positive result; we used 15 rotations for DenseNet.

Table 5. Linear regression model for the InceptionV3, ResNet, and DenseNet

InceptionV3		ResNet		DenseNet	
Adadelta	−0.087335	Adadelta	−0.0723572	Adadelta	−0.113673
Adagrad	0.211924	Adagrad	0.235464	Adagrad*	0.288506
Adam*	0.259773	Adam*	0.272921	Adam	0.273938
rmsprop	0.179444	rmsprop	0.236486	rmsprop	0.256848
SGD	0.0855492	SGD	0.0128875	SGD	−0.0506198
height_shift_range*	0.0307315	height_shift_range	−0.022101	height_shift_range	−0.000724389
rotation_range*	4.64182E-05	rotation_range	−2.23799E-05	rotation_range*	0.000409661
shear_range*	0.0205741	shear_range	−0.0668904	shear_range	−0.0186237
width_shift_range	−0.0625221	width_shift_range	−0.018865	width_shift_range	−0.0320386
zoom_range*	0.0604625	zoom_range	−0.0664451	zoom_range	−0.0299814
Bias	0.649355	Bias	0.685401	Bias	0.654998

3.3   Comparisons of the Three DL Algorithms

After selecting the optimizer and setting the appropriate parameter configurations of the data augmentation methods, we replicated the three algorithms on the same data set for 30 runs. We list the average accuracy result and the saved model size in Table 6. In terms of average accuracy, DenseNet was demonstrated to be the best algorithm, followed by ResNet. However, ResNet offers the largest model size, whereas the model size of DenseNet is approximately 28 MBs. Hence, DenseNet might be more suitable for solving the CHD ultrasound image classification problem.

Table 6. Average accuracy of the compared algorithms

Algorithm	N	Minimum	Mean	Maximum
InceptionV3	30	0.86667	0.90667	0.96667
ResNet	30	0.83333	0.89667	0.96667
DenseNet	30	0.9	0.96111	1

Table 7 lists the average epoch of each algorithm. Based on the early-stopping criterion, the algorithm run was stopped when the system exhibited no improvement for 75 epochs. In other words, DenseNet might converge near 40, and InceptionV3 required 70 epochs. Therefore, DenseNet does not need extensive computational effort during model training. Because of the high solution quality of DenseNet, it required few epochs to converge, thereby making it suitable for use in ultrasound image classification.

Table 7. Average epoch of the compared algorithms

Algorithm	N	Minimum	Mean	Maximum
InceptionV3	30	105	142.07	150
ResNet	30	104	124.23	150
DenseNet	30	101	114.8	150

4   Empirical Results of All VSD Type for a Larger Problem Size

We obtained 45 ultrasound videos from Kaohsiung Veterans General Hospital. The videos were converted to PNG figures and de-anonymized. The total number of images was 1390. The numbers of figures for Normal, VSDType1, VSDType2, and VSDType4 were 930, 80, 142, and 138, respectively. These figures were divided into training, validation, and testing categories. The image arrangement is depicted in Fig. 5.

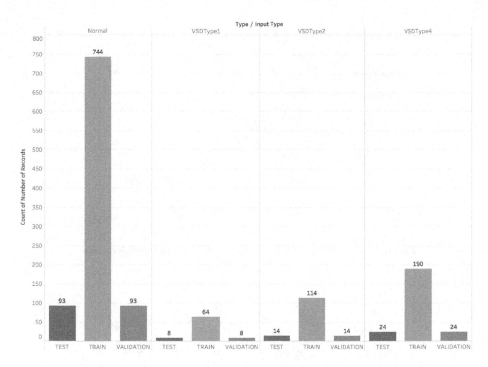

Fig. 5. Larger data set of the normal and three VSD types

Because experimental results revealed DenseNet to be the most accurate algorithm, we compared DenseNet with Google AutoML. DenseNet ran for 150 epochs with an early-stopping criterion; it stopped at 83 epochs. Figure 6 presents the confusion matrix of DenseNet. The accuracy in determining the Normal type was excellent. However, the result for VSDType4 was markedly unsatisfactory. Evidently, the recognition of Type 4 VSD could be improved. The overall accuracy, precision, and recall of DenseNet were 85%, 92%, and 75%, respectively.

Subsequently, we employed Google AutoML to solve the classification problem. The training time was set as 6 h, and we selected the highest prediction quality in the edge computing model. Figure 7 illustrates the confusion matrix of Google AutoML after 6 h of training. The performance of Google AutoML in classifying Normal, VSDType1, and VSDType2 was acceptable; in particular, the classification of VSDType1 was the most accurate. However, Google AutoML could not accurately classify VSDType4. This might be because DL algorithms present difficulty in classifying VSDType4. The corresponding average accuracy, precision, and recall are 88%, 91%, and 85%, respectively, which were better than the values obtained for DenseNet. Hence, we considered Google AutoML to be a good benchmark for the study of CHD ultrasound image classification.

Fig. 6. Confusion matrix of DenseNet for a large problem size

True Label \ Predicted Label	Normal	VSDType1	VSDType2	VSDType4	Recall
Normal	98%	1%	-	-	99%
VSDType1	11%	89%	-	-	89%
VSDType2	20%	-	80%	-	80%
VSDType4	66%	1%	-	33%	33%
Precision	82%	86%	100%	100%	

Fig. 7. Confusion matrix of Google AutoML for the larger data set

True Label \ Predicted Label	Normal	VSDType1	VSDType2	VSDType4	Recall
Normal	97%	-	1%	2%	97%
VSDType1	-	100%	-	-	100%
VSDType2	7%	-	93%	-	93%
VSDType4	50%	-	-	50%	50%
Precision	87%	100%	93%	86%	

5   Conclusions

Few researchers have studied the performance of deep learning frameworks, optimization algorithms, and data augmentation methods in CHD ultrasound image classification. In this study, we compared the well-known algorithms InceptionV3, ResNet, and DenseNet. Data augmentation is not necessarily suitable for this problem and should be used with care, if at all. Although data augmentation might be useful in some other image classification problems, some methods cause negative results when we utilize the linear regression to discover the relationship between the data augmentation method and classification accuracy.

After selecting the most suitable optimization algorithm and parameter settings for each DL framework, we replicated the three selected algorithms for 30 runs. We found that Adam exhibited higher performance for InceptionV3 and

ResNet and was slightly lower performance for DenseNet than Adagrad. In terms of the classification accuracy on average, DenseNet was superior to the others. In addition, the model size of DenseNet was approximately 28 MBs for solving the CHD ultrasound classification problem. Thus, DenseNet is recommended for solving the CHD ultrasound classification problem. We also evaluated an improved DenseNet with Google AutoML. Google AutoML was more efficient for larger problems, although both algorithms failed to accurately classify VSD Type 4. Hence, future research should investigate methods to improve the classification results.

Acknowledgments. The data used in this study are restricted by the Research Ethics Review Committee of the Kaohsiung Veterans General Hospital with the number 19-CT8-10(190701-2) to protect participant privacy. We thank the Ministry of Science and Technology for supporting this research with ID MOST 107-2221-E-230-007 and MOST 108-2221-E-230-004.

References

1. Avendi, M., Kheradvar, A., Jafarkhani, H.: A combined deep-learning and deformable-model approach to fully automatic segmentation of the left ventricle in cardiac mri. Med. Image Anal. **30**, 108–119 (2016)
2. Bridge, C.P., Ioannou, C., Noble, J.A.: Automated annotation and quantitative description of ultrasound videos of the fetal heart. Med. Image Anal. **36**, 147–161 (2017)
3. Carneiro, G., Nascimento, J.C.: Combining multiple dynamic models and deep learning architectures for tracking the left ventricle endocardium in ultrasound data. IEEE Trans. Pattern Anal. Mach. Intell. **35**(11), 2592–2607 (2013)
4. Carneiro, G., Nascimento, J.C., Freitas, A.: The segmentation of the left ventricle of the heart from ultrasound data using deep learning architectures and derivative-based search methods. IEEE Trans. Image Process. **21**(3), 968–982 (2012)
5. Carvalho, J., et al.: Isuog practice guidelines (updated): sonographic screening examination of the fetal heart. Ultrasound Obstet. Gynecol. **41**(3), 348–359 (2013)
6. Chen, H., Zheng, Y., Park, J.-H., Heng, P.-A., Zhou, S.K.: Iterative multi-domain regularized deep learning for anatomical structure detection and segmentation from ultrasound images. In: Ourselin, S., Joskowicz, L., Sabuncu, M.R., Unal, G., Wells, W. (eds.) MICCAI 2016. LNCS, vol. 9901, pp. 487–495. Springer, Cham (2016). https://doi.org/10.1007/978-3-319-46723-8_56
7. Gao, H., Liu, Z., van der Maaten, L., Weinberger, K.Q.: Densely connected convolutional networks. In: Proceedings of the IEEE Conference on Computer Vision and Pattern Recognition (2017)
8. Ghesu, F.C., et al.: Marginal space deep learning: efficient architecture for volumetric image parsing. IEEE Trans. Med. Imag. **35**(5), 1217–1228 (2016)
9. He, K., Zhang, X., Ren, S., Sun, J.: Delving deep into rectifiers: surpassing human-level performance on imagenet classification. In: Proceedings of the IEEE International Conference on Computer Vision, pp. 1026–1034 (2015)
10. He, K., Zhang, X., Ren, S., Sun, J.: Deep residual learning for image recognition. In: Proceedings of the IEEE Conference on Computer Vision and Pattern Recognition, pp. 770–778 (2016)

11. Krizhevsky, A., Sutskever, I., Hinton, G.E.: Imagenet classification with deep convolutional neural networks. In: Pereira, F., Burges, C.J.C., Bottou, L., Weinberger, K.Q. (eds.) Advances in Neural Information Processing Systems, vol. 25, pp. 1097–1105. Curran Associates, Inc. (2012)

12. LeCun, Y., Bengio, Y., Hinton, G.: Deep learning. Nature **521**(7553), 436–444 (2015)

13. Moradi, M., Guo, Y., Gur, Y., Negahdar, M., Syeda-Mahmood, T.: A cross-modality neural network transform for semi-automatic medical image annotation. In: Ourselin, S., Joskowicz, L., Sabuncu, M.R., Unal, G., Wells, W. (eds.) MICCAI 2016. LNCS, vol. 9901, pp. 300–307. Springer, Cham (2016). https://doi.org/10.1007/978-3-319-46723-8_35

14. Nascimento, J.C., Carneiro, G.: Multi-atlas segmentation using manifold learning with deep belief networks. In: 2016 IEEE 13th International Symposium on Biomedical Imaging (ISBI), pp. 867–871 IEEE (2016)

15. Pézard, P., et al.: Influence of ultrasonographers training on prenatal diagnosis of congenital heart diseases: a 12-year population-based study. Prenat Diagn. **28**(11), 1016–1022 (2008)

16. Poudel, R.P.K., Lamata, P., Montana, G.: Recurrent fully convolutional neural networks for multi-slice MRI cardiac segmentation. In: Zuluaga, M.A., Bhatia, K., Kainz, B., Moghari, M.H., Pace, D.F. (eds.) RAMBO/HVSMR -2016. LNCS, vol. 10129, pp. 83–94. Springer, Cham (2017). https://doi.org/10.1007/978-3-319-52280-7_8

17. Sundaresan, V., Bridge, C.P., Ioannou, C., Noble, J.A.: Automated characterization of the fetal heart in ultrasound images using fully convolutional neural networks. In: 2017 IEEE 14th International Symposium on Biomedical Imaging (ISBI 2017), pp. 671–674. IEEE (2017)

18. Szegedy, C., et al.: Going deeper with convolutions. In: Proceedings of the IEEE Conference on Computer Vision and Pattern Recognition, pp. 1–9 (2015)

19. Wu, M.H., Chen, H.C., Lu, C.W., Wang, J.K., Huang, S.C., Huang, S.K.: Prevalence of congenital heart disease at live birth in Taiwan. J. Pediatrics **156**(5), 782–785 (2010)

20. Yeh, S.J., et al.: National database study of survival of pediatric congenital heart disease patients in Taiwan. J. Formos. Med. Assoc. **114**(2), 159–163 (2015)

A Neural Network Model for Lead Optimization of MMP12 Inhibitors

Tewodros M. Dagnew[1]([envelope]) [iD], Claudio Silvestri[1,2], Debora Slanzi[1,3] [iD], and Irene Poli[1] [iD]

[1] European Centre for Living Technology (ECLT), Ca' Foscari University of Venice, Venice, Italy
{tewodros.dagnew,silvestri,debora.slanzi,irenpoli}@unive.it
[2] Department of Environmental Sciences, Informatics and Statistics, Ca' Foscari University of Venice, Venice, Italy
[3] Department of Management, Ca' Foscari University of Venice, Venice, Italy

Abstract. Lead Optimization is a complex process, whereby a large number of interacting entities give rise to molecular structures whose properties should be optimized in order to be considered for drug development. We will study molecular systems that are characterized by high dimensionality and dynamically interacting networks with the goal of discovering the optimal molecules with respect to the set of essential properties. Currently, the research involves the screening and the identification of molecule with desirable properties from large molecule libraries. Lead Optimization is a multi-objective optimization problem. The classical approaches involving in-vitro laboratory analysis are time consuming and very expensive. To address this problem, we propose in this paper an in-silico approach: Lead Optimization based on Neural Network (NN) model in order to help the chemist in the lab experimentation by requiring a small set of real laboratory tests. We propose and estimate a predictive network model to derive a simultaneous optimal multi-response property following a single and multi-objective optimization procedure. We adopt different architectures in this study and we compare our procedure with other state-of-the-art method showing the better performance of our approach.

1 Introduction

The general structure of a drug can be represented as a combination of small molecule fragments that interact with a pharmacological target of interest and have a therapeutic effect on a particular disease. Most current practices in drug discovery research involve the screening of large sets of chemical libraries composed of thousands or millions of compounds, with the aim of identifying candidate molecule/s with suitable characteristics [16]. These selected molecules with desirable properties are referred as lead molecules. In this work we consider the following relevant properties that molecules should satisfy: Activity, Solubility, Safety, Lipophilicity (cLogP) and Molecular Weight. The basic idea of drug discovery is in fact to identify the molecules that present these specific properties

© Springer Nature Switzerland AG 2021
A. Del Bimbo et al. (Eds.): ICPR 2020 Workshops, LNCS 12664, pp. 323–335, 2021.
https://doi.org/10.1007/978-3-030-68799-1_23

and that can bind themselves to a particular target protein achieving the drug likeness characteristics [6]. In drug design cycle, analysis of quantitative structure to property relationship (QSPR) is a common approach [4]. There are several ways of molecular feature representation as described in [11] where the authors used a 3D molecular representation with deep 3D-convolutional neural networks.

Lead molecule optimization is then the process of identifying the molecule, or a small set of molecules, with desirable properties in a large set. Usually this identification process involves the analysis of molecular systems where each molecule is described by an extremely large set of fragment features, whose presence and interaction can lead to optimal property values. The research in this filed shows that in-silico methods can be effective and generate better results than in vitro results, thereby saving time and complex real lab trials [1].

In this research, we wish to identify the lead molecules from a set of 'n' molecules by estimating and optimizing with a neural network model, five molecular properties i.e.: Activity, Solubility, Safety, cLogP and Molecular Weight (MW). The optimal molecules in our study refer to molecules having the highest Activity, Solubility and Safety values while having the lowest cLogP and MW values. According to the common practice in order to discover these molecules, we should conduct high throughput screening experiments, which are very expensive and time-consuming procedures. It could then be very helpful to develop in silico approaches that can save resources and also be effective and efficient.

Machine learning techniques can provide a methodological way that can map the molecule features to their respective response properties. This approach can address this problem of optimization in high dimensional settings discovering lead molecules and reducing the experiment costs and lab time drastically.

In this paper we address this optimization problem by building a procedure based on neural networks models. Networks models have been particularly successful in solving several classification and regression problems in different research fields [5] and particularly in drug design in different stages of the cycle and contexts [7,9,10,12,13].

For the optimization problem we evaluate different network topologies, different activation functions and learning approaches and several training schemes with the aim to achieve the final neural architecture able to perform the task in a satisfactory way. We then select a particular network model for prediction and employ it in identifying lead molecules in a set of a large molecule library.

Analyzing data from molecular systems we deal with small high dimensional data set (i.e., $p >> n$, where p is the number of features and n is the number of tested molecules). This problem structure has been recently addressed from statistical perspective by developing predictive statistical models with an evolutionary approach [2,14,15]. Authors in [15], tackled this problem by designing an evolutionary procedure based on statistical models. With this method the authors identify the lead molecules from a set of 2500 molecules characterized by 22750 fragments (features), applying also some dimension reduction techniques. The method called Evolutionary Design for Optimization (EDO) works as follows: 1) initially, a small set of molecules is selected in a random way building a first-generation of candidate lead molecules; 2) a predictive statistical model is built on the selected molecules to predict the response of the remained molecules in the molecules space. 3) a second generation of molecules is then selected according to their best predicted values (ranking these values in either ascending or descending way depending on the optimization task). 4) the process is iterated by going back to step 2 for building few generations. The lead molecules are selected as the best molecules from each iteration. Authors in [2] address the lead optimization process by using the same strategy both to design experiments and to model data in a reagent based analysis. With the same goal of lead optimization, authors in [14] developed genetic algorithms to solve the aforementioned problem of lead optimization of MMP-12 inhibitors (Matrix metalloproteinase-12 (MMP-12)) that is an enzyme that in humans is encoded by the MMP12 gene. It is mainly involved in the inflammatory process of chronic obstructive pulmonary disease (COPD).

In this work we propose a neural network model for lead optimization, where the chemist can identify the lead candidate molecule by conducting laboratory experiments only on a very small set of the molecules (approximately much less than 10% of the data). The goal and the novelty of this paper is to build a particular neural network model able to find the optimal values of MMP-12, conducting a very small number of tests and thus with small investments of resources and limited negative impact on the environment. From the estimated model we can also learn the molecular fragment features which deeply affect the molecule properties. The study is conducted on the same dataset considered in [2,14,15] to make easy evaluations and comparisons.

Our contribution is also in deriving a multi-objective lead optimization procedure to identify the molecules with best properties values in a simultaneously way. For comparison with other current approaches we also build and train a network model for single-objective optimization. The rest of the paper is organized as follows: Sect. 2 describes Lead Optimization Problem, Sect. 3 describes the proposed methods, Sect. 4 presents the experimental results of our study, after describing the dataset used in the analysis and the evaluation criteria. Finally, in Sect. 5 we draw conclusions and describe future directions of our research.

2   Lead Optimization Problem

Lead optimization is a multi-objective optimization problem whose solution consists in a set of molecules with the best values for some selected molecular properties (e.g., low Molecular Weight or high Activity with respect to a specific therapeutic target).

The input of the analysis is a library of candidate molecules described by binary vectors. Each molecule may contain a set of fragments and their presence is represented using a vector of binary variables. Each element of the binary vector is either 0 or 1 indicating the presence/absence of the fragment associated with the element position in the molecule. The presence of fragments in the input dataset can thus be represented as a binary matrix, where rows corresponds to molecules and columns corresponds to fragments, also named features. Figure 1 shows graphically the fragments found in each molecule of the library used in this study, as described in detail in Sect. 4.

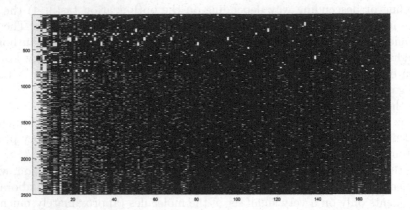

Fig. 1. The 2500 molecules with 175 fragment dataset. Each point represents the presence/absence of a fragment in a molecule.

Molecules are associated with vectors containing numerical molecular properties whose values can be determined paying a cost (for laboratory analysis). In this study the properties we consider are the pharmacological Activity at the target protein, the Solubility, the Safety, the Lipophilicity (cLogP) and the Molecular Weight (MW).

Given a large library of candidate molecules and a small set of property vectors associated to some of the molecules, the lead optimization problem consists in finding a small set of molecules from the library exhibiting molecular property values that are either optimal or in the region of optimality (top/bottom percentiles for each single molecular property) with respect to the complete library. The values of the properties for the selected optimal molecules are in general unknown, but all the molecules from the library can be part of the solution, also the ones for which the properties were measures and used as an input.

3   Lead Optimization Based on Predictive Neural Network Models

In this work we propose two methods for the solution of the lead optimization problem, both based on neural network (NN) models. The network models are trained as molecule properties predictors on a small set of molecules and then used on the complete molecule library to predict the unknown properties of other molecules.

Given a set of candidate molecules $\mathcal{C}$, a function **prop** that gives the properties **prop**(c) associated with a candidate molecule $c \in \mathcal{C}$, a budget M representing the maximal allowed number of evaluations of the function `prop` (to account for the cost of molecule examination), and a number of candidates to use for training T, the methods we propose operate as follow:

1. Choose a random subset $C' \subset C$ containing T candidates for training the model ($|C'| = T$);
2. Get the properties for the training set C', by evaluating (T times) the function `prop`: $P' = \{(c, \texttt{prop}(c)) | c \in C'\}$;
3. Design and train a neural network model using C' and the associated set of physically measured properties P';
4. Predict the properties of candidates in $C \setminus C'$;
5. Rank the candidates according to the predicted values (see later for details);
6. Build C'' as the set of the best $M - T$ ranked molecules;
7. Get the properties for candidates in C'', by evaluating ($M - T$ times) the function `prop`: $P'' = \{(c, \texttt{prop}(c)) | c \in C''\}$;
8. return the set of candidates $C' \cup C''$ that is expected to contain the lead molecules and their physically measured properties $P' \cup P''$

We observe that the output contains both the selected candidates as well as the ones that were randomly selected. The rational behind this is that any molecule for which the actual properties are known is useful to select the lead ones and we already payed the cost for measuring the properties of the candidates used for training. Thus the result contain a total of M candidates (and their actual properties), corresponding to the maximal number of physical measurement allowed by the budged, that will be re-ranked to select the lead molecules. Please note that the accurate prediction of properties values is not relevant, as long as the ranking of predicted values is consistent with the one based on actual values. This allow us to train our neural network models using sample with low cardinality, which usually are not suitable for obtaining accurate predictions.

The two methods we propose share the same outline sketched above and are based on multilayer perceptrons. The architecture of the corresponding NN models are presented in Fig. 2. Exponential linear unit activation is used in the hidden layers. During the training, the objective is to minimize the mean squared error and 10% of the training set is used as a validation to optimize the parameters. The training is set to run for 100 epochs exploiting the early stopping rule and adjusting the learning rate using the plateau callbacks function of Keras library [3]. The latter callback adjusts the learning rate if there is no improvement in the learning while the first call back terminates the learning if the weight gradient changes are below a threshold. The training procedure minimized the root mean squared error using a batch size of 1 molecule. `RMSprop` is used as an optimizer [17]. The responses are scaled from 0 to 1.

We tested two approaches for Lead optimization that solve two slightly different problems. The first one try to find lead molecules that are optimal for all properties, whereas the second is used when optimization is focused only on one of the properties.

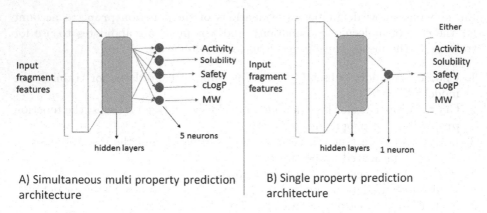

Fig. 2. The proposed neural network architecture for lead molecule optimization.

Lead Optimization Using Simultaneous Multiple Property Prediction (SMPP). In this case the goal is to predict all the properties for a specific molecule given the set of contained fragments. In the graphical representation of the NN model in Fig. 2A we observe that there is an output neuron for each molecular property that we learn to predict during the training phase. The figure refer to the dataset described in Sect. 4, so the network has 5 output neurons that corresponds to Activity, Safety, Solubility, cLogP and Molecular Weight.

Lead Optimization Using Single Property Prediction (SPP). This is a subcase of SMPP, but considering only a molecular property has implications that go beyond the fact that there is just one output neuron. In this case, indeed, all the hidden neurons are dedicated to the prediction of a single value and we expect to have a more accurate prediction than in SMPP when using the same number of neurons. Figure 2B shows the simplified network configuration used in this case.

In both cases, the software is implemented in Python3 using the Keras machine learning library and Jupiter notebook.

4   Results and Discussion

In this section we report on the performance of the proposed methods, *SMPP* and *SPP*, comparing them with those of the Evolutionary Data Optimization (EDO) [15] state-of-the-art method. Before describing the details of the comparison, we introduce the evaluation criteria that will be used and we describe the dataset used in this analysis, consisting of 2500 molecules each with 22750 fragments then reduced to 175.

4.1   Evaluation Metrics

As evaluation criteria for the reliability of the method in selecting lead molecules, we consider; Region of optimality (RegOp) and Optimum value (Opt).

1. **Region of optimality**: This metric checks if the output vector of molecules holds molecules within a region of optimality. The response optimality threshold for dataset is as follows: Activity $\geq$ 7.5, Solubility $\geq$ −2.415, Safety $\geq$ 3, clogP $\leq$ 4.5 and Molecular Weight $\leq$ 339.3.
2. **Optimum value**: This metric checks if the output vector contain a molecule with the optimum value. The optimum values for the response of the molecules in dataset are the following: Activity = 8, Solubility = −1.766, Safety = 3.6262, clogP = −2.505 and Molecular Weight = 291.3.

Since the properties of all molecules are available for the dataset we use in our analysis, it is possible to simulate several different runs, by choosing each time different training molecules to prove the robustness of the proposed method. To this end, we repeat the experiments 1000 times to estimate the degree of accuracy and we report the number of times the above criteria are satisfied by the solution. So when we report 1000 for the optimum value criteria, this means that in each trial the method is able to identify the best molecules.

4.2   Dataset

The analysis is carried on a dataset of 2500 molecules. Each molecule is composed of a set of fragments represented as a vector of binary variables indicating the presence/absence of each considered fragment. In Fig. 1 we describe the dataset matrix. The original feature dimension of the molecules is 22750, representing the number of fragments composing the molecules. Adopting Formal Concept Analysis for feature reduction as in [8], the resulting feature number is 175, and on this set we will develop our study.

These molecules satisfy a set of properties among which we consider the following most relevant for the optimization problem: the pharmacological Activity (Y1) at the target protein, the Solubility (Y2), the Safety property (Y3) and structural properties, such as the Lipophilicity (Y4) and the Molecular Weight (Y5). In particular, the pharmacological Activity at the target protein, defined as the capacity to produce physiological or chemical effects by the binding of a compound to the therapeutic target, will be denoted by Activity. The Solubility of a compound is the capacity to dissolve in a liquid and it will be denoted by Solubility. The Safety refers to the capacity of the chemicals to avoid undesirable effects. The Lipophilicity is the capacity of the compound to dissolve into lipid structures, and it will be denoted by cLogP. Finally, the Molecular Weight describe the compound size and will be denoted by MW. For a more detailed explanation of these biological concepts we refer to [6], which describes the influence of each of compound property on ADME and Safety. In our analysis we will consider these properties as the experimental response variables and their values will be the target of our optimization study.

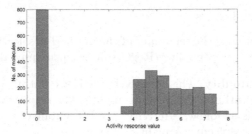

Fig. 3. Activity response (Y1) distribution histogram. The objective is to identify the molecules on the right tail of the distribution that have the highest Activity values.

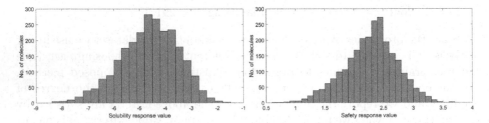

Fig. 4. Solubility response (Y2), and Safety response (Y3) distribution histograms. The objective is to identify the molecules on the right tail of the distribution that have the highest Solubility/Safety values.

Some summary statistics of this dataset are as follows:

- **Activity (Y1)** takes 8 as a maximum value, which corresponds to the optimal value. The 99-th percentile of the response variable distribution is 7.5. The target is the maximization of Y1. There are 24 molecules out of 2500 have Activity response ≥ 7.5. The histogram of this response distribution is depicted in Fig. 3.
- **Solubility (Y2)** takes -1.766 as a maximum value, which corresponds to the optimal value. The 99-th percentile of the response variable distribution is -2.415. The target is the maximization of Y2. There are 26 molecules out of 2500 have Solubility response ≥ -2.415. The histogram of this response distribution is depicted in Fig. 4.
- **Safety (Y3)** takes 3.6262 as a maximum value, which corresponds to the optimal value. The 99-th percentile of the response variable distribution is 3.2309. The target is the maximization of Y3. There are 25 molecules out of 2500 have Safety response ≥ 3.2309. The histogram of this response distribution is depicted in Fig. 4.
- **cLogP (Y4)** takes -2.505 as a minimum value, which corresponds to the optimal value. The 1-th percentile of the response variable distribution is 0.033. The target is the minimization of Y4. There are 25 molecules out of 2500 have cLogP response ≤ 0.033. The histogram of this response distribution is depicted in Fig. 5.

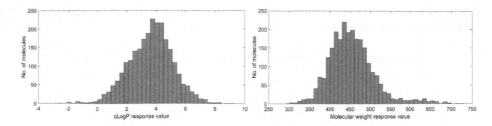

Fig. 5. cLogP response (Y4) and Molecular Weight response (Y5) distribution histograms. The objective is to identify the molecules on the left tail of the distribution that have lowest cLogP/MW values.

– **MW (Y5)** takes 291.3 as a minimum value, which corresponds to the optimal value. The 1-th percentile of the response variable distribution is 339.3. The target is the minimization of Y5. There are 25 molecules out of 2500 have MW response $\leq$ 339.3. The histogram of this response distribution is depicted in Fig. 5.

4.3   Results of Lead Optimization Using SMPP

A neural network is trained with 200 molecules each with 175 fragments and the model is used to predict the properties for all the other candidate molecules in the set. Then the 50 molecules exhibiting the best values and the 200 molecules used during training are considered as lead candidates and the missing molecular properties are retrieved (in real-world cases this would imply actual measurement). The fully connected architecture takes 175 molecular fragments as an input with two hidden layers (100 and 50 neurons) and a final dense output layer with 5 neurons (i.e., each neuron outputs Activity, Solubility, Safety, cLogP and Molecular Weight). The exponential linear unit is used as an activation function in the hidden layers except the last layer. The training procedure minimized the root mean squared error using a batch size of 1 molecule. RMSprop is used as an optimizer [17]. The responses are scaled from 0 to 1. Table 1 shows the results. In the column contining the results of our method, there are two additional evaluation metrics denoted as optimal-2 and optimal-3, indicating how many times the second and third optimal molecules respectively are discovered by our method. We compared our result with [15], where they used a combination of linear models in an iterative and evolutionary approach. Our results outperform the competitor in optimization of Solubility, Safety and Molecular Weight properties with a huge margin. To improve the result in the optimization of Activity and cLogP we design a different neural network architecture (SPP).

Moreover, we analyze the optimization procedure by assessing the behavior of the prediction on the best 50 lead candidate molecules (i.e., for Activity, Solubility and Safety; the best molecules are the ones with highest response values, whereas for Lipophilicity and Molecular Weight the best molecules are

Table 1. Performance of the model for **lead optimization using simultaneous multiple property prediction-SMPP** (1000 repetitions).

Molecule property	Evaluation metrics	SMPP	EDO
Activity (max)	1st, 2nd, 3rd-Opt	883, 979, 986	**916**
	RegOp	1000	1000
Solubility (max)	1st, 2nd, 3rd-Opt	**984**, 990, 991	912
	RegOp	1000	1000
Safety (max)	1st, 2nd, 3rd-Opt	**990**, 996, 997	467
	RegOp	1000	1000
clogP (min)	1st, 2nd, 3rd-Opt	824, 831, 997	**918**
	RegOp	1000	1000
Molecular weight (min)	1st, 2nd, 3rd-Opt	**994**, 1000, 1000	887
	RegOp	1000	1000

the ones with lowest response values). In Fig. 6 we can observe that the ranking based on the average response prediction is consistent with the one based on actual molecular properties (ground truth) for all of the five molecular responses.

4.4 Results of Lead Optimization Using SPP

With the aim of increasing the performance obtained with SMPP on the optimization of Activity and cLogP, we adopt a neural network architecture for Single Objective optimization using single property prediction. Once again, we used 175 fragments as an input with the same setting as the previous experiment and we searched for the best performing architecture. For predicting and optimizing Activity we found the architecture with 3 hidden layers with 200, 50, 25 neurons and 1 neuron as an output to be optimal. For cLogP, instead, the optimal architecture was found to have 3 hidden layers with 100, 50, 25 neurons and 1 neuron as an output. Table 2 shows the results.

Table 2. Performance of the model for **lead optimization using single property prediction-SPP** (1000 repetitions).

Molecule property	Performance evaluation	SPP	EDO
Activity (max)	Opt	**922**	916
	RegOp	1000	1000
clogP (min)	Opt	**972**	918
	RegOp	1000	1000

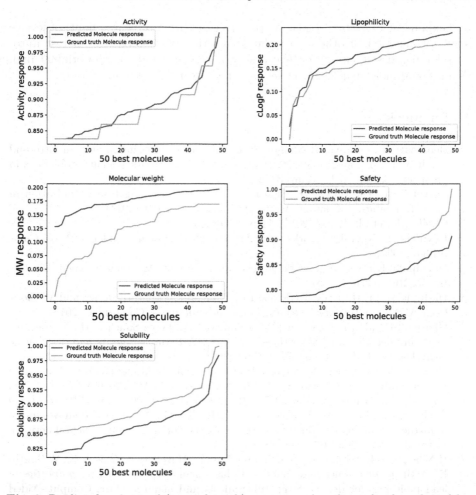

Fig. 6. Predicted and actual (ground truth) property values for molecules selected by SMPP. Since the goal is to select the best molecules, the numerical values of predictions are not significant as long as the ordering of predicted and actual values is the same.

5   Conclusion and Future Directions

We propose an in-silico lead optimization approach based on a neural network-based model that provides good results in both single and multi-objective molecular property optimization.

Results show that our method outperforms the state-of-the-art approach when we optimize Solubility, Safety and Molecular Weight properties using the proposed Simultaneous Multiple Property Prediction neural network architecture. For cLogP and Activity molecular properties, our method outperforms the others using the Single Property Prediction architecture. Our method could help screening candidate lead molecules using an in-silico approach, limiting laboratory tests to one tenth of the molecule library and reducing laboratory time and expensive laboratory trials. We plan to test this approach in multi-objective lead optimization of other molecular properties. We would like to test auto-encoder based self-supervised feature extraction methods on the original high dimensional molecular fragment feature space to reduce the feature set.

References

1. Benfenati, E., Gini, G., Hoffmann, S., Luttik, R.: Comparing in vivo, in vitro and in silico methods and integrated strategies for chemical assessment: problems and prospects. Altern. Lab. Anim. **38**(2), 153–166 (2010)
2. Borrotti, M., De March, D., Slanzi, D., Poli, I.: Designing lead optimisation of mmp-12 inhibitors. Comput. Math. Methods Med. **2014** (2014)
3. Chollet, F., et al.: Keras (2015). https://keras.io
4. Devillers, J.: Neural Networks in QSAR and Drug Design. Academic Press, Cambridge (1996)
5. Devulpalli, K.: Neural networks for classification and regression. Biomed. Biostat. Int. J. **2**(6), 00046 (2015)
6. Di, L., Kerns, E.H.: Drug-like Properties: Concepts, Structure Design and Methods from ADME to Toxicity Optimization. Academic press, Cambridge (2015)
7. Ghasemi, F., Mehridehnavi, A., Perez-Garrido, A., Perez-Sanchez, H.: Neural network and deep-learning algorithms used in QSAR studies: merits and drawbacks. Drug Discov. Today **23**(10), 1784–1790 (2018)
8. Giovannelli, A., Slanzi, D., Khoroshiltseva, M., Poli, I.: Model-based lead molecule design. In: Rossi, F., Piotto, S., Concilio, S. (eds.) WIVACE 2016. CCIS, vol. 708, pp. 103–113. Springer, Cham (2017). https://doi.org/10.1007/978-3-319-57711-1_9
9. Grebner, C., Matter, H., Plowright, A.T., Hessler, G.: Automated de-novo design in medicinal chemistry: which types of chemistry does a generative neural network learn? J. Med. Chem. **63**, 8809–8823 (2020)
10. H Nilar, S., B Lakshminarayana, S., Ling Ma, N., H Keller, T., Blasco, F., W Smith, P.: Artificial neural network analysis of pharmacokinetic and toxicity properties of lead molecules for dengue fever, tuberculosis and malaria. Curr. Comput.-Aided Drug Des. **12**(1), 52–61 (2016)
11. Jiménez-Luna, J., et al.: DeltaDelta neural networks for lead optimization of small molecule potency. Chemical Science **10**(47), 10911–10918 (2019)
12. Kadurin, A., et al.: The cornucopia of meaningful leads: applying deep adversarial autoencoders for new molecule development in oncology. Oncotarget **8**(7), 10883 (2017)
13. Kwon, Y., Shin, W.H., Ko, J., Lee, J.: AK-score: Accurate protein-ligand binding affinity prediction using the ensemble of 3D-convolutional neural network. Int. J. Mol. Sci. **21**(22), 8424 (2020)
14. Pickett, S.D., Green, D.V., Hunt, D.L., Pardoe, D.A., Hughes, I.: Automated lead optimization of mmp-12 inhibitors using a genetic algorithm. ACS Med. Chem. Lett. **2**(1), 28–33 (2011)
15. Slanzi, D., Mameli, V., Khoroshiltseva, M., Poli, I.: Multi-objective optimization in high-dimensional molecular systems. In: Pelillo, M., Poli, I., Roli, A., Serra, R., Slanzi, D., Villani, M. (eds.) WIVACE 2017. CCIS, vol. 830, pp. 284–295. Springer, Cham (2018). https://doi.org/10.1007/978-3-319-78658-2_21

16. Tautermann, C.S.: Current and future challenges in modern drug discovery. In: Heifetz, A. (ed.) Quantum Mechanics in Drug Discovery. MMB, vol. 2114, pp. 1–17. Springer, New York (2020). https://doi.org/10.1007/978-1-0716-0282-9_1
17. Tieleman, T., Hinton, G.: Lecture 6.5-rmsprop: divide the gradient by a running average of its recent magnitude. COURSERA: Neural Netw. Mach. Learn 4(2), 26–31 (2012)

An Empirical Analysis of Integrating Feature Extraction to Automated Machine Learning Pipeline

Hassan Eldeeb$^{(\boxtimes)}$, Shota Amashukeli, and Radwa El Shawi

Data Systems Group, University of Tartu, Tartu, Estonia
{hassan.eldeeb,shota.amashukeli,radwa.elshawi}@ut.ee

Abstract. Machine learning techniques and algorithms are employed in many application domains such as financial applications, recommendation systems, medical diagnosis systems, and self-driving cars. They play a crucial role in harnessing the power of Big Data being produced every day in our digital world. In general, building a well-performing machine learning pipeline is an iterative and complex process that requires a solid understanding of various techniques that can be used in each component of the machine learning pipeline. Feature engineering (FE) is one of the most time-consuming steps in building machine learning pipelines. It requires a deep understanding of the domain and data exploration to discover relevant hand-crafted features from raw data. In this work, we empirically evaluate the impact of integrating an automated feature extraction tool (`AutoFeat`) into two automated machine learning frameworks, namely, `Auto-Sklearn` and `TPOT`, on their predictive performance. Besides, we discuss the limitations of `AutoFeat` that need to be addressed in order to improve the predictive performance of the automated machine learning frameworks on real-world datasets.

Keywords: Automated machine learning · Automated feature engineering

1 Introduction

In recent years, there has been an enormous interest in deep learning systems across the industry, government, and academia. A central driver is that deep learning techniques learn task-specific representations of input data [1,12], eliminating FE, which used to be the most tedious task. However, deep learning has significant upfront costs: these methods need massive training instances to learn from, which is sometimes challenging. Although great progress has been made on the interpretability of deep learning models, machine learning practitioners still find it hard to trust such complex models, especially in critical domains such as self-driving cars and medical diagnosis systems [19]. To avoid these shortcomings, explicit FE techniques are highly needed to create a new better set of features

© Springer Nature Switzerland AG 2021
A. Del Bimbo et al. (Eds.): ICPR 2020 Workshops, LNCS 12664, pp. 336–344, 2021.
https://doi.org/10.1007/978-3-030-68799-1_24

that may lead to sufficiently accurate predictions on real-world data while providing transparent models with a high acceptance rate amongst machine learning practitioners. In general, FE is a complex task requiring highly skilled data scientists with strong machine learning and statistics backgrounds to extract useful patterns from raw data. Top data scientists reported that FE is the most time-consuming step. For example, in Grupo Bimbo Inventory Demand competition, the winners reported that 95% of their time was spent in the FE step and only 5% is for modelling[1]. Since the number of data scientists can not scale with the growing number of machine learning applications, it becomes unrealistic to perform FE for these applications manually. This motivates the need for automatic FE which is a sub-problem of the automated machine learning problem (AutoML) [6,17,22].

FE is an unarguably valuable phase of the machine learning pipeline. However, to the best of our knowledge, the current open-source AutoML frameworks address it from a high level of abstraction. These frameworks [3,15,18] produce basic feature preprocessors by applying techniques such as principal component analysis (PCA), and Linear Discriminant Analysis (LDA), without creating derived features.

Most state-of-the-art automatic FE techniques, in contrast to AutoML tools, follow one of three main approaches to obtain useful features. In the first approach, which is *generate-and-select*, an exhaustive feature pool is generated, and selection performed on the whole feature set [2,10,13,16]. In the second approach, a reinforcement learning technique is used to solve the FE problem by adopting Monte-Carlo tree search [5], Q-learning [11], or deep Q-learning [21]. In the third approach, a meta-learning based algorithm is trained offline over a collection of datasets before being used online to recommend the promising features for the input dataset [9,11,14]. Another line of research focuses on automatically constructing features from relational databases through deep feature synthesis [8]. Some other studies consider performing FE during training by introducing operations such as feature cross [20].

This paper reports empirical research to demonstrate the effectiveness of integrating one of the automated featured engineering tools to the state-of-the-art AutoML frameworks. In this work, we chose AutoFeat [7] as, to the best of our knowledge, it is the only general-purpose open-source library for automated FE and selection. The main contributions of this paper are as follows:

- Empirically studying the effectiveness of combining the AutoFeat with two AutoML frameworks after disabling the feature preprocessing stage of the AutoML. In particular, we compare the performance of two AutoML frameworks, with their feature preprocessors or with AutoFeat feature preprocessor, to the baseline of the performance of a pipeline without any feature preprocessing.
- Identifying the advantages and limitations of AutoFeat and exploring improvements that could be made.

---

[1] https://medium.com/kaggle-blog/grupo-bimbo-inventory-demand-winners-interview-clustifier-alex-andrey-1e3b6cec8a20.

- As ensuring repeatability is one of the main goals of this work, we provide access to the source codes and the detailed results of our study [2].

The remainder of the paper is organized as follows: after recapitulating the related work in Sect. 1, we provide an overview of the different AutoML techniques and feature extraction technique considered in this study in Sect. 2. Section 3 describes our experimental setup details in terms of experimental design and the used datasets. The results of our experiments are presented in Sect. 4, before we conclude the paper in Sect. 5.

2   Reference Tools

In this section, we give an overview on the automated FE tool and the AutoML frameworks considered in this study.

2.1   Automated Feature Engineering Tool

AutoFeat [7] is a python library that provides linear regression and classification models, inheriting scikit-learn style, with automated feature generation and selection capabilities. AutoFeat is a general-purpose library for a single table datasets. Since neural networks and deep learning can be viewed as a linear classifier operating on better, but not explainable, features, the authors assume that generating informative and explainable features could be used with a linear model to achieve a comparable result. To transform the raw features into the new informative features, they follow the *generate-and-select* technique. Firstly, the tool generates all the possible combinations of applying predefined nonlinear operators (e.g., $log(x), \sqrt{x}, 1/x, x^2, x^3, sin(x), e^x$) on the input features, and then these derived features are combined using the four main mathematical operators $(+, -, *,$ and $/)$. Secondly, these features are normalized, and a linear model is trained over them to select those with high coefficients.

2.2   AutoML Frameworks

Auto-Sklearn [3] has been implemented on top of ScikitLearn[3], a popular Python machine learning package [4]. Auto-Sklearn uses SMAC as a Bayesian optimization technique for the problem of algorithm selection and hyperparameters optimization. Both meta-learning and ensemble methods are integrated to the vanilla SMAC optimization [3].

TPOT [15] framework is a different solution that has been implemented on top of Scikit-Learn. It is based on genetic programming, exploring the search space through the mutation phase, and exploiting the best-performing pipelines during the cross-over phase.

---

[2] https://github.com/DataSystemsGroupUT/auto_feature_engineering.
[3] https://scikit-learn.org/.

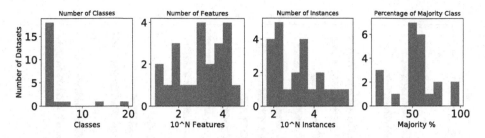

Fig. 1. Histogram of the main characteristics of the 22 datasets.

3 Experimental Setup

3.1 Experiment Design

The experiment has been mainly designed to answer the question of whether it will be useful to divide the pipeline generation process into two phases and solve each one independently. Typically, in the first phase, the feature representation is enhanced using `AutoFeat`. In the second phase, the model is selected and its hyperparameters tuned using an `AutoML` framework. Hence, we introduce three setups:

1. Combined setup: the setup to be investigated where `AutoFeat` is used for the FE phase, followed by `AutoML` for model selection and tuning.
2. `AutoML` setup: the regular setup which `AutoML` produces a full pipeline, including FE phase.
3. Baseline: `AutoML` is only used to select the model (classifier) and tune its hyperparameters, with no FE preprocessors.

The three setups have been experimented on a Linux based machine with 256 GB of RAM and 64 vCPUs. The time consumed by `AutoFeat` is always added to the time budget of `AutoML` in the second and third setups.

3.2 Datasets

Twenty-two datasets have been examined through each design setup. We selected these datasets to cover different characteristics like the number of instances, features, and the majority class percentage. Figure 1 summarizes the main characteristics, and the detailed properties are in the experiment repository.

4 Results and Discussion

We run `AutoFeat` with 0, 1, and 2 feature generation (FG) steps, followed by a feature selection (FS) step. For `Auto-Sklearn`, Fig. 2 shows the `F1 Score` of the three setups over different datasets, ordered by the number of features. `AutoFeat` is used with only feature selection step (0 steps of FG). The combined setup

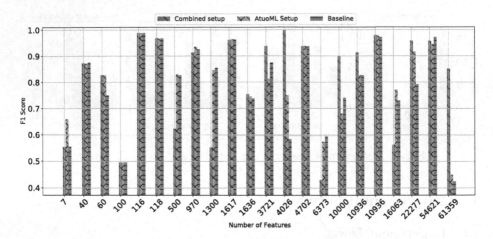

Fig. 2. Setups of combing `AutoFeat` with `Auto-Sklearn`. Each dataset denoted by number of features of raw data.

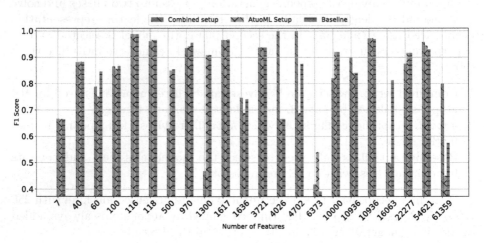

Fig. 3. Setups of combing `AutoFeat` with `TPOT`.

with only FS has a fluctuating performance difference, where it improved the F1 Score of six datasets but decreased performance for the remaining four. Similar performance scores are also obtained using TPOT, as shown in Fig. 3.

Figure 4 shows the effect of adding more time, i.e. 6 times the original time budget, to the AutoML tool (Auto-Sklearn) to search for a better pipeline compared to adding a single step of FG using AutoFeat. Generally, it has mitigated the big performance drops of the four datasets from Fig. 2. Additionally, the additional FG step added a minor improvement to the FS step while being comparable in performance with the AutoML setup having extended time budget (Fig. 4 and Fig. 5).

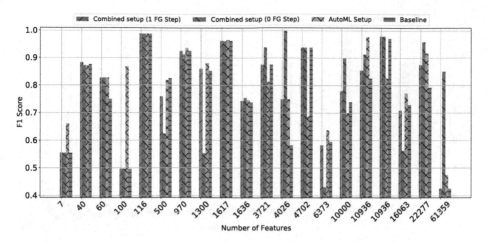

Fig. 4. Impact of adding a FG step AutoFeat compared to adding more time to Auto-Sklearn.

From these four figures, TPOT performs slightly better than Auto-Sklearn since, in contrast to Auto-Sklearn, TPOT can build a dynamic pipeline structure with multiple processing components [15]. While on both frameworks, increasing the time budget improves the performance. The combined setup mostly outperforms the pure AutoML solution with the same time budgets, but it could not beat it with an increased time budget for AutoML. That is, the current AutoML techniques scale better on larger time budgets compared AutoFeat with more iterations. Moreover, generating new features has led to improved performance scores compared to only FS. Hence, we conjecture that generating features from deeper FE steps will boost performance.

It worth mentioning that the technique used in FE phase has the largest impact over performance score compared to the effect of the technique used for model selection and hyperparameters tuning. That is, the two figures of Auto-Sklearn are almost the same as the two figures of TPOT.

We tried to run AutoFeat with two steps of FG, but none of the datasets had a successful run within 24 h. The majority have memory errors before this time limit. In theory, after each FG step, the number of produced features is kN^2, where k is the number of operators, N is the number of input features, and N^2 is all feature combination, assuming that we have only binary operators, e.g., $+, -, *, /$. That is, only two features are used with each operator. So, the memory complexity is $O(kN^{2^i})$. The memory complexity is double exponential of the number of the steps (i). For example, one of the used datasets (tumor_C[4]) occupies almost 1.8 MB of RAM with 7130 features. After the first FG step, it occupies about $1.8 * 7130 = 12.8$ GB with 7130^2 features, neglecting the effect of k. After the second step, it would occupy $1.8 * 7130^3 = 652$ PB, only if such

---

[4] https://www.openml.org/d/1107.

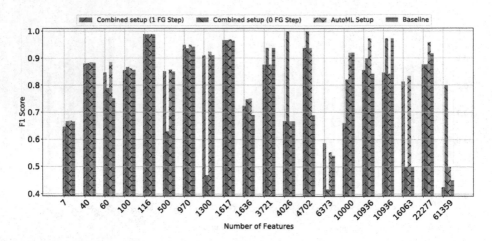

Fig. 5. Impact of adding a FG step `AutoFeat` compared to adding more time to TPOT.

RAM exists. The time complexity is $O(kN^{2^i})$ as well. Therefore, it would take a very long time with $i = 2$, as recommended by the authors of `AutoFeat`. This finding reveals a new challenge and opportunity to improve the performance of `AutoFeat`.

5 Conclusion and Future Work

We investigated the effectiveness of automatically obtaining new features for the `AutoML` frameworks, as an alternative to the regular pick and trial method currently used in their FE phase. Generally, pure FS does not clearly improve the pipeline performance score; however, a single step of FS has a minor improvement. Due to the limitations of the current `AutoFeat` techniques, it is practically impossible to generate complex features from deeper FG steps, i.e., more than one step, for realistic datasets. On the other hand, there is no doubt that we live in the era of big data, and we are witnessing an enormous surge in the amount of data generated daily from different sources. These generated data calls for distributed automatic FE frameworks that address the current techniques' shortages.

Acknowledgement. This work is funded by the European Regional Development Funds via the Mobilitas Plus programme (grant MOBTT75).

References

1. Bengio, Y.: Deep learning of representations: looking forward. In: Dediu, A.-H., Martín-Vide, C., Mitkov, R., Truthe, B. (eds.) SLSP 2013. LNCS (LNAI), vol. 7978, pp. 1–37. Springer, Heidelberg (2013). https://doi.org/10.1007/978-3-642-39593-2_1
2. Fan, W., et al.: Generalized and heuristic-free feature construction for improved accuracy. In: Proceedings of the 2010 SIAM International Conference on Data Mining, pp. 629–640. SIAM (2010)
3. Feurer, M., Eggensperger, K., Falkner, S., Lindauer, M., Hutter, F.: Auto-sklearn 2.0: The next generation (2020). arXiv preprint arXiv:2007.04074
4. Feurer, M., Klein, A., Eggensperger, K., Springenberg, J., Blum, M., Hutter, F.: Efficient and robust automated machine learning. In: Advances in Neural Information Processing Systems, pp. 2962–2970 (2015)
5. Gaudel, R., Sebag, M.: Feature selection as a one-player game (2010)
6. He, X., Zhao, K., Chu, X.: Automl: A survey of the state-of-the-art. arXiv preprint arXiv:1908.00709 (2019)
7. Horn, F., Pack, R., Rieger, M.: The `autofeat` python library for automated feature engineering and selection. In: Cellier, P., Driessens, K. (eds.) ECML PKDD 2019. CCIS, vol. 1167, pp. 111–120. Springer, Cham (2020). https://doi.org/10.1007/978-3-030-43823-4_10
8. Kanter, J.M., Veeramachaneni, K.: Deep feature synthesis: towards automating data science endeavors. In: 2015 IEEE International Conference on Data Science and Advanced Analytics, DSAA 2015, Paris, France, 19–21 October 2015, pp. 1–10. IEEE (2015)
9. Katz, G., Shin, E.C.R., Song, D.: Explorekit: automatic feature generation and selection. In: 2016 IEEE 16th International Conference on Data Mining (ICDM), pp. 979–984. IEEE (2016)
10. Kaul, A., Maheshwary, S., Pudi, V.: Autolearn–automated feature generation and selection. In: 2017 IEEE International Conference on data mining (ICDM), pp. 217–226. IEEE (2017)
11. Khurana, U., Samulowitz, H., Turaga, D.: Feature engineering for predictive modeling using reinforcement learning. arXiv preprint arXiv:1709.07150 (2017)
12. LeCun, Y., Bengio, Y., Hinton, G.: Deep learning. Nature **521**(7553), 436 (2015)
13. Markovitch, S., Rosenstein, D.: Feature generation using general constructor functions. Mach. Learn. **49**(1), 59–98 (2002)
14. Muller, K.R., Mika, S., Ratsch, G., Tsuda, K., Scholkopf, B.: An introduction to kernel-based learning algorithms. IEEE Trans. Neural Netw. **12**(2), 181–201 (2001)
15. Olson, R.S., Moore, J.H.: Tpot: a tree-based pipeline optimization tool for automating machine learning. In: Proceedings of the Workshop on Automatic Machine Learning (2016)
16. Piramuthu, S., Sikora, R.T.: Iterative feature construction for improving inductive learning algorithms. Exp. Syst. Appl. **36**(2), 3401–3406 (2009)
17. Shawi, R.E., Maher, M., Sakr, S.: Automated machine learning: state-of-the-art and open challenges. CoRR abs/1906.02287 (2019). http://arxiv.org/abs/1906.02287

18. Thornton, C., Hutter, F., Hoos, H., Leyton-Brown, K.: Auto-WEKA: combined selection and hyperparameter optimization of classification algorithms. In: KDD (2012). https://doi.org/10.1145/2487575.2487629
19. Tonekaboni, S., Joshi, S., McCradden, M.D., Goldenberg, A.: What clinicians want: contextualizing explainable machine learning for clinical end use (2019). arXiv preprint arXiv:1905.05134
20. Wang, R., Fu, B., Fu, G., Wang, M.: Deep & cross network for ad click predictions. In: Proceedings of the ADKDD 2017, pp. 1–7 (2017)
21. Zhang, J., Hao, J., Fogelman-Soulié, F., Wang, Z.: Automatic feature engineering by deep reinforcement learning. In: Proceedings of the 18th International Conference on Autonomous Agents and MultiAgent Systems, pp. 2312–2314 (2019)
22. Zöller, M.A., Huber, M.F.: Benchmark and survey of automated machine learning frameworks (2019)

Input-Aware Neural Knowledge Tracing Machine

Moyu Zhang, Xinning Zhu$^{(\boxtimes)}$, and Yang Ji

Key Laboratory of Universal Wireless Communications, Ministry of Education, Beijing
University of Posts and Telecommunications, Beijing, China
{zhangmoyu,zhuxn,jiyang}@bupt.edu.cn

Abstract. Knowledge Tracing (KT) is the task of tracing evolving knowledge
state of each student as (s)he engages with a sequence of learning activities
and can provide personalized instructions. However, exiting methods such as
Bayesian Knowledge Tracing (BKT) and Deep Knowledge Tracing (DKT) either
cannot capture the relationship among different concepts or lack of interpretability.
Although Knowledge Tracing Machines (KTM) makes up for these shortcomings,
it only uses a linear function to model students' knowledge states, which cannot
capture more information contained in each feature. To solve above problems,
this work introduces a novel model called Input-aware Neural Knowledge Tracing Machine (INKTM) which can enhance the interpretability to some extent and
capture more complex structure information of real-world data to improve prediction performance. Unlike standard FM-based methods that focus on the feature
interactions, our model focuses more on the information contained in each feature itself and retains all 2-order feature interactions. By converting weights of
each feature to a multidimensional vector, our model can use the vectors to learn
a unique attention weight of each feature in different instances by an attention
network, so as to highlight important features and then enhance interpretability.
At last, we input re-weighted features to a deep neural network to capture the
non-linear and complex inherent structure of data. Experiment results show our
model can consistently outperform existing models in a range of KT datasets.

Keywords: Knowledge Tracing · Factorization Machine · Deep Learning

1 Introduction

Up to now, computer-aided education (CAE) systems have been widespread. With the
help of these systems, we can obtain a large amount of exercises logs of students answering problems which are labelled with some underlying knowledge concepts (KCs). A
KC can be a skill or a concept. For example, when a student is given a problem such as "3
$\times\, 2 + 7$", he has to master the skills of addition and multiplication at the same time. And
the goal of knowledge tracing (KT) is to trace the knowledge state of students based on
their past exercise. KT has been critical for both personalized learning and teaching. It
not only can help students choose what they need, but help teachers understand students'
mastery level of each KC. Therefore, KT require interpretable feedback.

© Springer Nature Switzerland AG 2021
A. Del Bimbo et al. (Eds.): ICPR 2020 Workshops, LNCS 12664, pp. 345–360, 2021.
https://doi.org/10.1007/978-3-030-68799-1_25

So far, there have been lots of researches about student modelling such as Bayesian Knowledge Tracing (BKT) [1] and Deep Knowledge Tracing (DKT) [2]. BKT can't model complex knowledge states transitions, and DKT has long been criticized for its parameters not being explainable. Although other variants have appeared, such as Dynamic Key-value memory network (DKVMN) [4] which enhances the interpretability to some extent, they still can't handle the issue of generalizing while dealing with sparse data. And the real-word data are always sparse as students tend to answer few problems.

Because Factorization Machines (FM) [5] is the most common and effective method for sparse data and then J.-J.Vie et al. proposed Knowledge Tracing Machines (KTM), which uses FM to estimate students' knowledge states [6]. By reconstructing input features, KT task is transformed from a time series prediction problem to a general binary classification problem in machine learning. KTM not only can handle the issue of highly sparsity data, also uses the number of students' correctly and wrongly answer exercises to predict, which can be more interpretable. For example, generally speaking, the more students' practice, the higher probability they answer problems correctly.

However, KTM only uses FM to model students' exercise, which is a linear function. Therefore, based on KTM, J.-J.Vie applied Deep Factorization Machine (DeepFM) [7] for KT, which is designed to learn high-order and 2-order feature interactions at the same time, but found it may performed worse than KTM [8]. The reason may be the input of models is sparse and multi-field categorical in KT. For example, if one problem is involved with multiple KCs, we can think of these KCs as belonging to the skill field, and in which, each KC can be regarded as a feature of the skill field. And we find that once one problem involves multiple KCs, DeepFM tends to lose some 2-order feature interactions in the same field, such as the interactions of KCs in skill field and we will discuss in detail in Sect. 3. Besides, there are researches have pointed out high-order feature interactions are not necessarily [9–11].

Therefore, it is quite challenging to further enhance the interpretability and improve the performance of KTM. To address these challenges, the present work introduces a new model called Input-aware Neural Knowledge Tracing Machine (INKTM) that combines the best of three worlds: the ability to enhance interpretability of KTM and differentiate the contribution of each feature, and the ability to capture non-linear and complex inherent structure of real-world data and the ability to be more generic.

Actually, in KT, the contribution of each field in different instances to the prediction can be different. For example, problems may have different difficulty-levels, and some problems are hard to solve no matter how many times students have attempted. And these problems will play a more important role in forecasting than attempts. We then use a vector which can contain more information than scalars, instead of the original one-dimensional weight to represent each feature. And we take advantage of this vector to learn a unique weight for each feature in different instances by an attention network to highlight their different contributions. The attention network not only can help improve the prediction performance, but also can enhance interpretability of our model. As mentioned above, the input of KTM has interpretability to some extent, but it's quite abstract because we can't make sure the contributions of each field in the input to the prediction. Our model INKTM can outputs the attention scores of each field, and it's easier for us to find which field is more important and understand the relationship between input and output. Finally, we adopt a deep neural network (DNN) to capture the complex inherent structure of real-world data.

Since INKTM can retain all information of 2-order feature interactions and mainly highlight the different contributions of each field, it avoids introducing unimportant information like high-order feature interactions which have been pointed out that sometimes are not necessarily, and then it is a quite generic model. To the best of our knowledge, INKTM is the first FM-based model to convert the one-dimensional weight to a vector and apply several neural networks on this vector so as to capture more useful information of each feature. In INKTM, we set two kinds of vectors to learn contributions of each field itself and the 2-order feature interactions, respectively.

Our main contributions can be summarized as follows:

- We proposed a novel neural network model INKTM that distinguishes the contributions of each feature to prediction by using an attention network and deep neural networks to capture the complex and non-linear structure of real-word data.
- Our model introduces a new kind of vector to capture more information, which is used to learn the unique weight for each field and can enhance the interpretability. Our model points out that not only features interactions matter for predicting, each feature itself can also make important contributions to the prediction.
- Our model is quite generic and can consistently outperform several famous models in KT field on three real-world datasets.

2 Related Work

In KT field, BKT is the most popular model. But it assumes knowledge state as a binary variable and can't handle the situation when one problem is related with multiple KCs. DKT leverages Long Short-Term Memory (LSTM) to model a student's knowledge state, but it lacks interpretability. And, several researches [3, 12, 13] have shown that factor analysis models which are really simple, like Item Response Theory and additive factor model could match the performance of DKT.

2.1 Factor Analysis

Item Response Theory (IRT) is a classical model in the educational literature. Rasch came up with his own model, which is regarded as the basis of IRT [14]. The IRT model is the simplest model for factor analysis:

$$y_{ij} = \theta_i - d_j \tag{1}$$

where θ_i stands for the ability of student i and d_j measures the difficulty of problem j. Although IRT is simple, it can perform better than DKT even without temporal features for the latter has many parameters to estimate and easily overfit [3]. Multidimensional Item Response Theory (MIRT) is the multidimensional version of IRT, but its performance has not improved because it's always hard to train [15].

Because IRT ignores the learning process, an additive factor model (AFM) was proposed [16, 17], which takes the number of exercises into account and then it can capture the dynamic learning process of student:

$$y_{ij} = \sum_{k \in KC(j)} \beta_k + \gamma_k N_{ik} \tag{2}$$

where N_{ik} stands for the number of times student i practice questions involved skill k. In this paper, we note $KC(j)$ the sets of skills involved with question j. Based on AFM, the performance factor analysis model (PFA) [18] regards positive attempt and negative attempt as two separate situations to offer higher sensitivity to student performance.

2.2 Knowledge Tracing Machine

Because students always interact with few problems, and it is difficult to extract useful information due to highly dispersed distribution of each sample. Therefore, J.-J.Vie et al. proposed Knowledge Tracing Machines (KTM), which uses FM to predict as below:

$$y_{ij} = \mu + \sum_{i=1}^{N} w_i x_i + \sum_{1 \leq i < j \leq N} \langle V_i, \ V_j \rangle x_i x_j \tag{3}$$

where N is the number of features in each event, μ is a global bias, w_i and $V_i \in R^k$ both are weights for feature i. KTM is a family of models including IRT, AFM and PFA as shown in [6]. However, KTM only uses Bayesian Factorization Machines (BFM) [19] which is a linear structure and can't capture more information of input.

2.3 DeepFM for Knowledge Tracing

J.-J.Vie tried to apply DeepFM for KT, which was proposed to learn both low-order and high-order feature interactions by means of two components, FM and DNN component. DeepFM has a shared input to these two components, where FM component can model order-2 feature interactions, and DNN component aims to learn high-order feature interactions. All parameters are trained jointly for combined prediction model:

$$y(x) = y_{FM} + y_{DNN} \tag{4}$$

However, because DeepFM has a shared input, and when one filed contains multiple features, the FM component will lose some information such as interaction in the same field, while FM uses all features to interact with each other. As a result, DeepFM can't achieve better performance than BFM. We will discuss this in the next section in detail.

3 Proposed Model

Usually, KT is formulated as a supervised sequence learning problem. Suppose we have users $U = [u_1, u_2, \cdots, u_m]$ and problems $V = [v_1, v_2, \cdots, v_n]$, where m and n denote the number of students and questions, respectively. Students can attempt one question several times and may learn between different attempts. We can obtain an interaction

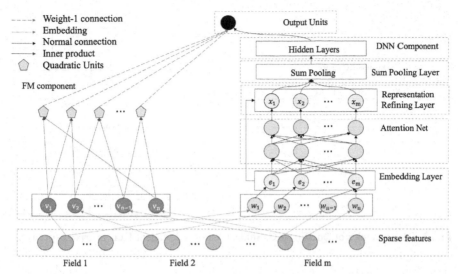

Fig. 1. The architecture of INKTM. The number of feature field and non-zero feature value in this example is m and n, respectively.

Fig. 2. An example of encoding features including student, problem, skills, wins and fails.

pair $x = (u, v)$ and the corresponding outcome $y \in \{0, 1\}$ (1 denotes correct). Given a student's past interactions pairs $X = [x_1, x_2, \cdots, x_t]$ and outcomes of each interaction. $Y = [y_1, y_2, \cdots, y_t]$, we want to predict outcome y_{t+1} for interaction pair x_{t+1}.

We think that although students and exercises are all important for prediction, their contributions can be different. Therefore, based on KTM, we design a novel model to learn a unique weight for each field, and capture non-linear and complex inherent structure of real-world data. Figure 1 shows our model structure, and in this section, we will describe the implementation of our model in detail.

3.1 Encoding and Embedding Layer

We can transform features into high-dimensional sparse features via one-hot encoding. For example, assume that we observed student 2 answered problem 3 which involves KCs 1, 2 and 3. The encoding of this instance is shown in Fig. 2:

Considering that students learn in the process of doing exercises, we choose the number of successful and failed attempts involved a KC to represent the opportunity for mastering it. In order to construct features related with attempts, we need to count the number of students answered problems involved with a specific KC as below:

$$w_{i,j,t} = \sum_{t=0}^{T} \delta\left(n_{i,j,t} = 1\right) \tag{5}$$

$$f_{i,j,t} = \sum_{t=0}^{T} \delta\left(n_{i,j,t} = 0\right) \tag{6}$$

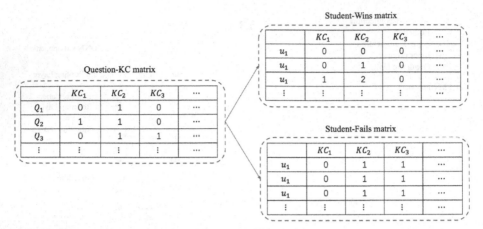

Fig. 3. The examples of Question-KC, Student-Wins and Student-Fails matrix, respectively.

where $n_{ij,t}$ denotes the answer of student i attempts exercise involved KC j at time t, and if answer is correct, $n_{ij,t}$ will be 1, otherwise it will be 0. $\delta(\cdot)$ denotes if condition within the function occurs, then it will be 1, 0 otherwise. $w_{ij,T}$ represents the number of correct attempts until time T, and $f_{ij,T}$ is the number of failed attempts.

We can use a Question-KC matrix to record the relationships among KCs and questions. A Question-KC matrix $O \in \{0, 1\}^{Q \times K}$, where Q denotes the number of questions and K is the number of KCs. If KC j is related with question i, O_{Q_i, K_j} is 1. Assume at time T, we get a set of observed, chronologically ordered triplets (1, 3, 0) (it means student u_1 attempted question 3 and answered it wrongly), (1, 1, 1), (1, 2, 1), (1, 3, 0). From the Question-KC matrix in Fig. 3, we can find that question 1 is related with KC 2, question 2 involves KCs 1, 2, and question 3 involves KCs 2, 3.

Figure 3 shows the Student-Wins matrix and Student-Fails matrix, which record the number of successful and failed attempts of student u_1 about each KC, respectively. For example, u_1 firstly failed answered question 3, and in the Student-Wins matrix, the counts of each KC remain unchanged, we only add the count by 1 of corresponding KCs in Student-Fails matrix like KC 2 and KC 3. At last, student u_1 has correctly answered exercises involved KC 1 once, KC 2 twice and KC 3 zero times. Similarly, student u_1 has wrongly answered exercises involved KC 1 zero times, KC 2 and KC 3 once. The encoding of features of attempts in this instance is shown in Fig. 2, which will be concatenated with the one-hot vector mentioned above as the input of INKTM:

Because students usually interact with few problems, the input is always sparse, and we need to compress them to a low-dimensional vector. Different from the traditional FM models, our model contains two embedding parts. Because one-dimensional weight may only capture a part of information of each feature, we convert the weight in FM to a vector to obtain more information. We refer w_i as feature vectors and refer V_i as embedding vectors. As Fig. 1 shows, V_i is utilized to learn 2-order feature interaction and w_i is used to learn a unique weight for each feature in different instances.

Because the questions always involve multiple KCs in reality, we need to sum feature vectors in the same field to make sure input vectors have the same length as Fig. 4 shows:

$E = [e_1, e_2, \cdots, e_m]$, where m denotes the number of fields, n stands for the number of non-zero values in input features, and $e_i \in R^d$ denotes the sum of feature vectors of the filed i. Especially, we only sum feature vectors w_i and keep embedding vectors v_i as they are. This way can retain more important low-level interactions information than DeepFM. Because the features in one field are multivalent, as [20] mentioned that the sum of feature embedding in the same field can be regard as the embedding of this field. And as a result, even the feature length and the number of instances is various, their embeddings are of the same length. Assume there are 2 fields, and embedding vectors V_1, V_2, V_3 belong to field 1, V_4, V_5 belong to field 2. Let $e_1 = V_1 + V_2 + V_3$, $e_2 = V_4 + V_5$. And then, the 2-order feature interactions will be $e_1, e_2 = V_1 + V_2 + V_3$, $V_4 + V_5$ in DeepFM, where, denotes the inner product operation. While in FM, the feature interactions are $\sum_{1 \leq i < j \leq 5} V_i, V_j$. Therefore, DeepFM truly will loses the information of feature interactions in the same field such as V_1, V_2, which our method can retain so as to be a quite generic model.

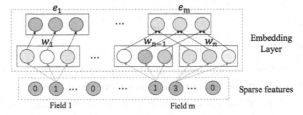

Fig. 4. The process of feature vectors embedded.

3.2 Attention Net

The attention mechanism allows different features to contribute differently to the target when compressing them to a single representation. As for KT task, features may have different effects on prediction. For example, if the difficulty-levels of problems are low, the features of attempts may account for a greater proportion of predicting. What's more, the unique attention scores of each field can make the model more interpretable. The attention scores can tell us which feature is more important for prediction, and the outcome of model will be more interpretable. Therefore, we propose to employ attention mechanism on the input feature field vectors e to learn unique weights:

$$h_1 = \sigma_1(W_1 E + b_1)$$
$$h_l = \sigma_l(W_l h_{l-1} + b_l) \tag{7}$$

where W_l, b_l, σ_l and h_l are the weight matrix, bias vector, activation function and the output of the l-th layer, respectively. And then we use the output vector of the last hidden layer h_l to learn a unique factor for each feature field in the input E:

$$a'_e = h_l H, H \in R^{t \times m} \tag{8}$$

$$a_{e,i} = m \times \frac{\exp\left(a'_{e,i}\right)}{\sum_{j=1}^{m} \exp\left(a'_{e,j}\right)} \tag{9}$$

The attention scores are normalized through the softmax function. And we add a real number m before the softmax function, which denotes the number of fields. So, the sum of all scores is equal to m, which can retain the probabilistic explanation of attention network and can emphasis the contributions of important fields to prediction.

3.3 Representation Refining and Sum Pooling Layer

After obtaining the outputs from attention layer, we use them to refine the input field vectors. The input is the input feature field vector $e_i \in R^d$ and attention scores $a_{e,i}$ from the attention layer. Formally, redefinition and sum process are as follows:

$$x_{e,i} = a_{e,i}e_i \tag{10}$$

$$X' = \sum_{i=1}^{m} x_{e,i} \tag{11}$$

3.4 Hidden Layers

Because FM still belongs to the family of linear models according to (3), and some researches mentioned out that linear method is sub-optimal due to the lack of learning process [21]. Therefore, in order to capture the non-linear and complex inherent structure of real-world data, we adopt DNN to model pooling vectors from previous layer:

$$\begin{aligned} d_1 &= \sigma_1\left(W_1^d X' + b_1^d\right) \\ d_l &= \sigma_l\left(W_l^d d_{l-1} + b_l^d\right) \end{aligned} \tag{12}$$

At last, we use a weight vector to transform d_l to the final prediction score:

$$f_1(x) = h^T d_l \tag{13}$$

3.5 Prediction Units

Assume embedding size is k and the number of features are n, we can reformulate (14) to reduce the runtime of FM from time complexity O (n^2) to O (kn):

$$f_2(x) = \sum_{i=1}^{n} \sum_{j=i+1}^{n} \langle V_i, V_j \rangle x_i x_j = \frac{1}{2} \sum_{f=1}^{k} \left[\left(\sum_{j=1}^{n} v_{x,j,f} x_j \right)^2 - \sum_{j=1}^{n} v_{x,i,j}^2 x_j^2 \right] \tag{14}$$

Finally, all parameters are trained jointly for the prediction function:

$$\hat{y}_x = w_0 + f_1(x) + f_2(x) \tag{15}$$

3.6 Optimization

At last, we can estimate model parameters by minimizing the binary cross-entropy loss. In order to prevent overfitting, we employ dropout and L_2 regularization:

$$L_c = \sum_{i=1}^{S} y_i logp(x_i) + (1 - y_i) \log(1 - p(x_i)) + \lambda ||\Phi||^2 \tag{16}$$

4 Experiments

4.1 Datasets

ASSISTments2009[1]. This dataset is described in [22] and we use the latest updated "skill-builder" dataset. Records without skill names are discarded in the preprocessing. After preprocessing, there are 3127 students, 17708 problems and 123 skills. We obtain 278425 observed instances in total. What need to point out is some problems only involve 1 skill while some involve 4 skills. The dataset is sparse as the sparsity is 0.005.

ASSISTments2012 **(See footnote 1).** ASSISTmentss2012 dataset contains 24750 students, 52976 problems, 265 skills, and 2692889 observed instances. It's worth mentioning that in this dataset, each problem only involves one skill. Besides, although the number of records in this dataset is larger than ASSISTments2009, it's sparser with a sparsity of 0.002.

Bridge Algebra2006[2]. Bridge to Algebra 2006–2007 stems from the KDD Cup 2010 EDM Challenge [23]. There are 1130 students, 129255 problems and 493 skills. We obtain in total 1816935 observed instances. some problems involve multiple KCs in this dataset. It's also a sparse dataset with a sparsity of 0.012.

We excluded concept tags that are unnamed to avoid noise for three datasets. The complete characteristics of three datasets are summarized in Table 1. Sparsity is equal to the total number of instances divided by the product of number of user and problems. And in this paper, we use two commonly used metrics for KT: AUC (Area Under the ROC curve) and ACC (Accuracy). ACC measures the accuracy of classifier classification results. And AUC is insensitive to class imbalance problem and can provides a robust metric for binary prediction evaluation.

4.2 Baselines

We compare INKTM with several famous models of the EDM literature including IRT, MIRT, AFM, PFA, KTM and DeepFM, which have been introduced in Sect. 2.

KTM [6]. It is learned using the MCMC Gibbs sampler implementation of libFM in C++ [24] which denotes BFM. LibFM is an official implementation of FM released by Rendle, and MCMC is highly praised by the author because it can achieve better performance than other optimizer such as Adagrad [25] without tuning parameters.

---

[1] https://sites.google.com/site/assistmentsdata/.

[2] https://pslcdatashop.web.cmu.edu/KDDCup/downloads.jsp.

Table 1. Statistics of the evaluation datasets.

Datasets	Users	Problems	Skills	Instances	Skills per problem	Sparsity
ASSISTments2009	3127	17708	123	278425	1.197	0.005
ASSISTments2012	24750	52976	265	2692889	1.000	0.002
Bridge Algebra2006	1130	129255	493	1816935	1.004	0.012

In this paper, we used the codes of DeepFM provided by Alibaba on GitHub. We also implemented our method using Tensorflow. And for a fair comparison, all methods use Adagrad (learning rate: 0.01) except KTM, which uses MCMC. As far as we know, many researches compare their models with libFM optimized with SGD. In fact, it's not really fair. Because their models always are optimized with Adagrad which is better than SGD. And we suggest that the future research should also compare with the best performance of libFM, rather than only the performance of being optimized with SGD.

In this paper, we performed 5-fold cross validation for all datasets. For each fold, 20% of the entries are held out as a test set and the remaining are used as the training set. For all experiments, the training batch sizes are set to 2048.

4.3 Student Performance Prediction

Because traditional models in KT field can't perform well with high embedding sizes for they're hard to train, and then we set the embedding size to 20 as other researches did [6] for fair comparison. We list the average test results of the evaluation measures in Table 2. The highest scores are denoted in bold. As Table 2 shows, our model can predict student performance more accurately in all datasets which shows the generalization ability. On the ASSISTments2009 dataset, the INKTM model achieves the average test AUC of 80.43%. KTM produces an average AUC of 79.05%, and DeepFM produces an AUC value of 77.97%. Other models perform worse. On ASSISTments2012 dataset, the test AUC of INKTM is 77.24%, which is better than 77.01% for KTM, 76.97% for DeepFM. With regard to Bridge Algebra2006 dataset, KTM gains the AUC of 78.26% and DeepFM obtains an AUC of 79.61%. However, our implemented INKTM leads to an AUC of 80.24%, outperforming all previous models. In order to seek the best embedding size of INKTM, KTM and DeepFM, we perform some experiments on different datasets. The results are listed in Table 3.

In conclusion, our model can perform better than other methods across all datasets, particularly on the Asistment2009 dataset whose problems are related with more skills, which demonstrates that INKTM can retain more useful information especially when knowledge components are multiple. And in reality, very few exercises are related to only one skill or concept. That's why DeepFM can't achieve the best performance in KT. Although DeepFM can capture high-order feature interactions, it will lose a part of.

2-order feature interactions information. While INKTM mainly highlights the contributions of important features and retain all 2-order feature interactions information. This way can effectively reduce noise interference and reinforce the generalization of INKTM. Therefore, INKTM consistently plays a positive role in three datasets.

Table 2. Overall performance of all models on three datasets with embedding size of 20.

Model	Embedding size	ASSISTments2009		ASSISTments2012		Bridge algebra2006	
		ACC	AUC	ACC	AUC	ACC	AUC
INKTM	20	**0.7648**	**0.8043**	**0.7538**	**0.7724**	**0.8492**	**0.8024**
KTM	20	0.7498	0.7905	0.7527	0.7701	0.8402	0.7826
DeepFM	20	0.7411	0.7797	0.7524	0.7697	0.8472	0.7961
IRT	0	0.7310	0.7651	0.7457	0.7602	0.8401	0.7692
MIRT	20	0.7309	0.7648	0.7456	0.7602	0.8402	0.7692
PFA	0	0.6997	0.7092	0.7106	0.6691	0.8392	0.7520
	20	0.7005	0.7091	0.7086	0.6689	0.8390	0.7518
AFM	0	0.6698	0.6316	0.6995	0.6108	0.8355	0.7112
	20	0.6701	0.6312	0.6995	0.6103	0.8355	0.7104

Table 3. Comparison of multidimensional models on datasets with different embedding sizes.

Model	Embedding size	ASSISTments2009		ASSISTments2012		Bridge algebra2006	
		ACC	AUC	ACC	AUC	ACC	AUC
INKTM	32	0.7709	0.8109	**0.7554**	**0.7736**	0.8494	0.8029
	64	0.7738	0.8162	0.7537	0.7725	**0.8497**	**0.8033**
	128	**0.7852**	**0.8287**	0.7524	0.7718	0.8496	0.8012
	256	0.7850	0.8277	0.7519	0.7710	0.8494	0.8011
KTM	32	0.7502	0.7910	0.7531	0.7707	0.8407	0.7844
	64	0.7507	0.7922	0.7540	0.7713	**0.8419**	**0.7862**
	128	0.7519	0.7923	0.7545	0.7717	0.8409	0.7861
	256	**0.7526**	**0.7930**	**0.7551**	**0.7729**	0.8402	0.7831
DeepFM	32	0.7414	0.7805	**0.7539**	**0.7713**	0.8475	0.7967
	64	0.7415	0.7815	0.7523	0.7701	0.8479	0.7976
	128	0.7435	0.7829	0.7512	0.7684	**0.8479**	**0.7984**
	256	**0.7456**	**0.7836**	0.7470	0.7645	0.8473	0.7961

What's more, from Table 2, we can easily observe that although IRT has been extended multidimension, the prediction performance has not been improved for MIRT has more parameters to learn which makes it become hard to train. That's the reason why MIRT are not frequently encountered in the EDM literature. Besides, PFA and AFM always perform worse than IRT. Because the PFA and AFM only consider the attempts and neglect the different abilities of students and problems have different difficulty. These results explain why INKTM achieves the best performance for INKTM can capture all useful features and learn their different contributions.

Table 4. The efficacy of each component on ASSISTments 2009 with embedding size of 256.

Method	ACC	AUC
INKTM	0.7860 ± 0.0020	0.8277 ± 0.0027
INKTM(the embedding size of w is set to 1)	0.7771 ± 0.0021	0.8131 ± 0.0023
INKTM(remove the attention component)	0.7814 ± 0.0024	0.8220 ± 0.0026
INKTM(remove the DNN component)	0.7788 ± 0.0041	0.8146 ± 0.0038

Table 5. The attention scores of each feature field on ASSISTments2009.

Instances	Model	e_1(user)	e_2(problem)	e_3(skill)	e_4(wins)	e_5(fails)	$|\bar{y} - y|$
1	KTM	e_1	e_2	e_3	e_4	e_5	0.07
	INKTM	$1.135 * e_1$	$1.293 * e_2$	$0.970 * e_3$	$0.824 * e_4$	$0.777 * e_5$	0.04
2	KTM	e_1	e_2	e_3	e_4	e_5	0.11
	INKTM	$1.134 * e_1$	$1.290 * e_2$	$0.971 * e_3$	$0.825 * e_4$	$0.779 * e_5$	0.05
3	KTM	e_1	e_2	e_3	e_4	e_5	0.15
	INKTM	$1.135 * e_1$	$1.291 * e_2$	$0.970 * e_3$	$0.825 * e_4$	$0.778 * e_5$	0.09
4	KTM	e_1	e_2	e_3	e_4	e_5	0.39
	INKTM	$1.136 * e_1$	$1.294 * e_2$	$0.970 * e_3$	$0.823 * e_4$	$0.776 * e_5$	0.24
5	KTM	e_1	e_2	e_3	e_4	e_5	0.11
	INKTM	$1.137 * e_1$	$1.298 * e_2$	$0.969 * e_3$	$0.822 * e_4$	$0.774 * e_5$	0.04

4.4 Ablation Study

In this section, we conduct ablation studies to investigate the effectiveness of some important components of our proposed model.

Feature Vectors Analysis

To further demonstrate the necessity of converting the weight w_i to a vector, we set its dimension size to 1 (converted back to a weight) and 256, respectively. The Table 4 shows the results on ASSISTments2009 dataset. INKTM denotes w_i with embedding size of 256, and we can observe that if we set the embedding size of w_i to 1, the AUC will degrade from 82.77% to 81.31%. The reason is that vectors can represent more information through multiple dimensions than the original one-dimensional weight.

DNN Component Analysis

INKTM leverages a DNN to capture more complex structure of data. Therefore, we remove the DNN component to demonstrate the effectiveness of DNN. From Table 4, we can see that if we remove this component, the AUC will degrade from 82.77% to 81.46%, which verifies that non-linear structure can capture more information.

Attention Mechanism Analysis
To provide more insights into the effect of attention component, we remove it from our model INKTM. As Table 4 shows, the AUC becomes 82.26%, which is lower than 82.77%. And in order to better demonstrate the effect of the attention network, we select several test examples to show attention scores of each feature field. Table 5 shows attention net can improve performance, where $\bar{y}$ is prediction value, and y is target value.

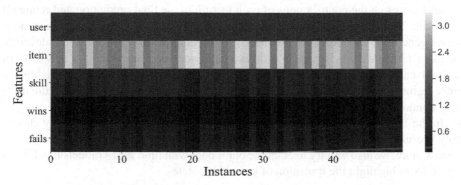

Fig. 5. Heatmap of the attention scores of features on Bridge Algebra2006.

In fact, attention network can also make model more explainable for the attention weight can tell us which feature is important. And then it will be easier for us to understand the prediction results. For example, if the attention score of item field is high and the student answered wrongly, we can think the reason may be the problem is too difficult. And from Fig. 5, we can easily find that user and problem are always given greater scores which means they contribute more to the final prediction on ASSISTments2009. As Table 2 shows that IRT performs only slightly worse than other state-of-the- art models, which only uses user and problem features to predict. The experiment results also demonstrate the reliability of the attention network.

What's more, we find the attention scores of each field are various in different datasets. As Fig. 5 shows, the attention scores of problems on Bridge Algebra2006 are quite high. However, from Table 5 we can observe the attention scores of each field are similar on ASSISTments2009. We believe that this is mainly due to the differences in users and problems on the platforms collected by each dataset. For example, if the difficulty-levels of problems in one dataset are similar, problem may get low attention scores because the final outcome will depend more on other features, like attempts.

5 Conclusion

This work proposes a novel model called INKTM which convert the weight of each feature to a new kind of vectors to effectively learn a unique weight for each field in different instances through an attention network and capture more complex structure of datasets by deep neural networks. INKTM not only can achieve the best performance among several famous models in KT, but also can enhance the interpretability of KTM by giving each field a unique attention weight.

What's more, compared with standard FM-based methods which only focus on the feature interactions and easily lose useful 2-order feature interactions information, our model focuses on the contributions of each feature to the final prediction and retain all important 2-order feature interactions information. In this way, our model can achieve strong generalization. Our work provides new research possibilities for future developments of FM models by retaining 2-order interactions and introducing a new kind of vector to capture more information. The experiment results also demonstrate that higher-order feature interactions are not really necessarily for prediction, and each feature itself truly contains some useful information.

In the future, we are particularly interested in exploring how to effectively choose truly useful 2-order feature interactions to further improve the accuracy of prediction. What's more, we also will try to combine our model with time series models like LSTM and RNN to highlight the transition of knowledge states.

Acknowledgements. This work is supported by the Beijing Education Reform Project "Research and Practice of AI-Driven New Cultivation Mode for Undergraduate Students Majored in Information and Communication" and "Research on the Evaluation Index of Self-learning in Mass Entrepreneurship and Innovation Education" and "Research on the Indicators of Freshmen's Autonomous Learning and Learning Situation Oriented to Three Cross Mass Entrepreneurship and Innovation".

References

1. Corbett, A.T., Anderson, J.R.: Knowledge tracing: modeling the acquisition of procedural knowledge. User Model. User-Adap. Inter. **4**(4), 253–278 (1994)
2. Piech, C., et al.: Deep knowledge tracing. In Advances in Neural Information Processing Systems (NIPS), pp. 505–513 (2015)
3. Wilson, K. H., et al.: Estimating student proficiency: deep learning is not the panacea. In Neural Information Processing Systems, Workshop on Machine Learning for Education, p. 3 (2016b)
4. Zhang, J., Shi, X., King, I., Yeung, D.-Y.: Dynamic key-value memory networks for knowledge tracing. In: Proceedings of the 26th international conference on World Wide Web, pp. 765–774. International World Wide Web Conferences Steering Committee (2017)
5. Rendel, S.: Factorization machines. In: Proceedings of 10th IEEE International Conference on Data Mining, pp. 995–1000 (2010)
6. Vie, J.-J., Kashima, H.: Knowledge tracing machines: factorization machines for knowledge tracing. In: Proceedings of the 33th AAAI Conference on Artificial Intelligence, pp. 750–757 (2019)

7. Guo, H., Tang, R., Ye, Y., Li, Z., He, X.: DeepFM: a factorization-machine based neural network for CTR prediction. In: Proceedings of the 26th International Joint Conference on Artificial Intelligence, pp. 1725–173 (2017)
8. Vie, J.-J.: Deep factorization machines for knowledge tracing. In: Proceedings of the Thirteenth Workshop on Innovative Use of NLP for Building Educational Applications, pp. 370–373 (2018)
9. Chen, C., Zhang, M., Ma, W., Liu, Y., Ma, S.: Efficient non-sampling factorization machines for optimal context-aware recommendation. In: Proceedings of The Web Conference, pp. 2400–2410 (2020)
10. Xiao, J., Ye, H., He, X., Zhang, H., Wu, F., Chua, T.-S.: Attentional factorization machines: learning the weight of feature interactions via attention networks. In: Proceedings of the 26th International Joint Conference on Artificial Intelligence, pp. 3119–3125 (2017)
11. Yu, Y., Wang, Z., Yuan, B.: An input-aware factorization machine for sparse prediction. In: Proceedings of the 28th International Joint Conference on Artificial Intelligence, pp. 1466–1472 (2019)
12. Xiong, X., Zhao, S., Inwegen, E. V., Beck, J.: Going deeper with deep knowledge tracing. In: Proceedings of the 9th International Conference on Educational Data Mining, pp. 545–550 (2016)
13. Wilson, K.H., Karklin, Y., Han, B., Ekanadham, C.: Back to the basics: bayesian extensions of IRT outperform neural networks for proficiency estimation. In: Proceedings of the 9th International Conference on Educational Data Mining, pp. 539–544 (2016a)
14. Rash, G.: Probabilistic Models for Some Intelligence and Attainment Tests. Danish Institute for Educational Research, Copenhagen (1960)
15. Desmarais, M.C., Baker, R.S.: A review of recent advances in learner and skill modeling in intelligent learning environments. User Model. User-Adap. Inter. 22(1–2), 9–38 (2012)
16. Cen, H., Koedinger, K., Junker, B.: Learning factors analysis–a general method for cognitive model evaluation and improvement. In: International Conference on Intelligent Tutoring Systems, pp. 164–175 (2006)
17. Cen, H., Koedinger, K., Junker, B.: Comparing two IRT models for conjunctive skills. In: International Conference on Intelligent Tutoring Systems, pp. 796–798 (2008)
18. Pavlik, P.I., Cen, H., Koedinger, K.R.: Performance factors analysis: a new alternative to knowledge tracing. In: Proceedings of the 14th International Conference on Artificial Intelligence in Education, pp. 531–538 (2009)
19. Freudenthaler, C., Schmidt, T.L., Rendel S.: Bayesian factorization machines. In: Proceedings of the NIPS Workshop on Sparse Representation and Low-rank Approximation (2011)
20. Lian, J., Zhou, X., Zhang, F., Chen, Z., Xie, X., Sun, G.: xDeepFM: combining explicit and implicit feature interactions for recommender systems, In: Proceedings of the 24th ACM SIGKDD International Conference on Knowledge Discovery & Data Mining, pp. 1754–1763 (2018)
21. Yu, Y., Wang, Z., Yuan, B.: CFM: convolutional factorization machines for context-aware recommendation. In: Proceedings of the 28h International Joint Conference on Artificial Intelligence, pp. 3926–3932 (2019)
22. Feng, M., Heffernan, N., Koedinger, K.: Addressing the assessment challenge with an online system that tutors as it assesses. User Model. User-Adap. Inter. 19(3), 243–266 (2009)
23. Stamper. J., Niculescu. A., Ritter, S., Gordon, G., Koedinger, K.: Algebra I 2005–2006 and Bridge to Algebra 2006–2007. https://pslcdatashop.web.cmu.edu/KDD--Cup/downloads.jsp

24. Rendle, S.: Factorization machines with libFM. ACM Trans. Intell. Syst. Technol. **3**(3), 1–22 (2012). article. 57
25. Duchi, J., Hazan, E.: Adaptive subgradient methods for online learning and stochastic optimization. J. Mach. Learn. Res. **12**, 2121–2159 (2011)

Towards Corner Case Detection by Modeling the Uncertainty of Instance Segmentation Networks

Florian Heidecker$^{(\boxtimes)}$, Abdul Hannan, Maarten Bieshaar, and Bernhard Sick

Universität Kassel, Wilhelmshöher Allee 73, 34121 Kassel, Germany
{florian.heidecker,a.hannan,mbieshaar,bsick}@uni-kassel.de
www.ies.uni-kassel.de

Abstract. State-of-the-art instance segmentation techniques currently provide a bounding box, class, mask, and scores for each instance. What they do not provide is an epistemic uncertainty estimate of these predictions. With our approach, we want to identify corner cases by considering the epistemic uncertainty. Corner cases are data/situations that are underrepresented or not covered in our data set. Our work is based on Mask R-CNN. We estimate the epistemic uncertainty by extending the architecture with Monte-Carlo dropout layers. By repeatedly executing the forward pass, we create a large number of predictions per instance. Afterward, we cluster the predictions of an instance based on the bounding box coordinates. It becomes possible to determine the epistemic position uncertainty for the bounding boxes and the classifier's epistemic class uncertainty. For the epistemic uncertainty regarding the bounding box position and the class assignment, we provide a criterion for detecting corner cases utilizing the model's epistemic uncertainty.

Keywords: Epistemic uncertainty estimation · Corner case detection · Object detection · Machine learning · Automated learning

1 Introduction

As we move more and more towards autonomous driving, the classification of objects and the perception of the environment is of great importance. The accuracy, reliability, and robustness of the machine learning algorithms used are also of great importance and must be improved continuously. As we see, a wide range of skills is required to master this task.

With this goal in mind, there is a need for different methods. An essential part of this is the detection of objects and their location [19,20,24] in an image. Additionally, there is the classification of objects, which in most current approaches is done at the pixel level and is known as semantic segmentation

---

F. Heidecker and A. Hannan—Contributed equally to this work.

© Springer Nature Switzerland AG 2021
A. Del Bimbo et al. (Eds.): ICPR 2020 Workshops, LNCS 12664, pp. 361–374, 2021.
https://doi.org/10.1007/978-3-030-68799-1_26

[22,25]. The combination of both methods is referred to as instance segmentation [9,18], which can separate and classify objects, from now on called instances, in the image data at the same time. Existing methods are powerful, but to make them even more robust and reliable, we present below our approach to epistemic uncertainty estimation and show how it can be used to detect corner cases. Corner cases are instances/samples, as shown in Fig. 1, which are difficult to classify, not covered by the training dataset, or underrepresented, and therefore represent a considerable learning advance for the model This learning advance can be explained by the fact that by iteratively adding more data, the situations covered by the model increase, and for situations where the results were previously poor, more training data is now available.

Fig. 1. The object detection and classification of this image is a real challenge for neural networks. If we only look at the red or yellow framed area, a model would almost certainly choose a motorcycle and, in the other case, a car. If we look at the whole vehicle (blue frame), we have to assume that the model is very uncertain so that we can consider this picture as a corner case. [23]

Our approach deals with the uncertainty modeling of Mask R-CNN [9] in a post-processing use case, for which we extend the architecture of the model by Monte-Carlo (MC) dropout layers, which allows us to model the epistemic uncertainty. Now consider for the sake of simplicity the classification part of the Mask R-CNN. The softmax layer at the output of the networks provides as output a probabilistic distribution for the individual classes (car, pedestrian, etc.). However, it does not provide information about the confidence in these probabilities, which can lead to overconfident predictions. It can only capture the aleatoric uncertainty, which is the uncertainty in the data, but not the epistemic uncertainty. The epistemic uncertainty describes the uncertainty of the model parameters [12]. For estimating this uncertainty, we use the MC dropout technique [8] in our approach. By implementing the MC dropout technique, we sample bounding boxes. However, it is not always clear which of the bounding boxes belongs to which instance. The subsequent clustering of the sampled bounding boxes allows for a prediction of the uncertainty. Besides the bounding box describing the location, the classifier provides a class probability for each bounding box. The clustering also allows us to determine the epistemic uncertainty for the classification. By modeling these two uncertainties, we can now implement a criterion to detect corner cases in the data.

As already mentioned, Fig. 1 shows a corner case example where we illustrate our approach of uncertainty estimation in an object detection use case. The vehicle looks like a car from the perspective of the yellow rectangle. From the perspective of the red rectangle, it is more likely to be classified as a motorcycle. However, from the perspective of the blue rectangle, the model type should be very uncertain. Since the uncertainties for both the bounding box and the class can be assumed to be very high, it should be possible to detect the scenario as a corner case based on these uncertainties.

The main contributions of this article are:

- We extend the architecture of Mask R-CNN by Monte-Carlo dropout layers to describe its epistemic uncertainty.
- We model two different epistemic uncertainties—first, the position uncertainty of the bounding boxes, and second, the class uncertainty of the classifier.
- We present an implementation for detecting corner cases based on epistemic uncertainties in an instance segmentation task.

The remainder of this article is structured as follows: In the following Sect. 2, we present the related work. In Sect. 3, we introduce the Mask R-CNN model, which we later modify according to our requirements. In Sect. 4, we introduce epistemic uncertainty modeling in neural networks. Afterward, we present our unique approach in Sect. 5 to estimate the epistemic uncertainty and to detect corner cases. In Sect. 6, we introduce the dataset and evaluate our results. This is followed by the conclusion and an outlook to future work in Sect. 7.

2   Related Work

Object detection, semantic segmentation, instance segmentation, region proposal networks, and many other methods based on deep neural networks deliver impressive results. The methods' output contains bounding boxes, object masks, instances, or the object class with the highest probability (softmax). However, what is missing is a prediction and distinction into aleatoric and epistemic uncertainty to evaluate how sure or uncertain the model is regarding the output.

To avoid the model output's uncertainty prediction, we can try to give the model only inputs, which can be processed with high reliability. Therefore out-of-distribution detection methods, e.g., [5,11,15,21] can be used. With such an out-of-distribution detection method, we obtain, in the best case, a precise prediction about whether the available input data is from the same distribution as the data used during model training. However, what we do not get is an indication/value that expresses the uncertainty of the model about its prediction.

It should also be possible to derive corner cases from misclassified and out-of-distribution examples whose detection is shown in [11].

Several methods use Bayesian Neural Networks (BNN) to estimate the uncertainties. For example, the work from [17] presents a method using prior networks for classification tasks that distinguish the distributional uncertainty, which arises due to data shift between training and test data, data uncertainty, and the model uncertainty. Another competitive approach to uncertainty estimation is the use of Monte-Carlo sampling, as found in Bayes by Backprop [3] and MC dropout [8].

On the other hand, ensembles [6] are widely used in machine learning and also used for uncertainty estimation [13,16]. An auspicious method is called Deep Ensembles [14], where an ensemble of neural networks is trained to estimate uncertainties.

A different approach is presented in [4], where they use mixture density networks for uncertainty estimation. They also show how uncertainty can decompose into explained and unexplained variances, which describes the ignorance about the model and noise.

In our approach, we use the MC dropout [8] technique for approximating the model uncertainty. To use this approximated uncertainty as an uncertainty measure for the object class and bounding box, we do not change the neural network's output layer in any way, but we use the predictions of several repetitions and cluster them. However, little research has been conducted in the field of modeling epistemic uncertainty in object detection tasks, i.e., bounding box regression, and object classification. Our approach goes beyond the state of the art by modeling the epistemic uncertainty with respect to the bounding box position and its dimensions as well as the actual object classification. With the knowledge gained from this, we try to detect corner cases not to exclude them

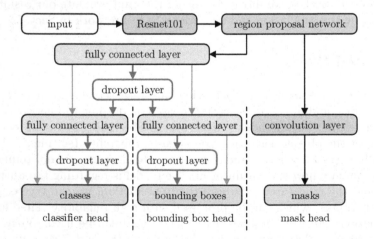

Fig. 2. The architecture of the Mask R-CNN model [9] and the modifications we introduced. We do not change the black connections of the Mask R-CNN architecture. The orange connections are replaced by the blue connections and the added MC dropout layers.

from the processes because the model is not capable like an out-of-distribution detection would; instead, we want to include them in future model training.

3   Mask R-CNN

To fully understand our approach and the changes made to the Mask R-CNN model, we give a brief introduction to this model here. For further details, we refer to [9].

In Fig. 2, the architecture of the original Mask R-CNN model is shown in black/orange arrows. The model takes an image as input and outputs for each instance, a bounding box, a class, and a mask via three different heads. First, the image passes the so-called backbone of the architecture. As a backbone, we use Resnet101 [10]. The features extracted by the backbone are then used by the Region Proposal Network (RPN) [20] to generate region boxes. In addition to generating region boxes, also called anchors, the RPN ranks the anchors according to the probability that they contain an instance. For this purpose, the network classifies the image data into the foreground and background and uses a regressor to optimize the bounding box around the instance. Then the topmost selected anchors are passed through fully connected layers of the classifier head to get the predicted class (e.g. car, truck, sidewalk) and the bounding box head to get a bounding box for each instance. In the mask head, the selected anchors are passed through convolution layers to get a mask for each instance. The difference between a bounding box and a mask is that the bounding box encloses the object with a box while the mask contains the predicted contour of the object.

4   Bayesian Uncertainty Modeling

Bayesian Neural Networks are a method to model the epistemic uncertainty of neural networks. We give a short introduction to BNNs (for further details see [3,8]).

In BNNs, we model the uncertainty regarding the weights $\mathbf{w}$ of a neural network using a probability distribution $p(\mathbf{w})$ over the weights. Given new training data $\mathbf{D}$, which comprises tuples of features and the respective targets, we update the prior distribution, i.e., we compute the posterior of the weights $p(\mathbf{w}|\mathbf{D})$. However, the computation is analytically not feasible. For this reason, we adopt the variational inference as a way to approximate this distribution. We search for a so-called variational distribution over the weights $\mathbf{w}$. We optimize this variational distribution to minimize the Kullback-Leibler divergence $KL(q(\mathbf{w})||p(\mathbf{w}|\mathbf{D}))$ between q and the posterior distribution $p(\mathbf{w}|\mathbf{D})$.

An equivalent way is the use of MC dropout because, as shown in [7], it can be interpreted as a variational approximation of the posterior distribution in a BNN. In our approach, we use the MC dropout technique as described in [8]. In each processing step during training, we drop a fixed percentage of random neurons in each layer and propagate the error back. In the testing setup, we pass the same image multiple times through the network each time with a fixed dropout rate. As a result, each time we draw different samples of $q(\mathbf{w})$, and this allows us to approximate the posterior distribution for that input by sampling.

5   Our Approach for Corner Case Detection

As mentioned before, our approach uses a Mask R-CNN as a basis. The functionality of the network does not change. However, we extend the architecture with dropout layers, as shown in Fig. 2, to estimate the epistemic uncertainty of the bounding box and classification path. In the following, we refer with the term uncertainty to the epistemic uncertainty. The figure also shows the mask generation path. However, we focus on the first path and leave this for future research. Furthermore, the design of our approach is not intended to be used for real-time applications, and we primarily intend to use it for the detection of corner cases in post-processing steps.

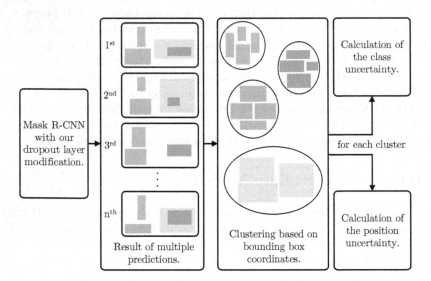

Fig. 3. Our Approach for uncertainty estimation via MC dropout.

5.1   Uncertainty Estimation

A visual description of our approach to the calculation of epistemic uncertainty is presented in Fig. 3. First, we train the modified model with the training dataset. After the training is completed, we use the test dataset's images and propagate each images through the network several times. In our case, each image is passed through the network 100 times. We could not see a better result with a larger number of repetitions. However, we have to keep in mind that the higher the number of repetitions, the longer the computing time, so 200 repetitions would double the runtime. Due to the dropout layers and a dropout rate of 10%, we get slightly different predictions each time, as indicated in Fig. 3 by the colored areas. Each of these predictions contains the bounding boxes found in the image and the classes of the instances. The number of predictions per input image is a hyperparameter and is adjustable. Besides the already mentioned sensitivity to the number of repetitions per image, the dropout rate also influences the result. A higher rate should generate more results with higher variance, but this influence needs further analysis.

To predict the epistemic uncertainty of the bounding boxes and the class, we first have to cluster the predictions as illustrated in Fig. 3. For the clustering, we use the bounding boxes' coordinates, so it becomes possible to cluster the bounding boxes of the same instance across all predictions based on the instance location. In the end, we get a cluster of slightly varying bounding boxes for each detected instance in the image. To cluster the bounding box coordinates, we use the Dirichlet process mixture models [2]. This method allows us to obtain the clusters and their co-/variances directly from the clustering process, which we can use to calculate the position uncertainty for the bounding boxes of this cluster/instance.

Due to the previous clustering, we know which bounding boxes belong to which instance. This allows us to calculate the class probabilities of each cluster, consisting of mean and variance for each class. The class with the highest mean value determines the class for the cluster and the instance.

5.2   Corner Case Detection

Based on the determined class uncertainty of the classifier and the position uncertainty of the bounding boxes, we present two example criteria for the detection of corner cases. The data from the detected corner cases can then be used to extend an existing training dataset and to improve the model iteratively by labeling the recorded data later and finetune the model afterward.

Nevertheless, before that, we have to detect and filter out corner cases. The iterative process we use, which iterates through each instance of each image, is shown in Algorithm 1. The decision about a possible corner case is based on the two criteria. If we detect a corner case, we save the instance/image and continue processing with the next image.

Algorithm 1. Corner Case Detection

---
    for each *image* **do**
        for each *instance* **in** *image* **do**
            if criterion(*class*) **or** criterion(*bbox*) **then**
                buffer.append(*instance*, *image*)
                break
            end if
        end for
    end for

First, we would like to discuss the definition of the criterion for class uncertainty. Let us take a look at Fig. 6b. This figure represents a single instance from the image Fig. 6a, the class label was decided in favor of the truck because of the greater mean value (red line). The variance of the value is not taken into account at all. However, the interquartile distance points out that the network is not sure about the prediction. The criterion considers the maximal class standard deviation $class_{std_{max}}$ of any class in each instance, and if the value is greater than a threshold t_{class}, we assume the instance to be a corner case:

$$\text{criterion}(class) = \begin{cases} 1 & \text{if } class_{std_{max}} > t_{class} \\ 0 & \text{else} \end{cases}. \tag{1}$$

Next, we consider the criterion for position uncertainty. For this, we first take a look at the red-colored instance in Fig. 5c. This bounding box, like all the others, is determined by its position, width, and height. These values are given in pixels by the output of the Mask R-CNN. Thus, the resulting position uncertainty is also given in pixels. This fact makes it challenging to define a threshold t_{bbox} for each of the four edges to decide if the present bounding box standard deviation $bbox_{std}$ of the position uncertainty $bbox$ is high or low. The reason for this is that a small standard deviation of pixels may be negligible for a large bounding box, but for a much smaller bounding box, it may be the opposite. Therefore, we scale the $bbox_{std}$ of the left/right edge according to the $bbox_{height}$ and the top/bottom edge by the $bbox_{width}$ respectively. In case of the left edge it means:

$$bbox_{std\_left_{norm}} = \frac{bbox_{std\_left}}{bbox_{height}}, \tag{2}$$

which makes the scaled $bbox_{std\_left_{norm}}$ independent of the size of the bounding box. Because the standard deviation of the bounding box is now normalized and a threshold value can be effectively applied:

$$\text{criterion}(bbox) = \begin{cases} 1 & \text{if } bbox_{std_{norm}} > t_{bbox} \\ 0 & \text{else} \end{cases}. \tag{3}$$

6   Experimental Setup and Result Analysis

First, we describe the dataset which we used in our investigations, and then we discuss the evaluation of our approach.

6.1   Dataset

For the evaluation of our approach, we use the publicly accessible driving dataset provided by Audi [1]. The dataset contains images, semantic segmentation, point clouds, and bounding boxes with labels. To record the data, Audi used several cameras and lidar sensors that cover all four sides of the vehicle. In our approach, we focus on the data from the front camera. The dataset contains data from three

Fig. 4. Ground truth example taken from the Audi driving dataset.

cities with different situations of road traffic and environmental conditions. Our training dataset consists of 65% randomly selected images, while the validation dataset consists of 15%, the remaining 20% constitutes the test dataset.

The dataset has 38 semantic segmentation categories. Because some of these categories are very similar, we reduce the categories. We use only eight classes, which are background, cars, trucks, pedestrians, bicycles, sidewalks/curbs, drive-able areas, and traffic signals. The reason for reducing the categories is that our focus is not on high detection accuracy or instance segmentation scoring, but on modeling the uncertainty for detecting corner cases. A sample figure with visualized ground truth is shown in Fig. 4.

(a) Ground truth.   (b) Model mask result.

(c) Model bounding box result.

Fig. 5. The withe dotted line in (c) represents the mean of the predicted bounding boxes, while the filled colored area represents two standard deviations of the corresponding edge.

6.2   Qualitative Evaluation

In the following, we carry out a qualitative evaluation of the results of our approach. For this purpose, we first address the position uncertainty of the bounding box and then the class uncertainty of the classifier.

Position Uncertainty. First, we consider the position uncertainty of the bounding boxes. Figure 5 shows an example. Each white dotted bounding box in the Fig. 5c shows the mean bounding box of all predicted bounding boxes for this particular instance. The color annotated areas of each edge of the bounding box reflect the positional uncertainty of the respective bounding box edge. For example, on the right edge of the yellow bounding box, which represents the sidewalk, the position uncertainty is very high. This circumstance is understandable because the network did not reliably recognize the sidewalk, as we compare Fig. 5b and the ground truth in Fig. 5a. The same also happens to the edges of the red bounding box, which represents the drivable area. Apart from that, the upper edge of the red bounding box indicates that the model does not recognize the farther away section of the road, as the uncertainty at this edge is also rather low. In comparison to this, the position uncertainty of the car is much smaller. Only the two edges that are close to the front and rear of the car have a slightly higher position uncertainty. This slightly higher position uncertainty is, in our understanding, due to the motion blur of dynamic objects.

Classifier Uncertainty. Next, we consider the class uncertainty deriving from the class distribution of each cluster, Fig. 6 shows an example. For visualization purposes, we do not show the position uncertainty in this figure. Instead, we show the colored bounding box, and the label followed by the class probability, and the standard deviation, which we consider to be the class uncertainty. In Fig. 6a, we see several vehicles whose class uncertainty mostly can be classified as rather low. The two cars in the background have a higher standard deviation, which makes sense because they are both slightly farther away. Only the orange car (van) on the other side of the median strip has a much higher standard deviation. The reason for this can be seen in the box plots shown in Fig. 6b. The class probabilities of the two classes car and truck are different, with the mean probability for truck outweighing the mean probability for car. However, the interquartile range for both classes is high, which indicates a high degree of class uncertainty in the network. Moreover, the figure shows that the interquartile distance is much smaller for other classes than car and truck. These small values mean that the network is sure that these objects do not belong to any of those classes. Next, we look at another example in Fig. 7 that also deals with class uncertainty. Figure 7a shows the original image in this example, while Fig. 7b shows the prediction of the network. For the two trucks and the driveable area, the class probabilities are above 0.97, while the standard deviations are small at 0.07, 0.004, and 0.001, respectively. These low values indicate that the network is confident about its prediction. For the object detected as a car (red), which is a

(a) The image shows the prediction with the name of the label, the mean value for that particular class, and the standard deviation, which we interpret as the class uncertainty.

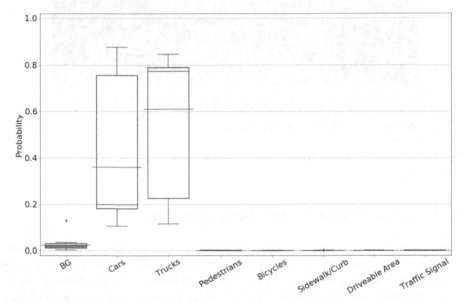

(b) The figure shows the class probabilities/box plots of the truck surrounded by an orange box in Fig. 6a. The red line describes the mean, and the blue line the median.

Fig. 6. This example shows the prediction of the model (a), and the class probabilities/box plots of one instance (b).

trailer with a car on top, the situation is slightly different. The class probability is relatively high, and the standard deviation is also quite low, but the object is not a car, it is just very similar to one. This example shows that the detection of corner cases is not perfect due to the uncertainties and further improvements are necessary.

(a) Original image.

(b) The image shows the prediction with the name of the label, the mean value for that particular class, and the standard deviation, which we interpret as the class uncertainty.

Fig. 7. This example shows the original image (a), and the prediction of the model (b) for some instances.

7   Conclusion and Future Improvements

In this article, we present an approach to estimate the epistemic uncertainty for an instance detection model. We calculate the position uncertainty of the bounding box and the class uncertainty of the classifier for each instance separately. We then use the estimated uncertainties to detect corner cases in large amounts of data and store them for later use. Our unique approach of modifying an instance detection model with MC dropout layers and combining it with clustering proved to be simple but yet powerful and effective for estimating epistemic uncertainty. But there are still some limitations to this approach.

A point that we want to examine in more detail is the bounding box prediction and the clustering based on that predictions. Currently, bounding boxes of two small instances close to each other are combined in one cluster because there are often not enough samples. We have to sample a lot more to be able to resolve the cluster into two. However, more samples mean more calculations and repetitions. In the future, we would like to include the predicted masks, which represent a more precise object outline based on the contour, in the clustering process.

Furthermore, in this article, we only estimated the position uncertainty of the bounding boxes and the class uncertainty of the classifier and used it to detect corner cases. The instance segmentation additionally provides masks for each object, whose epistemic uncertainty can also be approximated by an MC dropout layer. We want to use this epistemic uncertainty in our future work and integrate it into the approach presented here. We hope that masks will improve the clustering and the detection of corner cases.

It is also necessary to further analyze the presented criteria for corner case detection and adapt them accordingly. Another open point is how a quantitative evaluation for corner cases could look like since there is no ground truth. In the future, we will also have to investigate the difference between data-driven corner cases and corner cases that are obvious to humans.

Acknowledgment. This work results from the project KI Data Tooling (19A20001O) funded by BMWI (Deutsches Bundesministerium für Wirtschaft und Energie/German Federal Ministry for Economic Affairs and Energy) and the project DeCoInt^2 supported by the German Research Foundation (DFG) within the priority program SPP 1835: "Kooperativ interagierende Automobile", grant number SI 674/11-2.

References

1. Audi AG: Driving Dataset (2019). https://www.a2d2.audi/a2d2/en.html. Accessed 12 Apr 2020
2. Blei, D., Jordan, M.: Variational inference for dirichlet process mixtures. J. Bayesian Anal. **1**, 121–144 (2006)
3. Blundell, C., Cornebise, J., Kavukcuoglu, K., Wierstra, D.: Weight Uncertainty in Neural Network. In: Proceedings of the 32nd ICML, vol. 37, pp. 1613–1622. PMLR, Lille, France (2015)
4. Choi, S., Lee, K., Lim, S., Oh, S.: Uncertainty-aware learning from demonstration using mixture density networks with sampling-free variance modeling. In: 2018 IEEE ICRA, pp. 6915–6922. IEEE, Brisbane, QLD, Australia (2018)
5. DeVries, T., Taylor, G.W.: Learning Confidence for Out-of-Distribution Detection in Neural Networks arxiv:1802.04865v1 (2018)
6. Dietterich, T.G.: Ensemble methods in machine learning. In: Kittler, J., Roli, F. (eds.) MCS 2000. LNCS, vol. 1857, pp. 1–15. Springer, Heidelberg (2000). https://doi.org/10.1007/3-540-45014-9_1
7. Gal, Y.: Uncertainty in Deep Learning. Ph.D. thesis, University of Cambridge (2016)
8. Gal, Y., Ghahramani, Z.: Dropout as a Bayesian Approximation: Representing Model Uncertainty in Deep Learning. In: Proceedings of The 33rd ICML, vol. 48, pp. 1050–1059. JMLR.org, New York (2016)
9. He, K., Gkioxari, G., Dollar, P., Girshick, R.: Mask R-CNN. In: 2017 IEEE ICCV, pp. 2980–2988. IEEE, Venice, Italy (2017)
10. He, K., Zhang, X., Ren, S., Sun, J.: Deep Residual Learning for Image Recognition. In: 2016 IEEE CVPR, pp. 770–778. IEEE, Las Vegas, NV, USA (2016)
11. Hendrycks, D., Gimpel, K.: A Baseline for Detecting Misclassified and Out-of-Distribution Examples in Neural Networks arxiv:1610.02136v3 (2017)
12. Hüllermeier, E., Waegeman, W.: Aleatoric and Epistemic Uncertainty in Machine Learning: A Tutorial Introduction (2019)
13. Ilg, E., Cicek, O., Galesso, S., Klein, A., Makansi, O., Hutter, F., Brox, T.: Uncertainty estimates and multi-hypotheses networks for optical flow. In: ECCV, pp. 652–667. Munich, Germany (2018)

14. Lakshminarayanan, B., Pritzel, A., Blundell, C.: Simple and Scalable Predictive Uncertainty Estimation Using Deep Ensembles. In: Proceedings of the 31st NIPS, pp. 6405–6416. Curran Associates Inc, Red Hook, NY, USA (2017)
15. Liang, S., Li, Y., Srikant, R.: Enhancing The Reliability of Out-of-distribution Image Detection in Neural Networks arxiv:1706.02690v4 (2018)
16. Liu, J.Z., Paisley, J., Kioumourtzoglou, M.A., Coull, B.: Accurate uncertainty estimation and decomposition in ensemble learning. In: Proceedings of the 33rd NIPS, pp. 8952–8963. Curran Associates Inc, Vancouver, Canada (2019)
17. Malinin, A., Gales, M.: Predictive Uncertainty Estimation via Prior Networks. In: Proceedings of the 32nd NIPS, pp. 7047–7058. Curran Associates Inc, Red Hook, NY, USA (2018)
18. Pinheiro, P.O., Collobert, R., Dollár, P.: Learning to Segment Object Candidates. In: Proceedings of the 28th NIPS, vol. 2, pp. 1990–1998. MIT Press, Cambridge, MA, USA (2015)
19. Redmon, J., Divvala, S., Girshick, R., Farhadi, A.: You Only Look Once: Unified, Real-Time Object Detection. In: 2016 IEEE CVPR, pp. 779–788. IEEE, Las Vegas, NV, USA (2016)
20. Ren, S., He, K., Girshick, R., Sun, J.: Faster R-CNN: towards real-time object detection with region proposal networks. IEEE TPAMI 39(6), 1137–1149 (2017)
21. Shalev, G., Adi, Y., Keshet, J.: Out-of-Distribution Detection using Multiple Semantic Label Representations. In: Proceedings of the 32nd NIPS, pp. 7375–7385. Curran Associates Inc. (2018)
22. Shelhamer, E., Long, J., Darrell, T.: Fully convolutional networks for semantic segmentation. IEEE TPAMI 39(4), 640–651 (2017)
23. Sieman, R.: Strange Off-Road Dirt Bikes & Vehicles (2013), https://www.off-road.com/dirtbike/feature/strange-offroad-dirt-bikes-vehicles-53605.html?printable Accessed 12 Apr 2020
24. Uijlings, J.R.R., van de Sande, K.E.A., Gevers, T., Smeulders, A.W.M.: Selective search for object recognition. IJCV 104(2), 154–171 (2013)
25. Yuan, Y., Chen, X., Wang, J.: Object-Contextual Representations for Semantic Segmentation arxiv:1909.11065v2 (2019)

Intelligent and Interactive Video Annotation for Instance Segmentation Using Siamese Neural Networks

Jan Schneegans$^{(\boxtimes)}$, Maarten Bieshaar, Florian Heidecker, and Bernhard Sick

Intelligent Embedded Systems, University of Kassel,
Wilhelmshöher Allee 73, 34121 Kassel, Germany
{jschneegans,mbieshaar,florian.heidecker,bsick}@uni-kassel.de
www.ies.uni-kassel.de

Abstract. Training machine learning models in a supervised manner requires vast amounts of labeled data. These labels are typically provided by humans manually annotating samples using a variety of tools. In this work, we propose an intelligent annotation tool to combine the fast and efficient labeling capabilities of modern machine learning models with the reliable and accurate, but slow, correction capabilities of human annotators. We present our approach to interactively condition a model on previously predicted and manually annotated or corrected instances and explore an iterative workflow combining the advantages of the intelligent model and the human annotator for the task of instance segmentation in videos. Thereby, the intelligent model conducts the bulk of the work, performing instance detection, tracking, and segmentation, and enables the human annotator to correct individual frames and instances selectively. The proposed approach avoids the computational cost of online retraining by being based on the one-shot learning paradigm. For this purpose, we use Siamese neural networks to transfer annotations from one video frame to another. Multiple interaction options regarding the choice of the additional input data to the neural network, e.g., model predictions or manual corrections, are explored to refine the given model's labeling performance and speed up the annotation process.

Keywords: Instance detection · Tracking · Segmentation · Semi-supervised · Labeling · Video annotation · Object detection · One-shot learning

1 Introduction

Despite advances in transfer, semi- and unsupervised learning techniques, the training, testing, and evaluation of machine learning models requires vast amounts of labeled data [2]. Especially in the domain of image processing, i.e., object detection, tracking, and segmentation, the annotation of images is time-consuming, hard to automate, and thus very costly. It requires human annotators, often with

© Springer Nature Switzerland AG 2021
A. Del Bimbo et al. (Eds.): ICPR 2020 Workshops, LNCS 12664, pp. 375–389, 2021.
https://doi.org/10.1007/978-3-030-68799-1_27

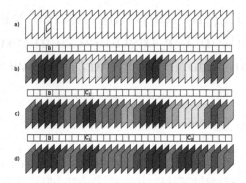

Fig. 1. The envisioned workflow using the intelligent annotation tool. The human annotator and the intelligent model work iteratively to annotate a video until reaching a
satisfactory annotation quality (shown in blue, darker is better). a) bounding box initialization (B) in an unlabeled video, b) initial prediction by the model, c) prediction
after the first correction (C_1) by the human annotator, d) prediction after the second
correction (C_2).

specialized proficiencies in specific domains, e.g., medical imaging, to thoroughly
examine each image, applying their knowledge about scenes, objects, and relationships between them. With rising applicability and use of machine learning models,
the demand for high-quality data increases. The automation of labeling processes
can alleviate this need. Even state-of-the-art machine learning models do not provide faultless and fully automated annotation capabilities. Thus, partial supervision by human annotators and machine-human interaction is still required, which
manifests in the use of tools assisting in the labeling process. This article proposes
a new methodology, which enables the human annotator to collaborate with an
intelligent model, i.e., a deep neural network. While the intelligent model handles
the bulk of the work, the human annotator provides corrections of image annotations. The intelligent model is a Siamese neural network that implements the one-
shot learning paradigm [7,9]. In the one-shot learning setting, we must correctly
make predictions given only a few training examples. With the aid of the Siamese
neural network, we exploit the temporal relationship between frames and annotations. In this way, we can quickly transfer annotations between frames without
the time-consuming retraining of the model.

Figure 1 depicts the envisioned workflow of annotating videos. We start with
an unlabeled sequence of video frames, where the human annotator provides
an initialization, e.g., a bounding box around an instance, marked as B. The
intelligent model now annotates the video. The annotation quality, e.g., bounding
box position or segmentation accuracy, here in blue, varies among the frames.
Hence, the human annotator monitors the predicted annotations and decides to
perform a correction, marked as C. The model supports the human annotator by
selecting frames or objects, e.g., based on confidence, before incorporating this
correction into its prediction, e.g., by using it as additional input. This process
iteratively improves the annotation quality around the predicted frames and is
repeated until reaching a satisfactory annotation quality.

The underlying intelligent model builds on the SiamMask network [17], which performs instance detection, tracking, and segmentation in a single model. SiamMask does not natively support incorporating additional inputs, i.e., the ability to process corrections made by a human annotator. To this end, we build a second model, i.e., interaction model, on top of SiamMask that can be conditioned on additional input and examine the effect of different combinations regarding the input data, e.g., human annotations. As the contour around an instance poses the most crucial part in the segmentation task, a pixel-wise weighting of the training loss emphasis those regions around the outline. Overall, we investigate three approaches with varying input data, i.e., additional feedback to the model, and a baseline without additional input data for comparison.

The main contribution of this article accounts to:

- The extension of the base SiamMask model to incorporate corrected annotations.
- A pixel-wise weighting of the training loss for training the interaction model, emphasizing the instance contour and distant instance parts.
- Evaluation of different combinations of conditioning input data to guide model inference without the need for online retraining.
- The intelligent video annotation itself, enabling human interaction and correction of annotations via brush and polygon utensils. We append a publicly available implementation of the proposed tool, including the pre-trained SiamMask networks[1].

The remainder of this article is structured as follows: Sect. 2 reviews previous works regarding intelligent and interactive instance segmentation tools. Section 3 shows how the underlying intelligent models extend the tool. This section also elaborates on the training procedure and describes the novel loss weighting. Afterward, Sect. 4 briefly describes the utilized dataset and metrics, followed by Sect. 5, which evaluates the approaches of incorporating corrected annotations. Section 6 further discusses qualitative results. Section 7 presents the conclusion and directions towards future work.

2   Related Work

The overall goal to reduce the labeling effort comes down to automating more parts of the labeling process while reducing user interaction. Methods for lowering video labeling time can be divided into two approaches: One that reduces the number of clicks per image and one that tries to reduce the number of images to annotate. The latter presents a semi-supervised labeling process.

Regarding a reduction in clicks per image, previous approaches [1,4] accelerate the annotation process by fitting a polygonal approximation around instances. While reducing the number of clicks for corrections and providing considerable time savings, this method produces less accurate annotations depending on the number of points. With this particular approach, also every frame is

---

[1] https://git.ies.uni-kassel.de/public_code/intelligent_video_annotation_tool.

annotated separately. Alternative approaches propose to mark foreground and background classes with rough strokes for segmentation [11] or single clicks for object detection [14].

Semi-supervised techniques, such as active learning, aim at minimizing the annotation time by intelligently selecting individual frames to annotate, thus, reducing the necessary annotator interactions [16]. Often, the first frame provides an initialization, requiring a single bounding box around an instance [17]. Based on this initial reference, the employed model then detects and extracts features from the marked instance and uses these to track and segment it throughout the remaining video. This approach reduces human interaction with the annotating system to a minimum. Still, it relies heavily on detecting, tracking, and segmentation parts of the model to function without errors. Otherwise, these errors propagate, and the labeling quality deteriorates quickly. In [5], instance tracking facilitates the propagation of annotations.

From the methodological point of view, one- or few-shot learning approaches are also closely related [7,9,18]. Here, the aim is to learn from a few labeled instances. Another related research field is object co-segmentation [15], defined as the joint segmentation of similar objects in multiple images or video frames. Both research fields are similar to the interactive and intelligent annotation approach that we aim at, where a segmentation, i.e., annotation, from one image, is transferred to other images. However, in the context of one-shot learning or co-segmentation, the focus is mostly on detection or segmentation and less on integrating the techniques into an interactive annotation process.

While there exist plenty of commercial platforms on the Internet to outsource the labeling effort to a human workforce, open tools for intelligent data labeling are scarce. [3] being one of the few approaches providing a holistic toolset for manual, semi-automatic, and automatic annotation.

3   Towards Intelligent Annotation

The intelligent model used in this work consists of two parts. SiamMask makes up the base model, which performs feature extraction, tracking, and segmentation of the selected instance, i.e., the chosen object to be tracked and segmented, throughout the video. The second part consists of a neural network, i.e., the interaction model, that facilitates additional inputs to incorporate manually corrected annotations. It takes the architecture of a U-Net [6]. The following gives a brief overview of the two utilized model architectures.

3.1   Base Model

SiamMask is a neural network that works in a semi-supervised manner, combining instance detection, tracking, and segmentation, in one architecture. The first detection step is performed by a ResNet model that extracts features in and around the given bounding box. The second part of the SiamMask model utilizes the extracted features in a region proposal network, predicting a proposed bounding box's position and size. Another separate part of the model is performing the segmentation task based on the extracted features. In this work, we use the detection and tracking parts for finding instances and cropping images to the area around the instance before segmentation. One of the approaches examined in this work will employ the segmentation part to create additional input to the interaction model.

The SiamMask model uses two input images: one annotated image, in which a bounding box marks an instance, and a second image, in which the instance is to be tracked and segmented. Based on the first image, features are extracted and used to search for the instance in the second image. Initially, the tracked instance features were extracted once from the first annotated frame of the video and then utilized in the remaining frames. We changed this so that the automated annotation process can be initialized from any frame in the video. The detection, tracking, and segmentation can then executed in both forward and backward directions. In the forward direction, features used for tracking and segmentation are extracted from the previous frame. And in the backward direction by the succeeding image frame.

Both methods have advantages and disadvantages: initializing once with the first image frame is faster because features are extracted only once. This approach might be more robust because errors in the bounding box prediction and subsequent feature extraction are not propagated, which is the major drawback of iterating over previous predictions. However, this method relies on a sound and robust feature representation of the annotated instance, which, in practice, is often lackluster, e.g., as the lighting condition change throughout a video, an instance is no longer recognized correctly. Features from the previous frame can capture such changes, but then the model is more prone to switching to another close and similar instance during tracking. This second method also enables a forward and backward pass, as explained above.

3.2 Interaction Model

To enable the human annotator to influence the segmentation process, further input to the model is needed. To this end, we extend the existing SiamMask model with a new network built on top of it. This network takes the form of a U-Net [6]. Figure 2 depicts the model architecture used in this work. Contrary to the original architecture, padding is applied during convolution so that there is no decrease in spatial dimensions towards the final output. In this way, we circumvented the possibility of cutting off parts of the segmented instances. Additionally, the downsampling layers apply average-pooling, upsampling layers consist of interpolation (factor two) followed by convolutions, and the nonlinearities take the form of ELU activation functions. The training procedure is restricted to the interaction model and uses the AdamW optimizer [10].

The interaction model was trained in an offline manner receiving additional inputs so that its predictions can later be conditioned interactively on manually

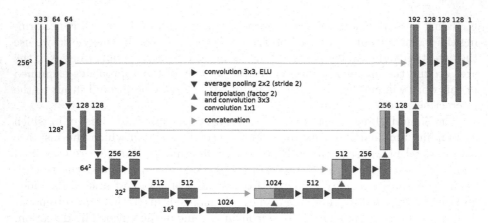

Fig. 2. The interaction model extends the base model to incorporate corrected image annotations. It is based on the U-Net architecture but differs in the spatial dimensions, up- and down-sampling layers, and the activation function.

corrected data. Thus no online training is required, as this would prohibit real-time interactions.

3.3   Weighted Binary Cross-Entropy Loss

As a loss function, the binary cross-entropy was computed on a per-pixel basis and then weighted to highlight three critical aspects that the model should focus on when comparing predictions to the binary target masks. These three aspects are:

a) the class balancing,
b) the distance from the instance contour,
c) and the distance from the instance center.

During instance segmentation, the goal is to assign each pixel to either a foreground or background class. The foreground class represents the instance. Typically, there are more background pixels than foreground pixels in an image. Therefore, these two classes should be weighted so that the model does not collapse to the prediction of only the background class. For this, we calculate the relative frequency of one of the classes (here foreground) as

$$\alpha = \frac{\#\ foreground\ pixels}{\#\ overall\ pixels}.$$

To emphasize the instance contour, as it is the crucial part of the predicted masks, pixels belonging to it should be weighted higher than the surrounding pixels. By placing a Gaussian kernel over the contour, the surrounding pixels are weighted by their distance to the nearest contour pixel. Previous work showed a performance increase with this technique [6]. The distance-based weighting

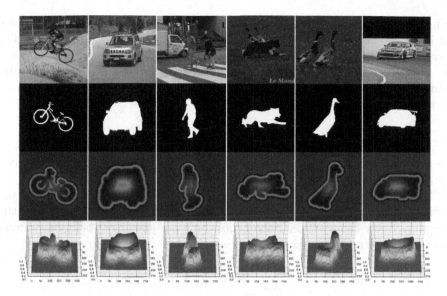

Fig. 3. Examples of the weighting function applied to the BCE loss. **First row**: original images; **second row**: ground truth annotations; **third and fourth row**: 2D and 3D representation of the weighting function, which focuses on the area around the contours and balances the emphasis on the foreground and background pixels according to the class frequencies.

around the contour is further enhanced by applying different kernel widths for the foreground and background classes. This way, pixels inside of the instance are weighted higher than those outside. Contrary to previous works, this does not set the weights for foreground pixels to one but reduces them towards the instance center, leading to an increased focus on the contour. In most cases, the central pixels of an image belong to the foreground class and are easier to classify. Together with the class balancing, the contour-based weighting function, so far, is given by the following equation

$$w_{contour}(\mathbf{x}) = \alpha + (1 - \alpha)\exp(\frac{-d(\mathbf{x})^2}{2\sigma^2}),$$

where $\mathbf{x}$ denotes a pixel in the image and $d(\mathbf{x})$ is the euclidean distance of a pixel to its nearest contour point. $w_{contour}(\mathbf{x})$ is calculated separately for the foreground and background classes with different $\alpha \in [0, 1]$ and σ depending on the image size. Here, we choose α and σ as 0.5 for the foreground class. For the background class, α is chosen according to the class balance of the particular image, and σ is fixed at 0.15. These values were approximated based on the image size of 256 × 256. Because the input image is cropped around the tracked instance, based on the predicted bounding box before performing the segmentation task, the probability of a pixel belonging to the foreground class decreases with the pixel's distance from the image center. As a result, current segmentation models tend to exclude extruding parts of an instance, e.g., hands or feet. To have a more substantial weighting for parts that are likely to be left out in the model prediction, the weighting is multiplied by an additional function, based on a pixel's distance to the image center. The image center approximates the instance center well because the instance is centered in the surrounding bounding box. This function takes the form of an upside-down 2-dimensional Gaussian centered in the image

$$w_{center}(\mathbf{x}) = 1 - \sigma \exp(-(x^2 + y^2)),$$

where x and y are the coordinates of pixel $\mathbf{x}$. Again corresponding to the image size, σ is set to 0.25.

The weighting approximates α for background pixels and tends towards one at the contour with a Gaussian transition in between. We are applying this weighting to the Binary Cross Entropy loss on a per-pixel basis before averaging, which leads to the Weighted Binary Cross-Entropy (WBCE) loss:

$$\text{WBCE}(\mathbf{X}, \mathbf{Y}) = \frac{1}{N} \sum_{\mathbf{x} \in \mathbf{X}, \mathbf{y} \in \mathbf{Y}} \text{BCE}(\mathbf{x}, \mathbf{y}) \cdot w_{contour}(\mathbf{x}) \cdot w_{center}(\mathbf{x})$$

where $BCE(\mathbf{x})$ is the Binary Cross-Entropy loss for one pixel, which is weighted and then averaged over all N pixels in an image I.

3.4   Approaches to Incorporate Corrections

We train the underlying model, which is the core component of the intelligent video annotation tool, to incorporate additional input data so that the human annotator can interactively condition the model on corrections. We examine three different approaches to incorporate the corrected annotations into the segmentation process and a baseline for comparison. Besides the original image at the current time step, colored masks of the current and previous time steps provide additional input to the models. These colored masks are created by stamping out the respective image with either the corrected annotations or the mask predicted by the base model. As such, the colored masks provide information about the shape of the tracked instance and its texture. Figure 4 illustrates the distinct approaches, which we further explain in the following.

Baseline. The baseline model provides a basis for comparison and receives no additional input besides the original image. Furthermore, it poses a more comparative approach than the SiamMask model's predictions, as it used the same model architecture as the other approaches.

Previous Annotation. The first approach utilizing additional input uses the corrected annotation of the previous frame. The current image and the additional input are concatenated and fed into the model. As in the other approaches, the additional input consists of a colored mask. During training, this mask is

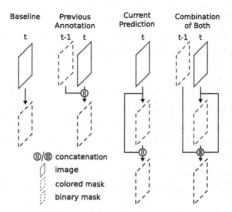

Fig. 4. Schematic of the three approaches to incorporate corrected user input, as well as the baseline.

given by the ground truth annotation to simulate a thorough human annotator correcting the base model's mistakes. However, this procedure is not optimal since the ground truth annotations are often not available during testing, where they are replaced with the model predictions. The hope is that these corrections propagate to the predictions in future time steps.

Current Prediction. In the second approach using additional input, the original image is first processed by the base model, whose prediction is then used to create a colored mask. This mask is concatenated with the original image and thus constitutes the interaction model's input. With this approach, we investigate whether we can improve the base model's predictions by stacking an additional model on top.

Combination of Both. The last approach combines the additional input from both the first and second approaches by concatenating them before feeding them to the U-Net model. Here, the goal is that the model learns to compare the instance representation of the previous and current frame, and thus, learn to recognize parts of the instance labeled in the previous frame and incorporate them into the current prediction. This approach might improve the performance because the previous instance is often not well aligned to the current one due to its or the camera's movement.

4   Evaluation Methodology

Annotators using the intelligent annotation tool are expected to provide corrections for some of the frames in a video so that the model can improve its predictions based on this feedback. We simulate the human annotator during training by using the existing ground truth annotations as corrections for every frame. Each of the four experiments named above was carried out three times, starting from three different initializations to ensure the starting point of optimization is not biased towards one particular experiment.

Dataset. The models were trained and tested on a small dataset, namely the publicly available parts of the DAVIS challenge [12,13], whereby the official validation set was further split into two to retain a separate test set. Overall the training, validation, and test sets consist of 60, 15, and 15 videos, each containing approximately 60 frames per video. The validation set was used for the hyperparameter search (learning rate) and to select the best performing model out of the training epochs. We report the following final results on the test set.

Metrics. First, the training results are reported based on the Weighted Binary Cross-Entropy loss for each approach and initialization. The second metric commonly used to evaluate image detection and segmentation tasks is the Intersection Over Union (IOU) metric. It measures the relative overlap of the predicted masks with the ground truth masks of an instance.

5   Evaluation

This Section presents the quantitative evaluation of the proposed approaches and compares them to the base model. The training of one model took roughly one day on an NVidia Tesla V100 graphics card.

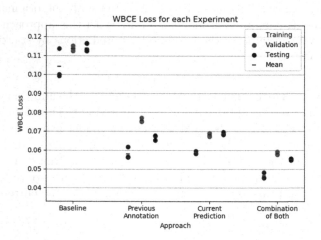

Fig. 5. The weighted binary cross entropy (WBCE) loss of each approach, showing all three training runs each with different initializations. All of the approaches utilizing additional information in form of predictions or correction achieve lower losses than the baseline approach.

5.1 Weighted Binary Cross-Entropy Loss

Figure 5 reports the WBCE loss after training, validation, and testing, respectively. One can see that the additional input in the form of the colored masks does improve the performance over the base model, which received no additional input besides the original image. Using the previous annotation or the current prediction end up at around the same loss, while utilizing both leads to better performances. We compare the original SiamMask network in the next part of this section, as the training loss of it is neither available nor directly comparable to the one used here.

5.2 Intersection over Union

The BCE loss is calculated over the model output after applying a sigmoid function, resulting in a probability measure with values in $[0, 1]$ for every pixel. To arrive at a binary mask, which can be used to calculate the IOU score, we apply a threshold. To choose an appropriate value for this threshold, the IOU was calculated for multiple values in $[0, 1]$ with step size 0.1. Figure 6 shows the IOU over these threshold values. A threshold of around 0.5 leads to peak performance. This is to be expected because the models were trained with the WBCE loss applied to the sigmoid output, which separates the foreground and background classes in the middle of its range, which is 0.5. However, it should be noted that the original SiamMask approach uses a threshold value of 0.3. A lower threshold value makes the predicted segmentations look less separated and extruding parts of the segmented instance are less likely to be excluded.

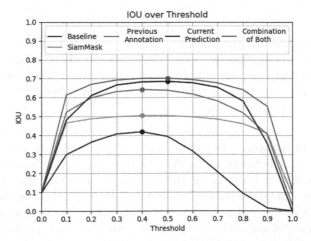

Fig. 6. Intersection Over Union for different threshold values applied to the predicted pixel-wise probabilities, only depicting each approach's best model. For comparison, we show the score of the SiamMask model without any added network.

As seen in Fig. 6, the best IOU of around 0.7 is reached by the approach using both previous and current colored masks. The approaches utilizing only one of those additional inputs follow closely. The SiamMask model peaks just above 0.5, leaving a gap to the additional stacked models. The baseline performs worst, with an IOU slightly above 0.4, and is, therefore, worse than the base SiamMask model. The approaches utilizing predicted or corrected annotation masks converged much faster than the baseline, which meets our expectations because the masks already provide a helpful guideline to the model.

6   Discussion

In this section, we briefly discuss qualitative examples pertaining to each of the approaches. Surprisingly, none of the approaches improves the segmentation performance of SiamMask qualitatively to a satisfactory extent, contrary to what the IOU scores suggest. Figure 7 provides a qualitative comparison of the different approaches before and after correcting the current and previous frames. The exemplary images are not part of the DAVIS datasets. The upper left quadrant depicts the SiamMask model's predictions, and the lower left quadrant depicts the corrected annotations of these frames. The right quadrants depict the predictions of each approach before and after correcting the additional input for them. The approach utilizing only the previous annotation is the only approach that shows signs of the desired effect of incorporating the correction into its prediction.

The following presents a more detailed discussion of the individual approaches at the example of a typical road traffic scene focused on a vulnerable road user, i.e., a cyclist.

Baseline. As expected, the baseline approach, not facilitating any additional input, performs worst, likewise to the previously calculated scores. Some limbs, like the head and legs, are not captured in the prediction. After correction, nothing changes, of course, because this approach does not facilitate any additional input besides the original image.

Previous Annotation. Before any corrections, the head starts not to be included in the predicted mask in the previous frame. Consequently, this approach also excludes it from the prediction. After correcting the previous frame to include the head/helmet, this approach does capture it in its prediction. Although parts of the background are now labeled as foreground, and as before, the shadowy part of the cyclist is falsely excluded from the prediction.

When the misalignment of the instance in the previous frame relative to the current one becomes too large, the model showed clear signs of confusion regarding the instance's position. Consequently, the predictions sometimes are torn between the two instances positions.

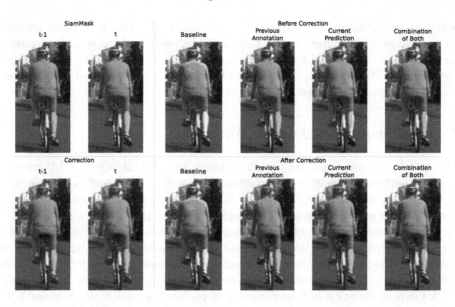

Fig. 7. Qualitative results of each approach before and after correction of two consecutive exemplary frames. Upper left quadrant: SiamMask predictions for current and previous frames. Lower left quadrant: corrected annotations of current and previous frames. Upper right quadrant: predictions of approaches before corrections. Lower right quadrant: predictions of approaches after corrections.

Current Prediction. This approach adheres too strictly to the base model's current prediction and does not manage to improve on it in this example. Accordingly, correcting the current frame fixes the prediction. However, the model does not appear to be capable of transferring the information to subsequent frames. This approach provides nothing more than unnecessary overhead compared to the predictions of the base model in this example.

Combination of Both. Combining the previous two approaches fall into the same trap as the approach utilizing the current prediction. It relies too heavily on the prediction of the base model and ignores the previous corrected frame in favor of the faulty prediction of the current frame. The results show that it is possible to incorporate corrected annotations into the segmentation process, but additional information can also be counterproductive. As seen in the latter approach, errors can propagate from the base model to the additional segmentation model without being corrected. In theory, the proposed augmentation with additional input leads to improved performance. However, in practice, the segmentation performance needs to improve by a fair bit to make it usable in an intelligent labeling tool. Especially when it comes to fine detail in the predicted masks; otherwise, the annotator needs to correct almost every frame of the video, as these details in the labeled data can be vital [8]. The corrections are currently only incorporated for a few frames before the model falls back to excluding instance parts again. Although seeming to work in some cases, the underlying model's segmentation performance proved so unreliable that the final results regarding the incorporation of annotator corrections remain inconclusive. We leave it as an open question whether improvements of the model's segmentation performance can alleviate some of the encountered hurdles.

7   Conclusion and Future Work

This article presents a novel approach to interactive and intelligent video annotation for instance segmentation. Our approach involves a Siamese neural network at its core to leverage the transfer of annotations from one frame to another. Our approach allows integrating feedback and corrections from human annotators without the necessity of time-consuming retraining of the complete neural network. In this context, we investigate and evaluate multiple approaches to incorporate feedback and human corrections. Our empirical experiments show that the usage of these correction increases the annotation quality considerably.

A possible direction of future work is to update the underlying Siamese neural network sporadically. In the approach followed in this article, the network is only trained once. The network remains unchanged during the annotation process, i.e., we only use it in the one-shot learning fashion to determine the similarity between instances. However, for the network update, the samples annotated and corrected through the annotation process can be utilized in a semi self-supervised manner. The human and machine in this interactive collaboration, thereby, act in a teacher and student relationship. This can be complemented by incorporating active learning techniques to implement an intelligent selection of unknown and new object instances. This includes using techniques to estimate the uncertainty or confidence of the network's predictions, e.g., using Bayesian neural networks. In this respect, the network might then explicitly ask for feedback or corrections where it is still uncertain. In this way, we can further increase the degree of interaction and, at the same time, substantially reduce the required annotation effort. Closely related to this is also the semi-supervised estimation of the labeling quality, i.e., how good the semi-automatically determined annotations are and whether we can automatically determine when to stop the annotation process. Furthermore, we also consider extending the approach to the concurrent annotation of multiple object instances. Moreover, we aim to extend the approach to other sensor modalities, e.g., LIDAR, and RADAR, for future work.

Acknowledgments. This work results from the project KI Data Tooling (19A20001O) funded by BMWI (German Federal Ministry for Economic Affairs and Energy), and the project DeCoInt^2 supported by the German Research Foundation (DFG) within the priority program SPP 1835: "Kooperativ interagierende Automobile", grant number SI 674/11-2.

References

1. Acuna, D., Ling, H., Kar, A., Fidler, S.: Efficient Interactive Annotation of Segmentation Datasets with Polygon-RNN++. In: Proceedings of the IEEE conference on Computer Vision and Pattern Recognition, pp. 859–868 (2018)
2. Asano, Y.M., Rupprecht, C., Vedaldi, A.: A critical analysis of self-supervision, or what we can learn from a single image. In: ICLR, pp. 1–16. Vienna, Austria (2020)
3. Bianco, S., Ciocca, G., Napoletano, P., Schettini, R.: An interactive tool for manual, semi-automatic and automatic video annotation. CVIU **131**, 88–99 (2015)
4. Castrejón, L., Kundu, K., Urtasun, R., Fidler, S.: Annotating object instances with a polygon-RNN. In: CVPR, pp. 4485–4493. Honolulu, HI, USA (2017)
5. Fagot-Bouquet, L., Rabarisoa, J., Pham, Q.: Fast and accurate video annotation using dense motion hypotheses. In: ICIP, pp. 3122–3126. Paris, France (2014)
6. Falk, T., et al.: U-Net: deep learning for cell counting, detection, and morphometry. Nature Methods **16**, 67–70 (2018)
7. Fei-Fei, L., Fergus, R., Perona, P.: One-shot learning of object categories. TPAMI **28**(4), 594–611 (2006)
8. Karras, T., Aila, T., Laine, S., Lehtinen, J.: Progressive growing of GANs for improved quality, stability, and variation. In: ICLR, pp. 1–26. Vancouver, BC, Canada (2017)
9. Koch, G., Zemel, R., Salakhutdinov, R.: Siamese neural networks for one-shot image recognition. In: ICML, pp. 1–8. Lille, France (2015)
10. Loshchilov, I., Hutter, F.: Decoupled weight decay regularization. In: ICLR, pp. 1–19. New Orleans, LA, USA (2019)
11. Nagaraja, N., Schmidt, F.R., Brox, T.: Video segmentation with just a few strokes. In: ICCV, pp. 3235–3243. Santiago, Chile (2015)
12. Perazzi, F., et al.: A benchmark dataset and evaluation methodology for video object segmentation. In: CVPR, pp. 724–732. Las Vegas, NV, USA (2016)
13. Pont-Tuset, J., Perazzi, F., Caelles, S., Arbeláez, P., Sorkine-Hornung, A., Van Gool, L.: The 2017 DAVIS challenge on video object segmentation. arXiv:1704.00675 (2017)
14. Subramanian, A., Subramanian, A.: One-click annotation with guided hierarchical object detection. arXiv:1810.00609 (2018)
15. Vicente, S., Rother, C., Kolmogorov, V.: Object cosegmentation. In: CVPR, pp. 2217–2224. Colorado Springs, CO, USA (2011)
16. Vondrick, C., Ramanan, D.: Video annotation and tracking with active learning. Adv. Neural Inf. Process. Syst. **24**, 28–36 (2011)
17. Wang, Q., Zhang, L., Bertinetto, L., Hu, W., Torr, P.: Fast online object tracking and segmentation: a unifying approach. In: CVPR, pp. 1328–1338. Salt Lake City, UT, USA (2018)
18. Wang, Y., Yao, Q., Kwok, J., Ni, L.M.: Generalizing from a few examples: a survey on few-shot learning. ACM Comput. Surv. **53**, 1–34 (2019)

Imputation of Rainfall Data Using Improved Neural Network Algorithm

Po Chan Chiu[1,2,3], Ali Selamat[1,2,4,5]([✉]), Ondrej Krejcar[5], and King Kuok Kuok[6]

[1] School of Computing, Faculty of Engineering, Universiti Teknologi Malaysia, 81310 Johor Bahru, Johor, Malaysia
aselamat@utm.my

[2] MagicX (Media and Games Center of Excellence), Universiti Teknologi Malaysia, 81310 Johor Bahru, Johor, Malaysia

[3] Faculty of Computer Science and Information Technology, Universiti Malaysia Sarawak, 94300 Kota Samarahan, Sarawak, Malaysia

[4] Malaysia Japan International Institute of Technology (MJIIT), Universiti Teknologi Malaysia Kuala Lumpur, Jalan Sultan Yahya Petra, 54100 Kuala Lumpur, Malaysia

[5] Faculty of Informatics and Management, University of Hradec Kralove, Rokitanského 62, 500 03 Hradec Kralove, Czech Republic

[6] Faculty of Engineering, Computing and Science, Swinburne University of Technology Sarawak Campus, 93350 Kuching, Sarawak, Malaysia

Abstract. Missing rainfall data have reduced the quality of hydrological data analysis because they are the essential input for hydrological modeling. Much research has focused on rainfall data imputation. However, the compatibility of precipitation (rainfall) and non-precipitation (meteorology) as input data has received less attention. First, we propose a novel input structure for the missing data imputation method. Principal component analysis (PCA) is used to extract the most relevant features from the meteorological data. This paper introduces the combined input of the significant principal components (PCs) and rainfall data from nearest neighbor gauging stations as the input to the estimation of the missing values. Second, the effects of the combination input for infilling the missing rainfall data series were compared using the sine cosine algorithm neural network (SCANN) and feedforward neural network (FFNN). The results showed that SCANN outperformed FFNN imputation in terms of mean absolute error (MAE), root means square error (RMSE) and correlation coefficient (R), with an average accuracy of more than 90%. This study revealed that as the percentage of missingness increased, the precision of both imputation methods reduced.

Keywords: Imputation · Missing rainfall data · Meteorological data · Principal component analysis (PCA) · Sine cosine algorithm neural network (SCANN)

1 Introduction

Rainfall is a critical component of the hydrological cycle. Numerous hydrological research areas, such as flood forecasting [1], flood risk assessment [2] and rainfall forecasting [3], require reliable and complete rainfall data series. However, hydrological

© Springer Nature Switzerland AG 2021
A. Del Bimbo et al. (Eds.): ICPR 2020 Workshops, LNCS 12664, pp. 390–406, 2021.
https://doi.org/10.1007/978-3-030-68799-1_28

data analysis is challenging due to the presence of missing rainfall data. For this reason, data imputation has attracted a great deal of attention from researchers to fill in the missing values with approximations. The traditional imputation approaches include listwise deletion [4], arithmetic mean and median imputation [5], and multiple imputations [6]. However, these methods are time-consuming and less accurate [7].

In recent years, numerous studies such as the Levenberg–Marquardt backpropagation algorithm [8], the Gaussian mixture model-based K-nearest neighbor (GMM-KNN) algorithm [9] and Bayesian principal component analysis (BPCA) [10] have been applied to impute the missing values in water resource engineering. Although artificial neural networks (ANNs) have been applied to treat the problem of missing data, ANNs tend to be trapped in local optima as it smoothly converges towards local minima rather than global minima. To overcome this, several novel approaches have been combined with ANNs to improve the performance of the estimation results. The sine cosine algorithm (SCA) is a metaheuristic optimization technique introduced by Mirjalili [11] to solve continuous optimization problems. SCA has been successfully applied in modal dimensional [12], short-term hydrothermal scheduling [13], support vector regression [14], and the traveling salesman problem [15]. To the best of the author's knowledge, there is no existing sine cosine algorithm neural network (SCANN) that focuses on missing rainfall data imputation.

Furthermore, the use of raw hourly rainfall data from nearest neighbor stations could be unreliable for the prediction of the missing data of the target station. The long dry periods contain long sequences of zero values at the beginning, middle, or end of the records, in which rain does not usually fall every hour. Modeling long dry rainfall periods poses challenges such as underestimation or overestimation of the length of long dry periods [16, 17]. As a result, a neural network is not able to estimate the missing rainfall value based on hourly rainfall datasets accurately. Hence, the hourly rainfall dataset needs to be combined with other non-precipitation data such as meteorological data for the estimation of missing rainfall data.

According to Kashiwao et al. [18], rainfall is caused by a variety of meteorological conditions, and the mathematical model for it is non-linear. The meteorological data have different units of measurement and accuracy. Thus, the meteorological data need to be pre-processed prior to imputation. Normalization is the most used approach. Yen [19] applied a mapminmax approach to normalize the meteorological parameters in the study. Meanwhile, Grange [20] proposed using a random forest machine learning algorithm for meteorological normalization to detect interventions in an air quality time series. According to Kashiwao et al. [18], the investigation into the method used to choose meteorological data is needed because suitable data can vary among prediction points due to the difference in the effect of conditions, such as altitude, ocean current, and airflow. For this reason, this paper uses principal component analysis (PCA) to extract the core relationships in the meteorological data. This study introduces the combined input of the significant principal components (PCs) and rainfall data from nearest neighbor gauging stations as the input for the estimation of missing rainfall values.

The contributions of this paper are the following:

- To propose a novel input structure that consists of the significant PCs and rainfall data from nearest neighbor gauging stations as the input to the missing data imputation.

- To introduce the proposed methodology of missing data imputation.
- To propose a sine cosine algorithm neural network (SCANN) imputation that focuses on treating the missing time series data.

2 Study Region

The selected study area for this study is Sungai Merang, or the Merang River gauge station, approximately 80 km from Kuching City, Sarawak, Malaysia. Sungai Merang is one of the five rainfall gauge stations in the Bedup River catchment, as shown in Fig. 1. Its nearest neighbor gauge stations over the basin are Bukit Matuh (BM), Semuja Nonok (SN), Sungai Busit (SB) and Sungai Teb (ST).

Fig. 1. Sungai Merang and its nearest neighbor gauging stations

3 Methodology

The proposed methodology employed in this study consists of two main phases, as shown in Fig. 2.

3.1 Phase 1: Data Preparation

In this study, the data preparation phase attempts to transform raw data into an understandable format prior to the missing data imputation. First, a subset of relevant features is identified from the datasets. In the works by [21–23], the studies have used historical rainfall data series from nearest neighbor stations to treat the problems of missing data. Therefore, this study identifies the nearest neighbor features as relevant features that contain very useful information for the missing data imputation. Consequently, the nearest neighbor features do not need to be pre-processed using PCA. However, the raw meteorological data must be pre-processed due to the variety of measurement units. During the pre-processing task, PCA is employed to extract the most relevant features from the meteorological data. By using PCA, a number of principal components (PC1, PC2, …,

PC10) are constructed. The significant principal components (PCs) are then integrated with the rainfall data from nearest neighbor gauging stations as the input to the missing data imputation. Hence, the output of this data preparation phase is the combined input of the significant PCs and rainfall data from the nearest neighbor stations.

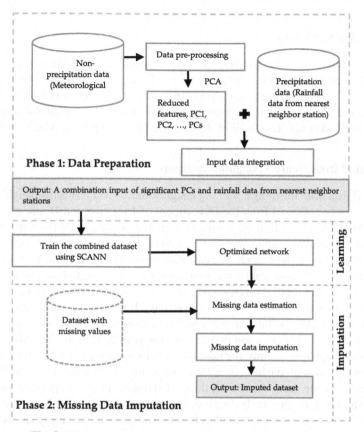

Fig. 2. The proposed methodology of missing data imputation.

3.2 Phase 2: Missing Data Imputation

The missing data imputation phase consists of two sub-phases, namely learning and imputation. In the learning sub-phase, the combined input from phase 1 will be used as an input to the neural network training. By using the SCANN, the neural network is trained to learn the complex and non-linear relationships between the features in the dataset. The output of the learning sub-phase is an optimized network with a set of optimal network weights and biases. Next, the imputation sub-phase involves missing data estimation using the optimized network. During the missing data imputation, the estimated missing data are imputed into the missing values in the dataset. Hence, the final output of this phase is the imputed database.

4   Material and Methods

This section begins with a brief description of meteorological and rainfall data from the nearest neighbor stations, followed by pre-processing data approach using PCA and the proposed imputation approach.

4.1   Meteorological Data

The meteorological data for Kuching station was acquired from the Malaysian Meteorological Department. In this study, ten types of meteorological data were collected: date, time, the pressure at mean sea-level (MSL), dry-bulb temperature, relative humidity, mean surface wind (direction), mean surface wind (speed), rainfall duration, rainfall amount and cloud cover. The detail of the data pre-processing is introduced in Sect. 4.3.

4.2   Rainfall from Nearest Neighbor Stations

The rainfall data from the Sungai Merang gauging station and its nearest neighbor gauging stations were collected from the Department of Irrigation and Drainage, Sarawak, as shown in Table 1. Overall, the correlation coefficients between the Sungai Merang station and each of the neighbor stations are greater than 0.8 and located within a radius range of 5 km. Since the Sungai Merang gauging station exhibits a high correlation coefficient with its nearest neighbor stations, the complete rainfall data series from the four neighbor stations of the corresponding hour, day, month and year are used to predict the missing values of Sungai Merang's rainfall data. Based on the availability of continuous and complete data (without missing values) for the five gauging stations, this study analyzed the observed hourly rainfall data from the year 2002 until 2003. With a sample size of 11,680 complete records, the neural networks were trained with a training length of 8180 and tested with datasets of 3500 records. In [24–27], the data were randomly deleted and removed from the testing datasets. Hence, for the preparation of missing values in rainfall data, this study employed a rate-based approach [28] in which 10%, 20%, 30%, 40%, and 50% were randomly removed from the testing datasets. In total, two sets of testing data were prepared for each percentage of the missingness. In this study, the missing data was categorized as missing completely at random (MCAR) [29] because the presence of missing rainfall data at the Sungai Merang gauge station is not affected by the data in that area or any nearby area.

4.3   Pre-processing Data Using Principal Component Analysis (PCA)

This study found that raw meteorological data could not be used as an input to estimate the missing data imputation due to the different units of measurement. For example, the values of mean surface wind (direction) are stored at 00°, 010°,..., 058°. These characters are considered noise in the data because the neural network could not understand and interpret those characters accurately. According to Kurita [30], principal component analysis (PCA) is a standard tool in modern data analysis that reveals hidden structure in the dataset and filters out the noise in data. PCA was proposed by Pearson [31] and

Table 1. The Sungai Merang gauging station and its nearest neighbor gauging stations.

Station name	Latitude	Longitude	Distance from Sg Merang (km)	Correlation coefficient
Sungai Merang	001 05 40	110 36 25	–	–
Bukit Matuh	001 03 50	110 35 35	3.88	0.8558
Semuja Nonok	001 06 25	110 35 50	2.10	0.8647
Sungai Busit	001 05 25	110 34 40	3.44	0.8676
Sungai Teb	001 03 15	110 37 00	4.37	0.8046

formalized by Hotelling [32]. Hence, PCA was used to pre-process the raw meteorological data in order to extract important information and to reduce the noise in the data. PCA reduces the number of meteorological features by constructing a new and smaller number of variables that capture a significant portion of the original meteorological features. The covariance and correlation between every pair of variables (meteorological features) were calculated based on the following equations [33, 34]:

$$cov(x, y) = \frac{\sum_{i=1}^{n}\left(x_i - \bar{x}\right)\left(y_i - \bar{y}\right)}{n - 1} \tag{1}$$

where $cov(x, y)$ indicates the covariance of the variables x and y, x_i is the independent variable of observations, y_i is the independent variable of observations and n is the number of data points in the observations.

$$r(x, y) = \frac{cov(x, y)}{s_x s_y} \tag{2}$$

where $r(x, y)$ is the correlation of the variables x and y, s_x is the sample standard deviation of the random variable x and s_y is the number of data points in the observations.

Then eigenvector and eigenvalue of the matrix are obtained using (4).

$$A = \begin{bmatrix} A_{1,1} & A_{1,2} & \cdots & A_{1,n} \\ A_{2,1} & A_{2,2} & \cdots & A_{2,n} \\ \vdots & \vdots & \ddots & \vdots \\ A_{n,1} & A_{n,2} & \cdots & A_{n,n} \end{bmatrix} \tag{3}$$

$$Av = \lambda v \tag{4}$$

The eigenvector v of each variable can be obtained by identifying the determinant of its characteristic polynomial as follows:

$$(A - \lambda I)v = 0 \tag{5}$$

The eigenvalue can be formulated using the following equation.

$$p(\lambda) = |A - \lambda I| \tag{6}$$

After these steps, the principal components (PC1, PC2, ..., PCs) can be determined. The first principal component accounts for the highest variance in the meteorological dataset, followed by the second principal component for the next highest variance. This continues until the total of the principal components is equal to the number of features in the meteorological dataset.

The last step is to compute the feature vector. A matrix M of dimensions n x d is represented as

$$M = \begin{bmatrix} f_{1,1} \, f_{1,2} \, f_{1,3} \, \cdots \, f_{1,d} \\ f_{2,1} \, f_{2,2} \, f_{2,3} \, \cdots \, f_{2,d} \\ f_{3,1} \, f_{3,2} \, f_{3,3} \, \cdots \, \cdots \\ \vdots \quad \vdots \quad \vdots \quad \ddots \quad \vdots \\ f_{n,1} \, f_{n,2} \, f_{n,3} \, \cdots \, f_{n,d} \end{bmatrix}$$

where f_{ij} is a reduced feature vector from n x n original data to size n x d, n is the number of data points in the observations and d is the number of principal components.

The final output of the PCA is combined with the raw rainfall data from the nearest neighbor gauging stations and then used as the input to the neural network for missing data imputation.

4.4 Imputation Approaches

This sub-section describes an existing neural network, the feedforward neural network (FFNN), followed by a detailed description of the proposed sine and cosine algorithm neural network (SCANN) imputation.

Feedforward Neural Network (FFNN)
The feedforward neural network (FFNN) model is the simplest type of ANN. The architecture of the FFNN network consists of p-many inputs (input neurons), a single hidden layer with q-many hidden neurons, and a single output. The number of inputs, p, is based on the number of cumulative principal components (PC1, PC2, ..., PC10) and raw rainfall data from nearest neighbor stations, as shown in Fig. 3. A simulation for estimation of the missing rainfall data using FFNN was carried out with ten neurons in the hidden layer. The activation functions for the hidden layer and output layer are tansig and purelin, respectively.

Sine Cosine Algorithm Neural Network (SCANN) Imputation Method
The sine cosine algorithm (SCA) is a metaheuristic optimization technique introduced by Mirjalili [11] to solve continuous optimization problems. One of the most significant advantages of SCA is its simplicity, as reported by Qu et al. [12]. SCA has fewer parameters that need to be fine-tuned compared to other algorithms. The capability of SCA in missing rainfall data imputation has not yet been explored. Hence, in this study, SCA is employed to train the feedforward neural network for infilling the missing rainfall data series.

In this sub-section, the sine cosine algorithm neural network (SCANN) imputation method is proposed to model the missing hourly rainfall data relationship in Sungai

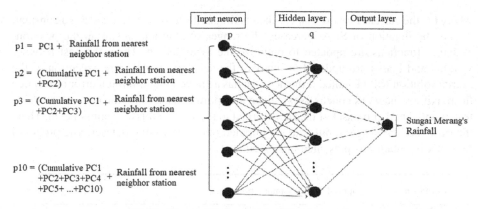

Fig. 3. The architecture of the FFNN network

Merang, Sarawak, Malaysia. Firstly, the network is trained using a feedforward neural network with the combination input of the significant principal components (PC1, ..., PCs) and nearest neighbor rainfall data. The network is used to identify and learn the relationships between features in the dataset. Then, the SCA is employed to optimize the search solutions by determining the optimal network weights and biases.

SCA starts the optimization process with a set of search solutions, X. The set of search solutions is initialized randomly and repeatedly evaluated by an objective function. The objective function indicates how much each search solution contributes to the value to be minimized in the optimization process. The objective function takes the following form [35]:

$$f_1(x) = \sum_{i=1}^{n} x_i^2$$

Next, the search solution is improved by the position-updating function in Eq. (7) [11]. The SCA updates the best solutions obtained and denotes it as a destination point, P.

$$X_i^{t+1} = \begin{cases} X_i^t + r1 \times \sin(r_2) \times \left| r_3 P_i^t - X_{i,}^t \right|, r_4 < 0.5 \\ X_i^t + r1 \times \cos(r_2) \times \left| r_3 P_i^t - X_{i,}^t \right|, r_4 \geq 0.5 \end{cases} \tag{7}$$

where X_i is the position vector of the current solution in the ith dimension, t is the current iteration, P_i is the destination solution and r_1, r_2, r_3, r_4 are random variables; the r_4 value is between 0 and 1.

As seen in (7), there are four parameters in SCA, namely r_1, r_2, r_3 and r_4. The parameter r_1 is the movement direction parameter that determines the region of the next solution, which is updated using (8). The parameter r_2 identifies the movement of forwards or outwards P_i within the value of 0 and 2π. Next, the parameter r_3 is the random weights of Pi with a value either less than 1 or greater than 1. The parameter r_4 is used to switch between the sine and cosine function.

$$r_1(t) = a \times \left(1 - \frac{t}{t_{max}} \right) \tag{8}$$

where t is the current iteration, t_{max} is the maximum iteration of SCA and a is a constant.

As the iteration of SCA increases, the ranges of sine and cosine in the position-updating functions are updated to optimize the local search. Then, the best network weights and biases are updated to improve the network model. The execution of the search solution will be halted if the network has achieved the minimum error or reached the maximum network epochs. Next, given the optimized network, the network model is tested with another dataset of the same format to predict the missing rainfall data. Then, the estimated missing rainfall data are imputed into the missing dataset. The proposed SCANN imputation is presented in Algorithm 1.

Algorithm 1: The proposed SCANN imputation

Begin
Do
Load training dataset
Initialize SCANN parameters as in Table 2
Initialize a set of random search agents (solutions) (X) and SCA parameters (r_1, r_2, r_3, r_4)

    Do
        Evaluate each of the search agents by the objective function
        Update the best solution obtained so far (P)
        Update the parameters r_1, r_2, r_3, and r_4
        Update the position of search agents using Equation (8)
    While (t < maximum number of iterations)
    Return the best solution (P) obtained as the global optimum solution
Track the best network into net
Update training state into net
While (MSE > the minimum error or E < maximum number of epochs)
Use the optimized net
Train the optimized net for another dataset of the same format
Output: Estimated missing rainfall data
Do
    Impute the estimated values into the missing value
While (there is missing value)
End
where X is the search solutions, P is the destination solution, r_1 is the movement direction parameter, r_2 identifies the movement of forwards or outwards P within the value of 0 and 2π, r_3 is the random weights of P (value less than 1 or greater than 1) and r_4 is the random variables (0 < r_4 <1).

Note: The algorithm in the dotted line box was adapted from Mirjalili [11].

In addition, different values of the parameters are introduced to the SCANN. The parameters are tuned based on the try and error method. The parameter settings are outlined in Table 2.

4.5 Performance Measures

The performances of the two imputation methods are measured by the mean absolute error (MAE), root mean square error (RMSE), and correlation coefficient (R).

Mean absolute error (MAE)

$$MAE = \frac{1}{N} \sum_{i=1}^{N} |O_i - T_i|$$

Table 2. The SCANN parameters

Parameters for SCANN	Value
Hidden layers, q	10
a	2
Search agents	30
Max number of epochs	1000
Max iteration of SCA	500

Rootmean square error (RMSE).

$$RMSE = \sqrt{\frac{\sum_{i=1}^{N} (O_i - T_i)^2}{N}}$$

The correlation coefficient (R).

$$R = \frac{\sum \left(T - \bar{T}\right)\left(O - \bar{O}\right)}{\sqrt{\sum \left(T - \bar{T}\right)^2 \left(O - \bar{O}\right)^2}}$$

where N is the total number of observations, 0 is the actual values of observations and T is the imputed values.

5 Results and Discussion

The proposed SCANN missing data imputation was compared with the FFNN missing data imputation using a combination input p of the significant PCs (cumulative PC) and rainfall data series from the nearest neighbor stations. Determining the number of significant PCs that represent the original data set is one key part in PCA. Works by [36] use eigenvalues to determine the significant number of PCs, while works by [37] uses eigenvectors to determine the significant number of PCs. In this paper, a different number of combination input p was introduced, from p1 to p10, to determine the significant input p to the neural network. The average result gave the minimum MAE and RMSE measures but the highest measure of R was chosen as the significant input p. For better evaluation of the proposed algorithm, we tested the imputation algorithms on two missing datasets. For each missing dataset, all the imputation algorithms were executed with 30 independent runs over each input p at different missing data rates (10%, 20%, 30%, 40%, and 50%). Table 3 shows the average values of the performance measures for SCANN and FFNN imputation respectively, over two missing datasets.

As seen in Table 3, among the input p values, the first and second input p (p1, p2) demonstrated good performances for predicting missing rainfall data. In particular, the SCANN imputation for p1 showed excellent performance in estimating the various

Table 3. Comparison of MAE, RMSE, AND R values using SCANN imputation and FFNN imputation at different percentages of missingness.

Input p	SCANN MAE (mm)						FFNN MAE (mm)					
	10%	20%	30%	40%	50%	Avg	10%	20%	30%	40%	50%	Avg
p1	**0.0718**	**0.1345**	**0.1626**	**0.2436**	**0.3010**	**0.1827**	0.1126	0.2169	0.2754	0.3981	0.4970	0.3000
p2	0.0866	0.1649	0.2137	0.3096	0.3878	0.2325	0.1198	0.2370	0.3155	0.4446	0.5724	0.3379
p3	0.1056	0.2102	0.2798	0.3998	0.5029	0.2996	0.1336	0.2711	0.3623	0.5138	0.6465	0.3855
p4	0.1161	0.2277	0.2962	0.4265	0.5293	0.3192	**0.1078**	**0.2124**	**0.2798**	**0.3970**	**0.4969**	**0.2988**
p5	0.1145	0.2211	0.2977	0.4198	0.5275	0.3161	0.1475	0.2862	0.3870	0.5329	0.6862	0.4080
p6	0.1133	0.2199	0.2961	0.4201	0.5224	0.3144	0.1526	0.3163	0.4244	0.5734	0.7624	0.4458
p7	0.1216	0.2411	0.3200	0.4451	0.5622	0.3380	0.1503	0.2976	0.4055	0.5705	0.7079	0.4264
p8	0.1420	0.2788	0.3814	0.5367	0.6647	0.4007	0.1478	0.2897	0.3909	0.5485	0.6897	0.4133
p9	0.1282	0.2496	0.3340	0.4684	0.5887	0.3538	0.1643	0.3451	0.4517	0.6104	0.7860	0.4715
p10	0.1232	0.2415	0.3219	0.4535	0.5679	0.3416	0.1662	0.3130	0.4244	0.5882	0.7424	0.4468

Input p	SCANN RMSE (mm)						FFNN RMSE (mm)					
	10%	20%	30%	40%	50%	Avg	10%	20%	30%	40%	50%	Avg
p1	0.8309	**1.1076**	**0.9515**	**1.3441**	**1.4733**	**1.1415**	1.0294	1.4542	1.3375	1.7405	1.9763	1.5076
p2	**0.8296**	1.1081	1.0136	1.3754	1.5341	1.1722	0.9943	1.5871	1.4953	1.8104	2.5168	1.6808
p3	0.8874	1.2137	1.1264	1.5138	1.7254	1.2933	1.0425	1.6286	1.5146	2.0156	2.3504	1.7106

(continued)

Table 3. (continued)

Input p	SCANN RMSE (mm)						FFNN RMSE (mm)					
	10%	20%	30%	40%	50%	Avg	10%	20%	30%	40%	50%	Avg
p4	0.9571	1.3050	1.1400	1.5457	1.7385	1.3373	**0.8706**	**1.2102**	**1.1029**	**1.4353**	**1.6376**	**1.2513**
p5	0.8765	1.1700	1.0636	1.4376	1.6171	1.2330	1.1862	1.7574	1.6776	1.8868	2.4270	1.7870
p6	0.8687	1.1576	1.0384	1.4248	1.5910	1.2161	1.1599	2.6579	2.5854	2.3485	3.9494	2.5402
p7	0.9720	1.6611	1.5517	1.5820	2.1131	1.5760	1.0032	1.4244	1.3650	1.7236	1.9201	1.4872
p8	1.0190	1.4121	1.3236	1.7326	1.9196	1.4814	1.0702	1.5126	1.3925	1.7734	2.0606	1.5618
p9	0.9328	1.2399	1.1273	1.5015	1.6951	1.2993	1.2187	2.8911	2.7826	2.2853	3.5474	2.5450
p10	0.9092	1.1791	1.0882	1.4686	1.6404	1.2571	1.0564	1.4835	1.3806	1.7666	2.0243	1.5423

Input p	SCANN R						FFNN R					
	10%	20%	30%	40%	50%	Avg	10%	20%	30%	40%	50%	Avg
p1	0.9510	0.9124	**0.9360**	**0.8739**	**0.8420**	**0.9031**	0.9200	0.8342	0.8628	0.7891	0.7109	0.8234
p2	**0.9518**	**0.9139**	0.9295	0.8700	0.8371	0.9005	0.9289	0.8610	0.8816	0.8064	0.7613	0.8478
p3	0.9439	0.8973	0.9134	0.8488	0.8117	0.8830	0.9253	0.8553	0.8721	0.8025	0.7543	0.8419
p4	0.9334	0.8746	0.9052	0.8320	0.7850	0.8660	**0.9467**	**0.8961**	**0.9140**	**0.8558**	**0.8080**	**0.8841**

(continued)

Table 3. (*continued*)

Input p	SCANN						FFNN					
	R						R					
	10%	20%	30%	40%	50%	Avg	10%	20%	30%	40%	50%	Avg
p5	0.9462	0.9023	0.9203	0.8554	0.8117	0.8872	0.9185	0.8645	0.8837	0.8177	0.7735	0.8516
p6	0.9467	0.9038	0.9245	0.8565	0.8148	0.8893	0.9200	0.8623	0.8769	0.8120	0.7642	0.8471
p7	0.9298	0.8632	0.8824	0.8261	0.7670	0.8537	0.9268	0.8554	0.8705	0.8079	0.7589	0.8439
p8	0.9289	0.8764	0.8968	0.8289	0.7858	0.8634	0.9169	0.8406	0.8681	0.7916	0.7233	0.8281
p9	0.9386	0.8885	0.9091	0.8408	0.7878	0.8730	0.8967	0.7946	0.8196	0.7522	0.6801	0.7886
p10	0.9421	0.8973	0.9170	0.8493	0.8087	0.8829	0.9189	0.8472	0.8675	0.7981	0.7266	0.8317

Note: The best results obtained are made bold.

percentages of missingness in terms of MAE, RMSE, and R. The proposed SCANN imputation achieved an average accuracy of more than 90%. The average MAE and RMSE measures of SCANN were 0.1827 mm and 1.1415 mm at p1, respectively. Meanwhile, the performances of FFNN increased as the input p decreased. It was observed that FFNN produced the best accuracy in total when input p4 was applied to the network. The average values of MAE, RMSE, and R measures for FFNN imputation were 0.2988 mm, 1.2513 mm, and 0.8841 at p4, respectively. Hence, input p1 and input p4 are the significant inputs for SCANN imputation and FFNN imputation, respectively, to achieve optimal imputation performance.

In general, the SCANN imputation has lower MAE and RMSE values but higher R values for all percentages of missingness than the FFNN imputation. This comparison revealed that the proposed SCANN imputation method outperformed FFNN imputation in treating the missing values in the dataset.

A closer inspection revealed that the values of MAE for both imputation methods linearly increased when the proportions of missing values increased. However, the values of RMSE linearly increased when the dataset had more than 30% missing values. This study supports the previous findings of Gill [38], Kim [39] and Ayilara [40] that the performance of imputation decreased when the proportion of missingness increased. According to Gill [38], the effect of missing data in information becomes very significant for hydrologic predictions as the percentage of missing data increases. The performances of ANN became unreliable because the ANN models may have converged to local optima when the amount of missing data increased significantly. On the other hand, the performances of the SCANN imputation are acceptable even though the proportion of missing values increased. The position-updating function of SCANN for searching the best destination solution tends to minimize the effect of local optima and consequently improve the accuracy of the missing rainfall data imputation. Hence, this study revealed that more missing rainfall data in the dataset results in a poorer model performance, which is consistent with previous research [38–40].

6 Conclusions

This study investigated the potential of using the combination input of significant PCs and rainfall data from nearest neighbor gauging stations for infilling missing rainfall data. The comparison of a different combination input in imputation was presented and evaluated using two imputation methods, SCANN and FFNN. With medium size data of 11,680 real-life records, the two methods were trained and compared at five different percentages of missingness under MCAR conditions (10%, 20%, 30%, 40%, and 50%). This study revealed that the two imputation methods were able to estimate the missing rainfall data series at an average accuracy of 88%. Hence, this finding suggests the use of significant PCs values and nearest neighbor station variables that have high correlation coefficients as the input to the missing rainfall data imputation. This study also concluded that the proposed SCANN imputation has a higher capability in treating missing values in the dataset than FFNN imputation in terms of MAE, RMSE, and R. By adopting the position-updating function, the proposed SCANN imputation successfully achieved better accuracy in missing data estimation as compared to the traditional approach.

Therefore, the results of the proposed SCANN imputation in this work support its use for infilling real-life missing rainfall data.

Acknowledgment. The authors would like to acknowledge the Malaysian Meteorological Department and Department of Irrigation and Drainage (DID), Sarawak, Malaysia, for providing the meteorological and rainfall data in this study. The authors sincerely thank Universiti Teknologi Malaysia (UTM) under Research University Grant Vot-20H04, Malaysia Research University Network (MRUN) Vot 4L876; Fundamental Research Grant Scheme (FRGS) Vot 5F073 and SLAI supported under Ministry of Higher Education Malaysia for the completion of the research. The work is partially supported by the SPEV project (ID: 2103–2020), Faculty of Informatics and Management, University of Hradec Kralove. We are also grateful for the support of Ph.D. students Jan Hruska and Michal Dobrovolny in consultations regarding application aspects from Hradec Kralove University, Czech Republic. The APC was funded by the SPEV project 2103/2020, Faculty of Informatics and Management, University of Hradec Kralove.

References

1. Muñoz, P., Orellana-Alvear, J., Willems, P., Célleri, R.: Flash-flood forecasting in an Andean mountain catchment—development of a step-wise methodology based on the random forest algorithm. Water **10**(11), 1519 (2018)
2. Szewrański, S., Chruściński, J., Kazak, J., Świąder, M., Tokarczyk-Dorociak, K., Żmuda, R.: Pluvial Flood Risk Assessment Tool (PFRA) for rainwater management and adaptation to climate change in newly urbanised areas. Water **10**(4), 386 (2018)
3. Kuok, K.K.: Parameter Optimization Methods for Calibrating Tank Model and Neural Network Model for Rainfall-runoff Modeling. Doctoral dissertation, Ph.D. thesis. Universiti Technology Malaysia (2010)
4. Mcdonald, R.A., Thurston, P.W., Nelson, M.R.A.: Monte Carlo study of missing item methods. Organizational Res. Methods **3**(1), 71–92 (2000)
5. McKnight, P.E., McKnight, K.M., Sidani, S., Figueredo, A.J.: Missing Data: A Gentle Introduction. Guilford Press (2007).
6. Lee, K.J., Carlin, J.B.: Multiple imputation for missing data: fully conditional specification versus multivariate normal imputation. Am. J. Epidemiol. **171**(5), 624–632 (2010)
7. Gao, Y., Merz, C., Lischeid, G., Schneider, M.: A review on missing hydrological data processing. Environ. Earth Sci. **77**(2), 1–2 (2018). https://doi.org/10.1007/s12665-018-7228-6
8. Mispan, M.R., Rahman, N.F.A., Ali, M.F., Khalid, K., Bakar, M.H.A., Haron, S.H.: Missing river discharge data imputation approach using artificial neural network. Methodology **25**, 20 (2015)
9. Chiu, P.C., Selamat, A., Krejcar, O.: Infilling missing rainfall and runoff data for sarawak, malaysia using gaussian mixture model based k-nearest neighbor imputation. In: Wotawa, F., Friedrich, G., Pill, I., Koitz-Hristov, R., Ali, M. (eds.) IEA/AIE 2019. LNCS (LNAI), vol. 11606, pp. 27–38. Springer, Cham (2019). https://doi.org/10.1007/978-3-030-22999-3_3
10. Lai, W.Y., Kuok, K.K.: A study on bayesian principal component analysis for addressing missing rainfall data. Water Resour. Manage **33**(8), 2615–2628 (2019). https://doi.org/10.1007/s11269-019-02209-8
11. Mirjalili, S.: SCA: a sine cosine algorithm for solving optimization problems. Knowl.-Based Syst. **96**, 120–133 (2016)

12. Qu, C., Zeng, Z., Dai, J., Yi, Z., He, W.: A modified sine-cosine algorithm based on neighborhood search and greedy levy mutation. Computational intelligence and neuroscience (2018)

13. Das, S., Bhattacharya, A., Chakraborty, A.K.: Solution of short-term hydrothermal scheduling using sine cosine algorithm. Soft Comput. **22**(19), 6409–6427 (2018)

14. Li, S., Fang, H., Liu, X.: Parameter optimization of support vector regression based on sine cosine algorithm. Expert Syst. Appl. **91**, 63–77 (2018)

15. Tawhid, M.A., Savsani, P.: Discrete Sine-Cosine Algorithm (DSCA) with Local Search for Solving Traveling Salesman Problem. Arab. J. Sci. Eng. **44**(4), 3669–3679 (2018). https://doi.org/10.1007/s13369-018-3617-0

16. Chandler, R.E., Isham, V.S., Leith, N.A., Northrop, P.J., Onof, C.J., Wheater, H.S.: Uncertainty in Rainfall Inputs. World Scientific/Imperial College Press, London (2011)

17. Stoner, O., Economou, T.: An Advanced Hidden Markov Model for Hourly Rainfall Time Series. arXiv:1906.03846 (2019)

18. Kashiwao, T., Nakayama, K., Ando, S., Ikeda, K., Lee, M., Bahadori, A.: A neural network-based local rainfall prediction system using meteorological data on the Internet: a case study using data from the Japan Meteorological Agency. Appl. Soft Comput. **56**, 317–330 (2017)

19. Yen, M.H., Liu, D.W., Hsin, Y.C., Lin, C.E., Chen, C.C.: Application of the deep learning for the prediction of rainfall in Southern Taiwan. Sci. Rep. **9**(1), 1–9 (2019)

20. Grange, S.K., Carslaw, D.C.: Using meteorological normalisation to detect interventions in air quality time series. Sci. Total Environ. **653**, 578–588 (2019)

21. Londhe, S., Dixit, P., Shah, S., Narkhede, S.: Infilling of missing daily rainfall records using artificial neural network. ISH J. Hydraulic Eng. **21**(3), 255–264 (2015)

22. Canchala-Nastar, T., Carvajal-Escobar, Y., Alfonso-Morales, W., Cerón, W.L., Caicedo, E.: Estimation of missing data of monthly rainfall in southwestern Colombia using artificial neural networks. Data Brief **26**, 104517 (2019)

23. Chiu, P.C., Selamat, A., Krejcar, O., Kuok, K.K.: Missing rainfall data estimation using artificial neural network and nearest neighbor imputation. In: Advancing Technology Industrialization Through Intelligent Software Methodologies, Tools and Techniques: Proceedings of the 18th International Conference on New Trends in Intelligent Software Methodologies, Tools and Techniques (SoMeT_19), 318, 132. IOS Press (2019)

24. Henry, A.J., Hevelone, N.D., Lipsitz, S., Nguyen, L.L.: Comparative methods for handling missing data in large databases. J. Vasc. Surg. **58**(5), 1353–1359 (2013)

25. Cheema, J.R.: Some general guidelines for choosing missing data handling methods in educational research. J. Mod. Appl. Stat. Meth. **13**(2), 3 (2014)

26. Zhu, P., Xu, Q., Hu, Q., Zhang, C., Zhao, H.: Multi-label feature selection with missing labels. Pattern Recogn. **74**, 488–502 (2018)

27. Hassani, H., Kalantari, M., Ghodsi, Z.: Evaluating the performance of multiple imputation methods for handling missing values in time series data: a study focused on East Africa. Soil-Carbonate-Stable Isotope Data. Stats. **2**(4), 457–467 (2019)

28. Oba, S., Sato, M.A., Takemasa, I., Monden, M., Matsubara, K.I., Ishii, S.: A Bayesian missing value estimation method for gene expression profile data. Bioinformatics **19**(16), 2088–2096 (2003)

29. Little, R.J., Rubin, D.B.: Statistical Analysis with Missing Data. Wiley, New York (2014)

30. Kurita, T.: Principal Component Analysis (PCA). In: Ikeuchi, K. (eds) Computer Vision. Springer, Boston (2014)

31. Pearson, K.: Principal components analysis. London, Edinburgh, Dublin Philos. Mag. J. Sci. **6**(2), 559 (1901)

32. Hotelling, H.: Analysis of a complex of statistical variables into principal components. J. Educ. Psychol. **24**, 417 (1933)

33. Smith, L.I.: A tutorial on principal components analysis (2002) https://www.cs.otago.ac.nz/cosc453/student_tutorials/principal_components.pdf. Accessed 03 Jan 2020
34. Khattree, R., Naik, D.N.: Multivariate Data Reduction and Discrimination with SAS Software. Cary, N.C., SAS Institute (2000)
35. Jamil, M., Yang, X.S.: A literature survey of benchmark functions for global optimisation problems. Int. J. Math. Modell. Numer. Optim. 4(2), 150–194 (2013)
36. Zuśka, Z., Kopcińska, J., Dacewicz, E., Skowera, B., Wojkowski, J., Ziernicka–Wojtaszek, A.: Application of the principal component analysis (PCA) method to assess the impact of meteorological elements on concentrations of particulate matter (PM10): a case study of the Mountain Valley (the Sącz Basin, Poland). Sustainability 11, 6740 (2019)
37. De Silva, C.C., Beckman, S.P., Liu, S., Bowler, N.: Principal component analysis (PCA) as a statistical tool for identifying key indicators of nuclear power plant cable insulation degradation. In: Proceedings of the 18th International Conference on Environmental Degradation of Materials in Nuclear Power Systems–Water Reactors, pp. 1227–1239. Springer, Cham (2019)
38. Gill, M.K., Asefa, T., Kaheil, Y., McKee, M.: Effect of missing data on performance of learning algorithms for hydrologic predictions: implications to an imputation technique. Water Resour. Res. 43(7) (2007)
39. Kim, T., Ko, W., Kim, J.: Analysis and impact evaluation of missing data imputation in day-ahead PV generation forecasting. Appl. Sci. 9(1), 204 (2019)
40. Ayilara, O.F., Zhang, L., Sajobi, T.T., Sawatzky, R., Bohm, E., Lix, L.M.: Impact of missing data on bias and precision when estimating change in patient-reported outcomes from a clinical registry. Health Quality Life Outcomes 17(1), 106 (2019)

Novelty Based Driver Identification on RR Intervals from ECG Data

Florian Heidecker[(⊠)], Christian Gruhl, and Bernhard Sick

Intelligent Embedded Systems, University of Kassel, Wilhelmshöher Allee 73,
34121 Kassel, Germany
{florian.heidecker,cgruhl,bsick}@uni-kassel.de
www.ies.uni-kassel.de

Abstract. We present an approach for driver identification, which is useful in many automotive applications such as safety or comfort functions. Driver identification would also be of great interest to other business models, such as car rental and car-sharing companies. The identification method is based on the driver's physiological state or rather his/her electrocardiogram (ECG) data. For this purpose, we have recorded ECG data of 25 people driving in a simulated environment. To identify a driver, we extend our existing novelty detection by aggregating local features over time. To do so, we extracted features and trained a Gaussian Mixture Model (GMM) to exploit localities present in the recorded sensor data. With novelty detection by aggregating local features, we are smoothing the noisy signal and reducing the dimensionality for further processing in a one-class SVM classification. Based on the output, a decision function decides whether the driver is unknown or well-known and if the driver is well-known, who of the known driver is it.

Keywords: Driver identification · RR-intervals · Novelty detection · One-class classification · Temporal data · Feature aggregation

1 Introduction

For most people, driving a car is a normal part of life. To further improve the adaptation of comfort functions and, in some cases, driver functions in a car, it is essential to identify the current driver. Driver identification is also economically valuable as it is an integral part of other commercial business models. Particularly for fleet managers of car-sharing and rental companies because, with a build-in driver identification, the company knows if an undeclared driver has driven the vehicle, which usually leads to an additional charge. Or someone is

The project "VitaB - Klassifizierung der Vtalparameter zur Individuellen vitalen und kognitiven Zustandsbestimmung des Menschen" (HA project no. 545/17–27) is financed with funds of LOEWE – Landes-Offensive zur Entwicklung Wissenschaftlich-ökonomischer Exzellenz, Förderlinie 3: KMU-Verbundvorhaben (State Offensive for the Development of Scientific and Economic Excellence).

© Springer Nature Switzerland AG 2021
A. Del Bimbo et al. (Eds.): ICPR 2020 Workshops, LNCS 12664, pp. 407–421, 2021.
https://doi.org/10.1007/978-3-030-68799-1_29

trying to steal the vehicle, but the identification of an undeclared person prevents it, which again is a useful function for all private owners, too.

In order to identify the driver, the algorithm has to characterize the driver. The characterization is based on his/her behavior or the physiological state and the features derived from it. Many publications for driver identification, for instance, [11,14], or [9] are using dynamic vehicle data such as steering rate, vehicle velocity, acceleration, pedal pressure, following distance to the leading car, and many more features to analyze the driving behavior. Instead of using vehicle data as features, [6] additionally uses heart rate and variance. In this article, we present an algorithm that uses a novelty detection technique for driver identification. Furthermore, the only requirement for the identification is the acquisition of the physiological state of the driver. As a database, we recorded electrocardiogram (ECG) data with shimmer sensors [13] of 25 different drivers while they had to drive through different scenarios sitting in front of a driving simulator (cf. Figure 1). We extract several features in the time and frequency domain from these collected data, such as heart rate or information entropy of Fourier coefficients. We use a trained Gaussian Mixture Model (GMM) to exploit localities present in the recorded sensor data (e.g., stressful situations are different from relaxed ones). We extend our existing novelty detection approach CANDIES [7] by aggregating local features over time, which smoothes the noisy signal and reduces dimensionality for further processing in the one-class SVM classification. The main contribution is the feature aggregation and the use of locally trained SVM for novelty detection. Based on the output of the one-class SVM, a decision function decides whether the driver is well-known or unknown. Besides, we want to find out whether the driver is well-known or unknown, but we also want to detect which of the known drivers is currently driving.

The remainder of this article is structured as follows. In Sect. 2, we present the related work. Section 3 describes in detail the data set and the data acquisition process. Section 4 describes our approach and is divided into feature extraction, novelty detection, and driver identification. In Sect. 5, we present the experimental evaluation of our described driver identification method. Finally, a conclusion and outlook for future work are given in Sect. 6.

Fig. 1. Experimental setup for data recording.

2   Related Work

The identification of a car driver is a challenging problem and, according to the literature, possible using different kinds of data. Different approaches use video data and dynamic vehicle data, which are related to the driver's operation signals such as steering rate, vehicle velocity, acceleration, pedal pressure, and many more. One of the first papers is from Wakita et al. [14]. They achieved an identification rate of 81% on drivers using a driving simulator. As features, they used some of the vehicle signals mentioned above and the distance to the vehicle in front. Miyajima et al. [11] extended the work of [14] using new features generated by spectral analysis and achieved a better result with an identification rate of 89.6% for a driving simulator. Many different approaches and models use this kind of feature and other vehicle data for driver identification. A good overview of the different approaches can be found in [6]. Furthermore, [6] uses ECG data in the form of heart rate and the heart rate variance in one of their data sets in addition to the vehicle signals. More common is the use of ECG data in driver distraction detection [5], monitoring [4], and also in stress detection [10] applications.

3   Data

The data we use to identify a person or, in this case, a car driver, are ECG records. A typical ECG signal of two heartbeats is shown in Fig. 2. The distance between two heartbeats is defined by the time between two beginnings of the heart chamber's contraction. The starting point of the chamber contraction (red vertical line) is the R peak and shown in Fig. 2. The time difference between two R peaks is therefore called the RR interval. These RR intervals are not constant. They vary over time. The quantification and analysis of these differences are called heart rate variability (HRV), and our starting point for all upcoming steps to identify a driver is based on her/his ECG data. Within our study, we collected ECG signals from 25 test persons. To collect the data, we used Shimmer sensors [13] together with the bipolar limb leads configuration. Each record has a duration of around 90 min. During the recording, all persons were driving a car in a simulator (cf. Figure 1) and wearing a virtual reality headset. The benefit of the simulator is that each test person has to drive under the same conditions. We can also control the traffic environment, and each test driver has to pass the same

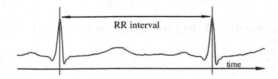

Fig. 2. ECG signal, each of the two red vertical lines is marking an R peak. (Color figure online)

scenarios. The outcome of these recordings is the vital parameters of each driver (ECG data), which characterize him/her. Based on this unique characteristic, we try to identify the driver. One thing that makes this task even more difficult is that our collected data is very noisy, e.g., due to the test subjects' movements during the ride.

4   Own Approach

The description of our approach is divided into three subsections. In the first subsection, we cover the feature extraction. Afterward, we describe the novelty detection process in detail, and last how we identify a driver and decide if an unknown or known driver is currently driving.

4.1   Feature Extraction

To extract the feature extraction from the RR intervals, we use a sliding window with two parameters. The first parameter is the window length and defines how many data points we use to calculate a set of features. The second parameter is the shift of the window to calculate the next set of features. The parameter N for the sliding window's length is dynamic because the RR intervals vary over time. Each window contains as many values that the sum of the individual RR intervals corresponds to one minute. Therefore a set of features takes all RR intervals from the last minute into account. The sliding window shift and the calculation of the next set of features is set to one second. Within each of these windows, we calculate the following eight features:

– The heart rate in beats per minute (HR)

$$HR = \frac{|X_t|}{60} \left[\frac{1}{s}\right] ,$$ (1)

– the standard deviation of the heart rate variability (HRV_{SD})

$$HRV_{SD} = \sqrt{Var(X_t)} ,$$ (2)

– the root-mean-squared difference of successive RR intervals (HRV_{RMSSD})

$$HRV_{RMSSD} = \sqrt{\frac{1}{N-1} \left(\sum_{i=1}^{N-1}(RR_{i+1} - RR_i)^2\right)} ,$$ (3)

– the ratio of the number of decreasing and increasing RR intervals (HRV_{DI})

$$HRV_{DI} = \frac{a_{increase}}{a_{decrease}}$$

$$\text{where} \quad \begin{cases} a_{increase} = count(R_i < R_{i+1}) \\ a_{decrease} = count(R_i > R_{i+1}) \end{cases} ,$$ (4)

– and the stress index (SI)

$$SI = \frac{n_D}{2D * (HRV_{max} - HRV_{min})} . \tag{5}$$

The SI value is a well-established heart rate variability value in the outer space medicine introduced by Baevskii [1]. It is a mathematical description of the RR interval histogram. n_D is the number of the modal value D whereas ($HRV_{max} - HRV_{min}$) describes the variability. Besides the analysis of the HRV in the time domain, the frequency domain is also an option. For this purpose, we used the Fourier transform to calculate the Fourier coefficients $C_t = (C_{t,1}, \cdots, C_{t,M})$ to get the following features:

– The energy of the Fourier transformation (fft_{en})

$$fft_{en} = \frac{1}{M} \sum_{i=1}^{M} |C_{t,i}|^2 , \tag{6}$$

– the information entropy of the Fourier coefficients ($fft_{inf-entr}$)

$$P_t = \frac{|C_t|}{\sum_{i=1}^{M} |C_{t,i}|} , \tag{7}$$

$$fft_{inf-entr} = \frac{-\sum_{i=1}^{M} P_t \log P_t}{\log N} , \tag{8}$$

– and the ratio of 33% of the coefficients with the lowest frequency compared to the 33% of the coefficients with the highest frequency ($L3rd - H3rd_{ratio}$)

$$L3rd - H3rd_{ratio} = \frac{\sum_{t=0}^{N/3} C_t}{\sum_{t=2N/3}^{N} C_t} . \tag{9}$$

After this feature extraction, we standardize the values so that the mean is 0, and the standard derivation is 1.

4.2 Novelty Detection

According to our definition, novelty detection is the task of identifying the emergence of a novel, i.e., unknown processes/driver, or a change in the characteristics of a known process/driver at runtime. In general, these novel processes are responsible for the emission of samples, which deviate substantially (based on an appropriate measure) from expected samples (i.e., the emissions of known processes/drivers). Commonly, those samples are called anomalies (often also outliers).

Our approach to novelty detection is CANDIES [7] (Combined Approach to Novelty Detection in Intelligent Embedded Systems), which is based on a probabilistic model and has two focal points. The first one is to detect agglomerations of anomalies in so-called low-density regions (LDR), regions where the estimated densities are low, and no regular samples are expected to be observed. The second focal point is towards the high-density regions (HDR). These are regions where the density is high and where more samples are expected to be observed. HDR novelties are detected by monitoring the distribution of currently observed samples, e.g., with a sliding window and comparing it against the expected (i.e., previously learned) distribution. If an agglomeration of anomalies is detected in LDR, this can be interpreted as a reliable indicator for the presence of an unknown process/driver – a novelty. Similar to that is the change of the HDR sample distribution, an indicator of another novelty type.

Two ideas support the mapping of the driver identification problem to novelty detection. First, the sample data from the known driver(s) is used to learn a probabilistic model (e.g., a Gaussian mixture model) representing the expected system behavior (that is, how the person's physiological state responds to various driving situations). This model will be used later as the decision base for CANDIES. Secondly, CANDIES is now applied to the sample data of other drivers who are considered unknown to the model. Since each driver's physiological state reacts differently to various driving situations (i.e., the samples are from different processes), sample data from the other drivers should result in the detection of novelties. Further, if multiple known drivers are using the car, the approach can detect who is currently driving.

Gaussian Mixture Models. As starting point for CANDIES, a probabilistic model is required which approximates the continuous density distribution of observed samples. Here, we rely on a Gaussian mixture model (GMM) which is the superposition, cf. Eq. (10), of multiple (here J) Normal distributions (also known as Gaussians) called components. Each component j has its own set of parameters: the mean vector $\boldsymbol{\mu}_j$ which describes its location and a covariance matrix $\boldsymbol{\Sigma}_j$ that describes its shape. To ensure that the GMM fulfills the requirements of a density (i.e., $\int P(\boldsymbol{x})\mathrm{d}\boldsymbol{x} = 1$) mixture coefficients π_j, with the constraints given in Eq. (11) and Eq. (12), are introduced for each component.

$$P(\boldsymbol{x}) = \sum_{j=1}^{J} \pi_j \cdot \mathcal{N}(\boldsymbol{x}|\boldsymbol{\mu}_j, \boldsymbol{\Sigma}_j) \tag{10}$$

$$0 \leq \pi_j \leq 1 \tag{11}$$

$$1 = \sum_{j=1}^{J} \pi_j \tag{12}$$

$$\gamma_{\boldsymbol{x}',j} = \frac{\pi_j \mathcal{N}(\boldsymbol{x}'|\boldsymbol{\mu}_j, \boldsymbol{\Sigma}_j)}{P(\boldsymbol{x}')} \tag{13}$$

The parameters of the GMM are estimated using our own implementation of Variational Bayesian Inference (VI) [2] (cf. the well known python library sklearn [12][1] for a comparable implementation). The probabilities $\gamma_{x',j}$ (cf. Eq. (13)) are called responsibilities and specify the degree to which a given sample x' "belongs" to a component j of the given GMM.

CANDIES [2]. The advancement of our two-stage novelty detection [8] (2SND) is to detect novelties in HDR. In CANDIES, samples are considered to be suspicious by the model if they reside in LDR. Based on a user-selected threshold of α, the model decides whether a sample is within HDR or LDR, which effectively controls how close a sample must be located to any of the components. To measure the distance, we use the Mahalanobis distance, cf. Eq. (14), of the component:

$$\Delta_j(x') = (x' - \mu_j)^T \Sigma_j^{-1}(x' - \mu_j) . \tag{14}$$

This procedure is the first stage of 2SND. Since we do not rely on LDR detection in this publication, the second stage of the LDR will be discussed only briefly. Suspicious samples get stored in a ring buffer. On the ring buffer, a DBSCAN-inspired non-parametric clustering is updated each time the LDR adds a new suspicious sample. When the data points in the buffer form a cluster and reach a certain size, the method reports a novelty.

Samples that are not considered suspicious and therefore located in the HDR region are linked to individual components of the GMM using the responsibilities mentioned above. This linking allows us to handle novelty detection differently for each component, thus exploiting each component's locality. In CANDIES, we monitor the distance distributions of recent samples affiliated with their respective component j, i.e., the distances $\Delta_j(x'))$, and compare it to the distribution observed with training samples using a χ^2-test with appropriate significance level. If the null hypothesis (observed samples are from the same distribution) is rejected, the method reports a novelty.

For driver identification, we used another approach for HDR detection. The distribution of sample distances is also the base in this approach, but utilizing the quantiles of the corresponding ecdf (empirical cumulative density function) in, what we call, QSpace.

QSpace. Especially in high dimensional spaces, SVM tends to require a large number of samples to yield satisfying results. Thus, we are interested in reducing the dimensionality but also in aggregating longer sequences of samples.

Our approach exploits the locality resulting from underlying GMM. Processed samples are affiliated with components based on their responsibilities $\gamma_{x',j}$, cf. Eq. (13). The image in Fig. 3a shows such a component with affiliated samples. The samples themselves are drawn from two distributions with identical mean μ, but with different diagonal covariances (blue ○: (1,1.5), orange △: (1,1)).

---

[1] sklearn.mixture.BayesianGaussianMixture.
[2] https://novelty-detection.de/p/mcandies.

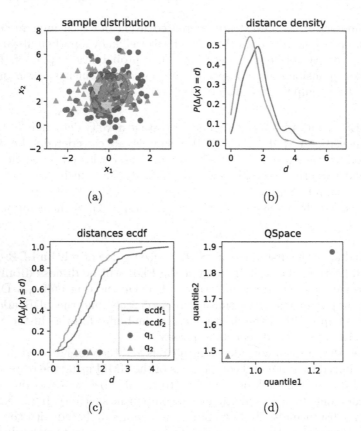

Fig. 3. Samples drawn from two distributions with identical mean, but different covariances (100 per distribution). Depicted are the density distributions of the distances of samples to mean, the corresponding ecdf, and the QSpace representations of both distributions.

A first dimension reduction is performed by calculating the Mahalanobis distance $\Delta_j(x')$ from each sample x to the mean μ_j of the component. Figure 3b depicts the resulting distribution of distances. By ascending ordering of the distances, we get an approximation of the cdf – the empirical cdf or ecdf, Fig. 3c. In contrast to other representations, this dimension reduction is lossy since we lose information about possible dependencies between the dimension in the data. The next step is to aggregate all samples represented by their ecdf to even fewer values. This approximation of the ecdf is made by a fixed number of quantiles Q. That means the ecdf gets divided into Q segments, and the area under the curve is the same for each segment. If we chose $Q = 3$ we need $2(Q-1)$ supporting points to segment the ecdf, since the first segment's domain is $(-\infty, q_1)$ and the last segment's domain (q_{Q-1}, ∞). Figure 3c shows the ecdf for both distributions, together with the supporting points for $Q = 3$. The supporting points themselves span a new vector space, which we call QSpace (depicted in Fig. 3d).

In this configuration ($Q = 3$) we can represent the original 100 2D samples (200 values) with only two values q_1 and q_2 in a 2-dimensional space.

By combining a sliding window with the QSpace aggregation, the method performs a distances based reduction, an approximation of the ecdf, and calculation of supporting points for each window. Through this procedure, we obtain a powerful tool to detect changing distributions. Consider Fig. 4a, where we applied QSpace to 2000 samples (1000 samples drawn from the earlier mentioned distributions). The Fig. 4b depicts the QSpace for $Q = 3$ for these samples. We can identify two closely located but distinctive clusters for the supporting points of both distributions.

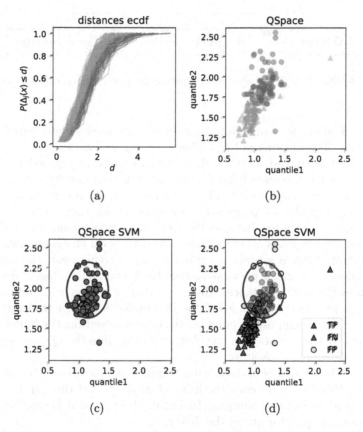

Fig. 4. Sliding window applied to synthetic data (2000 samples, window size set to 50 and step size to 10). Depicted are the ecdf for each window, the resulting QSpace with two quantiles, and a one-class SVM enclosing samples from the circle ○ distribution in the QSpace. The recall and precision for the novel class (triangle △) is 0.88 and 0.9, respectively.

Now, we can use a one-class νSVM[3] to perform novelty detection. The SVM decides whether the samples contained in a window come from the expected distribution or not. In Fig. 4c, the blue ○ samples represent the expected distribution, and the thick red line is the decision boundary. If we then use the same SVM to classify the supporting points of windows populated with samples from the second distribution, we achieve satisfying novelty detection rates with a recall of 0.88 and a precision of 0.9 (depicted in the most right image of Fig. 4d).

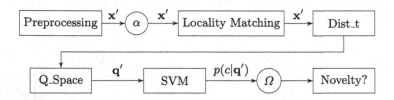

Fig. 5. Modified CANDIES execution pipeline for novelty detection in vital data.

Pipeline. Finally, we can build the core of our novelty detection for driver identification. The pipeline in Fig. 5 illustrates the various steps.

The GMM is trained with the driver's data to be recognized (well-known driver) and used as the underlying density model for our novelty detection. New, previously unseen samples (data of an unknown driver) are preprocessed. The preprocessing includes all previously mentioned steps, such as the calculation of RR intervals. Since we disabled the LDR detection facilities of CANDIES, the preprocessed samples x' are affiliated directly with the components of the driver's GMM. After a sample is affiliated with a component, the distance d' to the component's mean gets calculated. Each component has a ring buffer, which stores the last n distances, thus recreating a sliding window's behavior. Whenever a new distance is added to the ring buffer, a QSpace transformation is performed, i.e., the supporting points of the ecdf representing the buffer contents are updated. In the last stage, a one-class SVM classifies the QSpace supporting points and detects the novelties.

The one-class SVM is trained with training samples from the well-known driver (i.e., those used to learn the initial GMM) passed through the pipeline and then used as positive samples. In Fig. 6, the ecdf and QSpace are shown with training samples for one of the drivers.

If the one-class SVM classifies the current buffer content of the component as negative, it is assumed that it does not belong to the expected driver (unknown driver). To compensate for noisy data, i.e., frequent fluctuations between positive and negative classification, another buffer is used on the classified samples (denoted as Ω in the pipeline), which performs a majority voting.

---

[3] libsvm [3], rbf kernel, $\nu = 0.1, \gamma = 0.1$.

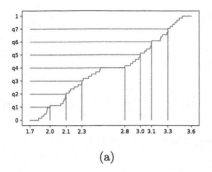 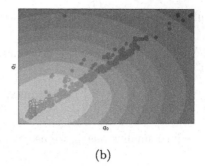

(a)   (b)

Fig. 6. Examples from evaluated driver data. (a) ecdf of one component's sliding window, segmented into eight quantiles, which leads to seven supporting points in the QSpace. (b) QSpace of the component from (a) with seven supporting points (only the first two dimensions are displayed). The green circles o are the training samples from the expected driver.

4.3   Driver Identification

Now we want to find out which car driver D_{curr} is currently driving. First, we have to train a novelty detection model with the driver data. This driver is now marked as well-known by the model. Whenever we want to recognize more than one driver, we have to train a second novelty detection model, one model for each driver d_i with $i \in 1...n$. Therefore we have the single and multi (well-known) driver scenario.

Single Driver Scenario. If we apply the model on the testing data, we get for each input sample, either 0 or 1. Zero means the input data does not match the model of the well-known driver, and we have an unknown driver. In the case of one, the well-known driver was recognized. Afterward, we store the predicted values in a ring buffer and calculate the relative frequency f_i of ones. If the relative frequency is equal or higher than a threshold t_i, the method detects the well-known driver. Whenever the relative frequency is below this threshold, the current car driver is unknown. The determination of the threshold is done during the training by predicting the training data and calculating the relative frequency.

Multi Driver Scenario. Whenever we have more than a single car driver, the threshold t_i, as mentioned above, is getting more important. Because in this case, we not only want to identify the expected driver, but we also want to decide which potential driver is driving. The first step is the same as with only one well-known driver. We store the predicted values in a driver-specific ring buffer and calculate the relative frequency f_i, but in this case, for multiple drivers. Afterward, we calculate the ratio per driver by how much the relative frequency is larger than the threshold value (cf. Eq. (15)). The final decision for multiple well-known drivers is made with Eq. (16):

$$d_i = \max\left(0, \frac{f_i - t_i}{1 - t_i}\right),$$

(15)

$$D_{curr} = \begin{cases} 1 + \arg\max((d_1, \cdots, d_i)) & \text{if } \sum d_i \neq 0 \\ 0 & \text{if } \sum d_i = 0 \end{cases}.$$

(16)

To reduce smaller peaks, we use a smoothing filter afterward.

5   Experimental Evaluation

To test our method, we use our 25 recorded data sets. As mentioned earlier, each of these data set contains data of around 90 minutes. We used 60% of the data belonging to one recorded driver to train the model. All types of situations passed through during the simulator drive are covered by this 60% of driving data. We use the remaining driver data to test whether the model recognizes the trained driver. Besides this, we used all other driver data sets to test whether the model can identify this as an unknown driver. For this reason, our test data is very imbalanced. Before we started the evaluation, we used a grid search to optimize the hyperparameter of our novelty detection model.

In Table 1, we show how our presented CANDIES method behaves on testing data if we train the model with one, two, or three drivers. With only one driver, we train the model once with each driver's data and test it as described before. At last, we averaged the results of the 25 cases. To test our CANDIES model on two drivers, we included all possible combinations resulting from 25 recorded drivers. That resulted in 300 possible combinations whose scores we as well averaged. We repeated the same procedure with three drivers, but in this case, we had 2300 possible combinations. Table 1 shows that the performance is dropping by adding a new driver. This circumstance affects all scores. However, the f1 scores are always better than if we only predicted the majority class $f1_{maj}$. On the other side, these results suggest that the presented method may not be usable for a larger group of well-known drivers.

Table 1. Performance of the CANDIES model depending on two and three expected drivers.

Well-known drivers	1	2	3
Classes	Unknown, driver 1	Unknown, driver 1/2	Unknown, driver 1/2/3
Precision	0.568	0.435	0.3783
Recall	0.766	0.640	0.5612
F1 score	0.598	0.477	0.4238
$f1_{maj}$	0.333	0.167	0.100

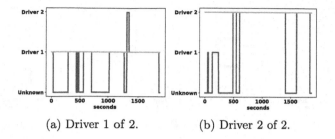

(a) Driver 1 of 2.   (b) Driver 2 of 2.

Fig. 7. The two presented results show the performance of driver identification in the case of two well-known drivers.

In Fig. 7, 8, and 9, we show examples of our driver identification with CAN-DIES. In all figures, the orange line describes the ground truth, while the blue line is the predicted car driver.

In Fig. 7 and 8, we can see the results of the CANDIES model trained with two or three drivers. In most cases, the driver identification method recognizes the well-known driver (driver 1, 2, and 3) correctly. Nevertheless, there are also many jumps and wrong detections, which may be related to the noisy data and similarities in the data between multiple drivers.

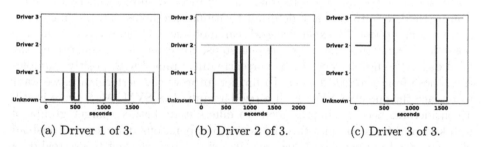

(a) Driver 1 of 3.   (b) Driver 2 of 3.   (c) Driver 3 of 3.

Fig. 8. The three presented results show the performance of driver identification in the case of three well-known drivers.

Fig. 9 shows three examples if we test the model with data from an unknown driver. The prediction results are perfect in some cases, e.g., Fig. 9a. These perfect identifications show that the well-known and unknown drivers have a different RR interval pattern, and therefore it is easy to separate them. However, there are also cases, e.g., Fig. 9c, where the well-known and unknown drivers are similar, and therefore the identification is challenging. Overall, the most common cause is like the one in Fig. 9b, where the model recognizes the unknown driver only for some short time ranges.

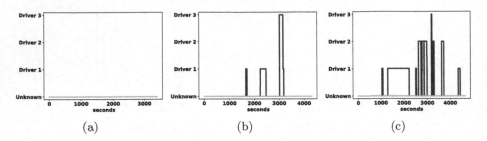

Fig. 9. The three presented results show the performance of driver identification in the case of three different unknown drivers.

6   Conclusion and Outlook

In this article, we have proposed a method for driver identification based on novelty detection. We modified our existing CANDIES approach, which relies on a GMM specific trained for an individual driver and uses our QSpace transformation to reduce dimensionality effectively. The final detection is performed with a one-class SVM. We could show that the developed method is feasible to implement driver identification tasks. Even though the data contains severe noise, we obtained promising results. In our future work, we will focus on the optimization of the novelty detectors, for instance, by implementing a calibration method to automatically determine various thresholds and parameters to make a computationally expensive grid search obsolete. Other improvements are possible by adjusting the preprocessing steps, for example, by smoothing the vital data. The novelty detector is currently only trained with positive samples (i.e., those from the real driver). To further improve the performance, we want to replace the one-class SVM with a regular multiclass SVM. Also, there is still a considerable need for improvement to differentiate between larger groups of well-known drivers. Besides this, we want to expand our data set, and instead of using data from people driving a car in a simulator, we want to use real data in the future.

References

1. Baevskii, R.M.: Analysis of heart rate variability in space medicine. Hum. Physiol. **28**(2), 202–213 (2002)
2. Bishop, C.M.: Pattern Recognition and Machine Learning (Information Science and Statistics). Springer, New York (2006)
3. Chang, C.C., Lin, C.J.: LIBSVM: A library for support vector machines. ACM TIST **2**(3), 27:1–27:27 (2011)
4. Dehzangi, O., Williams, C.: Towards multi-modal wearable driver monitoring: Impact of road condition on driver distraction. In: IEEE BSN, pp. 1–6. IEEE, Cambridge, MA, USA (2015)
5. Deshmukh, S.V., Dehzangi, O.: ECG-Based Driver Distraction Identification Using Wavelet Packet Transform and Discriminative Kernel-Based Features. In: IEEE SMARTCOMP, pp. 1–7. IEEE. Hong Kong (2017)

6. Ezzini, S., Berrada, I., Ghogho, M.: Who is behind the wheel? Driver identification and fingerprinting. J. Big Data **5**(1), 1–15 (2018)
7. Gruhl, C., Sick, B.: Novelty detection with CANDIES: a holistic technique based on probabilistic models. Int. J. Mach. Learn. Cyber. **9**(6), 927–945 (2018)
8. Gruhl, C., Sick, B., Wacker, A., Tomforde, S., Hähner, J.: A building block for awareness in technical systems: Online novelty detection and reaction with an application in intrusion detection. In: IEEE iCAST, pp. 194–200. IEEE, Qinhuangdao, China (2015)
9. Jafarnejad, S., Castignani, G., Engel, T.: Towards a real-time driver identification mechanism based on driving sensing data. In: IEEE ITSC, pp. 1–7. IEEE Yokohama, Japan (2017)
10. Keshan, N., Parimi, P.V., Bichindaritz, I.: Machine learning for stress detection from ECG signals in automobile drivers. In: IEEE Big Data, pp. 2661–2669. IEEE Santa Clara, CA, USA (2015)
11. Miyajima, C., et al.: Driver modeling based on driving behavior and its evaluation in driver identification. Proc. IEEE **95**(2), 427–437 (2007)
12. Pedregosa, F., et al.: Scikit-learn: machine learning in python. JMLR **12**(85), 2825–2830 (2011)
13. Shimmer: http://www.shimmersensing.com/. Accessed 27 Jan 2020
14. Wakita, T., et al.: Driver identification using driving behavior signals. In: IEEE ITSC, pp. 396–401. IEEE, Vienna, Austria (2005)

Link Prediction in Social Networks by Variational Graph Autoencoder and Similarity-Based Methods: A Brief Comparative Analysis

Sanjiban Sekhar Roy[1], Aditya Ranjan[1], and Stefania Tomasiello[2]([✉])

[1] School of Computer Science and Engineering, Vellore Institute of Technology, Vellore, India
[2] Institute of Computer Science, University of Tartu, Tartu, Estonia
stefania.tomasiello@ut.ee

Abstract. Link prediction is an emerging and fast-growing applied research area. In a network, it is possible to predict the next link which is going to be formed. The usefulness of link prediction modeling has been proved in several fields and applications, such as biomedicine, recommending systems, and social media. In this short paper, we discuss the potential of Variational Graph Autoencoder, by comparing the results so obtained against those by some similarity-based methods, such as Adamic-Adar, Jaccard coefficient, and Preferential Attachment.

1 Introduction

Social networks have gained growing attention. They represent a way to model communication or interactions among many people who may share common interests and backgrounds. In such evolving networks, where the addition and/or deletion of several links and vertices take place many times every day, it is interesting to predict the possible changes, e.g. for recommendations purposes. So social networks are dynamic entities, which can also be seen as a source of data for studying the interactions between the people in groups or communities. Such networks are formally modelled as a graph in which each vertex identifies a user and the edges represent the relation between them. The so-called egocentric social networks can be seen as a collection of connected egonets [1]. As a social network contains millions of nodes, it is very difficult to be analyzed and this suggests considering egonets. In egonets, the interactions of a node are only with its neighbors, based on social relationships [2]. The problem of link prediction is in a certain sense more difficult than some others in social networks, e.g. the prediction of positive and negative tweets [3]. Many methods have been investigated for the link prediction, as discussed in a recent survey [4]. For instance, in order to take into account the uncertainty in the formation of new links, there are some techniques based on fuzzy set theory [5–7] or on the belief function theory [8]. According to [4], there have not been works discussing extensively the use of Variational Graph Autoencoder (VGAE) after it has been introduced [9]. VGAE is an unsupervised framework which adapts the variational autoencoder scheme to graph-structured data.

© Springer Nature Switzerland AG 2021
A. Del Bimbo et al. (Eds.): ICPR 2020 Workshops, LNCS 12664, pp. 422–429, 2021.
https://doi.org/10.1007/978-3-030-68799-1_30

In this short paper, we present the performance of different methods for link prediction on egonets. In particular, here the VGAE has been considered for link prediction in different egonets and its performance compared against some local similarity-based methods (Adamic-Adar, Jaccard coefficient, and Preferential Attachment). Similarity-based approaches include local similarity-based methods and global network similarity-based methods. In the first class, the prediction is made by considering information about common neighbours and node degree, ignoring the global structure of the network. In the second class, the computation is performed by using entire topological information of a network. The computational cost of global methods is higher and seems to be infeasible for large networks [3].

The VGAE scheme that we used here has been adapted from [12]. In the encoding stage, the graph convolutional network (GCN) learns the distributions of input data from the network structure and node features, while in the decoding stage, the VGAE reconstructs the original input data and gets the prediction score.

The remaining part of this paper has been organized as follows: Sect. 2 briefly introduces the problem and the related works. The considered methods are briefly recalled in Sect. 3. Section 4 presents the experimental outcome, along with the experimental setup and data sets, metrics. Section 5 gives some conclusions.

2 Link Prediction: Problem Description and Related Works

Let G(V,E) be a graph of the network, where V is a vertex-set and E(i,j) the link-set. The link prediction problem can be formally summarized as follows [4]. Let G_{t0-t1} (V, E) be a graph representing a snapshot of a network during the time interval [t_0, t_1] and E_{t0-t1}, a set of links in that snapshot. The task of link prediction is to find set of links E_{t2-t3} during the time interval [t_2, t_3] where [t_0, t_1] $\leq$ [t_2, t_3]. In a very recent survey [3], the different techniques used for the link prediction problem have been reviewed, ranging from classical structural and probabilistic ones to recent network embedding methods, fuzzy models, and deep learning models. The authors classified the methods into similarity-based, probabilistic models, dimensionality reduction-based, entropy-based, and clustering-based. The authors also observed that local and quasi-local approaches perform well, usually. They also pointed out that although several link prediction methods have been explored in the literature, there are still some open issues, such as the best structural properties and how to deal with the large size of the network. In addition to the above-mentioned survey, it is the case to mention a couple of recent papers. In [10], the authors proposed a novel link prediction method in noisy networks, combining low-rank representation and non-negative matrix factorization for similarity matrix calculation. In [11], it has been discussed a linear model to integrate several types of local indices, by using two model averaging approaches, for hybrid link prediction.

3 Methodology

In this section, we briefly recall the methods which have been used for the comparative analysis.

3.1 Variational Graph Autoencoder

The VGAE is used to deal with graph structured data. It is a framework based on variational autoencoder which is influenced by the graph convolutional network (GCN) [9, 12]. The VGAE model includes two stages, encoding and decoding. In the encoding stage, it takes an adjacency matrix U and a feature matrix Y as input and gets a latent variable v as output, while in the decoding stage, the VGAE gets a reconstructed adjacency matrix according to the latent variable v. Without losing generality, let us consider the sub-network H_n. Let N be the number of nodes in H_n, U_n the adjacency matrix, and C the degree matrix of U_n. As mentioned before, Y denotes the feature matrix, and v the latent variable. By looking at the encoding stage, the VGAE includes two layers of GCN. The first layer of GCN generates a lower dimensional feature matrix Y′, with ReLU activation

$$Y' = \text{ReLU}(U'_n Y W_0) \tag{1}$$

with

$$U'_n = C^{-1/2}(U_n + I)C^{-1/2} \tag{2}$$

being I the identity matrix. The second GCN layer generates the data distribution as follows:

$$\mu = U'_n Y' W_m, \ \log \sigma = U'_n Y' W_s \tag{3}$$

Then, for ε following N(0, 1), the latent variable v can be obtained:

$$v = \mu + \sigma \varepsilon \tag{4}$$

The decoder is defined by an inner product and the output U^d_n is the reconstructed adjacency matrix:

$$U^d_n = R\left(v \ v^T\right) \tag{5}$$

where R is the sigmoid function. The loss function includes the binary cross-entropy between the target and the computed output [9, 12].

3.2 Similarity-Based Methods

The Adamic/Adar index was originally designed to predict links in social networks [13] and further adapted for the counting of common features [14]

$$A(i, j) = \sum_{u \in \Gamma(i) \cap \Gamma(j)} \frac{1}{\log|\Gamma(u)|} \tag{6}$$

Where $\Gamma(i)$ and $\Gamma(j)$ are neighbors of the node i and j respectively.

Jaccard's coefficient measures the probability that i and j have a common feature [15]:

$$J(i,j) = \frac{|\Gamma(i) \cap \Gamma(j)|}{|\Gamma(i) \cup \Gamma(j)|} \tag{7}$$

Preferential attachment is based on the idea that the probability that a new edge involves the node i is proportional to the current number of neighbors of i. This is a very basic prediction method [15]

$$P(i, j) = |\Gamma(i)| * |\Gamma(j)| \tag{8}$$

According to [4], this index shows the worst performance on most networks.

4 Numerical Experiments

In this section, any relevant information about the numerical experiments will be provided.

4.1 Data Set

The dataset is from Google+ social circle data, publicly available at https://snap.stanford.edu/data/ego-Gplus.html. This dataset consists of circles from Google+. It includes ego networks, circles and node features (profiles). The dataset has 107614 nodes, 13673453 edges as a whole and an average clustering coefficient of 0.4901. For the experiment purposes, two egonets from these datasets have been extracted for the link prediction. They consist of different circles. Table 1 shows the network statistics of the egonets which have been used. All the networks have a diameter which equals 2. The network in Fig. 1 has 658 nodes, while the one in Fig. 2 is the smallest network with an average node degree of 20.72.

Table 1. Network Statistics of egonets

Net ID	Diam	Avg degree	Num edges	Density	Transitivity	Nodes
Net_1	2	45.042	14819	0.06855	0.26912	658
Net_2	2	20.724	3720	0.05788	0.34064	359

4.2 Performance Metrics

The metrics that have been considered are Average Precision (AP), receiver operating characteristic (ROC). Additionally, Precision (P), Recall (R) and F-measure have also been calculated:

$$P = \frac{TP}{TP + FP} \tag{9}$$

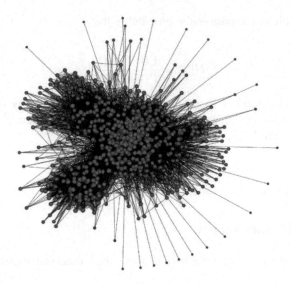

Fig.1. EGO-network 1

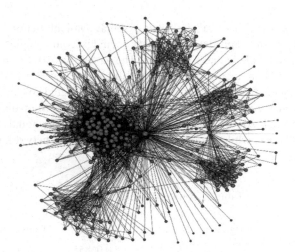

Fig. 2. EGO-network 2

$$R = \frac{TP}{TP + FN} \tag{10}$$

$$F1 = 2 * \frac{Precision * Recall}{Precision + Recall} \tag{11}$$

For the model to be trained, each graph has been divided into two parts: training set which consists of the links randomly selected from the graph and test set consisting of the remaining links.

4.3   Numerical Results

Two different egonets were taken from Google plus dataset for the application study. An incomplete network was created for the model to be trained, by randomly removing some of the links. The dimension of the latent variable was 32 and 16 for the first and second network respectively. The total number of epochs for each network is 300.

In Table 2, the behavior of the trained VGAE on the different networks for different epochs is reported. As one can observe, the ROC (here meant as area under the curve) and the average precision increase by increasing the number of epochs. The VGAE shows better accuracy on the smaller network.

Table 2. VGAE performance for different epochs

	Epoch 100		Epoch 200		Epoch 300	
	ROC	AP	ROC	AP	ROC	AP
Net_1	0.789	0.811	0.849	0.858	0.861	0.867
Net_2	0.858	0.874	0.893	0.901	0.898	0.910

A brief comparative analysis among Variational Graph Autoencoder (VGAE), Adamic-Adar, Jaccard's Coefficient, Preferential Attachment is shown in Tables 3, 4 and 5 The overall performance based on ROC and AP values, from Table 3, is ranked as Adamic-Adar>Preferential attachment>VGAE> Jaccard's coefficient. Different conclusion in terms of recall and F-measure. In fact, by looking at Table 4 and 5, the VGAE

Table 3. ROC and AP values for different approaches

	Net_1		Net_2	
	ROC	AP	ROC	AP
VGAE	0.861	0.867	0.898	0.910
Adamic-Adar	0.903	0.899	0.931	0.933
Jaccard coefficient	0.747	0.703	0.813	0.745
Preferential attachment	0.866	0.856	0.868	0.880

Table 4. Precision and Recall values for different approaches

	Net_1		Net_2	
	Precision	Recall	Precision	Recall
VGAE	0.767	0.678	0.797	0.700
Adamic-Adar	0.901	0.530	0.964	0.448
Jaccard coefficient	0.701	0.480	0.746	0.472
Preferential attachment	0.857	0.468	0.885	0.483

Table 5. F-measure for different approaches

	Net_1	Net_2
VGAE	0.719	0.745
Adamic-Adar	0.667	0.611
Jaccard coefficient	0.569	0.578
Preferential attachment	0.605	0.624

outperformed the other methods. It is worth mentioning that algorithms which show consistent recall and precision values are better algorithms with better stability.

5 Conclusions

In this short paper, we discussed a VGAE model for link prediction. A comparative study involving different link prediction approaches, by combining feature information and network topology has been carried out. The VGAE outperformed the other models in terms of recall and F-measure on two egonets, even though Adamic-Adar seems to perform better in terms of ROC value, meant as area under the curve.

References

1. Sen, P., et al.: Collective classification in network data. AI Mag. **29**(3), 93–106 (2008)
2. Tabourier, L., Libert, A.-S., Lambiotte, R.: Predicting links in ego-networks using temporal information. EPJ Data Sci. **5**(1), 1–16 (2016). https://doi.org/10.1140/epjds/s13688-015-0062-0
3. Roy, S.S., Biba, M., Kumar, R., Kumar, R., Samui, P.: A new SVM method for recognizing polarity of sentiments in twitter. In: Handbook of Research on Soft Computing and Nature-Inspired Algorithms, pp. 281–291. IGI Global (2017)
4. Kumar, A., et al.: Link prediction techniques, applications, and performance: a survey. Phys. A **553**, 124289 (2020)
5. D'Aniello, G., Gaeta, M., Reformat, M., Troisi, F.: Link prediction in signed social networks using fuzzy signature. In Proceedings - IEEE International Conference on Systems, Man and Cybernetics, pp. 2524–2529 (2019)
6. Reformat, M.Z., D'Aniello, G., Gaeta, M.: Knowledge graphs, category theory and signatures. In: Proceedings - 2018 IEEE/WIC/ACM International Conference on Web Intelligence, pp. 480–487 (2018)
7. Loia, V., Parente, D., Pedrycz, W., Tomasiello, S.: A granular functional network with delay: some dynamical properties and application to the sign prediction in social networks. Neurocomputing **321**, 61–71 (2018)
8. Mallek, S., Boukhris, I., Elouedi, Z., Lefevre, E.: Evidential link prediction in social networks based on structural and social information. J. Comput. Sci. **30**, 98–107 (2019)
9. Kipf, T., Welling, M. Variational graph auto-encoders. arXiv:1611.07308 (2016)
10. Chen, X., Tao, W., Xian, X., Wang, C., Yuan, Y., Ming, G.: Enhancing robustness of link prediction for noisy complex networks. Phys. A Stat. Mech. Appl. **555**, 124544 (2020)

11. Zhang, Q., Tong, T., Wu, S.: Hybrid link prediction via model averaging. Phys. A Stat. Mech. Appl. **556**, 124772 (2020)
12. Ding, Y., Tian, L. P., Lei, X., Liao, B., Wu, F. X. Variational graph auto-encoders for miRNA-disease association prediction. Methods (2020). in press
13. Adamic, L.A., Adar, E.: Friends and neighbors on the web. Soc. Netw. **25**(3), 211–230 (2003)
14. Liben-Nowell, D., Kleinberg, J.: The link-prediction problem for social networks. J. Am. Soc. Inform. Sci. Technol. **58**(7), 1019–1031 (2007)
15. Lichtenwalter, R.N., Lussier, J.T. , Chawla, N.V., New perspectives and methods in link prediction. In: Proceedings of the 16th ACM SIGKDD International Conference on Knowledge Discovery and Data Mining. ACM (2010)

A Hybrid Wine Classification Model for Quality Prediction

Terry Hui-Ye Chiu, Chien-Wen Wu, and Chun-Hao Chen(✉)

Department of Information and Finance Management, National Taipei University Technology,
Taipei 106, Taiwan
terry.h.chiu@gmail.com, {xcwwu,chchen}@ntut.edu.tw

Abstract. "Wine is bottled poetry" a quote from Robert Louis Stevenson shows the wine is an exciting and complex product with distinctive qualities that make it different from other products. Therefore, the testing approach to determine the quality of the wine is complex and diverse. The opinion of a wine expert is influential, but it is also costly and subjective. Hence, many algorithms based on machine learning techniques have been proposed for predicting wine quality. However, most of them focus on analyzing different classifiers to figure out what the best classifier for wine quality prediction is. Instead of focusing on a particular classifier, it motivates us to find a more effective classifier. In this paper, a hybrid model that consists of two classifiers at least, e.g. the random forest, support vector machine, is proposed for wine quality prediction. To evaluate the performance of the proposed hybrid model, experiments also made on the wine datasets to show the merits of the hybrid model.

Keywords: Machine learning · Decision tree · Random forest · Support vector machine · Wine quality prediction

1 Introduction

Wine has always been an essential part of the dinning culture in western countries. From the manufacturer point of aspect, understanding the quality of the wine and creating a steady production is an important goal for the industry. However, testing the quality of the wine is complex and diverse. The wine quality is evaluated in terms of subtlety and complexity, ageing potential, stylistic purity, varietal expression, ranking by experts, or consumer acceptance. By excluding the controllable object measures, the views of experts are very subjective because it may cause the most considerable influence on both winemakers and consumers [1].

Recording the steps of wine production procedure is to preserve the quality and knowledge of the whole winemaking process, and the collected information is the best tool to guarantee the wine quality. Currently, the wine industry has established the protected designation of origin (PDO) system [2] with the support of analytical chemistry and chemometric tools to obtain information related to a specific wine. With the improvement of technology both in software and hardware, winemakers started to use

© Springer Nature Switzerland AG 2021
A. Del Bimbo et al. (Eds.): ICPR 2020 Workshops, LNCS 12664, pp. 430–438, 2021.
https://doi.org/10.1007/978-3-030-68799-1_31

the collected data to improve the winemaking technique. Due to the high cost and lack of technological resources, it was difficult for most of the wine industries to classify the wines based on the chemical components. Many algorithms based on machine learning to assess the quality of wine have then been gained much attention for the wine industry to determine what attributes make a "good" wine that the consumers can satisfy with them. For instance, Yeo et al. focused on predicting the wine price using a machine learning technique by using past historical wine price data [3]. For wine production, Ribeiro et al. utilized the linear regression, neuron network and decision tree for predicting the wine vilification [4].

In 2009, Cortez et al. collected a wine quality dataset which consists of significant larger instances [6]. Then, three machine learning models, including multiple regression, support vector machine (SVM) and neuron network (NN), are trained using the collected wine dataset. It shows that SVM outperforms the other two methods, and indicates the importance of the correct setting of hyperparameters. Over the years, the wine dataset has been adopted in several studies with various methods such as SVM [7–10] , random forest (RF) [11–14], decision-tree-based algorithms [12,14], and NN [4,7,8] to predict the quality of the wine based on physiochemical characteristics in the wine.

However, the past literature mostly focused on using or comparing different machine learning models that can provide the best prediction result for specific datasets. To get a more effective classifier, in this paper, a hybrid model that consists of two classifiers at least, e.g. the random forest, support vector machine, is proposed for wine quality prediction. To evaluate the performance of the proposed hybrid model, experiments also made on the wine dataset to show the merits of the hybrid model.

2 Background Knowledge

Over the years, several different machines learning models are used to predict wine quality. The literature suggested that the LR and SVM provide better results than other models. In this section, the two commonly used classifiers are described.

2.1 The SVM Classifier

The support vector machine (SVM) [19] is a supervised machine learning model for solving a classification problem. The central concept of SVM is utilized the kernel function to find the hyperplane that can separate instances into categories. As mentioned earlier, the SVM has proven to be an effective classifier for wine quality prediction [7–10]. There are three hyperparameters in SVM that include penalty factor C, parameter gamma γ and kernel function in SVM. The parameter C is a regularization parameter that controls the trade-off between maximizing the margin and minimizing the training error. The gamma parameter defines how far the influence of a single training example reaches. The final parameter is a kernel; there are three different main types of kernels linear, poly and rbf, which may fit best with the different dataset. Hyperparameter tuning relies more on experimental results than theory, and therefore the best method to determine the optimal settings is by trial and error.

2.2 Random Forest

Random forest (RF) [20] is a supervised learning algorithm, but different from SVM used on both classification as well as regression. Several studies have shown that using random forest can provide a good prediction accuracy, and one study showed an extremely high accuracy rate [14]. In this study, we used for classification problems. Based on the name, RF constructed from different trees and more trees means more robust forest. In general, the RF algorithm creates different decision trees on random data samples and then gets the prediction from each of the trees. Next step is to use a voting technique to select the best solution. It has an advantage over other methods because it reduces the over-fitting by averaging the result. In this study, we will tune the RF model to provide the best result. The random forest has six hyperparameters: (1) No. of estimators (2) Maximum features (3) Maximum depth (4) Minimum samples split (5) Minimum samples leaf and (6) bootstrap. Tuning hyperparameters can improve the accuracy of the model. However, if evaluating the model only on the training set can lead to overfitting.

3 Proposed Hybrid Wine Classification Model

In this paper, the goal is to predict the quality of wine using the hybrid wine classification model that composes of multiple classifiers. The flowchart of the proposed model is shown in Fig. 1.

From Fig. 1, it first selects the classifiers from the machine learning models pool. The hyperparameter of selected classifiers is them determine by the randomized search method. The selected models are gathered as a hybrid classification model. Then, the input red and white data sets are used to train and test the model. This paper attempts to provide a hybrid wine classification model that produces the best performance. The pseudo-code of the proposed model is illustrated in Table 1(Algorithm 1).

In Table 1, the algorithm firstly selected at least two models (line 2). When the selection process is complete, the initial ranges of the hyperparameters associated with each model are set (line 3). For example, for SVM model (M_0), the hyperparameters as $p^m = \{$C: [1, 100, 10000], gamma: [0.1 0.01, 0.0001] and kernel: ['rbf', 'linear', 'poly']$\}$ are initially set. The hyperparameters for each model is different, and there is no specific appropriate range of value for any particular model. Therefore, the trial and error strategy is used in the algorithm in order to find the proper range of values that can provide a better model performance. For example, the hyperparameters are fitted to the model, and the model is then evaluated using the predefined criteria (mainly accuracy). Base on the performance, it decides whether the range values should be added or removed from the initial setting (line 6). After the modification process, we compare the performance of new setting against past settings (line 7). If the new setting indeed finds "optimal" solution, then the process of trial and error can be interrupted. Otherwise, it will continue the process until reaching the required iteration setting (lines 4–9). Continue using SVM as an example, the hyperparameter setting after the trial and error process is $p^m = \{$C: [500, 1000, 10000], gamma: [0.01 0.001, 0.0001] and kernel: ['rbf']$\}$. This range set of hyperparameters are passed to the next procedure. That is to find the best set of hyperparameter for SVM by using the random grid search method (line 11). After the tuning procedure, the selected models are merged to form a hybrid model M^{new} (line 16).

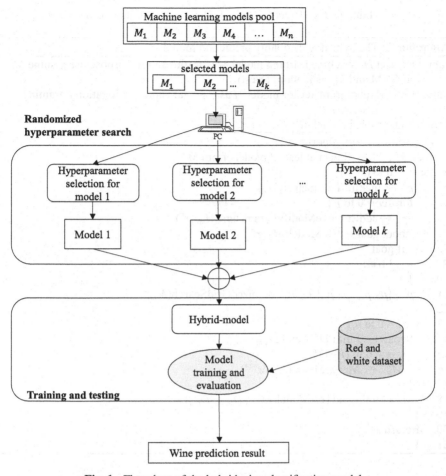

Fig. 1. Flowchart of the hybrid wine classification model

The model M^{new} is then trained and tested for n times with different training and testing data for each iteration (line 18). The criteria used to evaluate the models are accuracy, precision, recall and f1-score. From past studies, prediction techniques like SVM and RF are commonly used and have better results. Therefore, in this paper, we use those two models as the selected models and the dataset collected by [6] as our testing dataset. Experimental results are shown in the next section.

4 Experimental Evaluation

4.1 Dataset Description

The wine dataset from the UCI database [6] that consists of two sets of wine data (red and white) is used in this paper. The red wine contains 1599 instances, and white wine contains 4898 instances. Both datasets contain 11 physiochemical variables, including

Table 1. Pseudo-code of the hybrid wine classification model.

Algorithm 1: The hybrid wine quality prediction model
Input: Dataset D, Machine learning model M (for illustration purpose use assume M_0 as SVM and M_1 as random forest)
Parameters: Hyperparameter for model M as P_M, trial and error iteration j, training iteration n
Output: The hybrid model M^T

```
1. Procedure hybrid_algorithm(){
2. {M₀, M₁, ...} ← Select at least 2 models from M
3. for m = 0 to (i - 1){
4.     pᵐ ← pickInitParameters(Mₘ)
5.     for iter = 0 to j {
6.        pᵐ ← addRemoveModifyParameters(Mₘ, pᵐ)
7.        bool goal ← testModel(Mₘ, pᵐ)
8.        if goal
9.           break
10.    }
11.    pᵐ: {p₀, p₁, ..., pₘ} ← RandomizedGridSearch(Mₘ)
12. }
13. for i = 0 to n {
14.    Separate D into D_train and D_test
15.    for m = 0 to (i - 1){
16.        Mⁿᵉʷ ← MergeModel(Mᵢ, Mⁿᵉʷ)
17.    }
18.    Mᵢᵀ ← trainAndTestModel(Mⁿᵉʷ, D_train, D_test)
19. }
20. Return Mᵀ
21. }
```

fixed acidity, volatile acidity, citric acid, residual sugar, chlorides, free sulfur dioxide, total sulfur dioxide, density, pH, Sulphates, and alcohol. The output data (sensory data) is a quality rating from 0 (very bad) to 10 (excellent). We carefully examined the data to make sure no anomaly exists. As a result, for red wine, there are 24 duplicate instances, and 937 duplicate instances for white wine. However, those are not classified as an anomaly because all features and label values are precisely the same.

4.2 Performance Measure Metrics

The four criteria, including accuracy, precision, recall and f1-score, used to evaluate the performance of the model, are described in the following. The accuracy is to measure how often the classifier correctly classifies instances. The formulation is defined as follows:

$$accuracy = \frac{TP + TN}{TP + TN + FP + FN},$$
(1)

where TP (true positive) means the samples in the data which classified as belonging to the correct class, TN (true negative) means the samples in the data which classified as not belonging to the expected class correctly, FP (false positive) means the samples in the data which classified as belonging to the expected class incorrectly, and FN (false negative) means the samples in the data which classified as not belonging to the expected class, incorrectly.

In a multi-class condition, the micro and averaged accuracy, precision, and recall are always the same. Also based on past works, it is obvious the wine data is imbalanced, hence only using accuracy may not provide a clear picture. Therefore, the macro-averaging measurement for precision (macro-precision) and recall (macro-recall) are employed for a more detailed comparison.

The precision measures the percentage of the relevant results. When precision is one, it means the algorithm's prediction is perfect. The macro-precision will be lower than average precision for the model performing poorly on the rare classes even it performs well on specific classes. Since the measurement value can tell another story, hence it can be a complementary metric. Macro-averaging precision is performed by first computing the precision of each class, and then taking the average of all precisions.

$$Macro - PrecisionMP_{Classifier} = \frac{\sum_c P_c}{number of classes} \tag{2}$$

The recall measures the percentage of total relevant results correctly classified by the algorithm. When the recall is one, it means that all truly positive samples were predicted as the positive class. Similar to micro-precision, the value will be lower if rare classed performed poorly.

$$Macro - RecallMR_{classifier} = \frac{\sum_c R_c}{number of classes} \tag{3}$$

Accuracy is useful when the class distribution in the dataset is even, but F1-score is a better metric when the dataset has imbalanced classes. Hence, we also used it as a criterion to evaluate the performance of the model, and F1-score is defined as follows.

$$Macro - F1 F1_{classifier} = \frac{\sum_c F1_c}{number of classes} \tag{4}$$

4.3 Experimental Analysis

Since most of the past works mainly focus on accuracy, therefore, we compared the accuracy of the proposed model against others. In addition, because most works set the training and testing datasets ratio to 80/20, we also set the same ratio for comparison. For comparison, we included the work from Cortez et al. [6] and Apalasamy et al. [17] for performance comparison. The accuracy of each model is shown in Table 2.

In Table 2, the accuracies of the proposed model for red and white wines are 0.66 and 0.67. The results reveal that the proposed model performed slightly better than other models. Also, like most models, the performance of white wine is slightly higher than

Table 2. Comparison of different models in terms of accuracy.

Models	Accuracy	
	Red wine	White wine
Cortez et al. [6]	0.45	0.51
Apalasamy et al. [17]	0.62	0.65
Proposed model	**0.66**	**0.67**

Table 3. Result for different testing data ratio

Testing dataset size percentage	Red wine			White wine		
	10%	*20%*	*30%*	*10%*	*20%*	*30%*
Accuracy	0.69	0.71	0.66	0.70	0.68	0.67
Macro-precision	0.42	0.43	0.34	0.42	0.37	0.47
Macro-recall	0.41	0.38	0.32	0.41	0.34	0.35
Macro-F1 score	0.41	0.39	0.32	0.41	0.34	0.36

red wine. The experiments were then made to examine further the performance of the proposed model under different training and testing data ratio. The results of different testing size for red and white wine of the proposed model are shown in Table 3.

For red wine, the accuracy was at the highest when testing dataset ratio set at 20%. The macro-precision and recall are low across different ratio. It means the amount of false positive is very close or equal to a false negative. The F1 score for red wine gradually decreases with the increase in ratio. The white wine shows a different result, where the precision is always higher than recall. When the ratio was set at 10%, the accuracy and F1 score are at the highest. Also, the macro-precision and recall are at the closest. The low macro-F1 score for red and white wine indicates the data is highly skewed on some classes.

5   Conclusions and Future Work

This paper has proposed a hybrid wine classification model for quality prediction, which is unlike most past works focusing on which machine learning models provide the best performance in predicting wine quality. The proposed algorithm first selects n models from the given model pool. Then, the hyperparameters are then searched by the randomized search method. The models with acceptable performances are merged as the hybrid model. Experiments were done on the real dataset that contains red and white wines, and the results indicated the proposed hybrid model is effective in terms of accuracies when compared to other existing approaches. In the future, we will continue to design an algorithm that can be used to obtain both the hybrid models and hyperparameters for any wine dataset based on evolutionary algorithms.

References

1. Cardebat, J.-M., Livat, F.: Wine expert rating: a matter of taste? Int. J. Wine Bus. Res. **28**, 43–58 (2016)
2. Canizo, B.V., Escudero, L.B., Pellerano, R.G., Wuilloud, R.G.: 10 – Quality monitoring and authenticity assessment of wines: analytical and chemometric methods. In: Quality Control in the Beverage Industry, Grumezescu, A.M., Holban, A.M., (eds.), pp. 335–384. Academic Press (2019)
3. Yeo, M., Fletcher, T., Shawe-Taylor, J.: Machine learning in fine wine price prediction. J. Wine Econ. **10**(2), 151–172 (2015)
4. Ribeiro, J., Neves, J., Sanchez, J., Delgado, M., Machado, J., Novais, P.: Wine vinification prediction using data mining tools. In: International Conference on European Computing Conference, Tbilisi, Georgia (2009)
5. Andonie, R., Johansen, A.M., Mumma, A.L., Pinkart, H.C., Vajda, S.: Cost efficient prediction of Cabernet Sauvignon wine quality. In: IEEE Symposium Series on Computational Intelligence, pp. 1–8 (2016)
6. Cortez, P., Cerdeira, A., Almeida, F., Matos, T., Reis, J.: Modeling wine preferences by data mining from physicochemical properties. Decis. Support Syst. **47**(4), 547–553 (2009)
7. Gupta, Y.: Selection of important features and predicting wine quality using machine learning techniques. Procedia Comput. Sci. **125**, 305–312 (2018)
8. Lingfeng, Z., Feng, F., Heng, H.: Wine quality identification based on data mining research. Int. Conf. Comput. Sci. Educ. 358–361 (2017)
9. Bhattacharjee, S., Chaudhuri, M.R.: Understanding quality of wine products using support vector machine in data mining. Prestige Int. J. Manag. IT-Sanchayan **5**(1), 67–80 (2016)
10. Er, Y., Atasoy, A.: The classification of white wine and red wine according to their physicochemical qualities. Int. J. Intell. Syst. Appl. Eng. 23 (2016)
11. Trivedi, A., Sehrawat, R.: Wine quality detection through machine learning algorithms. In: International Conference on Recent Innovations in Electrical, Electronics & Communication Engineering, pp. 1756–1760 (2018)
12. Shaw, B., Suman, A.K., Chakraborty, B.: Wine quality analysis using machine learning. In: Mandal, J.K., Bhattacharya, D. (eds.) Emerging Technology in Modelling and Graphics. AISC, vol. 937, pp. 239–247. Springer, Singapore (2020). https://doi.org/10.1007/978-981-13-7403-6_23
13. Hu, G., Xi, T., Mohammed, F., Miao, H.: Classification of wine quality with imbalanced data. In: IEEE International Conference on Industrial Technology, pp. 1712–1217 (2016)
14. Aich, S., Al-Absi, A.A., Hui, K.L., Lee, J.T., Sain, M.: A classification approach with different feature sets to predict the quality of different types of wine using machine learning techniques. In: International Conference on Advanced Communication Technology, pp. 1–2 (2018)
15. Kumar, S., Agrawal, K., Mandan, N.: Red wine quality prediction using machine learning techniques. In: International Conference on Computer Communication and Informatics, pp. 1–6 (2020)
16. Mahima, G.U., Patidar Y., Agarwal, A., Singh, K.P.: Wine quality analysis using machine learning algorithms. In: The Micro-Electronics and Telecommunication Engineering, Lecture Notes in Networks and Systems (2020). https://doi.org/10.1007/978-981-15-2329-8_2
17. Appalasamy, P., Mustapha, A., Rizal, N., Johari, F., Mansor, A.: Classification-based data mining approach for quality control in wine production. J. Appl. Sci. **12**, 598–601 (2012)
18. Petropoulos, S., Karavas, C.S., Balafoutis, A.T., Paraskevopoulos, I., Kallithraka, S., Kotseridis, Y.: Fuzzy logic tool for wine quality classification. Comput. Electron. Agri. **142**, 552–562 (2017)

19. Vapnik, V.N.: An overview of statistical learning theory. IEEE Trans. Neural Netw. **10**(5), 988–999 (1999)
20. Liaw, A., Wiener, M.: Classification and regression by random forest (2007)
21. Bergstra, J., Bengio, Y.: Random search for hyper-parameter optimization. J. Mach. Learn. Res. **13**, 281–305 (2012)

A PSO-Based Sanitization Process with Multi-thresholds Model

Jimmy Ming-Tai Wu[1], Gautam Srivastava[2], Shahab Tayeb[3],
and Jerry Chun-Wei Lin[4(✉)]

[1] College of Computer Science and Engineering,
Sandong University of Science and Technology, Qingdao, Sandong, China
wmt@wmt35.idv.tw
[2] Department of Mathematics and Computer Science, Brandon University,
Brandon, Canada
SRIVASTAVAG@brandonu.ca
[3] Department of Electrical and Computer Engineering, California State University,
Fresno, USA
tayeb@csufresno.edu
[4] Department of Computer Science, Electrical Engineering and Mathematical
Sciences, Western Norway University of Applied Sciences, Bergen, Norway
jerrylin@ieee.org

Abstract. Earlier, many PPDM algorithms have been proposed to conceal sensitive items in a database in order to disclose sensitive itemsets. All prior techniques, however, ignored a crucial problem in setting minimum support thresholds. Thus, a new concept of minimal support for solving this issue is proposed in this paper. In compliance with a given threshold function, the proposed approach would set a tighter threshold for an object containing several items. Experimental results are then evaluated to show the performance of the traditional Greedy PPDM approach, GA-based PPDM approaches, and the proposed PSO-based algorithm with the new flexible and minimal support function.

Keywords: Data mining · PPDM · Partical swam optimisation · Multi-thresholds

1 Introduction

Data mining techniques [1] have been utilized in the last few decades in various domains and applications, in order to retrieve very large-scale useful and meaningful knowledge. Apriori [1] is called the fundamental algorithm of mining required patterns. The minimal support threshold is used for first detecting the set of frequent itemsets in a level-wise way. Several extensions [15,21] of knowledge discovery in the database (KDD) were then implemented to handle various scenarios and areas for revealing of different knowledge.

Although KDD techniques can be used to mine the useful information in the database, confidential/private information during the mining process [2] may also

© Springer Nature Switzerland AG 2021
A. Del Bimbo et al. (Eds.): ICPR 2020 Workshops, LNCS 12664, pp. 439–446, 2021.
https://doi.org/10.1007/978-3-030-68799-1_32

be identified or referred to from relevant information. Most algorithms were pro-grammed to conceal confidential information in an original database [3,10,22]. The sanitization processes applied evolutionary computation (EC)-based models [11,16,17,19,23] based on the pre-defined fitness functions generate an optimum set of transactions/items to hide the sensitive information. Such models, however, use a single threshold to determine how sensitive information in specific environments is hidden. But it is unreasonable in real applications. In actual implementations, it is not practical to use a single criterion to define the same importance of attributes. In this paper, we try to provide a framework for the security of privacy that can be used more appropriately in a specific environment to handle sensitive data based on the multi-thresholds model. The developed model is more suitable and reasonable for the real-life situations compared to the existing approaches.

2   Related Work

Genetic Algorithm (GA) was invented by John [7] at the beginning of 1970s. GA will represent a workable solution as a population on a chromosome. A predefined fitness function would be used to estimate each chromosome. In the whole process, GA must retain the best solution as the optimum global solution. GA is easily designed and usually works well in many applications compared to other EC algorithms, but the convergence speed is usually slower than that of other EC algorithms.

PSO was proposed by Kenny and Eberhart in 1995 as a population-based method [8]. In PSO, particles are used to solve problems, in which each particle has a speed that is a flying direction to the approximate solutions. PSO initializes particles first spontaneously. An iterative process of adaptation is then carried out afterward. A particle is replaced with the best value (*pbest*) by itself and the best global value (*gbest*) in conjunction with the predefined fitness function in every iteration. This evaluates and updates particles and their corresponding speeds using these two values. PSO was originally applied in numerous applications to find solutions to ongoing problems.

As the essential knowledge can be exposed from databases, sensitive and protected information that triggers users' privacy and safety risks can also be found during the mining process. Privacy-preserving data mining (PPDM) has been a big concern in recent years as it can reveal not only valuable information but can also mask sensitive information by sanitization progress [2,4,22]. Oliveira and Zaiane [20] provide various methods of sanitization, which can hide frequent itemsets by a heuristic framework. Then Lin et al. proposed a program for SPMF to provide many PPDM sanitization algorithms [13]. However, the PPDM is known as an NP-hard problem [3,22] and so meta-heuristic methods to find the near-best solutions are better offered. Lin *et al.* then proposed a variety of GA-based solutions to hide sensitive information by removing transactions, such as sGA2DT [11], pGA2DT [11] and cpGA2DT [12]. In order to illustrate the quality of the gene, a fitness function is often programmed to evaluate three side effects with the preset weights. There are also developed some extensions of PPDM evolutionary computation [6,14]. Several multi-objective PPDM models [19] were presented and discussed to balance the trade-off between side effects.

3   Background Knowledge

Let $A = \{A_1, A_2, \ldots, A_r\}$ be a finite set of r attributions in a identifiable dataset D. The dataset $D = \{C_1, C_2, \ldots, C_n\}$ is a set of records about identifiable information. Each attribution $A_q \in A$, $A_q = \left\{\emptyset, a_1^q, a_2^q, \ldots, a_{m_q}^q\right\}$ includes m_q values and 1 empty value. Each record $C_s \in D$, $C_s = \{c_1^s, c_2^s, \ldots, c_r^s\}$ such as $c_t^s \in A_t$. For simple expression of C_s, remove the value $\emptyset$ in C_s and set as $C_s = \{c_1'^s, c_2'^s, \ldots, c_u'^s\}$. It means there are u non-empty attribution values in C_s and C_s is a subset of $A_1 \cup A_2 \cup \cdots \cup A_r$. A minimum support threshold function $\delta()$ is given to reveal the high-frequency patterns in D. This function provides the different thresholds for the different lengths of patterns. Note that a pattern with longer length has a smaller threshold and a minimum value of minimum support δ_m will be set to avoid revealing a high-frequency pattern with low frequent.

Definition 1. Frequent Pattern: *A pattern p is a high-frequency pattern if and only if its support count higher or equal to the value of the minimum support threshold function multiplies the size of the dataset. It can be defined as:*

$$sup(p) \geq \delta\left(|p|\right) \times |D| \tag{1}$$

where $sup(p)$ is the support count of p and $\delta\left(|p|\right)$ is the minimum support count with the length of p. Finally, the set of frequent patterns in the dataset D can be denoted as $FPs = \{f_1, f_2, \ldots, f_k\}$

Definition 2. Sensitive Pattern: *A sensitive pattern in the dataset D, $SPs = \{s_1, s_2, \ldots, s_p\}$ is a subset of FPs ($SPs \subseteq FPs$) and each $s_i \in SPs$ needs to be hidden in the revealing information.*

Definition 3. Sanitized Database: *While a database D' is called as a sanitized database from the original database D, D' is generated by pruning some records from D.*

Problem Statement: PPDM is to build a sanitized database D' from D such that the support counts for all of the sensitive patterns $s_i \in SPs$ are less than the minimum support counts (calculated by minimum support threshold function), that is:

$$sup(s_i) < \delta\left(|s_i|\right) \times |D| \tag{2}$$

In this paper, the proposed algorithm applies a minimum support threshold function to adjust the value of the minimum support threshold. It will provide more strict thresholds to the high-frequency pattern with a longer length. The detail implementation is described in the following section.

4   Proposed Sanitization Algorithm

A fitness function is used to estimate the tours (solutions) generated by the selection process. The designed MPSO2DT typically follows the same fitness function as previous works [11], but introduces a rule to compare two tours that their fitness values remain equal. The used fitness function, which is defined as the weighted sum of the three side effects, is set as follows:

$$f(p) = w_1 \times \alpha + w_2 \times \beta + w_3 \times \gamma, \tag{3}$$

where α is the number of patterns for hiding failure; β is the number of patterns for missing cost, and γ is the number of patterns for artificial cost. The w_1, w_2 and w_3 are the weights that are set by the user for the relative value of each side effect. The w_1 is then often set to a number greater than the other two weighted numbers since the importance to hide the sensitive information is higher than others.

In the previous sensitive pattern hiding researches, a fixed minimum support threshold is set for any length of patterns. However, it is unreasonable for an identifiable information database. That is because a pattern is very easy to be identified to a specific individual or group when the length of the pattern is long. In the proposed MGA2DT, the minimum support threshold function is introduced to provide multiple minimum support thresholds for different lengths of patterns. It is to remove some records in a dataset in order to hide the sensitive information and keep the completeness of the dataset simultaneously. Each particle in the population indicates a possible solution, and would be evaluated by the fitness function. The size of a particle is set to an appropriate number of deleted transactions for hiding the sensitive itemset.

Definition 4. The appropriate number of transactions to be deleted is denoted as m, and is defined as the difference between the largest support count among all sensitive itemsets in SIs and the minimum support count, that is:

$$m = \left\lceil \frac{Max\_sup(s_i) - \delta \times |D|}{1 - \delta} \right\rceil \tag{4}$$

Definition 5. A particle p_i is an m dimensional vector, where each dimension represents a transaction to be deleted $T_q \in D^*$, and stores its TID. Note that a dimension may optionally contain the value **null**, representing no transaction.

The proposed MPSO2DT follows the design from the previous PSO2DT. Particles indicate a transaction sets to be removed in D^* for the hiding of sensitive information and represented as TID sets. The difference and union operations are applied for a particle with *pbest* and *gbest* to modify the velocity by Eq. 5. The proposed process calculates the difference between two particles and applies a union operator to generate a new particle.

$$v_i(t+1) = (pbest - x_i(t)) \cup (gbest - x_i(t)) \tag{5}$$

$$x_i(t+1) = rand(x_i(t), \boldsymbol{null}) + v_i(t+1) \tag{6}$$

Before performing the proposed PSO-based algorithm, the particle size, the size of the projected transactions, the original frequent itemsets will be first obtained by Eq. 4 and Apriori algorithm. A set of sensitive itemsets and a minimum support threshold function are also set before performing the sanitation process. The pseudo-code of the proposed method is shown in Algorithm 1. Note that the original frequent itemsets generated by Apriori also apply the dynamic minimum support threshold, which is defined as follows:

Definition 6. *To perform the multi-threshold model on the sensitive information, the equations are then used below as:*

$$\delta_t = F_s(L), and \tag{7}$$

$$\delta = max(\delta_t, S_u), \tag{8}$$

where F_s is the minimum support threshold function, S_u is the minimum value of the minimum threshold function, and L, the length of an itemsets. After that, δ is then returned that will be adopted to the designed MPSO2DT model for sanitization. After the designed algorithm is performed, the database is sanitized and the confidential information will be then hidden based on the developed multi-threshold model.

5   Experimental Results

The suggested MPSO2DT approach was compared in substantial experiments with pPSO2DT [18], GA-based sGA2DT, pGA2DT, cpGA2DT [11,12] and non-evolutionary Greedy sanitization approach [9]. Two datasets called chess and foodmart [5] are then used in the experiments for evaluation.

5.1   Fitness Value

Following the previous session, the related fitness values calculated by the three side-effects in terms of various minimum support thresholds are shown in Fig. 1. From the results, it is obvious to see that the proposed MPSO2DT has better performance than the other compared algorithms in terms of fitness value. This indicates that the lower side effects is then achieved by the designed model, and the sensitive information can be successfully hidden.

5.2   Runtime

To further compare the proposed MPSO2DT and the previous algorithms, the value of runtime is shown in Fig. 2. The performances of the runtime of MPSO2DT and previous PSO-based algorithm "pPSO2DT" are very similar but MPSO2DT spends a bit more runtime than pPSO2DT. That is because MPSO2DT and pPSO2DT are both applied the traditional PSO process but

---

Algorithm 1. MPSO2DT

Input:	D^*, the dataset of projected transactions; SIs, a set of sensitive patterns to be hidden; FI_s, the set of frequent patterns in D^*; F_s, the minimum support threshold function; S_u, the minimum value of the minimum threshold function; M, the size of population and N, the maximum value of iterations.
Output:	D', a sanitized database.

Termination conditions: the number of iterations achieves N.

1: $t = 1$ ▷ indicate the number of iteration
2: calculate the size of a particle as m ▷ by equation 4
3: **for** $i = 1 to M$ **do**
4:     **for** $j = 1 to m$ **do**
5:         $p_i(t) \leftarrow p_i(t) \cup rand(D^*)$
6:     **end for**
7: **end for**
8: **while** the number of iterations is not larger than N **do**
9:     calculate the fitness function for each particle ▷ by dynamic minimum support threshold
10:     **for** $i = 1$ to M **do**
11:         **if** $fitness(p_i(t)) \leq pbest_i$ **then**
12:             $pbest_i \leftarrow p_i(t)$
13:         **end if**
14:         **if** $fitness(p_i(t)) \leq gbest$ **then**
15:             $gbest \leftarrow p_i(t)$
16:         **end if**
17:         update $v_i(t+1)$ of the i-th particle   ▷ by equation 5
18:         update $p_i(t+1)$ of the i-th particle   ▷ by equation 6
19:     **end for**
20:     $t = t + 1$
21: **end while**
22: delete the transactions in $gbest$ from D^* generate D'
23: **return** the sanitized database D'

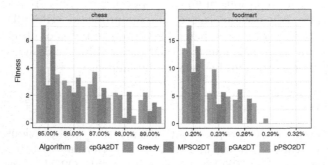

Fig. 1. Fitness w.r.t various minimum support thresholds.

the proposed method need handle the multi-thresholds process. Due to the simply process, the runtimes of Greedy algorithm are always shorter than the other compared algorithms. The runtimes of the two GA-based algorithms are a little more than the two PSO-based algorithms.

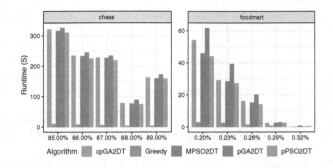

Fig. 2. Runtime w.r.t various minimum support thresholds.

6   Conclusions

In this paper, the proposed MPSO2DT is modified from the previous effective algorithm pPSO2DT and give a new concept of minimum support threshold function. It compared to the previous GA-based and Greedy approaches in two transaction databases to show its performance. Experimental results showed the proposed MPSO2DT could effectively obtain better fitness value and try to hide all sensitive information.

References

1. Agrawal, R., Srikant, R.: Fast algorithms for mining association rules. In: VLDB Conference, pp. 487–499 (1994)
2. Agrawal, R., Srikant, R.: Privacy-preserving data mining. In: ACN SIGMOD International Conference on Management of Data, pp. 439–450 (2000)
3. Atallah, M., Bertino, E., Elmagarmid, A., Ibrahim, M., Verykios, V.: Disclosure limitation of sensitive rules. In: Workshop on Knowledge and Data Engineering Exchange, pp. 45–52 (1999)
4. Dasseni, E., Verykios, V.S., Elmagarmid, A.K., Bertino, E.: Hiding association rules by using confidence and support. In: International Workshop on Information Hiding, pp. 369–383 (2001)
5. Fournier-Viger, P., et al.: The SPMF open-source data mining library version 2. In: Joint European Conference on Machine Learning and Knowledge Discovery in Databases, pp. 36–40 (2016)
6. Han, S., Ng, W.K.: Privacy-preserving genetic algorithms for rule discovery. In: International Conference on Data Warehousing and Knowledge Discovery, pp. 407–417 (2007)

7. Holland, J.H., et al.: Adaptation in natural and artificial systems: an introductory analysis with applications to biology, control, and artificial intelligence. MIT Press (1992)

8. Kennedy, J., Eberhart, R.: Particle swarm optimization. In: International Conference on Neural Networks, vol. 4, pp. 1942–1948 (1995)

9. Lin, C.W., Hong, T.P., Chang, C.C., Wang, S.L.: A greedy-based approach for hiding sensitive itemsets by transaction insertion. J. Inf. Hiding Multimed. Signal Process. **4**(4), 201–227 (2013)

10. Lin, C.W., Hong, T.P., Hsu, H.C.: Reducing side effects of hiding sensitive itemsets in privacy preserving data mining. Sci. World J. **2014**, 235837 (2014)

11. Lin, C.W., Hong, T.P., Yang, K.T., Wang, S.L.: The GA-based algorithms for optimizing hiding sensitive itemsets through transaction deletion. Appl. Intell. **42**(2), 210–230 (2015)

12. Lin, C.W., Zhang, B., Yang, K.T., Hong, T.P.: Efficiently hiding sensitive itemsets with transaction deletion based on genetic algorithms. Sci. World J. **2014**, 398269 (2014)

13. Lin, J.C.W., Fournier-Viger, P., Wu, L., Gan, W., Djenouri, Y., Zhang, J.: PPSF: an open-source privacy-preserving and security mining framework. In: IEEE International Conference on Data Mining Workshops, pp. 1459–1463 (2018)

14. Lin, J.C.W., Liu, Q., Fournier-Viger, P., Hong, T.P., Voznak, M., Zhan, J.: A sanitization approach for hiding sensitive itemsets based on particle swarm optimization. Eng. Appl. Artif. Intell. **53**, 1–18 (2016)

15. Lin, J.C.W., Pirouz, M., Djenouri, Y., Cheng, C.F., Ahmed, U.: Incrementally updating the high average-utility patterns with pre-large concept. Appl. Intell. **50**(11), 3788–3807 (2020)

16. Lin, J.C.W., Srivastava, G., Zahng, Y., Djenouri, Y., Aloqaily, M.: Privacy preserving multi-objective sanitization model in 6g IoT environments. IEEE Internet Things J. (2020)

17. Lin, J.C.W., Wu, J.M.T., Fournier-Viger, P., Djenouri, Y., Chen, C.H., Zhang, Y.: A sanitization approach to secure shared data in an IoT environment. IEEE Access **7**, 25359–25368 (2019)

18. Lin, J.C.W., Yang, L., Fournier-Viger, P., Wu, M.T., Hong, T.P., Wang, L.S.L.: A swarm-based approach to mine high-utility itemsets. In: Multidisciplinary Social Networks Research, pp. 572–581 (2015)

19. Lin, J.C.W., Zhang, Y., Zhang, B., Fournier-Viger, P., Djenouri, Y.: Hiding sensitive itemsets with multiple objective optimization. Soft. Comput. **23**(23), 12779–12797 (2019)

20. Oliveira, S.R., Zaiane, O.R.: Privacy preserving frequent itemset mining. In: International Conference on Privacy, Security and Data Mining, vol. 14, pp. 43–54 (2002)

21. Srivastava, G., Lin, J.C.W., Zhang, X., Li, Y.: Large-scale high-utility sequential pattern analytics in internet of things. IEEE Internet Things J. (2020)

22. Verykios, V.S., Bertino, E., Fovino, I.N., Provenza, L.P., Saygin, Y., Theodoridis, Y.: State-of-the-art in privacy preserving data mining. ACM SIGMOD Record **33**(1), 50–57 (2004)

23. Wu, T.Y., Lin, J.C.W., Zhang, Y., Chen, C.H.: A grid-based swarm intelligence algorithm for privacy-preserving data mining. Appl. Sci. **9**, 1–20 (2019)

Task-Specific Novel Object Characterization

Gertjan J. Burghouts[(✉)]

TNO, Hague, The Netherlands
gertjan.burghouts@tno.nl

Abstract. In an open world, a robot encounters novel objects. It needs to be able to deal with such novelties. For instance, to characterize the object, if it is similar to a known object, or to indicate that an object is unknown or irrelevant. We present an approach for robots to deal with an open world in which objects are encountered that are not known beforehand. Our method first decides whether an object is relevant for the task at hand, if it is similar to an object that is known to be relevant. Relevancy is determined from a task-specific taxonomy of objects. If the object is relevant for the task, then it is characterized through the taxonomy. The task determines the level of detail that is needed, which relates to the levels in the taxonomy. The advantage of our method is that it only needs to model the relevant objects and not all possible irrelevant and often unknown objects that the robot may also encounter. We show the merit of our method in a real-life experiment of a search and rescue task in a messy and cluttered house, where victims (including novelties) were successfully found.

Keywords: Open world · Novel objects · Task-specific taxonomy · Object characterization

1 Introduction

In an open world, a robot encounters novel objects. It needs to be able to deal with such novelties. For instance, to characterize the object, if it is similar to a known object, or to indicate that an object is unknown or irrelevant. The robot needs to be able to characterize objects that are relevant for the task. Typically, not all encountered objects are relevant. For instance, for a search and rescue task, the robot needs to find all humans, while all other objects can be ignored for the search. Therefore, we follow an approach of first assessing the relevancy of the (novel) object. If the object is relevant for the task, then it is characterized.

The advantage is that the robot only needs a limited knowledge of the world. Its world model is a limited taxonomy of task-specific objects. The taxonomy enables the robot to characterize an object in relation to the parents in the taxonomy. This is beneficial, because for various tasks the required level of detail may differ. For instance, for the search and rescue task, the robot needs to find all humans, whereas if there is very limited time it needs to give priority to a child (subcategory of human hence more specific). An example result of our method (Sect. 3), is shown in Fig. 1, where a novelty, i.e., the child, is found in a messy, cluttered house (Sect. 4) and it is characterized as relevant for the search and rescue task and as a subcategory of human.

© Springer Nature Switzerland AG 2021
A. Del Bimbo et al. (Eds.): ICPR 2020 Workshops, LNCS 12664, pp. 447–455, 2021.
https://doi.org/10.1007/978-3-030-68799-1_33

Bernard:~$ SNOW says: rescue! it is a human (child)

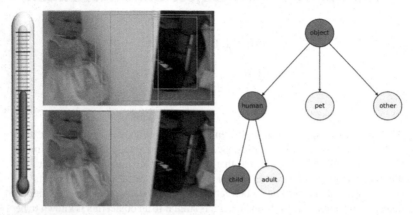

Fig. 1. Our method finds relevant objects for the task (red box) based on a task-specific taxonomy (right), in the midst of irrelevant objects (green boxes). (Color figure online)

Our approach does not require a model of all objects. For each of the relevant objects, it has learned a model before deployment, through a small labelled training set. Irrelevant objects are not modelled. Instead, they will be ignored during deployment by comparing them to the known objects and assessing that they are out of distribution.

Example applications of the proposed method are: robots operating in an open world and encountering novel objects; selecting the known and good quality products that pass by a camera in a production process and ignoring other objects that have not been produced properly; recognizing partially visible objects by their parts, of which the relations are modelled by a taxonomy.

Our method consists of four components that are detailed in Sect. 3: generic detection to localize candidate objects, a task-specific taxonomy that describes the relevant objects, a model to decide on relevancy for the task, and a model to characterize a relevant object.

The paper starts with describing related work (Sect. 2). The method is detailed in Sect. 3. Experiments are covered in Sect. 4, followed by conclusion (Sect. 5).

2   Related Work

Hierarchical novelty detection was proposed to simultaneously detect that an object is novel and describe it by its super-category [1]. This enabled an assessment such as 'this is a novel animal'. The notion that subclasses of a category are more similar to each other than subclasses of another category, was translated into a semantic distance which was applied as a loss function in a deep learning framework [2]. Hereby, the model learned to group together semantically related classes, based on a given, fixed taxonomy. This is especially important for a robot, because it may not always need to classify objects at the finest level of detail. Often, coarser classification suffices for the task at hand, e.g., classifying a child as a human is sufficient for a search and rescue robot, yet in some other cases it is important to know that it is a child. We will extend this framework [2] to learn task-specific taxonomies that may change when a robot needs to switch to another task.

Deciding whether a test object was from the training distribution is important to assess how capable a learned model is to predict the object's class. Typically an out of distribution analysis is performed to decide if a classifier is able to make a good prediction [3]. Recent progress is made by more advanced statistical modelling and generative models [4]. We aim to decide if a test object is sufficiently similar to the relevant classes from the task-specific taxonomy. To this end, we rely on out of distribution analysis, by comparing the test object to the statistics of the relevant samples.

None of the related works assess both the object's relevancy for the robot's task as well as characterizing the object even if it is a novelty. Combining these capabilities is essential for robotic vision and is the contribution of this paper.

3 Method

Our method consists of four components. To find candidate objects in the scene, we rely on a generic object detector from the state-of-the-art (3.1). Each candidate object is analyzed in more detail given the task at hand, which is specified by a task-specific taxonomy (3.2). The relevancy is assessed (3.3) before it is characterized using the taxonomy (3.4).

3.1 Detection of Common Objects

To find the potentially relevant objects in the scene, we consider a generic object detector, such that each candidate can be analyzed in more detail later. This approach is inspired by [5], where the detection and classification were decoupled. We adopt Retinanet [6]. It yields candidate regions in the image. For each candidate, a feature vector is extracted using ROI pooling on the final feature map of a Resnet50 backbone that was pretrained on Imagenet. The next steps of our method are detailed in the next subsections, and involve the assessment of relevancy of each candidate object and (if relevant) to characterize the object.

3.2 Task-Specific Taxonomy

Semantics are essential to deal with novelties, to assess whether a novel object looks like something that is important for the current task. For instance, when having seen several pets, a novel pet will similar to the known pets. Also, semantics are important for reporting object characteristics at the right level of detail. For instance, that adults and children are all humans who should be rescued. We capture these semantics by a taxonomy of objects. To classify objects, we train a deep learning model (CNN). Objects that are close together in the taxonomy are more semantically similar and should be assessed as such. Therefore, we adopt the semantic distance loss [2] that punishes confusions between very related objects less than very different objects. Semantic distance is measured by the distance to the common ancestor, normalized by the distance between leafs and root. This distance is used in the optimization of the model parameters, such that is able to discriminate better between semantically distinct objects.

3.3 Uncertainty and Relevancy for the Task

Most deep learning classification models employ a soft-max layer as the classifier. It is well-known that the soft-max outputs, i.e. the confidences, are over-confident. That means that even for an unknown object that is very dissimilar from the known objects, the confidence may be very high. That may lead to the unwanted result that an unknown, irrelevant object is considered to be very relevant for the task. To counter this, we correct the confidence by a measure of how well the object looks like the relevant, known objects. This measure is a ratio of the average of distances from the object to the n closest known objects (as an inter-distance measure), divided by the average of distances from those n known objects to their n closest known objects. In practice, the measure will be between 0 (the object is very different from known objects) to 1 (very similar to known objects). If the object is very different from known objects, it is unknown, and effectively the final confidence will be close to zero. Given that the known objects are all relevant, the confidence of a novel object will be high only when sufficiently similar to the known objects.

3.4 Characterizing Relevant Objects

For different tasks, reporting about the objects may be required at different levels of detail. For instance, for search and rescue all humans need to be found, but when almost no time is left, children may deserve priority (more detail needs to be reported than just that the object is human). The reporting level is coupled to the level in the taxonomy (Sect. 3.2). The classifier confidences (Sect. 3.3) are connected to respective nodes in the taxonomy. For classifiers at the finest level of detail, the trace through the taxonomy enables to relate the object to parent classes and coarser reporting levels.

4 Experiments

For search and rescue, no datasets of sensor data are available (to the best of our knowledge). We recorded a video dataset in a personal home. The head-mounted camera is moving through the house, as a proxy to a moving robot platform (e.g., a robot dog).

The goal is to find 'humans' (barbie toys), 'children' (doll toys) and 'pets' (fluffed toys). We consider toys as the objects of interest for this experiment, to demonstrate that we can quickly adapt to other kinds of objects of interest than standard everyday objects. That is, for real humans and animals, many datasets and pre-trained models are available already. For each object class, we collected 100 images from the Bing image search website. An impression of the train set is shown in Fig. 2.

The goal is to find the relevant objects including relevant novelties, i.e., a novel doll (child). The house is very messy hence cluttered with many distractor objects. It is essential not to be distracted by those objects. All of the clutter objects were not seen during training of the method, i.e., all clutter objects are novel. Some of the clutter objects are somewhat similar to objects of interest, e.g., a unicorn. This is to test whether the method can correctly ignore all irrelevant objects including the near-positives which serve as distractors and should not result in false positives.

Fig. 2. Train set for the experiments, our method requires only training samples for relevant objects and only a limited number of samples per class (here 100).

4.1 Demo Video on YouTube

The full capture of the video stream while moving through the house, and the localization of the relevant objects by our method, are available as a demo video on YouTube.

4.2 Visualization

Example visualizations of our method is shown in Fig. 1 (in the Introduction) and the figures in Sects. 4.3 and 4.4 (Fig. 3 and onwards). The visualization, at the top, shows a text line that describes the assessment of our robot (called 'Bernard') by our proposed method (called 'SNOW'). In this case: it is a child and hence it is relevant for the task, which is search and rescue. Therefore, SNOW says that the child indeed should be rescued. On the left, a thermometer is shown, indicating the level of certainty that the object is relevant. In the middle, the two images show respectively the candidate objects (top, green boxes) and the objects that are considered to be relevant (bottom, red boxes). On the right, the task-specific taxonomy is shown, with the highlighted nodes that show the characterization of an encountered (novel) object.

4.3 Finding the Relevant Objects

The goal is to find 'humans' (barbie toys), 'children' (doll toys) and 'pets' (fluffed toys) in the messy, cluttered house – and to ignore all the irrelevant clutter objects. Figure 3 shows the found objects: respectively, a child, a dog pet and another child. Each of the relevant objects has a high confidence, visualized by the thermometer which indicates high values. Each of the objects is correctly characterized, indicated by the highlighted taxonomy nodes.

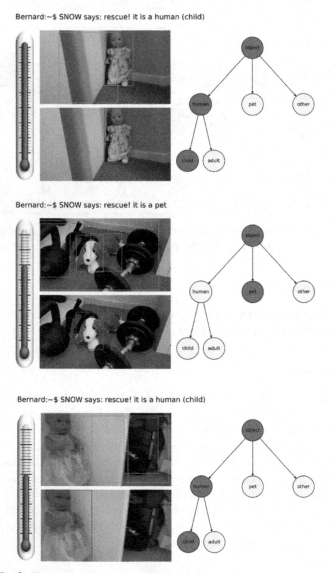

Fig. 3. Examples of relevant objects that have been successfully found.

Bernard:~$ SNOW says: rescue! it is a human (adult)

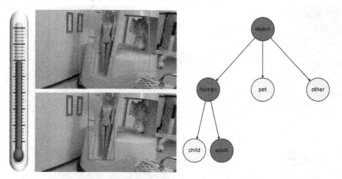

Fig. 4. A relevant *novel* object that has been successfully found.

The barbie doll in Fig. 4 is a novelty: none of the training images is a barbie doll with a bathing suit. It is correctly assessed as relevant for the task, and correctly characterized as an adult.

The false negative is shown in Fig. 5: the cat is in a novel pose and the confidence is intermediate (see the thermometer) but insufficient for a detection (i.e., no red box in the bottom image pane and no red nodes in the taxonomy).

Bernard:~$

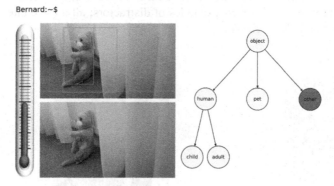

Fig. 5. False negative.

4.4 Ignoring the Irrelevant Objects

Since the model is only trained on a limited set of *relevant* objects, it is interesting to see how well our method is capable of assessing irrelevant objects as such. Figure 6 shows examples of clutter objects that are assessed to be irrelevant indeed (i.e., no red box in the bottom image pane and no red nodes in the taxonomy). The confidences are very low for these objects (see thermometers).

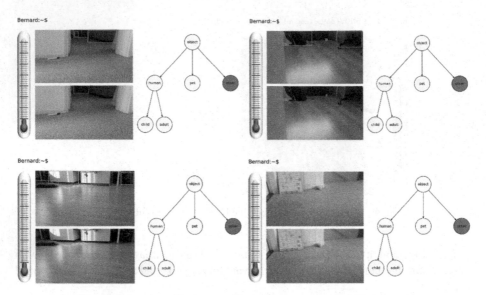

Fig. 6. Examples of irrelevant objects that were successfully ignored.

It is especially interesting how well the method is able to assess distractor objects as irrelevant, i.e., objects that are somewhat similar to the relevant objects such as other kinds of pet toys. Figure 7 shows examples of distractors; all these clutter objects are correctly assessed to be irrelevant.

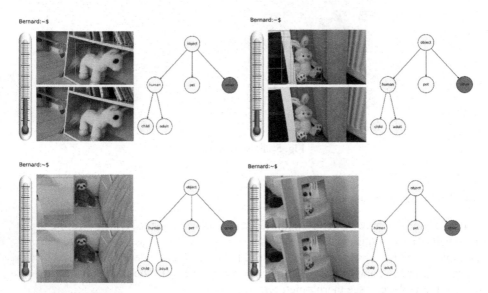

Fig. 7. Examples of irrelevant *distractor* objects that were successfully ignored.

4.5 Summary of the Results

In summary, the results are as follows. In total there are 36 objects. At a high threshold $>= 60\%$, 85% of the relevant objects can be found with no false positives. That is, 5 out of 6 relevant objects were found, including novel objects (novel child doll), 1 pet missed (1 false negative). At an intermediate threshold $>= 0.4$, 100% of the relevant objects can be found, at the cost of 3 false positives. That is, 6 out of 6 relevant objects found, 1 out of 3 distractors detected (1 false positive), 2 out of 26 other objects detected (2 false positives). This is achieved with training on a limited training set for each node of the taxonomy of relevant objects, and *no* training samples of any of the irrelevant objects.

5 Conclusions

We presented an approach for robots to deal with an open world in which objects are encountered that are not known beforehand. We proposed a method that first decides whether an object is relevant for the task at hand, if it is similar to a known object that is known to be relevant. Relevancy is determined from a task-specific taxonomy of objects. If the object is relevant for the task, then it is characterized through the taxonomy. The task determines the level of detail that is needed at this moment, and it relates directly to the levels in the taxonomy. The advantage of our method is that it only needs to model the relevant objects and not all possible irrelevant and often unknown objects that the robot may also encounter. This is essential for (semi-)autonomous systems that operate in a open, changing world. We showed its merit in a real-life experiment of a search and rescue task in a messy house, where the victims (including novelties) were successfully found.

Acknowledgement. This research was sponsored by the Appl. AI program at TNO, in the SNOW project.

References

1. Lee, K., Lee, K., Min, K., Zhang, Y., Shin, J., Lee, H: Hierarchical novelty detection for visual object recognition. In: CVPR (2018)
2. Barz, B., Denzler, J.: Hierarchy-based image embeddings for semantic image retrieval. In: IEEE Winter Conference on Applications of Computer Vision (WACV) (2019)
3. Lee, K. Lee, K., Lee, H., Shin, J.: A simple unified framework for detecting out-of-distribution samples and adversarial attacks. In: NeurIPS (2018)
4. Ren, J., et al: Likelihood ratios for out-of-distribution detection. In: NeurIPS (2019)
5. Singh, B., Li, H., Sharma, A., Davis, L.S.: R-FCN-3000 at 30fps: decoupling detection and classification. In: CVPR (2018)
6. Lin, T.Y., Goyal, P., Girshick, R., He, K., Dollár, L.S.: Focal loss for dense object detection. In: ICCV (2017)

IML - International Workshop
on Industrial Machine Learning

Workshop on Industrial Machine Learning (IML)

Workshop Description

IML is a forum for researchers and practitioners working on machine learning techniques, algorithms, and applications in the field of industrial and manufacturing data analysis. With the advent of Industry 4.0 paradigm, data has become a valuable resource, and very often an asset, for every manufacturer. Data from the market, from machines, from warehouses and many other sources are now cheaper than ever to be collected and stored. It has been estimated that in 2020 we will have more than 50B devices connected to the Industrial Internet of Things, generating more than 500ZB of data. With such an amount of data, classical data analysis approaches are not sufficient and only automated learning methods can be applied to produce value, a market estimated in more than 200B$ worldwide. Through the use of machine learning techniques manufacturers can use data to significantly impact their bottom line by greatly improving production efficiency, product quality, and employee safety. The introduction of ML in industry has many benefits that can result in advantages well beyond efficiency improvements, opening doors to new opportunities for both practitioners and researchers. Some direct applications of ML in manufacturing include predictive maintenance, supply chain management, logistics, quality control, human-robot interaction, process monitoring, anomaly detection and root cause analysis to name a few. This workshop drew attention to the importance of integrating ML technologies and ML-based solutions into the manufacturing domain, while addressing the challenges and barriers to meet the specific needs of this sector. Workshop participants had the opportunity to discuss needs and barriers for ML in manufacturing, the state-of-the-art in ML applications to manufacturing, and future research opportunities in this domain.

The first edition of the International Workshop on Industrial Machine Learning (IML) was held online on January 10th 2021. The workshop was a satellite event of the 25th International Conference on Pattern Recognition (ICPR).

We received over 20 submissions for reviews, from authors belonging to 11 distinct countries in 3 continents. After an accurate and thorough peer-review process, we selected 19 papers for presentation at the workshop; 15 papers were presented as spotlight and the best 4 papers as oral presentation. The review process focused on the quality of the papers, their scientific novelty and applicability to real-world industrial problems. The acceptance of the papers was the result of the reviewers' discussion and agreement. All the high quality papers were accepted, with a set of articles that represent an interesting mix of applications of machine learning techniques to quality control, visual inspection, energy consumption estimation, logistics, robotics, and fault detection. From a methodological point of view, articles focused both on the introduction of novel methods and neural networks architectures, evaluation of existing methods and datasets, presentation and discussion of case studies. As for the technologies, papers presented works related to additive manufacturing, laser cutting, and evaluation of edge computing platforms.

The program was completed by three invited talks given by experts in the field covering the main aspects of ML applications to industrial case studies, from predictive maintenance to design support systems and collaborative robotics.

We would like to thank the IML Program Committee, whose members made the workshop possible with their rigorous and timely review process. We would also like to thank ICPR for hosting the workshop and our industrial community, and the ICPR workshop chairs for the valuable help and support.

Organization

IML Chairs

Luigi Di Stefano   University of Bologna, Italy
Massimiliano Mancini   University of Tübingen, Germany
Vittorio Murino   University of Verona, Italy
     Huawei Technologies Co. Ltd., Ireland
     Istituto Italiano di Tecnologia, Italy
Paolo Rota   University of Trento, Italy
Francesco Setti   University of Verona, Italy

Program Committee

Martino Alessandrini   Datalogic, Italy
Carlos Beltran   Istituto Italiano di Tecnologia, Italy
Michael Bortz   Fraunhofer-ITWM, Germany
Paolo Bosetti   University of Trento, Italy
Marco Carletti   Embedded Vision Systems, Italy
Nicoló Carissimi   Istituto Italiano di Tecnologia, Italy
Valerio Carpani   FIZYR, The Netherlands
Fabio Cermelli   Politecnico di Torino, Italy
Samrjit Chakraborty   University of North Carolina, USA
Chiara Corridori   Deltamax Automazione, Italy
Daniele De Gregorio   EyeCan.ai, Italy
Giacomo De Rossi   University of Verona, Italy
Ahmad Delforouzi   Fraunhofer, Germany
Siddhartha Dutta   Technische Universität Kaiserslautern, Germany
Dario Fontanel   Politecnico di Torino, Italy
Francesco Fornasa   Embedded Vision Systems, Italy
Emanuele Frontoni   Univ. Politecnica delle Marche, Italy
Giovanni Gualdi   Deep Vision Consulting, Italy
Florian Kleber   TU Wien, Austria
Donato Laico   SACMI, Italy
Adriano Mancini   Univ. Politecnica delle Marche, Italy
Matteo Moro   University of Trento, Italy
Danilo Pau   STMicroelectronics, Italy
Nicola Piccinelli   University of Verona, Italy
Paolo Piccinini   Marchesini Group, Italy

Deep Learning Based Dimple Segmentation for Quantitative Fractography

Ashish Sinha[(⊠)] and K. S. Suresh[(⊠)]

Department of Metallurgical and Materials Engineering,
Indian Institute of Technology Roorkee, Roorkee, India
{asinha,ks.suresh}@mt.iitr.ac.in

Abstract. In this work, we try to address the challenging problem of dimple segmentation from Scanning Electron Microscope (SEM) images of titanium alloys using machine learning methods, particularly neural networks. This automated method would in turn help in correlating the topographical features of the fracture surface with the mechanical properties of the material. Our proposed, UNet-inspired attention driven model not only achieves the best performance on dice-score metric when compared to other previous segmentation methods when applied to our curated dataset of SEM images, but also consumes significantly less memory. To the best of our knowledge, this is one of the first work in fractography using fully convolutional neural networks with self-attention for supervised learning of deep dimple fractography, though it can be easily extended to account for brittle characteristics as well.

Keywords: Image segmentation · Machine learning · Fractography · Dimple fractures

1 Introduction

Titanium and it's alloys are used for making aircraft components and biomedical implants due to it's high strength and wear resistance. Hence, the material needs to be defect free else it can be detrimental [1–3].

Fracture patterns of metals in general and high-strength titanium and iron alloys in particular happens in stages undergoing deformation and stress accumulation. Deformation of metals is caused due to a sudden change in the external factors like temperature and pressure, and accelerated by the conjoined forces of load and corrosive actions of nature. This leads to an accumulation of pores on the surface and bulk of the material which coalesce with the grains (can be thought as domains in magnetic field) leading to the growth of the crack in a continuous fashion in the direction of application of load. Thus, the crack grows by thinning and breaking connections between the pores. The newly formed crack and prior history of external environmental conditions leave traces on the surface of the material in the form of dimples [4–6].

A. Sinha– Work done when the author was at Indian Institute of Technology Roorkee.

A. Del Bimbo et al. (Eds.): ICPR 2020 Workshops, LNCS 12664, pp. 463–474, 2021.
https://doi.org/10.1007/978-3-030-68799-1_34

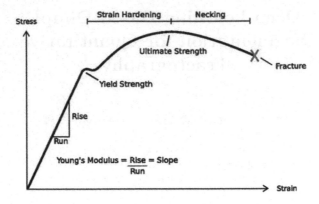

Fig. 1. Stress-Strain curve.[1]

The process of damage to a material can be depicted on stress-strain curve in Fig. 1 where fracture is the final stage of deformation. Administered by both the extrinsic (e.g. imposed loading, environmental conditions) and the intrinsic (microstructure) characteristics is the process of fracture of materials. The data regarding the affect of both intrinsic and extrinsic characteristics of the fracture process is contained in the surface of the fracture. To attain the topographic characterization of the surface, fractography is used. Fractographic analysis uses physics of solid body, material science, optic-digital methods to determine the causes of fracture. Earlier, parameter measurements of fracture surfaces were made manually or automatically but the software was positioned by an operator. A large variety of materials made the generalization difficult [7].

Many models of nucleation, growth, and coalescence of pores are known and their application under uncertain conditions leads to fracture of materials, the results of which are usually complicated to compare. The ponderous process of quantitative and qualitative assessment of fracture surface is executed manually by experienced and well-trained technicians, hence requiring significant labour. Quantitative estimations are likely to have more errors because there is a huge dependency on human factor. These shortcomings raise the need for automated methods of quantitative fractography.

In our work, we address this issue by exploring the application of various deep neural networks to segment deep dimples from SEM images of fractographs, which would further prove useful for the quantitative analysis of fractographs leading to reduced human error and labour. As part of this work, we have curated a dataset of 216 high-resolution SEM images to test our proposed model. We also propose a novel U-Net inspired model containing dual-attention residual blocks with dense connections. Extensive experiments and validation shows that this model performs significantly better than the baselines.

---

[1] https://www.instructables.com/Steps-to-Analyzing-a-Materials-Properties-from-its/

2   Background and Related Work

In the last century, metallurgists and material scientists have acquired, analysed and compared images of micro-structures in a progressive way. Most of the effort was put in to learn a defined material system and different classes of micro structures. If the features of micro structures are known then digital analysis techniques on images would be helpful in characterising, segmenting and comparing different structures with high resolution. Presently the analytics of microstructures is mostly focused on finding the relationship between the structure and properties by using the shape, size and appearance of the features of the material surface. These approaches have moved forward to more of a machine learning approach where the properties are found out by the choosing of correct algorithms. Some of the areas where machine learning techniques are used in metallurgy and materials science are: new material design, material property prediction, microstructure recognition, analysis of failure in a material, etc. [8–10].

2.1   Computer Vision in Quantitative Fractography

The main interest in fractography is due in finding the correlation between the features of the surface that is fractured and the environment or conditions that lead to its failure. For centuries, this has remained qualitative in nature. Mostly, scientists examine the Scanning Electron Microscope (SEM) images of samples for failure analysis due to it's higher magnification power. Therefore, quantitative analysis brings the potential for improving and understanding the mechanisms that control the fracture process and also determine the reliability of the models in the current material design system. With the latest advancement in the field of image analysis and moreover the availability of machine learning tools, it has become easier to automate the event of finding important features and information from the fractographs.

The use of computer vision methodologies by Hu et al. [11] and Chowdhary et al. [12] led to identify the images which contained dendritic morphology to classify if the the direction was longitudinal or transverse. Another use case was automatically measuring the volume of ferritic (iron) volume fraction from the binary phase structures of ferritic and austenite (a phase of iron). In the field of fractography, many successful attempts have been made to build automated models for quantitative fractography. Bastidas et al. used non-linear algorithms of machine learning (ANN and SVM) and combined it with texture analysis to classify the images into there modes: ductile sudden, brittle sudden and fatigue [7]. In the previous works by Konovalenko et al. [13] and Maruschak et al. [14], the authors have proposed methods for fractographic recognition, control and calculation of parameters of the dimples of based on neural networks. In [15], Konovalenko et al. trained 17 models of neural networks with various sets of hyper-parameters, then their speed and accuracy were evaluated and the optimal neural network was selected. A recent work by Tsopanidas et al. aims to quantify fracture surface for materials with brittle fracture characteristics [16]. Our work focuses primarily on ductile materials.

In our project, we use different variants of deep neural networks (DNNs) to segment deep dimples and benchmark the results, in ductile fracture materials on SEM images of titanium alloys curated by us; which can be further used to find the properties of fracture mechanism. We propose a fully convolutional U-Net [17] inspired deep neural network with position and channel based attention residual blocks with dense connections [18–20] and squeeze and excitation [21] block in the bottleneck layer for the segmentation of deep dimples.

Our work explores the application of deep learning methods in fractography, an active field of research in material science. Below, we briefly explain the terms necessary to better understand our work.

2.2   Fractography

Fractography is a technique to understand the causes of failures and also to verify theoretical failure predictions with real life failures. It can be used in forensics, for analyzing broken products which have been used as weapons, such as broken bottles. Thus, a defendant might claim that a bottle was faulty and broke accidentally when it impacted a victim of an assault. Fractography could show the allegation to be false and that considerable force was needed to smash the bottle before using the broken end as a weapon to deliberately the victim. In these cases, the overall pattern of cracking is important in reconstructing the sequence of events, rather than the specific characteristics of a single crack, since crack grows by coalescing with other grains in the microstructure of the metals.

2.3   Crack Growth

The initiation and continuation of crack growth is dependant on several factors such as bulk material properties, geometry of the body, geometry of the crack, loading rate, loading distribution, load magnitude, environmental conditions, time and microstructure. Cracks are initiated, and as the cracks grow, energy is transmitted to the crack tip at an energy release rate G, which is a function of the applied load, crack length and the geometry of the body. These cracks conjoined with external environmental factors leads to the formation of dimples. All solid materials, have an intrinsic energy release rate G_C, where G_C is referred to as the fracture energy or fracture toughness of the material. A crack will grow if $G \geq G_C$.

2.4   Microstructure

Microstructure is a very small scale structure of a material, defined as the structure of a prepared surface of a material as revealed under an optical microscope above 25x magnification. The microstructure of a material correlates strongly with the strength, toughness, ductility, hardness, wear resistance, etc. of the material. A microstructure's influence on the physical and mechanical properties of a material is governed by the different defects present or absent in the structure. These defects can take many forms but the primary ones are the pores. To acquire micrographs, both optical as well as electron microscopy is used, but electron microscopy is preferred for obtaining fractographs.

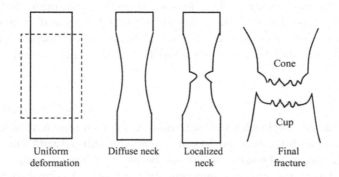

Fig. 2. Steps of ductile fracture under stress

2.5   Dimple Fracture

A dimple fracture is a type of material failure on a metal's surface that is characterized by the formation and collection of micro-voids along the granular boundary of the metal i.e. the fracture path. The occurrence of dimple fractures is directly proportional to increased corrosion rates. The material appears physically dimpled when examined under high magnification. There are three main types of dimple fractures:

- Shear fractures
- Tearing fractures
- Tensile fractures

All three of these fractures are characterized by tiny holes, known as microvoids, which are microscopically located in the interior of a piece of metal when under the force of an external load. The greater the load, the greater the proximity and the total gap volume of these voids. Generally, the merging of micro-pores in the material during deformation in the plastic region leads to such dimples as shown in Fig. 2. A scanning electron microscope can be used to examine a dimple rupture at a magnification of about 2500x.

3   Methodology

Here we discuss the fractographic analysis for detection of dimples in fracture metal surfaces using various deep learning models. We consider Titanium (Ti) alloys as our focus of discussion. In the next section we briefly discuss the previous methods used for segmentation tasks but which are new in this domain, and then we explain our proposed model.

3.1   Previous Approaches

U-Net

The U-Net [17] is mainly employed for bio-medical image segmentation. It has two parts: a contracting encoder and an expanding decoder. The encoder, a feature correction path, is a continuous stack of convolution and pooling layers; used for image identification. The decoder, a feature expanding path, is used for collecting exact localisation of fractures using transposed convolutions or deconvolutions. The model is an end-to-end fully convolutional network. The image of any size can be fed in the model as it lacks any densely connected layer.

U-Net++

This model uses dense block ideas of DenseNet [20] to improve upon U-Net [17].

This model architecture consists of an encoder sub-network and a decoder sub-network after it. The skip connections between each node comprises of three convolution layers and a deep dense convolution block. A convolution layer follows every concatenation layer. This convolution layer considers output from preceding convolution layers of the dense block and up-sampled output from lower dense block, and fuses them.

Mask R-CNN

Mask R-CNN [22] is the skeletal that assists in object detection and localization. It completes the task in two steps: scanning the image and generating proposals to point the probable locations of an object, classification on first step proposals, and generation of bounding boxes and masks.

3.2   Proposed Model

Like the U-Net architecture, our proposed network is divided into two parts, a contracting encoder and an expanding decoder. For the encoder part, we employ 5 layers of residual convolutional blocks with channel and spatial attention as proposed in [23] but instead of using attention blocks in parallel we find that using them as proposed in [18] gives better results in our case. We employ a dense connection [20] in each residual block used in the encoder of the model. Bottleneck blocks consists of 3 layers of dense connection of residual convolution blocks followed by a squeeze and excitation block [21]. The encoder part starts with a convolution of filter size 5 and stride 1 followed by a batch normalization layer [24] and a PRelu activation. The other convolutions have a filter size of 3.

In the decoder part, we make use of the residual convolutional blocks similar to that of the encoder and use parameter-free bilinear upsampling instead of transposed convolutional operations to reduce the number of trainable parameters [25]. The overall model architecture is shown in Fig. 3. Each upsampling block is followed by a channel and spatial attention block.

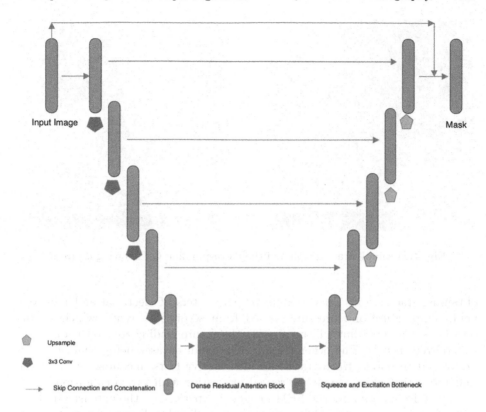

Input Image

Mask

Upsample

3x3 Conv

Skip Connection and Concatenation

Dense Residual Attention Block

Squeeze and Excitation Bottleneck

Fig. 3. Our proposed model

The goal of this work is not to propose a novel architecture but to establish a baseline for further development and also to show the application of deep learning models on an age old problem of dimple detection for quantitative fractography so as to determine the cause of fracture in materials which in turn will lead to the design of new and better materials. This work can be a reference for the material scientists to further explore the domain of fractography with machine learning.

4   Evaluation

4.1   Dataset

For conducting our experiments, we collected 216 high-resolution SEM images of Ti alloys (tested under various physical conditions) at 200x, 500x, 1000x and 2000x magnification. It is a very difficult process to obtain this kind of SEM data in such magnitude since the material has to be tested under different temperature and pressure to ensure generalization, as even a slight change in the testing environment can severely affect the fracture surfaces. The time scale for

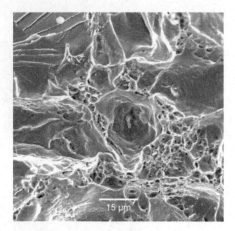 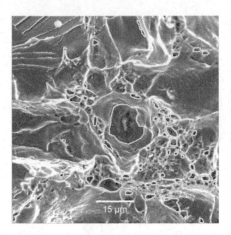

Fig. 4. A sample fractograph and the corresponding GT showing dimples

obtaining fractographs under extensive parameters of electrical and magnetic field, temperature and pressure, varied from seconds to years, which is why our dataset is small but will increase in the future leading to robust and more generalized models. The other reason for a small dataset being that the tests performed to obtain fractographs are destructive tests, meaning the material undergone the testing cannot be reused for testing with varied conditions.

Apart from obtaining the SEM images, we performed the extensive tedious task of annotating the SEM images with deep dimples. During annotation, we classify the areas of the SEM images that presented the most characteristic features of deep dimple (dark areas), while the areas with unclear classification or ambiguous features are labeled as background. As visible from Fig. 4, the dark, small regions are circled and overlapped on the SEM image to give the reader a better understanding of what deep dimples are. The larger surrounding region demarcated by the bright lines are the shallow dimples, but in this work we are primarily interested in the deep dimples, hence shallow dimples are not discussed. The annotated SEM images were then cropped into slices of 128 × 128px size to generate around 17786 images. This methodology of data preparation is popular in medical imaging domain involving whole-scale images (WSI) and the domain of satellite images. For the purpose of training our method, we used around 80% the total images for training our model, 10% for validating our model and 10% of the remaining dataset was reserved for testing our model. Since there is no overlap in the training, validation and testing datasets, we believe the model is able to achieve generalization.

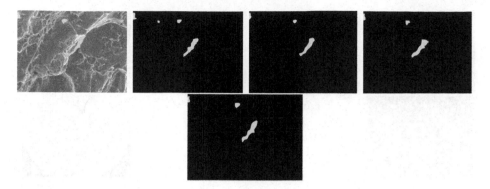

Fig. 5. Qualitative results. From Left to Right: GT, U-Net, UNet++, Attention Unet, (bottom) Res-Unet with Dual Attention **(ours)**

4.2 Experiments

In the experiments, we observed that training the model for a total 150 epochs was enough to reach convergence. The input dimensions of the image was a single channel 128×128 SEM and the output dimension was also a single channel 128×128 segmented mask. After extensive hyper-parameter tuning with optimizer, learning rates and weight decay, we found out that Adam [26] optimizer as an optimizer performs the best with a learning rate of 1e–4, a weight decay of 1e–6, $\beta_1 = 0.9$ and $\beta_2 = 0.999$. We use an exponential decay for learning rate after every 10 epochs. We used a weighted loss function consisting of dice score and binary cross-entropy loss, the weights were generated empirically with λ_1 = 1.25 for dice-score (L_{dice}) and $\lambda_2 = 0.95$ for binary cross-entropy loss (L_{bce}). The batch size for all the experiments were kept to 1 due to memory constraints of our GPU. The hyper-parameters were consistent across all the reported methods. We evaluate our model on the basis of dice score [27] and compare it to other state-of-the-art methods on semantic segmentation tasks. For a fair-comparison with our proposed model (DA ResU-Net), all the U-Net based models employ 5 layers similar to ours, and are trained from scratch.

$$L_{total} = \lambda_1 * (1 - L_{dice}) + \lambda_2 * L_{bce}$$

The code was written in PyTorch [28] and trained on a single Nvidia GeForce GTX 1080Ti. Each experiment took around 6–7 h.

$$DiceScore(DSC) = \frac{2 * |A \cap B|}{|A| + |B|}$$

where A and B are predicted segmentation map and ground truth, respectively

4.3 Results

We evaluate the performance of our proposed model and the baseline models on the widely used image segmentation metric of Dice-coefficient. The results of

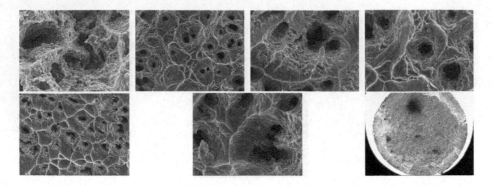

Fig. 6. Predictions of our model on SEM images.

Table 1. Quantitative results on microstructures of Ti alloys.

Method	Dice Score (DSC)	# Params
U-Net	0.68509 (±7.88%)	31.0M
U-Net++	0.73423 (±6.51%)	31.7M
U-Net w/Attention	0.81630 (±5.27%)	34.8M
ResU-Net w/Dual Attention (ours)	**0.86305** (±5.05%)	**20.9M**

various methods are tabulated in Table 1. It is visible from the quantitative and qualitative results Tables 5, 6 that our proposed model performs the best as compared to other previous established methods. Not only this our proposed model is memory efficient, and has around 14 million less parameters as compared to the second best performing model. After getting the segmentation results, the results can be analyzed to understand the materials inherent microstructure, grain boundary strength as well as the conditions leading to the failure of the material.

Quantifying the probability of occurrence of the mechanisms of ductile fracture is cumbersome and highly biased on the user, this makes the task at hand more challenging. Moreover, the size of the features on the fracture surface changes drastically from one material to the other. Since, simple hand-crafted features based model or deep learning based classification or object detection algorithms are not able to effectively tackle these challenges, we use semantic segmentation algorithms to classify every pixel in the SEM images which allows for the topographic characterization of the fracture surface, making this approach best suited for a fractographic analysis. This method tries to learn to classify the background pixels even though they do not follow a certain pattern, which leads to certain mis-classifications. The results presented herein can be improved, given a larger, annotated, training set. Depending on the area occupied by the deep dimples or shallow dimples (beyond the scope of this work) present in the SEM image, we can analyze if the material underwent brittle or ductile fracture and what mechanism did the material follow. Thus, this work aims to reduce the time and effort of material science researchers for fractographic analysis.

5   Conclusion

In this work, we have discussed the automated methods to identify fractures and obtain meaningful features which can help in avoiding the fractures while in service. We have presented an overview of the traditional as well as modern approaches involved in detecting of these defects on the microscopic scale. We also present new methods for the segmentation of deep dimples, which may serve as the first step towards categorizing the type of defect a material had by analyzing the segmented fractographic surface of the SEM image. This work presently focuses on deep dimple segmentation for ductile materials like Ti alloys, but can be easily extended to other kind of defects and other materials too. This is a robust method and the results can be improved provided the availability of large data. We would also like to explore unsupervised methods for dimple segmentation in the recent future. This work is an aim to foster machine learning and deep learning principles in automating the traditional methods applied in material science for fractography or material forensics. We hope to see better models in the future which can obtain the desired results with limited data and can lead to reduced human labour and material wastage, increased efficiency during production of new materials and so light-weight, that they can be deployed on small devices.

Acknowledgment. The authors would like to thank Sumit Yadav for his contribution on assisting with the annotatation of a few the SEM images. We are also very thankful to the free GPU service Colab by Google and Kaggle which was used for conducting extensive experiments for this research work apart from burning our own laptops.

References

1. Attar, H., Calin, M., Zhang, L., Scudino, S., Eckert, J.: Manufacture by selective laser melting and mechanical behavior of commercially pure titanium. Mater. Sci. Eng. A **593**, 170–177 (2014)
2. Ehtemam-Haghighi, S., Prashanth, K., Attar, H., Chaubey, A.K., Cao, G., Zhang, L.: Evaluation of mechanical and wear properties of tixnb7fe alloys designed for biomedical applications. Mater. Des. **111**, 592–599 (2016)
3. Kabashkin , I.V., Yatskiv, I.V.: Reliability and statistics in transportation and communication (2010)
4. Beachem, C., Yoder, G.: Elastic-plastic fracture by homogeneous microvoid coalescence tearing along alternating shear planes. Metall. Trans. 4(4), 1145–1153 (1973)
5. Kardomateas, G.: Fractographic observations in asymmetric and symmetric fully plastic crack growth. Scr. Metall. **20**, 609–614 (1986)
6. Merson, E., Danilov, V., Merson, D., Vinogradov, A.: Confocal laser scanning microscopy: The technique for quantitative fractographic analysis. Eng. Fract. Mech. **183**, 147–158 (2017)
7. Bastidas-Rodriguez, M., Prieto-Ortiz, F., Espejo, E.: Fractographic classification in metallic materials by using computer vision. Eng. Fail. Anal. **59**, 237–252 (2016)
8. Liu, Y., Zhao, T., Ju, W., Shi, S.: Materials discovery and design using machine learning. J. Materiomics **3**(3), 159–177 (2017)

9. Xie, T., Grossman, J.C.: Crystal graph convolutional neural networks for an accurate and interpretable prediction of material properties. Phys. Rev. Lett. **120**(14), 145301 (2018)

10. Popat, M., Barai, S.: Defect detection and classification using machine learning classifier. In: 16th World Conference on NDT, August. Citeseer (2004)

11. Hu, W., Wiliem, A., Lovell, B., Barter, S., Liu, L.: Automation of quantitative fractography for determination of fatigue crack growth rates with marker loads. In: 29th ICAF Symposium Nagoya (2017)

12. Chowdhury, A., Kautz, E., Yener, B., Lewis, D.: Image driven machine learning methods for microstructure recognition. Comput. Mater. Sci. **123**, 176–187 (2016)

13. Konovalenko, I., Maruschak, P., Chausov, M., Prentkovskis, O.: Fuzzy logic analysis of parameters of dimples of ductile tearing on the digital image of fracture surface. Proc. Engin **187**, 229–234 (2017)

14. Maruschak, P., Konovalenko, I., Chausov, M., Pylypenko, A., Panin, S., Vlasov, I., Prentkovskis, O.: Impact of dynamic non-equilibrium processes on fracture mechanisms of high-strength titanium alloy vt23. Metals **8**(12), 983 (2018)

15. Konovalenko, I., Maruschak, P., Prentkovskis, O., Junevičius, R.: Investigation of the rupture surface of the titanium alloy using convolutional neural networks. Materials **11**(12), 2467 (2018)

16. Tsopanidis, S., Moreno, R.H., Osovski, S.: Toward quantitative fractography using convolutional neural networks. Eng. Fract. Mech. **231**, 106992 (2020)

17. Ronneberger, O., Fischer, P., Brox, T.: U-Net: convolutional networks for biomedical image segmentation. In: Navab, N., Hornegger, J., Wells, W.M., Frangi, A.F. (eds.) MICCAI 2015. LNCS, vol. 9351, pp. 234–241. Springer, Cham (2015). https://doi.org/10.1007/978-3-319-24574-4_28

18. Woo, S., Park, J., Lee, J.-Y., Kweon, I.S.: Cbam: convolutional block attention module. In: Proceedings of the European Conference on Computer Vision (ECCV), pp. 3–19 (2018)

19. He, K., Zhang, X., Ren, S., Sun, J.: Deep residual learning for image recognition. In: Proceedings of the IEEE Conference on Computer Vision and Pattern Recognition, pp. 770–778 (2016)

20. Huang, G., Liu, Z., Weinberger, K., van der Maaten, L.: Densely connected convolutional networks. arxiv 2017. arXiv preprint arXiv:1608.06993

21. Hu, J., Shen, L., Sun, G.: Squeeze-and-excitation networks. In: Proceedings of the IEEE Conference on Computer Vision and Pattern Recognition, pp. 7132–7141 (2018)

22. He, K., Gkioxari, G., Dollár, P., Girshick, R.: Mask r-cnn. In: Proceedings of the IEEE International Conference on Computer Vision, pp. 2961–2969 (2017)

23. Fu, J.: Dual attention network for scene segmentation. In: Proceedings of the IEEE Conference on Computer Vision and Pattern Recognition, pp. 3146–3154 (2019)

24. Ioffe, S., Szegedy, C.: Batch normalization: Accelerating deep network training by reducing internal covariate shift. arXiv preprint arXiv:1502.03167 (2015)

25. De Fauw, J., et al.: Clinically applicable deep learning for diagnosis and referral in retinal disease. Nature Med. **24**(9), 1342–1350 (2018)

26. Kingma, D.P., Ba, J.: Adam: A method for stochastic optimization (2014)

27. Dice, L.R.: Measures of the amount of ecologic association between species. Ecology **26**(3), 297–302 (1945)

28. Paszke, A., et al.: Pytorch: an imperative style, high-performance deep learning library. In: Advances in Neural Information Processing Systems, pp. 8026–8037 (2019)

PaDiM: A Patch Distribution Modeling Framework for Anomaly Detection and Localization

Thomas Defard⬤, Aleksandr Setkov$^{(\boxtimes)}$⬤, Angelique Loesch⬤,
and Romaric Audigier⬤

Université Paris-Saclay, CEA, List, 91120 Palaiseau, France
thomas.defard@imt-atlantique.net,
{aleksandr.setkov,angelique.loesch,romaric.audigier}@cea.fr

Abstract. We present a new framework for Patch Distribution Modeling, PaDiM, to concurrently detect and localize anomalies in images in a one-class learning setting. PaDiM makes use of a pretrained convolutional neural network (CNN) for patch embedding, and of multivariate Gaussian distributions to get a probabilistic representation of the normal class. It also exploits correlations between the different semantic levels of CNN to better localize anomalies. PaDiM outperforms current state-of-the-art approaches for both anomaly detection and localization on the MVTec AD and STC datasets. To match real-world visual industrial inspection, we extend the evaluation protocol to assess performance of anomaly localization algorithms on non-aligned dataset. The state-of-the-art performance and low complexity of PaDiM make it a good candidate for many industrial applications.

Keywords: Anomaly detection · Anomaly localization · Computer vision

1 Introduction

Humans are able to detect heterogeneous or unexpected patterns in a set of homogeneous natural images. This task is known as anomaly or novelty detection and has a large number of applications, among which visual industrial inspections. However, anomalies are very rare events on manufacturing lines and cumbersome to detect manually. Therefore, anomaly detection automation would enable a constant quality control by avoiding reduced attention span and facilitating human operator work. In this paper, we focus on anomaly detection and, in particular, on anomaly localization, mainly in an industrial inspection context. In computer vision, anomaly detection consists in giving an anomaly score to images. Anomaly localization is a more complex task which assigns each pixel, or each patch of pixels, an anomaly score to output an anomaly map. Thus, anomaly localization yields more precise and interpretable results. Examples of

A. Del Bimbo et al. (Eds.): ICPR 2020 Workshops, LNCS 12664, pp. 475–489, 2021.
https://doi.org/10.1007/978-3-030-68799-1_35

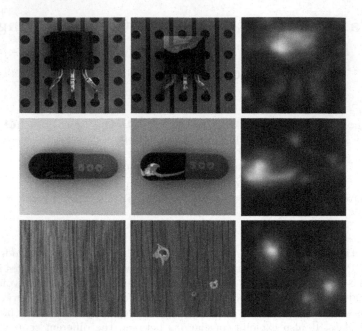

Fig. 1. Image samples from the MVTec AD [6]. *Left column*: normal images of Transistor, Capsule and Wood classes. *Middle column*: images of the same classes with the ground truth anomalies highlighted in yellow. *Right column*: anomaly heatmaps obtained by our PaDiM model. Yellow areas correspond to the detected anomalies, whereas the blue areas indicate the normality zones. Best viewed in color. (Color figure online)

anomaly maps produced by our method to localize anomalies in images from the MVTec Anomaly Detection (MVTec AD) dataset [6] are displayed in Fig. 1.

Anomaly detection is a binary classification between the normal and the anomalous classes. However, it is not possible to train a model with full supervision for this task because we frequently lack anomalous examples, and, what is more, anomalies can have unexpected patterns. Hence, anomaly detection models are often estimated in a one-class learning setting, *i.e.*, when the training dataset contains only images from the normal class and anomalous examples are not available during the training. At test time, examples that differ from the normal training dataset are classified as anomalous.

Recently, several methods have been proposed to combine anomaly localization and detection tasks in a one-class learning setting [7,9,30,31]. However, either they require deep neural network training [5,30] which might be cumbersome, or they use a K-nearest-neighbor (K-NN) algorithm [10] on the entire training dataset at test time [9,31]. The linear complexity of the KNN algorithm increases the time and space complexity as the size of the training dataset grows. These two scalability issues may hinder the deployment of anomaly localization algorithms in industrial context.

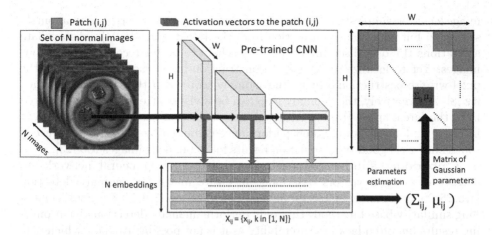

Fig. 2. For each image patch corresponding to position (i, j) in the largest CNN feature map, PaDiM learns the Gaussian parameters (μ_{ij}, Σ_{ij}) from the set of N training embedding vectors $X_{ij} = \{x_{ij}^k, k \in [\![1, N]\!]\}$, computed from N different training images and three different pretrained CNN layers. (Color figure online)

To mitigate the aforementioned issues, we propose a new anomaly detection and localization approach, named PaDiM for Patch Distribution Modeling. It makes use of a pretrained convolutional neural network (CNN) for embedding extraction and has the two following properties:

- Each patch position is described by a multivariate Gaussian distribution;
- PaDiM takes into account the correlations between different semantic levels of a pretrained CNN.

With this new and efficient approach, PaDiM outperforms the existing state-of-the-art methods for anomaly localization and detection on the MVTec AD [6] and the ShanghaiTech Campus (STC) [19] datasets. Besides, at test time, it has a low time and space complexity, independent of the dataset training size which is an asset for industrial applications. We also extend the evaluation protocol to assess model performance in more realistic conditions, *i.e.*, on a non-aligned dataset.

2   Related Work

Anomaly detection and localization methods can be categorized as either reconstruction-based or embedding similarity-based methods.

Reconstruction-based methods are widely-used for anomaly detection and localization. Neural network architectures like autoencoders (AE) [6,8,12,14], variational autoencoders (VAE) [16,18,28,30] or generative adversarial networks (GAN) [2,24,27] are trained to reconstruct normal training images only. Therefore, anomalous images can be spotted as they are not well reconstructed. At the image level, the simplest approach is to take the reconstructed error as an anomaly score [12] but additional information from the latent space [1,24], intermediate activations [15] or a discriminator [2,3] can help to better recognize anomalous images. Yet to localize anomalies, reconstruction-based methods can take the pixel-wise reconstruction error as the anomaly score [6] or the structural similarity [8]. Alternatively, the anomaly map can be a visual attention map generated from the latent space [18,30]. Although reconstruction-based methods are very intuitive and interpretable, their performance is limited by the fact that AE can sometimes yield good reconstruction results for anomalous images too [23].

Embedding similarity-based methods use deep neural networks to extract meaningful vectors describing an entire image for anomaly detection [4,5,25,26] or an image patch for anomaly localization [7,9,21,31]. Still, embedding similarity-based methods that only perform anomaly detection give promising results but often lack interpretability as it is not possible to know which part of an anomalous images is responsible for a high anomaly score. The anomaly score is in this case the distance between embedding vectors of a test image and reference vectors representing normality from the training dataset. The normal reference can be the center of a n-sphere containing embeddings from normal images [26,31], parameters of Gaussian distributions [17,25] or the entire set of normal embedding vectors [4,9]. The last option is used by SPADE [9] which has the best reported results for anomaly localization. However, it runs a K-NN algorithm on a set of normal embedding vectors at test time, so the inference complexity scales linearly to the dataset training size. This may hinder industrial deployment of the method.

Our method, PaDiM, generates patch embeddings for anomaly localization, similar to the aforementioned approaches. However, the normal class in PaDiM is described through a set of Gaussian distributions that also model correlations between semantic levels of the used pretrained CNN model. Inspired by [9,25], we choose as pretrained networks a ResNet [13], a Wide-ResNet [32] or an EfficientNet [29]. Thanks to this modelisation, PaDiM outperforms the current state-of-the-art methods. Moreover, its time complexity is low and independent of the training dataset size at the prediction stage.

3   Patch Distribution Modeling

3.1   Embedding Extraction

Pretrained CNNs are able to output relevant features for anomaly detection [4]. Therefore, we choose to avoid ponderous neural network optimization by only using a pretrained CNN to generate patch embedding vectors. The patch embedding process in PaDiM is similar to one from SPADE [9] and illustrated in Fig. 2. During the training phase, each patch of the normal images is associated to its spatially corresponding activation vectors in the pretrained CNN activation maps. Activation vectors from different layers are then concatenated to get embedding vectors carrying information from different semantic levels and resolutions, in order to encode fine-grained and global contexts. As activation maps have a lower resolution than the input image, many pixels have the same embeddings and then form pixel patches with no overlap in the original image resolution. Hence, an input image can be divided in a grid of $(i, j) \in [1, W] \times [1, H]$ positions where $W \times H$ is the resolution of the largest activation map used to generate embeddings. Finally, each patch position (i, j) in this grid is associated to an embedding vector x_{ij} computed as described above.

The generated patch embedding vectors may carry redundant information, therefore we experimentally study the possibility to reduce their size (Sect. 5.1). We noticed that randomly selecting few dimensions is more efficient that a classic Principal Component Analysis (PCA) algorithm [22]. This simple random dimensionality reduction significantly decreases the complexity of our model for both training and testing time while maintaining the state-of-the-art performance. Finally, patch embedding vectors from test images are used to output an anomaly map with the help of the learned parametric representation of the normal class described in the next subsection.

3.2   Learning of the Normality

To learn the normal image characteristics at position (i, j), we first compute the set of patch embedding vectors at (i, j), $X_{ij} = \{x_{ij}^k, k \in [\![1, N]\!]\}$ from the N normal training images as shown on Fig. 2. To sum up the information carried by this set we make the assumption that X_{ij} is generated by a multivariate Gaussian distribution $\mathcal{N}(\mu_{ij}, \Sigma_{ij})$ where μ_{ij} is the sample mean of X_{ij} and the sample covariance Σ_{ij} is estimated as follows :

$$\Sigma_{ij} = \frac{1}{N-1} \sum_{k=1}^{N} (\mathbf{x_{ij}^k} - \mu_{ij})(\mathbf{x_{ij}^k} - \mu_{ij})^{\mathrm{T}} + \epsilon I \tag{1}$$

where the regularisation term ϵI makes the sample covariance matrix Σ_{ij} full rank and invertible. Finally, each possible patch position is associated with a multivariate Gaussian distribution as shown in Fig. 2 by the matrix of Gaussian parameters.

Our patch embedding vectors carry information from different semantic levels. Hence, each estimated multivariate Gaussian distribution $\mathcal{N}(\mu_{ij}, \Sigma_{ij})$ captures information from different levels too and Σ_{ij} contains the inter-level correlations. We experimentally show (Sect. 5.1) that modeling these relationships between the different semantic levels of the pretrained CNN helps to increase anomaly localization performance.

3.3   Inference: Computation of the Anomaly Map

Inspired by [17,25], we use the Mahalanobis distance [20] $M(x_{ij})$ to give an anomaly score to the patch in position (i, j) of a test image. $M(x_{ij})$ can be interpreted as the distance between the test patch embedding x_{ij} and learned distribution $\mathcal{N}(\mu_{ij}, \Sigma_{ij})$, where $M(x_{ij})$ is computed as follows:

$$M(x_{ij}) = \sqrt{(x_{ij} - \mu_{ij})^T \Sigma_{ij}^{-1}(x_{ij} - \mu_{ij})} \qquad (2)$$

Hence, the matrix of Mahalanobis distances $M = (M(x_{ij}))_{1<i<W,1<j<H}$ that forms an anomaly map can be computed. High scores in this map indicate the anomalous areas. The final anomaly score of the entire image is the maximum of anomaly map M. Finally, at test time, our method does not have the scalability issue of the K-NN based methods [5,9,21,31] as we do not have to compute and sort a large amount of distance values to get the anomaly score of a patch.

4   Experiments

4.1   Datasets and Metrics

Metrics. To assess the localization performance we compute two threshold independent metrics. We use the Area Under the Receiver Operating Characteristic curve (AUROC) where the true positive rate is the percentage of pixels correctly classified as anomalous. Since the AUROC is biased in favor of large anomalies we also employ the per-region-overlap score (PRO-score) [7]. It consists in plotting, for each connected component, a curve of the mean values of the correctly classified pixel rates as a function of the false positive rate between 0 and 0.3. The PRO-score is the normalized integral of this curve. A high PRO-score means that both large and small anomalies are well-localized.

Datasets. We first evaluate our models on the MVTec AD [6] designed to test anomaly localization algorithms for industrial quality control and in a one-class learning setting. It contains 15 classes of approximately 240 images. The original image resolution is between 700×700 and 1024×1024. There are 10 object and 5 texture classes. Objects are always well-centered and aligned in the same way across the dataset as we can see in Fig. 1 for classes Transistor and Capsule. In addition to the original dataset, to assess performance of anomaly localization models in a more realistic context, we create a modified version of the MVTec AD, referred as Rd-MVTec AD, where we apply random rotation $(-10, +10)$ and random crop (from 256×256 to 224×224) to both the train and test sets. This modified version of the MVTec AD may better describe real use cases of anomaly localization for quality control where objects of interest are not always centered and aligned in the image.

For further evaluation, we also test PaDiM on the Shanghai Tech Campus (STC) Dataset [19] that simulates video surveillance from a static camera. It contains 274 515 training and 42 883 testing frames divided in 13 scenes. The original image resolution is 856×480. The training videos are composed of normal sequences and test videos have anomalies like the presence of vehicles in pedestrian areas or people fighting.

4.2   Experimental Setups

We train PaDiM with different backbones, a ResNet18 (R18) [13], a Wide ResNet-50-2 (WR50) [32] and an EfficientNet-B5 [29], all pretrained on ImageNet [11]. Like in [9], patch embedding vectors are extracted from the first three layers when the backbone is a ResNet, in order to combine information from different semantic levels, while keeping a high enough resolution for the localization task. Following this idea, we extract patch embedding vectors from layers 7 (level 2), 20 (level 4), and 26 (level 5), if an EfficientNet-B5 is used. We also apply a random dimensionality reduction (Rd) (see Sects. 3.1 and 5.1). Our model names indicate the backbone and the dimensionality reduction method used, if any. For example, PaDiM-R18-Rd100 is a PaDiM model with a ResNet18 backbone using 100 randomly selected dimensions for the patch embedding vectors. By default we use $\epsilon = 0.01$ for the ϵ from Eq. 1.

We reproduce the model SPADE [9] as described in the original publication with a Wide ResNet-50-2 (WR50) [32] as backbone. For SPADE and PaDiM we apply the same prepocessing as in [9]. We resize the images from the MVTec AD to 256×256 and center crop them to 224×224. For the images from the STC we use a 256×256 resize only. We resize the images and the localization maps using bicubic interpolation and we use a Gaussian filter on the anomaly maps with parameter $\sigma = 4$ like in [9].

We also implement our own VAE as a reconstruction-based baseline implemented with a ResNet18 as encoder and a 8×8 convolutional latent variable. It is trained on each MVTec AD class with 10 000 images using the following data augmentations operations: random rotation $(-2°, +2°)$, 292×292 resize, random crop to 282×282, and finally center crop to 256×256. The training is performed during 100 epochs with the Adam optimizer [16] with an initial learning rate of 10^{-4} and a batch size of 32 images. The anomaly map for the localization corresponds to the pixel-wise L2 error for reconstruction.

5   Results

5.1   Ablative Studies

First, we evaluate the impact of modeling correlations between semantic levels in PaDiM and explore the possibility to simplify our method through dimensionality reduction.

Inter-Layer Correlation. The combination of Gaussian modeling and the Mahalanobis distance has already been employed in previous works to detect adversarial attacks [17] and for anomaly detection [25] at the image level. However those methods do not model correlations between different CNN's semantic levels as we do in PaDiM. In Table 1 we show the anomaly localization performance on the MVTec AD of PaDiM with a ResNet18 backbone when using only one of the first three layers (Layer 1, Layer 2, or Layer 3) and when summing the outputs of these 3 models to form an ensemble method that takes into account the first three layers but not the correlations between them (Layers 1+2+3). The last row of Table 1 (PaDiM-R18) is our proposed version of PaDiM where each patch location is described by one Gaussian distribution taking into account the first three ResNet18 layers and correlations between them.

Table 1. Study of the anomaly localization performance using different semantic-level CNN layers. Results are displayed as tuples (AUROC%, PRO-score%) on the MVTec AD.

Layer used	All texture classes	All object classes	All classes
Layer 1	(93.1, 87.1)	(95.6, 86.5)	(94.8, 86.8)
Layer 2	(95.0, 89.7)	(96.1, 87.9)	(95.7, 88.5)
Layer 3	(94.8, 89.6)	(97.1, 87.7)	(95.7, 88.3)
Layer 1+2+3	(95.4, 90.7)	(96.3, 88.1)	(96.0, 89.0)
PaDiM-R18	(**96.3, 92.3**)	(**97.5, 90.1**)	(**97.1, 90.8**)

It can be observed that using Layer 3 produces the best results in terms of AUROC among the three layers. It is due to the fact that Layer 3 carries higher semantic level information which helps to better describe normality. However, Layer 3 has a slightly worse PRO-score than Layer 2 that can be explained by the lower resolution of Layer 2 which affects the accuracy of anomaly localization. As we see in the two last rows of Table 1, aggregating information from different layers can solve the trade-off issue between high semantic information and high resolution. Unlike model Layer 1+2+3 that simply sums the outputs, our model PaDiM-R18 takes into account correlations between semantic levels. As a result, it outperforms Layer 1+2+3 by 1.1p.p (percent point). for AUROC and 1.8p.p. for PRO-score. It confirms the relevance of modeling correlation between semantic levels.

Table 2. Study of the anomaly localization performance with a dimensionality reduction from 448 to 100 and 200 using PCA or random feature selection (Rd). Results are displayed as tuples (AUROC%, PRO-score%) on the MVTec AD.

	All texture classes	All object classes	All classes
Rd 100	(95.7, 91.3)	(97.2, 89.4)	(96.7, 90.5)
PCA 100	(93.7, 88.9)	(93.5, 84.1)	(93.5, 85.7)
Rd 200	(96.1, 92.0)	(97.5, 89.8)	(97.0, 90.5)
PCA 200	(95.1, 91.8)	(96.0, 88.1)	(95.7, 89.3)
all (448)	(**96.3, 92.3**)	(**97.5, 90.1**)	(**97.1, 90.8**)

Dimensionality Reduction. PaDiM-R18 estimates multivariate Gaussian distributions from sets of patch embeddings vectors of 448 dimensions each.

Decreasing the embedding vector size would reduce the computational and memory complexity of our model. We study two different dimensionality reduction methods. The first one consists in applying a Principal Component Analysis (PCA) algorithm to reduce the vector size to 100 or 200 dimensions. The second method is a random feature selection where we randomly select features before the training. In this case, we train 10 different models and take the average scores. Still the randomness does not change the results between different seeds as the standard error mean (SEM) for the average AUROC is always between 10^{-4} and 10^{-7}.

From Table 2 we can notice that for the same number of dimensions, the random dimensionality reduction (Rd) outperforms the PCA on all the MVTec AD classes by at least 1.3p.p in the AUROC and 1.2p.p in the PRO-score. It can be explained by the fact that PCA selects the dimensions with the highest variance which may not be the ones that help to discriminate the normal class from the anomalous one [25]. It can also be noted from Table 2 that randomly reducing the embedding vector size to only 100 dimensions has a very little impact on the anomaly localization performance. The results drop only by 0.4p.p in the AUROC and 0.3p.p in the PRO-score. This simple yet effective dimensionality reduction method significantly reduces PaDiM time and space complexity as it will be shown in Sect. 5.4.

Table 3. Comparison of our PaDiM models with the state-of-the-art for the anomaly localization on the MVTec AD. Results are displayed as tuples (AUROC%, PRO-score%)

Type	Reconstruction-based methods			Embedding similarity based methods			Our methods	
Model	AE simm [6–8]	AE L2 [6,7]	VAE	Student [7]	Patch SVDD [31]	SPADE [9]	PaDiM-R18-Rd100	PaDiM-WR50-Rd550
Carpet	(87, 64.7)	(59, 45.6)	(59.7, 61.9)	(-, 69.5)	(92.6, -)	(97.5, 94.7)	(98.9, 96.0)	(**99.1**, **96.2**)
Grid	(94, 84.9)	(90, 58.2)	(61.2, 40.8)	(-, 81.9)	(96.2, -)	(93.7, 86.7)	(94.9, 90.9)	(**97.3**, **94.6**)
Leather	(78, 56.1)	(75, 81.9)	(67.1, 64.9)	(-, 81.9)	(97.4, -)	(97.6, 97.2)	(99.1, 97.9)	(**99.2**, **97.8**)
Tile	(59, 17.5)	(51, 89.7)	(51.3, 24.2)	(-, 91.2)	(91.4, -)	(87.4, 75.9)	(91.2, 81.6)	(**94.1**, **86.0**)
Wood	(73, 60.5)	(73, 72.7)	(66.6, 57.8)	(-, 72.5)	(90.8, -)	(88.5, 87.4)	(93.6, 90.3)	(**94.9**, **91.1**)
All texture classes	(78, 56.7)	(70, 69.6)	(61.2, 49.9)	(-, 79.4)	(93.7, -)	(92.9, 88.4)	(95.6, 91.3)	(**96.9**, **93.2**)
Bottle	(93, 83.4)	(86, 91.0)	(83.1, 70.5)	(-, 91.8)	(98.1, -)	(**98.4**, **95.5**)	(98.1, 93.9)	(98.3, 94.8)
Cable	(82, 47.8)	(86, 82.5)	(83.1, 77.9)	(-, 86.5)	(96.8, -)	(**97.2**, **90.9**)	(95.8, 86.2)	(96.7, 88.8)
Capsule	(94, 86.0)	(88, 86.2)	(81.7, 77.9)	(-, 91.6)	(95.8, -)	(**99.0**, **93.7**)	(98.3, 91.9)	(98.5, 93.5)
Hazelnut	(97, 91.6)	(95, 91.7)	(87.7, 77.0)	(-, 93.7)	(97.5, -)	(**99.1**, **95.4**)	(97.7, 91.4)	(98.2, 92.6)
Metal Nut	(89, 60.3)	(86, 83.0)	(78.7, 57.6)	(-, 89.5)	(98.0, -)	(**98.1**, **94.4**)	(96.7, 81.9)	(97.2, 85.6)
Pill	(91, 83.0)	(85, 89.3)	(81.3, 79.3)	(-, 93.5)	(95.1, -)	(**96.5**, **94.6**)	(94.7, 90.6)	(95.7, 92.7)
Screw	(96, 88.7)	(96, 75.4)	(75.3, 66.4)	(-, 92.8)	(95.7, -)	(**98.9**, **96.0**)	(97.4, 91.3)	(98.5, 94.4)
Toothbrush	(92, 78.4)	(93, 82.2)	(91.9, 85.4)	(-, 86.3)	(98.1, -)	(97.9, **93.5**)	(98.7, 92.3)	(**98.8**, 93.1)
Transistor	(90, 72.5)	(86, 72.8)	(75.4, 61.0)	(-, 70.1)	(97.0, -)	(94.1, **87.4**)	(97.2, 80.2)	(**97.5**, 84.5)
Zipper	(88, 66.5)	(77, 83.9)	(71.6, 60.8)	(-, 93.3)	(95.1, -)	(96.5, 92.6)	(98.2, 94.7)	(**98.5**, **95.9**)
All object classes	(91, 75.8)	(88, 83.8)	(81.0, 71.4)	(-, 88.9)	(96.7, -)	(97.6, **93.4**)	(97.3, 89.4)	(**97.8**, 91.6)
All classes	(87, 69.4)	(82, 79.0)	(74.4, 64.2)	(-, 85.7)	(95.7, -)	(96.5, 91.7)	(96.7, 90.1)	(**97.5**, **92.1**)

5.2   Comparison with the State-of-the-art

Localization. In Table 3, we show the AUROC and the PRO-score results for anomaly localization on the MVTec AD. For a fair comparison, we used a Wide ResNet-50-2 (WR50) as this backbone is used in SPADE [9]. Since the other baselines have smaller backbones, we also try a ResNet18 (R18). We randomly reduce the embedding size to 550 and 100 for PaDiM with WR50 and R18 respectively.

We first notice that PaDiM-WR50-Rd550 outperforms all the other methods in both the PRO-score and the AUROC on average for all the classes. PaDiM-R18-Rd100 which is a very light model also outperforms all models in the average AUROC on the MVTec AD classes by at least 0.2p.p. When we further analyze the PaDiM performances, we see that the gap for the object classes is small as PaDiM-WR50-Rd550 is the best only in the AUROC (+0.2p.p) but SPADE [9] is the best in the PRO-score (+1.8p.p). However, our models are particularly accurate on texture classes. PaDiM-WR50-Rd550 outperforms the second best model SPADE [9] by 4.8p.p and 4.0p.p in the PRO-score and the AUROC respectively on average on texture classes. Indeed, PaDiM learns an explicit probabilistic model of the normal classes contrary to SPADE [9] or Patch-SVDD [31]. It is particularly efficient on texture images because even if they are not aligned and centered like object images, PaDiM effectively captures their statistical similarity accross the normal train dataset.

Additionally, we evaluate our model on the STC dataset. We compare our method to the two best reported models performing anomaly localization without temporal information, CAVGA-RU [30] and SPADE [9].As shown in Table 4, the best result (AUROC) on the STC dataset is achieved with our simplest model PaDiM-R18-Rd100 by a 2.1p.p. margin. In fact, pedestrian positions in images are highly variable in this dataset and, as shown in Sect. 5.3, our method performs well on non-aligned datasets.

Detection. By taking the maximum score of the anomaly maps issued by our models (see Section 3.3) we give anomaly scores to entire images to perform anomaly detection at the image level. We test PaDiM for anomaly detection with a Wide ResNet-50-2 (WR50) [32] used in SPADE and an EfficientNet-B5 [29]. The Table 5 shows that our model PaDiM-WR50-Rd550 outperforms every method except MahalanobisAD [25] with their best reported backbone, an EfficientNet-B4. Still our PaDiM-EfficientNet-B5 outperforms every model by at least 2.6p.p on average on all the classes in the AUROC. Besides, contrary to the second best method for anomaly detection, MahalanobisAD [25], our

Table 4. Comparison of our PaDiM model with the state-of-the-art for the anomaly localization on the STC in the AUROC%.

Model	CAVGA-RU [30]	SPADE [9]	PaDiM-R18-Rd100
AUROC score%	85	89.9	**91.2**

Table 5. Anomaly detection results (at the image level) on the MVTec AD using AUROC%.

Model	GANomaly [2]	ITAE [14]	Patch SVDD [31]	SPADE [9] (WR50)	Mahalano-bisAD [25] (EfficientNet-B4)	PaDiM-WR50-Rd550	PaDiM EfficientNet-B5
All textures classes	-	–	94.6	–	97.2	98.8	**99.0**
All objects classes	-	–	90.9	–	94.8	93.6	**97.2**
All classes	76.2	83.9	92.1	85.5	95.8	95.3	**97.9**

Table 6. Anomaly localization results on the non-aligned Rd-MVTec AD. Results are displayed as tuples (AUROC%, PRO-score%)

Model	VAE (R18)	SPADE (WR50)	PaDiM-WR50-Rd550
All texture classes	(54.7, 23.1)	(84.6, 75.6)	**(92.4, 77.9)**
All object classes	(65.8, 30.2)	(88.2, 65.8)	**(92.1, 70.8)**
All classes	(62.1, 27.8)	(87.2, 69.0)	**(92.2, 73.1)**

model also performs anomaly segmentation which characterizes more precisely the anomalous areas in the images.

5.3 Anomaly Localization on a Non-aligned Dataset

To estimate the robustness of anomaly localization methods, we train and evaluate the performance of PaDiM and several state-of-the-art methods (SPADE [9], VAE) on a modified version of the MVTec AD, Rd-MVTec AD, described in Sect. 4.1. Results of this experiment are displayed in Table 6. For each test configuration we run 5 times data preprocessing on the MVTec AD with random seeds to obtain 5 different versions of the dataset, denoted as Rd-MVTec AD. Then, we average the obtained results and report them in Table 6. According to the presented results, PaDiM-WR50-Rd550 outperforms the other models on both texture and object classes in the PRO-score and the AUROC. Besides, the SPADE [9] and VAE performances on the Rd-MVTec AD decrease more than the performance of PaDiM-WR50-Rd550 when comparing to the results obtained on the normal MVTec AD (refer to Table 3). The AUROC results decrease by 5.3p.p for PaDiM-WR50-Rd550 against 12.2p.p and 8.8p.p decline for VAE and SPADE respectively. Thus, we can conclude that our method seems to be more robust to non-aligned images than the other existing and tested works.

5.4 Scalability Gain

Time Complexity. In PaDiM, the training time complexity scales linearly with the dataset size because the Gaussian parameters are estimated using the entire

Table 7. Average inference time of anomaly localization in seconds on the MVTec AD with a CPU intel i7-4710HQ @ 2.50 GHz.

Model	SPADE (WR50)	VAE (R18)	PaDiM R18-Rd100	PaDiM-WR50-Rd550
Inference time (sec.)	7.10	0.21	0.23	0.95

training dataset. However, contrary to the methods that require to train deep neural networks, PaDiM uses a pretrained CNN, and, thus, no deep learning training is required which is often a complex procedure. Hence, it is very fast and easy to train it on small datasets like MVTec AD. For our most complex model PaDiM-WR50-Rd550, the training on a CPU (Intel CPU 6154 3 GHz 72th) with a serial implementation takes on average 150 s on the MVTec AD classes and 1500 s on average on the STC video scenes. These training procedures could be further accelerated using GPU hardware for the forward pass and the covariance estimation. In contrast, training the VAE with 10 000 images per class on the MVTec AD following the procedure described in Sect. 4.2 takes 2h40 per class using one GPU NVIDIA P5000. Conversely, SPADE [9] requires no training as there are no parameters to learn. Still, it computes and stores in the memory before testing all the embedding vectors of the normal training images. Those vectors are the inputs of a K-NN algorithm which makes SPADE's inference very slow as shown in Table 7.

In Table 7, we measure the model inference time using a mainstream CPU (Intel i7-4710HQ CPU @ 2.50 GHz) with a serial implementation. On the MVTec AD, the inference time of SPADE is around seven times slower than our PaDiM model with equivalent backbone because of the computationally expensive NN search. Our VAE implementation, which is similar to most reconstruction-based models, is the fastest model but our simple model PaDiM-R18-Rd100 has the same order of magnitude for the inference time. While having similar complexity, PaDiM largely outperfoms the VAE methods (see Sect. 5.2).

Memory Complexity. Unlike SPADE [9] and Patch SVDD [31], the space complexity of our model is independent of the dataset training size and depends only on the image resolution. PaDiM keeps in the memory only the pretrained CNN and the Gaussian parameters associated with each patch. In Table 8 we show the memory requirement of SPADE, our VAE implementation, and PaDiM, assuming that parameters are encoded in float32. Using equivalent backbone, SPADE has a lower memory consumption than PaDiM on the MVTec AD. However, when using SPADE on a larger dataset like the STC, its memory consumption becomes intractable, whereas PaDiM-WR50-Rd550 requires seven times less memory. The PaDiM space complexity increases from the MVTec AD to the STC only because the input image resolution is higher in the latter dataset as described in Sect. 4.2. Finally, one of the advantages of our framework PaDiM is that the user can easily adapt the method by choosing the backbone and the embedding size to fit its inference time requirements, resource limits, or expected performance.

Table 8. Memory requirement in Gb of the anomaly localization methods trained on the MVTec AD and the STC dataset.

Model	SPADE (WR50)	VAE (R18)	PaDiM R18-Rd100	PaDiM-WR50-Rd550
MVTec AD	1.4	0.09	0.17	3.8
STC	37.0	-	0.21	5.2

6 Conclusion

We have presented a framework called PaDiM for anomaly detection and localization in one-class learning setting which is based on distribution modeling. It achieves state-of-the-art performance on MVTec AD and STC datasets. Moreover, we extend the evaluation protocol to non-aligned data and the first results show that PaDiM can be robust on these more realistic data. PaDiM low memory and time consumption and its ease of use make it suitable for various applications, such as visual industrial control.

References

1. Abati, D., Porrello, A., Calderara, S., Cucchiara, R.: Latent space autoregression for novelty detection. In: 2019 IEEE/CVF Conference on Computer Vision and Pattern Recognition (CVPR), pp. 481–490 (2019)
2. Akcay, S., Atapour-Abarghouei, A., Breckon, T.P.: GANomaly: semi-supervised anomaly detection via adversarial training. In: Jawahar, C.V., Li, H., Mori, G., Schindler, K. (eds.) ACCV 2018. LNCS, vol. 11363, pp. 622–637. Springer, Cham (2019). https://doi.org/10.1007/978-3-030-20893-6_39
3. Akçay, S., Atapour-Abarghouei, A., Breckon, T.P.: Skip-ganomaly: skip connected and adversarially trained encoder-decoder anomaly detection. In: 2019 International Joint Conference on Neural Networks (IJCNN), pp. 1–8 (2019)
4. Bergman, L., Cohen, N., Hoshen, Y.: Deep nearest neighbor anomaly detection. In: arXiv, 2002.10445 (2020)
5. Bergman, L., Hoshen, Y.: Classification-based anomaly detection for general data. In: International Conference on Learning Representations (ICPR) (2020)
6. Bergmann, P., Fauser, M., Sattlegger, D., Steger, C.: Mvtec ad-a comprehensive real-world dataset for unsupervised anomaly detection. In: 2019 IEEE/CVF Conference on Computer Vision and Pattern Recognition (CVPR), pp. 9584–9592 (2019)
7. Bergmann, P., Fauser, M., Sattlegger, D., Steger, C.: Uninformed students: Student-teacher anomaly detection with discriminative latent embeddings. In: 2020 IEEE/CVF Conference on Computer Vision and Pattern Recognition (CVPR), pp. 4182–4191 (2020)
8. Bergmann, P., Löwe, S., Fauser, M., Sattlegger, D., Steger, C.: Improving unsupervised defect segmentation by applying structural similarity to autoencoders. In: Proceedings of the 14th International Joint Conference on Computer Vision, Imaging and Computer Graphics Theory and Applications (VISIGRAPP), vol. 5, VISAPP (2019)

9. Cohen, N., Hoshen, Y.: Sub-image anomaly detection with deep pyramid correspondences. In: arXiv 2005.02357 (2020)
10. Cover, T., Hart, P.: Nearest neighbor pattern classification. IEEE Trans. Inf. Theory **13**(1), 21–27 (1967)
11. Deng, J., et al.: Imagenet: a large-scale hierarchical image database. In: 2009 IEEE Conference on Computer Vision and Pattern Recognition (CVPR), pp. 248–255 (2009)
12. Gong, D., et al.: Memorizing normality to detect anomaly: Memory-augmented deep autoencoder for unsupervised anomaly detection. In: 2019 IEEE/CVF International Conference on Computer Vision (ICCV), pp. 1705–1714 (2019)
13. He, K., Zhang, X., Ren, S., Sun, J.: Deep residual learning for image recognition. In: 2016 IEEE Conference on Computer Vision and Pattern Recognition (CVPR), pp. 770–778 (2016)
14. Huang, C., Ye, F., Cao, J., Li, M., Zhang, Y., Lu, C.: Attribute restoration framework for anomaly detection. In: arXiv, 1911.10676 (2019)
15. Kim, K.H., et al.: Rapp: Novelty detection with reconstruction along projection pathway. In: 2020 International Conference on Learning Representations (ICLR) (2020)
16. Kingma, D.P., Welling, M.: Auto-encoding variational bayes. In: Proceedings of the 2nd International Conference on Learning Representations (ICLR) (2014)
17. Lee, K., Lee, K., Lee, H., Shin, J.: A simple unified framework for detecting out-of-distribution samples and adversarial attacks. In: Proceedings of the 32nd International Conference on Neural Information Processing Systems, NIPS 2018, pp. 7167–7177. Curran Associates Inc., Red Hook, NY, USA (2018)
18. Liu, W., et al.: Towards visually explaining variational autoencoders. In: 2020 IEEE/CVF Conference on Computer Vision and Pattern Recognition (CVPR), pp. 8639–8648 (2020)
19. Liu, W., Luo, W., Lian, D., Gao, S.: Future frame prediction for anomaly detection - a new baseline. In: 2018 IEEE/CVF Conference on Computer Vision and Pattern Recognition (CVPR), pp. 6536–6545 (2018)
20. Mahalanobis, P.: On the generalized distance in statistics. In: National Institute of Science of India (1936)
21. Napoletano, P., Piccoli, F., Schettini, R.: In: Sensors (Basel, Switzerland), vol. 18, p. 209 (2018)
22. Pearson, K.: On lines and planes of closest fit to systems of points in space. London, Edinburgh, Dublin Philosophical Mag. J. Sci. **2**(11), 559–572 (1901)
23. Perera, P., Nallapati, R., Xiang, B.: Ocgan: one-class novelty detection using gans with constrained latent representations. In: 2019 IEEE/CVF Conference on Computer Vision and Pattern Recognition (CVPR), pp. 2893–2901 (2019)
24. Pidhorskyi, S., Almohsen, R., Adjeroh, D.A., Doretto, G.: Generative probabilistic novelty detection with adversarial autoencoders. In: Proceedings of the 32nd International Conference on Neural Information Processing Systems, NIPS 2018, pp. 6823–6834. Curran Associates Inc., Red Hook, NY, USA (2018)
25. Rippel, O., Mertens, P., Merhof, D.: Modeling the distribution of normal data in pre-trained deep features for anomaly detection. In: arXiv, 2005.14140 (2020)
26. Ruff, L., et al.: Deep one-class classification. Proceedings of Machine Learning Research, vol. 80, pp. 4393–4402. PMLR, Stockholmsmässan, Stockholm Sweden (10–15 July 2018)
27. Sabokrou, M., Khalooei, M., Fathy, M., Adeli, E.: Adversarially learned one-class classifier for novelty detection. In: 2018 IEEE/CVF Conference on Computer Vision and Pattern Recognition, pp. 3379–3388 (2018)

28. Sato, K., Hama, K., Matsubara, T., Uehara, K.: Predictable uncertainty-aware unsupervised deep anomaly segmentation. In: 2019 International Joint Conference on Neural Networks (IJCNN), pp. 1–7 (2019)
29. Tan, M., Le, Q.: EfficientNet: rethinking model scaling for convolutional neural networks. In: Proceedings of Machine Learning Research, vol. 97, pp. 6105–6114. PMLR, Long Beach, California, USA (09–15 June 2019)
30. Venkataramanan, S., Peng, K.C., Singh, R.V., Mahalanobis, A.: Attention guided anomaly localization in images. In: European Conference on Computer Vision (ECCV), vol. 2020 (2020)
31. Yi, J., Yoon, S.: Patch svdd: Patch-level svdd for anomaly detection and segmentation. In: arXiv, 2006.16067 (2020)
32. Zagoruyko, S., Komodakis, N.: Wide residual networks. In: Richard C. Wilson, E.R.H., Smith, W.A.P. (eds.) Proceedings of the British Machine Vision Conference (BMVC). pp. 87.1-87.12. BMVA Press (September 2016)

Real-Time Cross-Dataset Quality Production Assessment in Industrial Laser Cutting Machines

Nicola Peghini[1]([✉])[iD], Andrea Zignoli[1][iD], Davide Gandolfi[2][iD], Paolo Rota[3][iD], and Paolo Bosetti[1][iD]

[1] Department of Industrial Engineering, University of Trento, Via Sommarive, 9, 38123 Povo, Trento, Italy
{nicola.peghini,andrea.zignoli,paolo.bosetti}@unitn.it
[2] ADIGE S.P.A., Via per Barco, 11, 38056 Levico Terme, Trento, Italy
lasertech@adige.it
[3] Department of Information Engineering and Computer Science, University of Trento, Via Sommarive, 9, 38123 Povo, Trento, Italy
paolo.rota@unitn.it
https://www.dii.unitn.it/en, https://www.blmgroup.com/en/home, https://www.disi.unitn.it/

Abstract. In laser cutting processes, cutting failure is one of the most common causes of faulty productions. Monitoring cutting failure events is extremely complex, as failures might be initiated by several factors, the most prominent probably being the high production speeds required by modern standards. The present work aims at creating and deploying a classifier able to assess the status of a production cutting quality in a real-time fashion. To this aim, multiple datasets were collected in different environmental conditions and with different sensors. Model inputs include photo-sensors and production parameters. At first, different algorithms were tested and rated by prediction ability. Second, the selected algorithm was deployed on a GPU embedded system and added to the current machine configuration. The final system can receive the input data from the sensors, perform the inference, and send back the results to the computer numerical control. The data management is based on a client-server architecture. The selected algorithm and hardware showed good performances despite multiple changes in the environmental conditions (domain adaptation ability) both in terms of prediction ability (accuracy) and computational times.

Keywords: Process monitoring · Domain adaptation · Time-series · Classification · Deep learning

1 Introduction

Production monitoring is the practice of producing and using actionable information on the status of a manufacturing process to improve production quality. Highly efficient coupling between data acquisition (sensors) and information

© Springer Nature Switzerland AG 2021
A. Del Bimbo et al. (Eds.): ICPR 2020 Workshops, LNCS 12664, pp. 490–505, 2021.
https://doi.org/10.1007/978-3-030-68799-1_36

management (data processing) can prevent production system failures, low quality, or inefficient production lines.

In recent years, in several industrial applications, data availability has been surpassing the capability of the data processing systems, therefore calling for a more efficient/informative data usage and more powerful processing units. The development of new deep learning algorithms for a wide variety of industrial applications, and the concomitant spread of newer, faster, and cheaper GPU units, expanded the frontiers of the production monitoring capabilities.

Laser cutting uses a high-power laser to slice materials (e.g. metal parts, sheets, tubes, etc.) and constitutes a relatively new technology in the vast world of industrial manufacturing. Laser cutting machines, by their nature, are equipped with a high degree of automation, as both the laser optics and the computer numerical control (CNC) that move all the machine axes are typically automatically controlled. During cutting, the laser beam melts and vaporizes the material, which is eventually carried out from the cutting kerf by a jet of gas (e.g. N_2).

Several parameters/variables (such as material, laser power, gas type/pressure, cutting speed, etc.) are used to control the cut through the processed metal, and to obtain an edge with a high-quality surface finishing. However, some uncontrollable environmental and material variables can unexpectedly affect the quality of the process. These include local material property changes (e.g. due to different material batches, preprocessing, welding, polishing, deposition, rusting, etc.), external temperature/humidity, defects in the optical fiber (i.e. dirt), and geometrical inaccuracies (inconsistencies or misalignment between the computer numerical control and the actual material shape). As a consequence, the laser cutting action might not always provide a satisfactory result, leaving portions of uncut material or undesired burr made of material correctly melted but not effectively evacuated. Bad processing results are often spotted only after the end of the whole process, leading to a lot of scrap material and time-wasting. This problem is even more exacerbated in case of unsupervised and fully automated cutting operations, e.g. during night shifts.

Being able to automatically and timely detect failures during the laser cutting operations can save a considerable amount of time, energy, and material during the production process. The control loop implemented in the CNC usually takes limited information into account, e.g. distance from the material surface from proximity sensors and axis positions from encoders. Photodiodes can be used to assess the status of the cutting operations, by measuring the intensity of the light reflected by the material surface at different wavelengths. The correct and reliable interpretation of the photodiode signals would help to complement and fusing the feedback information to be provided to the CNC. If this feedback information is timely enough and it is promptly available, then the CNC can be correctly controlled before the production process is compromised.

In a recent research work, Santolini et al. [18] shown that machine learning techniques (such as Gaussian Mixture and Hidden Markov Models) and deep learning techniques (such as Recurrent and Convolutional Neural Networks) can be used to classify the quality of the laser cutting process by fusing the sensors' information collected from the photodiodes, the laser beam source, and the CNC. However, the work of Santolini et al. [18] suffers from two major limitations: an accurate classification can be only obtained on a single material/thickness couple at a time (therefore requiring new supervised training for new materials and cutting conditions), and the time needed to retrieve the result of the inference is not suitable for real-time applications and industrial processes.

Therefore, the goal of this new research work was two-fold: to design a new classifier able to accurately detect laser cutting process failures in multiple materials without requiring custom-built labeled datasets, and to deploy the classifier and perform the inference within an extremely limited amount of time.

2   Problem Description

Similarly to Santolini et al. [18] the model must be able to make a distinction between three classes:

1. CUT: overall good cut production quality;
2. PLASMA: a condition where cutting still occurs but the quality is lower than the standard;
3. WELDING: a condition where the final part will still be attached to the scrap material either because the laser beam is not cutting through the material or the gas flow is not enough to evacuate all the molten material.

In laser cutting production lines welding conditions are just not acceptable, while plasma conditions -still suboptimal-might only be accepted in a narrow band of circumstances. These conditions can be assessed by visually inspecting the cut surface quality.

The system has to elaborate data coming from the different sensors of the laser cutting machine and predict the cut quality (i.e. the corresponding class) that occurred in the analyzed time frame. The sensors collect data every 25 μs, and a buffer characterizing a time frame of 20 ms is filled. Once the buffer is ready, the classifier predicts the cut quality. The data buffer is updated every 5 ms, after the first inference, by eliminating the oldest 200 sensors' measures and filling the buffer with the data of the most recent 5 ms. These time frames were set by previous analysis [17]. To reduce the latency of the system to the minimum, the communication of the data buffer and the inference process has to occur under 5 ms, so the prediction is available before the next buffer is sent to the inference server. This method will leave the system with a latency of 5 ms derived from the data acquisition process required to fill the buffer.

Since the classifier must perform under the 5 ms target time, the model needs to have an adequate number of parameters. In addition, the minimal accuracy should be set at the level of the state-of-the-art solutions, i.e. ~82% [18]. Model design and development therefore must consider a trade-off between model complexity and model computational performance. To further improve performance, since some sensors' data are redundant with one another, some signals are merged to preserve all the information but lower the input dimensions.

One of the big challenges of this project was that the two datasets were collected with two different sets of photodiodes, this resulted in different raw signals which represent the same conditions. The rest of the paper is structured as follows: Sect. 3 plots an overview of the scarce literature related to the topic of this work. Section 4 shows an extension to [18] comparing a few different architectures (i.e. Sect. 4.1), it also describes the adopted training procedure to align datasets acquired in different environmental conditions (i.e. Sect. 4.2). Section 5 describes the protocols and the hardware chosen for the deployment. In Sect. 6 we describe the dataset used for the experiments and we also present qualitative and quantitative results that are discussed in Sect. 7. Finally conclusions are drawn in Sect. 8.

3   Related Works

Machine Learning in Industrial Applications. Modern industrial technologies are starting to incorporate machine learning in nearly every aspect of the production process: from product inspection to quality control, from failure prediction to digitization of the industrial documentation. Importantly, machine learning enables predictive monitoring, with machine learning algorithms forecasting system failures, anomalies, breakdowns and poor production outcomes [5,13]. In particular, in laser cutting processes, machine learning algorithms (e.g. support vector machine and artificial neural networks) have been applied to classify the quality of the production process and to help defining the optimal production parameters [12,21]. Machine learning has also been adopted to predict bad cutting surface quality in the so-called *heat affected zones* [2] and in the dross attachment [7]. However, to the best of our knowledge, the reliability of the aforementioned models have been rarely assessed on the field. In addition, the models were not directly deployed on the industrial machines and they were not always tested for computational times. Clearly, the real-time assessment of the production quality in laser cutting production lines is still challenging [1,9].

Domain Adaptation. Despite the blowing scientific interest in the topic [4,15, 16,22], there are not many industrial projections of such amount of knowledge. A possible explanation is, as claimed in [14], industrial applications may have very different data distributions according to a multitude of different scenarios, depending on the intrinsic difference between sensors and the uncertain definition of a specific task. This is an observation that we share with the authors of [14] and we will briefly discuss it in Sect. 7. Differently to [14], however, we face a typical domain adaptation problem where the classes of the source dataset are the same as those in the target dataset, moreover we do not use adversarial learning but we try to align the distributions by considering separate statistics for different datasets.

Autonomous Driving is one of the most promising industrial application where domain adaptation is applied [19,23,24]. However, such task disposes of a multitude of large scale video datasets, in our case we only have a limited amount of self-produced data. Other works are worth to be cited but still related to generic problems such as NLP (e.g. recommendation systems [8]) or medical imaging [6].

Deployment of Deep Learning Models. Of great interest for the industrial applications, are the inference computational times of different models on different hardware solutions. For example, in [3], computational time and memory usage are reported for different model architectures and for different hardware solutions (e.g. NVIDIA Jetson TX1 and Titan X Pascal). When often times it is difficult to compare models across different disciplines (e.g. image recognition *vs* time-series classification), computational time and memory usage is known to be highly determined by model complexity.

4   Inference Model

4.1   Architectures

In the first part of the work, a pool of network architectures has been tested. Starting from well known models adopted in machine vision application a Residual Networks [10], an InceptionNet [20] and a DenseNet [11] network were investigated.

- **ResNet**: In brief, Residual Networks (ResNet) are a specific type of architecture that is configured in blocks (i.e. residual blocks) each composed by traditional convolutional, regularization and skip layers (that enables the network to learn residual information from each block). In traditional neural networks, the output of one layer represents the input of the next layer, and all the layers are trained sequentially. The residual block allows skipping a few layers using a skip-connection created by defining the output of the block as the sum of the "traditional" layer output with the identity of the block inputs. Therefore, if the block layers do not contribute to the training process, they are skipped and the next layer outside the ResBlock is fed with the block input itself.
- **InceptionNet**: The basic layer of the InceptionNet is based on filter concatenation of three different convolution layers and one max pooling layer. These layers instead of going deeper they operate at the same level and then the concatenation of these layers is the input of the following inception block. To reduce the computational time by minimising the input dimensions, an extra 1×1 convolution layer is added before the convolution layer. This network is quite deep and thus the problem of vanishing gradient can arise, for this reason, Inception Networks have multiple heads in different parts of the network where the tasks can be regressed. The loss will thus be a weighted sum of the auxiliary losses (calculated on the auxiliary classifiers) and the final loss (in traditional networks for classification the loss is computed only at the last layer).

- **DenseNet**: DenseNet connects each layer with the following layers in a feed-forward fashion. The fundamental layers of this network are the DenseBlock and the TransitionBlock. In the DenseBlock a convolution block is performed, each time that this block is called in the DenseBlock is receiving as an input the feature maps of the previous iterations and its own feature-maps will be used as input into all subsequent layers. The transition block is used to concatenate seamlessly the DenseBlock.

In Fig. 1, the structures of the aforementioned architectures are presented. However, in the present research work, the input is not bi-dimensional as expected in the traditional computer vision application (i.e. images), but it is a set of machine signals and sensors at a specific time, so the network has to be adapted to accept time-series input shapes.

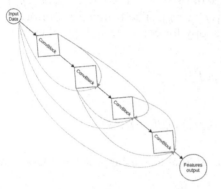

(a) Schematic representation of the DenseNet neural network characteristic layer.

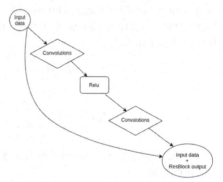

(b) Schematic representation of the ResNet neural network characteristic layer.

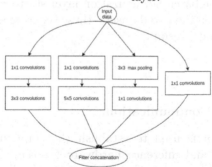

(c) Schematic representation of the InceptionNet neural network characteristic layer.

Fig. 1. Different neural network characteristic layers.

4.2 Adaptation

Maintaining a neural network for industrial applications that are inherently changing, updating and evolving technologies is quite challenging. For these particular applications, the bottle-neck consists of the large quantity of labelled data needed to re-train and tune the weights when the system is used under different conditions (i.e. different machine models, model updates, new materials, cutting parameters updates, environmental changes, etc.) with respect to the one used to collect the training dataset. This impediment arises the need to establish a starting point (model with labelled samples) from which it is possible to tune the weights using unlabelled data collected, for example, while setting up the machine. To this, inspired by the works of Carlucci et al. [4] and Roy et al. [16] a simultaneous training of a classification network using a labelled (source-dataset) and an unlabelled (target-dataset) was performed (Fig. 2).

In this work we consider a labelled source dataset $S = \{(x_1^s, y_1^s), \ldots, (x_n^s, y_n^s)\}$ and a non-labelled target dataset $T = \{x_1^t, \ldots, x_m^t\}$. The training is performed fusing two different losses, a sparse crossentropy for S (i.e. Eq. (1)) and entropy for T (i.e. Eq. (2)).

$$\mathcal{L}_s(\theta) = -\frac{1}{n} \sum_{i=1}^n \log f_s^\theta (y_i^s; x_i^s) \tag{1}$$

$$\mathcal{L}_u(\theta) = -\frac{1}{m} \sum_{i=1}^m \sum_{y \in \mathcal{Y}} f_t^\theta (y; x_i^t) \log f_t^\theta (y; x_i^t) \tag{2}$$

The two losses have then been fused in an unique loss using a scaling factor λ as shown in (3).

$$\mathcal{L}(\theta) = \mathcal{L}_s(\theta) + \lambda \mathcal{L}_u(\theta) \tag{3}$$

During training, the batch normalization layer statistics were computed in a distinct manner between the two datasets, these separate statistics have an high impact on the final model result.

5 Deployment

5.1 Client Server Communication

A client/server system is used to establish the communication between the machine side (*client*) and inference hardware (*server*). The *client* is responsible for reading, storing, and buffering the sensors' information (this is made possible by a dynamic library). The *server* receives the data in the buffer and invokes a dynamic library that takes care of the inference. Once the results of the inference have been computed, they are sent back to the *client*, which uses the information thereby contained to instruct the control loop. To make the communication robust, fast, and efficient, the high-performance asynchronous messaging queuing library ZeroMQ has been used.

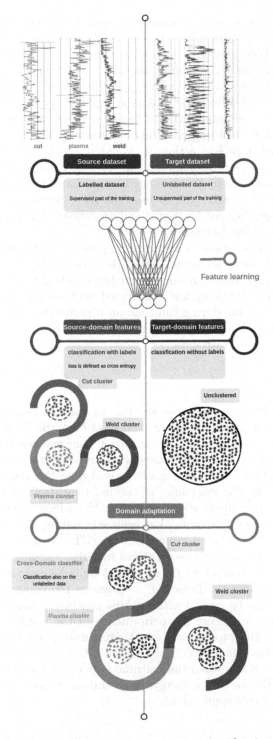

Fig. 2. Workflow for the simultaneous training of a classification network using a labelled (source-dataset) and an unlabelled (target-dataset)

The sensors' information is transferred from the *client* side in binary form while undergoing a serialisation and deserialisation process designed to minimise the cycle-time. This process is supervised by an error-detecting code (cyclic redundancy check CRC8). Therefore, the buffer sent from the *client* is composed by a header of 4 bytes indicating the buffer length, this allows to check if all the information needed from the sensors are gained correctly, then all the data themselves (sensors information) and finally a check byte (CRC8) to check if any communication error occurred. The inference result availability is checked by the *client* with a polling action; if available, the predicted class is read by the *client* and thus the control system can react accordingly. On the *server* side the results are processed by the dynamic library that takes care of the inference. The dynamic library was created to leave the *client/server* agnostic to the data buffer and the deployment hardware and workflow, this approach allows the deployment on multiple platforms.

In industrial applications is always good practice to be able to store logs of the process to intervene in case any issue arises. In this case, to avoid interfering with the process and create any undesirable lag, a publisher-subscriber setup was created. The *server*, if this option is activated, will publish the inference results and any error occurring during run-time on a dedicated thread, the industrial PC will then store the log elsewhere, this setup was created to avoid clogging up the inference hardware with data.

5.2 Hardware

The selected hardware needed to be suitable for the real-time task required, i.e.: cost-effective but still able to retrieve and send back the result of the inference within 5 ms; this upper limit in communication time was set to give enough time to the control loop to react and consequently modify the machine parameters. Some of the adjustments can be made in real-time but the majority of reactions are limited by the mechanics of the machine itself. A compact solution suitable for industrial applications (ZiggyBox^{TM}, which is based on NVIDIA® Jetson^{TM}TX2 computing module) has been selected for the deployment.

The inference is computed using the TensorRT^{TM}framework, which uses an ONNX file, open format built to represent machine learning models, to decode the model. This solution allows the system to be independent of the framework used to train the network. TensorRT^{TM}optimises the computational times for NVIDIA® platforms (equipped with ARM64 processing unit architectures).

With the goal to increase the performance of the hardware, automatize as much as possible the process and avoid any unnecessary communication and inference calculation time jitter, the Ubuntu GUI was deactivated. A start-up service (hardware starts when the machine is booted) was created, which activates all the CPUs, maximize energy usage and clock rate (power consumption is not an issue for this application).

6   Experiments

6.1   Datasets

Two experimental campaigns were performed to collect the data in different environmental conditions and constitute two different datasets (DS1 and DS2). The two data acquisition campaigns were performed on the same machine with different sensors (i.e. the set of photodiodes), different material batches and different layer thicknesses. The goal was to introduce variability in the datasets, which was used to test the reliability and the robustness of the system. The DS1 was collected in a cool environment ($\sim$18°C) during the winter season on the following combinations of materials and thicknesses: stainless steel (1.5, 2.5, 4.0 and 8 mm), construction steel (2.3, 3.0, 5.0 and 7 mm) and aluminium (2.0, 4.0 and 8 mm). The DS2 was collected in a warm environment ($\sim$24°C) during the summer season on the following combinations of materials and thicknesses: stainless steel (2.0, 3.0 and 6 mm), construction steel (1.8, 3.0 and 6 mm) and aluminium (1.5, 3.0 and 6 mm).

6.2   Training Procedure

Eight different training/testing sessions (S1–8, Table 1) were performed to assess the accuracy of the different model architectures. In S1, supervised (i.e. with labelled data) training and testing were both performed on DS1 normalised on max/min values found in DS1. In S2, supervised training was performed on DS1 and testing was performed on DS2, with DS1 and DS2 both normalised on max/min values found in DS1. In S3, supervised training was performed on DS1 and testing was performed on DS2, with DS1 and DS2 normalised on max/min values found in DS1 and DS2, respectively. In S4, supervised (i.e. with labelled data) training and testing were both performed on DS2 normalised on max/min values found in DS2. In S5, supervised training and testing were both performed on an unified dataset (DS1 + DS2), which was normalised on max/min values found in the same unified dataset. In S6, supervised training and testing were both performed on an unified dataset (DS1 + DS2), where DS1 and DS2 were normalised on the max/min values found in the correspondent datasets. In S7, supervised training was performed with DS1 and unsupervised (i.e. without labelled data) training was performed on DS2. Testing was performed on the unified dataset (DS1 + DS2), where DS1 and DS2 were normalised on the max/min values found in the correspondent datasets. In S8, unsupervised training was performed with DS1 and supervised training was performed on DS2. Testing was performed on the unified dataset (DS1 + DS2), where DS1 and DS2 were normalised on the max/min values found in the correspondent datasets.

6.3   Results

Model Accuracy. In Table 1 the model accuracy results are presented for the ResNet, which provided the best accuracy (the DenseNet accuracy was as high as

89.48% in the case of different normalization but decreased to 87.03% in the case of domain adaptation; the InceptionNet instead showed low accuracy (~75%) in the case of different normalization of the dataset and was not tested for domain adaptation abilities).

The high accuracy obtained in S1 was not clearly maintained during S2 (88.76% *vs* 23.61%). Apparently, normalisation only played a minor but meaningful role in this loss of accuracy, since in S3 the accuracy level was still very poor (40.50%). The accuracy provided in S4 (89.92%) confirmed that the ResNet was able to preserve a high prediction ability when the same dataset used for training was also used for testing. The accuracy levels obtained in S5 and S6 (70.44% *vs* 84.78%) revealed again the marginal but tangible contribution of the normalisation procedure. Interestingly, high levels of accuracy were maintained in the two sessions were the ResNet was tested for training adaptation abilities

Table 1. Combination of the different dataset and normalization criteria used during training and testing, accuracy on the test dataset, the second part relate to the experiments performed using the proposed domain adaptation strategy; all the results are relative to the ResNet model.

Session	Train.	Test	Norm. DS1	Norm. DS2	Acc.
S1	DS1	DS1	DS1	–	88.76%
S2	DS1	DS2	DS1	DS1	23.61%
S3	DS1	DS2	DS1	DS2	40.50%
S4	DS2	DS2	–	DS2	89.92%
S5	DS1 + DS2	DS1 + DS2	DS1 + DS2	DS1 + DS2	70.44%
S6	DS1 + DS2	DS1 + DS2	DS1	DS2	84.78%
S7	sDS1 + unsDS2	DS1 + DS2	DS1	DS2	89.92%
S8	unsDS1 + sDS2	DS1 + DS2	DS1	DS2	89.69%

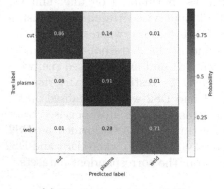

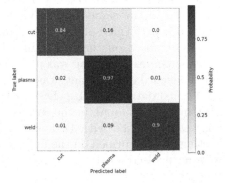

(a) Confusion matrix of S6

(b) Confusion matrix of S7 on testing subset derived from the target dataset

Fig. 3. Confusion matrices

(S7–S8). In these two last sessions, accuracy was as high as 89.92% and 89.69%. The confusion matrices reported in Fig. 3, revealed a tangible improvement when changing from the fully supervised to the partially-supervised approach.

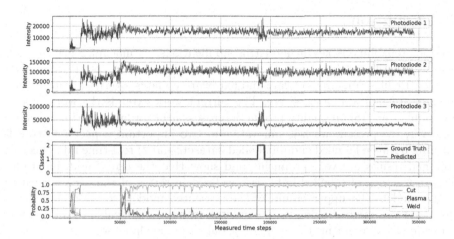

Fig. 4. Inference plot, the information reported in this plots are the sensors signals (network inputs), the labels and the prediction returned to the client in order to close the information loop and finally the network predictions.The sample is 6 mm thick aluminum.

To evaluate the network performances more practically the network predictions, labels, and sensor signals for the whole cut geometry are plotted together in Fig. 4. With these plots it is possible to appreciate the accuracy for the single geometry, this allows also to analyze the physical part and identify the critical situations. The value sent back to the client after the inference is performed is an integer representing the cutting class (cut quality) not the probabilities predicted of the network. But the probabilities are also reported: these are helpful to understand if the model is classifying the cut with high certainty or if the right or wrong prediction is a consequence of the maximization of the probabilistic prediction. The model tends to exclude the more distant class, this behaviour can be appreciated also from the confusion matrices where the top-right and bottom-left corners have near-zero probability value.

Deployment. The communication and inference times for 12 experiments are reported in Fig. 5. Each test is performed on 10000 inferences, thus corresponding to a process elapsed time of 50 s. The average time for both communication and inference was 2.3 ± 0.15 ms, well below the target time of 5 ms. In a few occasions (in the 0.02% of the total collected samples) the inference took longer to provide the results. The communication system must also deal with these outliers: if the result is not available for a set number of consecutive cycles the connection with the server is interrupted by the client and reestablished. This solution was selected because one missed step is acceptable but if multiple predictions are missed consecutively the system will crash and so the whole process must be automatically started again.

7 Discussion

Product reliability is key in industrial applications, as machines need to compensate for the different and often chaotic challenges that could occur across their entire lifespan, e.g.: software and hardware updates, overuse, wearing, environmental condition changes, etc. The current work extends the work done by Santolini et al. [18] on quality production assessment for laser cutting machines. However, the present work is innovative in two important aspects: 1) transfer learning between labelled and unlabelled datasets and 2) real-time inference. The obtained results are very promising and show that the use of additional data, despite the lack of labelled data, is beneficial in terms of the overall accuracy of the model. The upper bound of the presented models seems to be close to 90%, and this might be due to the annotation noise implacable to an operator-dependent disagreement. Unfortunately, only a unique label set was available, so such claim remains hypothetical and speculative. An additional observation is

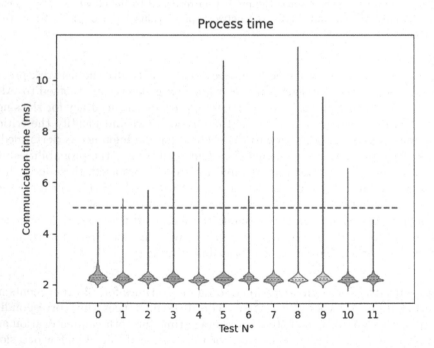

Fig. 5. The probability density of the data at different values (violin plot) is reported for a number of inference tests. Markers for the median of the data and indicating the interquartile ranges are reported. A red dashed line represents the upper limit to the computational time required (i.e. 5 ms). (Color figure online)

about the normalization strategy adopted during the training/testing sessions. It was noticed that model performance improved when per-dataset normalization was applied. This is reasonable, considering the different set of sensors employed in the two datasets: in the second campaign the adopted sensor appeared much more sensitive and they resulted in a more noisy signal spanning a slightly different range of magnitudes. In addition to the adaptation procedure, this paper also shows a comparison to a set of different well-known architectures, adapted to the specific task. Densenet and ResNet resulted to have similar accuracy, however, the latter has less parameters and therefore ensures a faster inference.

Finally, an important contribution of the study is the deployment of the model on the laser cutting machine. The complexity and potentially the maximal accuracy of the neural network was restrained by the maximal time allowed for computing the inference. As previously mentioned, the time limit of 5 ms has been set *a priori*. If the neural network was able to complete the inference within this time limit, then the CNC control loop could be close in due time. In the world of practice, with CNC controlling the laser beam at speeds as high as 30 m/min, an intervention time of 5 ms would allow sub-optimal cutting to occur for a length of about 0.5 mm (worst case scenario). Given this strict time constraint, a lot of energy has been invested in developing and optimizing a model-agnostic deployment framework, that allows deployment on multiple platforms and, thanks to the TensorRT^{TM} framework (i.e. NVIDIA® platforms only), to be independent of the training process. The final deployment has been made on a separated NVIDIA® hardware, to allow easy testing and better performance on industrial computers.

8   Conclusion

This work proved the feasibility of a neural network real-time deployment for an industrial application. In terms of technology readiness level, this research work does not only provide a proof of concept, but a component validated under controlled laboratory conditions. The neural network deployed here can complete the inference process at the ~89% of accuracy within 5 ms. The neural network has been proved to be robust against sensor and environmental changes and displayed the ability to adapt and transfer to new datasets without the need for an additional labelling campaign. This work opens up new avenues for future research in the field of partially and fully unsupervised models operating in industrial environments. However, to eliminating the labelling process, further testing is warranted.

Acknowledgment. The authors would like to thank Adige S.p.A. for the opportunity of working on such interesting research project, for their expert advice and collaborative flair. We would like to thank ProM Facility for providing computational resources and support that have contributed to these research results.

The project presented in this paper has been funded with the contribution of the Autonomous Province of Trento, Italy, through the Regional Law 6/99. Name of the granted Project: LT4.0.

References

1. Alippi, C., Bono, V., Piuri, V., Scotti, F.: Toward real-time quality analysis measurement of metal laser cutting. In: 2002 IEEE International Symposium on Virtual and Intelligent Measurement Systems (IEEE Cat. No. 02EX545), pp. 39–44. IEEE (2002)
2. Anicic, O., Jović, S., Skrijelj, H., Nedić, B.: Prediction of laser cutting heat affected zone by extreme learning machine. Opt. Lasers Eng. **88**, 1–4 (2017)
3. Bianco, S., Cadene, R., Celona, L., Napoletano, P.: Benchmark analysis of representative deep neural network architectures. IEEE Access **6**, 64270–64277 (2018)
4. Carlucci, F.M., Porzi, L., Caputo, B., Ricci, E., Bulo, S.R.: Autodial: automatic domain alignment layers. In: 2017 IEEE International Conference on Computer Vision (ICCV), pp. 5077–5085. IEEE (2017)
5. Das, S.K., Das, S.P., Dey, N., Hassanien, A.-E. (eds.): Machine Learning Algorithms for Industrial Applications. SCI, vol. 907. Springer, Cham (2021). https://doi.org/10.1007/978-3-030-50641-4
6. Dong, J., Cong, Y., Sun, G., Zhong, B., Xu, X.: What can be transferred: unsupervised domain adaptation for endoscopic lesions segmentation. In: Proceedings of the IEEE/CVF Conference on Computer Vision and Pattern Recognition, pp. 4023–4032 (2020)
7. Franceschetti, L., Pacher, M., Tanelli, M., Strada, S.C., Previtali, B., Savaresi, S.M.: Dross attachment estimation in the laser-cutting process via convolutional neural networks (CNN). In: 2020 28th Mediterranean Conference on Control and Automation (MED), pp. 850–855. IEEE (2020)
8. Glorot, X., Bordes, A., Bengio, Y.: Domain adaptation for large-scale sentiment classification: a deep learning approach. In: ICML (2011)
9. Halm, U., Arntz-Schroeder, D., Gillner, A., Schulz, W.: Towards online-prediction of quality features in laser fusion cutting using neural networks. In: Arai, K., Kapoor, S., Bhatia, R. (eds.) IntelliSys 2020. AISC, vol. 1250, pp. 346–359. Springer, Cham (2021). https://doi.org/10.1007/978-3-030-55180-3_26
10. He, K., Zhang, X., Ren, S., Sun, J.: Deep residual learning for image recognition. In: Proceedings of the IEEE Conference on Computer Vision and Pattern Recognition, pp. 770–778 (2016)
11. Huang, G., Liu, Z., Van Der Maaten, L., Weinberger, K.Q.: Densely connected convolutional networks. In: Proceedings of the IEEE Conference on Computer Vision and Pattern Recognition, pp. 4700–4708 (2017)
12. Jurkovic, Z., Cukor, G., Brezocnik, M., Brajkovic, T.: A comparison of machine learning methods for cutting parameters prediction in high speed turning process. J. Intell. Manuf. **29**(8), 1683–1693 (2018)
13. Larrañaga, P., Atienza, D., Diaz-Rozo, J., Ogbechie, A., Puerto-Santana, C.E., Bielza, C.: Industrial Applications of Machine Learning. CRC Press, New York (2018)
14. Li, X., Zhang, W.: Deep learning-based partial domain adaptation method on intelligent machinery fault diagnostics. IEEE Trans. Ind. Electron. (2020)
15. Munro, J., Damen, D.: Multi-modal domain adaptation for fine-grained action recognition. In: Proceedings of the IEEE/CVF Conference on Computer Vision and Pattern Recognition, pp. 122–132 (2020)
16. Roy, S., Siarohin, A., Sangineto, E., Bulo, S.R., Sebe, N., Ricci, E.: Unsupervised domain adaptation using feature-whitening and consensus loss. In: Proceedings of the IEEE Conference on Computer Vision and Pattern Recognition, pp. 9471–9480 (2019)

17. Santolini, G.: Deep Learning Models for Cut Interruption Detection in Laser Cutting Machines. Master's thesis, University of Trento (Department of Industrial Engineering), Trento (2019)
18. Santolini, G., Rota, P., Gandolfi, D., Bosetti, P.: Cut quality estimation in industrial laser cutting machines: a machine learning approach. In: Proceedings of the IEEE Conference on Computer Vision and Pattern Recognition Workshops (2019)
19. Shan, Y., Lu, W.F., Chew, C.M.: Pixel and feature level based domain adaptation for object detection in autonomous driving. Neurocomputing **367**, 31–38 (2019)
20. Szegedy, C., et al.: Going deeper with convolutions. In: Proceedings of the IEEE Conference on Computer Vision and Pattern Recognition, pp. 1–9 (2015)
21. Tercan, H., Al Khawli, T., Eppelt, U., Büscher, C., Meisen, T., Jeschke, S.: Improving the laser cutting process design by machine learning techniques. Prod. Eng. **11**(2), 195–203 (2017)
22. Tzeng, E., Hoffman, J., Saenko, K., Darrell, T.: Adversarial discriminative domain adaptation. In: Proceedings of the IEEE Conference on Computer Vision and Pattern Recognition, pp. 7167–7176 (2017)
23. Zhang, Y., David, P., Gong, B.: Curriculum domain adaptation for semantic segmentation of urban scenes. In: Proceedings of the IEEE International Conference on Computer Vision, pp. 2020–2030 (2017)
24. Zhao, S., et al.: Multi-source domain adaptation for semantic segmentation. In: Advances in Neural Information Processing Systems, pp. 7287–7300 (2019)

An Online Deep Learning Based System for Defects Detection in Glass Panels

Matteo Moro[1,2(✉)], Claudio Andreatta[2], Chiara Corridori[2], Paolo Rota[1], and Niculae Sebe[1]

[1] University of Trento, 38123 Trento, Italy
matteo.moro@alumni.unitn.it
[2] Deltamax Automazione Srl, Via di Spini, via Kufstein, 5, 38100 Trento, Italy
info@deltamaxautomazione.it

Abstract. Automated surface anomaly inspection for industrial application is assuming every year an increasing importance, in particular, deep learning methods are remarkably suitable for detection and segmentation of surface defects. The identification of flaws and structural weaknesses of glass surfaces is crucial to ensure the quality, and more importantly, guarantee the integrity of the panel itself. Glass inspection, in particular, has to overcome many challenges, given the nature of the material itself and the presence of defects that may occur with arbitrary size, shape, and orientation. Traditionally, glass manufacturers automated inspection systems are based on more conventional machine learning algorithms with handcrafted features. However, considering the unpredictable nature of the defects, manually engineered features may easily fail even in the presence of small changes in the environment conditions. To overcome these problems, we propose an inductive transfer learning application for the detection and classification of glass defects. The experimental results show a comparison among different deep learning single-stage and two-stage detectors. Results are computed on a brand new dataset prepared in collaboration with Deltamax Automazione Srl.

Keywords: Machine vision · Deep learning · Glass panels · Industrial application

1 Introduction

In automated industrial processes, the quality inspection of the product is one of the most important task for the assessment of the final product quality. Glass manufacturers have traditionally carried out the quality control manually, employing specialized workers trained to identify complex surface defects. This approach, however, implies serious timing and efficiency limitations. The automation of such process has been partially discussed in few machine vision works using traditional machine learning techniques such as: SVM, kNN and Random Forest (Tsai et al. [1], Paniagua et al. [2], He et al. [3], Bulnes et al. [4]). However, custom heuristics and extensive fine tuning in machine vision algorithms do not follow

© Springer Nature Switzerland AG 2021
A. Del Bimbo et al. (Eds.): ICPR 2020 Workshops, LNCS 12664, pp. 506–522, 2021.
https://doi.org/10.1007/978-3-030-68799-1_37

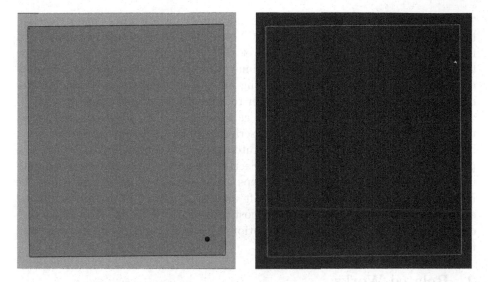

Fig. 1. Acquired images with different inspection lights, on the left a bright field acquired image, on the right a dark field acquired image.

the Industry 4.0 paradigm where fast adaptability and generalization are cornerstones (Oztemel et al. [5]). Data-driven algorithms such as deep learning algorithms allows to develop flexible solutions, shifting the focus from the definition of handcrafted features to the selection of a suitable deep learning model and the data processing.

This work aims at investigating different well known deep learning detection methods applied on our specific industrial application. This paper compares two different detection approaches, i.e., four *region-proposal based* models (Faster R-CNN, Mask R-CNN) and two *regression based* models (YOLO) in order to determine the more suitable approach for the detection task on glass panels images. In addition, we have defined a *custom pipeline* for detection and classification tasks performing a *channel shift analysis* with the purpose of defining the requirements for a viable early fusion approach with different illumination source, confirming the detection improvement as a result of the information increment given by the multi-channel acquisition. Selected models are compared, not only for the overall classification performance, but also for the computational cost. An extensive evaluation is performed on the selected models over a *novel dataset* acquired in collaboration with Deltamax Automazione Srl. The data are sampled over real world acquisition on working systems, this entails the opportunity to perform training and evaluation over non-synthetic data, but also involves a significant imbalance between defect classes given the different occurrence probability of each defect during glass manufacturing. On the studied domain, the two-stage model Mask R-CNN ResNet-50 represents a solid choice for glass defect detection task, reaching nearly 95% in detection with a feasible computational time of 36 ms per image. On the other hand, YOLO has shown a detection score of just

72%, however it presents the advantage of a significant inference time reduction, being 3x faster compared to Mask R-CNN ResNet-50. As already pointed out in other works (Jing and Dai [6], Liu et al. [7], Pham et al. [8]), it is possible to detect small objects on large scale images, however, unfortunately the small amount of data does not allow reaching an adequate detection score. Further investigation will be conducted in order to improve the current dataset.

The rest of the paper is presented as follows: Sect. 2 collects related works regarding surface detection. A brief description of the dataset and the currently implemented algorithm by Deltamax Automazione Srl is provided in Sect. 3. The proposed methodology is outlined in Sect. 4, where the overall system configuration and the main preprocessing steps are described. Section 5 provides the results obtained with the channel shift analysis together with the conducted experiments in order to establish the most suitable DNN model. The paper concludes with Sect. 6 with final considerations on the overall system proposed and future prospects.

2 Related Works

In the last decade, the advent of Deep Learning methods applied to the industrial context has brought the advent of Industry 4.0. The automation has been progressively outsourced to Machine Learning algorithms, in particular after the introduction of deep neural networks such as AlexNet (Krizhevsky et al.) [9] and pioneer industrial applications such as the work of Masci et al. [10]. These first applications have proven that it is possible to adopt neural networks algorithms to image detection and classification tasks. In particular the application of convolutional neural networks in automation context like surface defect detection outperforms any traditional machine vision approaches such as Random Tree Forest [11], Support Vector Machines [12], K-Nearest Neighbors algorithms [13]. In addition, deep neural network architectures do not require handcrafted features for the investigated domain, allowing to spare time and better generalize the problem with the unsupervised feature extraction.

In this paper we focus on automatic quality inspection for glass panels and there are only few related works in literature.

In the past, several works such as Zhao et al. [14] have used SVMs and classical machine vision algorithms to address the defect detection problem. These works rely on manual feature extraction, such as the foreground binarization, and assume very restrictive conditions in order to successfully work. However, the real world conditions are clearly different than the research environment in which these methods were developed.

Recently, Tabernik et al. [15] proposed a DNN architecture that allows to performs segmentation of the image surface and then passes the extracted features to the decision network in order to detect surface cracks. This approach has led to improved state of the art results, however the proposed method focuses only on crack detection and it performs inference evaluation over images of 1408 × 512 px with a time of 120 ms. However, this time-frame cannot satisfy the required standard for an online application.

A recent work by Park et al. [16] proposes a multi-DNN inspection system (4-DarkNet) for smartphone glasses that allows to exploit multi channel data for the detection and classification of several defect classes. This study permits to underline the importance of the image sources, in particular dark bright fields, described in Sect. 5.2. Although our work is related to this paper, the study proposed in Park et al. [16] does not provide a method for the image patch extraction, in fact the proposed regions are already generated by a proprietary algorithm of the smartphone manufacturer. Additionally each image channel is treated separately, in fact the proposed system needs to train 4 different models and then combines the prediction scores with a late fusion approach [17]. This approach is computationally expensive, giving as a result a system that is not applicable to large glass panels inspection.

The main purpose of our study is to provide a more general and computationally efficient application, exploiting the state of the art convolutional neural network architectures.

3   Problem Description

This section contains a brief description of the actual pipeline algorithm that is implemented on the Deltamax systems and an overview of the dataset employed for this work. The proposed solution must be able to analyze the entire panel in order to perform the defect detection within a particular time. Two main camera scanning speeds are given:

- *standard* production line velocity = 20 m/min, resolution = 0.2 mm/px: scan time for 800 px rows ≈ 0.5 s
- *demanding* production line velocity = 40 m/min, resolution = 0.1 mm/px: scan time for 800 px rows ≈ 0.12 s

3.1   Deltamax Detection System

The glass inspector system analyzes images of flat glass in order to spot defects such as scratches, marks and halos. The system relies on image analysis and computer vision algorithms that are well known in literature. The chain of processing elements consists of:

- noise removal algorithms, such as flat field correction, to cancel the effects of image artifacts caused by variations in the pixel-to-pixel sensitivity of the detector, by distortions in the optical path and by the non uniformity of the light source;
- blob extraction with adaptive threshold algorithms;
- blob features computation, the system characterizes the image regions with *geometrical* (height, width, bulkiness) and *color features* (contrast, gray level histograms);
- feature based blob classification with a fully connected multi layer perceptron.

Moreover, in order to identify and classify a specific kind of defects the system has an array of per-defect detectors followed by a weighted fusion to produce the final result.

3.2   CLEVER Glass Dataset

The dataset employed in this work is provided by Deltamax Automazione Srl[1] and it consists of a set of glass images, acquired by a group of line scan cameras, organized in arrays of 2 to 5 cameras. The images are acquired with 2 different inspection lights, in particular two illumination fields are exploited: bright-field and dark-field (Fig. 1). Bright-field imaging is the most typical technique used for microscopy illumination, it employs direct light, giving as a result a bright glass background with an emphasized dark defect. Instead in dark-field imaging illumination technique, the light does not go directly into the camera sensor, giving as a result a dark glass background with a bright defect. The dimensions of the glass sheets can be diverse, given the different application context: it can reach 1 m for the household appliance sector, between 2/2.5 m for the automotive sector, and up to 5 m for the architectural sector. The camera system resolution acquisition for 1 pixel is 0.1 mm, giving as a result for the largest glass panel to a 50000 pixels rows image. This implies a careful management of the computational resources for the adopted solution.

The entire dataset consists of 407 images with different resolution and visible in different lights conditions (Table 1) [18], for a total of 2410 labelled defects. For the experiments the dataset is partitioned in two groups, 90% for the training and 10% for the test, each partition is pseudo randomized with a seed that also determine and distinguish the trained models. It should be noted that the dataset images are acquired in two different modality: *single channel* and *multi channel*. In single channel mode the image is acquired only in bright field, meanwhile in multi channel mode the image is acquired both in bright and dark field. The dataset contains an heterogeneous range of defect sizes, starting from 2×2 px for 'dirt points' defects, up to 2000×2000 px for 'halo' defects. Furthermore, each defect presents in the dataset, has a segmented binary mask.

Despite the facts that there is not a standard in glass defects classification, we tried to define it. The main aims of defects classification for glass surface are:

– the assignment of a *severity index* of the defect based on defined quantities relating to the class (e.g. scratches length, bubble diameter)
– the distinction between *removable defects* with cleaning and *permanent defects*. This distinction is essential for systems that are not supervised by the online operator.

Glass defects can be very diverse from each other, we focus on aesthetic defects that are located inside the glass or on the surface. All the defects inside the glass (e.g. bubbles, inclusions) are permanent defects; some defects on the glass surface can be removed (e.g. dirt) although some are permanent (e.g. scratches). See Table 1.

The dataset is partitioned in 25 unbalanced defect classes, organized in 6 groups by defects similarity, for example 'scratch light' and 'scratch heavy' will belong to the same group class. Then, to each defect is assigned an *original class label* and a *group class label*. The list of original defects is presented in Table 1.

---

[1] http://www.deltamaxautomazione.it/.

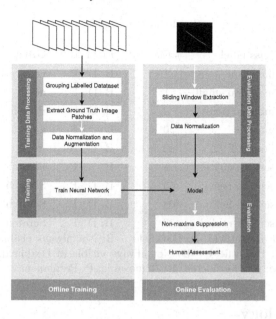

Fig. 2. Offline training and online evaluation sections.

Table 1. Deltamax automazione srl glass data

Class label	#Elements	Field	R/P	Group
'break'	3	B, D	P	nc
'bubble'	84	B, D	P	Bubble
'bubble hole'	13	B, D	P	Bubble
'bubble small'	77	B, D	P	Bubble
'chip'	1	B, D	P	nc
'coating'	55	B	P	nc
'dirt'	36	B, ~D	R	nc
'dirt halo'	122	~B, D	R	Halo
'halo'	121	~B, D	R	Halo
'halo points'	4	~B, D	R	Halo
'dirt p multi'	165	B, D	R	Dirt
'dirt point'	75	B, D	R	Point
'dirt point small'	412	B, D	R	Point
'dirt td'	42	B	R	nc
'dust'	17	B, D	R	nc
'glass id'	99	B, D	R	Glass ID

(Continued)

Table 1. *(Continued)*

Class label	#Elements	Field	R/P	Group
'inclusion'	1	B, D	P	nc
'inclusion small'	1	B, D	P	nc
'internal point'	2	B, D	P	Bubble
'mark'	26	B	R	nc
'point td'	42	B, D	R	Point
'scratch coating'	8	B	P	Scratch
'scratch heavy'	53	B, D	P	Scratch
'scratch light'	308	B, D	P	Scratch
'scratch multi'	38	B, D	P	Scratch

B: Brightfield, D: Darkfield, ~B: Not always visible
in Brightfield, ~D: Not always visible in Darkfield,
nc: not considered, R: Removable, P: Permanent

4   Methodology

In this section we provide an overview of the proposed pipeline for the offline
training and the online inference of the glass defects. In addition, the preprocessing will be outlined in detail as well as the data augmentation, given the
unbalanced nature of the dataset.

4.1   Overall System Configuration

Figure 2 shows the proposed approach, that is agnostic with respect to the considered DNN models. The overall system is composed of two main sections:
an *offline training* section and an *online evaluation* section. The offline training
section employs the labelled defect dataset; in the fist step the dataset defects are
grouped following the defined grouping rules, then each defect patch is extracted
and cropped, generating an image patch with the size of $s = 800 \times 800$ px. The
cropped patch is subsequently normalized and passed to the data augmentation
module. The augmented image patch is finally passed to the DNN architecture.

The chosen reference architectures in this work are Mask R-CNN and YOLO.
Mask R-CNN is a deep neural network that aims to solve instance segmentation
task in computer vision. It is composed by two main stages: the first stage
generates the proposal regions to inspect and detect the objects based on the
input dataset, the second stage aims to predict the class of the object, also giving
in output the bounding box and the binary mask. The main reason behind the
choice of this model relies on the fact that Mask R-CNN is the state of the
art deep learning architecture for the detection task. Furthermore our dataset is
provided with a binary segmentation mask that can be employed during training
in particular by this DNN model.

YOLO is a single-stage regression based algorithm (SSD). Single stage models consists of two parts: an SSD backbone model and an SSD head. In our study, the selected version of YOLO backbone is CSPNet [19]. In general, SSD backbones can be the same as for two-stage models but without the last fully connected classification layers that are replaced by the SSD head. During inference, there is no image ROIs selection, instead, the bounding boxes detection and the classes prediction are obtained for the image at once. This approach is more convenient for a real time application, compared to a two-stage detector. In Sect. 5 we compare the computational time and the accuracy of these two approaches.

In the last step the neural network classifier exploits all training examples to learn the model via backpropagation algorithm. The training section is not related to any runtime constraints since it runs in an offline mode.

During the online evaluation section, the inference on the tests samples is realized with a sliding window approach. The entire image I is partitioned in N patches W, where $W_1, W_2, ..., W_N$ are the investigated overlapping blocks with the same size of the training samples. Between W_i and W_{i+1} we perform a step of $s/2$ in each axes directions, allowing to detect the defects that could occur on the edge of W_i. The patch W_i is then normalized and passed to the learned model in order to detect potential glass defects. Each detected defect bounding box related to a window W_i is saved considering its relative position, therefore the local x-y coordinates are respectively summed with Δx and Δy in order to refer back to the entire image I coordinates system. Therefore, a non-maxima suppression algorithm is applied to the overall output list of bounding boxes, in order to reduce the multiple detection. A final human assessment step is performed in order to evaluate the accuracy of the model predictions.

4.2   Preprocessing

The image data preprocessing is one of the most important and challenging steps of glass surface defect detection. Once acquired, the overall image can be affected by non-uniform illumination, for this reason a background light correction algorithm is applied in order to mitigate this condition. Furthermore, in presence of an imperfect transportation system, the acquired image can present local deformations, for this case it is necessary to apply a local correction algorithm, in particular in the presence of serigraphy.

Given the diversity of the defects, it is necessary to carefully manage the bounding boxes that localize the defects, according to the deployed DNN architecture. The bounding boxes present in the dataset can reach dimensions that are bigger than the maximum dimension of the image patch. In order to avoid a down scaling of the defect, given also the presence of tiny defects, all the bounding boxes that are larger than 800 pixel, are split in sub-defects, maintaining the same original class and group label. Since the dataset is composed by single and multi channel images, the model must be able to learn features in both cases. For this reason, each pixel value x is scaled between 0 and 1, then each channel image x is normalized with its own mean and standard deviation. It should be noted that mean and standard deviation are previously calculated individually for each channel. In multi channel acquisitions, the segmentation binary mask could be slightly different, this depends on the visibility of the defect for each channel. In order to avoid information loss and, given the 2D requirement for the mask, the relative channel masks are combined together applying the union operation to the binary matrix.

The unbalanced nature of the dataset requires to filter out the classes that have a low number of items. All the training is performed only on sufficiently populated groups, this is done in order to avoid overfitting the model. A common preprocessing technique applied to the dataset involves the augmentation of the existing data with perturbed versions of the existing images. Data augmentation is then performed with the following transformations: scaling, rotations, warping, blurring, noise injection. This procedure is done in order to better generalize the defect detection performed by the neural network. This process allows to decrease the number of false positives and avoid to overfit the model.

5   Results

In this section we present the results produced with the dataset. All the quantitative results in this section are computed over the same class grouping, the reported results are the average of 5 repetitions for each experiment. The hyperparameters obtained during the experiments have been maintained for all the successive steps, in particular the choice of the order and distribution image channels parameters.

All the obtained results rely exclusively on the average detection and classification metrics, we avoided the most common metrics including the well-known average precision [20]. The main reason behind this choice relies on the nature of the detected objects. In particular, larger defect classes, such as "Glass ID", can achieves an average precision box of $AP_{75} = 39.4$, meanwhile small defects like "Point" only reach $AP_{75} = 15.2$. Despite this it is possible to notice in Table 3 that the "Point" group is one of the most accurately detected classes. In fact, the dataset is partially labeled, a consistent number of false positives are in fact, real defects. Secondly, the IoU score between small ground truth defects with the neural network proposed box can drops easily even with a slight misalignment. Therefore it is not possible to rely on the F1 score or Precision value. For these reasons, all the results and comparisons are based on average values of detection, classification and recall.

In this paper we present a list of experiments in order to determine which combination of backbones and head best fits for the glass defect detection task. The first main experiment presented in Sect. 5.1 aims at determining the best architecture configuration that provides better results in terms of detection and classification, also considering the time constraints presented in Sect. 3 for the production line. We present in detail the results obtained with the chosen architecture. Secondly, given the unusual image sources compared to the ImageNet pictures used to pre-train the neural networks weights, an experiment is proposed to propose the best channel fitting [21]. We then discuss the obtained quantitative results in Sect. 5.3. Section 5.4 presents the qualitative results and therefore the advantages of using multiple channels for the defect detection.

The experimental results were obtained on a single GPU NVIDIA GeForce RTX 2080Ti 11 GB with a CPU Intel(R) 32 Core(TM) i9-7960X @2.80 GHz, and 64 GB of RAM.

5.1   Architecture Selection

This experiment is meant to establish the most efficient and accurate architecture, so we have considered different two-stage and single-stage detectors, namely: Faster R-CNN ResNet-50 [22], Faster R-CNN ResNet-101, Mask R-CNN ResNet-50 [23], Mask R-CNN ResNet-101, YOLOv5-L and YOLOv5-X [24] with custom CSPResNet backbone [19]. The selected architecture must be capable to infer a block of N sliding windows approximately with the estimated acquisition time on the production lines. In this way, it is possible to analyze large glass panels without slowing down the production.

In this section are listed the DNN architectures that are taken into consideration for the glass defect detection, each architecture is publicly open implemented[2,3]. The hyper-parameters for Mask R-CNN and Faster R-CNN models have been set as follows: learning rate $= 10^{-5}$, weight decay $= 10^{-4}$, gamma $= 0.955$, using Adam optimizer. Each model has been trained for 100 epochs. For YOLOv5 models we have left the hyper-parameter unchanged, each model has been trained for 300 epochs. In Table 3 it is possible to observe the comparative results among different architectures.

The graph in Fig. 3 compares different batch size for the inference, each value is calculated over 7687 sliding windows with the size of 800×800 and an evaluation batch size of 16. Table 2 shows a comparison among the average detection

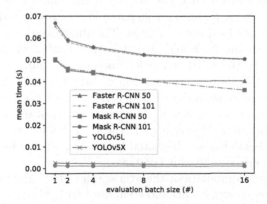

Fig. 3. Inference time for different inference batch size.

[2] https://github.com/facebookresearch/maskrcnn-benchmark.
[3] https://github.com/ultralytics/yolov5.

Table 2. Overall average detection and classification models comparison

Head	Backbone	Avg detection		Avg classification		Avg recall	Time*
		Mean	Std	Mean	Std		
YOLOv5-L	CSPResNet	65.08%	2.23	85.45%	3.56	64.35%	0.011
YOLOv5-X	CSPResNeXt50	72.16%	2.28	83.60%	2.76	68.97%	0.015
MasksRCNN	ResNet-50	93.82%	1.13	87.03%	2.89	81.63%	0.036
MasksRCNN	ResNet-101	93.67%	1.95	86.97%	3.99	81.41%	0.050
FasterRCNN	ResNet-50	93.18%	1.92	87.14%	3.04	81.10%	0.040
FasterRCNN	ResNet-101	94.18%	1.68	88.69%	3.35	83.78%	0.050

*time standard deviation: $SD = 10^{-5}$.

and classification for each tested neural network, giving also the average inference time using an evaluation batch size of 16, which has proven to be the most convenient value for the online inference.

In this section we also present the classification results obtained with Mask R-CNN ResNet-50 architecture. The results in Table 4 are the absolute values that are obtained with a single model, the last column displays the amount of *false positives* detected over the entire test set, meanwhile the last row shows the amount of *detected ground truth defects* with respect to the test set. Table 5 shows the average classification in percentage, calculated over 5 runs for the Group labels listed in Table 1. The overall average detection is 93.82% and the overall average classification is 87.03%.

5.2   Channel Shift Analysis

This analysis regards the fine tuning feasibility for the glass dataset, starting from the model pretrained on ImageNet. The first analysis is performed with Mask R-CNN ResNet-50, for each image patch W_i the *order* and the *distribution* of the bright and dark field are evaluated. The dimension of a single patch are $800 \times 800 \times 3$, it should be noted that 3 is the number of channels that are normally assigned to the RGB channels. Therefore in our case the 3 channels of W_i are filled with different combination of bright and dark field images. In Sect. 5 we refer to the image order with an array of 3×1 elements, each position of the array represents one channel. The array can contain 3 different values: '0' indicates the bright field channel, '1' the dark field channel, '–1' refers to the inverted dark field channel. It should be noted that the inverted channel is calculated with a center point value of 128.

For each channel combination, the average detection and classification percentage are estimated over 5 runs and computed with Mask R-CNN ResNet-50 architecture. In the first two rows in Table 6 we compare the result of the test performed with two different image order with the same distribution, in particular: 2 bright field and 1 dark field. Table 6 also presents the average detection and classification of different combination of bright field and dark field.

5.3   Discussion

In this section we discuss the obtained results described in the previous section. In the model selection test we have analyzed six DNN models, the results are shown in Table 2. Contrary to the expectations, our findings suggest that the average detection is similar for all the two-stage architectures, considering the classification values. Single-stage models such as YOLOv5-L and YOLOv5-X instead do not perform well during the detection task in particular for small defects such as "Point" and small "Scratch". Based on this fact, we do not consider single-stage models for further tests. Under the computational complexity, the models show a significant difference, in particular, comparing the results of the two most promising architectures, namely: Mask R-CNN ResNet-50 and Faster R-CNN ResNet-101. The average inference time for a single image patch is 0.0361s for Mask R-CNN ResNet-50 and 0.0501s for Faster R-CNN ResNet-101. Given the online inference constraints presented in Sect. 3, we have chosen Mask R-CNN ResNet-50 as our final architecture. There are evidences to suggest that inference performance increases with batch sizes that are power of 2, therefore we expect that with a batch size of 32 image patches the inference time performance can be additionally improved. Unfortunately the amount of memory of the NVIDIA GeForce RTX 2080Ti GPU does not allow such tests. Given the current results, we can estimate that for a production line with an array of 3 cameras, considering a step between each sliding window of $s = s/2$, it is possible to sustain the standard production velocity specified in Sect. 3. The previous model selection tests have fulfilled the expectation in terms of accuracy, these results have established the basis for the channel shift analysis study. This tests have the purpose to establish the feasibility to apply a standard image detector to glass images, in particular it is aimed to define the order and the distribution of the bright and dark fields. The results have highlighted that the channels order does not affect the detection, meanwhile the channels distribution has been proven to influence the detection value. On average, from the results listed in Table 6, it appears that the most promising image order is $[0, -1, -1]$.

The capability to train a model that can perform good prediction with both the channel acquisition methods (namely: single and multi channel acquisition), plays a vital role in the development of the glass defect detector, for this reason

Table 3. Average class detection comparison for the selected models

Method	Bubble	Dirt	Glass ID	Halo	Point	Scratch
YOLOv5-L	100%	39.08%	100%	44.34%	76.38%	39.24%
YOLOv5-X	100%	55.24%	100%	54.82%	80.18%	42.68%
MasksRCNN-50	100%	78.5%	100%	95.89%	91.82%	94.66%
MasksRCNN-101	97.64%	78.34%	100%	92.05%	93.34%	95.04%
FasterRCNN-50	98.89%	78.52%	100%	94.58%	90.24%	94.99%
FasterRCNN-101	98.89%	81.49%	100%	94.68%	93.76%	93.84%

Table 4. Detection and classification expressed in absolute values for a single Mask R-CNN RESNET-50 model model

	Bubble	Dirt	Glass ID	Halo	Point	Scratch	FP
Bubble	19	0	0	0	3	0	18
Dirt	0	5	0	2	1	1	119
Glass ID	0	0	7	0	0	0	7
Halo	0	0	0	18	0	0	55
Point	2	3	0	1	47	1	223
Scratch	0	3	0	0	1	44	136
Detection	21/21	11/13	7/7	21/21	52/55	50/53	

FP: False Positive.

Table 5. Average detection and classification confusion matrix example for a single Mask R-CNN RESNET-50 model

	Bubble	Dirt	Glass ID	Halo	Point	Scratch
Bubble	90.35%	2.5%	0%	0%	13.62%	0%
Dirt	0%	59.81%	0%	7.35%	0.84%	2.63%
Glass ID	0%	0%	100%	3.2%	0%	0%
Halo	0%	4.44%	0%	82.54%	0.43%	2.55%
Point	9.65%	19.73%	0%	0.95%	83.39%	0.4%
Scratch	0%	13.51%	0%	5.95%	1.72%	94.42%
Detection	100%	78.50%	100%	95.89%	91.82%	94.66%

much attention has been drawn to the early fusion criteria. Given the results for the domain adaptation and the groups profiling, we have compared the results of two models with the Mask R-CNN ResNet-50 architecture. Our experiments confirm that the addition of the dark field channel allows to detect more defects than the exclusive exploitation of the bright field. In particular this results is expected, given the nature of several defects, as specified in Table 1.

The detection results for Mask R-CNN ResNet-50 shown in Table 3 are in general, significantly positive, as is the classification performance shown in the confusion matrix in Table 5. The classification of the class "Dirt" is noticeably lower than the others. There are several possible explanations for this finding, in particular the nature of this class, in fact, "Dirt" and "Point" classes are both removable particles that only differ in the number and density of the particle itself. To the best of our knowledge, in order to mitigate this condition, further data collection is required.

Table 6. Channels order and distribution comparison, the model is a Mask R-CNN trained with a ResNet-50 backbone.

Channel order	Avg detection		Avg classification		Avg recall
	Mean	Std	Mean	Std	
[0, 1, 0]	92.88%	1.86	87.15%	2.23	78.81%
[0, 0, 1]	92.96%	1.73	86.40%	2.29	80.29%
[0, 1, 1]	93.80%	2.12	85.14%	3.62	79.81%
[0, 0,−1]	92.35%	1.74	85.04%	3.16	78.54%
[0,−1,−1]	93.82%	1.13	87.03%	2.89	81.63%

0: Bright field, 1: Dark field, −1: Inverted dark field

As stated for the results displayed in Table 4, the detection values are significantly high, nonetheless the number of false positives could bring to the conclusion that the selected model is not robust against noise or light variation. The qualitative results (presented in the next section, see Fig. 4) have shown that a large percentage of false positives are in fact unlabeled true defects. However, in several glass panels the detection of the class "Halo" has shown a consistent number of false positives, nonetheless this problem can be limited with an higher threshold for this specific class.

5.4   Qualitative Results: Single-Channel, Multi-channel

In this section we present the qualitative results obtained with Mask R-CNN ResNet-50 model. We compare the results obtained training a model only with the bright field channel to a model trained with multi-channel images along with single-channel bright field images. The results are compared over a single run for each method. For the sake of completeness we provide in Table 7 a numerical comparison between the two models. A significant qualitative confrontation can be done on the resulting images shown in Fig. 4 obtained with Mask R-CNN ResNet-50 architecture with the previously mentioned methods: single-channel model and multi-channel model. It should be noted that all the tests discussed in Sect. 5.2 concern only the multi-channel models. For each image sample we compare the results obtained with the two models.

In Fig. 4a it is possible to notice, on the left image, the detection performance of the single-channel model, meanwhile in the center and right images the detection of the multi-channel model. The comparison focus on the two ground truth "Halo" defects show the misdetection of the single-channel model. The crop shown in Fig. 4b depicts an illumination variation during the camera acquisition. The image on the top shows a *false positive* detected by the single-channel model, whereas the multi-channel model does not report the improper detection. Figure 4c shows on the left image, the detection of an "Halo" defect computed by the single-channel model. Although the detection made by the single-channel model is partially correct, in fact the inferred bounding box is not aligned with the ground truth. The center and the left images in Fig. 4c shows a more accurate detection of the "Halo" defect. It is also possible to notice that the ground truth "Point" is detected only by the single-channel model. It appears that multi-channel models are able to detect lightest "Halo" defects.

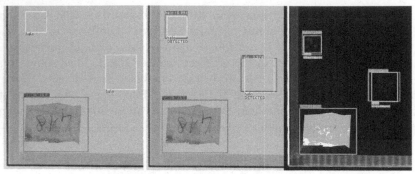

(a) On the left: single-channel model inference. In the center and in the right images: multi-channel model inference (enhanced dark channel)

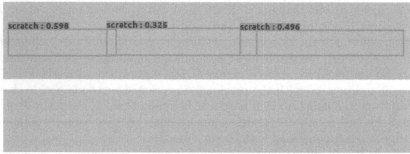

(b) False positive sample. Top image: single-channel model inference. Bottom image: multi-channel model inference

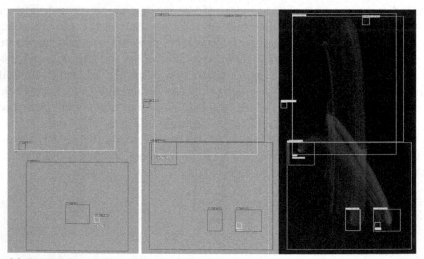

(c) On the left: single-channel model inference. In the center and in the right images: multi-channel model inference (enhanced dark channel)

Fig. 4. Qualitative comparison of meaningful detection between a model trained with single-channel images only and one trained with multi-channel + single-channel images

Table 7. Quantitative comparison of the detected ground truth bounding boxes for a model trained with single-channel images only and a model trained with multi-channel + single-channel images

Model	Detection
Mask R-CNN SCI	61.20%
Mask R-CNN MCI + SCI	95.30%

SCI: Single-Channel Images, MCI: Multi-Channel Images

6 Conclusion

The findings of this study indicate the possibility to apply deep learning techniques to the glass defect detection problem. The developed method allows to be more efficient compared to a traditional machine learning approach; handcrafted glass defects features do not satisfy the necessary level of generalization that can be provided instead with a deep learning approach. The obtained results have revealed promising prospects from the application point of view, in particular we have performed several experiments over a new dataset, providing results regarding the robustness relative to the environment condition variation and the adaptability of the proposed method to different image sources. We have presented the results of different architectures, underlying the detection and classification results. In conclusion we can assert that the developed application provides adequate performance. However, given the small sample size, caution must be taken and further analysis has to be done. From the viewpoint of future developments, the accuracy of classification could be further improved. Meanwhile, an improvement to the dataset, in particular the annotation of the image defects must be done with higher precision and efficiency.

References

1. Tsai, D., Chiang, I., Tsai, Y.: A shift-tolerant dissimilarity measure for surface defect detection. IEEE Trans. Industr. Inf. **8**(1), 128–137 (2012)
2. Paniagua, B., Vega-Rodríguez, M.A., Gomez-Pulido, J.A., Sanchez-Perez, J.M.: Improving the industrial classification of cork stoppers by using image processing and neuro-fuzzy computing. J. Intell. Manuf. **21**, 745–760 (2010) https://doi.org/10.1007/s10845-009-0251-4
3. He, Z., Sun, L.: Surface defect detection method for glass substrate using improved otsu segmentation. Appl. Opt. **54**(33), 9823–9830 (2015)
4. Bulnes, F.G., Usamentiaga, R., Garcia, D.F., Molleda, J.: An efficient method for defect detection during the manufacturing of web materials. J. Intell. Manuf. **27**(2), 431–445 (2014). https://doi.org/10.1007/s10845-014-0876-9
5. Oztemel, E., Gursev, S.: Literature review of industry 4.0 and related technologies. J. Intell. Manuf. **31**, 127–182 (2020)
6. Li, J., Dai, Y., Li, C., Shu, J., Li, D., Yang, T., Lu, Z.: Visual detail augmented mapping for small aerial target detection. Remote Sens. **11**, 14 (2018)

7. Liu, M., Wang, X., Zhou, A., Fu, X., Ma, Y., Piao, C.: Uav-yolo: Small object detection on unmanned aerial vehicle perspective. Sensors **20**, 2238 (2020)
8. Pham, M.-T., Courtrai, L., Friguet, C., Lefèvre, S., Baussard, A.: Yolo-fine: one-stage detector of small objects under various backgrounds in remote sensing images. Remote Sens. **12**, 2501 (2020)
9. Krizhevsky, A., Sutskever, I., Hinton, G.: Imagenet classification with deep convolutional neural networks. Neural Inf. Process. Syst. **25**, 1097–1105 (2012)
10. Masci, J., Meier, U., Ciresan, D., Schmidhuber, J., Fricout, G.: Steel defect classification with max-pooling convolutional neural networks. In: International Joint Conference on Neural Networks (IJCNN), pp. 1–6 (2012)
11. Saffari, A., Leistner, C., Santner, J., Godec, M., Bischof, H.: On-line random forests. In: IEEE International Conference on Computer Vision (ICCV), pp. 1393–1400 (2009)
12. Suykens, J., Vandewalle, J.: Least squares support vector machine classifiers. Neural Process. Lett. **9**, 293–300 (1999)
13. Fukunaga, K., Narendra, P.M.: A branch and bound algorithm for computing k-nearest neighbors. IEEE Trans. Comput. **C-24**(7), 750–753 (1975)
14. Zhao, J., Kong, Q.J., Zhao, X., Liu, J., Liu, Y.: A method for detection and classification of glass defects in low resolution images. In: International Conference on Image and Graphics, pp. 642–647 (2011)
15. Tabernik, D., Šela, S., Skvarč, J., Skočaj, D.: Segmentation-based deep-learning approach for surface-defect detection. J. Intell. Manuf. **31**(3), 759–776 (2019)
16. Park, J., Riaz, H., Kim, H., Kim, J.: Advanced cover glass defect detection and classification based on multi-dnn model. Manuf. Lett. **23**, 53–61 (2019)
17. Liu, D., Lai, K., Ye, G., Chen, M., Chang, S.: Sample-specific late fusion for visual category recognition. In: IEEE Conference on Computer Vision and Pattern Recognition, pp. 803–810 (2013)
18. Yao, M., Sharp, C., DeBrunner, L.S., DeBrunner, V.E.: Image processing for a line-scan camera system. In: Asilomar Conference on Signals, Systems and Computers, vol. 1, pp. 130–134 (1996)
19. Wang, C.Y.: et al.: Cspnet: a new backbone that can enhance learning capability of CNN. In: IEEE CVF Conference on Computer Vision and Pattern Recognition (CVPR) Workshops (2020)
20. Everingham, M., Van Gool, L., Williams, C., Winn, J., Zisserman, A.: The pascal visual object classes (voc) challenge. Int. J. Comput. Vis. **88**, 303–338 (2010)
21. Wilson, G., Cook, D.J.: A survey of unsupervised deep domain adaptation. ACM Trans. Intell. Syst. Technol. (TIST) **11**(5), 1–46 (2020)
22. Girshick, R.: Fast r-cnn. In: IEEE International Conference on Computer Vision (ICCV), pp. 1440–1448 (2015)
23. He, K., Gkioxari, G., Dollár, P., Girshick, R.: Mask r-cnn. In: IEEE International Conference on Computer Vision (ICCV), pp. 2980–2988 (2017)
24. Jocher, G., et al.: ultralytics/yolov5: v3.0 (2020)

Evaluation of Edge Platforms for Deep Learning in Computer Vision

Christoffer Bøgelund Rasmussen[1]([✉]) [iD], Aske Rasch Lejbølle[1] [iD],
Kamal Nasrollahi[1,2] [iD], and Thomas B. Moeslund[1] [iD]

[1] Department of Architecture, Design and Media Technology, Aalborg University,
Aalborg, Denmark
cbra@create.aau.dk
[2] Research Department of Milestone Systems, Copenhagen, Denmark
http://www.vap.aau.dk

Abstract. In recent years, companies, such as Intel and Google, have brought onto the market small low-power platforms that can be used to deploy and run inference of Deep Neural Networks at a low cost. These platforms can process data at the edge, such as images from a camera, to avoid transfer of large amount of data across a network. To determine which platform to use for a specific task, practitioners usually compare parameters, such as inference time and power consumption. However, to provide a better incentive on platform selection based on requirements, it is important to also consider the platform price. In this paper, we explore platform/model trade-offs, by providing benchmarks of state-of-the-art platforms within three common computer vision tasks; classification, detection and segmentation. By also considering the price of each platform, we provide a comparison of price versus inference time, to aid quick decision making in regard to platform and model selection. Finally, by analysing the operation allocation of models for each platform, we identify operations that should be optimised, based on platform/model selection.

Keywords: Edge platforms · Deep learning · Computer vision

1 Introduction

Within the years that followed 2012, researchers were focused on developing Deep Neural Networks (DNNs) that were accurate and generalised well. Each year, the top-1 error on the large ImageNet dataset, used within object classification, gradually decreased [3,10]. As other computer vision (CV) tasks gained more interest, such as object detection and semantic segmentation, accuracies on benchmark datasets would increase each year [12,35]. However, recently, focus has shifted towards more practical usage of DNNs. Today, lowering the network complexity while maintaining a high accuracy is largely prioritised. Novel architectures are developed that contain fewer parameters [17,30], and larger

© Springer Nature Switzerland AG 2021
A. Del Bimbo et al. (Eds.): ICPR 2020 Workshops, LNCS 12664, pp. 523–537, 2021.
https://doi.org/10.1007/978-3-030-68799-1_38

networks are quantified to speed up inference. The most common datatype for DNNs today is the 32-bit floating point (FP32), however, using quantification techniques, networks can operate on 16-bit floating point (FP16), or even 8-bit integers, with almost no loss in precision [22].

Following the trend within academia of developing DNNs, companies are developing hardware to run these networks. This hardware, furthermore, should be able to process incoming data with low latency. Several cloud solutions, offered by big companies, such as Amazon's Amazon Web Services (AWS) [4], have emerged that can train and run models online. Furthermore, Internet of Things (IoT) have resulted in products that require smaller and cheaper computers, which can run trained models at the edge. As a result of this demand, companies like Intel and NVIDIA have brought onto the market edge platforms that deploy and run network inference at limited costs [19,27]. These platforms can, for example, be integrated with a camera to process data directly at the source. In the last few years, several minor and large companies have brought onto the market their own edge platforms, combined with software packages to optimise pre-trained models before deployment. These platforms are able to run models within a variety of CV tasks, including object classification, detection and segmentation.

In this work, we evaluate edge platforms on common CV tasks, including object classification, object detection and semantic segmentation. We evaluate DNN models of different precision and complexity within each task, to show and compare inference timings between high-precision complex models and medium-precision/simple models when the batch size is varied. For better comparisons between platforms, we evaluate a high-end GPU and use it as reference. Furthermore, we calculate the number frames per second (FPS) based on the inference timings, and include the retail price of each platform to calculate an FPS cost. The FPS cost is a measure to identify the cost effectiveness of a certain platform/model combination, using a specific batch size. Additionally, comparing retail price and FPS, we propose a framework which aid the optimal platform/model selection, depending task and budget/speed requirements. Finally, we compare the distribution of DNN operations across platforms and models to identify the parts of a DNN on each platform that result in a higher FPS costs. These investigations allow us also to evaluate different CV tasks over the edge platforms and conclude on the best platform for a given use-case in terms of value for money, budget and FPS.

Previous works have studied models of different complexities [5,7,16], however, these publications aim to provide analyses of the speed/accuracy trade-off between models. On the other hand, works have been published that evaluates and compares different edge platforms [2,6], but these works do not take into consideration the price of different platforms. By including the price of the platforms, we are able to provide a simple and extensive overview of the FPS cost, which can be used by companies to select the optimal platform/model combination depending on their requirements, resources and the given CV task.

2   Related Work

2.1   Object Classification

In the last couple of years, works have been published that compare classification models and their performances. Canziani, Culurciello and Paszke [7] analysed inference time, power consumption and system memory utilisation for models of different complexity, depending on the batch size. However, all tests were performed on a single NVIDIA Jetson TX1, and only complex models were considered. Biano et al. [5] extended the work of [7] by including several additional CNNs, while also performing the evaluation on an NVIDIA Titan X GPU, but considered the same parameters. Meanwhile, Velasco-Montero et al. [33] evaluated models of different complexities, implemented in different frameworks, on a low-power Raspberry Pi 3 model B, and considered accuracy, throughput and power consumption to find a subset of optimal model/framework combinations for real-time deployment.

More recently, Almeida et al. [2] conducted an evaluation of several classification models, including those in [5], but also less complex models. Furthermore, they considered five different platforms, including an edge platform. Similarly to [5,7], they compared inference time and accuracy between models, but rather than having a single plot from all platforms, the comparison was performed per platform to identify differences and similarities between the platforms with respect to the handling of the networks. While the work provides insight on how to build up an architecture based on the platform, it does not consider the cost of using a certain platform.

2.2   Object Detection

Huang et al. [16] performed a comparison of three popular object detectors by changing the feature extractor, to analyse the change in accuracy/speed/memory trade-off. Liu et al. [24] presented a more extensive survey of object detectors, where less complex detectors were also considered. However, the survey does not include a speed/accuracy analysis between the presented detectors. To our knowledge, no published work compare speed/accuracy and price across several platforms, including edge platforms.

2.3   Semantic Segmentation

Few works have been published in benchmarking of semantic segmentation networks. Guo et al. [13] provide an overview of different architectures with the purpose of identifying strengths, weaknesses, and challenges of current work. A more general survey by Garcia-Garcia et al. [12] was published that presents the key ideas behind segmentation networks and provide an overview of previously proposed architectures with focus on, among other things, accuracy and efficiency. While they provide a comprehensive overview, they do not directly compare models.

2.4 Platform Benchmarks

Only few works compare performance of models of different complexities across difference platforms. Trindade et al. [32] evaluated two popular frameworks, Caffe [21] and TensorFlow [1] and compared performance, with respect to training time, between a GPU and NUMA CPU. A more extensive evaluation of frameworks was presented by Zhang, Wang and Shi [34] who performed the evaluation on different platforms where inference time, memory footprint and energy consumption was evaluated. Blouw et al. [6] measured inference time and energy consumption across different platforms with respect to batch size, and analysed the speed and energy cost per inference as a function of the network size. However, they only evaluated platforms on a single custom architecture. Finally, Pena et al. [28] focused on low-power devices, by evaluating object classification models and frameworks with respect to inference time and power consumption.

To our knowledge, only a single previous publication compares different platforms across different tasks, which is the aim this work. Ignatov et al. [18] considered mobile platforms containing chips that are manufactured by major chipset companies. The chips were evaluated in nine tests, including two image recognition tests using MobileNet and Inception V3, respectively, and a memory limitation test to identify the maximum allowed image size for inference before running out of memory. Instead, we perform evaluation of edge platforms across different common CV tasks, consider the retail price of the platforms, and analyse the consequence of DNN operations across platforms.

3 Platform Evaluation

This section presents an overview of our methodology for evaluating the edge platforms. Specifically, we present the evaluation procedure to ensure comparable results between platforms, choice of deep learning framework, and overview of selected models and platforms.

3.1 Model Overview

The choices for method and models are based upon differences in the complexity of feature extractors dependent on the difficulty of a given task together with their performance on leading benchmark challenges. For each of the three tasks covered in this survey, models at up to three different levels of complexity are evaluated. For all tasks, complexity is defined as the number of Giga Floating Point Operations (GLOPS). For simplicity, we adopt pre-trained networks available in the official TensorFlow [1] framework. An overview of the models for each task can be seen in Table 1.

Classification. We adopt MobileNetV1 [15] as the small, ResNet50 [14] as a medium, and InceptionResNetV2 [31] as the larger more complex network. An overview of the classification models described can be seen in the top portion in Table 1.

Table 1. Overview of models over the three tasks. Top-1 accuracy is based on the ImagenNet classification task [29]. mAP is based on the COCO detection task [23]. mIOU is based on the VOC 2012 segmentation task [11].

Model	Year	GFLOPS[a]	Top-1 [%]
MobileNetV1 [15]	2017	1.15	70.9
ResNet50 [14]	2015	6.97	75.2
InceptionResNetV2 [31]	2017	26.36	80.4
			mAP [%]
SSD MobileNetV1 [15]	2015	2.49	21
SSD InceptionV2 [16]	2017	9.63	24
			mIOU [%]
DeepLabV3 MobileNetV2 [30]	2018	17.69	75.32
DeepLabV3 Xception65 [8]	2017	354	82.20

[a] As measured in TensorFlow

Object Detection. For benchmarking object detection networks, we use the SSD [25] with the distinction between the complexity of the SSD networks being done by switching the feature extractor. The middle portion in Table 1 summarises our choices for the two feature extractors with varying complexity, namely, MobileNetV1 and InceptionV2.

Semantic Segmentation. We adopt DeepLabV3 [8] for evaluating semantic segmentation networks. An overview of model backbone choices for evaluation of DeepLabV3 is shown in the bottom portion of Table 1, which in this case is MobileNetV2 and Xception65.

3.2 Platform Overview

This section introduces the platforms evaluated across the various classification, object detection and segmentation models. An overview of some of the key specifications for the platforms can be seen in Table 2, covering the number of cores, clock frequency, memory, Thermal Design Power (TDP) and price. We include a CPU, the Intel i7-7700K, since a GPU solution, occasionally, may not be possible due to price or space restrictions. Further, we include two low-power edge devices, the Intel NCS and NCS2, that can perform inference of DNN models. The NCS devices have the form of USB sticks and must be connected to a host machine for inference. Additionally, we include an NVIDIA Jetson TX2, which requires more power, compared to the NCS devices, but is more powerful. Finally, we include a reference, NVIDIA GTX 1080, to which we can compare our evaluation of edge platforms.

Table 2. Overview of evaluated platforms, including the reference GTX 1080.

Platform	Cores	Clock Freq. (GHz)	Memory (GB)	TDP (W)	Price[a] ($)
i7-7700K	4	4.2	64	91	350
Intel NCS	12[b]	0.6	0.5	1	69
Intel NCS2	16[b]	0.7	0.5	1	75
NVIDIA GTX 1080	2560[c]	1.6	8	180	580
NVIDIA Jetson TX2	256[c]	1.3	8	7.5	560

[a] Price per 01/09/2020 [9] [b] SHAVE cores [c] CUDA cores

3.3   Evaluation Overview

In case of the TX2, models run in three settings; (1) in the standard TensorFlow format, (2) by maximising the clock speed on the TX2, (3) and by optimising the models with the TensorRT (TF-TRT) package [26], which transforms and optimises the models, for example by fusing layers, such as Convolution and ReLU. Additionally, the precision of the model is changed from FP32 to FP16, with minimal loss in accuracy. To run inference on the NCS and NCS2, models are converted to an Intermediate Representation (IR), consisting of an *xml* file to describe the model topology and a *bin* file containing model weights and biases. This is accomplished using the OpenVINO toolkit [20], developed by Intel. Similarly to TF-TRT, this is done by fusion of certain layers of the network, such as Convolution and BatchNormalisation or removing layers that are not used at test time, for example, the dropout layer. Likewise, the precision of the model is changed to FP16 in order to speed up inference and make the model compatible.

Evaluations are performed using TensorFlow 1.10.1 for most platforms. Additionally for the NCS and NCS2, OpenVINO 2018_R5 is used to optimise and run evaluation. However, TensorFlow 1.8 is used in case of TX2 as this is compatible with TensorRT 4.0.1, which is required to optimise models to TRT. To accelerate performance on TX2 and GTX 1080, we use CUDA 9.0 with CUDNN 7.0. The GTX 1080 and i7-7700k are evaluated on a machine containing 64 GBs of RAM, running Ubuntu 16.04, while NCS and NCS2 are evaluated on a machine consisting of an i7-6700HQ CPU @ 2.60 GHz and 16 GBs of RAM. In all cases, evaluations are executed in Python 3.5.2.

The evaluations are run on images from the ImageNet dataset [10]. N images are loaded, where N is the batch size, and resized accordingly to the input size of the model. For NCS and NCS2, the batch size corresponds to the number of sticks that are run in parallel, asynchronously. We run inference for 100 iterations and calculate the mean inference time per image based on the total inference time and batch size. We evaluate inference time using batch sizes {1, 2, 3, 4, 8, 16, 32, 64, 128}, in case of NCS and NCS2, we evaluate inference time using 1, 2, 3 and 4 sticks in parallel. The entire evaluation procedure is summarised in Algorithm 1.

Algorithm 1. Evaluation procedure

Require:*model_name, batch_size, platform, imagepath*
Ensure:*mean_inference_time*
1: $model \leftarrow load(model\_name)$
2: **if** $platform == tx2\ trt\ ||\ platform == NCS$ **then**
3:     $model \leftarrow convert\_model(model)$
4: **end if**
5: $images \leftarrow read\_images(batch\_size, imagepath)$
6: $i \leftarrow 0$
7: $total\_time \leftarrow 0$
8: **while** $i < 100$ **do**
9:     $start\_time \leftarrow time()$
10:     $run\_inference(model, images)$
11:     $inference\_time \leftarrow \frac{time() - start\_time}{batch\_size}$
12:     $i \leftarrow i + 1$
13:     $total\_time \leftarrow total\_time + inference\_time$
14: **end while**
15: $mean\_inference\_time \leftarrow \frac{total\_time}{100}$

---

4 Experimental Results

We perform experiments to conclude on the optimal model/platform selection within each task. Extensive plots are provided to aid selection based on platform price, inference time, and batch size. First, we plot the FPS cost in relation to the FPS of the difference platforms across the models of different complexities. The FPS cost is calculated as the retail price divided by the number of FPS for at given platform/model combination. Additionally, we plot price against FPS in order to show potential speeds based on concrete price points. Finally, we plot the top operation allocations for each platform to further understand the differences between the platforms. Since many of the plots show large numerical differences between platforms we plot values on a logarithmic scale.

4.1 Classification

Figure 1 shows the FPS cost for the three classification models over the platform variants. An increasing batch size is shown by an increasing diameter of the bubbles representing the platforms. It is clear that the NCS2 is the most cost friendly edge platforms between batch size 1 to 4. A difference in the NCS and NCS2 to the other platforms is the consistent costs over batch size as the number of sticks increases accordingly, whereas for the TX2 variants and GTX 1080 we see lower FPS cost as batch size increases. The i7 FPS cost does not change over batch sizes but does decrease as the number of cores is increased. The TX2 TRT does become competitive in comparison to the NCS2 for our medium complexity ResNet50 model at larger batch sizes.

Looking at the FPS for a given price point together with FPS cost in Fig. 2 we are able to determine the trade-off between model complexity, FPS and price

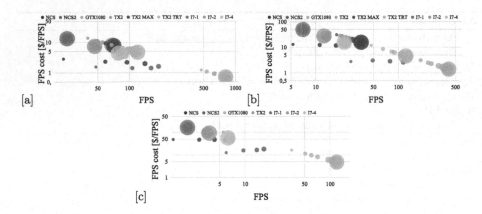

Fig. 1. FPS cost of classification models based on batch size and FPS. MobileNetV1 (a), ResNet50 (b) and InceptionResNetV2 (c).

budget for batch sizes 1 and 4. With these figures, if the budget is known for a deep learning system, it is possible to infer how complex a model can be run and at what speed. Additionally, in these figures we depict the three models and their complexity by the size of the bubble. The lowest complexity MobileNetV1 is shown by the smallest diameter and most complex InceptionResNetV2 by the largest. For batch size 1 in Fig. 2(a) and (c) the NCS and NCS2 are able to provide a relatively high FPS over the three model complexities for a low price point, furthermore, this is highlighted by the lower FPS cost. However, at batch size 4 the i7 CPU becomes more comparable in terms of price and, in the cases with less complex models and increased number of cores, have similar FPS to the NCS and NCS2.

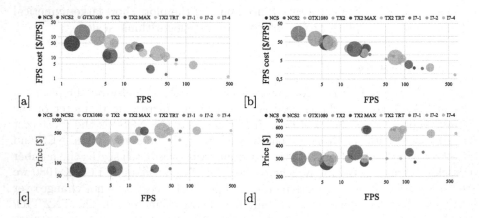

Fig. 2. Comparison of FPS cost (a) & (b) and retail price (c) & (d) based on FPS, for batch sizes one (a) & (c) and four (b) & (d). Small bubbles indicate MobileNetV1, middle-size bubbles indicate ResNet50, and large bubbles indicate InceptionResNetV2.

4.2 Object Detection

Again, we see in Fig. 3 that the NCS2 has the best FPS cost for the two SSD models, however, the NCS is competitive when MobileNetV1 is used as the backbone. In addition, despite the decreasing FPS cost as number of cores increase for the i7, or by increasing batch size and optimising for TX2 TRT, these variants are in general less viable for object detection purposes due to their overall FPS cost and lower FPS.

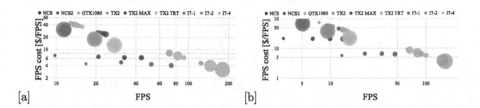

Fig. 3. FPS cost of SSD MobileNetV1 (a) and SSD InceptionV2 (b) based on batch size and FPS.

Regarding the concrete price point and FPS cost, Fig. 4 shows that the NCS2 is the best trade-off at batch size 1 at a lower price and high FPS, additionally, the FPS cost is comparable to that of the GTX 1080. At batch size 4 the i7 is more competitive, especially as the number of cores increases but still has a lower FPS than the NCS2 where it is able to obtain impressive amounts of FPS at almost 50 FPS with InceptionV2 and around 80 FPS with MobileNetV1. The FPS cost is again similar to that of the GTX 1080 for both SSD networks.

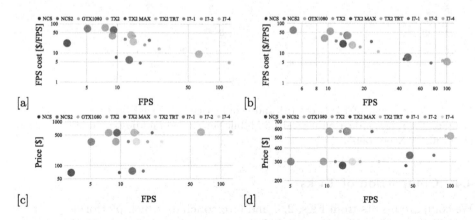

Fig. 4. Comparison of FPS cost (a) & (b) and retail price (c) & (d) based on FPS, for batch sizes one (a) & (c) and four (b) & (d). Small bubbles indicate SSD MobileNetV1 and middle-size bubbles indicate SSD InceptionV2.

4.3   Semantic Segmentation

It was only possible to run the DeepLabV3 models at batch size 1 due to memory constraints across the platforms. Figure 5 shows that none of the edge platforms could run the models near real-time. For the DeepLabV3 with MobileNetV2 the NCS had a considerably lower FPS cost compared to the other platforms. Whereas, with Xception65 NCS2 was the best but still at a high FPS cost.

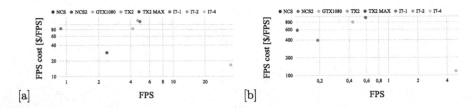

Fig. 5. FPS cost based of DeepLabV3 models based on batch size and FPS at batch size one. DeepLabV3 MobileNetV2 (a) and DeepLabV3 Xception65 (b).

Figure 6 shows that the NCS and NCS2 have a low price point but also a low FPS. Only the TX2 and TX2 MAX for the MobilenetV2 variant show promise with almost 5 FPS but at price point similar to the of the GTX 1080. FPS cost depends on the complexity of the model, for the lower complex DeepLabV3 with MobileNetV2 the NCS is the clear cheapest, whereas, with Xception65 as the backbone FPS cost is largely similar but with NCS2 as the cheapest option.

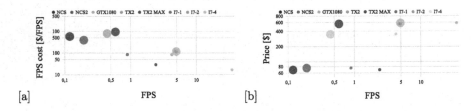

Fig. 6. Comparison of FPS cost (a) and retail price (b) based on FPS at batch size one. Small bubbles indicate DeepLabV3 MobileNetV2 and middle-size bubbles indicate DeepLabV3 Xception65.

4.4   Comparison of Tasks

We compare results from Figs. 2, 4 and 6 to conclude which platforms are more suited for specific tasks. If multiple NCS2 are combined, the platform is favourable in terms of both speed and price, to run classification or detection, independent of model complexity. Having a single NCS2, FPS performance on detection is still comparable to running TX2 TRT at batch size one, however, on classification TX2 TRT outperforms NCS2 in FPS. For both tasks, however, the FPS cost of NCS2 is still much lower. Nonetheless, on both classification and detection, TX2 TRT compares favourable to TX2 and TX2 MAX. Finally, segmentation is more suitable for the i7 or TX2, however, at a higher price compared to NCS2.

4.5   Inference Analysis

In order to understand more about the differences between the platforms we investigate the allocation of operations for the models. We use the TensorFlow profiler for the GTX 1080, TX2 and i7, whereas for the NCS and NCS2 we use the Deep Learning Workbench in OpenVINO. For each we visualise the operations as the top five and combine the remaining timings into one which we denote as *Other*. We only show the timings for the MobileNet variants from our three tasks in Figs. 7, 8 and 9 as similar trends were seen for the other backbones. The top-5 operations are largely the same for the GTX 1080 and TX2. We see that the TX2 TRT bundles a significant number of operations in *TRTEngineOp* for the classification model but not so much for the SSD variant. For all three tasks for the NCS and NCS2 a large portion is spent on the *Convolution* operation. Finally, the i7 is similar to that og the GTX 1080 and TX2 but does not show any type of convolution in the top-5.

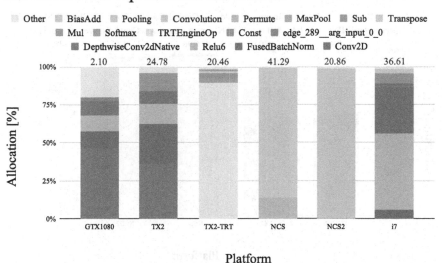

Fig. 7. Operation allocation for MobileNetV1. Numbers above bars indicate total time in ms.

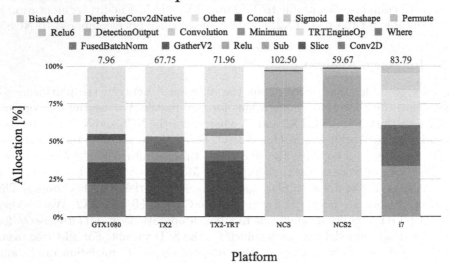

Fig. 8. Operation allocation for SSD MobileNetV1. Numbers above bars indicate total time in ms.

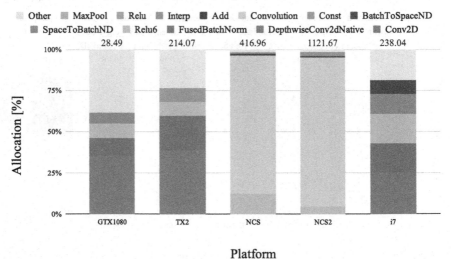

Fig. 9. Operation allocation for DeepLabV3 MobileNetV2. Numbers above bars indicate total time in ms.

5   Conclusion

In this work, we have evaluated different edge platforms within object classification, object detection and semantic segmentation. We have analysed the FPS cost together with batch size and budget to aid decision making of platform/model selection. Finally, we have analysed allocation of DNN operations. As a reference, all results of the edge platforms was compared with evaluations performed on a GTX 1080.

On classification, TX2 TRT is the optimal choice if a model runs at batch size one, and only speed is a requirement. However, if budget is limited, the NCS2 comes out as the better choice. Further, this is also the case for larger batch sizes, where the combination of multiple NCS2 is both cheaper and faster than compared to TX2, while being only slightly more expensive than the i7. For detection, a similar pattern is shown. However, at batch size one, differences between NCS2 and TX2 TRT in terms of FPS are much less, making the NCS2 favourable, independent of the number of sticks purchased. Finally, edge platforms are not yet suited for semantic segmentation, since only the GTX 1080 shows real-time inference timings. On the other hand, if real-time inference is not a requirement, either the NCS or NCS2 is the optimal choice in a strict budget, while TX2 is optimal in case of speed requirements.

Analysing the allocation of DNN operation across platform/model combinations, we have shown that several operations in detection and segmentation models should be made compatible with TensorRT to increase FPS, thus, reduce the FPS cost, while primarily *Convolution* and *Relu* operations should be optimised for NCS and NCS2 to speed up inference.

Acknowledgment. This work was funded by Innovation Fund Denmark under Grant 5189-00222B and 7038-00170B.

References

1. Abadi, M., et al.: Tensorflow: a system for large-scale machine learning. In: 12th USENIX Symposium on Operating Systems Design and Implementation (OSDI 2016), pp. 265–283 (2016). https://www.usenix.org/system/files/conference/osdi16/osdi16-abadi.pdf
2. Almeida, M., Laskaridis, S., Leontiadis, I., Venieris, S.I., Lane, N.D.: Embench: quantifying performance variations of deep neural networks across modern commodity devices. In: The 3rd International Workshop on Deep Learning for Mobile Systems and Applications, pp. 1–6 (2019)
3. Alom, M.Z., et al.: A state-of-the-art survey on deep learning theory and architectures. Electronics **8**(3), 292 (2019)
4. Amazon: Amazon web services (aws) (2020). https://aws.amazon.com/. Accessed 12 July 2020
5. Bianco, S., Cadene, R., Celona, L., Napoletano, P.: Benchmark analysis of representative deep neural network architectures. IEEE Access **6**, 64270–64277 (2018)
6. Blouw, P., Choo, X., Hunsberger, E., Eliasmith, C.: Benchmarking keyword spotting efficiency on neuromorphic hardware. arXiv preprint arXiv:1812.01739 (2018)

7. Canziani, A., Paszke, A., Culurciello, E.: An analysis of deep neural network models for practical applications. arXiv preprint arXiv:1605.07678 (2016)
8. Chen, L.C., Papandreou, G., Schroff, F., Adam, H.: Rethinking atrous convolution for semantic image segmentation. arXiv preprint arXiv:1706.05587 (2017)
9. Cosmic Shovel, I.: Amazon price tracker, amazon price history charts, price watches, and price drop alerts. https://camelcamelcamel.com/ (March 2019), accessed: 24 September 2020
10. Deng, J., Dong, W., Socher, R., Li, L.J., Li, K., Fei-Fei, L.: Imagenet: a large-scale hierarchical image database. In: IEEE Conference on Computer Vision and Pattern Recognition, pp. 248–255 (2009)
11. Everingham, M., Van Gool, L., Williams, C.K.I., Winn, J., Zisserman, A.: The pascal visual object classes (VOC) challenge. Int. J. Comput. Vision **88**(2), 303–338 (2010)
12. Garcia-Garcia, A., Orts-Escolano, S., Oprea, S., Villena-Martinez, V., Martinez-Gonzalez, P., Garcia-Rodriguez, J.: A survey on deep learning techniques for image and video semantic segmentation. Appl. Soft Comput. **70**, 41–65 (2018)
13. Guo, Y., Liu, Y., Georgiou, T., Lew, M.S.: A review of semantic segmentation using deep neural networks. Int. J. Multimed. Inform. Retrieval **7**(2), 87–93 (2018)
14. He, K., Zhang, X., Ren, S., Sun, J.: Deep residual learning for image recognition. In: Proceedings of the IEEE Conference on Computer Vision and Pattern Recognition, pp. 770–778 (2016)
15. Howard, A.G., et al.: Mobilenets: efficient convolutional neural networks for mobile vision applications. arXiv preprint arXiv:1704.04861 (2017)
16. Huang, J., et al.: Speed/accuracy trade-offs for modern convolutional object detectors. In: Proceedings of the IEEE Conference on Computer Vision and Pattern Recognition pp. 7310–7311 (2017)
17. Iandola, F.N., Han, S., Moskewicz, M.W., Ashraf, K., Dally, W.J., Keutzer, K.: Squeezenet: Alexnet-level accuracy with 50x fewer parameters and <0.5 mb model size. arXiv preprint arXiv:1602.07360 (2016)
18. Ignatov, A., et al.: Ai benchmark: running deep neural networks on android smartphones. In: Proceedings of the European Conference on Computer Vision, pp. 288–314 (2018)
19. Intel: Intel neural compute stick 2, September 2020. https://software.intel.com/en-us/neural-compute-stick. Accessed 6 Sept 2020
20. Intel: Openvino toolkit, December 2020. https://docs.openvinotoolkit.org/2018_R5/index.html. Accessed 3 Sept 2020
21. Jia, Y., et al.: Caffe: convolutional architecture for fast feature embedding. arXiv preprint arXiv:1408.5093 (2014)
22. Krishnamoorthi, R.: Quantizing deep convolutional networks for efficient inference: A whitepaper. arXiv preprint arXiv:1806.08342 (2018)
23. Lin, T.-Y., et al.: Microsoft COCO: common objects in context. In: Fleet, D., Pajdla, T., Schiele, B., Tuytelaars, T. (eds.) ECCV 2014. LNCS, vol. 8693, pp. 740–755. Springer, Cham (2014). https://doi.org/10.1007/978-3-319-10602-1_48
24. Liu, L., et al.: Deep learning for generic object detection: a survey. arXiv preprint arXiv:1809.02165 (2018)
25. Liu, W., et al.: SSD: single shot MultiBox detector. In: Leibe, B., Matas, J., Sebe, N., Welling, M. (eds.) ECCV 2016. LNCS, vol. 9905, pp. 21–37. Springer, Cham (2016). https://doi.org/10.1007/978-3-319-46448-0_2
26. NVIDIA: Accelerating inference in tf trt user guide, August 2019. https://docs.nvidia.com/deeplearning/frameworks/tf-trt-user-guide/index.html. Accessed 3 Sept 2019

27. NVIDIA: Jetson tx2 module (2020). https://developer.nvidia.com/embedded/jetson-tx2. Accessed 12 July 2020
28. Pena, D., Forembski, A., Xu, X., Moloney, D.: Benchmarking of CNNs for low-cost, low-power robotics applications. In: RSS 2017 Workshop: New Frontier for Deep Learning in Robotics, pp. 1–5 (2017)
29. Russakovsky, O., et al.: ImageNet Large Scale Visual Recognition Challenge. International Journal of Computer Vision (2015)
30. Sandler, M., Howard, A., Zhu, M., Zhmoginov, A., Chen, L.C.: Mobilenetv 2: Inverted residuals and linear bottlenecks. In: Proceedings of the IEEE Conference on Computer Vision and Pattern Recognition, pp. 4510–4520 (2018)
31. Szegedy, C., Ioffe, S., Vanhoucke, V., Alemi, A.A.: Inception-v4, inception-resnet and the impact of residual connections on learning. In: Proceedings of the AAAI Conference on Artificial Intelligence, pp. 4278–4284 (2017)
32. Trindade, R.G., Lima, J.V.F., Charão, A.S.: Performance evaluation of deep learning frameworks over different architectures. In: International Conference on Vector and Parallel Processing, pp. 92–104 (2018)
33. Velasco-Montero, D., Fernández-Berni, J., Carmona-Galán, R., Rodríguez-Vázquez, Á.: Optimum selection of DNN model and framework for edge inference. IEEE Access **6**, 51680–51692 (2018)
34. Zhang, X., Wang, Y., Shi, W.: PCAMP: performance comparison of machine learning packages on the edges. In: {USENIX} Workshop on Hot Topics in Edge Computing (HotEdge 18) (2018)
35. Zou, Z., Shi, Z., Guo, Y., Ye, J.: Object detection in 20 years: a survey. arXiv preprint arXiv:1905.05055 (2019)

BlendTorch: A Real-Time, Adaptive Domain Randomization Library

Christoph Heindl[1]([envelope])[ORCID], Lukas Brunner[1][ORCID], Sebastian Zambal[1][ORCID],
and Josef Scharinger[2]

[1] Visual Computing, Profactor GmbH, Steyr, Austria
{christoph.heindl,lukas.brunner,sebastian.zambal}@profactor.at
[2] Computational Perception, JKU, Linz, Austria
josef.scharinger@jku.at

Abstract. Solving complex computer vision tasks by deep learning techniques rely on large amounts of (supervised) image data, typically unavailable in industrial environments. Consequently, the lack of training data is beginning to impede the successful transfer of state-of-the-art computer vision methods to industrial applications. We introduce BlendTorch, an adaptive Domain Randomization (DR) library, to help create infinite streams of synthetic training data. BlendTorch generates data by massively randomizing low-fidelity simulations and takes care of distributing artificial training data for model learning in real-time. We show that models trained with BlendTorch repeatedly perform better in an industrial object detection task than those trained on real or photo-realistic datasets.

Keywords: Computer vision · Domain Randomization · Object detection

1 Introduction

Recent advances in computer vision depend extensively on deep learning techniques. With sufficient modeling capacity and enough (labeled) domain datasets, deeply learned models often outperform conventional vision pipelines [5,11,16]. However, providing large enough datasets is challenging within industrial applications for several reasons: a) costly and error-prone manual annotations, b) the odds of observing rare events are low, and c) a combinatorial data explosion as vision tasks become increasingly complex. If high capacity models are trained despite of low data quality, the likelihood of overfitting increases, resulting in reduced robustness in industrial applications [15].

In this work, we focus on generating artificial training images and annotations through computer simulations. Training models in simulations promise annotated data without limits, but the discrepancy between the distribution of training and real data often leads to poor generalizing models [19]. Increasing photo realism and massive randomization of low-fidelity simulations (Domain

© Springer Nature Switzerland AG 2021
A. Del Bimbo et al. (Eds.): ICPR 2020 Workshops, LNCS 12664, pp. 538–551, 2021.
https://doi.org/10.1007/978-3-030-68799-1_39

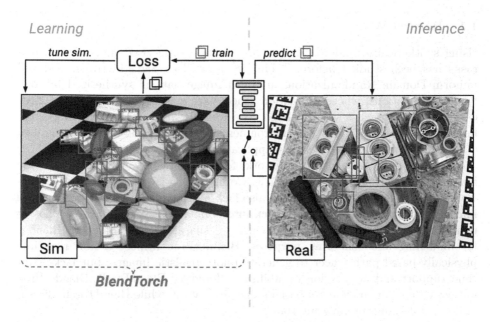

Fig. 1. We introduce BlendTorch, a real-time adaptive Domain Randomization library, for neural network training in simulated environments. We show, networks trained with BlendTorch outperform identical models trained with photo-realistic or even real datasets on the same object detection task.

Randomization) are two popular and contrary strategies to minimize the distributional mismatch. Recent frameworks focus on photorealism, but do not address the specifics of massive randomization such as: online rendering capabilities and adaptive simulations.

We introduce BlendTorch[1], a general purpose open-source image synthesis framework for adaptive, real-time Domain Randomization (DR) written in Python. BlendTorch integrates probabilistic scene composition, physically plausible real-time rendering, distributed data streaming, and bidirectional communication. Our framework integrates the modelling and rendering strengths of Blender [1] with the high-performance deep learning capabilities of PyTorch [12]. We successfully apply BlendTorch to the task of learning to detect industrial objects without access to real images during training (see Fig. 1). We demonstrate that data generation and model training can be done online in a single sweep, reducing the time required to ramp-up neural networks significantly. Our approach not only outperforms photo-realistic datasets, but on the same perception task, we also show that DR surpasses the detection performance compared to a real image data set.

---

[1] https://github.com/cheind/pytorch-blender.

1.1   Related Work

Using synthetically rendered images for training supervised machine learning tasks has been studied before. Tobin [18] as well as Sadeghi [13] introduced uniform Domain Randomization, in which image data is synthesized by low-fidelity simulations, whose simulation aspects are massively randomized. These earlier DR approaches were tailored to a specific application, which severely limits their reusability in other environments. BlendTorch is based on the idea of DR, but generalizes to arbitrary applications.

Recently, general purpose frameworks focusing on generating photo-realistic images have been introduced [3,14,17]. Compared to BlendTorch, these frameworks are not real-time capable and also lack a principled way to communicate information from model training back into simulation. The work most closely related to ours is BlenderProc [3], since we share the idea of using Blender for modelling and rendering purposes. BlenderProc utilizes a non real-time physically-based path tracer to generate photo-realistic images, but lacks real-time support and adaptation capabilities offered by BlendTorch. BlenderProc focuses on a configuration file based scene generation, while BlendTorch offers a more flexible programming interface.

Self-optimizing simulations by means of training feedback was introduced in Heindl [6] for the specialized task of robot keypoint detection. BlendTorch generalizes this idea to arbitrary applications and to real-time rendering.

1.2   Contributions

This paper offers the following contributions

1. BlendTorch, an adaptive, open-source, real-time domain randomization library that seamlessly connects modelling, rendering and learning aspects.
2. A comprehensive industrial object detection experiment that highlights the benefits of DR over real as well as photo-realistic training datasets.

2   Design Principles

BlendTorch weaves several ideas into a design that enables practitioners and scientists to rapidly realize and test novel DR concepts.

Reuse. Training neural networks in simulation using DR requires several specialized software modules. For a successful experiment, modeling and rendering tools as well as powerful learning libraries for deep learning are required. In the recent past, excellent open-source frameworks for the aforementioned purposes have emerged independently from each other. However, these tools are not interconnected and a basic framework for their online interaction is missing. BlendTorch aims to bring these separate worlds together as seamlessly as possible without losing the benefits of either of the software component.

Real-Time. As the complexity of visual tasks increases, the combinatorial variety of scene also increases exponentially. Several applications of DR separate the simulation from the actual learning process, because the slow image generation stalls model learning. However, offline data generation suffers from the following shortcomings: constantly growing storage requirements and the missing possibility of online simulation adaptation. BlendTorch is designed to provide real-time, distributed integration between simulation and learning that is fast enough not to impede learning progress.

Adaptability. The ability to adapt the simulation parameters during model training has already proven to be beneficial [6] before. Adaptability allows simulations to synchronize with the evolving requirements of the learning process in order to learn more efficiently. The meaning of adaptability is application-dependent and ranges from adjusting the level of simulation difficulty to the generation of adversarial model examples. BlendTorch is designed with generic bidirectional communication in mind, allowing application-specific workflows to be implemented quickly.

The remainder of this paper focuses on BlendTorch's reuse and real-time principles, while leaving a thorough discussion of its adaptability capabilities for future work.

3   Architecture

BlendTorch connects the modeling and rendering strengths of Blender [1] and the deep learning capabilities of PyTorch [12] as depicted in Fig. 2. Our architecture considers data generation to be a bidirectional exchange of information, which contrasts conventional offline, one-way data generation. Our perspective enables BlendTorch to support scenarios that go beyond pure data generation, including adaptive domain randomization and reinforcement learning applications.

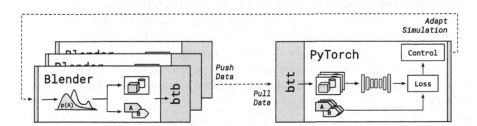

Fig. 2. BlendTorch overview. BlendTorch combines probabilistic (supervised) image generation in Blender [1] (left) with deep learning in PyTorch [12] (right) via a programmatic Python interface. Information is exchanged via a scalable network library (dotted lines). Implementation details are encapsulated in two Python subpackages `blendtorch.btb` and `bendtorch.btt`. An adaptation channel allows the simulation to be tuned to current training needs.

To seamlessly distribute rendering and learning across machine boundaries we utilize ZeroMQ [7] and split BlendTorch into two distinctive sub-packages that exchange information via ZMQ: `blendtorch.btb` and `bendtorch.btt`, providing the Blender and PyTorch views on BlendTorch.

A typical data generation task for supervised machine learning is setup in BlendTorch as follows. First, the training procedure launches and maintains one or more Blender instances using `btt.BlenderLauncher`. Each Blender instance will be instructed to run a particular scene and randomization script. Next, the training procedure creates a `btt.RemoteIterableDataset` to listen for incoming network messages from Blender instances. BlendTorch uses a pipeline pattern that supports multiple data producers and workers employing a fair data queuing policy. It is guaranteed that only one PyTorch worker receives a particular message and no message is lost, but the order in which it is received is not guaranteed. To avoid out-of-memory situations, the simulation processes will be temporarily stalled in case the training is not capable to catch up with data generation. Within every Blender process, the randomization script instantiates a `btb.DataPublisher` to distribute data messages. This script registers the necessary animation hooks. Typically, randomization occurs in `pre-frame` callbacks, while images are rendered in `post-frame` callbacks. Figures 3, 4 and 5 illustrates these concepts along with a minimal working example.

The train script `train.py` (Fig. 4) launches multiple Blender instances to simulate a simple scene driven by a randomization script `cube.blend.py` (Fig. 3). The training then awaits batches of images and annotations within the main training loop. The simulation script `cube.blend.py` randomizes the properties of a cube and publishes color images along with corner annotations. Color images and superimposed annotations are shown in Fig. 5.

4   Experiments

We evaluate BlendTorch by studying its performance within the context of an industrial 2D object detection task as follows. We train the same state-of-the-art object detection neural network, varying only the training dataset while keeping all other hyper-parameters fixed. We then evaluate each of the trained models on the same test dataset using the mean Average Precision (mAP) [4] metric that combines localization and classification performance. To avoid biases due to random model initialization, we repeat model learning multiple times for each dataset.

4.1   Datasets

For all our experiments we use the T-Less dataset [8] consisting of 30 industrial objects without significant texture or discriminating color which are captured from varying angles in semi-unstructured configurations. For reasons of presentation, we re-group the 30 classes into 6 super-groups as depicted in Fig. 6. Throughout our evaluation we distinguish four color image T-Less datasets that we describe next.

```
import bpy
from numpy.random import uniform
import blendtorch.btb as btb

def main():
        # Blender object
        cube = bpy.data.objects["Cube"]

        def pre_frame():
                # Randomize
                cube.rotation_euler = uniform(0,3.14,3)

        def post_frame():
                # Publish image + annotations
                pub.publish(
                        image=off.render(),
                        xy=off.camera.object_to_pixel(cube)
                )

        # Setup data publishing
        btargs, _ = btb.parse_blendtorch_args()
        pub = btb.DataPublisher(
                btargs.btsockets['DATA'],
                btargs.btid)

        # Setup rendering using def. camera
        off = btb.OffScreenRenderer(mode='rgb')

        # Setup animation and callbacks
        anim = btb.AnimationController()
        anim.pre_frame.add(pre_frame)
        anim.post_frame.add(post_frame)
        anim.play()

main()
```

Fig. 3. Simulation cube.blend.py

RealKinect. A real image dataset based on T-Less images taken with a Kinect camera (see Fig. 7a). It consists of 10^4 images grouped into 20 scenes with 500 images per scene. Images are taken in a structured way by sampling positions from a hemisphere. Each scene includes a variable number of occluders.

PBR. Is a publicly available photo-realistic synthetic image dataset generated by BlenderProc [3] in an offline step (see Fig. 7b). It consists of 5×10^4 images taken from random camera positions. PBR uses non parametric occluders which are inserted randomly.

```
from torch.utils import data
import blendtorch.btt as btt

def main():
        largs = dict(
                scene='cube.blend',
                script='cube.blend.py',
                num_instances=2,
                named_sockets=['DATA'],
        )
        # Launch Blender locally
        with btt.BlenderLauncher(**largs) as bl:
                # Create remote dataset
                addr = bl.launch_info.addresses['DATA']
                ds = btt.RemoteIterableDataset(addr)
                dl = data.DataLoader(ds, batch_size=4)

                for item in dl:
                        img, xy = item['image'], item['xy']
                        print('Received', img.shape, xy.shape)
                        # train with item ...

main()
```

Fig. 4. Training `train.py`

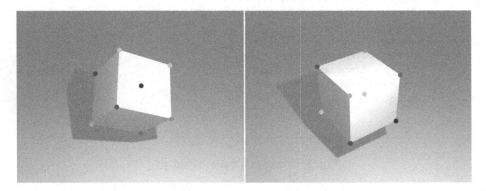

Fig. 5. Received images superimposed with annotations.

BlendTorch. This dataset corresponds to synthetic data generated by Domain Randomization BlendTorch (see Fig. 7c). The dataset consists of 5×10^4 images, whose generation details are given in Sect. 4.2.

BOP. A real image dataset of T-Less corresponding to the test dataset of the BOP Challenge 2020 [9] taken with a PrimeSense camera (see Fig. 7a). The dataset contains 10^4 images. We use this dataset solely for final evaluation of trained models.

4.2 BlendTorch Dataset

The BlendTorch dataset is generated as follows. For each scene we randomly draw a fixed number of objects according to the current class probabilities which might change over the course of training. Each object is represented by its CAD model. For each generated object, there is a chance of producing an occluder. Occluding objects are based on super-shapes[2], which can vary their shape significantly based on 12 parameters. These parameters are selected uniformly at random within a sensible range. Next, we randomize each object's position, rotation and procedural material. We ensure that initially each object hovers above a ground plane. Finally, we employ Blender's physics module to allow the objects to fall under the influence of gravity. Once the simulation has settled, we render N images using a virtual camera positioned randomly on hemispheres with varying radius. Besides color images, we generate the following annotations: bounding boxes, class numbers and visibility scores. Figure 7c shows an example output. Each image and annotation pair is then published through BlendTorch's data channels (refer to Figs. 2, 3, 4 and 5).

4.3 Training

Object detection is performed by CenterNet [21] using a DLA-34 [20] feature extracting backbone pre-trained on ImageNet [2]. The backbone features are inputs to multiple prediction heads. Each head consists of a convolution layer mapping to 256 feature channels, a rectified linear unit, and another convolution layer to have a task specific number of output channels. In particular, our network uses the following heads: a) one center point heatmap per class of dimension $6 \times \lfloor H/4 \rfloor \times \lfloor W/4 \rfloor$, and b) a bounding box size regression head of dimension $2 \times \lfloor H/4 \rfloor \times \lfloor W/4 \rfloor$, where (W, H) are width and height of input images.

During training, the total loss L_{total} is computed as follows

$$L_{total} = L_{hm} + 0.1 \cdot L_{wh}, \tag{1}$$

where L_{hm} is measured as the focal loss of predicted and ground truth center point heatmaps, L_{wh} is the L1 loss of regressed and true bounding box dimensions evaluated at ground truth center point locations.

We split the training dataset into training and validation data ($90/10\,\%$) and train for 2×10^5 images using Adam [10] ($lr = 1.25 \times 10^{-4}$, weight-decay $= 10^{-3}$). We perform the following augmentations independent of the dataset: scale to $512\,\text{px}$, random rotation, random horizontal flip, and color jitter. Regularization and augmentation are applied to all training sets to avoid skewing of results due to different dataset sizes.

Validation is performed every 1×10^5 images. The best model is selected based on the lowest total validation loss, as defined in Eq. 1. As mentioned above, each training session is repeated 10 times, and the best model in each run is stored for evaluation.

---

[2] https://github.com/cheind/supershape.

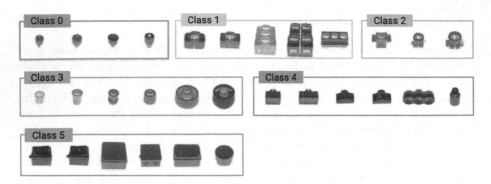

Fig. 6. T-Less dataset objects. 30 individual objects re-grouped into 6 super-classes in this work. Displayed in false colors for better visibility.

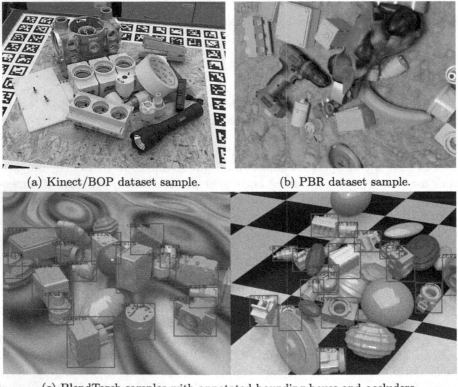

(a) Kinect/BOP dataset sample.   (b) PBR dataset sample.

(c) BlendTorch samples with annotated bounding boxes and occluders.

Fig. 7. Samples from datasets used throughout our work. Notice that PBR gives a much more realistic impression of the scene than BlendTorch using DR. Yet, we show DR is simpler to implement, generates images faster, and yields a better performing model.

4.4   Prediction

Objects and associated classes are determined by extracting the top-K local peak values that exceed a given confidence score from all center heatmaps. Bounding box dimensions are determined from the respective channels at center point locations. In Fig. 8 ground truths and predictions are depicted.

4.5   Average Precision

We assess the quality of each training dataset by computing the mean Average Precision (mAP) [4] from model predictions based on the unseen BOP test dataset (see Sect. 4.1) using the Intersection over Union (IoU) evaluation metric. The additional training runs per model allow us to compute the mAP with confidence. Unless otherwise stated, we report the mAP averaged over runs, error bars indicate ± 1 standard deviation (compare Table 1 and Fig. 9). We consider up to 25 model predictions that surpass a minimum confidence score of 0.1 during computation of the mAP.

Table 1 compares the average precision achieved for each training dataset over the course of 10 runs. BlendTorch outperforms photo-realistic and real image datasets. Although the BOP test dataset contains similar scenes to RealKinect, the characteristics of the cameras are different. Despite moderate model regularization, the network begins to overfit on the smaller RealKinect dataset, resulting in poor test performance. With BlendTorch generated images we slightly exceed the data provided by the photo-realistic PBR data set. However, BlendTorch trained models exhibit only half the standard deviation of all other models, making their performance more predictable in real world applications. This feature is shown clearly in Fig. 9 that compares the precision-recall behaviour of all three training datasets.

4.6   Runtime

We compare the time it takes to generate and receive synthetic images using BlendTorch. All experiments are performed on the same machine (Nvidia GeForce GTX 1080Ti) and software (Blender 2.90, PyTorch 1.60). Table 2 shows a runtime comparison of the data synthesis process for BlendTorch (ours) and the PBR data from BlenderProc [3]. Compared to photo-realistic rendering, BlendTorch creates images at interactive frame rates, even for physics enabled scenes involving millions of vertices.

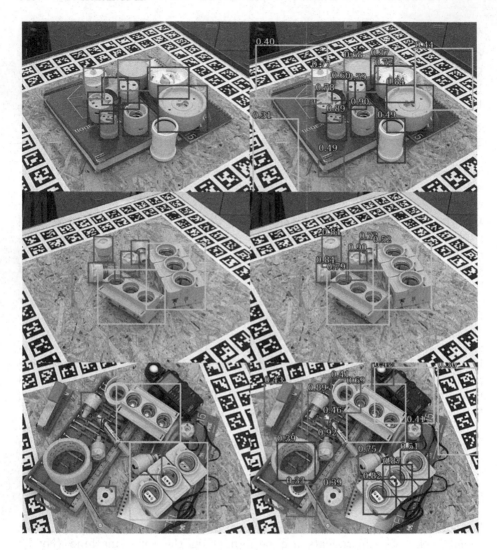

Fig. 8. Inference on the BOP test data done with a model trained on Blendtorch. Ground truths (left column) and predictions (right column) with color coded class affiliation (see Fig. 6). Predicted bounding boxes have the corresponding confidence score attached in the top left corner. A detection is drawn when the score exceeds the 0.3 confidence threshold. Real world applications would work with a higher threshold between 0.6 and 0.7, thus in row one and two examples of correctly classified scenes are given. (Color figure online)

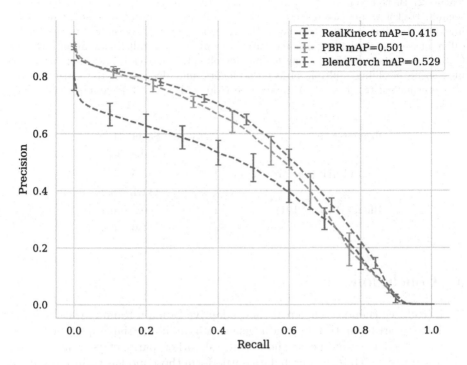

Fig. 9. Precision-recall curves as a result of evaluating object detection performance on varying training datasets. Despite the non-realistic appearance of domain randomized scenes in BlendTorch, our approach outperforms photo-realistic datasets and real image dataset at all IoU thresholds. We repeat the process 10 times and indicate the variations as $\pm 1\sigma$ error bars. Note that models trained with BlendTorch data have significantly less variability.

Table 1. Performance comparison. Average precision values over 10 training runs. Here mAP refers to the mean average precision computed by integrating over all classes, Intersection over Union (IoU) thresholds and all runs. σ_{mAP} denotes standard deviation of mAP measured over 10 runs. AP_{50} and AP_{75} represent average precision values for specific IoU-thresholds averaged over all runs and classes. Finally, mAP_{Ci} are average precision values averaged for specific classes averaged over all runs and thresholds. Except for σ_{mAP}, higher values indicate better performance.

Dataset	Overall AP				Per-Class mAP					
	mAP	σ_{mAP}	AP_{50}	AP_{75}	mAP_{C0}	mAP_{C1}	mAP_{C2}	mAP_{C3}	mAP_{C4}	mAP_{C5}
RealKinect	41.1	3.4	71.7	42.9	29.9	52.5	37.0	47.3	28.4	53.7
BlenderProc (PBR)	50.1	2.9	70.8	58.8	**40.3**	49.0	**59.4**	36.6	**59.2**	56.0
BlendTorch (ours)	**52.3**	**1.2**	**73.5**	**61.5**	34.0	**54.4**	56.3	**54.6**	55.3	**59.3**

Table 2. BlendTorch vs. photo-realistic rendering times. Timings are in frames per second. Higher values are better. All timings include the total time spent in rendering plus the time it takes to receive the data in training. We show timings for two different scenes, render engines and various numbers of parallel simulation instances. Cube represents a minimal scene having one object, whereas T-Less refers to a complex scene involving multiple objects, occluders and physics. Shared settings across all experiments: batch size (8), image size (640 × 512 × 3), and number of PyTorch workers (4).

Renderer	Instances	Scene [Hz]	
		Cube	T-Less
BlendTorch (renderer Eevee)	1	43.5	5.7
	4	111.1	18.2
BlenderProc (PBR)	1	0.8	0.4
	4	1.9	0.6

5   Conclusion

We introduced BlendTorch, a real-time, adaptive Domain Randomization library for infinite artificial training data generation in industrial applications. Our object detection experiments show that—all other parameters being equal—models trained with BlendTorch data outperform those models trained on photo-realistic or even real training datasets. Moreover, models trained with domain randomized data exhibit less performance variance over multiple runs. This makes DR a useful technique to compensate for the lack of training data in industrial machine learning applications. For the future we plan to explore the possibilities of adjusting simulation parameters with respect to training progress and to apply BlendTorch in the medical field, which exhibits similar shortcomings of real training data.

Acknowledgements. This work was supported by the strategic economic and research program "Innovatives OÖ 2020" of Upper Austria and by the European Union in cooperation with the State of Upper Austria within the project Investition in Wachstum und Beschftigung (IWB).

References

1. Blender Online Community: Blender - a 3D modelling and rendering package. Blender Foundation, Blender Institute, Amsterdam (2020). http://www.blender.org
2. Deng, J., Dong, W., Socher, R., Li, L.J., Li, K., Fei-Fei, L.: Imagenet: a large-scale hierarchical image database. In: 2009 IEEE Conference on Computer Vision and Pattern Recognition, pp. 248–255. IEEE (2009)
3. Denninger, M., et al.: Blenderproc: reducing the reality gap with photorealistic rendering

4. Everingham, M., Van Gool, L., Williams, C.K., Winn, J., Zisserman, A.: The pascal visual object classes (VOC) challenge. Int. J. Comput. Vision **88**(2), 303–338 (2010)
5. He, K., Zhang, X., Ren, S., Sun, J.: Deep residual learning for image recognition. In: Proceedings of the IEEE Conference on Computer Vision and Pattern Recognition, pp. 770–778 (2016)
6. Heindl, C., Zambal, S., Scharinger, J.: Learning to predict robot keypoints using artificially generated images. In: 24th IEEE International Conference on Emerging Technologies and Factory Automation (ETFA), pp. 1536–1539. IEEE (2019). https://doi.org/10.1109/ETFA.2019.8868243
7. Hintjens, P.: ZeroMQ: messaging for many applications. O'Reilly Media, Inc. (2013)
8. Hodan, T., Haluza, P., Obdržálek, Š., Matas, J., Lourakis, M., Zabulis, X.: T-less: An rgb-d dataset for 6d pose estimation of texture-less objects. In: 2017 IEEE Winter Conference on Applications of Computer Vision (WACV), pp. 880–888. IEEE (2017)
9. Hodaň, T., et al.: BOP challenge 2020 on 6D object localization. In: European Conference on Computer Vision Workshops (ECCVW) (2020)
10. Kingma, D.P., Ba, J.: Adam: a method for stochastic optimization. arXiv preprint arXiv:1412.6980 (2014)
11. Krizhevsky, A., Sutskever, I., Hinton, G.E.: Imagenet classification with deep convolutional neural networks. In: Advances in Neural Information Processing Systems, pp. 1097–1105 (2012)
12. Paszke, A., et al.: Pytorch: an imperative style, high-performance deep learning library. In: Advances in Neural Information Processing Systems, pp. 8026–8037 (2019)
13. Sadeghi, F., Levine, S.: Cad2rl: real single-image flight without a single real image. arXiv preprint arXiv:1611.04201 (2016)
14. Schwarz, M., Behnke, S.: Stillleben: realistic scene synthesis for deep learning in robotics. arXiv preprint arXiv:2005.05659 (2020)
15. Shorten, C., Khoshgoftaar, T.M.: A survey on image data augmentation for deep learning. J. Big Data **6**(1), 60 (2019)
16. Simonyan, K., Zisserman, A.: Very deep convolutional networks for large-scale image recognition. arXiv preprint arXiv:1409.1556 (2014)
17. To, T., et al.: NDDS: NVIDIA deep learning dataset synthesizer (2018). https://github.com/NVIDIA/Dataset_Synthesizer
18. Tobin, J., Fong, R., Ray, A., Schneider, J., Zaremba, W., Abbeel, P.: Domain randomization for transferring deep neural networks from simulation to the real world. In: 2017 IEEE/RSJ International Conference on Intelligent Robots and Systems (IROS), pp. 23–30. IEEE (2017)
19. Tremblay, J., et al.: Training deep networks with synthetic data: Bridging the reality gap by domain randomization. In: Proceedings of the IEEE Conference on Computer Vision and Pattern Recognition Workshops, pp. 969–977 (2018)
20. Yu, F., Wang, D., Shelhamer, E., Darrell, T.: Deep layer aggregation. In: Proceedings of the IEEE Conference on Computer Vision and Pattern Recognition, pp. 2403–2412 (2018)
21. Zhou, X., Wang, D., Krähenbühl, P.: Objects as points. arXiv preprint arXiv:1904.07850 (2019)

SAFFIRE: System for Autonomous Feature Filtering and Intelligent ROI Estimation

Marco Boschi[1]($\boxtimes$) [iD], Luigi Di Stefano[1] [iD], and Martino Alessandrini[2] [iD]

[1] Department of Computer Science and Engineering, University of Bologna, Viale Risorgimento 2, 40136 Bologna, Italy
`marco.boschi5@studio.unibo.it`, `luigi.distefano@unibo.it`
[2] Datalogic S.r.l., Via San Vitalino 13, 40012 Calderara di Reno, BO, Italy
`martino.alessandrini@datalogic.com`

Abstract. This work introduces a new framework, named SAFFIRE, to automatically extract a dominant recurrent image pattern from a set of image samples. Such a pattern shall be used to eliminate pose variations between samples, which is a common requirement in many computer vision and machine learning tasks.

The framework is specialized here in the context of a machine vision system for automated product inspection. Here, it is customary to ask the user for the identification of an anchor pattern, to be used by the automated system to normalize data before further processing. Yet, this is a very sensitive operation which is intrinsically subjective and requires high expertise. Hereto, SAFFIRE provides a unique and disruptive framework for unsupervised identification of an optimal anchor pattern in a way which is fully transparent to the user.

SAFFIRE is thoroughly validated on several realistic case studies for a machine vision inspection pipeline.

Keywords: Machine vision · Object detection · Pattern learning · Graph analysis · Anchor pattern identification · ROI detection

1 Introduction

Machine vision systems are key for the automation of the production line. Hereto, in a typical installation, a camera (or a set of cameras) is mounted on top of the conveyor transporting the manufactured items. The images are analyzed by a specific software which grades the "quality" of the product. Typical examples include: reading/verifying a printed text/barcode; identifying a pattern; measuring a part; verifying the absence of defects (flaw detection), and combinations of thereof. The output grading is then used to keep or discard that item for distribution.

© Springer Nature Switzerland AG 2021
A. Del Bimbo et al. (Eds.): ICPR 2020 Workshops, LNCS 12664, pp. 552–565, 2021.
https://doi.org/10.1007/978-3-030-68799-1_40

It is often the case that the content to be analyzed resides in a small sub-portion of the image (i.e., a *region of interest* (ROI)). As such, a necessary pre-processing step is the identification of such a region in the image. We call this process "ROI finding". Of note, ROI finding is not only required for speedup, but is often necessary for the meaningfulness of the following analysis. For instance, one might need to select the correct line of text before reading it, or to locate the specific mechanical part to be measured. In many cases, the inspected items are left free to assume an arbitrary pose relative to camera and, as such, the ROI has to be tracked by the automated software independently on each sample.

A typical way to address the ROI finding problem is by instructing the system with an *anchor pattern*. An anchor is a distinctive recurrent (i.e., present on all samples) image pattern (e.g., a specific mark on a printed label) identifying a local coordinate system where the ROI position is (as much as possible) constant. Such a pattern shall be detected by the machine vision system and used to position the ROI on a new sample. Importantly, the anchor is generally not located inside the ROI. Indeed, the focus of an inspection is commonly on regions which are expected to change.

In many machine vision software products, the anchor pattern is annotated manually by the user when training the system. Manual annotation is disadvantageous for several reasons:

1. It requires expertise: The user has to know what an anchor is and how to use it to configure the specific inspection (e.g., by instructing a template/pattern matching routine to locate the pattern on new samples);
2. Visual identification of a viable anchor pattern can be complex. For instance, the anchor pattern has to be isolated among many (e.g., for complex printed labels) or, possibly, the anchor is represented by a set of visual features which are distributed sparsely in the image;
3. It is time consuming: It requires accurate visual inspection of multiple items in the lot;
4. It is subjective: Different users might identify different anchors based on their expertise;
5. The anchor selected by a human is very unlikely to be the "best" anchor usable by the ROI selection algorithm.

Conditions (1), (2) and (3) are especially undesirable in a market which is constantly struggling to shift towards plug-and-play solutions.

This paper introduces a new framework, named SAFFIRE,[1] which overcomes the aforementioned limitations:

- The anchor is learned automatically from a set of training images;
- The framework requires only straightforward supervision which only implies a basic understanding of the task of interest. Namely, the user has to draw a ROI around the content of interest. Of note, this is considerably easier than identifying an anchor which can be uncorrelated with the specific task and distributed anywhere in the image;

---

[1] Pronounced as *sapphire*.

- The computed anchor is the best pattern for the ROI selection algorithm.

The proposed framework presents also several computational advantages against alternatives based on Deep Neural Networks (DNN), which are rapidly gaining popularity:

- SAFFIRE can be trained on a normal device within seconds (at most) while DNN-based systems would take minutes to hours and dedicated expensive hardware;
- SAFFIRE requires very few train samples (order of units) while DNN-based systems would require thousands;
- SAFFIRE is agnostic to the specific task while DNN-based solutions are task-specific (e.g., text detectors [5]);
- SAFFIRE can be trained incrementally, in that a new sample can be added seamlessly to the train set to refine the anchor pattern, while DNN-based solutions would require re-training the entire model (almost) from scratch;
- SAFFIRE outputs a lightweight anchor pattern which can be found on a new sample within tens of milliseconds. On the contrary, DNN-based systems output heavy models and inference operates at a slow rate;

The paper proceeds as follows: Related work is presented in Sect. 2; Sect. 3 describes the SAFFIRE algorithm, both at train-time and at run-time; Sect. 4 presents the evaluation setup while Sect. 5 presents the results. Section 6 elaborates on the results and Sect. 7 concludes the work.

2   Related Work

At its core, SAFFIRE exploits an *object detection* paradigm. Object detection is a well known problem in computer vision. Currently, the established approach is based on matching descriptors, i.e. compact representations of local image features, such as corner points, segments or blobs, extracted from the object of interest (see, e.g., [6]). Hereto, a range of techniques, such as the Generalized Hough Transform (GHT) or Random Sample Consensus (RANSAC) can be employed to retrieve instances of a *query* pattern in a new image [6].

Yet, applying the object detection framework as presented above requires the query pattern to be known. DNN-based approaches like YOLO [7] or Faster R-CNN [8] may be used to implicitly learn such pattern, but they do not represent a solution that fulfills the constraints of the addressed application: only very few training images are available, the training time should be quite short and the processing speed at run-time must be as high as possible with limited computational resources. In this context, we propose an original technique where the query pattern (i.e., the anchor) is learned autonomously from a set of images, satisfying the presented constraints without the employment of DNN-based solutions. Technically, this is done by finding the shortest path in a directed acyclic graph (DAG) where each path identifies a possible model candidate.

Moreover, different descriptors are known to perform differently on different kinds of images. For instance, SIFT is arguably the most popular approach

[6] for textured objects, although a number of alternative solutions have been also proposed, such as ORB [9] and SURF [1]. However, approaches based on segments [3] and combination of segments [10] have been shown to be more appropriate for texture-less objects.

The choice of the best descriptor for the specific dataset is typically the result of a trial-and-error process. On the contrary, SAFFIRE implements a selection logic where the features most suited to the specific data are computed at train time.

Finally, classic feature matching is used at run-time to retrieve the anchor model in a new sample and re-position the ROI. In the case of multiple instances, SAFFIRE selects the solution where the image content inside the ROI is the closest to that learned on the train set.

3   Materials and Methods

The processing workflow of SAFFIRE at train time and test time is represented in Fig. 1 and 2, respectively. A description of the pipeline follows.

3.1   Training Data and Reference Image

A set of N_{train} annotated train images $\{I_0, \dots, I_{N_{\text{train}}-1}\}$ is provided as input (Fig. 1a). One of such images, by convenience I_0, is adopted as *reference* image. Each image is annotated with an oriented rotated rectangular ROI R_i. Note that orientation is in general relevant for the task (for instance, it can disambiguate text reading direction).

We assume that all train images contain a distinctive *anchor* pattern mixed with a set of *noise* patterns which are irrelevant for the identification of the ROI. The ultimate goal of the training phase is therefore to retain the anchor pattern while filtering irrelevant patterns out.

In the toy example in Fig. 1a, the dashed rectangle denotes the ROI enclosing the "text" to be read. The highlighted corner is the origin of the ROI and indicates its orientation in that corners are ordered clockwise starting from the origin. The star represents the anchor pattern since its position relative to the ROI is stable. The triangle represents any other pattern in the image which cannot work as an anchor for this task.

3.2   Feature Extraction

A set of visual *features* is extracted from all train images (the red and yellow dots in Fig. 1b). Specifically, we use a well consolidated paradigm for object detection based on matching sets of visual landmarks using their local image descriptors [4]. We make use of multiple types of features to account for textured and texture-less patterns and the logic to choose the best feature type for the specific data is presented in Sect. 3.6.

With this in mind, we define the anchor pattern as the largest subset of features such that:

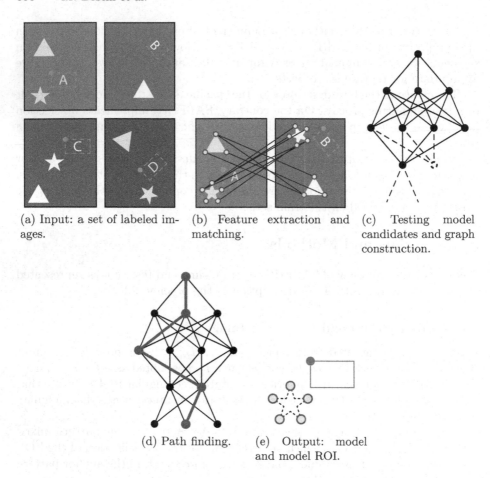

(a) Input: a set of labeled images.

(b) Feature extraction and matching.

(c) Testing model candidates and graph construction.

(d) Path finding.

(e) Output: model and model ROI.

Fig. 1. Train mode: build model and model ROI from train data. Please refer to the main text for a detailed explanation. (Color figure online)

- It is present in all of the train images;
- Its position in different train images can be related by a geometrical transformation. Hereto, although general transformations can be used, we made use of similarities (i.e., roto-translations with isotropic scaling) since they were appropriate for the geometry of the experimental setup and relevant to industrial applications;
- Its position correlates best with the position of the ROI. More quantitatively, the overlap between the ROI's is maximal when they are expressed in the anchor's coordinate system (see Sect. 3.3).

The procedure used to derive such an anchor pattern is described in the remaining of this Section.

The j-th feature on the i-th train image is denoted by f_i^j. A feature is geometrically defined by a position $\mathbf{x}_i^j$, a direction θ_i^j and a size l_i^j. The image content around a feature is accounted for by a *descriptor* $\mathbf{d}_i^j$. Descriptors shall be used to match features between images, as denoted by the blue connecting lines in Fig. 1b.

3.3   The Training Graph

For each feature f_0^j in the reference image, we find its 3 closest matches in the i-th train image. We use 3 nearest neighbours to allow good robustness to multiple instances and/or periodic patterns. We denote by f_i^k one of such matches. Matches with a large inter-descriptor distance are discarded by applying Lowe's ratio test [6]. Further filtering is applied by using the following geometrical considerations:

- Since in our setup scale is not expected to vary significantly, we exclude matches whose scale factor $s = l_0^j/l_i^k$ is not within the interval $[0.85, 1.15]$;
- Recalling our definition of an anchor pattern (see Sect. 3.2), we exclude matches which are not useful to superimpose the ROI's between samples. Hereto, we use the (unique) spatial transform between f_0^j and f_i^k to bring R_0 and R_i to a common origin, and we discard the match if the *intersection over union* (IoU) between the transformed ROI's is below 0.2.

Matches voting for the same spatial transform T_i^t are then aggregated into N_i clusters, with $t = 0, \ldots, N_i - 1$. Clustering is done by iterative application of the RANSAC algorithm [2]. At each iteration, RANSAC is applied to find a transform T_i^t and the "inliers" are removed from the list of matches. The procedure is repeated incrementally until not enough matches are left to estimate a transform.

As such, a directional acyclic graph (DAG), which we refer to as the *training graph* (see Fig. 1c), is created with a number of layers equal to the number of training images. Each layer contains N_i nodes, one for each cluster computed as above. Each node n_i^t in the i-th layer contains:

- A candidate transform T_i^t between I_0 and I_i;
- The indexes of the features in I_0 and I_i voting for T_i^t.

The root layer in the training graph is obtained by matching the reference image with itself. By construction, this gives a single node described by an identity matrix and the indexes of all features in I_0.

3.4   Search of Optimal Path in the Training Graph

As mentioned, the goal is to find the largest subset of features in I_0 which can be used as a common coordinate system maximizing the overlapping between the training ROI's. The problem is solved by searching for the shortest root-to-leaf path in the training graph (in red in Fig. 1d). The length (cost) of a path is defined as a function of two terms: the former depends on the number of overlaps along the path, which is computed from the intersection of the features voting for all the transforms in the path; the latter depends on the alignment and overlap between the aligned ROI's after bringing them all to a common coordinate system by means of the transformations along the path.

To measure ROI alignment, we use the *oriented IoU* (*oIoU*), a modified standard *intersection over union* (IoU) accounting for ROI orientation. Indeed, the classical IoU between two rectangular shapes does not distinguish 180° flips around the center. Specifically, for a node on the n-th layer, it is computed by using the set of spatial transformations on the intermediate path to first bring the n ROI's to a common coordinate system. The proposed oriented IoU is defined in such a way that its value for two flipped and perfectly overlapped rectangles is 0, while their standard IoU would be 1.

Having defined the cost of a path, we then use a tree-search algorithm to retrieve the optimal anchor pattern, the one corresponding to the path of least cost.

3.5   The Anchor

The best path is used to compute the final anchor model, which will be used at run-time to position the ROI on a new sample (see Fig. 1e). Such model is composed of four parts: i) a set of *model features*; ii) the associated *model descriptors*, iii) a *ROI model* R_m and iv) a *ROI descriptor*.

Model features are the subset of $\{f_0\}$ which vote for all the transforms in the best path. For what concerns the *model descriptors*, although one could use the descriptors computed on the reference image alone, this would result in a high bias to the specific choice of the reference image. As such, we defined descriptors which also account for the statistical variation of image content between train samples.

The *ROI descriptor* is used to keep track of the image content inside the ROI. It will be used to discriminate between multiple instances of the anchor model at run time, as described in Sect. 3.7. Let's note that a *global* image descriptor, i.e. accounting for average image properties such as mean image value or average contrast, is best suited for machine vision applications. Indeed, fine image details can change between samples (for instance, the content of a printed text). As for the model descriptors, the ROI descriptor is not tailored to a specific train image, but is rather based on the statistical distribution of image content between images.

3.6   Selection of the Best Feature Type

Since different features are best suited for different patterns (e.g., textured vs. texture-less objects), SAFFIRE employs an optimized feature selection strategy which chooses the feature-descriptor pair which performs best on the specific data. Both accuracy and speed are used as a selection criterion.

3.7   Testing

The trained model is used at run-time to position the ROI in a new unseen sample (see Fig. 2a) with a two-step process.

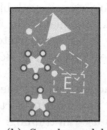

(a) Input: a test image.

(b) Search model in image.

(c) Remove bad matches with ROI content.

(d) Output: ROI on the new image.

Fig. 2. Test mode: find the trained model on unseen data. Please refer to the main text for a detailed explanation. (Color figure online)

First, a standard object detection paradigm is used to find instances of the trained model in the new image (Fig. 2b). Namely, features and descriptors of the selected type (see Sect. 3.6) are extracted on the test image and matched to the anchor pattern model (see Sect. 3.3). Using the GHT, as in [6], feature correspondences are turned into votes to an accumulation array (AA) with peaks in the AA representing possible instances of the anchor model in the test image. Hereto, a peak pruning procedure, based on the relative number of votes in each peak, is applied to keep only the most significant matches. For each valid instance of the anchor model, the corresponding spatial transform is used to position the ROI model in the new sample, as described in see Sect. 3.5.

With regard to Fig. 2, the test image contains two instances of the "star" anchor model, which presents circular symmetry. This could result in many matches (hence, ROI positions) with equal score, but only that containing the letter "E" is relevant. Such matches are displayed in Fig. 2b as yellow dashed boxes.

The following step consists therefore in discarding non-relevant matches (see Fig. 2c) by computing the content descriptor for each candidate ROI (see Sect. 3.5). The weighted ℓ-2 distance to the model ROI descriptor $\mathbf{d}_{\text{model}}$ is:

$$d(\mathbf{d}_{\text{test}}, \mathbf{d}_{\text{model}}, \mathbf{w}) = \sqrt{\sum_{i=0}^{47} w_i \left(\mathbf{d}_{\text{test}}(i) - \mathbf{d}_{\text{model}}(i)\right)^2} \qquad (1)$$

where $\mathbf{d}(i)$ denotes the i-th bin of $\mathbf{d}$. A weight $w_i \in [0, 1]$ assigns highest (respectively, lowest) weight to bins with the lowest (respectively, highest) variance. The ROI with the lowest distance is the one used to analyze the sample at run time, as depicted in Fig. 2c and Fig. 2d.

4   Evaluation Setup

4.1   Evaluation Datasets

The datasets used to validate SAFFIRE are presented in Fig. 3 with 2 samples from each alongside the total number of images available for each one. The datasets are not suited for a single task, but many can be applied: the main task is indeed reading the text to ensure it has been printed correctly, but also measurements (see Fig. 3i) and flaw detection (see Fig. 3c and Fig. 3e) can be carried out.

The images have been shuffled upon acquisition and the first three have been used for the training phase. The results presented in Sect. 5 are computed on the whole datasets, including the images used in training.

4.2   Evaluation Metrics

The search capabilities of the trained models are evaluated by comparing the computed ROI with the ground truth one by means of $oIoU$ between the two. For each model the individual $oIoU$'s are aggregated in a single plot showing the distribution of $oIoU$ over the samples in the dataset by using a box plot.

The box in each box plot extends from the first q_1 to the third quartile q_3 of the distribution and, inside it, the median is highlighted. Outside the box, the "whiskers" extend on the left to the lowest value in the distribution greater than $q_1 - 1.5 \cdot \text{IQR}$, with the interquartile range (IQR) defined as $q_3 - q_1$; similarly on the right to the highest value less than $q_3 + 1.5 \cdot \text{IQR}$. Samples outside the range of the whiskers are considered outliers and plotted as single points.

5   Results

5.1   Noise Patterns and Filtering

As detailed in Sect. 3.4, the path cost depends both on the number of overlaps of votes along the path and how well the ROI's are aligned, in the following we will show the importance of the second term.

Considering the StarCart dataset (see Fig. 3a), it can be seen that both the cart and the star have a somewhat static position with respect to the ROI, however the orientation changes in the case of the cart. We applied the training procedure using only segment based features due to the nature of the shapes, and report the results in Fig. 4. As depicted in Fig. 4a, considering only the overlaps results in taking the cart as the learned model due to the fact that its shape is formed by many more segments than the star is. Adding the term based on ROI $oIoU$ results instead in the desired model (see Fig. 4b).

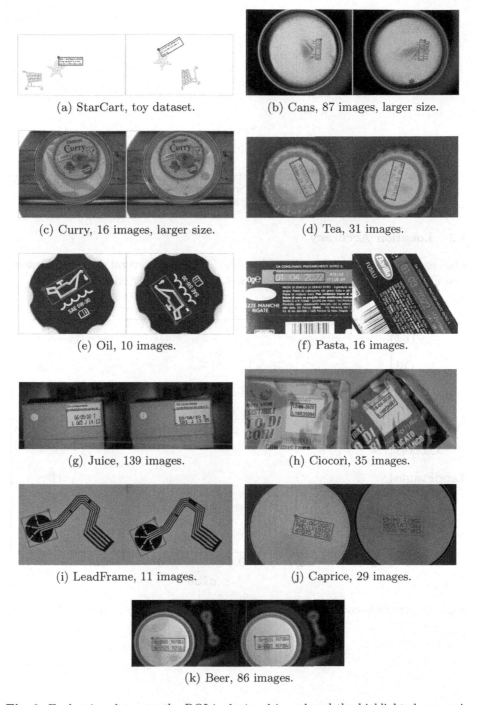

(a) StarCart, toy dataset.

(b) Cans, 87 images, larger size.

(c) Curry, 16 images, larger size.

(d) Tea, 31 images.

(e) Oil, 10 images.

(f) Pasta, 16 images.

(g) Juice, 139 images.

(h) Ciocorì, 35 images.

(i) LeadFrame, 11 images.

(j) Caprice, 29 images.

(k) Beer, 86 images.

Fig. 3. Evaluation datasets: the ROI is depicted in red and the highlighted corner is the origin. Real datasets also include the total number of images (Color figure online).

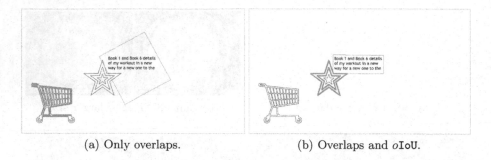

(a) Only overlaps.	(b) Overlaps and oIoU.

Fig. 4. Training over the StarCart dataset: the thick green lines are the segments of the model, the lighter ones delimit the ROI of said model. (Color figure online)

5.2 Location Accuracy

The search capabilities of SAFFIRE are evaluated over several different datasets (see Fig. 3) using oIoU as metric (see Sect. 4.2) and reported in Fig. 5.

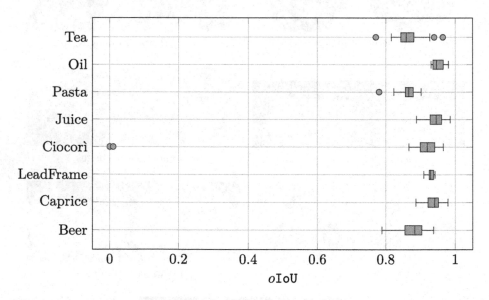

Fig. 5. Distributions of oIoU of computed and ground truth ROI over several datasets.

5.3   Computation Time

SAFFIRE has been implement to run on a Datalogic Memor 20,[2] powered by a Snapdragon 600 CPU. On this platform the train process takes just a few seconds and the search tens of milliseconds (with a few exceptions) as reported in Table 1. Search times are averaged over all images in the dataset and train times over 5 executions.

Table 1. Average times for training and search over several datasets.

		Tea	Oil	Pasta	Juice	Ciocorì	LeadFrame	Caprice	Beer
Train	**Time**	1.30 s	983.33 ms	8.63 s	3.13 s	3.83 s	896.10 ms	1.24 s	2.78 s
	Features	A	A	B	B	A	A	A	A
Search	**GHT**	8.62 ms	2.90 ms	744.38 µs	1.98 ms	774.97 µs	4.89 ms	1.11 ms	45.05 ms
	Total	37.05 ms	36.20 ms	106.21 ms	99.16 ms	35.28 ms	38.89 ms	32.32 ms	86.70 ms

During development, SAFFIRE has been tested also on a personal computer powered by an Intel Core i5 7267U CPU. The results over a different selection of datasets, which were mainly used during development, are reported in Table 2.

Table 2. Average times for training and search over a few datasets on the development platform.

		Cans	Curry	Beer
Train	**Time**	4.92 s	8.28 s	3.99 s
	Features	B	B	A
Search	**GHT**	19.40 ms	41.59 ms	26.53 ms
	Total	126.76 ms	216.23 ms	35.95 ms

6   Discussions

The results in Fig. 5 shows that SAFFIRE can obtain an high accuracy in identifying the ROI. From our experiments and by means of a qualitative evaluation of results, we found that even with $oIoU$ of about 0.85, localization of the ROI can be considered accurate. This is caused by using as model a ROI slightly larger than those used for training and as ground truth, due to how it is constructed. We also note some failures for Ciocorì denoted by the points around 0, likely caused by the reflective surface of the packaging (see Fig. 3h).

Table 1 indeed shows that the search time is fast, but by isolating the identification of the ROI, by means of the GHT and the ROI content descriptor to

---

[2] https://www.datalogic.com/eng/retail-manufacturing-transportation-logistics-healthcare/mobile-computers/memor-20-pd-869.html.

choose the right candidate, it emerges how our approach is extremely fast and most of the time is taken by feature extraction and descriptor matching. Indeed, the fastest features are preferred as SAFFIRE prefers choosing the ones with the least computational time even if this means sacrificing a bit of accuracy (see Sect. 3.6). The results in Table 2 are provided as a mean of comparison of running SAFFIRE on an non-embedded device and, as such, they lack some optimizations that led to the final results.

We also note that the SAFFIRE framework is extremely general and independent on the specific technical choices presented in this work:

- Any feature descriptor can be used;
- Other techniques can be used instead of RANSAC to compute the nodes of the training graph, such as the GHT;
- Arbitrary spatial transforms can be employed depending on the geometry of the problem;
- Any graph-search algorithm can be used to find the optimal model;
- Many image descriptors can be used to encode the ROI image content.
- The rectangular ROI in this study is not a limitation and can be replaced with arbitrary shapes with a defined notion of "direction", in order to simplify the analysis task following the ROI identification, e.g., to force text to be upright, with this definition even circles can be used.

The algorithm can also be adapted to work in a "fully unsupervised" fashion not requiring the provision of a ROI. This is simply achieved by adjusting the path cost in the graph to depend only on the number of overlaps, thus discarding the alignment of the ROI's. This way, SAFFIRE will look for the most stable pattern in the image sequence with the highest number of features. For instance, this could be sufficient to align images of similar items acquired in different positions, but care must be taken as it can results in cases like that presented in Sect. 5.1, in which relying only on feature overlaps results in a bad alignment of ROI's.

7   Conclusions

We have introduced a new algorithm, called SAFFIRE, for unsupervised identification of a stable anchor pattern from a set of images. Such an anchor is used to remove pose variations between samples before further processing. Data normalization is a fundamental pre-processing step for many computer vision and machine learning tasks.

The use of SAFFIRE was shown in the context of machine vision systems for industrial automation, where it represents a disruptive element enabling the implementation of an automatic ROI finding mechanism in an unsupervised fashion.

Intellectual Property. Patent application pending.

References

1. Bay, H., Tuytelaars, T., Van Gool, L.: SURF: speeded up robust features. In: Leonardis, A., Bischof, H., Pinz, A. (eds.) Computer Vision - ECCV 2006. Lecture Notes in Computer Science, pp. 404–417. Springer (2006). https://doi.org/10.1007/11744023_32

2. Fischler, M.A., Bolles, R.C.: Random sample consensus: a paradigm for model fitting with applications to image analysis and automated cartography. Commun. ACM **24**(6), 381–395 (1981). https://doi.org/10.1145/358669.358692

3. von Gioi, R.G., Jakubowicz, J., Morel, J.M., Randall, G.: LSD: a fast line segment detector with a false detection control. IEEE Trans. Pattern Anal. Mach. Intell. **32**(4), 722–732 (2010). https://doi.org/10.1109/TPAMI.2008.300

4. Krig, S.: Interest Point Detector and Feature Descriptor Survey, pp. 217–282. Apress, Berkeley (2014). https://doi.org/10.1007/978-1-4302-5930-5_6

5. Long, S., He, X., Yao, C.: Scene text detection and recognition: the deep learning era (2018)

6. Lowe, D.G.: Distinctive image features from scale-invariant keypoints. Int. J. Comput. Vis. **60**(2), 91–110 (2004). https://doi.org/10.1023/B:VISI.0000029664.99615.94

7. Redmon, J., Divvala, S., Girshick, R., Farhadi, A.: You only look once: unified, real-time object detection. In: 2016 IEEE Conference on Computer Vision and Pattern Recognition (CVPR), pp. 779–788 (2016). https://doi.org/10.1109/CVPR.2016.91

8. Ren, S., He, K., Girshick, R., Sun, J.: Faster R-CNN: towards real-time object detection with region proposal networks. IEEE Trans. Pattern Anal. Mach. Intell. **39**(6), 1137–1149 (2017). https://doi.org/10.1109/TPAMI.2016.2577031

9. Rublee, E., Rabaud, V., Konolige, K., Bradski, G.: ORB: an efficient alternative to SIFT or SURF. In: 2011 International Conference on Computer Vision, pp. 2564–2571 (2011). https://doi.org/10.1109/ICCV.2011.6126544

10. Tombari, F., Franchi, A., Di Stefano, L.: BOLD features to detect texture-less objects. In: 2013 IEEE International Conference on Computer Vision, pp. 1265–1272. IEEE (2013). https://doi.org/10.1109/ICCV.2013.160

Heterogeneous Feature Fusion Based Machine Learning on Shallow-Wide and Heterogeneous-Sparse Industrial Datasets

Zijiang Yang[1]([✉]), Tetsushi Watari[2], Daisuke Ichigozaki[2], Akita Mitsutoshi[2], Hiroaki Takahashi[2], Yoshinori Suga[2], Wei-keng Liao[1], Alok Choudhary[1], and Ankit Agrawal[1]

[1] Department of Electrical and Computer Engineering, Northwestern University, Evanston, USA
{zyz293,wkliao,choudhar,ankitag}@eecs.northwestern.edu
[2] Toyota Motor Corporation, Toyota, Japan

Abstract. Although machine learning has gained great success in industry, there are still many challenges in mining industrial data, especially in manufacturing domains. Because industrial data can be 1) shallow and wide, 2) highly heterogeneous and sparse. Particularly, mining on sparse data (i.e. data with missing features) is extremely challenging, because it is not easy to fill in some features (e.g. images), and removing data points would reduce the data size further. Thus, in this work, we propose a machine learning framework including transfer learning, heterogeneous feature fusion, principal component analysis and gradient boosting to solve these challenges and effectively develop predictive models on industrial datasets. Compared to a non-fusion method and a traditional fusion method on two real world datasets from Toyota Motor Corporation, the results show that the proposed method can not only maximize the utility of available features and data to achieve more stable and better performance, but also give more flexibility when predicting new unseen data points with only partial set of features available.(Code and data are available at: https://github.com/zyz293/FusionML.)

1 Introduction

Developing machine learning models on industrial datasets can be extremely challenging, especially in manufacturing domains: 1) Industrial dataset can be shallow and wide: Shallow means dataset is small because it is time-consuming to conduct an experiment or run a physics simulation in manufacturing domains. Wide means number of features is large due to the complexity and huge amount of parameters in experiment and physics simulation. Thus, the curse of dimensionality caused by shallow and wide data is usually a problem. 2) Industrial dataset is highly heterogeneous and sparse: Industrial data can have various types of data types, such as numerical data and image data. Taking experiments in materials science related manufacturing domain as an example, numerical data might contain composition and manufacturing processing information of

A. Del Bimbo et al. (Eds.): ICPR 2020 Workshops, LNCS 12664, pp. 566–577, 2021.
https://doi.org/10.1007/978-3-030-68799-1_41

materials, while image data can also be heterogeneous, such as different images obtained by different imaging techniques, or at different locations and different scales of the material sample. Thus, it is challenging to utilize such highly heterogeneous data, and the heterogeneity of industrial data could also cause the curse of dimensionality. Moreover, industrial data can be very sparse because there could be many missing features in the dataset, which can be missing attributes in numerical data and/or missing images in image data. However for machine learning, usually the input is required to have exactly same format of feature vector without missing values. There are usually two methods to handle missing features. One way is to fill in missing values before model training, but it is not easy to fill in missing images. The other way is to simply remove the data points with missing values, but that would further reduce the data size. Thus, it is desirable to somehow have the training process intelligently deal with the missing features so that the utility of available features/data can be maximized.

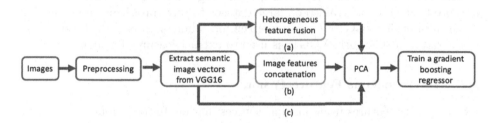

Fig. 1. The flowchart of (a) the proposed method where heterogeneous feature fusion is applied, and (b) baseline method where traditional feature fusion method is used by directly concatenating features. (c) non-fusion baseline method where single image is used as input.

In this work, we propose a data-driven methodology to overcome the above challenges and develop predictive models on industrial datasets, particularly focusing on image datasets (see Fig. 1(a)). In other words, a machine learning framework including transfer learning, heterogeneous feature fusion, principal component analysis (PCA) and gradient boosting is proposed to solve curse of dimensionality, handle data with missing images, and train predictive models on heterogeneous industrial data. More specifically, the reduced sets of image feature vectors are extracted from input images by transfer learning and principal component analysis, which solves the problem of the curse of dimensionality (i.e. the first challenge). The heterogeneous feature fusion would intelligently deal with missing input images and fuse heterogeneous image feature vectors to solve the second challenge. To evaluate the proposed method, we select a traditional fusion method in work [1] as baseline, which fuses the features by directly concatenating features from different data resources (see Fig. 1(b)). In addition, a non-fusion machine learning method, which takes single image as input is selected as another baseline (see Fig. 1(c)). We compare the performance of the proposed method and baseline methods on two real world datasets (collected by Toyota Motor Corporation). The results show that the proposed method can not only maximize the utility of available data to achieve more stable and better performance, but also give more flexibility when predicting new unseen data points with only partial set of features (i.e. images) available.

2   Background

2.1   Transfer Learning

Deep learning has shown its striking performance in various fields [2–4]. In general, large amount of reliable data is required to train a deep learning model from scratch. However, it is difficult or even impossible to collect such large amount of data in some industrial domains. Thus, transfer learning provides an alternative to take advantage of deep learning techniques on relatively small dataset. More specifically, the main concept of transfer learning is to reuse the learned features in a pre-trained model trained on large dataset to solve problems on similar but small dataset [5]. VGG16 [6] is one of the most famous pre-trained models for image data. In this work, VGG16 is used to extract features for images.

2.2   Heterogeneous Feature Fusion

By integrating features from multiple sources, feature fusion can be used to produce more accurate and useful information than that provided by single feature resource. However, traditional feature fusion is primarily used to fuse features from different models [6–8]. The proposed feature fusion in this work is different from existing methods as we focus on feature fusion of different images for a *single* data point, even if some of the images are missing. More specifically, unlike other feature fusion methods where the goal is to study the fusion of different features computed for the same input with different techniques or models, the proposed method is to fuse whatever features (i.e. image in this work) a data point might have into a uniform feature space so that the training set can be enlarged for model training. This is due to the fact that the input of machine learning model is usually required to have the same feature length. If there are missing features of a data point, we usually either fill in the missing features or simply remove that data point. However, it is not easy to fill in some features (e.g. image features), and removing data points would reduce the data size further. The proposed method can handle missing features (i.e. images features in this work) during training process so that the utility of available data could be maximized for model training. Thus, the proposed method could be very useful in industry where the dataset is shallow and wide, heterogeneous and sparse.

3   Method

3.1   Proposed Method

As mentioned in previous sections, the challenges of mining on industrial data is that the dataset is shallow and wide, heterogeneous and sparse. In traditional machine learning method, all the model inputs should have exactly same number of features, and missing feature values are not allowed. If more features are considered as input for the model (i.e. using multiple images as input in this work), the dataset becomes even smaller because data points without corresponding features (i.e. without corresponding images in this work) would be removed. Thus, it is of importance to design a machine learning model that can handle data points with missing features (i.e. missing images) so that utility of available data can be maximized. In other words, the model should be trainable on a dataset where data points can have different partial set of features (i.e. images) so that the size of training data can be enlarged. More importantly, such model also gives more flexibility when making predictions on the new data points where new data points also only have partial set of features (i.e. images) available. Thus, we propose a machine learning framework to achieve this purpose.

The flowchart of the proposed model is shown in Fig. 1(a), and the proposed method can be divided into five steps,

- Preprocessing: Image preprocessing, such as image resizing and rescaling, is necessary before feeding images into VGG16 to extract image feature vectors.
- Transfer learning: Because the dimensionality of images is much higher than the number of data points, curse of dimensionality would be a major problem for model training. Thus, we use transfer learning, heterogeneous feature fusion and PCA to extract low-dimensional image feature vectors v for input images of a data point and fuse them into a joint features space. More specifically, we applied VGG16 as image feature extractor by keeping all the convolutional blocks and removing fully connected layers. Image is fed into the network to compute the output of the last convolutional block (i.e. 512 feature maps), and global average pooling [9] is applied on each feature map individually to form a 1-D image feature vector v with 512 entries.
- Heterogeneous feature fusion (see Algorithm 1): Image feature vectors v of all the available input images of a data point are computed using transfer learning. Since the same pre-trained model is used for transfer learning, comparable features would be extracted even for different image types. In other words, the feature space of image feature vectors v for different input images are the same, which provides the opportunity to do further calculation even though input image types are different, some of which could even be missing. Thus, these image feature vectors v can be fused by computing feature-wise average for each feature in image feature vectors v of a data point. After fusion, a 1-D fused image feature vector F with 512 entries is obtained for each data point which represents all the input images of that data point. In other words, each data point would use a 1-D fused image feature vector F

to represent its input images. In this way, data points with different subset
of features can be used as input for a machine learning model.

- Principal component analysis: The dimensionality of fused features can still
  be high compared to the size of the data. Thus, PCA is applied to further
  reduce the dimensionality where the summation of explained variance of the
  selected principal components is ensured to be above 95%. In this way, after
  transfer learning, heterogeneous feature fusion and PCA, we can use a 1-D
  vector with about a couple of dozens of entries to represent multiple input
  images with tens of thousands of pixels.
- Model training: Finally, gradient boosting method is applied which takes such
  1-D vectors as input to train the predictive model.

Algorithm 1: Heterogeneous feature fusion

Select a data point i;
Initialize an empty vector, $feature\_vector\_list$;
for *the kth input image $img_{i,k}$ of data point i* **do**
    if $img_{i,k}$ *exists* **then**
        Compute image feature vector, $v_{i,k} = f1(img_{i,k})$;
        // $f1(.)$ denotes VGG16 feature extractor to compute image
            feature vector.
        Append $v_{i,k}$ to $feature\_vector\_list$;
    end
end
Compute fused image feature vector, $F = f2(feature\_vector\_list)$;
// $f2(.)$ computes feature-wise average for each feature of image
    feature vectors in feature_vector_list.
return F;

3.2   Baseline Methods

The method from work [1] is selected as the first baseline (referred as baseline 1),
which uses a low-level fusion approach by directly concatenating the extracted
features to construct a joint features space. As shown in Fig. 1(b), the training
process of baseline method is similar to that of the proposed method, except
that there is no heterogeneous feature fusion step. More specifically, after image
preprocessing, the low-dimensional image feature vector v is extracted using
transfer learning with VGG16 for each image of a data point. If multiple images
are used as input, the image feature vectors v of these images are concatenated
as a 1-D vector. Then, PCA is applied to further reduce the dimensionality, and
summation of explained variance of the selected principal components is ensured
to be above 95%. Finally, gradient boosting is used to train the predictive model.
In addition, a non-fusion baseline method (referred as baseline 2 as shown in

Fig. 1(c)) is also used where single image is used as input, and the training process is the same as the first baseline except that no fusion step is applied. Note that for baseline 1, only those data points for which all images are available can be used for modeling, while for baseline 2, only a single image can be used for each data point.

3.3   Evaluation Method

To evaluate the proposed method, we conduct a systematic experiment as shown in Algorithm 2. We first spilt dataset into training set and testing set. Then, we randomly select a portion of data points in training set (ranging from 1% to 99% with 1% increment). For each set of selected training data points, we carry out $N + 3$ experiments where N denotes the total number of images a data point has. The entire evaluation process is repeated for 30 times, and the averaged performance is reported.

- In the first $N - 1$ experiments, we randomly remove a fixed number of images (ranging from 1 to $N - 1$ with 1 increment) for each selected training data point. After processing, the proposed method is trained on processed training set, and evaluated on the same hold-out testing set.
- In the Nth experiment, we randomly remove a portion of images (any integer number between 1 to $N - 1$) for each selected training data point. Then, the proposed method is trained on processed training set (the resulting trained model is referred as $F(.)$) and evaluated on same hold-out testing set.
- In the $(N + 1)^{th}$ experiment, we randomly remove a portion of images (any integer number between 1 to $N-1$) for each data point in the testing set, and evaluate the trained proposed model in the Nth experiment $F(.)$ on processed testing set.
- In the $(N + 2)^{th}$ and $(N + 3)^{th}$ experiment, we remove the selected training data points from training set, because both baseline methods can not deal with data points with missing image feature vectors. Then both baseline methods are trained on reduced training set, and evaluated on the same hold-out testing set, respectively.

In this way, we can conduct a thorough comparison between the proposed method and both baseline methods. In addition, although cross validation is widely used to evaluate model's performance for relatively small datasets, we instead evaluate the model's performance on the same hold-out testing set for all the experiments. In this way, we can clearly compare and evaluate how the proposed method would perform when it is trained and evaluated on different number of data points with different number of missing images.

In this work, we use mean absolute percentage error (MAPE) as error metric,

$$MAPE = \frac{1}{N} \sum_{i=1}^{N} |\frac{y_i - \hat{y}_i}{y_i}| \times 100\% \tag{1}$$

Where N denotes the number of data points, y_i and $\hat{y}_i$ represent ground truth value and predicted value for the ith data point in the testing set, respectively.

Algorithm 2: Evaluation process

Split dataset into training set and testing set;
for *i% in the range from* 1% *to* 99% *with* 1% *increment* **do**

> Randomly select $i\%$ data points from training set;
>
> // First $N-1$ experiments where N denotes the total number of
>   images a data point has
>
> **for** *integer n in the range from* 1 *to* $N-1$ **do**
>
>> **for** *the* k^{th} *data point in selected training data points* **do**
>>
>>> | Randomly remove n images for the kth data point;
>>
>> **end**
>> Train the proposed method on processed training set, evaluate the
>>   performance on the same hold-out testing set, and save the results;
>
> **end**
>
> // The Nth experiments
>
> **for** *the* k^{th} *data point in selected training data points* **do**
>
>> Randomly select an integer n in the range from 1 to $N-1$;
>> Randomly remove n images for the kth data point;
>
> **end**
> Train the proposed method ($F(.)$) on processed training set, evaluate the
>   performance on the same hold-out testing set, and save the results;
>
> // The $(N+1)^{th}$ experiments
>
> **for** *the* k^{th} *data point in testing set* **do**
>
>> Randomly select an integer n in the range from 1 to $N-1$;
>> Randomly remove n images for the kth testing data point;
>
> **end**
> Evaluate the trained proposed method ($F(.)$) on processed testing set, and
>   save the results;
>
> // The $(N+2)^{th}$ and $(N+3)^{th}$ experiments
>
> Remove the selected training data points from training set;
> Train both baseline methods on reduced training set, evaluate the
>   performance on the same hold-out testing set, and save the results,
>   respectively;

end

---

4   Results and Discussion

4.1   Toyota Dataset No. 1

Toyota dataset No. 1 is collected by Toyota Motor Corporation [1], and it contains data of 116 material samples. Each sample has six Scanning Electron Microscopy (SEM) images, which is a combination of three different magnifications at two different locations of the material samples (see Fig. 2 for an example image). Four material properties are measured for each data point (referred as P1-P4). We randomly select 20% data points for testing, and the rest for training in each evaluation process. Two image preprocessing steps are applied. First, the label information at the bottom of SEM image is cropped since it does not contain any structure information of the material. Second, the SEM image is resized to 224×224.

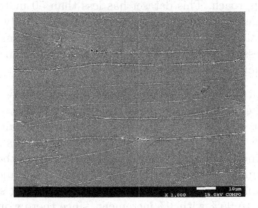

Fig. 2. An example of SEM image taken with $\times 1000$ magnification at the center location of material sample in Toyota dataset No. 1.

Figure 3(a)–(d) compares all the experiment results, except the $(N + 1)^{th}$ experiment for the four properties. Compared to baselines, the proposed method gives more stable and better performance irrespective of the ratio of data points with missing images and the number of available images of each data point. More specifically, when five images are removed for selected training data points (i.e. curve "fusion (5)"), the performance of the proposed method is not as good as when it is trained on processed training set with less images removed for selected training data points (although it still outperforms both the baselines). This might be because too much information gets hidden by removing five images for selected training data points and thus the proposed model is unable to make as accurate predictions as it does in other experiments.

Figure 3(e)–(h) compares the performance of the Nth to $(N + 3)^{th}$ experiments for all the four properties. "fusion (random) test (0)" curve shows the results of the Nth experiment, and "fusion (random) test (random)" curve represents the results of the $(N+1)^{th}$ experiment. We can observe that even though the proposed method predicts on testing set with random missing images, it gives stable and comparable performance as predicting on testing set without missing images, and it is significantly better than the baselines predicting on testing set without missing images. It is important to note that the traditional machine learning methods (i.e. baselines) cannot be used at all to make predictions on testing set with missing images.

4.2   Toyota Dataset No. 2

In addition, we also evaluate the proposed method on another dataset from Toyota Motor Corporation, the details of which cannot be shared due to confidentiality, but here we present the application of the proposed technique on this dataset at a high-level to illustrate another real-world scenario of shallow-wide and heterogenous-sparse dataset with actual missing images, and the usefulness of the proposed approach. This dataset has less than 50 material samples, and each data point has up to seven image types, of which only a subset is available for a given data point. The problem is to predict two material properties, and they are regression problems. To evaluate the performance of the proposed method, we conduct the experiments as below,

- We first tried all the possible combinations of input images for baseline 1 and baseline 2, and cross validation is used to select which combination of input images performs the best. Note that very few material samples (less than 5) have one of the seven image types, so this image type is not considered as input for baselines since that would significantly reduce the dataset.
- For best performing baseline model (best of baselines 1 and 2) of each property, we randomly select $i\%$ of its data points as testing set. Note that i is varied from 5% to 95% with 1% increment, since there would be only few or even no data points on both tails (i.e. between 1% to 5% and 95% to 99%). The training set consists of two parts, where one is the unselected part (i.e. $100\% - i\%$) of the dataset from baseline model, and the other part is rest of data that was removed due to missing images for baseline model. In this way, the proposed model can be evaluated on the same hold-out testing set as the baseline model for a fair comparison.
- The baseline model and the proposed method are trained and evaluated on the dataset splits from the previous step.
- The previous two steps are repeated 30 times, and the averaged results are reported.

Figure 4 shows that the proposed fusion method performs better than the baseline method for property No. 1, but the performance of both methods are significantly degraded when the ratio of data points in testing set reaches 85%. On the other hand, the performance of the proposed fusion method is pretty much comparable to the baseline method for property No. 2, with fusion method performing slightly worse (better) for testing set ratio below (above) 80%. Thus, the proposed fusion method can not only achieve comparable or even better performance compared to traditional machine learning methods, but also gives more

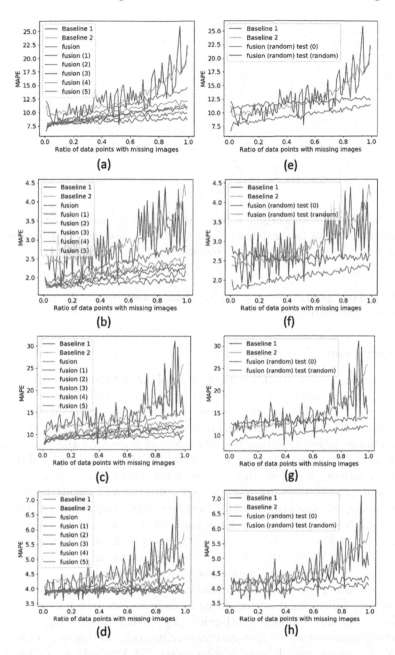

Fig. 3. Performance comparison between the proposed method and two baseline methods on Toyota dataset No. 1. (a) and (e), (b) and (f), (c) and (g), (d) and (h) shows the results comparison for property P1 to P4, respectively. Note that in plot (a)-(d), "fusion (i)" curves show the results of the first $N-1$ experiments where i indicates the number of images removed for selected training data points, and "fusion" curve represents the result of the Nth experiment. In plot (e)–(h), "fusion (random) test (0)" curve shows the results of the Nth experiment, and "fusion (random) test (random)" curve represents the results of the $(N+1)^{th}$ experiment.

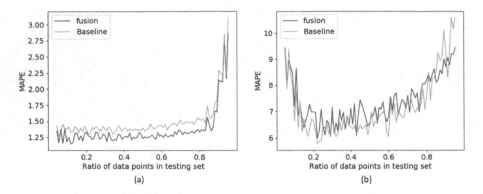

Fig. 4. Performance comparison between the proposed method and best baseline method on additional Toyota dataset No. 2. (a) and (b) present the performance comparison for Property No. 1 and No. 2, respectively.

flexibility when predicting new unseen data points even with partial features (i.e. images) available.

4.3 Discussion

From the experimental results on two industrial datasets from Toyota, we observe that the proposed fusion method is able to achieve more stable and better accuracy of predictive models. Further, the biggest advantage of the proposed method is that it gives more flexibility when making predictions on new unseen data points with only partial set of features (images), which is very important and practically useful in real-world scenarios from three aspects. (1) Maximizing the utility of data: Industrial datasets, especially those collected using expensive and time consuming experiments, usually contain missing features/images, which leads to low data utility when using traditional machine learning methods. The proposed method allows building predictive models using all the available data, so that the data utility can be maximized and more information could be learned, leading to more robust predictive models. (2) Reducing costs: Since the proposed models can be used to make predictions on data with missing images, they can be used to significantly reduce costs by screening the promising candidates. More specifically, experimentalists can first only collect the most important or easy-to-obtain images of the samples to make an initial prediction, and thus identify the most promising candidates for further investigation with higher-end processing steps. In this way, a tremendous amount of cost and man-hours could be saved. (3) Accelerating product design: Experiments are extremely expensive in terms of time, manpower, and money. Accurate and robust predictive models establishing the cause-effect relationships of science can in turn help to realize of inverse models of engineering (discovery and design) [10] by quickly identifying promising directions for future experiments. This combined with the significant resource savings can provide unprecedented opportunities for accelerating product design.

5   Conclusion

In this work, we propose a machine learning framework including transfer learning, heterogeneous feature fusion, principal component analysis, and gradient boosting to overcome the challenges when mining on industrial datasets. Compared to two baseline methods on two industrial datasets, the proposed fusion method was found to not only achieve more stable and superior performance, but also give more flexibility when making predictions on new unseen data points. Thus, the proposed method could be very useful and practical for building predictive models in real-world situations, since it can maximize the utility of available data, reduce costs, and accelerate product design.

Acknowledgment. This work was supported in part by Toyota Motor Corporation and NIST CHiMaD (70NANB19H005).

References

1. Yang, Z.: Data-driven insights from predictive analytics on heterogeneous experimental data of industrial magnetic materials. In: 2019 International Conference on Data Mining Workshops (ICDMW), pp. 806–813. IEEE (2019)
2. Agrawal, A., Choudhary, A.: Deep materials informatics: applications of deep learning in materials science. MRS Commun. **9**(3), 779–792 (2019). https://doi.org/10.1557/mrc.2019.73
3. Goodfellow, I., et al.: Generative adversarial nets. In: Advances in neural Information Processing Systems, pp. 2672–2680 (2014)
4. Yang, Z., et al.: Deep learning based domain knowledge integration for small datasets: illustrative applications in materials informatics. In: 2019 International Joint Conference on Neural Networks (IJCNN), pp. 1–8. IEEE (2019)
5. Gopalakrishnan, K., Khaitan, S.K., Choudhary, A., Agrawal, A.: Deep convolutional neural networks with transfer learning for computer vision-based data-driven pavement distress detection. Constr. Build. Mater. **157**, 322–330 (2017)
6. Simonyan, K., Zisserman, A.: Very deep convolutional networks for large-scale image recognition. arXiv preprint arXiv:1409.1556 (2014)
7. Lin, T.-Y., RoyChowdhury, A., Maji, S.: Bilinear CNN models for fine-grained visual recognition. In: Proceedings of the IEEE International Conference on Computer Vision, pp. 1449–1457 (2015)
8. Bodla, N., Zheng, J., Xu, H., Chen, J.-C., Castillo, C., Chellappa, R.: Deep heterogeneous feature fusion for template-based face recognition. In: 2017 IEEE Winter Conference on Applications of Computer Vision(WACV)
9. Lin, M., Chen, Q., Yan, S.: Network in network. arXiv preprint arXiv:1312.4400 (2013)
10. Agrawal, A., Choudhary, A.: Perspective: materials informatics and big data: realization of the "fourth paradigm" of science in materials science. APL Mater. **4**(053208), 1–10 (2016)

3-D Deep Learning-Based Item Classification for Belt Conveyors Targeting Packaging and Logistics

Ho-min Park[1,3](✉) , Byungkon Kang[2] , Arnout Van Messem[1,4] ,
and Wesley De Neve[1,3]

[1] Center for Biotech Data Science,
Ghent University Global Campus, Incheon, South Korea
{homin.park,arnout.vanmessem,wesley.deneve}@ghent.ac.kr
[2] Department of Computer Science,
The State University of New York Korea, Incheon, Korea
byungkon.kang@sunykorea.ac.kr
[3] IDLab, Department of Electronics and Information Systems, Ghent University,
Ghent, Belgium
[4] Department of Applied Mathematics, Computer Science and Statistics,
Ghent University, Ghent, Belgium

Abstract. In this study, we apply concepts taken from the fields of Artificial Intelligence (AI) and Industry 4.0 to a belt conveyor, a key tool in the packaging and logistics industries. Specifically, we present an item classification model built for belt conveyors, helping the conveyor control system to recognize items while minimizing its impact on the conveyor design and the movement of items. To that end, we followed a three-pronged approach. First, we converted a size measurement system into a 3-D shape reconstruction system by recycling a belt conveyor prototype developed in a previous study. Secondly, we transformed a scanned point cloud that varies in size, given the use of variable-length items, into a point cloud with a fixed size. Thirdly, we constructed three different end-to-end 3-D point cloud classification models, with the Dynamic Graph Convolutional Neural Network (DGCNN) model coming out on top when considering accuracy, response time, and training stability.

Keywords: 3-D object understanding · Data augmentation · Deep learning · Industry 4.0

1 Introduction

Interest in Industry 4.0 has recently been booming, resulting in attempts at cost-cutting by introducing information technology in various industrial areas, including production, logistics, and quality control [14,19]. One of the four core concepts of Industry 4.0 is "decentralized decisions" [4]. This means that each so-called Cyber-Physical System (CPS) [3] needs to be able to autonomously

© Springer Nature Switzerland AG 2021
A. Del Bimbo et al. (Eds.): ICPR 2020 Workshops, LNCS 12664, pp. 578–591, 2021.
https://doi.org/10.1007/978-3-030-68799-1_42

monitor and control an underlying physical system. To do so, a CPS typically generates a virtual copy of the physical world using its sensor system [1].

Along with this trend, belt conveyors, which are key to packaging and logistics, are also expected to evolve from simply moving around items to CPSs that recognize and track items. However, limitations are encountered when, for example, adding support for taking decentralized decisions as an add-on module to an already existing belt conveyor system, rather than having such a system designed from the ground up. Indeed, an add-on should, as much as possible, not interfere with the flow of execution of an already existing belt conveyor system and should ensure a near real-time response [16]. A number of efforts have been made to overcome the aforementioned limitations. In [16], for instance, a size measurement system is proposed that recognizes an item using Laser Curtain Sensors (LCS) and a rotary encoder.

The research and development effort presented in this paper is a follow-up to [16], starting from the prototype system presented in this previous study (see Fig. 1). However, in this paper, we aim at defining and solving the problem of item type classification, instead of solely obtaining size measurements for a particular item. In other words, we aim at scanning items transported by a conveyor belt and at subsequently classifying these items using state-of-the-art machine learning (ML) tools. To that end, the presented approach increases the number of sensor values obtained after upgrading one of the LCSs (that is, the LCS marked in blue in Fig. 1(b)), with this upgrade making it possible to not only measure the width of an item (a single value), but also to specify the start and end point of the width (two values). Taking advantage of the availability of more sensor values, this study then presents three contributions. First, we propose a method that reconstructs the 3-D shape of a sensed item. Next, we propose a data pre-processing method for ML-based classification that converts the 3-D reconstructed item shape into a point cloud that can be used as an input for a 3-D deep learning model. In this context, we construct three different kinds of 3-D deep learning models, evaluating their performance and their industrial applicability. Finally, using the results obtained, we provide suggestions on how and where to apply the proposed approach.

The paper is organized as follows. Section 2 introduces a number of item classification approaches for belt conveyor systems, as well as some basic knowledge of machine learning, deep learning, and 3-D deep learning. In Sect. 3, the different elements of our prototype system are described, including the hardware and the software used, as well as the way item measurement is performed. In Sect. 4, the process of going from sensor values to a final classification decision is explained in five steps. Finally, in Sect. 5 and Sect. 6, we discuss our experimental results, paying attention to a number of limitations of the proposed approach and also presenting directions for future research.

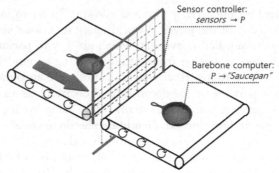

(a) Concept overview P denotes points generated by the sensor controller).

(b) Main direction.

(c) Opposite direction.

Fig. 1. Hardware prototype overview.

2   Background

2.1   Conveyor-Based Item Classification System

To the best of our knowledge, the number of studies related to ML-based classification of items in a belt conveyor setting is limited. In this context, [9] introduced an approach for classifying garbage using a 3-D camera and a robotic arm, applying an extremely randomized tree. The authors of [27] made use of nearest neighbours to classify different types of gears, disks, pulleys, and piston rods, whereas the authors of [13] leveraged nearest neighbors to classify various types of light bulbs.

2.2   Machine Learning

In artificial intelligence, ML corresponds to a set of methods that can detect patterns in data, and then uses the detected patterns to predict newly injected data. ML is usually divided into two main categories, supervised learning and unsupervised learning. When using supervised learning, the goal is to learn a mapping from known inputs x to known outputs y. In unsupervised learning, on the other hand, only the inputs x are given and the goal is then to find relevant patterns in the data without explicit knowledge of the output y.

For supervised learning, a dataset $\mathcal{D} = (x_i, y_i)_{i=1}^{n}$ consisting of pairs of inputs x and outputs y is given and the goal is to learn a mapping from x onto y. The inputs x are D-dimensional vectors, obtained from, for example, text, images, or a 3-D point cloud. The output y can either be discrete (that is, $y \in \{1, \ldots, C\}$, with C the number of different classes) or continuous. The former learning method is called classification and the latter is known as regression. In this paper, we focus on supervised learning, and classification in particular. This means that, given a scanned item x_i, we will train a function that maps x_i onto the corresponding output y_i, namely an item type.

In the simplest setting, each training input x_i is a D-dimensional vector of numbers that are called features or attributes (i.e., problem characteristics). In the real world, x_i can take the form of a complex structure, such as a sentence, an image, a time series, a point cloud, or a graph. To map such complex data directly, a large number of computational resources and a large amount of time may be required. Therefore, ML applications often apply manual feature extraction in practice, identifying and compressing essential information from complex-structured data and mapping this to a lower dimensional feature vector [15].

2.3 Deep Learning

Deep learning (DL), a type of ML, typically refers to the use of multi-layered feed-forward artificial neural networks, making it possible to automatically learn and classify features (end-to-end learning). These computational networks consist of artificial neurons, building on top of the perceptron model first introduced in [20], and where these artificial neurons correspond to mathematically simplified biological neurons. Moreover, like nerves, these artificial neurons combine signals or information from the previous stage, subsequently moving the outcome obtained to the next stage. Information thus only moves from the back to the front (feed-forward).

The goal of such a feed-forward artificial neural network is to approximate an ideal function f^*. For example, for a classifier, a feed-forward artificial neural network defines a mapping $y = f(x; \theta)$ and learns the parameter values θ that result in the best function approximation [2, p. 164–172]. Convolutional Neural Networks (CNNs) [10] are a specialized kind of artificial neural network, targeting the processing of data that are coming with a known grid-like topology. This includes, for example, images, which can be seen as 2-D grids of pixels. Starting from the 2012 ImageNet Large Scale Visual Recognition Challenge [8], CNNs have been shown to be tremendously successful in different application domains, and 2-D image analysis in particular [2, p. 326–335]. Similar to traditional feed-forward artificial neural networks, CNNs have trainable parameters (θ), making it possible to automatically extract the most important features from the given input data (feature learning). This is opposed to classical ML techniques, where a practitioner needs to manually specify features of interest (feature engineering).

2.4   3-D Deep Learning Models

The output of 3-D vision sensors, especially 3-D cameras and sensors using LiDAR (Light Detection and Ranging), is represented as a 3-D point cloud, coming with two characteristics. First, no matter how many points are transformed together, the positional relationship between two points (e.g., in terms of Euclidean distance) does not change. Secondly, this relationship does not depend on the order in which points are stored. When CNNs were first applied to 3-D point clouds, these neural networks could only deal with grid-like inputs, such as images or voxel-based representations. Therefore, a point cloud could not be used directly. Instead, a point cloud first had to be transformed into a grid-like form that a CNN could easily understand (e.g., after capturing 2-D surrounding pictures from a 3-D structure (multi-view) [21], the obtained data points would be placed into voxels [28]). Since CNNs have an excellent performance, these models also attained a good performance compared to ordinary ML. However, they cannot directly learn features from relationships between point pairs, with the transformation of data possibly also taking up a significant amount of time and computational resources. To remedy these shortcomings, PointNet [17] was released. This model is directly taking point clouds as input and suggests a method of embedding so that positional relationships are preserved. Over time, PointNet evolved into PointNet++ [18], which has now become a de facto baseline when evaluating new 3-D deep learning models [11, 25].

3   Experimental Setup

This section discusses various aspects of our experimental setup. First, we describe the hardware used by our prototype system, namely the belt conveyor and the sensors attached to it. Next, we introduce several software components. Finally, we detail the features of the items to be scanned.

3.1   Hardware Setup

The conveyor system used is the system presented in our previous study [16]. As depicted in Fig. 1, this prototype system consists of two conveyors parallel to each other with a spacing of 48 mm. Both conveyors are connected to one motor, rotating together at the same speed and in the same direction. In Fig. 1(c), the green arrow between the conveyors represents the motor. The rotational speed is measured using a rotary encoder, which is indicated by the red Θ in Fig. 1(a) and the yellow arrow in Fig. 1(b).

An item moves in the direction of the blue arrow (Fig. 1(a)) or the white arrow (Fig. 1(b) and Fig. 1(c)), and passes through the sensing range of the two LCSs installed in the gap between the two conveyors (marked with red and blue lines). Each LCS consists of a transmitter and a receiver: the transmitter emits laser signals and the receiver senses the area where the laser signals arrive. If part of an item is in-between both, a number of receiving areas will not be able to detect a laser signal. The sensor checks the area length and returns this information to the sensor controller. The red dashed lines denote the LCS used to measure the height of an item passing through. Similarly, the blue dotted lines denote the LCS used to measure the width of an item passing through. This LCS is located parallel to the drive shaft (motor and rotary encoder).

3.2 Software Setup

All sensors are controlled through B&R's X20CP1382 industrial computer (sensor controller). This computer can read sensor values every 10 ms. An automata-based control program [16] internally saves the sensor values when an item reaches the sensing area covered by the blue dotted lines, denoting the range of the LCS responsible for determining the width of an item. Once an item has passed completely, the control program stops saving sensor values. We implemented this approach using the Structured Text (ST) language [23].

As for the LCSs, Leuze's CML720i series was used. However, the LCS responsible for determining the width of an item, covering the blue dotted line area in Fig. 1, has been replaced with version CML720i-T05-640 A/L, which is the same type of LCS, but allowing for digital communication. In particular, this sensor can specify the start and the end position of an item in the sensing area, rather than a single value denoting all interrupted laser beams (as generated by the older sensor pair). To make this possible, an IO-Link [5] gender has been added to the sensor controller. Finally, the Autonics E40S6 Rotary Encoder was used to measure the number of rotations performed by the conveyor belt.

Since the sensor controller is an embedded computer, it comes with a number of computational constraints. For example, each operation must be completed within 10 ms, otherwise an error will occur in the system. In other words, it is difficult for the sensor controller to perform complex operations such as the ones needed for ML or DL. To compensate for this, a high-performance barebone computer was added that can classify items using sensor data that are stored in the sensor controller. The computer model used is ZOTAC's ZBOX MAGNUS EC52070D barebone computer with an Intel Core i5-8400T CPU and 32 GB of DDR4 RAM. Furthermore, a GeForce RTX 2070 graphics card with 8 GB of RAM was used.

3.3 Items Used

For evaluation purposes, we used a total of 8 items, as shown in Fig. 2: 3 types of boxes, 2 types of books, a saucepan, a cookie bucket, and a slipper. Each item was measured using a vernier caliper and a ruler, taking into account the criteria below.

– Item width: For an item on a flat surface, the item width is given by the shortest side of the minimum bounding rectangle surrounding the contact surface.

Table 1. Specification of the items used for our classification experiment. All values are in millimeter.

Item in Fig. 2	Width	Height	Length
A (Box type 1)	259.54	105.92	349.00
B (Box type 2)	257.48	213.58	347.00
C (Box type 3)	320.00	284.93	422.00
D (Red book)	162.01	46.22	239.10
E (Blue book)	209.23	48.12	237.08
F (Saucepan)	185.48	92.49	317.00
G (Bucket)	190.06	162.00	191.03
H (Slipper)	105.85	85.78	289.57

- Item length: For an item on a flat surface, the item length is given by the longest side of the minimum bounding rectangle surrounding the contact surface.
- Item height: For an item on a flat surface, the item height is given by the longest line segment among all line segments perpendicular to the ground, for all measured points belonging to the item.

Each item has a width in-between 105 mm and 320 mm, a length in-between 191 mm and 422 mm, and a height ranging from 46 mm to 284 mm. Table 1 summarizes our manual item measurements.

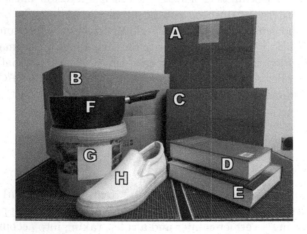

Fig. 2. Items used in our classification experiment. In alphabetical order, A to E are box-shaped items and F to H are special types of items [16].

4   Methods

In this section, we describe the processes associated with two major subsystems of our prototype belt conveyor, namely the sensor controller and the barebone computer. Referring to Fig. 1(a), we first introduce the process of acquiring and processing point clouds from the sensor controller, and we then discuss the training of various 3-D classification models using the processed point clouds and the barebone computer.

4.1   Dataset Acquisition from Sensor Controller

The prototype belt conveyor has a minimum speed of 0.1 m/s and can accelerate to up to 0.5 m/s [16]. Since the control system can read the values assigned to the sensor once every 10 ms, it is theoretically possible to acquire sensor values for an item once every 1 mm to 5 mm.

The acquired data are saved as shown in Table 2. The time slot, which is the first column in Table 2, indicates the sensor value acquisition time. The columns FIB and LIB refer to the values acquired by the blue LCS shown in Fig. 1(b). They denote the first and the last location of the area covered by an item, respectively. The column Height refers to the height of an item obtained through the red LCS in Fig. 1(c). The column Length contains the length of the item measured until that time.

The conveyor rotates thanks to force exerted by the shaft, which is a cylinder that transmits the turning force generated by the motor. Since the radius including the shaft and belt is 30 mm, once the shaft has rotated completely, the belt moves about 188.5 mm. At the same time, the rotary encoder generates 8,000 digital pulses per rotation. Therefore, it can be calculated that the belt moves by 0.023 mm per pulse. As a result, the length of an item is measured by counting the number of pulses from when the item is first detected, and where that number is subsequently multiplied by 0.023.

Given the values shown in Table 2, two points (FIB, Length, Height) and (LIB, Length, Height) can be generated for each time slot, following the order of (x, y, z). We can visualize these points in a 3-D space, as illustrated by Fig. 3(a). For each point, a line can be added that is drawn parallel to the z-axis, thus making it possible to produce an intuitive visualization.

Table 2. Data acquisition format used by the control system. Except for the time slot, all values are in millimeter.

Time slot (ms)	FIB	LIB	Height	Length
10	200	380	210	4.97
20	205	385	210	9.81
30	205	385	210	14.4
...	...	...	...	...

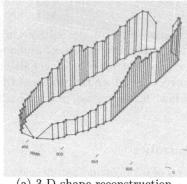

(a) 3-D shape reconstruction.   (b) 3-D triangular mesh creation.

(c) Fixed-size point cloud sampled
from a mesh.

Fig. 3. Point cloud sampling.

4.2   Point Cloud Equalization

Since our point generation method collects data as long as an item passes through the LCSs, the size of the stored data varies according to the length of the item and the speed of the belt conveyor. For example, if an item has a length of 100 mm, passing through at the maximum speed of 0.5 m/s, two points are created every 5 mm, resulting in a total of $20 \times 2 = 40$ points. Trivially, if the length of an item is 200 mm, the number of points is doubled.

However, the ML models need to receive inputs with the same fixed dimension. In particular, the final end-to-end 3-D point cloud classification model used in this study receives input data consisting of a fixed number of points, such as 512 or 1,024 points. Therefore, it is necessary to equalize the number of raw data points. To solve this problem, we transformed the scanned point cloud to a 3-D triangular mesh by connecting adjacent points with line segments as, for example, illustrated in Fig. 3(a) and Fig. 3(b). We then randomly selected a single triangle and performed single point cloud uniform random sampling within the selected triangle. This was repeated 512 times in order to extract the same number of points for each scan, leading to, for instance, the point cloud that can be seen in Fig. 3(c). For point sampling from mesh, we made use of PyntCloud, a Python library for working with 3-D point clouds [12].

4.3 Dataset Generation

In order to construct a DL model, as described in Sect. 4.4, we created a dataset for the purpose of training in three steps. First, each item in Fig. 2 was scanned 5 times, generating a point cloud using the aforementioned equalization approach. Secondly, all created point clouds were parallel translated so that each centroid coincides with the origin (0, 0, 0). Thirdly, we adopted data augmentation by using random rotations along the z-axis. As a result, 1,800 augmented point clouds were created for each of the eight different items, resulting in 14,400 point clouds in total. We used 80% of these point clouds for training (11,520 point clouds) and 20% for validation (2,880 point clouds).

4.4 Candidate Classification Models

In our classification experiment, we tested three end-to-end 3-D point cloud models. The characteristics of each model are as follows.

PointNet [17], the first model, is the pioneer of the end-to-end point cloud classification approach. For the design of this model, the authors used the property that symmetric functions can be used to extract invariant vectors from a point cloud. Compared to other models using multi-view images or voxels, Point-Net obtains a similar classification performance, but with a substantially lower running time complexity and a lower number of parameters.

PointNet++ [18], the second model, is an extended version of PointNet, addressing the shortcomings of the original model, including the loss of local features because of scale variance. In particular, to solve this latter shortcoming, the authors proposed to integrate the results obtained by different PointNets applied at various scales.

Lastly, Dynamic Graph CNN (DGCNN) [25] combines PointNet with several ideas taken from graph theory. In particular, DGCNN assumes the use of a graph that consist of vertices that represent points and edges that connect adjacent points, extracting features from points and edges using EdgeConv. Although PointNet [17] also considers neighboring points, DGCNN makes it possible to better reflect the geometric correlation between points during feature extraction.

4.5 Model Training Strategy

Since having 14,400 point clouds for eight item types is relatively insufficient, there is a high possibility of overfitting: although the model fits the training data very well, newly incoming data are, in general, not properly classified. Therefore, in this study, the transfer learning method of [22] is adopted to cope with this shortcoming. Using this method, a model is trained to capture the general

Table 3. Comparison of the effectiveness and the efficiency of the different models for item classification. For the validation set, the average accuracy over all epochs is given between brackets. The response time corresponds to the mean response time after 100 measurements (the standard deviation is given between brackets).

	Final training accuracy	Final validation accuracy	Model size (Mb)	Response time (ms)
PointNet	0.81	0.87 (0.79)	1.10	0.025 ($\pm$0.015)
PointNet++	0.97	0.79 (0.84)	5.77	0.453 ($\pm$0.001)
DGCNN	0.99	0.78 (0.83)	7.06	0.027 ($\pm$0.000)

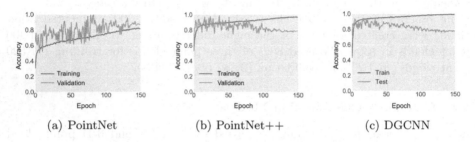

(a) PointNet   (b) PointNet++   (c) DGCNN

Fig. 4. Evolution of the training (blue) and validation (orange) accuracy for the selected 3-D classification models over all epochs. (Color figure online)

characteristics of 3-D data with a large dataset containing similar data items (pre-training), and the resulting model is then applied to a smaller dataset. We used ModelNet40 [26] for pre-training. This dataset contains 40 different items and consists of 10,000 3-D CAD models in total. We applied data augmentation to increase the size of the dataset to 1 million point clouds before performing pre-training. Similar to our own dataset, the augmented ModelNet40 dataset was split into a training set (80%) and a validation set (20%).

5   Experimental Results

In all experiments, Adam [6] was used as the optimization algorithm and cross-entropy was used as the loss function. For transfer learning, the candidate models were pre-trained over 250 epochs using the augmented ModelNet40 dataset. After pre-training, the model with the best validation accuracy was selected for re-training (fine-tuning) on our training dataset, using 150 epochs.

Table 3 shows our experimental results. The first and second column represent the final training and validation accuracy, respectively. For the validation set, we also provide the average accuracy over all epochs. We found that DGCNN has the highest training accuracy, closely followed by PointNet++, whereas PointNet obtained the highest validation accuracy. However, in terms of the average accuracy on the validation set, PointNet++ and DGCNN are again very similar, outperforming PointNet.

In addition, Table 3 shows model sizes and response times. The model size was measured in terms of the amount of storage needed, whereas the response time is the time needed for end-to-end classification, with the latter referring to the time between a model getting input and returning a label (type, class). In case of the response time, we performed measurements 100 times, subsequently obtaining the average and the standard deviation.

To get a more complete view on the model performance, we additionally investigated the evolution of the accuracy over all epochs for each model. The result is shown in Fig. 4. The vertical axis represents the accuracy and the horizontal axis indicates the number of epochs. Compared to the two other models, we can observe that PointNet comes with substantial fluctuations in the validation accuracy (orange line). Other than that, PointNet++ and DGCNN produce very similarly shaped training accuracy curves (blue line). However, since the curve of the validation accuracy of DGCNN has less variability than that of the other models (PointNet: 0.55–0.99, PointNet++: 0.75–0.99, DGCNN: 0.78–0.94), and since DGCNN comes with a faster response time than PointNet++, we finally opted to select DGCNN from the three candidate models.

6   Discussion and Conclusion

In this study, we presented an item classification model that works in combination with a belt conveyor. This model helps the belt conveyor control system with the recognition of items, while minimizing its impact on the conveyor design and item movement. To that end, we took three steps. First, we converted a previously proposed item size measurement system into a 3-D shape reconstruction system by recycling a belt conveyor prototype developed in a previous study [16], hereby only upgrading one LCS. Secondly, we proposed an equalization method that converts measured sensor values that vary in number according to item size into fixed-size point clouds that are suitable for training machine learning models. Finally, we trained three end-to-end 3-D point cloud classification models, selecting DGCNN as the best model when simultaneously considering accuracy, response time, and training stability.

We believe that the use of a 3-D deep learning approach in the context of a belt conveyor-based control system is of high interest, coming with significant industrial application potential. However, more experiments and analyses need to be conducted. For example, cross-validation [7] to assess the general effectiveness of each model has not been performed, nor has an in-depth analysis of model misclassifications using a confusion matrix been conducted [24].

Ultimately, we plan to put together an integrated measurement and classification system that covers the entire workflow, ranging from sensors to end users. We also plan to construct a more robust classification model for industrial applications. For instance, most products can currently only be distinguished from one another by making use of printed trademarks or barcodes, and not by the size and shape of an item. As an example, two shoe boxes can have the same size but their content may be different depending on the barcode attached.

To solve this problem, it may be of interest to apply multi-modal learning. This is a method that enables robust classification, even when dealing with items coming with highly varying characteristics, by obtaining additional data from various sensors, including 2-D cameras and barcode readers.

Acknowledgements. The research and development activities described in this paper were funded by Ghent University Global Campus (GUGC) and MOTIE (Ministry Of Trade, Industry, and Energy), Korea, under the Advanced Technology Center programme supervised by KEIT (Korea Evaluation Institute of Industrial Technology) (10077285, Development of a smart and automated packaging system for variable-type and variable-size products).

References

1. Chen, B., Wan, J., Shu, L., Li, P., Mukherjee, M., Yin, B.: Smart factory of industry 4.0: key technologies, application case, and challenges. IEEE Access **6**, 6505–6519 (2018)
2. Goodfellow, I., Bengio, Y., Courville, A., Bengio, Y.: Deep Learning, vol. 1. MIT press Cambridge (2016)
3. Gronau, N., Grum, M., Bender, B.: Determining the optimal level of autonomy in cyber-physical production systems. In: 2016 IEEE 14th International Conference on Industrial Informatics (INDIN), pp. 1293–1299 (2016)
4. Hermann, M., Pentek, T., Otto, B.: Design Principles for Industrie 4.0 scenarios. In: 2016 49th Hawaii International Conference on System Sciences (HICSS), pp. 3928–3937 (2016)
5. International Electrotechnical Commission: IEC 61131–9: IEC 61131–9:2013 Programmable controllers - Part 9: Single-drop digital communication interface for small sensors and actuators (SDCI) (2013). https://webstore.iec.ch/publication/4558. Accessed 21 Sept 2020
6. Kingma, D.P., Ba, J.: Adam: a method for stochastic optimization. arXiv preprint arXiv:1412.6980 (2014)
7. Kohavi, R.: A study of cross-validation and bootstrap for accuracy estimation and model selection. In: Proceedings of the 14th International Joint Conference on Artificial Intelligence (1995)
8. Krizhevsky, A., Sutskever, I., Hinton, G.E.: ImageNet classification with deep convolutional neural networks. In: Pereira, F., Burges, C.J.C., Bottou, L., Weinberger, K.Q. (eds.) Advances in Neural Information Processing Systems, pp. 1097–1105. Curran Associates, Inc. (2012)
9. Kujala, J.V., Lukka, T.J., Holopainen, H.: Classifying and sorting cluttered piles of unknown objects with robots: a learning approach. In: 2016 IEEE/RSJ International Conference on Intelligent Robots and Systems (IROS), pp. 971–978. IEEE (2016)
10. LeCun, Y., et al.: Backpropagation applied to handwritten zip code recognition. Neural Comput. **1**(4), 541–551 (1989).https://doi.org/10.1162/neco.1989.1.4.541
11. Lu, H., Shi, H.: Deep learning for 3D point cloud understanding: a survey. arXiv preprint arXiv:2009.08920 (2020)
12. MATATA., H.: Welcome to pyntcloud! - pyntcloud 0.1.3 documentation. https://pyntcloud.readthedocs.io/en/latest/. Accessed 15 Nov 2020

13. Mezei, A., Tamás, L., Bușoniu, L.: Sorting objects from a conveyor belt using active perception with a POMDP model. In: 2019 18th European Control Conference (ECC), pp. 2466–2471. IEEE (2019)
14. Mohamed, M.: Challenges and benefits of industry 4.0: an overview. Int. J. Supply Oper. Manage. **5**(3), 256–265 (2018). https://doi.org/10.22034/2018.3.7, http://www.ijsom.com/article_2767.html
15. Murphy, K.P.: Machine Learning: A Probabilistic Perspective, pp. 1–2. MIT press, Cambridge (2012)
16. Park, H., Van Messem, A., De Neve, W.: Item measurement for logistics-oriented belt conveyor systems using a scenario-driven approach and automata-based control design. In: 2020 IEEE 7th International Conference on Industrial Engineering and Applications (ICIEA), pp. 271–280 (2020)
17. Qi, C.R., Su, H., Mo, K., Guibas, L.J.: PointNet: deep learning on point sets for 3D classification and segmentation. In: Proceedings of the IEEE Conference on Computer Vision and Pattern Recognition (CVPR), pp. 652–660, July 2017
18. Qi, C.R., Yi, L., Su, H., Guibas, L.J.: PointNet++: deep hierarchical feature learning on point sets in a metric space. In: Guyon, I., Luxburg, U.V., Bengio, S., Wallach, H., Fergus, R., Vishwanathan, S., Garnett, R. (eds.) Advances in Neural Information Processing Systems, vol. 30, pp. 5099–5108. Curran Associates, Inc. (2017)
19. Rojko, A.: Industry 4.0 concept: background and overview. Int. J. Interact. Mob. Technol. (iJIM) **11**(5), 77–90 (2017). https://online-journals.org/index.php/i-jim/article/view/7072
20. Rosenblatt, F.: The perceptron: a probabilistic model for information storage and organization in the brain. Psychol. Rev. **65**(6), 386 (1958)
21. Su, H., Maji, S., Kalogerakis, E., Learned-Miller, E.: Multi-view convolutional neural networks for 3D shape recognition. In: Proceedings of the IEEE International Conference on Computer Vision (ICCV), pp. 945–953, December 2015
22. Tan, C., Sun, F., Kong, T., Zhang, W., Yang, C., Liu, C.: A survey on deep transfer learning. In: Kůrková, V., Manolopoulos, Y., Hammer, B., Iliadis, L., Maglogiannis, I. (eds.) ICANN 2018. LNCS, vol. 11141, pp. 270–279. Springer, Cham (2018). https://doi.org/10.1007/978-3-030-01424-7_27
23. Tiegelkamp, M., John, K.H.: IEC 61131-3: Programming industrial automation systems, vol. 14. Springer (1995). https://doi.org/10.1007/978-3-642-12015-2
24. Visa, S., Ramsay, B., Ralescu, A.L., Van Der Knaap, E.: Confusion Matrix-based Feature Selection. vol. 710, pp. 120–127 (2011). https://openworks.wooster.edu/facpub/88
25. Wang, Y., Sun, Y., Liu, Z., Sarma, S.E., Bronstein, M.M., Solomon, J.M.: Dynamic graph CNN for learning on point clouds. ACM Trans. Graph. **38**(5), October 2019. https://doi.org/10.1145/3326362
26. Wu, Z., et al.: 3D ShapeNets: a deep representation for volumetric shapes. In: Proceedings of the IEEE Conference on Computer Vision and Pattern Recognition, pp. 1912–1920, June 2015
27. Zhang, Y., Li, L., Ripperger, M., Nicho, J., Veeraraghavan, M., Fumagalli, A.: Gilbreth: a conveyor-belt based pick-and-sort industrial robotics application. In: 2018 Second IEEE International Conference on Robotic Computing (IRC), pp. 17–24. IEEE (2018)
28. Zhou, Y., Tuzel, O.: VoxelNet: end-to-end learning for point cloud based 3D object detection. In: 2018 IEEE/CVF Conference on Computer Vision and Pattern Recognition, pp. 4490–4499 (2018). https://doi.org/10.1109/CVPR.2018.00472

Development of Fast Refinement Detectors on AI Edge Platforms

Min-Kook Choi[1] and Heechul Jung[2(✉)]

[1] hutom, Seoul, Republic of Korea
mkchoi@hutom.io
[2] KNU, Daegu, Republic of Korea
heechul@knu.ac.kr

Abstract. Refinement detector (RefineDet) is a state-of-the-art model in object detection that has been developed and refined based on high-end GPU systems. In this study, we discovered that the speed of models developed in high-end GPU systems is inconsistent with that in embedded systems. In other words, the fastest model that operates on high-end GPU systems may not be the fastest model on embedded boards. To determine the reason for this phenomenon, we performed several experiments on RefineDet using various backbone architectures on three different platforms: NVIDIA Titan XP GPU system, Drive PX2 board, and Jetson Xavier board. Finally, we achieved real-time performances (approximately 20 fps) based on the experiments on AI edge platforms such as NVIDIA Drive PX2 and Jetson Xavier boards. We believe that our current study would serve as a good reference for developers who wish to apply object detection algorithms to AI edge computing hardware. The complete code and models are publicly available on the web (link).

Keywords: Object detection · AI edge platform · Refinement detector

1 Introduction

In recent years, the performance of object detection has dramatically improved due to the emergence of object detection networks that utilize the structure of CNNs [9]. Owing to this improvement, CNN-based object detectors have potential practical applications such as video surveillance [6], autonomous navigation [14], machine vision [17], and medical imaging [5]. Several industries have been making efforts to implement these technological advancements in conjunction with industrial applications.

Object detection methods using the CNN structure are broadly classified into two types: The first is a learning method by classifying class information, which locates and classifies objects in an image using a one-stage training method, by including them into one network stream and dividing the encoded features into different depth dimensions. Representative models of this one-stage network

© Springer Nature Switzerland AG 2021
A. Del Bimbo et al. (Eds.): ICPR 2020 Workshops, LNCS 12664, pp. 592–606, 2021.
https://doi.org/10.1007/978-3-030-68799-1_43

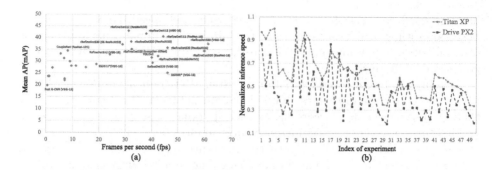

Fig. 1. Experiment results on MS-COCO dataset with Refinement Detectors. (a) Speed (fps) versus accuracy (mAP) on MS-COCO. Out models (red color) have a balanced speed and accuracy compared to the existing real-time object detection network-oriented models (purple color). Both the speed and accuracy of the proposed model were measured on the NVIDIA Titan XP. Details of the performance measurements are described in Sect. 4. (b) Speed (fps) variation based on Titan XP and Drive PX2 systems with the different experimental settings. X-axis represents the experiment number as shown in Table 1, and Y-axis represents speed (fps) normalized by the largest fps among 50 experiments for each system. Each experiment showed that the variation in inference speed on the edge board was much more sensitive to the training setup and architecture choices. (Color figure online)

are YOLO [18], single shot multibox detector (SSD) [8], SqueezeDet [14], and RetinaNet [26]. The advantage of these networks is that the encoder of the head part leading to the network output is constructed relatively intuitively, and the inference speed is high, while the accuracy of the bounding box regression, which specifies the exact position of the object, is generally lowered.

The second strategy is a two-stage learning method in which a network that classifies approximate positions and classes of objects is searched to obtain a more precise object location using a structure in which an independent encoder within the head is separated. These two-stage net-works have a region proposal network (RPN) that can more accurately determine the existence of objects in the head of the network. R-CNN [19] is the first model using this two-stage learning approach and Faster R-CNN [21], R-FCN [22], Deformable ConvNets [23], and Mask-RCNN [24] are some of the advanced versions. The advantage of the two-stage model is greater accuracy of the bounding box regression compared with the one-stage model. However, owing to its relatively complicated head structure, the inference and training time is longer than the one-stage model.

Therefore, models with modified overall detection network structures have been proposed to compensate for the disadvantages of the one-stage and the two-stage models [16,25]. [25] proposed a lightweight head structure to achieve an inference speed close to that of the one-stage models. Conversely, [16] suggested a model that uses the advantage of the two-stage model while maintaining the overall architecture of the one-stage model. In recent years, due to the increasing research efforts, the performance of object detection using CNN has improved dramatically in terms of accuracy and speed; however, the proposed algorithms still have various practical limitations in real-time applications in terms of accuracy and inference speed, especially in AI edge devices using embedded platforms.

Herein, we present several variations of RefineDet (refinement detector) [16], which is one of state-of-the-art object detection networks. To demonstrate how their performances vary according to their parameter settings and backbone architectures, we compare the inference performances between *on high-end GPU* and *on embedded platforms*. Unlike the inference on high-end GPU systems, obtaining the optimal parameters and backbone architectures is considerably important in embedded systems. Figure 1 shows the performance variation according to parameter settings and backbone architectures. Particularly, models tested on Drive PX2 exhibit a higher speed variance than high-end GPU systems, implying that hyper-parameters must be carefully selected when embedding or deploying models using the RefineDet architecture into embedded platforms. The fastest model in high-end GPU systems is not the fastest in embedded platforms.

Meanwhile, to achieve real-time inference speed with limited resources, we examined the limitations of the existing networks and applied various lightweight backbone structures to the head structure of RefineDet [16] by constructing optimal intermediate layers for an effective training. The utilized backbone CNNs include VGG [10] and shallow ResNet [3], which are the most widely used neural networks in image recognition. ResNeXt [15], Xception [2], and MobileNet [1,12], which are known for their high computation efficiency compared with layers of the same depth, were used; SENet [4] was applied using the "reweighting by local encoding" structure to generate punchy convolutional feature maps. To confirm the effect of feature generation on the real-time performance of the head structure, extensive comparative experiments were performed by applying head structures with a reduced channel depth for the head structure of RefineDet. Finally, we verified and confirmed the real-time performance on the NVIDIA Drive PX2 and the Jetson Xavier edge platforms using the ResNet18-based RefineDet model of reduced number of parameters and specific NMS parameters (see Table 1).

The major technical contributions of this study can be summarized as follows: 1) We present real-time detection models for AI edge platforms using a modified head structure of RefineDet. 2) The object detection network using the latest lightweight backbone with connections of the RefineDet head structure is extensively compared and analyzed using the MS-COCO 2017 detection dataset [7]. 3) Performance tests and model analysis were performed on NVDIA Drive PX2 and Jetson Xavier to achieve a balanced performance in terms of accuracy and speed for real-time detection on edge platforms. In addition, we introduced some issues related to the object detection models on the embedded GPU and summarized the concerns that must be addressed for achieving high real-time performance. We hope that this study will be useful for the development of real-time applications using diverse CNN-based architectures on AI edge platforms.

2   Baseline Architecture - RefineDet

We used the basic structure of a single-shot refinement object detector (RefineDet) [16] as the baseline architecture, which is derived from a one-stage learning structure of a single-shot multibox detector (SSD) [8]. The architectural weakness of the localization regressor of the one-stage detector was compensated using an anchor refinement module (ARM) branch, which functions similarly as the region proposal network in a two-stage detector. RefineDet minimizes the combined objective with two modules from the modified head structure of the former SSD architecture [16].

$$L(p_i, x_i, c_i, t_i) =$$

$$\frac{1}{N_b}\Big(\sum_{i=0}(L_b(p_i, [l_i \geq 1])) + \sum_{i=0}[l_i \geq 1]L_r(x_i, g_i)\Big)$$

$$+\frac{1}{N_m}\Big(\sum_{i=0}(L_m(c_i, l_i)) + \sum_{i=0}[l_i \geq 1]L_r(t_i, g_i)\Big),$$

(1)

where the objective's index i denotes the index of the anchor in the input mini-batch; objective input a is the inferenced output for the binary classifier to determine the objectness; x is the inferenced output of the regressor for the prediction of the bounding box location; l denotes the class vector of the ground truth label to obtain loss information, and g denotes the ground truth size and location. Furthermore, another objective input c is the predicted confidence value for the multi-class inferenced output of the object on the feature map; t is the regressor output for the location and size of the multi-class object; N_b and N_m denote the number of positive anchors entering each loss term; L_b is defined as a cross-entropy log loss for binary classification; and L_m is defined as a softmax loss for multi-class classification. The bracket indicator function $[l_i \geq 1]$ denotes the condition when the positive anchor is true.

The connection of the head structure of RefineDet to the backbone CNNs is divided into two branches. Among them, the ARM module is a type of region proposal classifier that learns to minimize the binary classification loss of objectness and supports multiclass inference with loss generation by the backpropagation of object existence and location information. The other branch comprises an object detection module (ODM) that is used to deduce the predicted confidence and localization bounding box information for multiclass objects. The entire network is trained such that the balanced loss for both branches is minimized. To utilize the concept of a feature pyramid through information coding between the upper and lower-layer features, [16] proposed a connection structure of a transfer connection block (TCB) and intermediate layers. To verify the efficiency of the network, the most widely known structure of CNNs, i.e., VGG-16 and ResNet-101, are used as the backbone for the training and evaluation results [16].

3   Combining Light-Weight Backbone CNNs

We combined state-of-the-art convolutional blocks of modified layers and structures with the lightweight backbone of the RefineDet architecture. Furthermore, we applied feature encoding blocks according to each model to the intermediate layer in the ODM branch for an efficient loss propagation between the head structure of RefineDet and the backbones. We applied VGG-16, ResNet-18, ReNeXt-26, ResNeXt-50, SE-ResNeXt-50, Inception-SENet, MobileNetV1, MobileNetV2, and Xception to the learning architecture. Furthermore, we used the feature encoding block of each backbone model as the intermediate layers. All backbone architectures were initialized through pre-training using ImageNet data [11]. To evaluate the tradeoff between inference accuracy and speed enhancement of the detection networks, the capability of the feature pyramid was validated. Hence, we verified the original (256) and reduced (128) channel depths of the TCB and intermediate layers, and we tested additional convolutional layers. Herein, the RefineDet model with reduced channel depth is known as the "reduced RefineDet (rRefineDet)."

VGG-16 [10]. Based on the VGG-16 model proposed in [10], the $fc6$ and $fc7$ layers were transformed into the convolution layers, $conv\_fc6$ and $conv\_fc7$ respectively, through the subsampling parameters as shown in [16]. To combine with the head structure of RefineDet, the subsequent layers including the last pooling layer of VGG-16 were removed and convolutions $conv6\_1$ and $conv6\_2$ were added to the top as additional convolution layers. As in [16], L2 normalization was used for the intermediate layers and some convolution layers.

ResNet-18 [3]. The authors of [16] applied the ResNet-101 architecture as the baseline backbone to improve the accuracy of RefineDet. In our study, RefineDet was learned using ResNet-18 pretrained by ImageNet for real-time inference as the backbone architecture. For the high-level feature encoding of backbone CNNs, the $res6$ block was added after the $res5$ block (similar to VGG-16), and the intermediate layer in the ODM branch used the residual encoding block with a channel depth of 256 in ResNet-18. For other parameter settings such as batch normalization (BN) and activations, learning was performed under the same conditions as those for ResNet-101 in [16].

ResNeXt-26 and 50 [15]. ResNeXt is a CNN structure that uses group convolution to improve the efficiency of computation for aggregated residual transformations using identity mapping. According to [15], its computational efficiency is higher than that of ResNet with the same depth, and it performed better for ImageNet data even though it uses only a few weight parameters. To effectively combine with the RefineDet head for ResNeXt-26, the outputs of $resx4$, $resx6$, and $resx8$ were used as the ARM and ODM outputs among eight $resx$ blocks. Furthermore, $resx9$ was used to process high-level features similar to those of ResNet. We added the features to the upper layer, and it could learn simultaneously by scratching. Regarding ResNeXt-50, $resx7$, $resx13$ and $resx16$ blocks of 16 $resx$ blocks were combined with the head of RefineDet, and $resx17$ was added to the upper layer. Regarding the intermediate layer, each output feature comprised the input for the feature pyramid encoded by one *resx* block.

SE-ResNeXt-50 and Inception [4]. SENet is a CNN configured to allow channel reweighting of the convolutional feature by applying a squeeze and excitation (SE) module to the output of the convolutional layer. The SE branch facilitates transformation in the depth dimension through channel-wise 1D encoding of the processed output feature. In our work, ResNeXt-50 and Inception towers with SE module were combined with the head of RefineDet and the performances of these models were verified. For SE-ResNeXt-50, the squeeze and excitation module was applied to all convolutional layers except the *conv*1 layer. To combine the heads, the output features of *conv*3_4, *conv*4_6, and *conv*5_3 layers were used as inputs to the AMR and ODM, and *conv*6 blocks were added to the top layer. The channel depth of the intermediate layer was 256, and that of the final output depth of the *conv*6 block was 256. Inception-SENet comprised 10 inception blocks after the first convolution layer (*conv*1), max pooling, and the second convolution layer (*conv*2). Among the feature outputs obtained when Inception-SENet was used as the backbone CNN structure, *inception*_3b, *inception*_4d, *inception*_5b and RefineDet head were connected and the *inception*_6 block was added to the top layer. The final output of the *inception*_6 block had a channel depth of 256, and the intermediate layer 256, which was the same as that of SE-ResNeXt-50.

Xception [2]. Xception applies a depth-wise separable convolution to the inception tower to re-evaluate the learning efficiency using CNNs of the same structure as InceptionV3 [13]. The authors of [2] proposed a network structure with a higher accuracy and better inference speed than InceptionV3, based on the fact that the "extreme" structure of Inception is almost equivalent in operation to the depth-wise separable convolution. To combine this Xception structure with the head of RefineDet, we used the outputs of *xception*11, *xception*12, and *conv*4_2 as inputs to the ARM and ODM branches and added an additional *xception* block (*xception*13) with a channel output depth of 256 to the top layer. Fur-thermore, we applied an *xception* block with a channel depth of 256 in the intermediate layer.

MobileNetV1 [1] **and V2** [12]. In MobileNetV1, the cost of computation is minimized by applying depthwise separable convolutions such as in [2] to reduce the operation cost from a typical convolution layer structure. In MobileNetV2, the depthwise separable convolution, which affects the computation time, is utilized as is, and the linear bottleneck structure is applied to minimize the burden on performance degradation. For MobileNetV1, we used the outputs of *conv*4_1, *conv*5_5, and *conv*6 to connect them to the head of RefineDet and added a convolution block with a channel depth of 512 to the top layer. Considering that the number of parameters of the backbone CNN structure is relatively smaller than that of the other architectures, we attempted to maintain the depth of the intermediate layers at the same feature volume as that of a basic convolutional block. Depth-wise separable convolution was applied to the top convolution block and

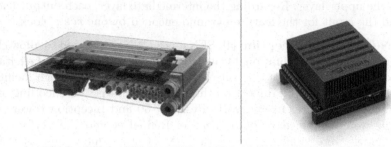

Fig. 2. NVIDIA Drive PX2 (left) and Jetson Xavier (right). Both AI edge devices support embedded GPU based parallel computing with Pascal (Drive PX2) and Volta (Jetson Xavier) architectures.

intermediate layer. Regarding MobileNetV2, the outputs of $conv3\_2$, $conv4\_7$ and $conv6\_4$ were used as the feature outputs for the coupling with head, and the $conv7$ block was added at the top. The output channel depth of $conv7$ was 96 after narrowing down and the intermediate layer was of the same size.

4   Evaluation Environments

Dataset Preparation. We used the MS-COCO 2017 object detection dataset [7] to evaluate the performances of various RefineDet networks. The dataset comprises 80 object classes, approximately 120,000 training images ($train17$), 5,000 validation images ($val17$), 80,000 test images ($test17$), and 120,000 unlabeled images ($unlabeled17$). For the evaluation of the proposed models, all models were trained with $train17$. Moreover, $val17$ and $test$-$dev17$ were used as test sets, separately. For the quantitative evaluations of $val17$ and $test$-$dev17$, the mean average precision (mAP) value was used as the evaluation criterion. The final mAP was calculated as the mean value obtained from a range of intersection over union (IoU) with $[0.5 : 0.05 : 0.95]$.

Plaforms. We used Titan XP, Drive PX2, and Jetson Xavier as our testing platform. Both platforms were developed for AI edge computing, and specifically for DrivePX2, they are designed for application to autonomous vehicles. We performed training on the Titan XP, Drive PX2, and Jetson Xavier systems. Drive PX2 is based on the Tegra X2 SoC board and contains 12 CPU cores: eight of A57 and four of Denver. The Pascal architecture of the GPU processor is based on the 16FinFET process and supports UART, CAN, LIN, FlexRay, USB, and 1 or 10 Gbit Ethernet communication (see Fig. 2). Jetson Xavier is a recently released embedded architecture based on a 512-Core Volta GPU with tensor cores and an 8-core ARM CPU. This architecture has 16 GB 256-bit of LPDDR4x memory and 32 GB of eMMC 5.1 flash storage and supports (2x) NVDLA DL accelerator engines.

Training. For training each model quickly, we used the pretrained weights as the initialization from ImageNet. The top layers of the convolution blocks, intermediate layers, and TCB of the RefineDet structure for the feature pyramid were initialized to a random Gaussian distribution with $\sigma = 0.01$. For a fair performance evaluation of the proposed models, the same parameters related to the detection model (anchor size, IoU with ground truth box, types of training data augmentation, etc.) were used except for the modification of the backbone and head structures. In addition, to confirm the role of the last convolutional block for high-level feature processing and the TBC for the feature pyramid, the channel depth of each feature map was divided into two cases of 128 and 256. The total learning duration was 120 epochs, and stochastic gradient descent was used as an optimizer. The base learning rate was started at 0.001, and a drop rate of 0.1 to 84 and 108 epochs was applied. The weight decay was 0.0005 and the momentum was 0.9. All the learning was performed in the Caffe environment using the Python interface. The depthwise separable convolution utilized the implementation code in the Caffe environment available to the public and the original code for the customizing layer of the SE module. In order to evaluate the performance of the model under various conditions, we trained different models according to the feature depth and input image size for TCB and top layers.

Testing. We performed testing process using *val*17 and *test-dev*17 datasets for each model. Additionally, we used several non-maxima suppression (NMS) parameters to demonstrate the effectiveness of the NMS according to the system. rRefineDet is a training model that focuses on the inference speed and limits the size of the feature depths of the TCB, intermediate layer, and top layer. As a reference, the evaluation results of MS-COCO *test-dev*14 data are included to show the results of original four RefineDet models reported in [16]. Unlike our experiments, the experiments reported in the original RefineDet paper were performed on a Titan X GPU. Therefore, a lower fps was indicated compared with our experiments even when the same model and parameter settings were used. The average for every inference speed was obtained based on 5,000 input images (10 for warm-up) (Table 2).

5   Results and Discussions

Changing Backbone Architecture. As shown in Table 1, changing the structure of the feature connection blocks between the head and backbone affected the performance. Models with VGG-16, ResNet-18 as the backbone structure demonstrated superior inference speed compared with the ResNeXt-26, MobileNetV1, and V2 models. Particularly, VGG-16 was the fastest architecture on Titan XP, but ResNet-18 was the fastest on Drive PX2. By combining the head structure of the existing object detection networks and CNN models, a balanced performance can be achieved, which is applicable to various fields. Moreover, the quantitative performance can be enhanced by setting a different input size for the image

Table 1. Performance change in different output feature depths from top layers of the high-level convolutional block, TBC and intermediate layer and difference in inference speed according to the NMS parameter on each platform. rRefineDet is a model that minimizes the feature output depth of the top convolutional blocks, TBC, and intermediate layers. NMS parameters in the NMS column include the maximum number of candidate bounding boxes for the NMS input, the maximum number of predicted bounding boxes according to the NMS output, and the confidence threshold for the final output. Every model was trained using *train*17 and tested on the Drive PX2 embedded platform. *td* means *test-dev*. The performance results of the first four models were extracted from the original RefineDet paper [16]. Red, blue, and green colors represent the 1st, 2nd, and 3rd fastest models among 50 experiments, respectively.

Exp. No.	Model	Backbone	NMS parameters	mAP (*val*17)	mAP (*td*17)	fps (Titan XP)	fps (Drive PX2)
[16]	RefineDet320	VGG-16	(1000, 500, 0.01)	–	29.4 (*td*14)	40.3 (Titan X)	–
[16]	RefineDet320	VGG-16	(1000, 500, 0.01)	–	33	24.1 (Titan X)	–
[16]	RefineDet320	ResNet-101	(1000, 500, 0.01)	–	32	9.3 (Titan X)	–
[16]	RefineDet320	ResNet-101	(1000, 500, 0.01)	–	36.4	5.2 (Titan X)	–
1	rRefineDet320	VGG-16	(400, 200, 0.1)	31.8	36.1	60.2	18.8
2	rRefineDet320	VGG-16	(1000, 500, 0.01)	32.7	37.2	56.5	10.9
3	RefineDet320	VGG-16	(400, 200, 0.1)	32.2	36.4	61.3	16.7
4	RefineDet320	VGG-16	(1000, 500, 0.01)	33.7	37.3	62.1	9.7
5	rRefineDet512	VGG-16	(400, 200, 0.1)	34.4	40.7	38.0	8.9
6	rRefineDet512	VGG-16	(1000, 500, 0.01)	35.5	41.7	40.3	5.8
7	RefineDet512	VGG-16	(400, 200, 0.1)	35.4	40.9	35.5	8.2
8	RefineDet512	VGG-16	(1000, 500, 0.01)	36.4	42.0	33.9	5.6
9	rRefineDet320	ResNet-18	(400, 200, 0.1)	31.0	33.5	52.9	21.6
10	rRefineDet320	ResNet-18	(1000, 500, 0.01)	31.7	34.3	50.3	8.9
11	RefineDet320	ResNet-18	(400, 200, 0.1)	32.4	34.2	59.9	19.6
12	RefineDet320	ResNet-18	(1000, 500, 0.01)	33.6	35.1	56.2	9.5
13	rRefineDet512	ResNet-18	(400, 200, 0.1)	35.8	40.2	44.4	13.6
14	rRefineDet512	ResNet-18	(1000, 500, 0.01)	36.2	40.5	40.7	6.2
15	RefineDet512	ResNet-18	(400, 200, 0.1)	36.6	40.3	35.7	12.2
16	RefineDet512	ResNet-18	(1000, 500, 0.01)	37.6	41.5	39.1	6.4
17	rRefineDet320	MobileNetV1	(400, 200, 0.1)	26.2	30.0	50.8	18.7
18	rRefineDet320	MobileNetV1	(1000, 500, 0.01)	27.2	30.8	46.9	6.9
19	RefineDet320	MobileNetV1	(400, 200, 0.1)	28.2	31.3	45.7	17.0
20	RefineDet320	MobileNetV1	(1000, 500, 0.01)	30.0	32.0	40.5	4.5
21	rRefineDet320	MobileNetV2	(400, 200, 0.1)	26.7	30.8	41.5	13.9
22	rRefineDet320	MobileNetV2	(1000, 500, 0.01)	27.5	31.1	38.8	7.1
23	RefineDet320	MobileNetV2	(400, 200, 0.1)	28.5	32.2	37.2	14.7
24	RefineDet320	MobileNetV2	(1000, 500, 0.01)	29.2	33.0	39.7	6.7
25	rRefineDet320	Inception-SENet	(400, 200, 0.1)	33.1	32.5	40.3	11.5
26	rRefineDet320	Inception-SENet	(1000, 500, 0.01)	33.2	28.3	40.3	7.3
27	RefineDet320	Inception-SENet	(400, 200, 0.1)	34.2	35.0	31.3	9.2
28	RefineDet320	Inception-SENet	(1000, 500, 0.01)	35.2	35.8	32.1	6.1
29	rRefineDet512	Inception-SENet	(400, 200, 0.1)	37.4	41.8	21.6	4.7
30	rRefineDet512	Inception-SENet	(1000, 500, 0.01)	37.7	42.1	21.1	3.9
31	rRefineDet320	SEResNeXt-50	(400, 200, 0.1)	35.2	37.2	28.9	9.9
32	rRefineDet320	SEResNeXt-50	(1000, 500, 0.01)	36.1	38.2	27.9	7.3
33	rRefineDet320	ResNeXt-26	(400, 200, 0.1)	30.5	34.9	36.0	11.8
34	rRefineDet320	ResNeXt-26	(1000, 500, 0.01)	31.3	35.7	29.9	6.8
35	RefineDet320	ResNeXt-26	(400, 200, 0.1)	32.1	35.7	32.7	11.0
36	RefineDet320	ResNeXt-26	(1000, 500, 0.01)	32.8	36.6	33.8	6.9
37	rRefineDet512	ResNeXt-26	(400, 200, 0.1)	34.4	39.5	25.1	6.9
38	rRefineDet512	ResNeXt-26	(1000, 500, 0.01)	35.4	40.5	25.2	4.6
39	RefineDet512	ResNeXt-26	(400, 200, 0.1)	35.2	40.3	24.8	6.4
40	RefineDet512	ResNeXt-26	(1000, 500, 0.01)	36.2	41.3	24.0	4.7

(continued)

Table 1. (*continued*)

Exp. No	Model	Backbone	NMS parameters	mAP (*val*17)	mAP (*td*17)	fps (Titan XP)	fps (Drive PX2)
41	rRefineDet320	Xception	(400, 200, 0.1)	34.6	37.2	37.9	10.9
42	rRefineDet320	Xception	(1000, 500, 0.01)	35.0	38.0	36.0	6.1
43	RefineDet320	Xception	(400, 200, 0.1)	34.9	37.8	35.8	10.4
44	RefineDet320	Xception	(1000, 500, 0.01)	35.7	38.9	33.8	5.2
45	rRefineDet320	ResNeXt-50	(400, 200, 0.1)	35.6	37.6	32.3	10.2
46	rRefineDet320	ResNeXt-50	(1000, 500, 0.01)	36.1	38.4	31.3	7.4
47	RefineDet320	ResNeXt-50	(400, 200, 0.1)	36.7	38.0	29.8	9.6
48	RefineDet320	ResNeXt-50	(1000, 500, 0.01)	37.6	38.4	28.1	7.0
49	rRefineDet512	ResNeXt-50	(400, 200, 0.1)	36.9	42.5	21.1	5.4
50	rRefineDet512	ResNeXt-50	(1000, 500, 0.01)	37.8	43.5	20.6	4.0

Table 2. Inference speed of NVIDIA Drive PX2 vs. Jetson Xavier (Jetson-X). Index of experiments is the same as that in Table 1.

Exp. No.	Model	Backbone	fps (Drive PX2)	fps (Jetson-X)
1	rRefineDet320	VGG-16	18.8	13.6
2	rRefineDet320	VGG-16	10.9	10.7
3	RefineDet320	VGG-16	16.7	12.5
4	RefineDet320	VGG-16	9.7	11.4
5	rRefineDet512	VGG-16	8.9	6.6
6	rRefineDet512	VGG-16	5.8	6.0
7	RefineDet512	VGG-16	8.2	7.6
8	RefineDet512	VGG-16	5.6	6.9
9	rRefineDet320	ResNet-18	21.6	**22.1**
10	rRefineDet320	ResNet-18	8.9	**18.5**
11	RefineDet320	ResNet-18	**19.6**	17.0
12	RefineDet320	ResNet-18	9.5	16.6
13	rRefineDet512	ResNet-18	13.6	11.9
14	rRefineDet512	ResNet-18	6.2	11.5
15	RefineDet512	ResNet-18	12.2	11.2
16	RefineDet512	ResNet-18	6.4	11.1

and applying the deformable operation to a specific convolutional layer according to the available computing resources. In addition, it is clear that using the improved convolutional block such as feature renormalization of the SE module and the inverted residual structure of MobileNetV2 in a specific head and backbone structure helps to prevent speed degradation and improve accuracy.

Effectiveness of NMS Parameters. As shown in Table 1, a slight adjustment of the NMS parameter resulted in a considerable improvement in the inference speed at the expense of a low accuracy. Furthermore, unexpected computational bottlenecks appeared when applying a state-of-the-art algorithm based on CNNs in a Drive PX platform. Layer-wise inference testing was performed on all layers

Table 3. Accuracy of state-of-the-art detectors using the MS-COCO dataset. The order is sorted by accuracy and speed. *td* means *test-dev*. We note that even on the same architecture, performance differences may occur due to different settings.

Model	Backbone	Training	mAP (data)
Fast R-CNN [20]	VGG-16	*train*14	19.7 (*td*14)
Faster R-CNN [21]	VGG-16	*trainval*14	21.9 (*td*14)
R-FCN [22]	ResNet-101	*trainval*14	29.9 (*td*14)
Def. R-FCN [23]	ResNet-101	*trainval*14	34.5 (*td*14)
Def. R-FCN [23]	Inception-ResNet	*trainval*14	37.5 (*td*14)
umd_det [30]	ResNet-101	*trainval*14	40.8 (*td*14)
G-RMI [31]	Ensemble	*trainval*14 (32k)	41.6 (*td*14)
SSD321 [8]	ResNet-101	*trainval*14 (35k)	28.0 (*td*14)
DSSD321 [33]	ResNet-101	*trainval*14 (35k)	28.0 (*td*14)
SSD513 [8]	ResNet-101	*trainval*14 (35k)	31.2 (*td*14)
YOLOv2 [18]	Darknet-19	*trainval*14 (35k)	31.6 (*td*14)
RetinaNet500 [26]	ResNet-101	*trainval*14 (35k)	32.0 (*td*14)
DSSD513 [33]	ResNet-101	*trainval*14 (35k)	33.2 (*td*14)
RetinaNet800 [26]	ResNet-101-FPN	*trainval*14 (35k)	36.4 (*td*14)
RefineDet320 [16]	VGG-16	*trainval*14 (35k)	29.4 (*td*14)
RefineDet320 [16]	ResNet-101	*trainval*14 (35k)	32.0 (*td*14)
RefineDet512 [16]	VGG-16	*trainval*14 (35k)	33 (*td*14)
RefineDet512 [16]	ResNet-101	*trainval*14 (35k)	36.4 (*td*14)
RefineDet320	MobileNetV2	*train*17	33.0 (*td*17)
RefineDet320	ResNet-18	*train*17	35.1 (*td*17)
RefineDet320	Inception-SENet	*train*17	35.8 (*td*17)
RefineDet320	ResNeXt-26	*train*17	36.6 (*td*17)
RefineDet320	VGG-16	*train*17	37.2 (*td*17)
RefineDet320	Xception	*train*17	38.0 (*td*17)
rRefineDet320	SEResNeXt-50	*train*17	38.2 (*td*17)
RefineDet512	ResNeXt-26	*train*17	41.3 (*td*17)
RefineDet512	ResNet-18	*train*17	41.5 (*td*17)
RefineDet512	VGG-16	*train*17	42.0 (*td*17)
rRefineDet512	Inception-SENet	*train*17	42.1 (*td*17)
rRefineDet512	ResNeXt-50	*train*17	43.5 (*td*17)

of the RefineDet architecture to analyze the cause. We discovered that the computation times of all layers operating on the embedded GPU increased linearly compared with the number of high-end GPU cores, but the post-processing for the bounding box occupied most of the bottlenecks. Because the model is designed to process bounding box filtering related to the NMS operation of the post-processing layer, the CPU operations on a certain computing platform causes a severe performance degradation compared with high-end GPU systems rich in CPU computing resources. Table 1 shows that performance degradation owing to the CPU operation bottleneck can cause a significant performance degradation not only in embedded environments, but also in general computing resources. Furthermore, Table 1

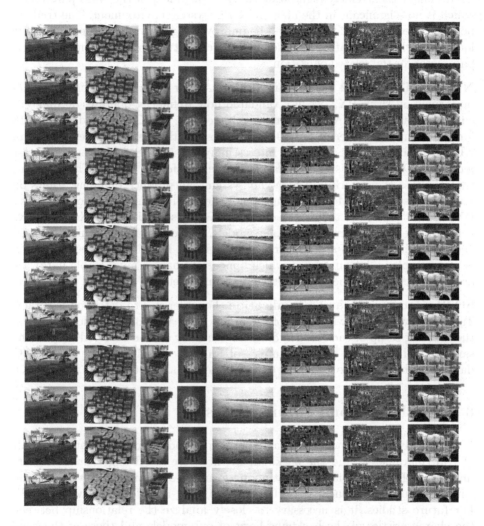

Fig. 3. Qualitative results on the MS-COCO *val*17. We denote each model as (model, backbone, size, nms parameter): from the top (RefineDet, MobileNetV2, 320, 1000), (rRefineDet, MobileNetV2, 320, 400), (rRefineDet, ResNet18, 320, 400), (RefineDet, ResNet18, 512, 1000), (rRefineDet, VGG16, 320, 400), (RefineDet, VGG16, 512, 1000), (rRefineDet, ResNeXt26, 320, 400), (RefineDet, ResNeXt26, 512, 1000), (rRefineDet, ResNeXt50, 320, 400), (rRefineDet, ResNeXt50, 512, 1000), (rRefineDet, Inception-SENet, 320, 400), and (rRefineDet, Inception-SENet, 512, 1000). Confidence threshold is set to 0.6 for better visualization. Best viewed in color.

shows the comparison results of the inference speed on various platforms according to the model structure and NMS parameters. To reduce the burden of the CPU operations, the input bounding box for NMS is set to a maximum of 400, and the confidence threshold for the output to the NMS is increased to 0.1. Although the accuracies are not significantly different, the performance gain in the Drive PX2 environment is extremely high owing to the adjustment of the NMS parameter, which is not significant in the high-end GPU system environment. The tradeoff between speed and performance can be improved significantly by identifying the location of the computational burden in the architecture and adjusting the hyperparameter for the related operation.

NVIDIA Drive PX2 vs. Jetson Xavier. As our experiments were primarily performed on Drive PX2, we also validated how RefineDet operates on different edge platforms. First, we select two architectures of VGG-16 and ResNet-18 for the comparison, because those backbones are the fastest models on Titan XP and Drive PX2, respectively. Interestingly, ResNet-18, which is the fastest model for Drive PX2, is also the fastest model for Jetson Xavier. This confirms that the fastest network in a high-end system can produce different results on an embedded board. Meanwhile, although the hardware specifications of Drive PX2's GPU are better than those of Jetson Xavier's GPU, Jetson Xavier shows similar or better performances in some models. We conjecture that this is because Jetson Xavier contains a deep learning accelerator. Finally, the speed variation is more stable in Jetson Xavier, which appears to have a more stable hardware architecture for a balanced computation resource between CPU and GPU cores.

SOTA Comparison and Visualization. For reference to detection accuracy, Table 3 shows mAPs with state-of-the-art models. We note that depending on the training setup of each model, different accuracy can be achieved despite the same model. Figure 3 shows examples of inference results on the MS-COCO dataset with confidence threshold 0.6 for bonding box visualization.

6   Conclusions and Future Works

In this study, we conducted several experiments and discovered that the choice of NMS parameters and the backbone architecture in AI edge hardware were important. Results indicated real-time performances (approximately 20 fps) on embedded platforms such as NVIDIA Drive PX2 and Jetson Xavier boards. For future studies, it is necessary to closely analyze the relationship between the characteristics of the backbone layer of our models and those of the head structure. Hence, it is necessary to perform an ablation study on each backbone structure and TBC layer or intermediate layer to analyze the importance of each relation about connecting to the head. Furthermore, we intend to apply half or mixed precision techniques such as TensorRT to obtain better optimizations on platforms with limited resources.

References

1. Howard, A.G., et al.: MobileNets: efficient convolutional neural networks for mobile vision applications. arxiv:1704.04861 (2017)
2. Chollet, F.: Xception: deep learning with depthwise separable convolutions. In: Proceedings of CVPR (2017)
3. He, K., Zhang, X., Ren, S., Sun, J.: Deep residual learning for image recognition. In: Proceedings of CVPR (2016)
4. Hu, J., Shen, L., Sun, G.: Squeeze-and-excitation networks. In: Proceedings of CVPR (2018)
5. Jin, A., et al.: Tool detection and operative skill assessment in surgical videos using region-based convolutional neural networks. In: Proceedings of WACV (2018)
6. Jung, H., Choi, M.K., Jung, J., Lee, J.H., Kwon, S., Jung, W.Y.: ResNet-based vehicle classification and localization in traffic surveillance systems. In: Proceedings of CVPRW (2017)
7. Lin, T.Y., et al.: Microsoft COCO: common objects in context. In: Proceedings of ECCV (2014)
8. Liu, W., et al.: SSD: single shot multibox detector. In: Proceedings of ECCV (2017)
9. Liu, L., Ouyang, W., Wang, X., Fieguth, P., Liu, X., Pietikainen, M.: Deep learning for generic object detection: a survey. arxiv:1809.02165 (2018)
10. Liu, W., Rabinovich, A., Berg, A.C.: ParseNet: looking wider to see better. In: Proceedings of ICLR (2016)
11. Russakovsky, O., et al.: ImageNet large scale visual recognition challenge. Int. J. Comput. Vis. **115**(3), 211–252 (2015)
12. Sandler, M., Howard, A., Zhu, M., Zhmoginov, A., Chen, L.-C.: MobileNetV2: inverted residuals and linear bottlenecks. In: Proceedings of CVPR (2018)
13. Szegedy, C., Vanhoucke, V., Ioffe, S., Shlens, J., Wojna, Z.: Rethinking the inception architecture for computer vision. In: Proceedings of CVPR (2016)
14. Wu, B., Wan, A., Iandola, F., Jin, P.H., Keutzer, K.: SqueezeDet: unified, small, low power fully convolutional neural networks for real-time object detection for autonomous driving. In: Proceedings of CVPR (2017)
15. Xie, S., Girshick, R., Dollar, P., Tu, Z., He, K.: Aggregated residual transformations for deep neural networks. In: Proceedings of CVPR (2017)
16. Zhang, S., Wen, L., Bian, X., Lei, Z., Li, S.Z.: Single-shot refinement neural network for object detection. In: Proceedings of CVPR (2018)
17. Yao, Y., et al.: End-to-end convolutional neural network model for gear fault diagnosis based on sound signals. Appl. Sci. **8**(9), 1584 (2018)
18. Redmon, J., Farhadi, A.: YOLO 9000: better, faster, stronger. In: Proceedings of CVPR (2017)
19. Girshick, R., Donahue, J., Darrell, T., Malik, J.: Rich feature hierarchies for accurate object detection and semantic segmentation. In Proceedings of CVPR (2014)
20. Girshick, R.: Fast R-CNN: towards real-time object detection with region proposal networks. In: Proceedings of ICCV (2015)
21. Ren, S., He, K., Girshick, R., Sun, J.: Faster R-CNN: towards real-time object detection with region proposal networks. In: Proceedings of NIPS (2015)
22. Dai, J., Li, Y., He, K., Sun, J.: R-FCN: object detection via region-based fully convolutional networks. In: Proceedings of NIPS (2016)
23. Dai, J., et al.: Deformable convolutional networks. In: Proceedings of ICCV (2017)
24. He, K., Gkioxari, G., Dollar, P., Girshick, R.: Mask R-CNN. In: Proceedings of ICCV (2017)

25. Li, Z., Peng, C., Yu, G., Zhang, X., Deng, Y., Sun, J.: Light-head R-CNN: in defense of two-stage object detector. arXiv: 1711.07264 (2017)
26. Lin, T.Y., Goyal, P., Girshick, R., He, K., Dallor, P.: Focal loss for dense object detection. In: Proceedings of ICCV (2017)
27. Shrivastava, A., Gupta, A., Girshick, R.: Training region-based object detectors with online hard example mining. In: Proceedings of CVPR (2016)
28. Bell, S., Zitnick, C.L., Bala, K., Girshick, R.: Inside-outside net: detecting objects in context with skip pooling and recurrent neural networks. In: Proceedings of CVPR (2016)
29. Zhu, Y., Zhao, C., Wang, J., Zhao, X., Wu, Y., Lu, H.: CoupleNet: coupling global structure with local parts for object detection. In: Proceedings of ICCV (2017)
30. Bodla, N., Singh, B., Chellappa, R., Davis, L.S.: Improving object detection with one line of code. In: Proceedings of ICCV (2017)
31. Huang, J., et al.: Speed/accuracy tradeoffs for modern convolutional object detectors. In: Proceedings of CVPR (2017)
32. Kong, T., Sun, F., Yao, A., Liu, H., Lu, M., Chen, Y.: RON reverse connection with objectness prior networks for object detection. In: Proceedings of CVPR (2017)
33. Fu, C.-Y., Liu, W., Ranga, A., Tyagi, A., Berg, A.C.: DSSD: deconvolutional single shot detector. arXiv:1701.06659 (2017)

Selecting Algorithms Without Meta-features

Martin Lukac[1]()[iD], Ayazkhan Bayanov[1], Albina Li[1], Kamila Abiyeva[1,2], Nadira Izbassarova[1], Magzhan Gabidolla[1,3], and Michitaka Kameyama[4]

[1] Department of Computer Science, Nazarbayev University,
Nur-Sultan 010000, Kazakhstan
martin.lukac@nu.edu.kz
[2] Nanyang Technological University, Singapore, Singapore
[3] University of California, Merced, USA
[4] University of Ishinomaki Senshu, Ishinomaki, Japan

Abstract. The algorithm selection has been successfully used on a variety of decision problems. When the problem definition is structured and several algorithms for the same problem are available, then meta-features, that in turn permit a highly accurate algorithm selection on a case-by-case basis, can be easily and at a relatively low cost extracted. Real world problems such as computer vision could benefit from algorithm selection as well, however the input is not structured and datasets are very large both in samples size and sample numbers. Therefore, meta-features are either impossible or too costly to be extracted. Considering such limitations, in this paper we experimentally evaluate the cost and the complexity of algorithm selection on two popular computer vision datasets VOC2012 and MSCOCO and by using a variety task oriented features. We evaluate both dataset on algorithm selection accuracy over five algorithms and by using a various levels of dataset manipulation such as data augmentation, algorithm selector fine tuning and ensemble selection. We determine that the main reason for low accuracy from existing features is due to insufficient evaluation of existing algorithms. Our experiments show that even without meta features, it is thus possible to have meaningful algorithm selection accuracy, and thus obtain processing accuracy increase. The main result shows that using ensemble method, trained on MSCOCO dataset, we can successfully increase the processing result by at least 3% of processing accuracy.

Keywords: Algorithm selection · Computer vision · Features augmentation

1 Introduction

Algorithm selection [26] is a well studied topic in almost every field of computing and under various forms. In this paper we focus on the so called per-instance selection approach [14] that in this paper we will refer to as Case-By-Case (CBC) selection. In the CBC task, a problem is given by a mapping $M : I \rightarrow O$ and by an associated accuracy function $\mathcal{A}(o, t)$, the target is to select from a set

© Springer Nature Switzerland AG 2021
A. Del Bimbo et al. (Eds.): ICPR 2020 Workshops, LNCS 12664, pp. 607–621, 2021.
https://doi.org/10.1007/978-3-030-68799-1_44

of algorithms $A = \{a_1, \ldots, a_k\}$ such that $\arg\max_i A(a_i(i), t)$. That is, for each instance of the problem $i \in I$ the algorithm with highest accuracy score of processing is selected.

The CBC selection was applied very successfully to many decision making problems. The most successful result wise is the SatZilla [30] where authors applied algorithm selection to the SAT solving competition [22]. Other applications of algorithm selection include scheduling problems, Traveling Salesman Problem and other type of constrain satisfaction decision making [2,3,11,17,25, 30]. Several software suites were implemented including the ASLib [3], AutoFolio [20], Flexfolio [19], etc. however the main feature that these approaches are exploiting are meta-features extraction, portfolio algorithm selection and feature hierarchy. Such strong limitations do not allow general purpose algorithms to be selected using these methods and tools.

For general purpose machine learning based algorithms, AutoML focuses on the configuration and design of Convolutional Neural Networks (CNN) using data and CNN structures. One of the auto-ml approaches combines the GPU acceleration of DL, CNN, Big Data (BD) and Reinforcement Learning (RL) into a meta approach for designing and recombining algorithms for certain applications. While searching the meta-space of algorithms and configurations with RL is an appealing approach, the extremely large data and time to obtain solution is in most of cases unrealistic and not achievable [23,27–29,31]. The main focus of the AutoML approach in general is however rarely focused on CBC selection and rather the design and configuration targets dataset as a whole.

Recently the algorithm selection was also applied to real world problems such as computer vision or image processing [1,4,21]. Some success was also obtained in more advanced tasks of computer vision such as scene understanding and semantic segmentation [15,21,24]. However, for the more advanced tasks a system based approach using heuristics is required [21,24].

The lack of meaningful meta-features was addressed previously in [21] where a heuristic approach was proposed using an iterative processing. In this approach the authors extract semantic meta-information after an initial processing by a feature based algorithm selector. The extracted semantic information is used to select algorithms based on relational, object and scene level attributes.

Because of the difficulty of effectively applying algorithm selection to real-world (unstructured) problems, we investigate the problem of algorithm selection without meta-features applied to real-world problems: we study the hardness of predicting the best algorithm from features used for the task itself. Instead of searching for meta-features (such as statistical co-occurrence of features across data samples, relative positions across samples, etc.) we experimentally determine the requirements of successful algorithm selection by only fine tuning the basic algorithm selection approach. The purpose of this study is to determine how feasible it is to use existing features for a fast and easy to set-up meta system that can be used for the purpose of either performance optimization or algorithm and training data evaluation.

We investigate the accuracy and hardness of algorithm selection in two computer vision tasks: classification and semantic segmentation. We estimate the accuracy of algorithm selector using various features selection, machine learning parameters adjustment, synthetic data generation, data sub-sampling and ensemble selectors. Additionally we also evaluate the algorithm selection with higher level regional features and semantic annotations. Finally we also test an ensemble and dynamical ensemble with sub-sampling approach to algorithm selection.

We show that in the studied problem, the CBC algorithm selection is effective only on the highest granularity of selection: a dynamic ensemble with local evaluation of accuracy. Also, the algorithm selection with reasonable accuracy $\approx 68\%$ is obtained only when enough samples are available. The positive observation is that at already a low level of algorithm selection accuracy the task performance is better than any of the used algorithms for processing. However single direct algorithm selection approach is not effective, and only using meta-ensemble approach results in improved processing accuracy.

This paper is structured as follows. Section 2 provides the necessary background into the algorithm selection and related topics. Section 3 describes the data set used and Sect. 4 presents the individual experimental settings. Section 5 discusses the results. Section 6 concludes the paper.

2 Background

Let $\mathbb{A} = \{a_0, \ldots, a_{k-1}\}$ be a set of algorithms, $\mathbb{I} = \{i_0, \cdots, i_{n-1}\}$ be the set of input images of size $X \times Y$ and $\mathbb{L} = \{l_0, \ldots, l_{l-1}\}$ be a set of labels. Let there be a set of labels: $\mathbb{S} = \{s_0, \ldots, s_{n-1}\}$. The set $\mathbb{S}$ contains a set of sets: each element $\mathbf{s}_j \in \mathbb{S}$, is a set of labels $\mathbf{s}_j = \{s_j(0,0), \ldots, s_j(x-1, y-1)\}$ for image i_j, such that $(x, y) = (0, 0), \ldots, (x-1, y-1)$. Each element of $s_j(x, y) \in \mathbf{s}_j$ represents the label for each pixel $p_j(x, y) \in i_j$ of an input image of size $X \times Y$.

2.1 Semantic Segmentation and Object Detection

For semantic segmentation and object detection, each algorithm assigns label to each pixel of the input image such as shown in Eq. 1

$$s_j^i(x, y) = a_i(i_j(x, y)) \tag{1}$$

with $s_j^i(x, y) \in \mathbb{L}$, and $i_j(x, y)$ is a pixel located at coordinates x, y in image $i_j \in \mathbb{I}$.

In semantic segmentation, each pixel can have a different label. In general to measure the accuracy of semantic segmentation several measures exists such as Jaccard index [13] (Intersection of Union) or the f-measure [8]: here we use the Intersection of Union (IOU). Let $s_j^i(x, y)$ and $s_j(x, y)$ be the label for pixel generated by algorithm a_i and the desired label from the ground truth respectively. As shown in Eq. 2 the IOU is the ratio of correctly labeled pixels over the number of all pixels that have been labeled a) correctly as label $s_j^i(x, y) = s_j(x, y)$ called *true positive (TP(x,y))*, b) incorrectly as label $l_k = s_j^i(x, y)$ while $s_j(x, y) = l_j$ (called *false positive (FP(x,y))*) and c) incorrectly as label $l_j = s_j^i(a, b$ while $s_j(x, y) = l_k$ (called *false negative (FN(x,y))*).

$$m_s(a_i, i_j) = \sum_{j=1}^{n} \sum_{a=0,b=0}^{x,y} \frac{TP(x, y)}{TP(x, y) + FP(x, y) + FN(x, y)} \tag{2}$$

The $m_s(a_i, i_j)$ will be referred to in this paper for simplicity as **Semantic Segmentation Accuracy (SSA)**.

2.2   Algorithm Selection (AS)

The algorithm selection starts by extracting a set of features F_j from the input image I_j. The features F_j and (if available) additional information ξ_j is used as input to the algorithm selection mechanism. The selection outputs the identifier a_j for a single algorithm which is then used to process the input image and generate output result.

The algorithm selection be defined as a multi-class classification/decision problem. Let there be a set $T = \{t_0, \ldots, t_{n-1}\}$, be the set of algorithms with highest SSA for each input image:

$$t_j = arg\ max_k\ m_s(a_k, i_j) \tag{3}$$

Then the average accuracy of any algorithm selector can be computer as:

$$\hat{s}_c(\mathbb{A}, \mathbb{I}) = \frac{\sum_{j=0}^{n-1} \sum_{x=0}^{X} \sum_{y=0}^{Y} (f(i_j(x, y)) == t_j)}{n \times X \times Y} \tag{4}$$

where $f()$ - algorithm selection function. The $\hat{s}_c(\mathbb{A}, \mathbb{I})$ will be referred to in the paper as Mean Algorithm Selection Accuracy (MASA).

3   Data Set for Algorithm Selection

To evaluate the accuracy of algorithm selection we used two computer vision problems: semantic segmentation and object detection. In particular, we used validation set of the VOC2012 [10] and MSCOCO [18]. The two datasets have been chosen for several reasons. First, VOC2012 is a relatively small data set compared to MSCOCO and thus the intent was to evaluate the problem on a data size dependent manner. Second, VOC2012 is older than MSCOCO and thus was evaluated using larger set of algorithms including algorithms based not on deep learning. We wanted to evaluate the contribution of non deep learning algorithms as compared to the state of the art. Comparing the task results on such two different data sets allows us to better understand the problem difficulty.

The reason for using the validation dataset is two folds: a) the algorithms were designed and learned on the training dataset and thus the results can be strongly biased due to learning convergence and b) the validation data set allows to directly evaluate the task accuracy as in most cases the test data is not provided with the ground truth.

The VOC2012 validation data set contains 1441 images and the validation data set of MSCOCO contains 40000 images, For the algorithm selection experiments each data set was divided into a training (80%) and testing (20%) subsets. The resulting datasets contain 1152(32000) images, and test set of 289(8000) images respectively for VOC2012 (MSCOCO). The VOC2012 dataset consists of 20 classes in 4 categories: Person, Animal, Vehicle, and Indoor. The MSCOCO dataset has 80 categories of objects.

Each data set contains the input images and the results of the evaluated algorithms with corresponding accuracy. For VOC2012 we used five different algorithms that do not include only the state of the art neural network approaches. These algorithms are A1 [16], A2 [5], A3 [12], A4 [7] and A4 [7]. The average SSA of each of the algorithms evaluated on the whole VOC2012 validation dataset are shown in Table 1.

Table 1. The Average SSA of each of used semantic segmentation algorithms

A1	A2	A3	A4	A5
48.473%	47.048%	67.637%	50.089%	69.873%

3.1   Ideal Algorithm Selection

To determine how the algorithm selection can improve a computer vision tasks we perform a synthetic experiment for semantic segmentation. We measure the semantic segmentation accuracy (SSA) as a function of algorithm selection accuracy (MASA). For each label s_j^i we rank each label-wise segmentation using the m_s^j score. Then, we construct a resulting semantic segmentation by selecting from each label s_j probabilistically according to the m_s^j. Let $\theta \in [0, 1]$ represent the accuracy of virtual algorithm selection, the s^1 and s^0 represent all best and all non-best results of semantic segmentation for label image I. Then, randomly select for each label s_j from s^1 and from s^0 using θ. Repeat the process k times to obtain average SSA.

Table 2 shows the results of the semantic segmentation for each object class at various MASA values. The first column in Table 2 indicates the object class, second to last columns shows semantic segmentation accuracy (SSA) at decreasing MASA threshold (θ).

The second column, $\theta = 1$ means that for each object class on each image, the algorithm with highest SSA for each object class present in the image is chosen for segmentation. The column with $\theta = 0.7$ shows the result of semantic segmentation accuracies for each class: 70% of times the algorithm with highest SSA is chosen for a particular class to make segmentation. For the remaining 30% of times, the algorithm for a class segmentation is chosen randomly among the other remaining four algorithms excluding the best one from the pool.

Table 2. Per class segmentation accuracies over all the images in VOC2012 with 100%–50% average selection accuracy (MASA)

Label	θ			
	1	0.9	0.7	0.5
Background	94.649%	94.649%	94.426%	92.753%
Aeroplane	87.363%	87.363%	87.363%	84.237%
Bicycle	39.996%	39.996%	39.838%	38.693%
Bird	90.811%	90.811%	88.266%	86.394%
Boat	81.360%	81.360%	81.360%	77.778%
Bottle	80.827%	80.827%	80.795%	78.496%
Bus	92.474%	92.474%	91.532%	90.605%
Car	87.655%	87.655%	87.655%	86.164%
Cat	92.404%	92.404%	92.312%	88.786%
Chair	53.704%	53.704%	51.137%	48.948%
Cow	91.060%	91.060%	91.060%	84.677%
Diningtable	79.888%	79.888%	79.895%	76.061%
Dog	89.636%	89.636%	89.730%	86.591%
Horse	87.995%	87.995%	87.578%	84.298%
Motorbike	84.106%	84.106%	83.647%	80.161%
Person	85.143%	85.143%	85.048%	80.867%
Pottedplant	73.632%	73.632%	73.632%	63.464%
Sheep	88.885%	88.885%	86.556%	79.957%
Sofa	75.154%	75.154%	74.969%	64.473%
Train	90.041%	90.041%	90.041%	83.006%
Tvmonitor	83.409%	83.409%	83.409%	76.555%
Average SSA	82.390%	82.390%	81.869%	77.76%

Interestingly the resulting SSA of the algorithm selection with $\theta = 0.5$ (50% MASA) results in average 77.76% SSA: which is higher than the top accuracy among the five algorithms (69.873%, Table 1). Thus, even with a relatively low MASA the resulting average SSA is higher than the SSA of algorithm with the highest SSA, A5.

For the MSCOCO dataset a different set of also five algorithms was used. The reason for using different algorithms for the MSCOCO dataset is because the target processing is used for object detection. Note, that for object detection we will also be using the SSA measure as it cab be directly used to evaluate object detection task. Table 3 shows information about average IoU scores of all detection algorithms (rows 2 to 7) and the highest possible average IoU score (row 1), using algorithm selection at $\theta = 1$ MASA. Mask R-CNN outperformed all individual algorithms, achieving average IoU score of 73.56% while the algorithm selection approach with 100% MASA results in even higher IoU 83.36%.

Table 3. Average IoU scores of object detection algorithms for MSCOCO (averaged across 5 runs).

Algorithm	Average IoU score
AS with $\theta = 1$	**83.36%**
RetinaNet	58.32%
SSD	30.99%
Mask R-CNN	**73.56%**
MultiPath Network	11.75%
R-FCN	37.85%
YOLO	62.37%

Observe that in the case of object detection (Table 3) the impact of algorithm selection is the increase of up to 10% of IoU while in semantic segmentation (Table 2) the impact of algorithm selection is much higher. The main reason is due to the fact that the largest amount of errors in object detection is miss detection resulting in many objects not being recognized at all by any algorithm.

4   Experiments

4.1   Experiments on VOC2012

The first set of experiments focuses on AS based on average SSA for the whole image and using the VOC2012 dataset.

In order to have an accurate algorithm selection, we need to obtain distinctive feature set [21] from the available fatures extractors. Note, that we do not look for specific meta features as this would require a heavy pre-processing. We evaluated the following features for the experiments on VOC2012:

- Feature set 1a: features obtained from the fourth convolutional layer of AlexNet (c4).
- Feature set 1b: features obtained from the fifth convolutional layer of AlexNet (c5).
- Feature set 1c: features obtained by concatenating output of the convolutional layer four to the output of the convolutional layer five of AlexNet (c4 + c5).
- Feature set 2: features obtained from ResNet18.
- Feature set 3: features obtained by concatenating the output of the last convolutional layer of VGG16
- Feature set 4: features obtained by concatenating the output of the last convolutional layer of GoogLeNet
- Feature set 5: features obtained by concatenating the output of the last convolutional layer of DenseNet
- Feature set 6: visual bag of words using SIFT descriptors.

Experiments with Features and Data Augmentation. The first set of experiments were conducted using Feature sets 1a-1c, which are extracted from pretrained AlexNet. The classification algorithm used at the early stage is SVM because it is a good choice whenever the number of instances is less than the number of features. Moreover, since the number of train instances is 1152, and is low compared to the number of features in Feature set-1a, Feature set-1b (both have 43264 features), and Feature set-1c (86528 features), we applied feature selection techniques such as XGBoost and PCA. The results of classification accuracy after using XGBoost is 38.75%, and 34.25% when PCA is applied to Feature set-1 to reduce the number of features to 289.

We conducted several experiments using uncompressed and non reduced features extracted from AlexNet. The best results on the classification of algorithm selection on semantic segmentation are obtained using Feature set-3 (concatenation of the fourth and the fifth layers of AlexNet) with RBF kernel in SVM, which resulted in 43.6% of classification accuracy. To determine if the features from AlexNet are suitable we also performed feature extraction using ResNet18, GoogLeNet, Densenet and VGG16. The features extracted using from each of the CNNs consist of 512, 1024, 2200, and 512 features respectively features. The accuracy of the classification results of the experiment with Feature set-2 and SVM is 34.6%. For GoogLeNet, Densenet and VGG16 features the relative accuracies of MASA were 35%, 35.5% and 34.5% respectively.

During some analysis on the dataset, we observed that most instances of the test set were predicted to belong to algorithm A4 and algorithm A5. The main reason for such a classification result is high class imbalance, which can be observed on the histograms Figs. 1a, b, and d. The distribution of samples across the different classes both in train and test sets are the same. We have very big number of samples classified to A4 and A5; therefore, all the samples in the test set are predicted to A4 and A5.

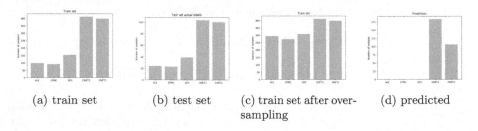

(a) train set   (b) test set   (c) train set after over-   (d) predicted
                                          sampling

Fig. 1. Best Algorithms distribution in the dataset. y-axis - number of samples, x-axis - a_i

To evaluate the impact of the class imbalance on the MASA, we evaluated different techniques for over sampling, synthetic data generation and data subsampling.

For oversampling we evaluated the following approaches. Sample data copy. This approach increases the amount of the instances from the minority class by copying (sampling) them. The result of this approach is show in row 8 of Table 4.

The generation of synthetic samples of the under-represented class using Synthetic Minority Class Oversampling Technique (SMOTE) [6]. This approach creates new data samples by creating them from existing samples by altering some of the existing data sample features. The result of this approach is showin in rows 5 and 7 of Table 4.

Next data augmentation used is Gaussian Mixture of Models (GMM). The GMM model is a technique to approximate arbitrary data distribution by fitting a set of Gaussian kernels onto the data. Using this approach we used the training set of samples to build the GMM model. Then the model was sampled for a total of 3000 samples per algorithm. The average accuracy using this method resulted in $MASA = 36\%$ shown in row 15 of Table 4.

One of the difficulty that can be related to the low results in MASA, can in fact be based on the nature of the used features for algorithm selection. The used features are obtained from pre-trained neural networks: high quality features effective for a task do not however guarantee enough information for distinguishing the different tool's that use these features.

Therefore, we evaluated the Feature set-6 formed using the visual bag of words on SIFT descriptors. The SIFT feature descriptors are 128 dimensional vectors. The results of this experiments in MASA are shown in rows 12 and 13 in Table 4.

Table 4. Summary of experiments showing MASA as a function of different features combination.

Method	Accuracy
AlexNet (c4), SVM	28.02%
AlexNet (c5) + PCA, SVM	34.26%
AlexNet (c4 + c5), SVM	**43.6%**
AlexNet (c4 + c5) + SMOTE, SVM	39.45%
ResNet18, SVM	34.6%
ResNet18 + SMOTE, SVM	20.41%
AlexNet (c4 + c5), Oversampling, SVM	39.1%
GoogLeNet, SVM	35%
Densenet, SVM	35.5%
VGG16, SVM	37%
SIFT, ANN	37%
SIFT, SVM	35.64%
AlexNet (c4 + c5), Two stage SVM	35.6%
All features, GMM, SVM	36%

Region Attributes and Semantic Labels. The algorithm selection using only processing oriented is quite limited in MASA. Therefore we consider the possibility to add meta features extracted from the first pass of processing as originally shown in [21]. These meta-features are related to objects detected within each region and are collected as a set of regional attributes. Also due to the fact that AlexNet features performed in a most satisfactory manner, we decided to use AlexNet features for most of the following experiments. The region attributes are extracted from a grayscale image (RPG). The regional attributes contain the following properties: area, centroid, eccentricity, euler number, extent, orientation, perimeter, solidity and equivalent diameter.

For all ground truth instances, the aforementioned attributes were extracted from numerous regions and 50 highest values were normalized and converted to histograms. Additionally, for each object category, using attribute information of instances belonging to it, histograms representing all categories, were calculated.

Table 5 shows the results of experiments where image features, context attributes (CA) (ground truth semantic labels of each object), and region attributes are used. For evaluation the SVM selection is compared to the Gradient Boosting Method.

Table 5. Summary of Experiments for determining the impact of Context Attributes (CA) and Regional Properties extracted from gray image (RPG) on MASA.

Configuration	SVM prediction	Gradient boosting
CA	37.60%	38.76%
AlexNet (c4)	37.98%	37.98%
AlexNet (c4), RPG	39.10%	38.40%
AlexNet (c4), CA	41.09%	40.31%
AlexNet (c4), RPG, CA	43.41%	**44.96%**

4.2   Experiments on MSCOCO

The VOC2012 dataset proved to be difficult with respect to algorithm selection. Therefore, a much larger dataset was considered for evaluation. To determine if the weak results on VOC2012 were due to the approach or due to the small amount of samples, the instance-based algorithm selection was also applied to MSCOCO validation set. We focus on the more successful selection previously described, in MSCOCO we perform algorithm selection in the task of object detection, using image features and object attributes. For each object class, an average set of attributes is calculated. This is obtained by standard averaging. Then, given an object instance, the attributes extracted from the image are compared to averaged attributes per object class. The distance measures between histogram of an instance and histograms of all object categories were calculated and category with the smallest distance was chosen. For more robustness, several distance measures were used and evaluated.

Therefore the dataset contained the following data fields: target object instance attributes, and seven closest object attributes calculated using seven different standard similarities such as euclidean distance, Manhattan distance, etc.

Table 6 displays algorithm selection accuracies (first column) and average IoU score (second column) for 7 classifiers, namely, K-Nearest Neighbors, Support Vector Machine, Decision Tree, Random Forest, Multi-layer Perceptron, AdaBoost and Gaussian Naive Bayes classifiers. The best algorithm selection accuracy of 42.69% was obtained by AdaBoost classifier, but its corresponding average IoU score of 71.59% was not the highest. Better average IoU scores of 71.87%, 72.03% and 72.06% were achieved by Random Forest, Multi-layer Perceptron and Decision Tree classifier respectively, although their algorithm selection accuracies were less than AdaBoost's.

Table 6. Accuracy of algorithm selection and corresponding average IoU scores averaged across 5 runs.

	KNN	SVM	Decision tree	Random forest	MLP	AdaBoost	Gaussian
MASA	37.65%	41.45%	42.20%	42.48%	41.43%	**42.69%**	39.83%
Average IOU	67.33%	70.71%	**72.06%**	71.87%	72.03%	71.59%	68.04%

4.3 Dynamic Ensemble Approach

In order to leverage the advantage of each algorithm individually, we determined that a more dynamic approach of selection is required. For this we used the dynamic ensemble (DE) approaches [9]. A dynamic ensemble is a method where from an ensemble of classifiers one or more selectors for a particular task is selected on a CBC basis. This makes it a good candidate for single instance algorithm selection. The novelty of this approach in the algorithm selection problem is that we are not going to select adaptively algorithms for semantic segmentation but the selection will be done on a set of heterogeneously trained selectors.

To build a DE we consider two different approaches. The first approach considers to train k one-vs-j (with $j = 1, \ldots, k-1$) algorithm selectors on perfectly balanced subset of training data. Therefore for each pair of target algorithms, different number of selectors can be obtained. The second approach only builds l selectors for all combinations of one-vs-j but using the whole training dataset (FDE). Then during inference, given a test sample, the distance to all training samples is calculated and k-closest neighbors are stored. Then only algorithm selectors that have classification accuracy higher than θ on the k-nearest neighbors are kept [9] and used for determining the result.

The experiments are summarized in Table 7.

Columns two to seven show various configurations of the dynamic ensemble. The second to fifth columns shows the matching selector criteria k: the number of closest training sample points on which the selected selector is evaluated.

Table 7. Results of using dynamic ensembles for algorithm selection

Data Set	Measure	DE Parameterization (k nearest neighbors)					
		5	10	20	40	FDE	E
VOC2012	MASA	35.7%	36%	38.7%	35%	**40%**	36%
	IOU	65%	64.3%	66%	65%	**68.5%**	66%
MSCOCO	MASA	38%	41%	40%	37%	**60%**	43%
	IOU	68%	70%	68.5%	66.9%	**76.5%**	72%

The sixth column shows the results of the DE that trained the ensemble using the second approach (FDE). Finally the column entitled E, shows the result of an ensemble approach, where the selector is selected without the use of closest neighbors information. The used selectors are XGBoost, SVM, AdaBoost, RandomForest, KNN and MLP.

The main observation is that similarly to single algorithm selector the dynamic ensemble and the ensemble method did not work for the data from the VOC2012. On the other hand the results from MSCOCO are showing a quite different trend. While the sub-sampling approach using ensemble method is better than s.

5   Results and Discussion

The experiments described in Table 4 to 7 show that for single algorithm selection the SVM classifier is the most accurate. The maximal accuracy of 44.96% (VOC2012) is far from an average random accuracy obtained by random selector resulting in $\approx 22\%$. Additionally the experiments demonstrated that features from AlexNet seem to be more effective than deeper features from ResNet18, GoogLeNet or others. This is interesting because in general these models has a higher classification accuracy compared to AlexNet.

For MSCOCO the single algorithm selection was most accurate with AdaBoost. However in both cases, single algorithm selection was not accurate enough to outperform the best algorithm in the SSA.

The most striking observation is the fact how the MASA changed on MSCOCO vs VOC2012 when using the DE classifier. The possible conclusion drawn from such a large difference is that VOC2012 is very undersized for the algorithm selection problem. While the algorithms can perform relatively well on the VOC2012 dataset (and being trained on it) the task of algorithm selection requires more extensive evaluation. In fact how much training is required for accurate algorithm selection is still an open question [14]. However more interestingly, one can look at the success rate of algorithm selection as a proof of learning and validation. Consider a machine learning task $T : I \rightarrow O$ that is developed the standard training and validation dataset. While the validation dataset provides a statistical averaged measure of learning, validation dataset does guarantee an algorithms generalization after large enough amount of points are selected only. The algorithm selection can give a direct estimation of how the data (and the learning procedure) is significant for the generalization. The main difference between applying selection and simple validation data evaluation, is the observation that learning selection will allow to precisely determine what areas of feature space should be evaluated in order to determine the each algorithm performance assessment. Naturally, if meta-features are available, the generalization cam be improved by considering.

As a consequence, the proposed algorithm selection was able to surpass Mask R-CNN in terms of detection accuracy (76.5% vs 73.56% respectively) using the ensemble approach only for the MSCOCO dataset. The reason for the positive result is because the MSCOCO dataset is better in representing individual algorithms behavior.

6   Conclusion

In this paper we evaluated various learning approaches to the problem of algorithm selection in computer vision problems without the usage of specifically crafted meta-features. We showed that given enough of sample data, algorithm selection can be accurate enough to increase the processing accuracy. In practice however the ideal set of features at a reasonable computational remains to be found.

A direct extension of this work is a context based algorithm selection and an evaluation based study that is inline with the current research in algorithm selection [14].

Acknowledgment. This work was funded by the FCDRGP research grant entitled *LFC: Intention Estimation: A Live Feeling Approach* from Nazarbayev University with reference number 240919FD3936.

References

1. Aguiar, G.J., Mantovani, R.G., Mastelini, S.M., de Carvalho, A.C., Campos,G.F., Junior, S.B.: A meta-learning approach for selecting image segmentation algorithm. Pattern Recogn. Lett. **128**, 480–487 (2019). https://doi.org/10.1016/j.patrec.2019. 10.018, http://www.sciencedirect.com/science/article/pii/S0167865519302983
2. Ali, S., Smith, K.: On learning algorithm selection for classification. Appl. Soft Comput. **6**, 119–138 (2006)
3. Bischl, B., et al.:ASlib: a benchmark library for algorithm selection. Artif. Intell. **237**, 41–58 (2016). https://doi.org/10.1016/j.artint.2016.04.003, http:// www.sciencedirect.com/science/article/pii/S0004370216300388
4. Bosch, M., Gifford, C., Dress, A., Lau, C., Skibo, J.: Improved image segmentation via cost minimization of multiple hypotheses. In: T.-K., Kim, Zafeiriou, G.B.S., Mikolajczyk, K. (eds.) Proceedings of the British Machine Vision Conference (BMVC), pp. 7.1–7.12. BMVA Press, September 2017. https://doi.org/10. 5244/C.31.7
5. Carreira, J., Li, F., Sminchisescu, C.: Object recognition by sequential figure-ground ranking. Int. J. Comput. Vis. **98**(3), 243–262 (2012)

6. Chawla, N., Bower, K.W., Hall, L., Kegelmayer, W.: SMOTE: synthetic minority over-sampling technique. J. Artif. Intell. Res. **16**, 321–357 (2002)
7. Chen, L., Papandreou, G., Kokkinos, I., Murphy, K., Yuille, A.: Semantic image segmentation with deep convolutional nets and fully connected CRFs. CoRR abs/1412.7062 (2014). http://arxiv.org/abs/1412.7062
8. Chinchor, N.: MUC-4 evaluation metrics. In: Proceedings of the 4th Conference on Message Understanding. MUC4 1992, pp. 22–29. Association for Computational Linguistics, USA (1992). https://doi.org/10.3115/1072064.1072067
9. Cruz, R.M.O., Sabourin, R., Cavalcanti, G.D.C.: Dynamic classifier selection: recent advances and perspectives. Inf. Fusion **41**, 195–216 (2018)
10. Everingham, M., Van Gool, L., Williams, C.K.I., Winn, J., Zisserman, A.: The PASCAL visual object classes (VOC) challenge. Int. J. Comput. Vis. **88**(2), 303–338 (2010)
11. Gunawan, A., Lau, H.C., Misir, M.: Designing and comparing multiple portfolios of parameter configurations for online algorithm selection. In: Festa, P., Sellmann, M., Vanschoren, J. (eds.) Learning and Intelligent Optimization - 10th International Conference, LION 10, 29 May – 1 June 2016, Ischia, Italy, Revised Selected Papers. Lecture Notes in Computer Science, vol. 10079, pp. 91–106. Springer (2016). https://doi.org/10.1007/978-3-319-50349-3_7
12. Hariharan, B., Arbeláez, P., Girshick, R., Malik, J.: Simultaneous detection and segmentation. In: European Conference on Computer Vision, pp. 297–312 (2014). https://doi.org/10.1007/978-3-319-10584-0_20
13. Jaccard, P.: Étude comparative de la distribution florale dans une portion des alpes et des jura. Bulletin del la Société Vaudoise des Sciences Naturelles **37**, 547–579 (1901)
14. Kerschke, P., Hoos, H.H., Neumann, F., Trautmann, H.: Automated algorithm selection: survey and perspectives. Evol. Comput. **27**(1), 3–45 (2019). https://doi.org/10.1162/evco_a_00242, pMID: 30475672
15. Kim, Y., Jang, T., Han, B., Choi, S.: Learning to select pre-trained deep representations with Bayesian evidence framework. CoRR abs/1506.02565 (2015). http://arxiv.org/abs/1506.02565
16. Ladicky, L., Russell, C., Kohli, P., Torr, P.: Graph cut based inference with co-occurrence statistics. In: Proceedings of the 11th European Conference on Computer Vision, pp. 239–253 (2010). https://doi.org/10.1007/978-3-642-15555-0_18
17. Leyton-Brown, K., Nudelman, E., Andrew, G., Mcfadden, J., Shoham, Y.: A portfolio approach to algorithm selection. In: IJCAI, vol. 3, pp. 1542–1543 (2003)
18. Lin, T., et al.: Microsoft COCO: common objects in context. CoRR abs/1405.0312 (2014). http://arxiv.org/abs/1405.0312
19. Lindauer, M., Hoos, H., Hutter, F.: From sequential algorithm selection to parallel portfolio selection. In: LION (2015)
20. Lindauer, M., Hoos, H.H., Hutter, F., Schaub, T.: AutoFolio: an automatically configured algorithm selector. J. Artif. Int. Res. **53**(1), 745–778 (2015)
21. Lukac, M., Abdiyeva, K., Kim, A., Kameyama, M.: Reasoning and algorithm selection augmented symbolic segmentation. In: Intelligent Systems Conference (2017)
22. van Maaren, H., Franco, J.: The International SAT Competition Web Page (2002). http://satcompetition.org/
23. Mnih, V., et al.: Playing atari with deep reinforcement learning. CoRR abs/1312.5602 (2013). http://arxiv.org/abs/1312.5602
24. Murdock, C., Li, Z., Zhou, H., Duerig, T.: Blockout: dynamic model selection for hierarchical deep networks. In: Proceedings of the IEEE Conference on Computer Vision and Pattern Recognition (CVPR), June 2016

25. Muñoz, M.A., Sun, Y., Kirley, M., Halgamuge, S.K.: Algorithm selection forblack-box continuous optimization problems: a survey on methods and challenges. Inf. Sci. **317**, 224 – 245 (2015). https://doi.org/10.1016/j.ins.2015.05.010, http://www.sciencedirect.com/science/article/pii/S0020025515003680

26. Rice, J.: The algorithm selection problem. Adv. Comput. **15**, 65118 (1976)

27. Rusu, A.A., et al.: Progressive neural networks. CoRR abs/1606.04671 (2016). http://arxiv.org/abs/1606.04671

28. Wang, J.X., et al.: Learning to reinforcement learn. CoRR abs/1611.05763 (2016). http://arxiv.org/abs/1611.05763

29. Wang, Z., de Freitas, N., Lanctot, M.: Dueling network architectures for deep reinforcement learning. CoRR abs/1511.06581 (2015). http://arxiv.org/abs/1511.06581

30. Xu, L., Hutter, F., Hoos, H.H., Leyton-Brown, K.: SATzilla: portfolio-based algorithm selection for SAT. J. Artif. Intell. Res. **32**, 565–606, July 2008. https://doi.org/10.1613/jair.2490

31. Zoph, B., Le, Q.V.: Neural architecture search with reinforcement learning. CoRR abs/1611.01578 (2016). http://arxiv.org/abs/1611.01578

A Hybrid Machine Learning Approach for Energy Consumption Prediction in Additive Manufacturing

Yixin Li[1(✉)], Fu Hu[1], Jian Qin[2], Michael Ryan[1], Ray Wang[3], and Ying Liu[1]

[1] High-Value Manufacturing Research Group, School of Engineering, Cardiff University, Cardiff, UK
{liy248,huf4,ryanm6,LiuY81}@cardiff.ac.uk
[2] School of Aerospace, Transport and Manufacturing, The Welding Engineering and Laser Processing Centre, Cranfield, Bedford, England
j.qin@cranfield.ac.uk
[3] Unicmicro (Guangzhou) Co., Ltd, Guangzhou, China
ray.wang@unicmicro.com

Abstract. Additive manufacturing (AM), as a fast-developing technology for rapid manufacturing, offers a paradigm shift in terms of process flexibility and product customisation, showing great potential for widespread adoption in the industry. In recent years, energy consumption has increasingly attracted attention in both academia and industry due to the increasing demands and applications of AM systems in production. However, AM systems are considered highly complex, consisting of several subsystems, where energy consumption is related to various correlated factors. These factors stem from different sources and typically contain features with various types and dimensions, posing challenges for integration for analysing and modelling. To tackle this issue, a hybrid machine learning (ML) approach that integrates extreme gradient boosting (XGBoost) decision tree and density-based spatial clustering of applications with noise (DBSCAN) technique, is proposed to handle such multi-source data with different granularities and structures for energy consumption prediction. In this paper, four different sources, including design, process, working environment, and material, are taken into account. The unstructured data is clustered by DBSCAN so to reduce data dimensionality and combined with handcrafted features into the XGBoost model for energy consumption prediction. A case study was conducted, focusing on the real-world SLS system to demonstrate the effectiveness of the proposed method.

Keywords: Additive manufacturing · Energy consumption · Modelling · Machine learning · Multi-source data

1 Introduction

AM is often referred to as a 3D printing technology and defined as a process of adding materials layer by layer to fabricate products based on 3D model data [1]. Compared

© Springer Nature Switzerland AG 2021
A. Del Bimbo et al. (Eds.): ICPR 2020 Workshops, LNCS 12664, pp. 622–636, 2021.
https://doi.org/10.1007/978-3-030-68799-1_45

with conventional manufacturing techniques, AM provides the feasibility for complex-shaped parts [2]. Besides, AM allows a short time to fabricate products from a concept, and it shows the significance of processing and improving material properties of products with low material consumption [3]. Therefore, AM is applied in the variety of industrial and medical usages, such as aerospace, automobile and dental equipment [4].

However, the growing environmental concerns on sustainability, particularly on energy consumption, have been considered [5]. Therefore, improving energy management in AM systems is urgent. So far, more and more researchers and manufacturers have increased their attention towards this aspect. AM has the potential to achieve a larger yield of product, resulting in an increasing amount of energy consumed. The eco-design is necessary at an early stage in the manufacturing system, supporting designers and manufacturers in energy management, decision-making and improvement in the process [6]. However, it is challenging to improve energy management, as the subsystems of AM will generate numerous data during the entire process. As investigated by Ahuett-Garza and Kurfess [7], the production process of an AM system consists of six stages, including 1) conversion, 2) positioning and orientation, 3) adding support structure, 4) slicing, 5) fabrication and 6) post-processing. The variety of data sources leads to the complexity of processing data due to the different dimensions and structures of the data in energy consumption prediction. According to Qin et al. [8], four categories of data related to design, process, working environment and material are considered, which are also known as the multi-source data. The multi-source data contains valuable information that uncovers the correlations between the selected features and energy consumption. Compared with traditional manufacturing techniques, the complexity of AM systems is very challenging for energy consumption analysis. However, the application of IoT technology can perform real-time monitoring from multiple processes. This makes a single model not suitable for this situation. Therefore, this paper proposes a hybrid ML method for predictive modelling of energy consumption.

The method is integrated by using supervised and unsupervised learning, which are proposed as XGBoost and DBSCAN, respectively. The data were obtained from an SLS system with different sources, where energy consumption is affected. The primary function of the DBSCAN algorithm is to get informative data and simultaneously reduce data dimensionality, combining selected features into the XGBoost model for energy consumption prediction. Root mean squared error (RMSE) and the model correlation coefficient (MCC) is used to assess the model performance.

Section 2 reviews the factors that influence energy consumption in AM systems and existing studies of ML techniques to establish predictive modelling in various AM system under different scenarios. Section 3 describes the detailed framework supporting the adoption of DBSCAN and XGBoost. Section 4 demonstrates the outcome of the proposed methodology in the specific SLS system. Section 5 concludes the paper.

2 Literature Review

2.1 Analysis of Energy Consumption in AM Systems

AM has promoted a new manufacturing pattern that is involved in small-batch manufacturing with customisation, satisfying customers' demand [9]. According to different working principle and material supplies, some typical AM techniques include electron beam melting (EBM) [10], fused deposition modelling (FDM) [6], selective laser sintering (SLS) [11] and selective laser melting (SLM) [12].

Table 1. The list of different AM systems in terms of working principles, material supply and energy consumption

AM system	Working principles	Material supply	Unit Energy consumption (kW/h)
EBM	Utilisation of high-intensive electron beam to melt the material powder	Ti-6Al-4V, 316L stainless steel	17–49.2
FDM	The nozzle of printer extrudes fused thermoplastic material	Acrylonitrile Butadiene Styrene (ABS), Polycarbonate (PC)	23.1–346.4
SLS	Laser sinters the material powder	Polyamide, nylon	14.2–40.0
SLM	Laser melts the material powder	Ti-6A-4, 316L stainless steel	23.1–163.3

Noticeably, different working principles and types of supplied material used in AM systems play a vital role in affecting the energy consumption according to Table 1. Other impact factors from process, design, working environment and material are also considered. Many researchers have investigated those impact factors in different AM systems. For instance, an investigation was given by Paul and Anand [13], who conducted numerical studies, demonstrating the impact factors, including layer thickness and part orientation. Baumers et al. [14, 15] conducted experiments to determine the four relevant factors during the AM process by comparing two different working platform. In their research, the processing stage, scanning, recoating, and Z-height consumed energy. Peng [16] focused on the process of 3D printing and broadened the analysis in terms of primary and secondary energy. Primary energy referred to direct energy consumption such as material form and properties, while secondary energy highly depended on in-process energy consumption. Watson and Taminger [2] computed an energy consumption model to illustrate the flow of energy consumption, considering the life cycle from feedstock to the end of product life. Differently from other authors, they highlighted that the transportation distance influenced the energy consumed in the AM system. Liu et al. [17] investigated the machine and process level in AM system that could have a significant impact on energy consumption. Furthermore, they also concentrated on material characteristics from the micro-level, which has an indirect impact on energy consumption.

For the investigations and studies above, it is found that various factors from different sources have a significant impact on the energy consumption of the AM systems. These contribute to a better understanding of applying the predictive model in practice.

With regards to different scenarios, it is essential to set up specific models which are progressive to the specific task. The following contents in this section briefly review the machine learning (ML) for predictive modelling in different AM systems.

2.2   Machine Learning for Predictive Modelling in AM System

In general, AM systems are complicated manufacturing systems, including many subsystems and the sensors and processors which generate massive amounts of data with various features and types under the IoT environment [9]. Hence, the time cost of preprocessing often makes up the largest proportion of time used during the entire prediction tasks. With consideration of the redundancy and irrelevance of some data, it is crucial to extract the relevant information by advanced data analytics. Machine learning (ML) and deep learning (DL) show merit to tackle the issue.

In the experiments by Bhinge et al. [18], data were collected from a FANUC controller and using a high-speed power meter to handle process data and power time-series data, respectively. The work applied the Gaussian Process regression model to cope with the small amount of data with high dimensionality. The model showed the significance of illustrating the correlation between features and energy usages in machine tools, which helps with depicting the tool path of the target machine for energy management.

Another application of ML and DL in anomaly detection was achieved by Zhang et al. [19]. A high-speed camera captured the images from the SLM system. Firstly, the conducted work was to apply principal component analysis on feature selection, reducing data dimensionality. Secondly, the image data were combined into the support vector machine and convolutional neural network (CNN) to classify the image data according to features with 90.1% and 92.7% of accuracy, respectively.

Other researchers benefited from ML techniques for the feature selection and extraction in AM systems. For instance, Wu et al. [20] conducted a numerical study to apply data integration method at the feature level, to process signals received by monitoring as the input, which was applied in ML models, followed by predicting the surface roughness of builds in the AM system. Some contributions were investigated in [8] and [21]. In [8], Qin et al. proposed a hybrid ML and DL approaches to deal with unstructured data with different features, types and dimensions in a complex AM system. The advantage is that it achieves considerable information compression. In [21], the focus of this paper was to integrate heterogeneous data to uncover the hidden knowledge with correlations between different features to help designers make decisions.

3   Methodology

This section will elaborate on the proposed methodology, targeting the multi-source data from an SLS system by using DBSCAN to cluster the unstructured data, which is then integrated into the XGBoost model to predict the energy consumption. The main stages of the experiment consist of three main stages, i.e. 1) data sensing and collection, 2) the hybrid ML-based approach for predictive modelling and 3) model validation and evaluation.

3.1 Multi-source Data Sensing and Collection

The original data is collected from the target AM system, where the data can be categorised into four different types. They are operation process, material, working environment and design. In Fig. 1, process data stem from the parameter settings collected from the SLS machine, such as the measured values from the dispenser, recoater speed and the laser power used in sintering, which relies on the experience and knowledge of the operators. With regards to material data, it depends on the material itself. In this case, the type of material is known, referring to two kinds of nylon powder. Design data is the data collected from computer-aided design (CAD) models created by designers, often including design parameters for each layer [5], which are often determined at the beginning stage of the entire process. The working environment can be monitored by sensors and data stored in the conditional monitoring files for the illustration. This kind of data source collected from the working environment by an IoT platform is considered as the layer level from real-time monitoring. Some monitoring files can demonstrate these data to better comprehend the structure of the data.

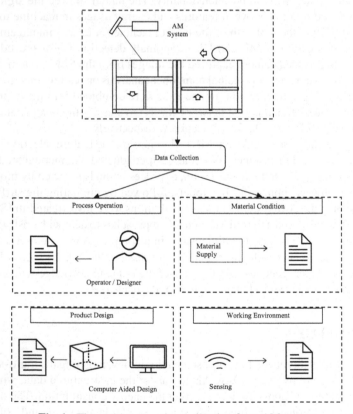

Fig. 1. The multi-source data collection from AM system

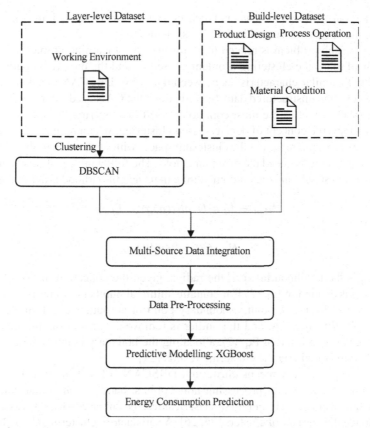

Fig. 2. The framework of proposed methodology for energy consumption prediction

Figure 2 demonstrates the framework of the proposed methodology in pre-processing and modelling. The entire process is divided into three stages, corresponding to their respective roles. At the first stage, the input data are collected from the SLS system and categorised into four types of datasets according to their sources. The working environment data contains different quantities in each separate file, which is essential for integrating these data with using DBSCAN to unify the structure of layer-level data. Secondly, the dimensionality of the integrated dataset is reduced through DBSCAN clustering, and this dataset is fed into the XGBoost decision tree. Finally, the energy consumption is obtained, and the performance of the XGBoost model is evaluated using RMSE and MCC.

3.2 The Hybrid ML Approach for Energy Consumption Prediction

Advanced data analysis and ML methods show the ability to predict the model. ML is usually divided into supervised learning, unsupervised learning and reinforcement learning according to the learning method [22]. This work uses a hybrid method that combined unsupervised learning and supervised learning, i.e. unsupervised learning aims to integrate different dimensional datasets, while supervised learning is utilised to predictive the energy consumed in the SLS system.

The clustering problem is related to an unsupervised learning problem. According to predefined rules, the clustering problem is used to find the uncovered patterns to be classified with similar characteristics between data [23]. DBSCAN is a data clustering algorithm targeting unstructured data. Specifically, DBSCAN used a density-based clustering approach, which is the most commonly used in clustering the spatial data. This algorithm adopts the concept of density-based clustering, which requires the number of points lied in a specific region of the clustering space, with minimum numbers of objects (*MinPts*) and it should exceed the given threshold. The following equations demonstrate the nature of DBSCAN and the random point p in its neighbourhood is defined in Eq. (1)

$$N_{Eps} = \{q \in D / dist(p, q) < Eps\} \tag{1}$$

$$N_{Eps}(P) > MinPts \tag{2}$$

where *Eps* is the neighbourhood of the radius, given the collection of objects D. The core point P is defined in Eq. (2) if it contains minimal numbers of points.

In other words, a core point, a boundary point or an outlier is determined by two indicators: *MinPts* and *Eps*, and the outlier is removed. The algorithm connects core points under the condition of Eq. (2), allocating the boundary point to the closest core point and finally obtaining the clustering results [24].

When comparing to k-means clustering, DBSCAN is faster in terms of clustering speed and more effective in processing noise points, handling abnormal data, and in exploring spatial clusters of random shapes. Besides, the unbiased-shaped clusters do not need to divide the number of clusters [25, 26]. A satisfactory clustering algorithm needs to have the following characteristic: 1) to determine knowledge from inputs, especially for the large datasets, 2) capable of finding arbitrary shaped clusters and 3) efficient to handle large datasets [27]. The working environment data is collected layer by layer over thousands of data in separate files with various types from the entire process, as a consequence of large data volume and heterogeneity. Therefore, DBSCAN is expected to tackle the issues. Furthermore, this algorithm was applied at the beginning, and it demonstrates the mean values which can be a representative of the entire cluster. These values can be combined into design-relevant datasets on the build-level, in order to unify the format of each working environment data file.

XGBoost refers to a tree-based ensemble learning using tree algorithm, proposed by Chen and Guestrin [28]. This boosting method is an effective ML method. XGBoost uses regression trees ensembles which have the same decision rules as the decision tree (DT), containing one score in each leaf value. Two aspects allow it to be distinguished from other tree boosting machine. Firstly, XGBoost has a different objective function. For each regression tree, this ensemble method accumulates the sum of scores as the prediction value for all tree. Assuming there are k trees, the output for tree ensemble is defined as follow:

$$\hat{y}_i = \sum_{k=1}^{K} f_x(x_i), f_x \in \mathcal{F} \tag{3}$$

Moreover, the objective function, which is the sum of training loss and complexity of the trees to control overfitting, is defined as follow:

$$Obj = \sum_{i=1}^{n} l(y_i, \hat{y}_i) + \sum_{k}^{K} \Omega(f_k), f_x \in \mathcal{F} \tag{4}$$

$$\Omega(f_k) = \gamma T + \frac{1}{2}\lambda\|w\|^2 \tag{5}$$

where $\hat{y}_i$ is the predicted value of the model, y_i stands for the i th feature label, f_k represents the k th tree model, T is the number of nodes and w is the collection of score combinations. In the reduction of the objective function, predicted value adds a new function f in each iteration. This additive training defines a new objective function to optimise and search for a new tree model.

Another difference is the division of nodes. There are four proposed splitting algorithms from Chen and Guestrin's work. XGBoost adopts (1) basic exact greedy algorithm, (2) approximate algorithm, (3) weighted quantile sketch and (4) sparsity-aware split finding methods. Among these four split finding algorithms, algorithm (2) and (3) is to solve the problem that the data fails to load into memory at once or algorithm (1) is not distributed efficiently. The XGBoost approach calculates the gain of each feature in parallel and chooses the feature with the largest information gain to split.

XGBoost provides an idea for processing sparse data and enables the handling of instance weights in tree learning. Compared with the traditional tree model, it shows the merits of regularisation in controlling the model complexity and reducing the variance of the model to avoid overfitting. This model is used to predict the energy consumption in the SLS system. By targeting this specific task, XGBoost integrates the weak learner to form a stronger learner to increase the accuracy. In addition, the sparsity-aware split finding method of XGBoost can process the missing values in the combined datasets. Also, it increases the learning rate effectively by controlling the model complexity, which is important when dealing with large datasets.

3.3 Validation of Prediction Model

Various subsystems consume electric power in the SLS system [29]. The consumption is noted as E_U, referring to the unit energy consumption in kWh/kg and it is:

$$E_U = \frac{E_T}{M_T} \tag{6}$$

where E_T is the total energy consumed in the AM system and M_T means the total weight of fabricated products.

The performance of the predictive model XGBoost decision tree can be evaluated by RMSE and MCC [29–31]. RMSE identifies the actual value (a_t) and the predicted value (p_t). The low value of RMSE determines the high accuracy of the model, which is given by:

$$RMSE = \sqrt{\frac{1}{N}\sum_{t=1}^{N}(p_t - a_t)^2} \tag{7}$$

$$MCC = \frac{S_{PA}}{\sqrt{S_P S_A}} \tag{8}$$

$$S_{PA} = \frac{\sum_i (p_i - \bar{p})(a_i - \bar{a})}{n - 1}; S_P = \frac{\sum_i (p_i - \bar{p})^2}{n - 1}; S_A = \frac{\sum_i (a_i - \bar{a})^2}{n - 1} \tag{9}$$

In Eq. (8) and (9), MCC reveals the correlations between the predicted and actual data obtained from the model, where $\bar{p}$ is the mean value of predicted data, and $\bar{a}$ is the mean value of the entire data.

4 Case Study

The case study was based on an SLS machine (EOS P700) using nylon powder (PA2200 and PA3200GF) to create builds. The data was collected from 2016 to 2018. The working environment data has different quantities, and pre-processing is considered.

4.1 Experimental Setup

Data Description. The data collected from the SLS system are divided into four categories, including process, material, working environment and design. These four datasets obtained from different sources can be distinguished from their data structure.

By inspecting the datasets, the variety of data types is taken into accounts, such as time-series data (build date), nominal data (material types) and numeric data (data collected from the four described sources). Different types of data sources make it complicated to analyse data directly, while data pre-processing is critical before predictive modelling. In details, the layer- and build-level data are allocated based on four kinds of datasets.

For instance, in the working environment dataset, data has different dimensions because each parameter or feature was collected layer by layer, indicating the different height of prints. With regards to the design data, combining with information of material supply and operation process, this combined dataset consists of build date, process parameters, material supply and unit energy consumption of each build. Figure 3 illustrates the statistical distribution of energy consumption. It can be observed that most of the values are located in the range of 200 to 400 kWh/kg. However, the energy consumption of each build showed the difference.

The multi-source data makes it complicated to model directly with these heterogeneous data. By observing the collected datasets, 31 attributes are recorded in the corresponding datasets. These datasets are allocated into two new classes: layer-level and build-level, which layer-level datasets include working environment data and build-level data constitute design, process operation and material. For parameters that remain constant during the process, such as scan speed, the material type and the height of builds, they can be classified as build-level data, corresponding to process, material and product design data, respectively. Conversely, the working environment data are collected from each layer where the information tends to be different during the manufacturing process.

In addition, the informative data monitored and collected from the working environment are placed into layer-level datasets, such as the chamber temperature and the oxygen content (Table 2).

Table 2. Data categories in terms of sources and types.

Data types	Data sources	Data attributes
Build-level data	Process operation	Dispenser values, Scan speed, Recoater speed, etc.
	Product design	Number of builds, Build height, Filling degree, etc.
	Material	Type of material supply
Layer-level data	Working environment	Chamber temperature, Frame temperature, Oxygen level, scanner temperature, etc.

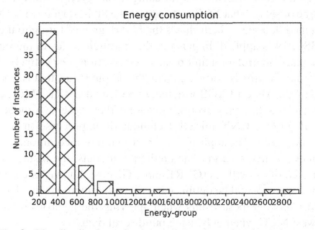

Fig. 3. The distribution of unit energy consumption from datasets.

Data Pre-processing. Prior to the establishment of the predictive model, this stage is used to minimise the data noise and outliers from the datasets. Some instances or features containing massive missing values, they were replaced by mean values or removed from the whole dataset. Considering the complexity and heterogeneity of working environment data, DBSCAN provides a clustering method to select a specific value which can represent the cluster, targeting the working environment data. This is utilised to select the valuable feature to conduct training from existing features, simultaneously reducing the dimensionality to unify the datasets. Differently from DL techniques, ML techniques need labelled and handcrafted features. This method aims to extract a series of the most representative data point from the clusters, which is also constructive to dimensionality reduction.

Model Setups. The proposed methodology is to use XGBoost decision tree to predict energy consumption. Three other algorithms are adopted to the prediction task as benchmarks to compare the performance. These are SVR with linear kernel, gradient boosting regression tree (GBRT) and convolutional neural network (CNN). By comparison, the prediction performance of the proposed model is more convincing.

Model Validation. The model adopts five-cross validation for testing. XGBoost and benchmarks can be evaluated by RMSE and MCC.

4.2   Results and Discussion

There are three ML and one DL algorithms adopted in energy consumption prediction of the case study. The working environment data was trained then combined with other types of datasets into the proposed ML technique and three other benchmarks. After applying XGBoost, RMSE, and MCC to determine the performance of the model, the effectiveness of the proposed approach is demonstrated. Figure 4 illustrates the comparison between the four algorithms with using layer-level, build-level and combined datasets. It can be observed that XGBoost appears the highest value of MCC, 0.708 when applying combined datasets, which shows the best degree of fitting to the experimental data when DBSCAN is applied. In general, this coefficient lies in the range of -1 to 1, representing negative correlation and positive correlation, respectively. It can be seen that all variables are positively correlated with the output value, that is, energy consumption. Followed by the MCC of XGBoost, that of SVR and GBRT's MCC obtains similar values (0.669 and 0.676, respectively), indicating that these two models are suitable for prediction. Regarding CNN, this DL technique often processes image data and performs classification tasks. The application of regression is uncommon. In addition, when applying the multi-source data into the predictive models, the results show that some of the performance of algorithms (GBRT and XGBoost, known as ensemble methods, based on the tree learner) will be optimised, while other methods yield a slight decrease in MCC. SVM yields the best results when applying the layer-level datasets, while it obtains the lowest MCC when only using build-level datasets.

RMSE describes the error in the models more intuitively and refers to the loss function from the regression analysis. Figure 5 demonstrates the comparison of the RMSE for each model. For the RMSE of XGBoost (130.783 kWh/kg), which is within an acceptable range, it measures the deviation between actual and predicted values. This value shows the lowest RMSE when applying the entire datasets. After that, the RMSE of CNN changes dramatically and has the largest value at 231.958 kWh/kg. For this neural net-based algorithm, the best application of industry is associated with pattern recognition or classification. The adoption of multi-source data will affect the performance of predictive models. Figure 6 is the comparison between test data and predicted data, which shows a similar trend for energy consumption. As a result, it can be observed that some outliers affected the final results. When utilising more heterogeneous data and combining them into XGBoost regression tree, the pattern of data fluctuates. However, the trend of predicted data and original still demonstrate a big gap as more data enters, which may be caused by irrelevant features from the datasets and can be solved by collecting data to create new features.

By analysing the nature of different predictive models, ensemble methods (GBRT and XGBoost) show the merits in prediction, and statistic-based algorithm (SVM), also has

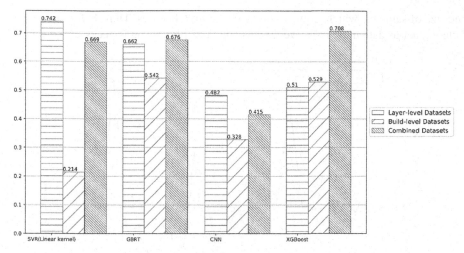

Fig. 4. Comparison of MCC of XGBoost and benchmarks.

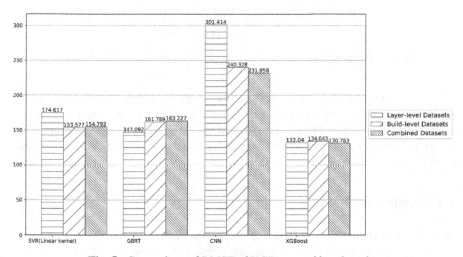

Fig. 5. Comparison of RMSE of XGBoost and benchmarks.

good learning performance. CNN, as the neural-net-based algorithm, has applicability in processing pattern and mere applications in the regression or prediction. Consequently, it has the lowest outcome in MCC and the highest values in RMSE. Another pattern can be observed by the integration of datasets. The combined dataset is determined to influence the prediction performance of each model. The consideration of the overfitting problem is the most essential. Complicated training models may lead to the overfitting of the training data. Furthermore, as one of the challenges of ML, the quality of data will influence the performance of models. If training data contains many noise and outliers, the performance of the model will be affected. In some situation, it depends on the size of samples, which requires the representative features. Thus, before employing heterogeneous data into predictive models, data pre-processing is necessary.

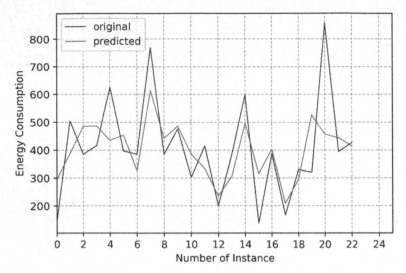

Fig. 6. The prediction result between predicted values and original values.

This hybrid ML approach has presented better performance among the other three algorithms and connected the target and input with high dimensionality when using combined datasets. For tackling the real-world issue regarding handling heterogeneous data, a single learner cannot be adopted, while the integration of DBSCAN and XGBoost suits the case.

5 Conclusions

AM is a comprehensive manufacturing technique embedded with various technologies, which currently employed across many different industries. Manufacturers and researchers are beginning to focus on the sustainability of AM and have started to optimise the process in terms of the energy aspect through data-driven approaches. However, different subsystems generate this data which makes it complicated to handle data, meaning a single predictive model does not adapt to this complicated situation. Hence, a hybrid ML-based approach is proposed.

The experiment firstly adopted the DBSCAN method to select informative and representative data, simultaneously reducing data dimensionality, where the data was clustered and combined into the predictive model. Secondly, this paper applied a tree-based ensemble learning technique, XGBoost, and used it to predict the energy consumption of the SLS system. The evaluation matrix of XGBoost demonstrated a significance in the performance in dealing with heterogeneous data in a complex SLS system. The performance of XGBoost outperformed the other benchmarks, demonstrating the highest MCC and lowest RMSE compared with other algorithms, which shows that XGBoost greatly improves the degree of fitting and accuracy of the model. XGBoost, as an ensemble learning algorithm, provides a feasible approach to predict energy consumption in a complex AM system.

However, the predicted and original value show a significant gap after prediction. This may be affected by interfering and irrelevant information from the raw datasets. Feature extraction will be considered in future work, which is expected to be implemented in the data pre-processing stage. This could improve the modelling performance when the model is built on the data from a large volume.

References

1. ASTM International: Standard Terminology for Additive Manufacturing Technologies. ASTM International, West Conshohocken, PA (2012)
2. Watson, J.K., Taminger, K.M.B.: A decision-support model for selecting additive manufacturing versus subtractive manufacturing based on energy consumption. J. Clean. Prod. **176**, 1316–1322 (2018)
3. Wang, L., Alexander, C.A.: Additive manufacturing and big data. Int. J. Math. Eng. Manage. Sci. **1**(3), 107–121 (2016)
4. Kellens, K., Baumers, M., Gutowski, T.G., Flanagan, W., Lifset, R., Duflou, J.R.: Environmental dimensions of additive manufacturing: mapping application domains and their environmental implications. J. Ind. Ecol. **21**, S49–S68 (2017)
5. Yang, Y., Li, L., Pan, Y., Sun, Z.: Energy consumption modeling of stereolithography-based additive manufacturing toward environmental sustainability. J. Ind. Ecol. **21**, S168–S178 (2017)
6. Freitas, D., Almeida, H.A., Bártolo, H., Bártolo, P.J.: Sustainability in extrusion-based additive manufacturing technologies. Prog. Addit. Manuf. **1**(1–2), 65–78 (2016). https://doi.org/10.1007/s40964-016-0007-6
7. Ahuett-Garza, H., Kurfess, T.: A brief discussion on the trends of habilitating technologies for Industry 4.0 and Smart manufacturing. Manuf. Lett. **15**, 60–63 (2018)
8. Qin, J., Liu, Y., Grosvenor, R.:Multi-source data analytics for AM energy consumption prediction. Adv. Eng. Inf. **38**, 840–850 (2018)
9. Wang, Y., Lin, Y., Zhong, R.Y., Xu, X.: IoT-enabled cloud-based additive manufacturing platform to support rapid product development. Int. J. Prod. Res. **57**(12), 3975–3991 (2019)
10. Frazier, W.E.: Metal additive manufacturing: a review. J. Mater. Eng. Perform. **23**(6), 1917–1928 (2014). https://doi.org/10.1007/s11665-014-0958-z
11. Ma, F., Zhang, H., Hon, K.K.B., Gong, Q.: An optimisation approach of selective laser sintering considering energy consumption and material cost. J. Clean. Prod. **199**, 529–537 (2018)
12. Sing, S.L., et al.: Direct selective laser sintering and melting of ceramics: a review. Rapid Prototyp. J. **23**(3), 611–623 (2017)
13. Paul, R., Anand, S.: Process energy analysis and optimisation in selective laser sintering. J. Manuf. Syst. **31**(4), 429–437 (2012)
14. Baumers, M., Tuck, C., Bourell, D.L., Sreenivasan, R., Hague, R.: Sustainability of additive manufacturing: measuring the energy consumption of the laser sintering process. Proc. Inst. Mech. Eng. Part B J. Eng. Manuf. **225**(12), 2228–2239 (2011)
15. Baumers, M., Tuck, C., Wildman, R., Ashcroft, I., Rosamond, E., Hague, R.: Transparency Built-in. J. Ind. Ecol. **17**(3), 418–431 (2013)
16. Peng, T.: Analysis of energy utilization in 3D printing processes. Procedia CIRP **40**, 62–67 (2016)

17. Liu, Z.Y., Li, C., Fang, X.Y., Guo, Y.B.: Energy consumption in additive manufacturing of metal parts. Procedia Manuf. **26**, 834–845 (2018)
18. Bhinge, R., Park, J., Law, K.H., Dornfeld, D.A., Helu, M., Rachuri, S.: Toward a generalised energy prediction model for machine tools. J. Manuf. Sci. Eng. **139**(4), 1–2 (2017)
19. Zhang, Y., Hong, G.S., Ye, D., Zhu, K., Fuh, J.Y.H.: Extraction and evaluation of melt pool, plume and spatter information for powder-bed fusion AM process monitoring. Mater. Des. **156**, 458–469 (2018)
20. Wu, D., Wei, Y., Terpenny, J.: Predictive modelling of surface roughness in fused deposition modelling using data fusion. Int. J. Prod. Res. **57**(12), 3992–4006 (2019)
21. Hu, F., Liu, Y., Qin, J., Sun, X., Witherell, P.: Feature-level data fusion for energy consumption analytics in additive manufacturing. In: 2020 IEEE 16th International Conference on Automation Science and Engineering (CASE), pp. 612–617 (2020)
22. Qiu, J., Wu, Q., Ding, G., Xu, Y., Feng, S.: A survey of machine learning for big data processing. EURASIP J. Adv. Signal Process. **1**, 2016 (2016)
23. Jiang, H., Li, J., Yi, S., Wang, X., Hu, X.: A new hybrid method based on partitioning-based DBSCAN and ant clustering. Expert Syst. Appl. **38**(8), 9373–9381 (2011)
24. Fong, S., Rehman, S.U., Aziz, K., Sarasvady, S.: DBSCAN: Past, Present and Future, pp. 232–238 (2014)
25. Birant, D., Kut, A.: ST-DBSCAN: an algorithm for clustering spatial-temporal data. Data Knowl. Eng. **60**(1), 208–221 (2007)
26. Sander, J., Ester, M., Kriegel, H.P., Xu, X.: Density-based clustering in spatial databases: the algorithm GDBSCAN and its applications. Data Min. Knowl. Discov. **2**(2), 169–194 (1998)
27. Tran, T.N., Drab, K., Daszykowski, M.: Revised DBSCAN algorithm to cluster data with dense adjacent clusters. Chemom. Intell. Lab. Syst. **120**, 92–96 (2013)
28. Chen, T., Guestrin, C.: XGBoost: a scalable tree boosting system. In: Proceedings of 22nd ACM SIGKDD International Conference on Knowledge Discovery Data Mining, pp. 785–794 (2016)
29. Qin, J., Liu, Y., Grosvenor, R.: Data analytics for energy consumption of digital manufacturing systems using internet of things method. In: 2017 13th IEEE Conference on Automation Science and Engineering (CASE), Xi'an, pp. 482–487 (2017)
30. Han, J., Pei, J., Kamber, M.: Data Mining: Concepts and Techniques. Elsevier, Amsterdam (2011)
31. Taylor, J.: Introduction to error analysis, the study of uncertainties in physical measurements (1997)

Bias from the Wild Industry 4.0: Are We Really Classifying the Quality or Shotgun Series?

Riccardo Rosati[1(✉)], Luca Romeo[1], Gianalberto Cecchini[2], Flavio Tonetto[2], Luca Perugini[3], Luca Ruggeri[3], Paolo Viti[3], and Emanuele Frontoni[1]

[1] Department of Information Engineering,
Marche Polytechnic University, Via Brecce Bianche 12, 60131 Ancona, Italy
`r.rosati@pm.univpm.it`
[2] Sinergia Srl, Viale Goffredo Mameli 44, 61121 Pesaro (PU), Italy
[3] Benelli Armi Spa, Via della Stazione 50, 61029 Urbino (PU), Italy

Abstract. The traditional data quality control (QC) process was usually limited by the high time consuming and high resources demand, in addition to a limit in performance mainly due to the high intrinsic variability across different annotators. The application of Deep Learning (DL) strategies for solving the QC task open the realm of possibilities in order to overcome these challenges. However, not everything would be a bed of roses: the inability to detect bias from the collected data and the risk to reproduce bias in the outcome of DL model pose a remarkable and unresolved point in the Industrial 4.0 scenario. In this work, we propose a Deep Learning approach, specifically tailored for providing the aesthetic quality classification of shotguns based on the analysis of wood grains without running into an unwanted bias. The task as well as the collected dataset are the result of a collaboration with an industrial company. Although the proposed DL model based on VGG-16 and ordinal categorical cross-entropy loss has been proven to be reliable in solving the QC task, it is not immune to those who may be unwanted bias such as the typical characteristics of each shotgun series. This may lead to an overestimation of the DL performance, thus reflecting a more focus on the geometry than an evaluation of the wood grain. The proposed two-stage solution named Hierarchical Unbiased VGG-16 (HUVGG-16) is able to separate the shotgun series prediction (shotgun series task) from the quality class prediction (quality task). The higher performance (up to 0.95 of F1 score) by the proposed HUVGG-16 suggests how the proposed approach represents a solution for automatizing the overall QC procedure in a challenging industrial case scenario. Moreover, the saliency map results confirm how the proposed solution represents a proof of concept for detecting and mitigated unwanted bias by constraining the network to learn the characteristics that properly describe the quality of shotgun, rather than other confound characteristics (e.g. geometry).

Keywords: Quality control · Bias · Deep learning

A. Del Bimbo et al. (Eds.): ICPR 2020 Workshops, LNCS 12664, pp. 637–649, 2021.
https://doi.org/10.1007/978-3-030-68799-1_46

1   Introduction

Nowadays, most of the vision instruments for the automation of quality control procedure available on the market focus on quantitative and deterministic analyses (e.g. dimensional control, an inspection of the roughness of materials, etc.), but there is no software instrument that allows the modeling and generalization of all those qualitative analyses that are still executed by highly specialized technicians. Thus, the traditional data quality control process was usually limited by the high time consuming and high resources demand, in addition to a limit in performance mainly due to the high intrinsic variability across different annotators [6]. It is therefore not surprising that the quality control task has quickly established itself as an industrial 4.0 use case. Implementing industrial IoT and vision systems to monitor the health and quality of the instrumentation/products/materials enables manufacturers to support the technicians during the process while reducing resource costs, intrinsic variability and improving productivity.

The application of Deep Learning (DL) strategies for solving the quality control task open the realm of possibilities in order to overcome these challenges. However, not everything would be a bed of roses. Although the potential of DL to learn the discriminatory patterns is very relevant and the range of applications is almost limitless, the inability to detect bias from the collected data and the risk to reproduce bias in the outcome of the DL model pose a remarkable and unresolved point in the Industrial 4.0 scenario. Accordingly, the holy grail of DL is to learn from examples and generalize to situations never seen before. These examples are usually provided by humans and humans are biased by our nature. Examples of bias situations in the application of predictive models could range from sentiment analysis, image classification, job recruiting, adaptive chatbot, recidivism assessment, gender classification and language translation [1]. Research direction aims to detect, rate and mitigate bias of the learning process. However, in the DL scenario, the inability to detect bias situations is mainly due to the lack of understanding of the reason for the algorithm outcome. In fact, usually, DL algorithms are conceived as black-box models, which not always lead to interpret and understand how the algorithm provides the decision rule.

Starting from the lesson we have learned by designing and developing a DL methodology, specifically tailored for providing the aesthetic quality classification of shotguns based on the analysis of wood grains, the aim of the paper is to propose a best practice and a proof of concept for detecting and mitigating unwanted bias in the collected data and in the outcome of DL model, in the Industry 4.0 scenario.

Thus the main contributions of the work in the computer vision and pattern recognition field are summarised below:

- the collection of annotated real dataset composed of 1902 images specifically tailored to solve the aesthetic quality classification of shotguns based on the analysis of wood grains. Each image, displaying a different view of the shotgun is properly annotated by a high specialized technician;

- the proposing of DL approach based on VGG-16 and ordinal categorical cross-entropy loss in a novel and challenging Industry 4.0 application, i.e. the quality control classification task;
- an in-depth analysis of the DL model to show the risk to reproduce the presence of possible bias in the collected data;
- the proposing of two-stage solution named Hierarchical Unbiased VGG-16 (HUVGG-16) for mitigating the detected bias.

2   Related Work

The Industry 4.0 revolution was triggered by the increasing availability of data, high-computing power and large storage capacity. These conditions have led the ML and DL appealing solutions in different industrial areas such as predictive maintenance [5,14], decision support system [13] and quality control [4,7,19]. Quality-control is a growing area in Industry 4.0 and an important step in every production system. Recently, the increasing amount of data in this scenario lied the foundation for combining DL and machine vision for performing texture classification thus helping to detect production issues and classify the quality of the product [3,16].

Standard DL approaches were employed in [11] to replace costly quality control procedures based on visual inspection during the welds mass production scenario with the aim to improve the defect detection accuracy. They mainly focused on the collection of balanced database and image pre-processing. Deep Neural Network (DNN), Deep Belief Network (DBN) and restricted Boltzmann machine are standard DL architectures that were applied in [18] to perform a visual inspection process in the printing industry 4.0 by using as input a high-resolution optical quality control camera. Similarly, in [12] a standard deep learning strategy fed by images acquired by a camera placed over the assembly line was implemented to predict the quality-control in a smart factor prototype. Differently, the authors in [2] employed as predictors one key-quality index and different process parameters monitored by the control instruments. They applied a DNN consisting of a DBN in the bottom and a regression layer on the top to predict the quality prediction of a complex manufacturing process.

Recently, a DL strategy was adopted for detecting geometric inaccuracy of the laser-based additive manufacturing process [8]. They combined the output of a Convolutional Neural Network (CNN) and the output of an Artificial Neural Network for analyzing the thermal images and include relevant process/design parameters respectively. The overall network was trained to predict the final pointwise distortion prediction. Also in [19] the authors proposed a CNN solution to automatically extract discriminative features of the images for defect detection and at the same time by ensuring a high processing speed which guarantees real-time detection.

The main differences with our work lie in the (i) different application of DL methodology we propose in unexplored and challenging quality-control application (i.e. we are interested to classify the aesthetic quality classification of shotguns based on the analysis of wood grains) and the (ii) different goal we aim to solve (i.e. the detection and mitigation of any unwanted bias in this scenario). The proposed solution to this task and challenges provides the main contribution of the presented paper.

3   Dataset

The commercial classification of wooden stocks is defined in categories ranging from grade 1, which indicates almost veinless wood, up to grade 5 with a very twisted and variegated grain pattern. Each different type of shotgun series manufactured by the company is equipped with a stock belonging to a specific grade class. Today the QC is mainly based on the evaluation of the human eye. The dataset refers to a real industrial case study of Benelli Armi Spa. The detention and conservation are regulated by an agreement between Benelli Armi Spa and Università Politecnica delle Marche.

Table 1. The aesthetic quality level of the collected dataset. All the grades are reported in ascending order together with the relative number of stocks.

Label	# Stocks
1	165
2^-	148
2	212
2^+	177
3^-	168
3	250
3^+	198
4^-	208
4	270
4^+	106

The collected dataset is composed of both left and right side images belonging to 951 different shotguns, for a total of 1902 images with a size of 1000×500 pixels. The stocks have been classified into 4 main grades (1, 2, 3, 4) and their relative minor grades ($2^-, 2^+, 3^-, 3^+, 4^-, 4^+$), resulting into 10 different classes as reported in Table 1. Figure 1 shows an example of stock for each of the 10 classes. The images were acquired with a high-definition RGB camera placed in the top-view configuration. During the annotation procedure, a highly specialized technician accurately inspects the item and assigns the labels of the stock using a custom data annotation platform (see Fig. 2).

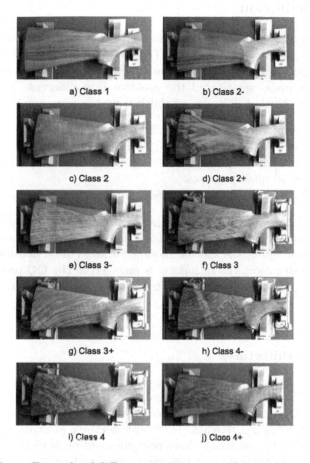

Fig. 1. Example of different stock for each of the 10 classes.

Fig. 2. Custom data annotation platform. The stock is placed in the box where the RGB camera and an industrial lamp are mounted. The annotation software allows the operator to capture the image and to record the grade.

4   Task Definition

The task that we aim to solve is the prediction of the aesthetic quality of shotguns based on the analysis of wood grains. Thus, our inputs are represented by the collected images, while outputs are the 10 different quality classes. Following companies' demands, we divided the problem into different QC tasks sorted according to the level of difficulty established by the company itself:

(a) pediction of middle quality classes: $Y = \{1, 2, 3, 4\}$;
(b) prediction of meta quality classes: $Y = \{1^*, 2^*, 3^*, 4^*\}$,
    where $1^* = \{1\}, 2^* = \{2^-, 2, 2^+\}, 3^* = \{3^-, 3, 3^+\}, 4^* = \{4^-, 4, 4^+\}$;
(c) pediction of all quality classes:
    $Y = \{1, 2^-, 2, 2^+, 3^-, 3, 3^+, 4^-, 4, 4^+\}$;

where X and Y are respectively the input and the output space. We are interested to learn an agnostic model that classifies the quality classes without being given information about the specific shotgun series. This is because, in the real-industrial case situation, the technician is engaged in the QC procedure within the information of the series of the shotgun is not known a priori.

5   Classification Model

5.1   CNN Architectures

The proposed classification tasks are based on the fine-tuning strategy of state-of-the-art CNNs for image classification, i.e. AlexNet [10], VGG16 [17] and ResNet50 [9]. A transfer learning approach was used to fine-tune the networks on ImageNet pre-trained weights. These architectures are chosen for two main reasons: i) they achieved competitive performances on ImageNet challenge, ii) they are relatively simple (i.e. not too deep), allowing to obtain low-level features for fine-tuning. For all the networks, the last fully-connected layer was modified from 1000 to K neurons, where K is the dimension of output space for each task as defined in Sect. 4.

5.2   Loss Functions

We solved each multi-class classification task a), b) and c) independently by considering respectively i) a standard Categorical Cross-Entropy (CCE) ii) a standard Categorical Cross-Entropy with a target vector which encourages an Ordinal Structure (CCE-OS) and iii) an ordinal categorical cross-entropy (OCCE).

6   Experimental Procedure

6.1   Training Settings

All the networks considered were fed with stock images resized to 224×224 pixels in order to match the ImageNet input dimension. The mean value was removed from each image. We adopted a mini-batch stochastic gradient descend (SGD) optimizer. We have explored the best batch size, the initial learning rate and the momentum in the range $\{32, 64, 128\}$, $\{1 \cdot 10^{-4}, 1 \cdot 10^{-3}, 1 \cdot 10^{-2}\}$, $\{0.8, 0.9\}$ respectively. For each task, we validated these hyperparameters in a separate validation set using a grid-search approach. The number of epochs was set to 30. The best-selected hyperparameters for the quality task are respectively 64, $1 \cdot 10^{-3}$ and 0.8 for batch size, initial learning rate and momentum. The dataset was split by a startified hold-out procedure, i.e. using 60% of images as training, 20% as validation and 20% as test. Images belonging to the same shotgun ID were maintained in the same set. This checking was performed to ensure that the algorithm may be able to generalize across different unseen shotgun stocks. Due to the small dimension of the dataset, data augmentation was performed on-the-fly on the training set, applying horizontal flip, rotation and zoom to original images. To cope with the slight unbalance of the dataset, class weights were computed for weighting the loss function. All the experiments were performed using TensorFlow 2.0 and Keras 2.3.1 frameworks on Intel Core i7-4790 CPU 3.60 GHz with 16 GB of RAM and NVIDIA GeForce GTX 970.

7   Results

In Sect. 7.1, we reported the results for solving the task a) b) and c). Taking into account that the experimental results demonstrated the higher effectiveness of OCCE with respect to CCE-OS, we decided to report the classification performance related to standard CCE and OCCE. Afterward, we described how we have detected unwanted bias for solving this task in Sect. 7.2. Finally, we propose our strategy and related results to mitigate the detected bias in Sect. 7.3.

7.1   Classification Performance

Table 2 shows the classification performance of VGG-16, ResNet50 and AlexNet for solving task a), b) and c) using the standard CCE loss on the test set.

For each task, the VGG-16 overcomes the ResNet50 and AlexNet in terms of Accuracy, Recall, Precision and F1. Hence, we show in Table 3 the performance of VGG-16 using the CCE and O-CCE as loss functions for solving tasks a), b) and c).

The O-CCE loss is more reliable for solving all tasks. The performance of the model decreases according to an increase in the difficulty of the task. Figure 3 shows the confusion matrices of the best performing model (VGG-16 Ω-CCE) for solving task a), task b) and c).

7.2   Bias Detection

Starting from the classification performance extracted in Sect. 7.1, we go further by analyzing any possible unwanted bias that may influence the classification performance. In particular, the bias may be unknown and embedded in

Table 2. Classification performance of VGG-16, ResNet50 and AlexNet for solving task a), b) and c) using the standard CCE loss on the test set. The best performing model in terms of F1 is reported in bold for each task.

Model	Accuracy	Precision	Recall	F1
Task a				
VGG16	**0.96**	**0.95**	**0.96**	**0.95**
ResNet50	0.92	0.92	0.93	0.92
AlexNet	0.95	0.94	0.95	0.94
Task b				
VGG16	**0.91**	**0.87**	**0.90**	**0.88**
ResNet50	0.88	0.82	0.87	0.84
AlexNet	0.89	0.84	0.87	0.85
Task c				
VGG16	**0.63**	**0.61**	**0.62**	**0.60**
ResNet50	0.58	0.56	0.58	0.55
AlexNet	0.60	0.58	0.58	0.56

Table 3. Classification performance of VGG-16 for solving task a), b) and c) using the CCE and O-CCE loss on the test set. The best performing model in terms of F1 is reported in bold for each task.

Loss	Accuracy	Precision	Recall	F1
Task a				
CCE	0.96	0.95	0.96	0.95
O-CCE	**0.96**	**0.96**	**0.96**	**0.96**
Task b				
CCE	0.91	0.87	0.90	0.88
O-CCE	**0.92**	**0.88**	**0.92**	**0.90**
Task c				
CCE	0.63	0.61	0.62	0.60
O-CCE	**0.65**	**0.64**	**0.65**	**0.63**

the dataset/images. In this scenario, following the company's suggestion, we have pointed out different possible bias factors: shotgun series, the instant of time (hour of the day) where the QC is carried out, stock sale id, production time (minutes). Considering the Cramer's correlation we have analyzed how these possible bias factors are correlated with respect to the quality classes $Y = \{1, 2^-, 2, 2^+, 3^-, 3, 3^+, 4^-, 4, 4^+\}$. Table 4 shows that the specific shotgun series is bounded to a specific quality class. Different shotgun series have different exclusive characteristics, i.e. size, shape, color, polishing, plastic insert and other specific treatments. The Cramer's analysis found that the shotgun series is

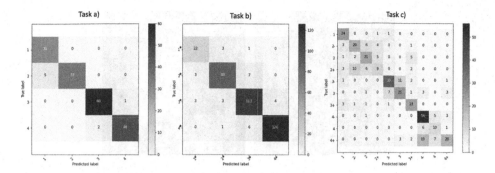

Fig. 3. Confusion matrices of the best performing VGG-16 models with O-CCE loss for solving task a), b), c) respectively.

significantly correlated $(0.67, p < .05)$ with respect to the ground-truth quality classes Y. As a consequence, the shotgun series represents a bias in the VGG-16 model.

Table 4. QC classes for each shotgun series

Shotgun series	Code	1	2-	2	2+	3-	3	3+	4-	4	4+
ACCADEMIA GR3+	2	0	0	0	0	0	24	85	9	2	0
RAFFAELLO 2013 GR3	3	0	0	0	11	130	9	0	0	0	0
RAFFAELLO 2013 GR2	4	0	32	149	26	14	3	0	0	0	0
ANNIVERSARY 50° GR4	6	0	0	0	0	0	1	19	101	252	61
828 NIKEL GR3+	8	1	0	0	0	11	176	14	0	0	0
828 CAL20 GR3+	9	0	0	1	0	2	35	70	12	2	0
MONTFELTRO EUROPE GR1	10	151	19	18	4	7	1	0	0	0	0
MONTEFELTRO CLARO GR2	11	3	79	27	105	4	0	0	0	0	0
ARGO E GR4	12	0	0	0	0	0	0	0	8	2	18
FRANCHI EUROPE GR2	13	10	18	17	31	0	0	0	0	0	0
ANNIVERSARY 50° CAL.20 GR4	14	0	0	0	0	0	1	3	35	1	14
RAFFAELLO LIMITED EDITION CAL.12 GR4	15	0	0	0	0	0	0	7	43	11	13

The detected bias is also demonstrated by the high significant Cramer's correlation $(0.70, p < .05)$ found between the VGG-16 prediction of task c) and the shotgun series. This fact is also confirmed by exploring the saliency map of the VGG-16 (see Fig. 4, left side) according to the approach proposed by [15]. The most discriminative pattern of the network is placed on the shotgun edge, thus reflecting a more focus on the geometry than an evaluation of the wood grain.

7.3   Bias Mitigation

Our objective is to classify the quality classes without being given information about the specific shotgun series. Thus, the model should be able to classify the quality levels independently from the geometry and the shotgun series.

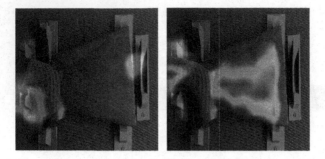

Fig. 4. Saliency maps of a grade 2 stocks belonging to Raffaello series. Left: VGG-16 model trained for task c); right: VGG-16 sub-network trained for solving the quality task on "Raffaello" series.

To achieve this objective we propose a hierarchical network approach, named HUVGG-16 in order to mitigate the bias due to shotgun series. It is designed to learn separately two-stage hierarchical networks that are able to predict respectively the shotgun series (shotgun series task) and the quality classes (quality task). The single network of the first stage predicts all the shotgun series. Based on the shotgun series predicted, we assign a specific second stage sub-network for classifying the quality classes. Each sub-network is conceived to predict only the quality classes associated with respect to the shotgun macro-series. Each macro-series is defined according to the company's knowledge by aggregating shotgun series which have the same geometrical characteristic. Figure 5 shows in detail the workflow of the proposed approach.

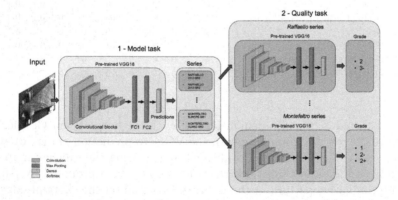

Fig. 5. Workflow of the proposed bias mitigation approach HUVGG-16. The first VGG-16 is conceived to learn the shotgun series (model task) while each sub-networks is specialized to classify the quality classes (quality task) for each shotgun macro-series. Each macro-series is defined according to the company's knowledge by aggregating shotgun series which have the same geometrical characteristic.

The HUVGG-16 approach is shown in Fig. 5. The first VGG-16 is conceived to learn the shotgun series (model task) while each sub-network is specialized to classify the quality classes (quality task) for each shotgun series. For solving the both tasks we employed the O-CCE loss.

The Accuracy, Precision, Recall and F1 of HUVGG-16 for solving the model task are respectively: 0.97, 0.98, 0.97, 0.97.

Figure 6 shows the confusion matrix of HUVGG-16 for solving the quality task on Raffaello and Montefeltro series. In particular, we focus only on the most numerous quality classes (i.e., 2 and 3- for Raffaello series and 1, 2+ and 2- for Montefeltro series). For Raffaello series the Accuracy, Recall, Precision and F1 are respectively: 0.95, 0.94, 0.96, 0.95; for Montefeltro series the Accuracy, Precision, Recall and F1 are respectively: 0.83, 0.85, 0.84, 0.80. Accordingly, the extracted saliency maps are constrained to focus on wood grains rather than the geometrical edges (see Fig. 4 right side). Thus, this strategy allows to alleviate the bias by separating the two task and providing the prediction of quality classes for each shotgun macro-series model.

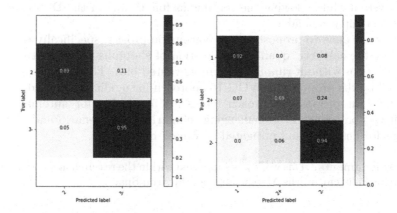

Fig. 6. Confusion matrices of quality classification task of HUVGG-16: Left: "Raffaello" series [codes 3,4]; right: "Montefeltro" series [codes 10,11].

8   Conclusions

Today the QC on the various types of shotgun stocks is mainly based on the evaluation of the human eye, following that the results are affected by a high inter and intra-subject variability. The introduction of DL, based on VGG-16 and ordinal categorical cross-entropy loss, for the automation of aesthetic QC of wooden stock, has been proven to be effective in order to automatize and standardize the overall QC process. The integration of the trained network in the QC platform, allows the reduction of QC timing from minutes to tenths of a second, allowing substantially faster business productivity and eliminating bottlenecks in business processes. The deep understanding of the overall procedure and the strict collaboration with the company lead to detect and avoid an unwanted bias in the QC classification procedure. The analysis confirms how the presence of bias in the dataset and in the learning procedure could be a disruptive finding that may lead to an overestimation of the performance. The proposed HUVGG-16 solution based on hierarchical networks allows to mitigate the detected bias by learning the characteristics that properly describe the quality of shotgun, rather than other confound characteristics. Taking into account that the DL model should focus on the wood part we are currently considering the inclusion of a segmentation step prior the classification of gradining. Additionally, we are currently testing the generalization power of the proposed DL approach in a different company's production chain for supporting the QC process of the technician in a different environment and operating condition. Future work may also be addressed to integrate the proposed approach into a QC serverless platform where the predicted quality class is obtained by ingestion event. The technician may trigger a cloud function that could invoke the DL model to provide the inference. This setting may ensure the high scalability of the system while allowing the continuous fine-tuning of the DL model based on the new input available.

Our work aims to propose a deep learning approach, specifically tailored for providing the aesthetic quality classification of shotguns based on the analysis of wood grains without running into an unwanted bias. The higher performance obtained by the proposed HUVGG-16 approach for quality class prediction suggests how the proposed approach represents a solution for automatizing the overall QC procedure in a challenging industrial case scenario and a proof of concept for detecting and mitigated unwanted bias.

Acknowledgments. This work was supported within the research agreement between Università Politecnica delle Marche and Benelli Armi Spa.

References

1. Internet trolls turn a computer into a nazi. https://www.propublica.org/article/machine-bias-risk-assessments-in-criminal-sentencing. Accessed 02 July 2020
2. Bai, Y., Li, C., Sun, Z., Chen, H.: Deep neural network for manufacturing quality prediction. In: 2017 Prognostics and System Health Management Conference (PHM-Harbin), pp. 1–5 (2017)
3. Basu, S., et al.: Deep neural networks for texture classification–a theoretical analysis. Neural Netw. **97**, 173–182 (2018)
4. Usuga Cadavid, J.P., Lamouri, S., Grabot, B., Pellerin, R., Fortin, A.: Machine learning applied in production planning and control: a state-of-the-art in the era of industry 4.0. J. Intell. Manufact. **31**(6), 1531–1558 (2020). https://doi.org/10.1007/s10845-019-01531-7
5. Calabrese, M., et al.: Sophia: an event-based iot and machine learning architecture for predictive maintenance in industry 4.0. Information **11**(4), 202 (2020)

6. Dai, W., Yoshigoe, K., Parsley, W.: Improving data quality through deep learning and statistical models. In: Latifi, S. (ed.) Information Technology - New Generations. AISC, vol. 558, pp. 515–522. Springer, Cham (2018). https://doi.org/10.1007/978-3-319-54978-1_66

7. Escobar, C.A., Morales-Menendez, R.: Machine learning techniques for quality control in high conformance manufacturing environment. Adv. Mech. Eng. **10**(2), 1687814018755519 (2018). https://doi.org/10.1177/1687814018755519

8. Francis, J., Bian, L.: Deep learning for distortion prediction in laser-basedadditive manufacturing using big data. Manufact. Lett. **20**, 10–14 (2019). https://doi.org/10.1016/j.mfglet.2019.02.001,http://www.sciencedirect.com/science/article/pii/S221384631830172X

9. He, K., Zhang, X., Ren, S., Sun, J.: Deep residual learning for image recognition. In: Proceedings of the IEEE Conference on Computer Vision and Pattern Recognition, pp. 770–778 (2016)

10. Krizhevsky, A., Sutskever, I., Hinton, G.E.: Imagenet classification with deep convolutional neural networks. In: Advances in Neural Information Processing Systems, pp. 1097–1105 (2012)

11. Muniategui, A., de la Yedra, A.G., del Barrio, J.A., Masenlle, M., Angulo, X., Moreno, R.: Mass production quality control of welds based on image processing and deep learning in safety components industry. In: Cudel, C., Bazeille, S., Verrier, N. (eds.) Fourteenth International Conference on Quality Control by Artificial Vision, vol. 11172, pp. 148–155. International Society for Optics and Photonics, SPIE (2019)

12. Ozdemir, R., Koc, M.: A quality control application on a smart factory prototype using deep learning methods. In: 2019 IEEE 14th International Conference on Computer Sciences and Information Technologies (CSIT) vol. 1, pp. 46–49 (2019)

13. Romeo, L., Loncarski, J., Paolanti, M., Bocchini, G., Mancini, A., Frontoni, E.: Machine learning-based design support system for the prediction of heterogeneous machine parameters in industry 4.0. Expert Syst. Appl. **140**, 112869 (2020). https://doi.org/10.1016/j.eswa.2019.112869, http://www.sciencedirect.com/science/article/pii/S0957417419305792

14. Sakib, N., Wuest, T.: Challenges and opportunities of condition-based predictive maintenance: a review. Procedia CIRP **78**, 267–272 (2018)

15. Selvaraju, R.R., Cogswell, M., Das, A., Vedantam, R., Parikh, D., Batra, D.: Gradcam: visual explanations from deep networks via gradient-based localization. In: Proceedings of the IEEE International Conference on Computer Vision, pp. 618–626 (2017)

16. Simon, P., Uma, V.: Deep learning based feature extraction for texture classification. Procedia Comput. Sci. **171**, 1680–1687 (2020)

17. Simonyan, K., Zisserman, A.: Very deep convolutional networks for large-scale image recognition. arXiv preprint arXiv:1409.1556 (2014)

18. Villalba-Diez, J., Schmidt, D., Gevers, R., Ordieres-Meré, J., Buchwitz, M., Wellbrock, W.: Deep learning for industrial computer vision quality control in the printing industry 4.0. Sensors **19**(18), 3987 (2019). https://doi.org/10.3390/s19183987

19. Wang, T., Chen, Y., Qiao, M., Snoussi, H.: A fast and robust convolutional neural network-based defect detection model in product quality control. Int. J. Adv. Manufact. Technol. **94**(9–12), 3465–3471 (2018)

Machine Learning for Storage Location Prediction in Industrial High Bay Warehouses

Fabian Berns[1]([✉]), Timo Ramsdorf[2], and Christian Beecks[1]

[1] University of Münster, Münster, Germany
{fabian.berns,christian.beecks}@uni-muenster.de
[2] Technical University of Berlin, Berlin, Germany
timo.ramsdorf@campus.tu-berlin.de

Abstract. Global trade and logistics require efficient management of the scarce resource of storage locations. In order to adequately manage that resource in a high bay warehouse, information regarding the overall logistics processes need to be considered, while still enabling human stakeholders to keep track of the decision process and utilizing their non-digitized, domain-specific, expert knowledge. Although a plethora of machine learning models gained high popularity in many industrial sectors, only those models that provide a transparent perspective on their own inner decision procedures are applicable for a sensitive domain like logistics. In this paper, we propose the application of machine learning for efficient data-driven storage type classification in logistics. In order to reflect this research problem in practice, we used production data from a warehouse at a large Danish retailer. We evaluate and discuss the proposed solution and its different manifestations in the given logistics context.

Keywords: Machine learning · Decision tree · Storage location prediction · ABC classification · Industry 4.0

1 Introduction

Globalization has become one of the major driving factors in global logistics. In order to enable the distribution of goods on a global scale and deliver them to customers all over the world, industrial high bay warehouses have become the major backbone of this development. Thus, an efficient and effective management of storage locations as the key and scarce resource of these warehouses is no trivial question. An ideal solution for storage location prediction in a production-scale warehouse needs to consider all the information available, while still enabling human stakeholders to keep track of the decision process and utilizing their non-digitized, domain-specific, expert knowledge.

In recent years, machine learning methods have been the answer to many analytical challenges, where no exact solution was available and a large amount

T. Ramsdorf—This research has been conducted, while the author was affiliated with University of Münster.

of attributes needed to be considered. While neural networks and deep learning methods gained high popularity in non-sensitive application domains, the intricate task of explaining and interpreting their inner workings and their opaque decision process inhibited its application in more sensitive domains [1,10]. Since we regard warehouse logistics as a sensitive domain due to its key position at the intersection of global trade, only explainable machine learning models are considered for the research problem of *Machine Learning for Storage Location Prediction in Industrial High Bay Warehouses* tackled in this paper.

In this paper, we propose the application of supervised machine learning for efficient data-driven storage type classification in logistics. To this end, we make use of logistic data provided by a Warehouse Management System (WMS) and show how to formalize the problem of storage type classification, i.e. ABC classification [12,24], by means of different domain-specific features. Our performance analysis on real-world logistic data indicates that our approach based on machine learning is able to outperform classical manual ABC decision making. We verified this claim via an in-depth simulation of a large-scale warehouse and millions of warehouse movements.

To summarize, we include the following contributions:

- We conceptualize and define a logistic data model from a mathematical perspective for the exemplary warehouse used for this research.
- We propose a decision-tree-based solution for the problem of ABC classification in high bay warehouses.
- We evaluate and discuss the performance of different manifestations of our solution in the given logistics context.

The remainder of this paper is structured as follows: In Sect. 2, we outline related work. In Sect. 3, we propose our approach for efficient storage location prediction based on machine learning. In Sect. 4, we report and discuss the results of our performance analysis. Section 5 finally concludes our paper with on outlook on promising future research directions.

2   Related Work and Background

In this section, we describe the two main building blocks our topic is concerned with and their respective related work, namely (High Bay) Warehouse Management and Storage Location Forecasting in particular as well as Machine Learning and Data Mining.

Van den Berg [2] defines a warehouse in general as *"a temporary place to store inventory and as a buffer in supply chains"*. As such, a warehouse can be seen as a central, static component of logistics, which facilitates *"the movement of goods from suppliers to customers, meeting demand in a timely and cost-effective manner"* [2,17]. A high bay warehouse in particular is characterized by its storage, retrieval and conveyor systems, as well as the actual high bay storage [21]. The latter is organized in racks separated by aisles, which accommodate transport routes for a Storage Retrieval Machine (SRM) each. Those enable the storage and retrieval of goods organized by Transport Units (TUs) from the racks to conveyor belts and vice versa. In order to provide a physical interface for TU storage and retrieval, these conveyor belts manage the TU transport between SRMs and dedicated picking areas [21].

The concept of high bay warehouses is designed to act as a blueprint for storing very large amounts of goods. Subsequently, dedicated software systems, called WMS, are needed to keep record of the storage capacity and storage location of TUs, as well as to track and control the movement of goods and TUs within a warehouse [12]. In order to account for the latter requirement, a WMS needs to automatically interact with the physical system[1] it is controlling by means of Programmable Logic Controllers (PLCs) [12]. Finally, a WMS controls all the inbound and outbound logistics of a high bay warehouse. While its way of determining storage locations is often described as *"chaotic"* (cf. [12,16]), dedicated systems are frequently employed to optimize transport routes of SRMs and conveyor belts and thus improve on energy and time efficiency. One common method to account for the latter is to sort goods into three classes (A, B, C) indicating how often they are needed and thus their respective TUs are stored and retrieved from the warehouse. Transport routes of frequently requested goods (class A) need to be short, while the overall warehouse efficiency is not seriously harmed by longer routes for rarely requested goods (class C) [11,12,17,24]. Due to the respective class names, this strategy is called *ABC strategy* [24]. Since these classes are still often determined by means of simple empirical and statistical heuristics, the application of machine learning and data mining techniques in this area of intralogistics is still an open area of research (cf. [13]).

In general, machine learning is an interdisciplinary field at the intersection of computer science, mathematics and cognitive science, that aims to design models adapting their inner workings based on prior knowledge and experience, both usually represented by data [7,23]. These adapted models are used to infer information about unknown data for knowledge representation [23] and analytical tasks such as classification [7,14], regression [7,15] and pattern recognition [6]. While some of these methods, such as Decision Trees [10] or Gaussian Process Models [3,4,8], are considered to be explainable by nature, extra effort is needed in order to make the internal processes of other models such as neural networks [15,20] transparent, interpretable and traceable [1,10]. Thus, the latter group of methods proves harder to apply in sensitive domains, where accountability is key [1].

3   Methods and Results

In this section, we propose our approach for efficient data-driven ABC classification based on machine learning. For this purpose, we first introduce the utilized logistic data model, which is based on *movement orders*. Since the retention period is the key indicator for assessing the correctness of a classification in

---

[1] For the sake of simplicity, we intentionally generalized the intricacies of a WMS and left out detailed concepts like the Material Flow System (MFS) [12].

hindsight, we need to further break down movement orders into two categories – *infeeds* and *outfeeds* – and their interrelation – denoted by means of the function *io* (cf. Subsect. 3.1). Since our given logistic dataset is not accompanied with appropriate ABC classes per movement order, Subsect. 3.2 is dedicated to calculate that information in hindsight. Subsection 3.3 outlines both the used feature directly retrieved from the given dataset as well as those constructed based on prior non-digitized knowledge from domain experts. Finally, we present training and testing of the used machine learning model – i.e. decision tree – in Subsect. 3.4.

3.1 Logistic Data Model

The exemplary logistic data model used in this paper encompasses 12 million entries and was recorded over a period of two years between 12/14/2017 and 12/14/2019. Every single entry represents a so-called *movement order*, which registers a command issued by the WMS to the physical actuators of the high bay warehouse. Every command refers to an individual TU and its transfer from one position in the warehouse to another. Formally, each *movement order* $m = (t, id^{\text{TU}}, id^{\text{a}}) \in M$ is a tuple comprising a timestamp t, which indicates the time of movement, an identifier id^{TU}, which denotes the corresponding TU, and an identifier id^{a} of the article that has been moved by the TU.

We define two non-disjoint subsets of all movement orders M dependent on the source and target locations of the movements, respectively. Let $M_i \subset M$ include all *infeeds*, i.e. movements $m_i \in M_i$ whose *target-locations* are in the high bay, and let $M_o \subset M$ include all *outfeeds*, i.e. movements $m_o \in M_o$ whose *source-locations* are in the high bay. According to these definitions both subsets must not be pairwise disjoint, because there are relocations, that are movements with source and target location in the high bay. The splitting of a relocation in outfeed and infeed is possible here, because the storage location search is triggered by relocations and infeeds, i.e. there is no difference regarding this investigation.

The retention period per movement order plays a key role in determining the correct ABC class in hindsight. Therefore, its appropriate, domain-specific representation is key. Since the warehouse of interest is operated only with limited capacity or not at all during holidays and weekends, commonly used UNIX-style timestamps distort the actual situation especially for orders issued right before these non-business days. Thus, we use the index of the list of all movement orders as timestamp t per movement order m.

In order to conveniently link an infeed of a TU with the succeeding outfeed, we define the function *io*. Each of these infeed-outfeed pair represents one storing process of a TU, thus having the same id^{TU}. Since a TU can be stored multiple times, it is important to pair them with respect to time. Only the earliest outfeed after an infeed builds a pair with it. Formally the pairs are given by the following function *io*, that returns the corresponding outfeed $m_o = (t_o, id_o^{\text{TU}}, id_o^{\text{a}})$ for a given infeed $m_i = (t_i, id_i^{\text{TU}}, id_i^{\text{a}})$:

$$io(m_i) = \underset{m_o \in M_o | t_o > t_i \wedge id_i^{TU} = id_o^{TU}}{\arg \min} (t_o) \qquad (1)$$

As can be seen in the definition above, the function io thus returns the earliest outfeed m_o for a given infeed m_i from the same TU.

3.2 Retrieving ABC Classes for Training Data

In this subsection, we outline how to calculate the ABC classes per infeed m_i in hindsight based on retention time $r(m_i)$ (cf. Eq. 2). In order to do so, we define marginal retention times, which separate infeeds belonging to two adjacent classes. The given warehouse comes with predefined capacities for the three classes. We sort all infeeds by storage time in ascending order and determine marginal retention times so that the resulting classification matches warehouse capacity per class.

$$r(m_i) = t_o - t_i(t_o, id_o^{TU}, id_o^a) = io(m_i) \qquad (2)$$

In particular, we assign class A to unprocessed infeeds having the smallest storage time, until the sum of mapped storage times matches the proportion of the class A locations of the high bay capacity. Then class B is filled similarly and all other infeeds (with the longest storage times) are mapped to class C. Thus, TUs stored for a shorter period of time are assigned to a higher class and vice versa. The following table shows the high bay capacities of the different classes in the example warehouse in comparison to the proportion of infeeds that are labeled with this class.

Table 1. High bay capacities

ABC-Class	A	B	C
High bay capacity	33,8%	21,0%	44,2%
Labeled infeeds	91,7%	4,4%	3,9%

Table 1 illustrates those predefined warehouse capacities and the number of infeeds per class resulting from our retrospective class assignment. Since A-classified items are usually stored for a much shorter period of time, these dominate the dataset with a proportion of over 90%. Thus, the resulting full dataset (movement order + assigned classes) is biased towards class A.

3.3 Extending Feature Set

Prior to the establishment of machine-learning-based ABC classification in the given exemplary warehouse, ABC classification was based on non-digitized knowledge and experience from domain experts supported by simple statistical performance indicators, such as the marginal retention times mentioned above. In order to benefit from their expertise, we enhance the existing set of *basic* features gathered (i.e. movement orders and further meta data of TUs) by means of *extended* features, reflecting particular domain knowledge. The following basic features have proven to be effective: Weight, Volume, Calenderweek, Day in month, Load Unit Type (e.g. euro pallet) and the Source Location Group. The three extended features, that have shown to be effective, are explained in the remainder of this subsection.

The first extended feature, that is able to improve the classification accuracy, is called *number of outfeeds* n_O. For a given infeed m_i, the number of outfeeds $n_O(m_i)$ indicates how many outfeeds of the respective same article id id^a happened over the last 30 d. Equation 3 defines the feature *number of outfeeds* given an infeed m_i its timestamp t_i and article id id_i^a as well as timestamp t_o and article id id_o^a of outfeeds $m_o \in M_o$.

$$n_O(m_i) = \sum_{m_o \in M_o} \mathbb{1}_{t_i - 30\,d < t_o < t_i \wedge id_o^a = id_i^a} \tag{3}$$

The second feature is based on an article's stock currently available in the warehouse. If there are many TUs of the same article in stock, the logistic system has to decide which specific stock item will be reserved for a new outfeed request. Commonly, TUs with an older so-called *FIFO-time* t_{FIFO} will be preferred, i.e. a newly stored TU will probably stay longer in the warehouse, if the current stock of its article is older. In general, articles, which are already represented by a high number of TUs in the warehouse, are less likely to be picked on short notice than scarce articles.

$$t_{FIFO}(m_i) = min(\{t \mid (t, id^{TU}, id^a) \in M_i, t < t_i, id^a = id_i^a\}) \tag{4}$$

The third extended feature that we examine in the scope of this paper is named *classification variance per infeed* $var_c(m_i)$ respectively per its article id id_i^a. This feature reflects, whether the infeeds of the same article are commonly assigned to the same class or whether class assignment varies a lot. Thus, we utilize the common statistical variance measure and apply it to the class assignment for all the infeeds representing the same article. In order to numerically represent the assigned ABC class $a \in \{A, B, C\}$ of an arbitrary infeed $m \in M_i$ we utilize function $c(m), c : \{A, B, C\} \rightarrow \{1, 2, 3\}$.

$$var_c(m_i) = var(\{c(m) \mid (t, id^{TU}, id^a) \in M_i, t < t_i, id^a = id_i^a\}) \tag{5}$$

Although the aforementioned extended features are manually designed, they epitomize the required domain knowledge for this particular warehouse. In what follows, we will present the implementation of our machine learning pipeline before we detail and discuss the impact of our proposed solution.

3.4   Implementation / Decision Tree

In this final part of the results section, we present the used machine learning method as well as its training and testing processes. Given the particular warehouse scenario of this paper, explainability of the model predictions and interpretability of the decision process is a major aspect. Therefore, we choose to use decision trees [10] in this particular scenario, since they deliver explainable model structures, whose inner workings are transparent to data scientists as well as to non-experts in this field [18,19].

In order to foster broad application and reproduction of the results given in this paper, we utilize a standard data mining tool to implement, train and test the chosen machine learning model. One of the prominent tools in that area is Konstanz Information Miner (KNIME) [5,9,22]. KNIME facilitates to build a strong structured workflow of a data mining pipeline. The latter encompasses data preprocessing, filtering, evaluation, and the actual application of a data mining or a machine learning method. We designed such a workflow with a common structure, which is illustrated in Fig. 1.

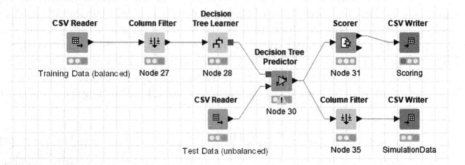

Fig. 1. KNIME-Workflow representing Training and Testing Pipeline for the used Decision Tree Architecture

The purpose of this workflow is to evaluate the performance of the different features mentioned above. First a balanced version (for training) and an unbalanced version (for testing) of the previously created data is imported via CSV Reader nodes. The evaluation is implemented with a so-called scorer node, which compares the predicted classes with the previously calculated correct classes. The other branch makes an export with predicted class labels for a further simulation.

The defined features are tested with the KNIME-Workflow above. The baseline feature set encompasses *Number of outfeeds* n_O and *Source Location Group* only. To measure the improvement of the new features, we determined the accuracy of the classification in the KNIME workflow with a feature set including the two old features and the tested feature and compared it to the accuracy of a classification with the two old features only. Table 2 shows the absolute accuracy and the improvement for each of the mentioned features.

Table 2. Accuracy of Decision Tree given different Feature Selections

Feature	Accuracy	Improvement
Baseline	72.7%	0.00%
Load Unit Type	73.0%	0.41%
Day in Month	73.3%	0.83%
Classification Variance	73.5%	1.10%
Weight	74.1%	1.93%
Calenderweek	74.2%	1.50%
Volume	74.3%	2.06%
FIFO-time	74.5%	2.48%
All Features Combined	74.6%	5.23%

To sum up, we outlined the initial situation and data model of the warehouse of interest for this paper. The assignment of appropriate class labels per data record was discussed and extended features based on existing, non-digitized knowledge have been proposed. Finally, we described the used machine learning method – i.e. decision tree – and justified the corresponding feature selection.

4 Evaluation and Discussion

In this section we will evaluate and discuss the theoretically proposed solution on two different abstraction levels. On the first level, we determine the accuracy of the classification, i.e. the difference between predicted storage class and correct storage class in hindsight. The second abstraction level focuses on the benefits of the proposed solution in the logistic context. For that purpose we will create a simulation for the logistic processes and measure its performance with the original movement orders of the example warehouse.

The classification accuracy of the decision tree is a directly calculated output of the KNIME workflow introduced above. The results depend on many factors, especially on the selected set of features. To compare the following values it is important to know the accuracy of the currently implemented solution. Using the legacy, manual ABC classification previously used for our exemplary warehouse only 67.2% of the infeeds are correctly classified. By applying the new forecasting proposal the accuracy increased to 74.6%.

In reality, the ABC classification is just one criterion to determine storage location search along many others. There are obvious physical restrictions, e.g. that only empty locations can be selected and that height of each TU ought to match the height of the respective storage location. Moreover, there are further restrictions for logistic strategies, e.g. pair building of TUs with the same article. Given these further restriction a thorough simulation is needed to evaluate, the influence of newly predicted ABC values on storage location search.

We implemented a simulation, that manages the occupancy of all high bay locations. For each infeed in the archived dataset it runs a storage location search, books the pallet to the selected high bay location and executes the corresponding outfeed later. It is developed in a flexible and modular manner, so that different ABC classifier can be used: either the retrospectively correct ABC values, the legacy classifier as well as our decision-tree-based classifier. During all of these processes of storage, retrieval and relocations the simulation measures the distance, that the SRMs have to move for each transportation in the high bay aisles (including the empty moves between transportation). In doing so, the distance traveled by SRMs is the defined quality measure. Since improving efficiency and efficacy of the high bay warehouse and therefore its main moving components, i.e. SRMs is the main objective of our research in this paper, we aim to minimize that distance.

Given this quality measure, we are now able to run an evaluation of our newly proposed solution for ABC classification considering all real-world constraint. Thus, we can also verify the earlier assumptions underlying our machine learning model design (cf. Section 3). In Subsect. 3.1 we assumed, that a UNIX-style timestamp is outperformed a "timestamp" reflecting the index of each movement order, since the latter time keeping is not distorted by warehouse operation during holidays and weekends. In order to verify this, we ran two simulations with the correctly calculated ABC-values, i.e. there is no classifying forecasting in both runs. In the first run the ABC-values are calculated with UNIX-style timestamps and the values for the second run are calculated with index per movement order. These two simulations resulted in 939 million movement units for a UNIX-style timestamps and 934 million units for the index-based solution. Thus, the index-based approach produces a small improvement.

In order to provide a baseline for further evaluation of our machine-learned ABC classifier, we first evaluated how the legacy ABC classifier and the retrospectively correct ABC classification affect the quality measure of distance by movement units. Table 3 illustrates these results and shows, how the legacy system poses a clear advantage over having no ABC classification whatsoever. Still, there is still potential for improvement, as the retrospectively correct ABC classification shows indicates. Furthermore, the simulation showed, that the definition of correct ABC-classes for training data indeed leads to shorter path lengths in the warehouse.

Given the baseline performance defined by Table 3 and the chosen set of features defined by Table 2, we evaluate our decision-tree-based approach for different preprocessing strategies: unbalanced dataset and two balancing approaches. Balancing approaches are needed, since class A infeeds are highly overrepresented in the dataset (cf. Subsect. 3.2) and therefore a classifier can achieve high accuracy by just labeling any input class A. In this case, there will be a bad logistic performance, because there will be items with a long-term storage, that occupy high bay locations in the front of the aisle. Table 4 illustrates that an unbalanced approach is not favorable in practice despite its high accuracy. Though balancing approaches achieve lower accuracy, they result in shorter distances traveled and are thus more efficient.

Table 3. Baseline Simulation results Different ABC Classifiers

Simulation	Movement Units	Improvement
No ABC-strategy	ca. 1085 Mio	−6,0%
Common ABC-strategy	ca. 1024 Mio	0%
Definition of correct ABC-classes	ca. 934 Mio	8,8%

Table 4. Results for Balancing Methods

Method	Accuracy	Movement Units
Unbalanced	82,8%	ca. 1024 Mio
Balanced by deleting	74,6%	ca. 987 Mio
Balanced by multiplying	75,9%	ca. 989 Mio

Even if accuracy decreased, the balancing improved the logisitic performance. According to the illustrated performance in terms of distance, i.e. movement units, it does not make much of a difference, whether the balancing is implemented with deleting or amplifying samples. Still, *balancing by deleting* leads to fewer training samples and shorter runtime to build the decision tree.

5   Conclusion

In this paper, we introduced the problem of ABC classification in industrial high bay warehouses and proposed an explainable machine learning solution based on decision trees. Therefore, we described the exemplary warehouse used for this research and the respective dataset retrieved on-site this production facility. Furthermore, the given data was described from a mathematical perspective in order to provide a clear foundation for applying the given analytical method. Finally, we evaluated and discussed the latter's performance by its classification accuracy and by means of an in-depth simulation of millions of warehouse movements.

As future work, we plan to drive organizational change that prepares stakeholders to accept machine learning models of higher accuracy such as neural networks in the process of ABC classification in a production-scale warehouse. Given this change, we plan to investigate a larger set of machine learning models, replace the given discrete classification problem with a more flexible continuous one. Finally, we aim to transfer the knowledge gained over the course of this project to other warehouse facilities in order to further foster the advantages of machine learning methods in logistic systems.

References

1. Adadi, A., Berrada, M.: Peeking inside the black-box: a survey on explainable artificial intelligence (xai). IEEE Access **6**, 52138–52160 (2018)
2. van den Berg, J.P.: Highly Competitive Warehouse Management: An Action Plan for Best-in-Class Performance. Management Outlook Publications, Buren (2012)
3. Berns, F., Beecks, C.: Automatic gaussian process model retrieval for big data. In: CIKM. ACM (2020)
4. Berns, F., Schmidt, K., Bracht, I., Beecks, C.: 3CS algorithm for efficient gaussian process model retrieval. In: 25th International Conference on Pattern Recognition (ICPR) (2020)
5. Berthold, M.R., et al.: KNIME - the konstanz information miner: version 2.0 and beyond. SIGKDD Explor. **11**(1), 26–31 (2009)
6. Bishop, C.: Pattern Recognition and Machine Learning. Springer, New York (2006)
7. Clarke, B.: Principles and Theory for Data Mining and Machine Learning. Springer, Dordrecht New York (2009)
8. Duvenaud, D., Lloyd, J.R., Grosse, R., Tenenbaum, J.B., Ghahramani, Z.: Structure discovery in nonparametric regression through compositional kernel search. In: Proceedings of the 30th International Conference on International Conference on Machine Learning, pp. III-1166–III-1174. ICML'13 (2013)
9. Dwivedi, S., Kasliwal, P., Soni, S.: Comprehensive study of data analytics tools (Rapidminer, Weka, R tool, Knime). In: 2016 Symposium on Colossal Data Analysis and Networking (CDAN), pp. 1–8 (2016)
10. Gilpin, L.H., Bau, D., Yuan, B.Z., Bajwa, A., Specter, M., Kagal, L.: Explaining explanations: an overview of interpretability of machine learning. In: 2018 IEEE 5th International Conference on Data Science and Advanced Analytics (DSAA), pp. 80–89. IEEE (2018)
11. Grupp, B.: Aufbau einer integrierten Materialwirtschaft, chap. 3.10.3 ABC-Analyse der Lagerartikel. Forkel, Wiesbaden (2003)
12. Hompel, M.: Warehouse Management : Automation and Organisation of Warehouse and Order Picking Systems. Springer, Berlin New York (2007)
13. Kartnig, G., Grösel, B., Zrnić, N.: Past, state-of-the-art and future of intralogistics in relation to megatrends. FME Trans. **40**(4), 193–200 (2012)
14. Kotsiantis, S.B., Zaharakis, I.D., Pintelas, P.E.: Machine learning: a review of classification and combining techniques. Artif. Intell. Rev. **26**(3), 159–190 (2006)
15. Prieto, A., et al.: Neural networks: an overview of early research, current frameworks and new challenges. Neurocomputing **214**, 242–268 (2016)
16. Quintanilla, S., Pérez, A., Ballestín, F., Lino, P.: Heuristic algorithms for a storage location assignment problem in a chaotic warehouse. Eng. Optim. **47**(10), 1405–1422 (2015)
17. Richards, G.: Warehouse Management: A Complete Guide to Improving Efficiency and Minimizing Costs in the Modern Warehouse. Kogan Page Publishers, London (2017)
18. Shalaeva, V., Alkhoury, S., Marinescu, J., Amblard, C., Bisson, G.: Multi-operator decision trees for explainable time-series classification. In: Medina, J., Ojeda-Aciego, M., Verdegay, J.L., Pelta, D.A., Cabrera, I.P., Bouchon-Meunier, B., Yager, R.R. (eds.) IPMU 2018. CCIS, vol. 853, pp. 86–99. Springer, Cham (2018). https://doi.org/10.1007/978-3-319-91473-2_8
19. Shulman, E., Wolf, L.: Meta decision trees for explainable recommendation systems. In: AIES, pp. 365–371. ACM (2020)

20. Szegedy, C., Toshev, A., Erhan, D.: Deep neural networks for object detection. In: Advances in Neural Information Processing Systems, pp. 2553–2561 (2013)
21. Triebig, C., Credner, T., Fischer, P., Leskien, T., Deppisch, A., Landvogt, S.: Agent-based simulation for testing control software of high bay warehouses. In: Eymann, T., Klügl, F., Lamersdorf, W., Klusch, M., Huhns, M.N. (eds.) Multiagent System Technologies, pp. 229–234. Springer, Berlin Heidelberg (2005)
22. Warr, W.A.: Scientific workflow systems: pipeline pilot and KNIME. J. Comput. Aided Mol. Des. **26**(7), 801–804 (2012)
23. Wojtusiak, J.: Machine learning. In: Seel, N.M. (ed.) Encyclopedia of the Sciences of Learning, pp. 2082–2083. Springer, US, Boston, MA (2012)
24. Zajac, P.: The construction and operation of modern warehouses. The Energy Consumption in Refrigerated Warehouses. E, pp. 1–21. Springer, Cham (2016). https://doi.org/10.1007/978-3-319-40898-9_1

A Deep Learning-Based Approach for Automatic Leather Classification in Industry 4.0

Giulia Pazzaglia(✉), Massimo Martini, Riccardo Rosati, Luca Romeo, and Emanuele Frontoni

Dipartimento di Ingegneria Dell'Informazione (DII), Università Politecnica Delle Marche, Via Brecce Bianche 12, 60131 Ancona, Italy
giuliapazzaglia.94@gmail.com, {m.martini,r.rosati}@pm.univpm.it, {l.romeo,e.frontoni}@univpm.it

Abstract. Smart production is trying to bring companies into the world of industry 4.0. In this field, leather is a natural product commonly used as a raw material to manufacture luxury objects. To ensure good quality on these products, one of the fundamental processes is the visual inspection phase to identify defects on leather surfaces. A typical exercise in quality control during the production is to perform a rigorous manual inspection on the same piece of leather several times, using different viewing angles and distances. However, the process of the human inspection is expensive, time-consuming, and subjective. In addition, it is always prone to human error and inter-subject variability as it requires a high level of concentration and might lead to labor fatigue. Therefore, there is a necessity to develop an automatic vision-based solution in order to reduce manual intervention in this specific process.

In this regard, this work presents an automatic approach to perform leather and stitching classification. The main goal is to automatically classify the images inside of a new dataset called LASCC (Leather And Stitching Color Classification) dataset. The dataset is newly collected and it is composed of 67 images with two different colors of leathers and seven different colors of stitching. For this purpose, Deep Convolutional Neural Networks (DCNNs) such as VGG16, Resnet50 and InceptionV3 have been applied to LASCC dataset, on a sample of 67 images.

Experimental results confirmed the effectiveness and the suitability of the approach, showing high values of accuracy.

Keywords: Deep learning · Industry 4.0 · Classification

1 Introduction

The evolution of digital technologies is placing companies in front of a potential paradigm shift characterized by greater interconnection and cooperation between

This work was not supported by any organization.

systems, people and information. This technological mix of automation, information, connection and programming is leading to the birth of the *fourth industrial revolution*, also known as **Industry 4.0**. Now the goal is to interconnect manufacturing with the economy as strongly indicated by the development programs of Germany [17].

In the manufacturing field, the concept of real physical systems is currently evolving into high-level information technologies. The main objectives in this context are the development of engineering processes, through which factories will have the capabilities of self-awareness, self-prediction, self-comparison, self-reconfiguration and self-maintenance [19].

Smart manufacturing refers to a new manufacturing system where manufacturing machines are fully connected through wireless networks, monitored by sensors, and controlled by advanced computational intelligence to improve product quality, system productivity, and sustainability while reducing costs. Recent technologies such as IoT, Cloud Computing, Cyber Physical System (CPS) provide key support to advance modern manufacturing [23, 37, 38, 40].

Deep learning (DL) has attracted much attention as a development of computational intelligence [24, 25]. By mining knowledge from aggregated data, DL methods are a fundamental tool in automatically learning from data and making decisions, allowing to transform manufacturing into highly optimized smart structures. The advantages are reducing operating costs, keeping up with changing consumer demand, improving productivity and reducing downtime, gaining better visibility and extracting more value from the operations for globally competitiveness [6, 26].

In the field of smart manufacturing, one of the most important sector is the manufacture inspection, which consists in detecting product defects: in fact, the main existing problem in this task is that current inspection systems cannot guarantee good performances while keeping the processing efficiency and reducing inter-subject variability. The traditional methods, such as filters or feature-based classification approaches, are simple but not so effective in all the scenarios. Then, the DL-based methods turned up, bringing greatly improved analysis and recognition abilities, by automatizing the overall procedure [34].

In this regard, this work is focused precisely in the field of industry 4.0. In fact, the aim is to classify the images inside of the new **LASCC dataset**[1] (*Leather And Stitching Color Classification*) composed by images of leather in two colors, black and alcantara, with stitching in seven different colors. For this purpose, three state-of-the-art CNNs have been used: VGG16 [32], ResNet50 [11] and InceptionV3 [36].

The overview of the proposed method is shown in Fig. 1, which represents the framework of the project: there are four phases which include images acquisition, preprocessing, training of CNN and classification.

---

[1] LASCC Dataset is available upon request at the following link: https://vrai.dii.univpm.it/content/lascc-dataset.

To briefly summarize, the main contributions of this work are:

– A challenging new dataset of leather images, hand-labeled with ground-truth;
– A demonstration that DL architectures can be applied for classification of leather and stitching images;
– A comparison of DCNNs performances for image classification.

The paper is organized as follows: Sect. 2 provides a description of the DL approaches adopted for Industry 4.0, and in particular for leather classification. Section 3 gives details on the proposed approach, which is the main core of this work, describing in detail the dataset and the various methods used. In Sect. 4, a comparative evaluation of our approach is offered. Finally, in Sect. 5, conclusions and discussion about future directions for this field of research are drawn.

2   Related Works

Until now, research in the field of the leather industry has mainly focused on identifying defects and classifying the type of leather itself. For the defect localization and segmentation tasks, one of the pioneer research works is conducted by Lovergine et al. [22]. They detect and determine the defective areas using a black and white CCD camera. Then, a morphological segmentation ([9] and [18]) process is applied on the collected images to extract the texture orientation features of the leather.

One of the most recent works in this field is what we can find in [2]: Bong et al. employed several image processing algorithms to extract the image features and identify the defect's position on the leather surface; the extracted features are evaluated on an SVM [3] classifier.

Jawahar et al. [12] proposed a wavelet transform to classify the leather images. They adopt the *Wavelet Statistical Features* (WSF) [1] and *Wavelet Co-occurrence Features* (WCF) [13] as feature descriptors. A binary SVM with Gaussian kernel [14] is exploited to differentiate the defective and non defective leather sample.

On the other hand, Pistori et al. [28] used Gray-level Co-Occurrence Matrix (GLCM) to extract the features of the images. Ridge estimators and logistic regression are adopted to learn the normalized Gaussian radial basis functions. They are then clustered by SVM, Radial Basis Functions networks (RBF) [5] and Nearest Neighbours (KNN) [7] as classifiers.

Another leather detection work is carried out by Pereira et al. [27]: a combination of features from GLCM, Local Binary Pattern (LBP) and Pixel Intensity Analyzer (PIA) are formed; then, classifiers such as KNN, MLM (Minimal Learning Machine), ELM (Extreme Learning Machine), and SVM are adopted independently and tested on the features extracted.

There are very few works in the literature exploiting DL techniques in analyzing the leather types. One of the recent works that applied a pre-trained Convolutional Neural Network (CNN) was conducted by Winiarti et al. [39], where they aims to realise an automatic leather grading system. At this stage, it classifies the type of leather on tanning leather images. It uses the first seven layers of AlexNet [16] to extract features from the images and then classifies the images using linear SVM. Using approximately 3000 leather sample images from the five types of leather (i.e., monitor lizard, crocodile, sheep, goat, and cow skin), they achieved a 99.9% classification accuracy.

In the engineering practice, it frequently occurs that designers, final or intermediate users have to roughly estimate some basic performance or specification data on the basis of input data available at the moment, which can be time-consuming. There is the need for a tool that will fill the missing gap in the optimization problems in engineering design processes, by making use of the advances in the artificial intelligence field. The work proposed by Romeo et al. first in [31] and then extended in [30], aims to fill this gap by introducing an innovative Design Support System (DesSS), originated from the Decision Support System (DSS), for the prediction and estimation of machine specification data.

Instead, the work carried out by Liong et al. [21] propose a method using both Convolutional Neural Network and Artificial Neural Networks as the feature extractors and classifiers to differentiate the defective/ non-defective leather images. Afterwards they developed their work in [20] where an instance segmentation DL model, namely, Mask Region-based Convolutional Neural Network (Mask R-CNN [10]), is utilized to develop a robust architecture to evaluate the test dataset and then the details of the defective regions is transferred to a robotic arm to automatically mark the boundary of the defect area.

Unlike the methods described above, this work aims to directly classify the colors of the seams on pieces of leather without focusing on defect detection. Different from other related work that employed feature extraction and standard ML technique we propose the application of DL for solving this challenging task. In particular, the Network used for the classification are deep neural networks such as VGG16 [32], ResNet50 [11] and InceptionV3 [36] .

3   Materials and Methods

In this section, the framework and the dataset used for evaluation are explained. The framework is depicted in Fig. 1.

In particular, it comprises four phases: *Data Acquisition*, *Preprocessing*, model building (*Networks*) and prediction (*Predicted class*). The framework is comprehensively evaluated on LASCC dataset, a dataset specifically collected for this work.

3.1   LASCC Dataset

LASCC dataset contains images of leather of two different colors (black leather and dark alcantara), with seams of seven different colors (red, dark blue, light blue, yellow, black, gray and gold).

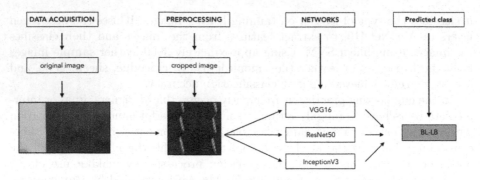

Fig. 1. Framework - The structure of the work carried out can be divided into 4 phases: the first phase can be identified by the *data acquisition*, images of leathers with stitching of different colors (in this case black leatherand red and light blue stitching); the second is a *preprocessing* phase, as the images have been cropped to be able to insert them as input to the network (in this case cropped image with black leather and light blue stitching); the third phase was that of *training the networks,* in which three different types of networks were used (VGG16, ResNet50, InceptionV3); the last phase is that of the *test,* in which the networks produce the label of the past image with input. (in this case the predicted label is BL-LB, i.e. Black leather and light blue stitching). (Color figure online)

The dataset consists of 67 images and it is divided in 8 classes as follows:

- 8 images of black leather with red and light blue stitching
- 7 images of black leather with dark blue and yellow stitching
- 7 images of black leather with black and grey stitching
- 8 images of black leather with grey and gold stitching
- 8 images of dark alcantara leather with red and light blue stitching
- 9 images of dark alcantara leather with dark blue and yellow stitching
- 9 images of dark alcantara leather with black and grey stitching
- 11 images of dark alcantara leather with grey and gold stitching

The dataset acquisition process is shown in the Fig. 2. The images within the dataset could be of different types: some had a single seam, others had a double seam; furthermore, these seams could be 7 different colors as described above.

In order to classify the color of stitching in leather pieces, the images within each class were split in

- 70% Train (the sample of data used to fit the model),
- 20% Validation (the sample of data used to provide an unbiased evaluation of a model fit on the training dataset while tuning model hyperparameters),
- 10% Test (the sample of data used to provide an unbiased evaluation of a final model fit on the training dataset).

The percentage of each split was maintained for each class. Furthermore, we have not balanced the dataset as the imbalance between the various classes is very low.

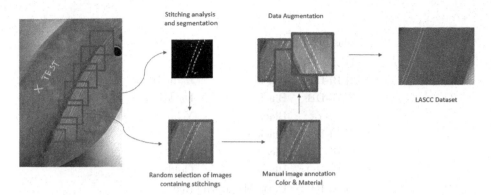

Fig. 2. Dataset acquisition process. From raw initial images, stitching analysis and segmentation was performed. Then the images that contained the stitching were randomly selected and once collected they were manually annotated for stitching color and leather material and color. At this point the data augmentation was carried out which consisted of rotations, horizontal and vertical symmetries and changes in brightness. In this way the LASCC dataset was created.

Lastly, a new division of the dataset was carried out: the images were then divided into 14 classes (2 colors of the leather × 7 colors of the seams), as described in the Table 1.

Data augmentation was also performed in the training images. In fact, from the original images, 224 × 224 crops were made in order to make the images accessible to the network. Furthermore, techniques such as rotations, vertical and horizontal symmetries and brightness changes were performed.

3.2 Deep Learning Models

In this subsection, the DL model used for the training phase are described.

For the classification of the images in LASCC dataset, three different DL networks are used:

- VGG16 [32],
- ResNet50 [11],
- InceptionV3 [36].

These networks have been chosen because they are already present in the state of the art and they are also the networks usually used for the classification task. Furthermore, the network architectures are all very different: we train a simple network (VGG16), an intermediate network (ResNet50) and a deep and heavier network (InceptionV3).

All these networks are pretrained on ImageNet [8], a dataset of over 15 million labeled high-resolution images belonging to roughly 22,000 categories. Therefore, the networks take images of fixed size 224 × 224 RGB as input: thus a preprocessing phase was required in which the initial images were resized.

Table 1. 14 Class

Class	Leather	Stitching
1) Bl-Rs	Black	Red
2) Bl-LBs	Black	Light Blue
3) Bl-DBs	Black	Dark Blue
4) Bl-Ys	Black	Yellow
5) Bl-Bs	Black	Black
6) Bl-GRs	Black	Grey
7) Bl-GOs	Black	Gold
8) Al-Rs	Dark Alcantara	Red
9) Al-LBs	Dark Alcantara	Light Blue
10) Al-DBs	Dark Alcantara	Dark Blue
11) Al-Ys	Dark Alcantara	Yellow
12) Al-Bs	Dark Alcantara	Black
13) Al-GRs	Dark Alcantara	Grey
14) Al-Gos	Dark Alcantara	Gold

The VGG16 Network was introduced by Simonyan and Zisserman in [32]. This network is characterized by its simplicity, using only 3 × 3 convolutional layers stacked on top of each other in increasing depth. Reducing volume size is handled by max pooling. Two fully-connected layers, each with 4,096 nodes are then followed by a softmax classifier composed of 1000 neurons used to classify the images according to the ImageNet dataset classes [8].

ResNet50 was first introduced by He et al. in [11]. This Network is based on the VGG network with a shortcut connection between the input of the convolutional block and the output. Even though ResNet is much deeper than VGG, the model size is actually substantially smaller due to the usage of global average pooling rather than fully-connected layers.

Inception micro-architecture was first introduced by Szegedy et al. in [35]. The goal of this network is to act as a "multi-level feature extractor" by computing 1 × 1, 3 × 3, and 5 × 5 convolutions within the same module of the network; the output of these filters are then stacked along the channel dimension and before being fed into the next layer in the network. The Inception V3 architecture included in the Keras core comes from the later publication by Szegedy et al. [36] which proposes updates to the inception module to further boost ImageNet [8] classification accuracy.

3.3 Performance Evaluation

In this subsection, the metrics used for the evaluation of all networks are presented.

They are:

- **Precision**, also called *positive predictive value*

$$Precision \ = \ \frac{TP}{TP + FP} \qquad (1)$$

- **Recall**, also known as *sensitivity*

$$Recall \ = \ \frac{TP}{TP + FN} \qquad (2)$$

- **F1-score**

$$F1 - Score \ = \ \frac{Precision * Recall}{Precision + Recall} \qquad (3)$$

- **Accuracy**

$$Accuracy \ = \ \frac{TP + TN}{TP + FP + TN + FN} \qquad (4)$$

[29, 33].

In all the previous equations:

- TP is True Positive,
- FP is False Positive,
- TN is True Negative,
- FN is False Negative.

as described in [4].

Lastly, Categorical Cross Entropy was chosen as loss function, that is the standard loss function used in multi-class classification tasks, and Softmax as the activation function.

4   Results and Discussions

In this section, the results of the experiments conducted on LASCC dataset are reported. For the training phase, we used three types of Convolutional Networks: VGG16, ResNet50 and InceptionV3.

All three different types of networks were pre-trained on ImageNet [8].

In any case, Adam [15] was used as an optimizer with a learning rate of 0.001 (10^{-3}), a batch size of 32 was chosen and the training was done for

- 20 epochs in case of VGG16 and ResNet50,
- 100 epochs in case of InceptionV3.

The experimental results are shown in Fig. 3. In the case of InceptionV3, 100 epochs were used because 20 epochs were very few to obtain satisfactory results. In fact, observing the graphs in Fig. 3(f) and 3(c), which represents the accuracy and the loss function of the InceptionV3 network, it can be seen that

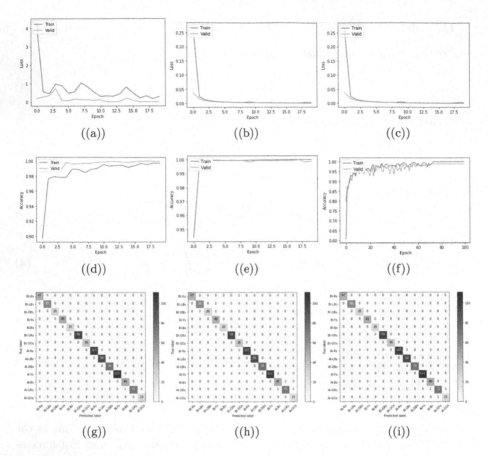

Fig. 3. Experiments Results - Fig. (3(a)) represent the loss function of VGG16, Fig. (3(b)) represent the loss function of ResNet50, Fig. (3(c)) represent the loss function of InceptionV3 - Fig. (3(d)) represent the accuracy values of VGG16, Fig. (3(e)) represent the accuracy values of ResNet50, Fig. (3(f)) represent the accuracy values of InceptionV3 - Fig. (3(g)) represent the confusion matrix of VGG16, Fig. (3(h)) represent the confusion matrix of ResNet50, Fig. (3(i)) represent the confusion matrix of InceptionV3

the trend of the curves begins to stabilize once it reaches 80 epochs. However, in all cases, the training time is 15 s per epoch and early stopping is used to avoid overfitting.

The parameters of the networks are shown in Table 2. The training and validation loss and accuracy are shown in Fig. 3. The final test accuracy of VGG16, ResNet50 and InceptionV3 is respectively 0.999, 0.999 and 0.991 (see Table 4). Table 3 shows the results of the classification for each class in terms of Precision, Recall and F1-Score.

Taking into account the confusion matrices of VGG16 and ResNet50 (Figs. (3(g)) (3(h))) it is possible to infer that both VGG16 and ResNet50 misclassifies an image Dark Alcantara leather - Grey stitching as an image Dark Alcantara leather - Black stitching. Inceptionv3 network (Fig. (3(i)), instead, misclassified 8 samples, 3 of these in the classification of the images Dark Alcantara leather - Grey stitching. Therefore, it can be deduced that the images Al-GRs are the most difficult to classify for all the networks.

Lastly, it can be stated that for LASCC dataset VGG16 and ResNet50 networks obtain the best result. The difference between the VGG16 and ResNet50 is the number of epochs needed to reach an accuracy of 0.999% (see Fig. 3(b) and Fig. 3(e)): so in this case, it can be assert that ResNet50 is the network that offers the greatest result in the shortest time.

InceptionV3 also gives an excellent result, but not up to the previous two networks, as it needs almost 100 epochs (see Fig. 3(c) and Fig. 3(f)) to reach an accuracy of 0.991%.

Table 2. Parameters of networks

Network	Total Parameters	Trainable Parameters	Non-trainable Parameters
VGG16 pretrained	40.420.174	25.705.486	14.714.688
VGG16 from skratch	134.317.902	134.317.902	0
ResNet 50 pretrained	23.616.398	28.686	23.587.712
ResNet 50 from skratch	23.616.398	23.563.278	53.120
InceptionV3 pretrained	21.831.470	28.686	21.802.784
InceptionV3 from skratch	21.831.470	21.797.038	34.432

Table 3. Results - Table 3(a) represents the values of Precision, Recall and F1-Score of VGG16 Network; Table 3(b) represents the values of Precision, Recall and F1-Score of ResNet50 Network; Table 3(c) represents the values of Precision, Recall and F1-Score of InceptionV3 Network

Class	Pre	Rec	F1	Class	Pre	Rec	F1	Class	Pre	Rec	F1
Bl-Rs	1.000	1.000	1.000	Bl-Rs	1.000	1.000	1.000	Bl-Rs	1.000	1.000	1.000
Bl-LBs	1.000	1.000	1.000	Bl-LBs	1.000	1.000	1.000	Bl-LBs	1.000	1.000	1.000
Bl-DBs	1.000	1.000	1.000	Bl-DBs	1.000	1.000	1.000	Bl-DBs	1.000	1.000	1.000
Bl-Ys	1.000	1.000	1.000	Bl-Ys	1.000	1.000	1.000	Bl-Ys	1.000	1.000	1.000
Bl-Bs	1.000	1.000	1.000	Bl-Bs	1.000	1.000	1.000	Bl-Bs	0.963	1.000	0.981
Bl-GRs	1.000	1.000	1.000	Bl-GRs	1.000	1.000	1.000	Bl-GRs	1.000	1.000	1.000
Bl-GOs	1.000	1.000	1.000	Bl-GOs	1.000	1.000	1.000	Bl-GOs	1.000	1.000	1.000
Al-Rs	1.000	1.000	1.000	Al-Rs	1.000	1.000	1.000	Al-Rs	0.981	1.000	0.991
Al-LBs	1.000	1.000	1.000	Al-LBs	1.000	1.000	1.000	Al-LBs	0.989	0.989	0.989
Al-DBs	1.000	1.000	1.000	Al-DBs	1.000	1.000	1.000	Al-DBs	0.987	0.974	0.980
Al-Ys	1.000	1.000	1.000	Al-Ys	1.000	1.000	1.000	Al-Ys	1.000	1.000	1.000
Al-Bs	0.980	1.000	0.990	Al-Bs	0.980	1.000	0.990	Al-Bs	0.980	0.980	0.980
Al-GRs	1.000	0.987	0.993	Al-GRs	1.000	0.987	0.993	Al-GRs	0.973	0.961	0.967
Al-Gos	1.000	1.000	1.000	Al-Gos	1.000	1.000	1.000	Al-Gos	1.000	0.967	0.978
(a)				(b)				(c)			

Table 4. Accuracy Evaluation.

	VGG16	ResNet50	InceptionV3
Accuracy	0.999%	0.999%	0.991%

5   Conclusions and Future Works

Deep learning provides advanced analytics and offers great potentials to smart manufacturing during this period of Industry 4.0 development. By unlocking the unprecedented amount of data into actionable and insightful information, DL gives decision-makers new visibility into their operations, as well as real-time performance measures and costs. One of the most popular in the field of smart manufacturing is the manufacture inspection, which is to detect the defects of the products. To facilitate the progress of smart manufacturing, various DL architectures such as Convolutional Neural Networks are used.

This project can be seen as a preliminary proof of concept of the application of DL based approach for automatizing the leather classification in the Industry 4.0 scenario. In fact, within this field, it is particularly concerned with the classification of images, inside of LASCC Dataset, which represents leather pieces in which stitching of different colors could be found. Using the Convolutional Networks at the state of the art such as VGG16, ResNet 50 and InceptionV3, excellent results in a short time are obtained, reaching up to 0.999% accuracy. The impact of these results is relevant in order to automatize the overall leather classification by reducing human error and inter-subject variability. In this work, the variants "single stitching" and "double stitching" have been merged and not classified.

Inside of the LASCC dataset, we also find images without stitching, and images of this type placed in different classes could create ambiguities on the network. Although the impact of this result could be potentially limited by the low amount of samples, we are currently increasing the sample size of the dataset by increasing the number of images, the number of classes and cleaning up any ambiguities. In particular, the number of classes could be increased, by adding other classes, such as double stitching or single stitching. In this context, as a further interesting future direction, a hierarchical DL based approach may be designed to learn top-level and bottom level classes. Finally, detection techniques could be adopted, as the mapping between the acquired crop and the original image is available.

References

1. Arivazhagan, S., Ganesan, L.: Texture classification using wavelet transform. Pattern Recogn. Lett. **24**(9–10), 1513–1521 (2003)
2. Bong, H.Q., Truong, Q.B., Nguyen, H.C., Nguyen, M.T.: Vision-based inspection system for leather surface defect detection and classification. In: 2018 5th NAFOSTED Conference on Information and Computer Science (NICS), pp. 300–304. IEEE (2018)

3. Boser, B.E., Guyon, I.M., Vapnik, V.N.: A training algorithm for optimal margin classifiers. In: Proceedings of the Fifth Annual Workshop on Computational Learning Theory, pp. 144–152 (1992)
4. Brismar, J., Jacobsson, B.: Definition of terms used to judge the efficacy of diagnostic tests: a graphic approach. AJR Am. J. Roentgenol. **155**(3), 621–623 (1990)
5. Broomhead, D.S., Lowe, D.: Radial basis functions, multi-variable functional interpolation and adaptive networks. Technical report, Royal Signals and Radar Establishment Malvern (United Kingdom) (1988)
6. Calabrese, M., et al.: Sophia: an event-based iot and machine learning architecture for predictive maintenance in industry 4.0. Information **11**(4), 202 (2020)
7. Dasarathy, B.V.: Nearest neighbor (nn) norms: Nn pattern classification techniques. IEEE Computer Society Tutorial (1991)
8. Deng, J., Dong, W., Socher, R., Li, L.J., Li, K., Fei-Fei, L.: Imagenet: a large-scale hierarchical image database. In: 2009 IEEE Conference on Computer Vision and Pattern Recognition, pp. 248–255. IEEE (2009)
9. Giardina, C., Dougherty, E.: Morphological Methods in Image and Signal Processing, Prentice-Hall, Inc., New Jersey (1988)
10. He, K., Gkioxari, G., Dollár, P., Girshick, R.: Mask r-cnn. In: Proceedings of the IEEE International Conference on Computer Vision, pp. 2961–2969 (2017)
11. He, K., Zhang, X., Ren, S., Sun, J.: Deep residual learning for image recognition. In: Proceedings of the IEEE Conference on Computer Vision and Pattern Recognition, pp. 770–778 (2016)
12. Jawahar, M., Babu, N.C., Vani, K.: Leather texture classification using wavelet feature extraction technique. In: 2014 IEEE International Conference on Computational Intelligence and Computing Research, pp. 1–4. IEEE (2014)
13. Jobanputra, R., Clausi, D.A.: Texture analysis using gaussian weighted grey level co-occurrence probabilities. In: First Canadian Conference on Computer and Robot Vision, 2004. Proceedings, pp. 51–57. IEEE (2004)
14. Keerthi, S.S., Lin, C.J.: Asymptotic behaviors of support vector machines with gaussian kernel. Neural Comput. **15**(7), 1667–1689 (2003)
15. Kingma, D.P., Ba, J.: Adam: A method for stochastic optimization. arXiv preprint arXiv:1412.6980 (2014)
16. Krizhevsky, A., Sutskever, I., Hinton, G.E.: Imagenet classification with deep convolutional neural networks. In: Advances in Neural Information Processing Systems, pp. 1097–1105 (2012)
17. Lasi, H., Fettke, P., Kemper, H.G., Feld, T., Hoffmann, M.: Industry 4.0. Bus. Inform. Syst. Eng. **6**(4), 239–242 (2014)
18. Lee, C., Wong, S.: A mathematical morphological approach for segmenting heavily noise-corrupted images. Pattern Recogn. **29**(8), 1347–1358 (1996)
19. Lee, J., Kao, H.A., Yang, S., et al.: Service innovation and smart analytics for industry 4.0 and big data environment. Procedia Cirp. **16**(1), 3–8 (2014)
20. Liong, S.T., Gan, Y., Huang, Y.C., Yuan, C.A., Chang, H.C.: Automatic defect segmentation on leather with deep learning. arXiv preprint arXiv:1903.12139 (2019)
21. Liong, S.T., et al.: Efficient neural network approaches for leather defect classification. arXiv preprint arXiv:1906.06446 (2019)
22. Lovergine, F.P., Branca, A., Attolico, G., Distante, A.: Leather inspection by oriented texture analysis with a morphological approach. In: Proceedings of International Conference on Image Processing, vol. 2, pp. 669–671. IEEE (1997)
23. Lu, Y., Xu, X., Xu, J.: Development of a hybrid manufacturing cloud. J. Manufact. Syst. **33**(4), 551–566 (2014)

24. Paolanti, M., Frontoni, E.: Multidisciplinary pattern recognition applications: a review. Comput. Sci. Rev. **37**, 100276 (2020)
25. Paolanti, M., Pietrini, R., Mancini, A., Frontoni, E., Zingaretti, P.: Deep understanding of shopper behaviours and interactions using rgb-d vision. Mach. Vis. Appl. **31**(7), 1–21 (2020)
26. Paolanti, M., Romeo, L., Felicetti, A., Mancini, A., Frontoni, E., Loncarski, J.: Machine learning approach for predictive maintenance in industry 4.0. In: 2018 14th IEEE/ASME International Conference on Mechatronic and Embedded Systems and Applications (MESA), pp. 1–6. IEEE (2018)
27. Pereira, R.F., Dias, M.L., de Sá Medeiros, C.M., Rebouças Filho, P.P.: Classification of failures in goat leather samples using computer vision and machine learning
28. Pistori, H., Paraguassu, W.A., Martins, P.S., Conti, M.P., Pereira, M.A., Jacinto, M.A.: Defect detection in raw hide and wet blue leather. In: Proceedings of the International Symposium Computing IMAGE, p. 355 (2018)
29. Powers, D.M.: Evaluation: from precision, recall and f-measure to roc, informedness, markedness and correlation (2011)
30. Romeo, L., Loncarski, J., Paolanti, M., Bocchini, G., Mancini, A., Frontoni, E.: Machine learning-based design support system for the prediction of heterogeneous machine parameters in industry 4.0. Expert Syst. Appl. **140**, 112869 (2020)
31. Romeo, L., Paolanti, M., Bocchini, G., Loncarski, J., Frontoni, E.: An innovative design support system for industry 4.0 based on machine learning approaches. In: 2018 5th International Symposium on Environment-Friendly Energies and Applications (EFEA), pp. 1–6. IEEE (2018)
32. Simonyan, K., Zisserman, A.: Very deep convolutional networks for large-scale image recognition. arXiv preprint arXiv:1409.1556 (2014)
33. Sokolova, M., Lapalme, G.: A systematic analysis of performance measures for classification tasks. Inform. Process. Manag. **45**(4), 427–437 (2009)
34. Sturari, M., Paolanti, M., Frontoni, E., Mancini, A., Zingaretti, P.: Robotic platform for deep change detection for rail safety and security. In: 2017 European Conference on Mobile Robots (ECMR), pp. 1–6. IEEE (2017)
35. Szegedy, C., et al.: Going deeper with convolutions. In: Proceedings of the IEEE Conference on Computer Vision and Pattern Recognition, pp. 1–9 (2015)
36. Szegedy, C., Vanhoucke, V., Ioffe, S., Shlens, J., Wojna, Z.: Rethinking the inception architecture for computer vision. In: Proceedings of the IEEE Conference on Computer Vision and Pattern Recognition, pp. 2818–2826 (2016)
37. Wang, L., Törngren, M., Onori, M.: Current status and advancement of cyber-physical systems in manufacturing. J. Manufact. Syst. **37**, 517–527 (2015)
38. Wang, P., Gao, R.X., Fan, Z.: Cloud computing for cloud manufacturing: benefits and limitations. J. Manufact. Sci. Eng. **137**(4) (2015)
39. Winiarti, S., Prahara, A., Murinto, D.P.I., Ismi, P.: Pretrained convolutional neural network for classification of tanning leather image. Network (CNN), vol. 9, no. 1 (2018)
40. Wu, D., Rosen, D.W., Schaefer, D.: Cloud-based design and manufacturing: status and promise. In: Cloud-Based Design and Manufacturing (CBDM), pp. 1–24. Springer, Berlin (2014)

Automatic Viewpoint Estimation for Inspection Planning Purposes

Siddhartha Dutta[1,2]($\boxtimes$) (iD), Markus Rauhut[2] (iD), Hans Hagen[1] (iD),
and Petra Gospodnetić[1,2] (iD)

[1] Department of Computer Science,
Technische Universität Kaiserslautern, 67663 Kasierslautern, Germany
s_dutta18@cs.uni-kl.de
[2] Department of Image Processing,
Fraunhofer ITWM, 67663 Kaiserslautern, Germany
petra.gospodnetic@itwm.fraunhofer.de

Abstract. Viewpoint estimation is an important aspect of surface inspection and planning. Typically viewpoint estimation has been done only with the 3D model and not with the actual object. This, therefore, limits the flexibility of using the actual object in inspection planning. In this work, we present a novel pipeline that can efficiently estimate viewpoints from live camera images. The pipeline can be used for different sized industrial objects to efficiently estimate the viewpoint. The achieved results are real-time and the method is easily generalizable to different objects. The presented method is based on a 3D model of the object, which is self-supervised and requires no manual data annotation or real images to be used as an input. The complete solution, together with documentation and examples is available in the public domain for testing. https://gitlab.itwm.fraunhofer.de/dutta/real-time-pose-est/.

Keywords: Viewpoint estimation · Inspection planning · 6D pose estimation

1 Introduction

Introduction of Industry 4.0 in production lines means a wider adoption of automation, enhancement using smart and autonomous systems and overall orientation towards product customization. Product customization means that a single production line should be capable of producing different products, with minimal overhead between changes. Every manufactured product must be examined and visual inspection systems are a common approach to inspection automation. Design of visual inspection systems needs proper positioning of the camera relative to the object, is an important part of the design process. To address the above problem we present a novel pipeline which enables the engineer who is designing a visual inspection system, to rotate the object in front of the camera and obtain its pose relative to the camera. That way desired positions can be easily determined, recorded, and used in further design.

© Springer Nature Switzerland AG 2021
A. Del Bimbo et al. (Eds.): ICPR 2020 Workshops, LNCS 12664, pp. 675–691, 2021.
https://doi.org/10.1007/978-3-030-68799-1_49

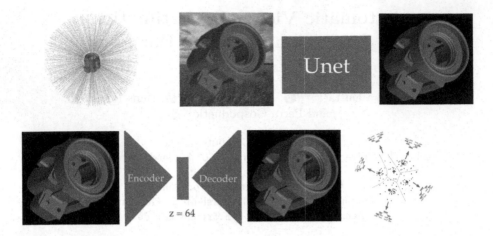

Fig. 1. (a) For each viewpoint an image is rendered and used in training. (b) Training example of Unet. (c) Rectification Network learns only to encode the model in a latent vector. (d) Orientation feature space clustering. For each cluster, a multivariate regression model using decision tree is trained.

Typically viewpoints are expressed as a combination of right, up, forward, and position vectors. These vectors represents the camera axes and the distance from the object. This is analogous to a 6D pose, which is essentially a rotation and a translation vector from the object's frame of reference. In the following work, we only compute these later vectors. An equivalent conversion can be easily made to right, up, forward and position vectors based on whether it is a right or left-handed coordinate system.

Classical pose estimation approaches (*e.g.* [16]) use tracking based methods that rely on localization of handcrafted targets or natural texture of the object. Such approaches are however very sensitive to noise and do not work well in varying light conditions. Deep learning has shown significant advancements in solving this problem and can accurately estimate pose of an object. Our approach is inspired from the works of [10,21]. The main idea of the presented approach is to segment out the object from the scene and then estimate the pose on the segmented object. It is done because the background does not provide any relevant information in estimating the pose. To start with, we employ a RetinaNet [13] to give us a bounding box for the object. The image from the bounding box is then resized to a fixed size and passed to a Unet [20] for segmentation. The task of the Unet is to segment the background and replace the original model in the image with a synthetic view. However, like most reconstruction approaches, the generated synthetic view is blurry. Therefore we introduce a Rectification Network to process the output of Unet. It is an autoencoder based network (with a bottleneck layer) which serves a dual purpose. Firstly, it works on the output of the Unet to rectify the image, and secondly, it only focuses on the features of the object and not any background information. We compared our reconstruction generated by our

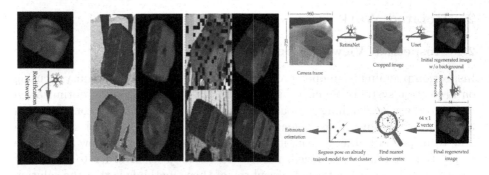

Fig. 2. (a) Figure shows the Rectification Network applied to Unet's output on Object 5 on T-LESS [8] dataset. (b) The first two columns shows the regeneration achieved through AAE [21], the third and fourth column shows the regeneration achieved after using our Rectification Network. Our network is able to reconstruct finite features and sharper images. (c) Orientation estimation run-time pipeline.

networks to the ones generated by [21], and as seen in Fig. 2(b), our model was able to generate a superior reconstruction. Once a considerable reconstruction is achieved, orientation and translation are estimated as follows. For each image in the training set, using the Rectification Network the value of the bottleneck vector ($\mathbf{z}$) is extracted. We use this value to create a dictionary of $\mathbf{z}$ vectors to its orientation in quaternion. Using the mini-batch k-means clustering algorithm, the dictionary is clustered into n different clusters. A separate multivariate decision regression tree is trained for each cluster with the computed $\mathbf{z}$ vector as input and quaternion values as regressed output. Finally, only the decision tree models and the cluster centers are saved in another dictionary to be used during runtime. At runtime, the latent vector ($\mathbf{z}$) generated from Rectification Network is compared to the saved cluster centers to find the nearest cluster. Once the cluster is recognized the trained decision tree is used to regress the orientation of the test image. The translation is estimated using a bounding box approach similar to [9]. The entire pipeline during the test time can be seen in Fig. 2(c). Training data generation for these networks are done automatically using the 3D model of the object. This has been described in Sect. 3.1.

In short, our contributions are as follows:

(1) Designing a novel approach for viewpoint estimation for industrial inspection that is automatic, scalable and time-efficient. This approach is configurable easily as per needs, and doesn't need manually annotated data.
(2) Part of this approach can also be applied to state-of-the-art approaches like [10,21] to improve their results further. We show that by generating better image reconstruction and compare the accuracy with the approach used by these methods.
(3) We make the code available publicly for further testing and development in the community.

2   Related Work

2.1   Inspection Viewpoint Selection

Inspection planning is an open problem within the research community focusing on inspection systems. So far it has been mostly observed from an optimization point of view, researching various approaches to deal with the complexity of the planning problem. [6,15] focused on finding a single optimal solution on how to inspect an object, given its 3D model. In simple terms, how should the acquisition system be designed, relative to the object, in order to inspect it completely. The most noticeable drawback of those methods is that the solution behaves in a manner of a black box, offering no possibility for the engineer to influence or evaluate results. Recent works of *Gospodnetic et al.* [4,5] are oriented more towards an interactive aspect of inspection planning, enabling the expert to control, oversee, and evaluate part of the process. They present an interactive 3D visualization tool intended to be used by an expert for evaluating automatic inspection planning results or manually determining good viewpoints to place a camera, relative to an object. The viewpoints are created on a predefined distance from the selected surface point, and the angle of the viewpoint can be subsequently adjusted. A drawback of such an approach is that there is no possibility to create a viewpoint that does not originate from the object surface.

2.2   Pose from RGB/RGB-D Data

Traditional approaches (*e.g.* [16]) use keypoint localisation and match them to solve the *PnP* problem. For this approach to work, local descriptors are needed and must be invariant to illumination, scale, and orientation. The descriptors can be either handcrafted features (*e.g.*) fiducials) or object texture. However, the use of descriptors typically demands the use of high-resolution images to achieve reliable tracking. *Rambach et al.* [18] use *Good Features To Track*, which requires a realistic 3D model and frame-to-frame tracking. Recently, CNN based approaches (*e.g.* [11,17]) have shown promising results using labeled real images for pose prediction. However, getting accurate labeling of real images is both time consuming and challenging. *Sundermeyer et al.* [21] and *Kehl et al.* [10] addressed this issue using a 3D model to generate accurately labelled pose data. They discretize the orientation space and make the pose estimation a classification task. Depth data can be used to further refine the pose estimation. *Xiang et al.* [25] fuse the depth data as an additional 4th input channel to the CNN. *Wang et al.* [24] use features predicted from RGB image with depth data and produce a pose estimate for each pixel. To the best of our knowledge, all of the approaches considered only RGB images and made no comments on performance in case of grayscale images or objects of different sizes.

Fig. 3. Shows the training images for RetinaNet. Random backgrounds from Pascal VOC dataset is used to render synthetic images of the model.

3   Approach

We use 6D pose which consists of rotation in quaternions $R = (q_x, q_y, q_z, q_w) \in \mathbb{R}^4$ and translation $t = (t_x, t_y, t_z) \in \mathbb{R}^3$. Both rotation and translation are determined relative to the camera position.

3.1   Data Generation

Deep learning networks require a lot of labeled training data to achieve good performance. Manually creating such a data set for pose estimation purposes requires a complex setup and is a very time-consuming task. Therefore, the automation of data generation and labeling processes is highly desired. For that purpose, we use a 3D model of an object to render synthetic images to be used for training. To cover the whole object like a sphere, thousands of viewpoints are created and are evenly distributed but are randomly rotated (see Fig. 1(a)). Additionally 25% viewpoints are created for validation. A fine sampling of viewpoints is encouraged to ensure more training data for a robust estimate. Once viewpoints are created, they are used to render the object for two different purposes - (1) to train the RetinaNet and (2) to train Unet, Rectification Network, and multivariate Decision Trees. An arbitrary number of viewpoints is randomly picked from the viewpoint list and rendered against random background images from Pascal VOC [3] dataset and the bounding box is calculated (see Fig. 3). The distance at which the model is rendered is varied in a few hundred mm (*e.g.*, 100–500 mm) randomly. The final images are also augmented with gaussian blur and random noises. The resulting image with the bounding box details is saved to train the RetinaNet.

For the Unet and the Rectification Network, rendering distance is kept fixed and images are generated for all the viewpoints. Similar to (1) above, the object is rendered against both black and random background, and the image is cropped containing only the object. Finally, the image is resized to a suitable size for training. All images rendered on random backgrounds are augmented in order to ensure that varying illumination and noise in the actual image do not affect the Unet reconstruction. To address this we add random noise, gaussian blur, and contrast normalization to the input images for the Unet.

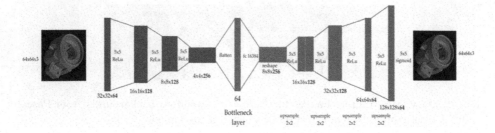

Fig. 4. Detailed architecture of the Rectification Network - a fully convolutional autoencoder network, with a bottleneck layer of 64.

For Unet training, clean images with a black background are considered as reconstruction targets and the corresponding augmented images on a random background as input. The Rectification Network uses the same clean images for input and for target reconstruction (Fig. 1(b)). This way it is ensured that Unet helps in background removal and the Rectification Network focuses only on the important features which can be used for pose estimation. All the images and their orientation in quaternions are saved in the disk to be used in training.

3.2   Architecture

The presented solution has five parts. (1) Object identification using RetinaNet (2) Object extraction using Unet (3) Regeneration of missing synthetic features and get a latent representation using Rectification Network (4) Multivariate Decision Trees for orientation regression (5) Projective distance estimation using bounding boxes.

Object Identification in the Scene. We used RetinaNet [13] for object detection in the scene, however, any object detection network(*e.g.,* YOLO [19]) should suffice.

Object Extraction Using Unet. For the first step of object segmentation, we use a modified Unet. Originally Unet was developed for bio-medical segmentation purposes but has also been extremely successful in many complex segmentation problems. It is a fully convolutional neural network used for semantic segmentation. The Unet consists of two parts - contraction path and expansion path. The contraction part downsamples the image, whereas the expansion path upsamples the image. We modify the expansion part with Inception-ResNet-v2 [22] to get improved segmentation. A large number of feature channels are used in upsampling which leads to the propagation of contextual information. This makes the network a kind of symmetric and leads to a U-shaped structure.

Regeneration of Missing Synthetic Features and Get a Latent Representation Using Rectification Network. To realise this step we use the Rectification Network. It is a fully convolutional autoencoder that takes an image input and recreates the same image. Originally, autoencoders were designed for dimensionality reduction techniques. As can be seen from Fig. 1(c) (detailed structure in Fig. 4), it consists of an Encoder f_E and a Decoder f_D, both of which are connected through a bottleneck layer (also known as latent vector). The training objective is for the recreated output $\hat{x}$ to be the same as input ($\mathbf{x} \in \mathbb{R}^n$), obtained through a lower-dimensional bottleneck ($\mathbf{z} \in \mathbb{R}^d$, where $d << n$). In this learning process, the most important features are captured in the bottleneck $\mathbf{z}$.

$$\mathbf{z} = f_E(\mathbf{x}) \tag{1}$$

$$\hat{\mathbf{x}} = f_D(\mathbf{z}) = f_D(f_E(\mathbf{x})) \tag{2}$$

The loss per training image is calculated pixel wise using binary cross-entropy function:

$$L_{ce}(\hat{x}) = \sum_k (x_k \log \hat{x_k} + (1 - x_k) \log(1 - \hat{x_k})) \tag{3}$$

Multivariate Decision Trees for Orientation Regression. Decision Trees (DTs) which can be used for both classification and regression problems are a non-parametric supervised learning method. The aim is to create a model learned from data features to predict the target variable. DTs are simple to interpret and understand, efficient in computation, and are able to handle multi-output problems. If there is no correlation between a k dimensional output vector, designing k independent models for each one of the k outputs can solve the problem. However, in rotation space, *e.g.*, while predicting quaternions, the individual values are likely to be correlated. Therefore, building a single model that predicts all of the k output values is the best possible approach. It not only reduces the training time but also increases the generalization accuracy. There are various DT algorithms known, and the one that has been used here is Classification and Regression Trees (CART) [1,14]. It works on constructing binary trees using the feature and threshold that yields the largest information gain at each node.

Consider training vectors $\mathbf{x_i} \in \mathbb{R}^n$, $i = 1, 2, ...$ and output vector $\mathbf{y} \in \mathbb{R}^l$, the idea of a DT is to recursively partition the space until similar target variables are grouped together. For any arbitrary node m whose data is Q, the split $\theta = (j, t_m)$ consists of a feature j and threshold t_m. This partitions the data into $Q_{left}(\theta) = (x, y)|x_j \leq t_m$ and $R_{right}(\theta) = Q \backslash Q_{left}(\theta)$.

An impurity function H is used to calculate the impurity m. The choice of H is based either classification or regression problem.

$$G(Q, \theta) = \frac{n_{left}}{N_m} H(Q_{left}(\theta)) + \frac{n_{right}}{N_m} H(Q_{right}(\theta)) \tag{4}$$

Solving for $\theta^* = argmin_\theta G(Q, \theta)$. In regression scenario, for node m which represents a region R_m and N_m observations, Mean Squared Error is used to find future splits [1].

$$\hat{y} = \frac{1}{N_m} \sum_{i \in N_m} y_i \tag{5}$$

$$H(X_m) = \frac{1}{N_m} \sum_{i \in N_m} (y_i - \hat{y}_m)^2 \tag{6}$$

Fig. 5. Shows translation estimation using bounding boxes.

Projective Distance Estimation Using Bounding Boxes. Translation can be estimated similar to the method described by [9, 21]. Consider two bounding boxes (BB) of two objects at different distances but at same orientation, one for the object in the test image and one for the synthetic image , as seen in Fig. 5. Using a pinhole model, the distance for test model t_z can be estimated by the ratio of the detected bounding box diagonal to synthetic bounding box (BB) diagonal as:

$$t_z = \frac{BB_{synthetic}}{BB_{detected}} \times t_{fixed} \tag{7}$$

For t_x and t_y, the pixel co-ordinates of the center u_x, u_y of the bounding box is calculated and the following is applied:

$$\left[t_x t_y t_z \right]^T = K^{-1} t_z \left[u_x u_y 1 \right]^T \tag{8}$$

where K is the intrinsic matrix of the camera in use.

3.3   Training Process

Within the automated training pipeline, the application generates training images for RetinaNet, Unet, and Rectification Network as configured. It further trains these networks. After successful training, all the clean images (*i.e.*, the training images of the Rectification Network) are re-passed to the trained Rectification Network to create a codebook of bottleneck feature vectors. Since the orientation details of the object rendered in each image are known, it is possible to create a dictionary which has feature vectors as keys and the corresponding quaternions as values. Using mini-batch k-means clustering algorithm all of the feature vectors are clustered into n different clusters. The number of clusters n, is arbitrary and can be different based on the object. Finally, it creates a multivariate decision trees and trains it on each cluster with the feature vector as input and the orientation in quaternion, as output. The decision trees and the cluster centers are stored for use in runtime.

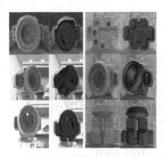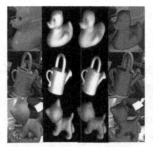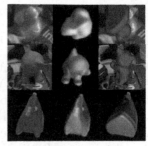

Fig. 6. (a) Shows the approach performing on grayscale images and RGB images on a 3D printed model (Object 12 in T-LESS dataset) on live camera. The left image shows the actual image and in the right, the estimated pose overlayed on the model. (b) Shows the approach applied to various test images of the LineMOD dataset for different objects. The first image is the scene crop of the object, the second image is the target reconstruction (generated from the 3D model rendered with the ground truth data provided with the dataset), the third image is the reconstructed output and the final image is the estimated pose calculated and overlayed on top of the original image. (c) Shows the failed cases. In all failed cases, the regenerated model (in the center) can be seen as incorrect which resulted in wrong orientation estimation. (Color figure online)

3.4   Runtime

At runtime, for each frame, RetinaNet detects the object and provides the bounding box details. Using these details the image is cropped and passed to the Unet to generate a blurry synthetic image of the actual object. The blurry synthetic image is then passed to the Rectification Network to recreate the missing features and also produce the latent vector (z) for the image. We find the nearest cluster using the Squared Euclidean distance of the z-vector from the cluster centers. Squared Euclidean distance was used because it is very easy to compute in runtime and also the distance grows non-linearly which can be used to put a maximum threshold. Once the nearest cluster is identified we use the pre-trained decision tree on this cluster to regress the orientation in quaternions. The steps can be seen in Fig. 2(a). After the orientation is regressed, we render the object in the scene from a fixed distance with the regressed orientation and calculate the synthetic bounding box. The translation is then estimated as described in Sect. 3.2. One important thing to note is, the accuracy of the translation is dependent on the accuracy of the object identification network. Networks like YOLO can provide fast detections, where as RetinaNet provides more accurate detections.

4   Experiments and Results

4.1   Experimental Setup

We implemented the solution in Python 3.6 using Keras and Tensorflow 1.15 on NVIDIA Titan V GPU. We evaluated it on LineMOD dataset [7] and visually tested it on T-LESS [8] objects on a Basler acA2500-20gc camera with resolution 960×720 pixels. For each object, 324000 viewpoints were created. Among them, 25000 viewpoints were selected randomly to train the RetinaNet. The rendering distance for RetinaNet training was varied from (100–500) mm randomly and for Unet and Rectification Network it was kept fixed at 300 mm. Unet and Rectification Network were trained in parallel with all the viewpoints. The final image to train these two networks was resized to 64×64. Both of these networks used *binary cross entropy* as loss function and *adam* as optimiser with learning rate as 1×10^{-3}. The size of latent vector $\mathbf{z}$ was set to 64. We train the Unet for 10 epochs and the Rectification Network for 20 epochs which took around 6 and 3.5 hours respectively. Further we clustered orientation space into $n = 16384$ clusters. In our experiments we saw good orientation estimation from $n \geq 4096$ onward. Within the application the above mentioned parameters can be configured easily to work with different objects and suite the required environment. Testing of the application was done in a *only* CPU based system using Intel Core i7-8700 CPU @ 3.20GHz x 12 cores, 16GB RAM running Ubuntu 18.04 OS.

4.2   Testing Using Different Objects

Evaluation Criteria. We evaluate the model in two ways (1) IoU score (2) Rotation error. The IoU score is the overlap between projections of the 3D model of given ground truth and the projections of the predicted pose. A pose is considered correct if the overlap is larger than 0.5, as used in [23]. In rotation error, we calculate the difference of rotation of the detected pose and the given ground-truth pose in degrees. We do not calculate the translation error here as the translation is based on the accuracy of the object identification network. We show cumulative error up to $10°$ of error.

Evaluation on Objects of LineMOD Dataset. We evaluated our approach on different objects of the LineMOD train dataset. The dataset has 15 different objects, 13 of which are present in Table 1. We intentionally use the train set because, firstly, we do not expect clutter and occlusion in an industrial inspection environment and the test set has a lot of clutter around the main object. Secondly, the seen objects in the test images are quite small, which discourages defect inspection. Thirdly, we train the model completely on synthetic data and the model hasn't seen the training data from the dataset. The results for the individual model can be seen in Table 1. We also achieve nearly 100% results when evaluated using the IoU metric. We do not evaluate the remaining two models - Cup and Bowl, as they do not have enough features to accurately estimate rotation. To the best of our knowledge LineMOD is the only publicly available dataset that has 3D model same to the actual object in the scene which is a requirement for our method.

Testing on T-LESS Objects. We 3D printed Object 5 and 12 from the T-LESS dataset and tested it on a live camera. Some of the tested images are present in Fig. 6. In our experiments, we repainted the object color of the 3D model so that it has a similar color that of the real object. The same can be seen in Fig. 2(a).

Table 1. Test on LineMOD objects: rotation error in degrees. Not evaluated on objects Cup and Bowl.

Objects / Rot. error	$\leq 5°$	$\leq 6°$	$\leq 7°$	$\leq 10°$
Ape	28.5	36.1	39.0	73.9
Benchvise	16.1	21.5	26.6	50.5
Camera	31.3	39.3	42.8	79.6
Can	21.8	28.2	30.3	59.0
Cat	31.0	36.5	39.3	76.8
Driller	8.3	12.4	15.6	29.8
Duck	27.7	32.1	34.2	66.4
Eggbox	13.4	17.1	18.9	37.6
Glue	10.5	12.7	15.0	30.2
Holepuncher	32.4	37.8	40.2	75.5
Iron	3.4	5.8	7.4	14.5
Lamp	17.4	25.5	33.8	55.0
Phone	24.1	30.6	34.4	62.6
Mean	20.7	25.8	29.0	54.7

4.3   Grayscale Images

Grayscale cameras are frequently used for industrial computer vision solutions for various reasons. The most prominent two are 1) smaller data footprint: 3 channels in RGB compared to only one in grayscale and 2) resolution: grayscale images use every pixel on the sensor, while RGB cameras, depending on their color-separation mechanism, may have reduced resolution because each pixel is dedicated for a specific color. Since the presented solution is aimed to ease the inspection planning process in determining good camera positions to inspect an object, it is desirable that it can be used with the camera which should be used in the inspection later on (regardless if it is grayscale or RGB camera). For using in grayscale mode, the RetinaNet and the Unet are trained on grayscale images. We use the already created dataset, however during the Unet training we convert the colored input image to grayscale in runtime. The Unet is made to learn to recreate a colored synthetic model. The rest of the pipeline remains the same.

We notice there is a drop in accuracy of estimated pose for the same object when tested in RGB. Figure 6(a) shows the results when tested in grayscale input.

4.4   Effect of Size of Bottleneck Layer

We tested our approach to select the appropriate size of the bottleneck layer (z). The average error on the train set and well as the test set for the Object Cat in LineMOD dataset was calculated. We started with length of $z = 32$ as the initial size and then moved up till 512. For the test set, the error decreases from 32 to 64 and doesn't show considerable decrease thereafter. Whereas for the train set the error doesn't see any significant drop. While the increase in the size of z can increase accuracy, it means a higher computation cost at runtime. Hence we set our bottleneck value to 64. Figure 7(d) shows the graph. This parameter can be easily configured.

Table 2. Average rotation error Nearest Neighbour vs Regression

Object / Error (degrees)	Ours 16384 clusters	Nearest Neighbour $k = 1$
Ape	11.79	**11.56**
Benchvise	**126.69**	127.41
Camera	9.31	**8.36**
Can	25.52	24.36
Cat	**7.54**	7.33
Driller	26.86	**24.99**
Duck	128.15	**127.78**
Eggbox	**57.32**	57.89
Glue	30.54	**30.31**
Holepuncher	12.72	**12.35**
Iron	**45.04**	45.38
Lamp	11.47	**9.66**
Phone	**10.54**	10.57
Mean	38.73	**38.3**

4.5   Effect of Cluster Numbers on the Accuracy and Performance

The purpose of clustering is to divide the orientation space into similar object views. Figure 7(a) and (b) shows the orientation space of Object 12 in T-LESS dataset divided into 4096 clusters. We rendered the first two values of the first 9 clusters out of 4096 clusters. As can be seen, k-means divides the space into similar views of objects. However, in some clusters the object views rotated by 180° are kept together, suggesting that there are not enough differentiating features for both of the views to separate them. While dividing the orientation space into more clusters creates finer divisions, but the trade-off lies in doing

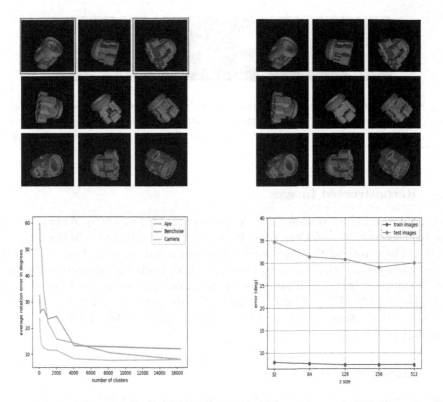

Fig. 7. Figure (a) and (b) show how the mini-batch k-means algorithm divides the orientation space of Object 12 in the T-LESS dataset. Showing the first two images from the first 9 clusters out of 4096 clusters. Each cluster by itself has a unique view of the object, however, within some clusters, the object has been rotated by $180°$ (marked in pink), which suggests that there weren't enough differentiating features to keep them in separate clusters. This behavior can result in incorrect regression of pose for objects with fewer features. (c) Shows the average rotation error on increasing the cluster number on some LineMOD objects. The error sees a fall with an increase in clusters and then saturates. (d) The figure shows how the rotation error varies with the change in z size.

more calculations in runtime to find the nearest cluster. We find the average rotation error on 50 random images of Object 1 (Ape), Object 2 (Benchvise), and Object 3 (Camera) of the LineMOD dataset and plot the error with the number of clusters (see Fig. 7(c)). The error gradually decreases with the cluster size, which as expected, is due to finer divisions of the orientation space. We also compare the average rotation error between the Nearest Neighbour approach (used by [10,21]) and ours. We achieve comparable error and also outperform in some cases in the same time being time efficient. Details are in Table 2.

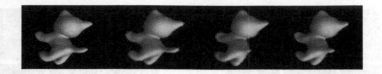

Fig. 8. Shows the reconstruction achieved using different networks used. From left to right, the first image is the target reconstruction, final reconstruction done after using Unet, FPN and Linknet respectively. Significant difference can be seen in the tail of the Cat.

4.6 Reconstructed Images

The final reconstructed object can be seen in Fig. 2(a) and 6(b). As can be seen, the Rectification Network can generate a very close resemblance to the target 3D model. The clarity of the reconstructed images serves as an indicator of a good orientation estimation. A wrong reconstruction leads to incorrect orientation estimation(see Fig. 6(c)). We further see the reconstruction target when the Unet is replaced with FPN [12] and Linknet [2]. We find that Unet has a superior reconstruction compared to the other segmentation models used. This can be seen in Fig. 8.

4.7 Inference Time

For real-time inference, we tested our solution in *only* CPU and recorded the inference time for each component. The details are present in Table 3. The maximum time is taken by Unet and estimating the translation component. As per our understanding, Unet can be replaced with a simpler architecture. This is because, for the industrial scenario the environments are simpler *i.e.*, without any occlusion and clutter unlike LineMOD environment and hence an efficient segmentation can be achieved easily. FPN and Linknet achieve slightly inferior result than Unet. For translation estimation per frame, a synthetic model is rendered at runtime and the bounding box is calculated. This process takes a considerable time (see column *Pose est.* in Table 3). We also tested our cluster-based regression approach to the Nearest Neighbour ($k = 1$) based approach. Our approach stands out 10 times faster. This is because we do not compare the whole codebook of 324000 vectors for each frame to get the right orientation instead regress it from the appropriate decision tree.

Table 3. Time taken per RGB image in LineMOD dataset

Time taken (ms)	Unet	Rec.n/w	Pose est	Avg. time
CPU (12 cores)	35 ms	4 ms	47 ms	86 ms

4.8   Failure Cases

Live Testing on T-LESS Object. We observed more failure cases during live testing of 3D printed objects from the T-LESS dataset compared to testing on LineMOD dataset objects. The reason is majorly because the objects in T-LESS are approximately symmetrical, *e.g.*, Object 5, and 12 as seen in Fig. 2(a and b). Moreover, the absence of enough unique features within the model doesn't produce a clear separation of views during clustering (Fig. 7(a and b)). This results in different identical views present in the same cluster which leads to a wrong estimate. The accuracy also drops with grayscale images.

LineMOD Dataset. Most failed cases in the LineMOD dataset are due to blur, specularity, or occlusion (when tried on the test dataset of LineMOD). Also, small image resolution (640 × 480) makes it difficult to identify unique features. Some of the failed cases can be seen in Fig. 6(c). In all failed cases, the reconstructed model can be seen as drastically dissimilar, which resulted in poor pose estimation.

5   Conclusion

Finding appropriate inspection viewpoints is a crucial part of visual inspection planning systems. We presented a pipeline that can instantly detect the pose of a given object from RGB or grayscale images. This entire pipeline is automated and doesn't require any manual intervention. The approach is flexible and is configurable to work with different inspection objects and environments. The solution avoids the storage of a large data in its memory, which makes it scalable with growing data. Around this approach, we make the pipeline available in the public domain for further testing and future developments.

Acknowledgement. The authors would like to thank Dr. Thomas Weibel for his valuable inputs and the reviewers for their useful comments. We would also like too thank the Department of Image Processing, Fraunhofer ITWM for funding this project.

References

1. Buitinck, L., et al.: API design for machine learning software: experiences from the scikit-learn project. In: ECML PKDD Workshop: Languages for Data Mining and Machine Learning, pp. 108–122 (2013)
2. Chaurasia, A., Culurciello, E.: Linknet: exploiting encoder representations for efficient semantic segmentation (2017)
3. Everingham, M., Winn, J.: The pascal visual object classes challenge 2012 (voc2012) development kit. Pattern Analysis, Statistical Modelling and Computational Learning, Technical Report, vol. 8 (2011)
4. Gospodnetic, P., Mosbach, D., Rauhut, M., Hagen, H.: Flexible surface inspection planning pipeline. In: 2020 6th International Conference on Control, Automation and Robotics (ICCAR), pp. 644–652 (2020)

5. Gospodnetic, P., Rauhut, M., Hagen, H.: Surface inspection planning using 3D visualization. In: LEVIA 2019: Leipzig Symposium on Visualization in Applications (2019)

6. Gronle, M., Osten, W.: View and sensor planning for multi-sensor surface inspection. Surf. Topogr. Metrol. Prop. **4**(2), 024009 (2016)

7. Hinterstoisser, S., et al.: Gradient response maps for real-time detection of textureless objects. IEEE Trans. Pattern Anal. Mach. Intell. **34**(5), 876–888 (2011)

8. Hodaň, T., Haluza, P., Obdržálek, Š., Matas, J., Lourakis, M., Zabulis, X.: T-less: an RGB-D dataset for 6D pose estimation of texture-less objects. In: IEEE Winter Conference on Applications of Computer Vision (WACV), pp. 880–888 (2017)

9. Kehl, W., Manhardt, F., Tombari, F., Ilic, S., Navab, N.: SSD-6D: making RGB-based 3D detection and 6D pose estimation great again. In: Proceedings of the IEEE International Conference on Computer Vision (ICCV), pp. 1521–1529 (2017)

10. Kehl, W., Milletari, F., Tombari, F., Ilic, S., Navab, N.: Deep learning of local RGB-D patches for 3D object detection and 6D pose estimation. In: Leibe, B., Matas, J., Sebe, N., Welling, M. (eds.) ECCV 2016. LNCS, vol. 9907, pp. 205–220. Springer, Cham (2016). https://doi.org/10.1007/978-3-319-46487-9_13

11. Kendall, A., Grimes, M., Cipolla, R.: Posenet: a convolutional network for real-time 6-DoF camera relocalization. In: IEEE International Conference on Computer Vision (ICCV) (2015)

12. Kirillov, A., Girshick, R., He, K., Dollár, P.: Panoptic feature pyramid networks. In: IEEE Conference on Computer Vision and Pattern Recognition (CVPR), pp. 6399–6408 (2019)

13. Lin, T., Goyal, P., Girshick, R., He, K., Dollár, P.: Focal loss for dense object detection. In: IEEE International Conference on Computer Vision (ICCV), pp. 2980–2988 (2017)

14. Loh, W.Y.: Classification and regression trees. Wiley Interdisc. Rev. Data Min. Knowl. Discovery **1**(1), 14–23 (2011)

15. Mohammadikaji, M., Bergmann, S., Irgenfried, S., Beyerer, J., Dachsbacher, C., Worn, H.: Inspection planning for optimized coverage of geometrically complex surfaces. In: 2018 Workshop on Metrology for Industry 4.0 and IoT, pp. 52–67 (2018)

16. Pagani, A.: Modeling reality for camera registration in augmented reality applications. KI-Künstliche Intelligenz **4**(28), 321–324 (2014)

17. Rad, M., Lepetit, V.: Bb8: a scalable, accurate, robust to partial occlusion method for predicting the 3D poses of challenging objects without using depth. In: IEEE International Conference on Computer Vision (ICCV), pp. 3828–3836 (2017)

18. Rambach, J., Pagani, A., Schneider, M., Artemenko, O., Stricker, D.: 6DoF object tracking based on 3D scans for augmented reality remote live support. Computers **7**(1), 6 (2018)

19. Redmon, J., Divvala, S., Girshick, R., Farhadi, A.: You only look once: unified, real-time object detection. In: IEEE Conference on Computer Vision and Pattern Recognition (CVPR) (2016)

20. Ronneberger, O., Fischer, P., Brox, T.: U-Net: convolutional networks for biomedical image segmentation. In: Navab, N., Hornegger, J., Wells, W.M., Frangi, A.F. (eds.) MICCAI 2015. LNCS, vol. 9351, pp. 234–241. Springer, Cham (2015). https://doi.org/10.1007/978-3-319-24574-4_28

21. Sundermeyer, M., Marton, Z., Durner, M., Brucker, M., Triebel, R.: Implicit 3D orientation learning for 6D object detection from RGB images. In: European Conference on Computer Vision (ECCV) (2018)

22. Szegedy, C., Ioffe, S., Vanhoucke, V., Alemi, A.A.: Inception-v4, inception-resnet and the impact of residual connections on learning. In: Thirty-First AAAI Conference on Artificial Intelligence (2017)

23. Tekin, B., Sinha, S.N., Fua, P.: Real-time seamless single shot 6D object pose prediction. In: IEEE Conference on Computer Vision and Pattern Recognition (CVPR), pp. 292–301 (2018)

24. Wang, C., et al.: Densefusion: 6D object pose estimation by iterative dense fusion. In: Proceedings of the IEEE Conference on Computer Vision and Pattern Recognition (CVPR), pp. 3343–3352 (2019)

25. Xiang, Y., Schmidt, T., Narayanan, V., Fox, D.: Posecnn: a convolutional neural network for 6D object pose estimation in cluttered scenes. In: Robotics: Science and Systems (RSS) (2018)

Localisation of Defects in Volumetric Computed Tomography Scans of Valuable Wood Logs

Davide Boscaini[1]([⊠])[iD], Fabio Poiesi[1][iD], Stefano Messelodi[1][iD], Ayman Younes[2], and Donato A. Grande[2]

[1] Fondazione Bruno Kessler, Povo, TN, Italy
{dboscaini,poiesi,messelod}@fbk.eu
[2] Meccanica del Sarca S.p.A., Dro, TN, Italy
{a.younes,d.grande}@sarca.it

Abstract. We present a novel pipeline to efficiently localise defects in volumetric Computed Tomography (CT) scans of valuable wood logs. We couple a 2D detector applied independently on each scan slice with a multi-object tracking approach processing detections along the scan direction to localise the defects in 3D. Our solution is designed to meet the real-time requirements of modern production lines, to optimise the wood sawing operations for high-quality final products and to reduce wood waste as well as carbon footprints. We effectively embedded our defect localisation algorithm in the Meccanica del Sarca S.p.A.'s production pipeline achieving a reduction of their economic loss by 7% compared to the previous years.

1 Introduction

Defects in precious wood logs are one of the most important cause of big economic loss in wood industries, producing also unnecessary carbon footprints. Meccanica del Sarca S.p.A., an Italian company that deals with the mechanical process of precious wood logs, estimated that they lost about 405K Euro in 2017, 430 K Euro in 2018, and 424 K Euro in 2019 due to wood waste. Usually, the log production process starts with an operator that inspects the external surface of a log searching for, or predicting, internal defects. On the one hand, the inspection of the log surface may not provide enough evidence to suggest the presence of internal defects. On the other hand, a visible defect on the log surface may suggest to discard the whole wood log despite being undamaged internally. The two main factors that lead to the economic loss are the incorrect classification of defected logs (i.e. material waste) and the late identification of defects during the production process (i.e. time waste). Therefore, the need for a automated system to accurately detect defects in the early stage of the log production process is key. If a defect is detected in the early production stage, it is possible to optimally plan the sawing operations in order to avoid defects and to minimise waste.

Wood log interiors can also be analysed using Computed Tomography (CT) [7] (Fig. 1). CT produces a set of image slices that are typically used by

© Springer Nature Switzerland AG 2021
A. Del Bimbo et al. (Eds.): ICPR 2020 Workshops, LNCS 12664, pp. 692–704, 2021.
https://doi.org/10.1007/978-3-030-68799-1_50

Old pipeline:

| Raw wood log | → | Drying and Shaving | → | Defect detection (1st round) | → | Mechanical processing by chip removal | → | Carved wood log | → | Defect detection (2nd round) |

New pipeline:

| Raw wood log | → | Drying and Shaving | → | Computed Tomography | → | Automatic defect detection | → | Mechanical processing by chip removal | → | Carved wood log |

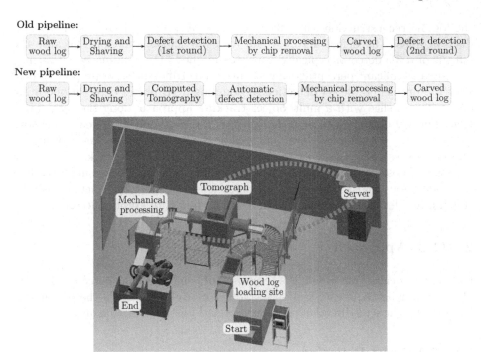

Fig. 1. Current wood log processing pipeline. Top row: previous Meccanica del Sarca S.p.A.'s pipeline where defects were found manually by experts that visually inspected the external surface of the wood logs before and after their mechanical processing. Bottom row: current wood processing pipeline where internal parts of the logs are acquired through Computed Tomography and the defect detection is performed automatically using our deep-learning based algorithm. Data flow is represented by blue arrows. (Color figure online)

operators to verify the presence of defects. This job is laborious and in certain circumstances not well defined. There is wide scope for subjective interpretation of what a defect is, which may lead to annotation mistakes. Elements of ambiguity may be related to the management of adjacent defects. There can be regions containing several small and spread defects, that for some experts it could be regarded as a single defect. The concept of proximity, considering the difficulty in defining defect boundaries, can be subjective.

The problem of wood defect detection has already been tackled in literature [1,3,11,13–15]. Proposed methods used algorithms based on Artificial Neural Networks [11], edge-based image analysis [1], Support Vector Machine classification applied to colour features [3] or, more recently, based on deep-learning, i.e. Faster-RCNN [13]. Because the CT scan of a wood log produces its 3D reconstruction, 3D detection methods could in principle be used to detect its internal defects [4]. However, we decided not to use 3D object detectors for two reasons. The operator would need to wait the end of the CT scan to examine the log 3D reconstruction, thus affecting the real-time requirements of the production lines. The defect may have an irregular shape, therefore, because the output of a 3D detector typically is an axis-aligned 3D bounding box, this may include a significant percentage of good material. Differently, we propose to deal with the problem of defect detection by coupling a 2D object detector operating on each scan slice with a multi-object tracking approach along the scan direction. The 2D object detector is modelled as a end-to-end deep neural network that takes the CT slices as input and generates the bounding boxes coordinates and categories of likely defects as output. The multi-object tracking acts as a refinement post-processing to linearly interpolate bounding boxes in the case of miss-detections and to prune outlier detections. We perform tracking using a tracking-by-detection algorithm formulated as bipartite graph matching.

2   Our Approach

Our defect localisation pipeline is composed of a module that detects defects as 2D bounding boxes for each scan slice, and of a module that tracks these bounding boxes along the scan direction using current and past information only.

2.1   2D Defect Detection

We base our detection module on the Single Shot MultiBox Detector (SSD) that was proposed for object detection in RGB images [8]. Figure 2 depicts a block diagram of the proposed architecture. We denote our model with Φ_Θ, where Θ is the collection of its learnable parameters. Φ_Θ is composed of three main modules: the feature extraction core, the classification head and the regression head. The main difference between Φ_Θ and the original SSD model [8] is in the classification and regression heads: we use less skip connections (two instead of six) and we place them at shallower levels (54×54 feature map locations instead of 38×38, and 27×27 instead of 19×19, respectively). These modifications are motivated by the fact that the defects we are aiming to detect are, on average, much smaller than the typical object size the original SSD model was developed for. Anticipating the skip connections allow us to extract feature maps at a higher resolution grid.

Ψ_Ω is the feature extraction module with Ω its learnable parameters. Ψ_Ω takes a mini-batch of b CT images of size 440×440 in input and outputs features at two depth levels, $\mathbf{h}^1 = \Psi^1_{\Omega^1}(\mathbf{x})$ of size $(b, 512, 54, 54)$ and $\mathbf{h}^2 = \Psi^2_{\Omega^2}(\mathbf{h}^1)$ of size $(b, 1024, 27, 27)$, where $\Psi_\Omega = \Psi^2_{\Omega^2} \circ \Psi^1_{\Omega^1}$. The precision of the input CT images is set to 16-bit in order not to lose possibly relevant information. The early layers of Ψ_Ω implements a VGG-16 backbone [12] truncated before the classification layer and endowed with Batch Normalization [5], and are followed by three custom layers: MaxPool2d, Conv2d(512, 1024) + ReLU, and Conv2d(1024, 1024) + ReLU.

We associate a default bounding boxes of different aspect ratios, called anchors, to each feature map cell of $\mathbf{h}^1, \mathbf{h}^2$. For each anchor the classification

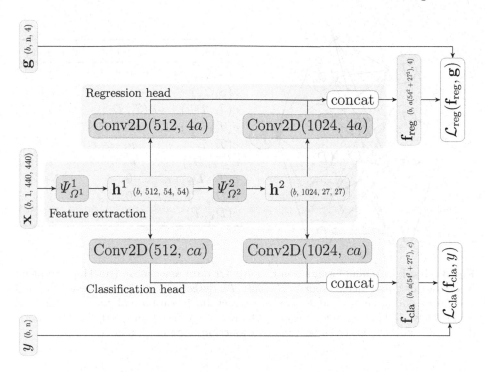

Fig. 2. Block diagram of our SSD model applied to CT scans. Colour key. Grey: input/output tensors. Blue: parametric layer. White: non-parametric layer. (Color figure online)

head predicts c scores that indicates the presence of a class instance inside it. The regression head predicts the offsets to apply to the 4 anchor coordinates to fit the ground-truth bounding box.

A classification loss $\mathcal{L}_{\mathrm{cla}}$ measures the consensus between the output of the classification head $\mathbf{f}_{\mathrm{cla}}$ and the ground-truth class y. Similarly, a regression loss $\mathcal{L}_{\mathrm{reg}}$ measures the error between the output of the regression head $\mathbf{f}_{\mathrm{reg}}$ and the offsets between the anchors and the ground-truth bounding boxes $\mathbf{g}$. Without loss of generality, we configure our detector for binary classification: a bounding-box can contain either a defect or not. We use the cross entropy loss as classification loss,

$$\mathcal{L}_{\mathrm{cla}}(\mathbf{f}_{\mathrm{cla}}, k) = -\log \frac{\exp \mathbf{f}_{\mathrm{cla}}(k)}{\sum_k \exp \mathbf{f}_{\mathrm{cla}}(k)},$$

and the smooth L1 loss, also known as Huber loss [2], as regression loss,

$$\mathcal{L}_{\mathrm{reg}}(\mathbf{f}_{\mathrm{reg}}, \mathbf{g}) = \mathrm{SmoothL1}(\mathbf{f}_{\mathrm{reg}}, \mathbf{g}).$$

The Huber loss is defined as a squared L2 norm if the absolute error falls below 1 and as an L1 norm otherwise. During training we optimise a linear combination of the two losses, i.e. $\mathcal{L} = \mathcal{L}_{\mathrm{cla}} + \lambda \mathcal{L}_{\mathrm{reg}}$. In our experiments we set $\lambda = 0.1$ by cross-validation.

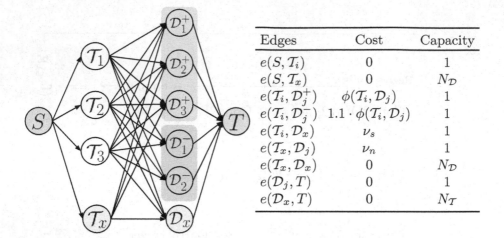

Edges	Cost	Capacity
$e(S, \mathcal{T}_i)$	0	1
$e(S, \mathcal{T}_x)$	0	$N_{\mathcal{D}}$
$e(\mathcal{T}_i, \mathcal{D}_j^+)$	$\phi(\mathcal{T}_i, \mathcal{D}_j)$	1
$e(\mathcal{T}_i, \mathcal{D}_j^-)$	$1.1 \cdot \phi(\mathcal{T}_i, \mathcal{D}_j)$	1
$e(\mathcal{T}_i, \mathcal{D}_x)$	ν_s	1
$e(\mathcal{T}_x, \mathcal{D}_j)$	ν_n	1
$e(\mathcal{T}_x, \mathcal{D}_x)$	0	$N_{\mathcal{D}}$
$e(\mathcal{D}_j, T)$	0	1
$e(\mathcal{D}_x, T)$	0	$N_{\mathcal{T}}$

Fig. 3. Graph matching formulation to solve our data association (tracking) problem. Nodes in this graph include source (S), sink (T), detections ($\mathcal{D}_j$), track states ($\mathcal{T}_i$), and two proxy nodes, i.e. $\mathcal{T}_x$ and $\mathcal{D}_x$, to allow track initialisation and track termination, respectively. $\mathcal{D}_j^+$ is a strong detection, $\mathcal{D}_j^-$ is a weak detections and $\mathcal{D}_j$ is a generic detection. $\phi(\mathcal{T}_i, \mathcal{D}_j)$ is the association cost function between $\mathcal{T}_i$ and $\mathcal{D}_j$.

2.2 Multi-defect Tracking

We filter out and associate detections across slices using online tracking. Given a set of $N_{\mathcal{D}}$ detections computed on the current slice we aim to associate them to the set of $N_{\mathcal{T}}$ tracks that are computed on the previous slice. Let $\mathcal{T}_i$ be the ith track and $\mathcal{D}_j$ be the jth detection. We formulate the tracking problem as a bipartite graph matching and solve it using Minimum Cost Flow [9]. We define our graph as $\mathcal{G} = (N, E)$, where N represents the set of nodes and E the set of edges. Tracking states and detections are the nodes of $\mathcal{G}$. Nodes in this graph include source (S), sink (T), detections ($\mathcal{D}_j$), track states ($\mathcal{T}_i$), and two proxy nodes, i.e. $\mathcal{T}_x$ and $\mathcal{D}_x$, to allow track initialisation and track termination, respectively. Specifically, our tracking algorithm initialises, associates and terminates tracks based on costs. Initialisation and termination costs are hyper-parameters set by us. The association cost is a function that depends on the position and on the bounding-box size between a detection and its predicted track state. Let $\phi(\mathcal{T}_i, \mathcal{D}_j)$ be the association cost function between $\mathcal{T}_i$ and $\mathcal{D}_j$. $\phi(\cdot)$ is a linear combination of distance between the centres, widths and heights of last bounding box of $\mathcal{T}_i$ and the bounding box of $\mathcal{D}_j$. We denote $e(n_1, n_2)$ as the edge between node n_1 and node n_2. Each edge is characterised by the cost and a capacity. Figure 3 shows the graph formulation of our tracking algorithm.

The solution of this graph is a one-to-one association between track states and detections. In our graph formulation we embed the weak and strong detection tracking model proposed in [10]. Strong detections are detections with confidence above the threshold δ^+, weak detections are detections with confidence between

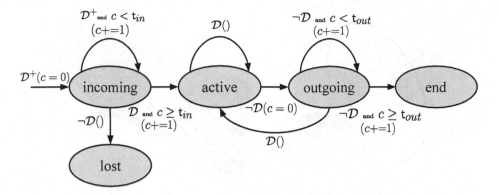

Fig. 4. Track state management. Incoming, lost, active, outgoing and end are the possible states of each track. A new track always starts in the incoming state. Key. input: $\mathcal{D}$ ($\neg\mathcal{D}$) detection is (not) associated to the track. Edge label: input [**and** condition] (action). c: counter.

the threshold δ^- and δ^+. These thresholds should be set such that $\delta^+ > \delta^-$. Strong detections are used for track initialisation and for tracking, whereas weak detections are used for tracking only. Let $\mathcal{D}_j^+$ be a strong detection, $\mathcal{D}_j^-$ be a weak detection and $\mathcal{D}_j$ be a generic detection. For each slice processed by the detector we solve the graph matching problem. For each track we build a state machine to manage its evolution. Figure 4 shows our state machine. An unassociated strong detection triggers the initialisation of a new track, which begins from the incoming state. If this track is successfully associated to a number of t_{in} consecutive strong detections, it goes in the active state. Subsequent associations in the active state use both strong and weak detections. In the case of a miss-detection, i.e. an unsuccessful association, the track goes in the outgoing state. If the association fails for more than t_{out} consecutive slices the track ends. Otherwise if the track is associated to a new detection in subsequent slices before reaching t_{out}, the track state goes back to the active state and the slices where miss-detections occurred are filled with linearly interpolated bounding boxes. The linear interpolation uses the information from the associated bounding boxes.

3   Experimental Results

3.1   Data Acquisition and Normalization

CT scans of the wood logs were captured at the Meccanica del Sarca S.p.A.'s premises using a MiTO tomograph developed by Microtec s.r.l., the world leading wood scanning solutions provider. During the CT acquisition a wood log is transported by a conveyor belt moving at a constant velocity. An X-ray scanner moves with a spiral pattern opposite to the conveyor belt direction to scan the log (Fig. 5, left). A spin of the X-ray scanner produces a 2D slice of the 3D wood

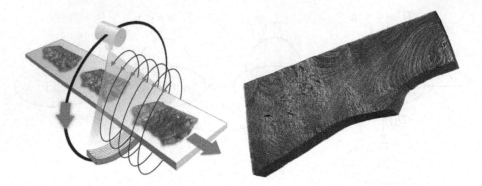

Fig. 5. Computed Tomography (CT) acquisition process and a wood log 3D reconstruction from its CT scans.

log. Multiple 2D image slices are captured while the logs moves on the conveyor belt. The collection of all slices form the 3D structure of the wood log (Fig. 5, right). The tomograph acquisition accuracy is set to 0.5 mm for both the 2D and 3D scan directions. This means that a voxel of the 3D reconstruction of the wood log corresponds to a cube of material of size 0.5 mm.

Slices are stored as signed 16-bit single-channel 2D images. Each pixel of the image contains an intensity value that is expected to be correlated with the material density. However, Fig. 6 shows that in practice the histogram of the intensities of the pixels of a CT scan is bimodal. The first mode corresponds to air (intensity approx. 0) plus noise related to acquisition artifacts. The second mode corresponds to the actual wood material. To focus our analysis on the wood material we normalise each image between 200 and 1200.

3.2   Datasets

We collected a dataset of 175 CT scans of wood logs of various sizes. Each scan contains between 412 to 1144 slices, for a total of 149,237 CT images. Each CT scan is composed of about 1000 slices, where each slice may contain up to tens of defects. We split the data in training and testing sets with proportions of 85% and 15%, respectively. The training set contains 150 CT scans, i.e. 127,954 CT images. The test set contains 25 CT scans, i.e. 21,283 CT images.

Experts from Meccanica del Sarca S.p.A. carefully annotated the wood defects by drawing axis-aligned bounding boxes for each slice. The annotation process was particularly difficult, time consuming and prone to errors. The difficulty was due to the fact that experts were used to identify defects by inspecting the surface of the wood log. They had to acquire some experience to understand how to identify defects from the CT slices. To speed-up this annotation process, instead of asking the experts to annotate every slice we proposed to annotate every 5–10 slices and then to linearly interpolate the missing bounding boxes. We discovered that there is wide scope for subjective interpretation of what a

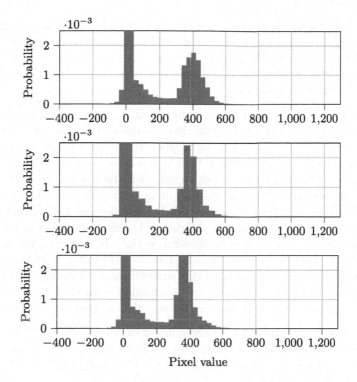

Fig. 6. Histograms of three CT scans that show the distribution of intensity values. The left-hand side mode represents empty regions (i.e. air), while the right-hand side mode represent the region with material.

defect is, which often led to annotation mistakes. To reduce these mistakes we had several iterations with the experts from Meccanica del Sarca S.p.A. to review and correct the annotated bounding boxes. We experimentally confirmed that accurate annotations lead to a considerable improvement in the performance of our deep-learning based defect detection algorithm.

3.3 Experimental Setup

Training. We train our detector for 200 epochs using the Stochastic Gradient Descent (SGD) as optimiser. We set SGD to an initial learning rate of 1e − 2. We decrease the learning rate by a factor of 0.75 every 20 epochs. We use a weight decay with a factor of 1e − 03 for regularisation. We train on an NVidia GeForce GTX 1080 with 8 GB RAM.

Inference. At inference time we apply Non-Maximum Suppression (NMS) [8]. NMS filters out bounding-box predictions that have a confidence lower than t_{conf} and that have an overlap bigger than t_{NMS}. We choose $t_{\mathrm{conf}} = 0.5$ (or 0.05) and $t_{\mathrm{NMS}} = 0.5$ in our experiments.

Pretraining. We use a pretrained version of VGG-16 layers on the ImageNet dataset for the feature extraction module. The other layers are trained from scratch. Although ImageNet is a RGB-image dataset, which is quite different from our CT scan dataset, we found that this pretraining enables us to achieve better performances in fewer epochs.

Data Augmentation. We use data augmentation to avoid overfitting on the training set. Specifically, we used (i) horizontal and vertical flipping of the image, each with a probability of 0.5 to be applied to the input image, (ii) random cropping also with probability set to 0.5, followed by a resizing of the crop region to the resolution of the input image, (iii) data normalisation using the mean μ and standard deviation σ of training data, i.e. $\mu \approx 0.074, \sigma \approx 0.012$. As far as random cropping is concerned, let h, w be the height and width of the input image, and $\bar{h}, \bar{w}$ be the height and width of the crop region, and $\bar{y}, \bar{x}$ be the top-left coordinates of the crop region. In our experiments, we set $\bar{h} \sim \mathcal{U}(0.3h, h), \bar{w} \sim \mathcal{U}(0.3w, w), \tilde{y} \sim \mathcal{U}(100, h - \bar{h})$, and $\tilde{x} \sim \mathcal{U}(50, w - \bar{w})$. If the cropped region has an aspect ratio smaller than 0.5 or bigger than 2 the hyper-parameters described above are re-sampled until this condition is met. The bounding boxes whose centres are outside the cropped region are discarded.

Detector Configuration. We set the number of anchors to $a = 9$. The anchors cover different aspect ratios, obtained by multiplying the width and height of the feature map cell by the following factors: $(1, 1)$, $(2, 2)$, $(3, 3)$, $(2, 1)$, $(1, 2)$, $(4, 2)$, $(2, 4)$, $(6, 3)$, and $(3, 6)$.

Tracking Configuration. The tracking hyper-parameters are set through cross-validation using the validation set: $t_{in} = 0$, $t_{out} = 3$, $\delta^+ = 0.5$, $\delta^- = 0.05$.

3.4   Analysis of the Results

We quantify the localisation performance using Precision, Recall and F1 Score [6]. We assess the quality of our pipeline with and without the tracking module activated, and by performing an ablation study on the key parameters.

Table 1 shows the results obtained using $t_{eval} = 0.25$, i.e. a bounding box is considered correctly estimated if its overlap with the ground-truth bounding box is greater than $1/4$ of its area. The first row of Table 1 shows the detection performance using bounding boxes predicted with a confidence greater than 0.5. Although the results show that we can achieve a high precision with this configuration (Prec. $\approx 90\%$), the number of missed defects is high (Rec $\approx 55\%$). This results in a F1-score of 68.4. The second row of Table 1 shows the detection results when the confidence threshold is lowered to 0.05. As expected, this affects the precision (Prec $\approx 40\%$), whereas the recall score increases to $\approx 80\%$. Figure 7 shows some examples of detection results. The latter is a suitable case to postprocess with tracking given its ability in filtering out false positives and to interpolate miss-detections. The third row of Table 1 shows how tracking effectively improves Precision, despite worsening Recall. We empirically observed that there are several situations where detections with a confidence above δ^+

Table 1. Quantitative comparison results in terms of True Positives (TP), False Negatives (FN), False Positives (FP), Precision (Prec.), Recall (Rec.) and F1-score (F1).

Method	TP	FN	FP	Prec	Rec	F1
SSD ($t_{conf} = 0.5$)	2475	1971	312	88.8	55.7	68.4
SSD ($t_{conf} = 0.05$)	3724	722	5832	39.0	**83.8**	53.2
Our ($t_{conf} = 0.05$)	3203	1243	1449	**68.9**	72.0	**70.4**

Table 2. Ablation on the data augmentation. Hyper-parameters used: $\lambda = 1e - 01$, learning rate $= 1e - 03$, weight decay $= 1e - 04$, $t_{conf} = 0.5$, $t_{eval} = 0.5$. Results reported are best values achieved on the test set.

Data augm.	TP	FN	FP	Prec	Rec	F1
data norm.	2075	2371	986	67.8	46.7	55.3
data norm. + crop	2018	2428	514	**79.7**	45.4	57.8
data norm. + crop + flip	2291	2155	999	69.6	**51.5**	**59.2**

are not enough consistent over consecutive slices to become tracks, thus deemed false positives and filtered out by tracking. Figure 8 shows examples of tracking results where false-positive detections are prunned and miss-detections are corrected.

Table 2 reports the ablation study on different data transformation configurations that we used for the training of our detector. We obtained the best performance in terms of F1-score by combining data normalisation with random cropping and image flipping. We found that data augmentation is key to avoid overfitting. Table 3 reports the ablation study using different tracking parameters. We can observe that a high value of the threshold to decide between strong

Table 3. Ablation on the tracking hyper-parameters. First row show the upper bound we can reach in recall.

t_{in}	t_{out}	δ^+	δ^-	TP	FN	FP	Prec	Rec	F1
0	3	0.05	0.05	3792	654	7142	34.7	85.3	49.3
0	3	0.1	0.05	3700	746	4608	44.5	83.2	58.0
0	5	0.1	0.05	3721	725	5029	42.5	83.7	56.4
1	5	0.1	0.05	3692	754	4303	46.2	82.0	59.1
0	3	0.4	0.05	3328	1118	1820	64.7	74.9	69.4
0	3	0.5	0.05	3203	1243	1449	68.9	72.0	**70.4**
1	5	0.5	0.05	3202	1244	1453	68.8	72.0	**70.4**
0	3	0.6	0.05	3020	1426	1260	70.6	67.9	69.2
0	3	0.7	0.05	2932	1514	1061	73.4	66.0	69.5

Fig. 7. Examples of detection results. Our detector can handle defects at different scales. Some detected regions are difficult to judge whether they are a defected, e.g. second row - second and third figure. There are then extreme cases where regions at the border of the log are detected as defects. Bounding-box key. Green: ground-truth. Red: estimated detection.

Fig. 8. Examples of tracking results. The first and second rows show how tracking can effectively filter out spurious detections. The last row shows the ability of tracking in interpolating miss-detections (top-left bounding boxes). Bounding-box key. Green: ground-truth. Red: estimated detection. Blue: estimated tracking state. (Color figure online)

and weak detections have a positive effect on the Precision, but that it affects the Recall. Based on the application at hand, a user can tune these parameters to increase or decrease the sensitivity of the approach in order achieve the desired output quality.

4 Conclusions

We presented a deep-learning based algorithm to localise defects in volumetric Computed Tomography scanned wood logs. We showed how to perform 3D defect localisation via 2D object detection and multi-object tracking in order to meet the real-time requirements of the production line. We trained our models on annotations made by experts that deal with the wood production industry. Although the annotation process seems straightforward for deep-learning engineers, we experienced several challenges in instructing the experts. Annotation accuracy is critical to deploy reliable data-driven algorithms on a real production lines. We experimentally showed that we achieved promising localisation performance, which especially contributed to reduce the company's economic loss by 7% compared to the previous years. This experience helped us understand that greater effort must be put into the creation of intuitive mechanisms for data annotation and into comprehensive protocols to localise well-defined defects.

Acknowledgments. This work has been developed within a collaboration between FBK and Meccanica del Sarca S.p.A. and funded by the "Programma operativo FESR 2014–2020 della Provincia di Trento".

References

1. Bhandarkar, S., Luo, X., Daniels, R., Tollner, E.: Detection of cracks in computer tomography images of logs. Pattern Recogn. Lett. **26**, 2282–2294 (2005)
2. Girshick, R.: Fast R-CNN. In: Proceedings of the IEEE International Conference on Computer Vision (ICCV) (2015)
3. Gu, I.Y.H., Andersson, H., Vicen, R.: Wood defect classification based on image analysis and support vector machines. Wood Sci. Technol. **44**, 693–704 (2010). https://doi.org/10.1007/s00226-009-0287-9
4. Gwak, J., Choy, C., Savarese, S.: Generative Sparse Detection Networks for 3D Single-shot Object Detection. arXiv:2006.12356, June 2020
5. Ioffe, S., Szegedy, C.: Batch normalization: accelerating deep network training by reducing internal covariate shift. In: Proeedings of the International Conference on Machine Learning (ICML) (2015)
6. ISO 5725-1: Accuracy (trueness and precision) of measurement methods and results, Part 1: General principles and definitions, International Organization for Standardization (1994)
7. Kak, A.C., Slaney, M.: Principles of computerized tomographic imaging. IEEE Press, the Institute of Electrical and Electronics Engineers Inc., New York (1999)
8. Liu, W., et al.: SSD: single shot MultiBox detector. In: Leibe, B., Matas, J., Sebe, N., Welling, M. (eds.) ECCV 2016, Part I. LNCS, vol. 9905, pp. 21–37. Springer, Cham (2016). https://doi.org/10.1007/978-3-319-46448-0_2

9. Mehlhorn, K.: Algorithms on Graphs. Data Structures and Algorithms. Graph Algorithms and NP-Completeness. Monographs in Theoretical Computer Science. An EATCS Series, vol. 2. Springer, Heidelberg (1984). https://doi.org/10.1007/978-3-642-69897-2

10. Sanchez-Matilla, R., Poiesi, F., Cavallaro, A.: Online multi-target tracking with strong and weak detections. In: Hua, G., Jégou, H. (eds.) ECCV 2016, Part II. LNCS, vol. 9914, pp. 84–99. Springer, Cham (2016). https://doi.org/10.1007/978-3-319-48881-3_7

11. Sarigul, E., Abbott, A., Schmoldt, D.: Progress in analysis of computed tomography (CT) images of hardwood logs for defect detection. In: Proceedings of the Tenth International Conference on Scanning Technology and Process Optimization in the Wood Industry (2003)

12. Simonyan, K., Zisserman, A.: Very deep convolutional networks for large-scale image recognition. In: Proceedings International Conference on Learning Representations (ICLR) (2015)

13. Urbonas, A., Raudonis, V., Maskeliunas, R., Damasevicius, R.: Automated identification of wood veneer surface defects using faster region-based convolutional neural network with data augmentation and transfer learning. Appl. Sci. **22**, 4898 (2019)

14. Yuhan, Q., Zhou, Y., Xu, J., Ge, Z.: Development of a wood computed tomography imaging system using a butterworth filtered back-projection algorithm. For. Prod. J. **68**, 147–156 (2018)

15. Zhao, P., Wang, C.K.: Hardwood Species Classification with Hyperspectral Microscopic Images. J. Spectro. (2019)

Image Anomaly Detection by Aggregating Deep Pyramidal Representations

Pankaj Mishra[ID], Claudio Piciarelli[(✉)][ID], and Gian Luca Foresti[ID]

Università degli Studi di Udine, via delle Scienze 206, 33100 Udine, Italy
mishra.pankaj@spes.uniud.it,
{claudio.piciarelli,gianluca.foresti}@uniud.it

Abstract. Anomaly detection consists in identifying, within a dataset, those samples that significantly differ from the majority of the data, representing the normal class. It has many practical applications, e.g. ranging from defective product detection in industrial systems to medical imaging. This paper focuses on image anomaly detection using a deep neural network with multiple pyramid levels to analyze the image features at different scales. We propose a network based on encoding-decoding scheme, using a standard convolutional autoencoders, trained on normal data only in order to build a model of normality. Anomalies can be detected by the inability of the network to reconstruct its input. Experimental results show a good accuracy on MNIST, FMNIST and the recent MVTec Anomaly Detection dataset.

1 Introduction

Anomaly detection is an application-driven problem, where the task is to identify the novelty of samples which exhibit significantly different characteristics with respect to an predefined notion of normal class. A system which can perform such task autonomously is highly in demand and its applications range from video surveillance to defect detection, medical imaging, financial transactions etc.

Only recently the topic of image anomaly detection has been investigated in the field of deep learning. The vast majority of the proposed methods rely on some for of encoding-decoding scheme, e.g. by using autoencoders, in order to train a network to reconstruct normal data [5]. The assumption is that the network is unable to correctly reconstruct anomalous images, which can be identified by direct comparison of the original and reconstructed image. However, current methods generally do not address the problem at different scales. Moreover, the comparison is often based on trivial pixel-by-pixel comparison, which is not necessarily the best approach to evaluate image similarity [3,14,24,30,35]. Finally, many papers are evaluated on toy datasets only, e.g. MNIST, which are not explicitly studied for anomaly detection problems.

This work was partially funded by Beantech srl.

A. Del Bimbo et al. (Eds.): ICPR 2020 Workshops, LNCS 12664, pp. 705–718, 2021.
https://doi.org/10.1007/978-3-030-68799-1_51

In order to address these aspects, we propose a reconstruction-based pyramidal network, which uses deep autoencoders for anomaly detection. The idea is to use several parallel autoencoders with different scaling factors in order to catch features at different scale levels. To the best of our knowledge, our work is the first advocating for such multi-level design for anomaly detection. Moreover, we adopted a more sophisticated anomaly score which performs better than vanilla MSE loss adopted in many works. Finally, we tested the proposed method on the MVTec dataset [7], which is explicitly studied for anomaly detection systems.

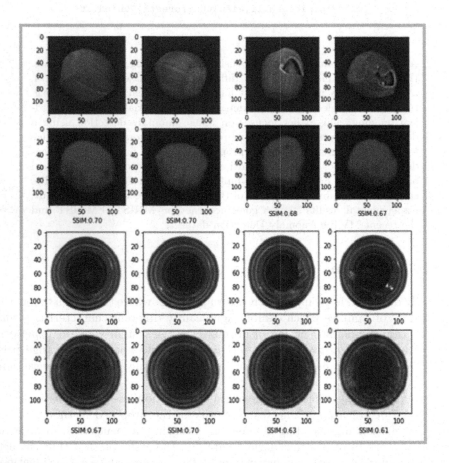

Fig. 1. Examples from the MVTec dataset. The first two columns show the original and reconstructed images for normal objects (hazelnuts and glass bottles). The last two columns show the same results for anomalous images (broken hazelnuts, defective bottles).

2   Related Work

Anomaly detection has been studied in many practical application fields, such as industrial inspection of manufactured products [15], detection of anomalous network activity in intrusion detection systems [2], medical image analysis for tumor detection [5], structural integrity check in hazardous or inaccessible environments [25], traffic analysis [26], fault-prevention in industrial sensing systems [13].

Anomaly detection can be addressed as a standard supervised binary classification problem, however in this case extreme dataset imbalance must be explicitly addressed, as in [27], because the amount of anomalous samples is typically very limited in real-world scenarios. In [19] the authors propose a transfer learning strategy to deal with class imbalance.

Most of the proposed works adopt a semi-supervised strategy: they learn a model for the normal class and try to compute a dissimilarity measure to identify the anomalies [3,11,12,33]. This approach is semi-supervised since it requires a labeled training set consisting of normal data only, although is often improperly described in literature as unsupervised.

Some of recent proposed models exploited and relied on either parametric [9,29] or non parametric [1,4] approaches for density estimations of latent space for anomaly detection. Parametric models, majorly traditional machine learning techniques, uses the Gaussian density estimation techniques and Gaussian Mixture Models are recent trends with deep learning methods [32]. However, remembering an event or an instance of a particular class implies adoption of dominant features at latent space, either by a dictionary of normal prototype - as commonly the adopted methods of sparse coding approaches [10] - or by remembering the features space as in the graph based techniques [18] or the most recently used deep autoencoders [6,8,17,21]. All these methods tries to estimate the remembered latent features of normal class, uses either of the density estimation approaches by minimising the log likelihood loss and expects higher log likelihood loss for an anomalous sample. However, deep autoencoders have limited capacity and are not able to capture the causal factors that generates an image and are relevant to anomaly detection job. And this limits such methods.

In recent times number of works have also been done using the Generative Adversarial Network (GAN), but none of them were directly developed for anomaly detection on images. Usually, they use MSE (Mean Squared Error) or L1-norm between the pixels of ground truth and the reconstructed images, which is not how humans perceive the similarity between two images. Another problem with GAN based techniques is to find the latent vector that recovers the input image after passing through the generator [3]. Moreover, the GAN base procedures are time consuming or consist of complex multi-step training steps or are often not stable.

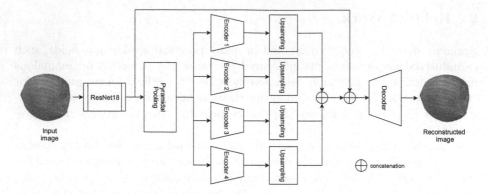

Fig. 2. Proposed network architecture. The network consists of the first levels of a Resnet18 network for feature extraction, followed by a pyramidal pooling layer that feeds 4 encoders, each one connected to an upsampling layer with shared weights, and a final decoder. The features obtained form the four upsampling layers are concatenated with the output features of resnet18.

3   Proposed Model

Following a global trend in deep anomaly detection, we propose a reconstruction-based approach. The basic idea is to find a low-dimension feature representation of the input image that captures its fundamental properties (the *causal factors*, as named in some works) from which the image can be reconstructed. The network thus has an encoder-decoder structure, as in standard autoencoders, which ideally models an identify function, but passing through a dimensionality-reduction bottleneck after the encoding part. The main idea is that the network, when trained on normal data, learns a mapping from input space to the low-dimensional latent space which is suitable only for normal data. If anomalous data are fed as an input to the network, their reconstruction should be poor in quality, and thus the anomalies can be detected by image comparison with the original input.

Compared to other [1,3,14,31] similar works, our main contribution consists in the addition of a pyramidal level in the network structure, in order to extract features at different resolution scales. This way we increase the chances to extract features at a scale level in which the anomaly is particularly evident. Another improvement consists in the way the input and reconstructed images are compared. Most methods rely on a trivial MSE loss that, when applied to image comparison context, consists in a simple pixel-by-pixel comparison. We instead propose a high-level perceptual loss, that better models visual similarity between images.

3.1   Network Architecture

We propose a network to learn the manifold of the normal class by analyzing the feature representation at different scale levels using pyramidal pooling. This way the network can better find meaningful features that describe the image content at different scales, and a consequence will perform better at detecting anomalies of different sizes.

Figure 2 shows the schematic diagram of our proposed novel network. The main components of the network include:

- **Resnet18** - A pre-trained Resnet18 network (trained over imagenet dataset) is being used for deep feature extraction from images. Only the first four layers of the network have been used. The basic idea is that the network can extract generic low-level features that are meaningful for many different types of images.
- **Pyramidal Pooling Layer** - The pyramidal pooling layer thus scales the input features at different magnification levels, thus increasing the possibility that features relevant for the anomaly detection task are actually extracted. The layer takes input from the Resnet 18 block and applies an average pooling such that the output will have a unit width and height. Then it uses a convolutional layer to reduce the channel features at different scales respectively 1, 2, 3, and 6, followed by a batch normalization and ReLU activation layers. The outputs are respectively fed into four encoders.
- **Encoders** - We use four encoders, which receive their input from the pyramidal pooling layer. Each encoder is composed of a sequence of three convolutional layers which reduce the input to a feature vector in $\mathbb{R}^8$.
- **Up-sampling Layer** - The upsampling blocks are composed of two linear layers and are used to upsample the latent features of the encoders from $\mathbb{R}^8$ to $\mathbb{R}^{512}$. All the upsampling blocks share the same network weights.
- **Decoder** - The decoder layer takes as input the concatenated features from upsampling layers and the output of the Resnet layer. Decoder uses 4 transposed convolutional layers to reconstruct the sample, thus giving in output an image of the same size of the input image.

Each network layer is followed by a batch normalization layer and uses the ReLU activation function except for the last layer, where a sigmoid activation forces the pixel values to be in the range $[0, 1]$. Tables 5 and 6 show the full structure of the encoder, upsampling and decoder layers.

3.2 Objective and Losses

In order to train the network we adopted a reconstruction-based approach, in which the network output is requested to be similar to the input. If the training is successful, it means that the low-dimension latent feature space in which the input is mapped after the encoders efficiently describes the visual properties of normal images. We assume that the same features will not be able to reconstruct anomalous images, which will then be identified by their higher loss. The network is trained using the following two losses:

- *Reconstruction loss*: It's a MSE loss computed between the input and the reconstructed image, i.e. $\frac{1}{WH}\|X - \hat{X}\|_2^2$, where X is the input and $\hat{X}$ is the output of the network (reconstructed image). This is a pixel-by-pixel image comparison, widely adopted in other anomaly detection works. However, because of its intrisic pixel-level independence assumption, it fails at modeling high-level visual features.

- *Perceptual loss*: Perceptual loss [16] is a more sophisticated loss trying to catch the high-level perceptual and semantic differences between images, rather than relying on a low-level, pixel-based comparison as in the standard reconstruction loss. It is a MSE loss computed between the high-level image features obtained by a pre-trained VGG11 network using its first four layers. The loss is defined as $\frac{1}{WHC}\|F(X) - F(\hat{X})\|_2^2$, where F is the transformation function applied through the trained four layers of VGG11 network, and W, C, H are the size of the resulting feature map. The trained network is only used for the calculation of loss and the weights are not updated during training. VGG11 network is used specifically as it has a very simple network structure without any maxpool layer, which makes it ideal candidate for this work.

the proposed objective function minimizes the total loss L:

$$L = \frac{1}{WH}\|X - \hat{X}\|_2^2 + \lambda\frac{1}{WHC}\|F(X) - F(\hat{X})\|_2^2 \tag{1}$$

where X is the input image, $\hat{X}$ is the network output, and F is the non-linear function computed by the first four layers of a pre-trained VGG11 network. λ is a weighing factor between two losses, all the experiments discussed in Sect. 4 are obtained with $\lambda = 1$.

4   Experimental Results

We tested the proposed model on the publicaly available standard datasets like MNIST [20] and FMNIST [34]. Although this dataset was not initially meant to be used in anomaly detection tasks, it has been widely adopted in literature to show the ability of the system to discriminate between one class, considered normal, and the other ones, considered anomalies. In addition to this we also tested our model over the recently published real-world anomaly detection dataset by MVTec [7], which contains more realistic data.

- **MNIST**: *MNIST* dataset consists of 60,000 28 × 28 gray images of hand written digits, grouped in 10 classes. The gray images were converted to RGB images and then passed through the network. For training, one class has been considered as the normal class while the remaining classes are considered as anomaly. Results are averaged over several runs in which each one of the original classes has been chosen as normality model.
- **Fashion MNIST**: *FMNIST* dataset composed of 60,000, 28 × 28 grey images of clothes from an online clothing store. The imaegs were standardised similar to the MNIST before passing to the network. Table 2 shows our network performs better in compare to other state of the art methods like GPND [28] and OCGAN [24]. The respective ROC curve score were used for the performance measurement.

– **MVTec** : MVTec dataset contains 5354 high-resolution color images of different texture and object categories (see Fig. 1). It contains normal and anomalous images (*70 different types of anomalies*) from real world products. Since gray-scales are quite common in industrial uses, it has 3 object categories (*zipper, screw and grid*) available solely in single channel images. As the original image sizes were large, the images were resized to *120x120* pixels before passing it to the proposed network. This size has been chosen as it maintains the structural integrity of the images such that anomalies are still visible by human eye.

Training is started by initializing the weights of the network using orthogonal initialization except the resent block, which was pretrained on imagenet, and the VGG11 block, which was pretrained on imagenet and it was kept fixed. The architectural hyper-parameters details are shown in Table 1.

Table 1. Training hyperparameters.

Adam learning rate	0.0001
Weight decay	0.0001
Batch size	120
Epochs	600

Table 3 shows the achieved results on the MNIST dataset. Tests have been done considering one of the classes as normal and using the remaining ones as anomalies, this has been done for each one of the 10 classes. The achieved results have been compared with standard methods such as one-class support vector machines and kernel density estimators, as well as with deep learning approaches such as denoising autoencoders, variational autoencoders [17], Pix-CNN [23] and Latent Space Autoregression [1]. The comparative results have been taken from [1]. Performance is measured using the Area Under ROC Curve (AUC) metric. As it can be seen, the proposed method achieves the best result on 6 out of 10 classes, and it has the best average result, at the same level of LSA. Table 2 shows the achieved results on the FMNIST dataset. Test is done similar to MNIST approach. The achieved ressults are compared with the other state-of-the art methods like GPND [28] and OCGAN [24]. Permoance is measured using AU ROC curve, averaged over 10 classes. The proposed methods performs at par and even better to the compared methods.

Table 4 shows the results on the MVTec dataset over all the 16 categories, comprising both textures (carpet, grid, leather, etc.) and objects (bottle, cable, capsule, etc.). Our results are compared with other deep learning anomaly detection algorithms such as autoencoders with L2 norm loss and structural similarity loss [8], the GAN-based approach AnoGAN [31], and CNN feature dictionary [22]. The comparative results have been taken from [7]. Performance is measured again using True Positive Rate (TPR) and True Negative Rate (TNR).

The study follows same method with that of the compared state of the art. The proposed method achieves the best results on 9 out of 16 categories, and it reaches the best average result.

Table 2. ROC AUC for anomaly detection using FMNIST. We report the average value for the network for all the classes.Results taken from [24,28]

Network	Average AU ROC
GPND	0.933
OCGAN	0.924
Ours	**0.936**

Table 3. AUC results of anomaly detection using MNIST. Each row shows the normal class on which the model has been trained. Comparative results takedn from [1]

Class	OC SVM	KDE	DAE	VAE	Pix CNN	GAN	LSA	Deep SVDD	Ours
0	0.988	0.885	0.991	0.994	0.531		0.993	0.98	**0.995**
1	**0.999**	0.996	**0.999**	**0.999**	0.995		**0.999**	0.0.997	**0.999**
2	0.902	0.710	0.891	**0.962**	0.476		0.959	0.917	0.941
3	0.950	0.693	0.935	0.947	0.517		**0.966**	0.919	**0.966**
4	0.955	0.844	0.921	**0.965**	0.739		0.956	0.949	0.960
5	0.968	0.776	0.937	0.963	0.542		0.964	0.885	**0.972**
6	0.978	0.861	0.981	**0.995**	0.592		0.994	0.983	0.992
7	0.965	0.884	0.964	0.974	0.789		0.980	0.946	**0.993**
8	0.853	0.669	0.841	0.905	0.340		**0.953**	0.0.939	0.895
9	0.955	0.825	0.960	0.978	0.662		0.981	0.0.965	**0.989**
Mean	0.95	0.81	0.94	0.97	0.62		**0.97**	0.948	**0.97**

5   Ablation Study

Here we propose a set of ablation studies, in which we removed some parts of the network and the whole setup is retrained in order to see the influence of those parts on the network performance. First we try to study the effect of number of encoders. We started it with one encoder and then successively tested the cases with 2, 4, and 6 encoders. We found that the proposed model performed better in terms of anomaly detection and reconstruction capabilities with 4 and 6 autoencoders, where the performances of the two cases are substantially the same. Ablation studies have been done on the MVTec dataset ('bottle' and 'carpet', one from product and one from texture category) so that complexity of the learning domain (accuracy) and the reconstruction capacity (SSIM) can be tested. All the tests have been done with the hyperparameters kept constant (Table 1). The result can be seen in Fig. 3 and 4.

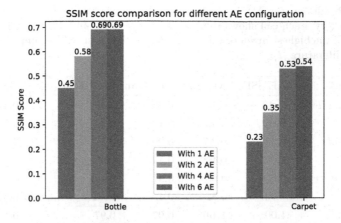

Fig. 3. SSIM comparison results for different AE configuration.

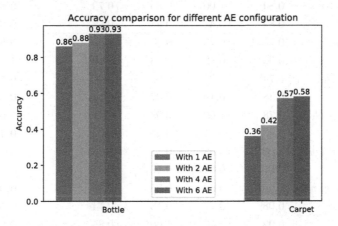

Fig. 4. Accuracy comparison results for different AE configuration.

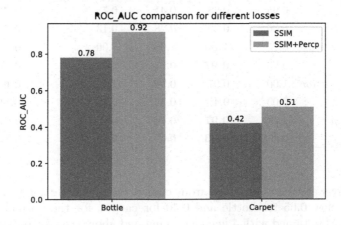

Fig. 5. ROC AUC comparison for different losses.

Table 4. Results on the MVTec dataset. Each row shows the results achieved on a specific category. Each cell shows the best TNR (bottom) and TPR (top) values. The method with the highest mean of the two values is shown in bold. Comparative results taken from literature [7].

Class	AE SSIM	AE (L2)	Ano Gan	CNN Feat. Dict.	Ours
Carpet	**0.43**	0.57	0.82	0.89	0.42
	0.90	0.42	0.16	0.36	0.72
Grid	0.38	**0.57**	0.90	0.57	0.86
	1.00	**0.98**	0.12	0.33	0.53
Leather	0.00	0.06	0.91	**0.63**	0.62
	0.92	0.82	0.12	**0.71**	0.625
Tile	1.00	**1.00**	0.97	0.97	0.44
	0.04	**0.54**	0.05	0.44	0.85
Wood	0.84	1.00	0.89	0.79	**0.85**
	0.82	0.47	0.47	0.88	**0.95**
Bottle	0.85	0.70	0.95	1.00	**0.84**
	0.90	0.89	0.43	0.06	**1.00**
Cable	0.74	0.93	0.98	0.97	**0.58**
	0.48	0.18	0.07	0.24	**0.89**
Capsule	0.78	1.00	0.96	0.78	**0.62**
	0.43	0.24	0.20	0.03	**0.74**
Hazelnut	1.00	0.93	0.83	0.90	**0.90**
	0.07	0.84	0.16	0.07	**0.89**
Metal nut	1.00	0.68	0.86	0.55	**0.98**
	0.08	0.77	0.13	0.74	**0.55**
Pill	0.92	1.00	1.00	0.85	**0.76**
	0.28	0.23	0.24	0.06	**0.62**
Screw	0.95	0.98	0.41	0.73	**0.73**
	0.06	0.39	0.28	0.13	**0.71**
Toothbrush	0.75	**1.00**	1.00	1.00	0.8
	0.73	**0.97**	0.13	0.03	0.92
Transistor	1.00	0.97	0.98	1.00	**0.60**
	0.03	0.45	0.35	0.15	**0.89**
Zipper	**1.00**	0.97	0.78	0.78	0.64
	0.60	0.63	0.40	0.29	0.82

While average SSIM obtained from configuration having 1 and 2 encoders remained below 0.65 for bottle and 0.40 for carpet, for the normal class, the average SSIM obtained with 4 encoders remained above 0.68 for bottle and 0.53 for carpet. Also, accuracy has been tested to choose the best model configuration.

Table 5. Encoders and Up-sampling layer architecture: in, out, k, s, p means in‑channel, out‑channel, kernel, stride and padding respectively. 'mf' is multiplying factor which is calculated as 0.5*(output height of Resnet18 features)

Encoder1	Encoder 2	Encoder 3	Encoder 4	Upsampling
Conv2d in:512,out:16, k:3,s:1,p:1	Conv2d in:256,out:16, k:3,s:1,p:1	Conv2d in:170,out:16, k:3,s:1,p:1	Conv2d in:85,out:16, k:3,s:1,p:1	**Linear in:8,out:128**
Batch norm ReLU	*Batch norm ReLU*	*Batch norm ReLU*	*Batch norm ReLU*	*Batch norm ReLU*
Conv2d in:16,out:8, k:3,s:1,p:1	**Conv2d in:16,out:8, k:3,s:1,p:1**	**Conv2d in:16,out:8, k:3,s:1,p:1**	**Conv2d in:16,out:8, k:3,s:1,p:1**	Linear in:128, out:512*mf^2
Batch norm ReLU	*Batch norm ReLU*	*Batch norm ReLU*	*Batch norm ReLU*	*ReLU*
Conv2d in:8,out:8, k:1,s:1	Conv2d in:8,out:8, k:1,s:1	Conv2d in:8,out:8, k:1,s:1	Conv2d in:8,out:8, k:1,s:1	

Table 6. Decoder structure

MNIST, FMNIST	Mvtech
ConvTranspose2d in:64,out:16:k:5,s:1p:1	ConvTranspose2d in:1024,out:16:k:3,s:2p:1
Batch norm ReLU	*Batch normReLU*
ConvTranspose2d in:16,out:32:k:5,s:1	ConvTranspose2d in:16,out:32:k:3,s:2p:1
Batch norm ReLU	*Batch norm ReLU*
ConvTranspose2d in:32,out:32:k:6,s:1	ConvTranspose2d in:32,out:32:k:4,s:2
Batch norm ReLU	*Batch norm ReLU*
ConvTranspose2d in:32,out:32:k:6,s:1	ConvTranspose2d in:32,out:3:k:4,s:2,p:1
Batch norm ReLU	*Tanh*
ConvTranspose2d in:32,out:3:k:5,s:1	
Tanh	

The accuracy with configuration having 1 and 2 encoders remained below 85% and 36% for bottle and carpet respectively in compare to 93% and 57% obtained with 4 encoder configuration for bottle and carpet respectively. Distinctively the study showed that by adding more AE, SSIM and accuracy didn't improve much. As with the 6 autoencoder configuration the results did not improve much and remained pretty much same. Hence, we choose 4 autoencoders configuration for our further studies.

We also tried to study the effect of perceptual loss (3), over the model performance. To measure the system performance two model has been trained in following configuration: a) MSE loss only ($\lambda = 0$); b) MSE + Perceptual loss. The netowrk was trained with 4 autoencoder configuration and constant hyperparameters (see Table 1) for 'Bottle' and 'Carpet'. The results as measured in terms of the AUC and can be refered in Fig. 5. As it can be seen, the introduction of the perceptual loss greatly enhanced the system performances.

6   Conclusions

In this paper we proposed a deep pyramidal network for anomaly detection. Anomalies are identified by means of a network that encodes normal images in a low-dimensional latent space and then reconstructs them, ideally modeling an identity function. Since the network is trained on normal data only, its fails at reconstructing anomalous images, which can be detected by an image similarity loss. The main contributions of this work consist in the usage of a multi-scale pyramidal approach that extract latent features at different resolutions, and the usage of a high-level perceptual loss to better compare images at feature level, rather than at pixel level. We also found that the proposed model worked best for product images (bottle, capsule, etc.), while in the case of texture images (carpet, grid, etc.) it can be further improved. Moreover, differing from many works that have been evaluated on basic datasets only such as MNIST and FMNIST, we also tested the proposed network on MVTec, a real-world dataset of defective products. Achieved results are promising and often outperform other state-of-the-art methods.

References

1. Abati, D., Porrello, A., Calderara, S., Cucchiara, R.: Latent space autoregression for novelty detection. In: Proceedings of the IEEE Conference on Computer Vision and Pattern Recognition, pp. 481–490 (2019)
2. Ahmed, M., Mahmood, A.N., Hu, J.: A survey of network anomaly detection techniques. J. Netw. Comput. Appl. **60**, 19–31 (2016)
3. Akcay, S., Atapour-Abarghouei, A., Breckon, T.P.: GANomaly: semi-supervised anomaly detection via adversarial training. In: Jawahar, C.V., Li, H., Mori, G., Schindler, K. (eds.) ACCV 2018, Part III. LNCS, vol. 11363, pp. 622–637. Springer, Cham (2019). https://doi.org/10.1007/978-3-030-20893-6_39
4. Ambrogioni, L., Güçlü, U., van Gerven, M.A., Maris, E.: The ernel mixture network: A nonparametric method for conditional density estimation of continuous random variables. arXiv preprint arXiv:1705.07111 (2017)
5. Antonie, M.L., Zaïane, O.R., Coman, A.: Application of data mining techniques for medical image classification. In: Proceedings of the Second International Conference on Multimedia Data Mining, MDMKDD 2001, pp. 94–101. (2001)
6. Baldi, P.: Autoencoders, unsupervised learning, and deep architectures. In: Proceedings of ICML Workshop on Unsupervised and Transfer Learning, pp. 37–49 (2012)
7. Bergmann, P., Fauser, M., Sattlegger, D., Steger, C.: MVTec AD-a comprehensive real-world dataset for unsupervised anomaly detection. In: Proceedings of the IEEE Conference on Computer Vision and Pattern Recognition, pp. 9592–9600 (2019)

8. Bergmann, P., Löwe, S., Fauser, M., Sattlegger, D., Steger, C.: Improving unsupervised defect segmentation by applying structural similarity to autoencoders. In: International joint Conference on Computer Vision, Imaging and Computer Graphics Theory and Applications (2019)
9. Bishop, C.M.: Mixture Density Networks. Aston University, Birmingham (1994)
10. Cai, Q., Pan, Y., Yao, T., Yan, C., Mei, T.: Memory matching networks for one-shot image recognition. In: Proceedings of the IEEE Conference on Computer Vision and Pattern Recognition, pp. 4080–4088 (2018)
11. Chalapathy, R., Chawla, S.: Deep learning for anomaly detection: A survey. CoRR abs/1901.03407 (2019). http://arxiv.org/abs/1901.03407
12. Chandola, V., Banerjee, A., Kumar, V.: Anomaly detection: a survey. ACM Comput. Surv. **41**(3), 151–1558 (2009)
13. Chen, P., Yang, S., McCann, J.A.: Distributed real-time anomaly detection in networked industrial sensing systems. IEEE Trans. Ind. Electron. **62**(6), 3832–3842 (2015)
14. Deecke, L., Vandermeulen, R., Ruff, L., Mandt, S., Kloft, M.: Image anomaly detection with generative adversarial networks. In: Berlingerio, M., Bonchi, F., Gärtner, T., Hurley, N., Ifrim, G. (eds.) ECML PKDD 2018, Part I. LNCS (LNAI), vol. 11051, pp. 3–17. Springer, Cham (2019). https://doi.org/10.1007/978-3-030-10925-7_1
15. Huang, S.H., Pan, Y.C.: Automated visual inspection in the semiconductor industry: a survey. Comput. Ind. **66**, 1–10 (2015)
16. Johnson, J., Alahi, A., Fei-Fei, L.: Perceptual losses for real-time style transfer and super-resolution. In: Leibe, B., Matas, J., Sebe, N., Welling, M. (eds.) ECCV 2016, Part II. LNCS, vol. 9906, pp. 694–711. Springer, Cham (2016). https://doi.org/10.1007/978-3-319-46475-6_43
17. Kingma, D.P., Welling, M.: Auto-encoding variational bayes. In: International Conference on Learning Representations (2014)
18. Klushyn, A., Chen, N., Kurle, R., Cseke, B., van der Smagt, P.: Learning hierarchical priors in VAEs. In: Advances in Neural Information Processing Systems, vol. 32, pp. 2866–2875. Curran Associates, Inc. (2019). http://papers.nips.cc/paper/8553-learning-hierarchical-priors-in-vaes.pdf
19. Kumagai, A., Iwata, T., Fujiwara, Y.: Transfer anomaly detection by inferring latent domain representations. In: Advances in Neural Information Processing Systems, pp. 2467–2477 (2019)
20. LeCun, Y., Bottou, L., Bengio, Y., Haffner, P., et al.: Gradient-based learning applied to document recognition. Proc. IEEE **86**(11), 2278–2324 (1998)
21. Mishra, P., Piciarelli, C., Foresti, G.L.: A neural network for image anomaly detection with deep pyramidal representations and dynamic routing. Int. J. Neural Syst. **30**(10), 2050060 (2020)
22. Napoletano, P., Piccoli, F., Schettini, R.: Anomaly detection in nanofibrous materials by CNN-based self-similarity. Sensors **18**(1), 209 (2018)
23. Van den Oord, A., Kalchbrenner, N., Espeholt, L., Vinyals, O., Graves, A., et al.: Conditional image generation with pixelCNN decoders. In: Advances in Neural Information Processing Systems, pp. 4790–4798 (2016)
24. Perera, P., Nallapati, R., Xiang, B.: Ocgan: One-class novelty detection using GANs with constrained latent representations. In: Proceedings of the IEEE Conference on Computer Vision and Pattern Recognition, pp. 2898–2906 (2019)
25. Piciarelli, C., Avola, D., Pannone, D., Foresti, G.L.: A vision-based system for internal pipeline inspection. IEEE Trans. Ind. Inf. **15**(6), 3289–3299 (2018)

26. Piciarelli, C., Micheloni, C., Foresti, G.L.: Trajectory-based anomalous event detection. IEEE Trans. Circuits Syst. Video Technol. **18**(11), 1544–1554 (2008)
27. Piciarelli, C., Mishra, P., Foresti, G.L.: Image anomaly detection with capsule networks and imbalanced datasets. In: Ricci, E., Rota Bulò, S., Snoek, C., Lanz, O., Messelodi, S., Sebe, N. (eds.) ICIAP 2019, Part I. LNCS, vol. 11751, pp. 257–267. Springer, Cham (2019). https://doi.org/10.1007/978-3-030-30642-7_23
28. Pidhorskyi, S., Almohsen, R., Doretto, G.: Generative probabilistic novelty detection with adversarial autoencoders. In: Advances in Neural Information Processing Systems, pp. 6822–6833 (2018)
29. Qin, X., Cao, L., Rundensteiner, E.A., Madden, S.: Scalable kernel density estimation-based local outlier detection over large data streams. In: EDBT, pp. 421–432 (2019)
30. Ruff, L., et al.: Deep one-class classification. In: Dy, J., Krause, A. (eds.) Proceedings of the 35th International Conference on Machine Learning. Proceedings of Machine Learning Research, vol. 80, pp. 4393–4402. PMLR, Stockholmsmässan, Stockholm Sweden (2018)
31. Schlegl, T., Seeböck, P., Waldstein, S.M., Schmidt-Erfurth, U., Langs, G.: Unsupervised anomaly detection with generative adversarial networks to guide marker discovery. In: Niethammer, M., Styner, M., Aylward, S., Zhu, H., Oguz, I., Yap, P.-T., Shen, D. (eds.) IPMI 2017. LNCS, vol. 10265, pp. 146–157. Springer, Cham (2017). https://doi.org/10.1007/978-3-319-59050-9_12
32. Viroli, C., McLachlan, G.J.: Deep Gaussian mixture models. Stat. Comput. **29**(1), 43–51 (2019)
33. Wulsin, D., Blanco, J., Mani, R., Litt, B.: Semi-supervised anomaly detection for EEG waveforms using deep belief nets. In: 2010 Ninth International Conference on Machine Learning and Applications, pp. 436–441 (2010)
34. Xiao, H., Rasul, K., Vollgraf, R.: Fashion-mnist: a novel image dataset for benchmarking machine learning algorithms. arXiv preprint arXiv:1708.07747 (2017)
35. Zhou, C., Paffenroth, R.C.: Anomaly detection with robust deep autoencoders. In: Proceedings of the 23rd ACM SIGKDD International Conference on Knowledge Discovery and Data Mining, KDD 2017, pp. 665–674. ACM, New York (2017)

Fault Detection in Uni-Directional Tape Production Using Image Processing

Somesh Devagekar[1] (ID), Ahmad Delforouzi[2](✉) (ID), and Paul G. Plöger[3] (ID)

[1] Department of Computer Science, Hochschule Bonn-Rhein-Sieg, Bonn, Germany
`somesh.devagekar@smail.inf.h-brs.de`
[2] Institute for Algorithms and Scientific Computing Fraunhofer SCAI,
Bonn, Germany
`ahmad.delforouzi@scai.fraunhofer.de`
[3] Department of Computer Science, Hochschule Bonn-Rhein-Sieg,
Bonn, Germany
`Paul.Ploeger@h-brs.de`
`https://www.scai.fraunhofer.de/`, `https://www.h-brs.de`

Abstract. The quality of uni-directional tape in its production process is affected by environmental conditions like temperature and production speed. In this paper, computer vision algorithms on the scanned images are needed to be used in this context to detect and classify tape damages during the manufacturing procedure. We perform a comparative study among famous feature descriptors for fault candidate generation, then propose own features for fault detection. We investigate various machine learning techniques to find best model for the classification problem. The empirical results demonstrate the high performance of the proposed system and show preference of random forest and canny edges for classifier and feature generator respectively.

Keywords: Unidirectional thermoplastic composites · Quality control · Feature extraction · Machine-learning · Object detection

1 Introduction

Unidirectional Tapes (UD-Tapes) are thermoplastic composite comprised of reinforced fibers usually of the type glass, carbon, or natural fibers, and prove increased reinforcement when it comes to composite markets. Also, the cost for the production of two-dimensional preforms is 30% less than any other woven fabric-based organic composites [1]. These properties of a TPC, promise an alternative option to reinforced sheets along with a minimum scrap rate and a decreased production cycle time. Today, with numerous applications in lightweight series production, an evolution of a more mature supply chain has begun. Thus, it is adopted in many applications of the car industry like central floor, door panel, and wheel rim. The production of these tapes is influenced by manufacturing processes thus providing an enormous challenge to process controls such as quality assurance. The poor quality of the tapes lowers the quality

© Springer Nature Switzerland AG 2021
A. Del Bimbo et al. (Eds.): ICPR 2020 Workshops, LNCS 12664, pp. 719–732, 2021.
https://doi.org/10.1007/978-3-030-68799-1_52

of the final product. Automatic systems are needed to be developed to control the quality of UD-tape manufacturing. Advanced vision-based technologies allow a better understanding of the production process. The concept of machine-based vision begins with individual features analyzed and accordingly foresees the end value product. The data gathered from a UD-Tape helps to understand the debilitating defects, e.g., local fiber deviations, porosities, dents.

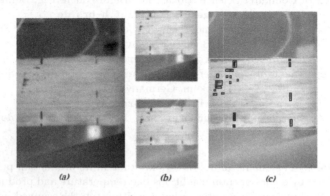

(a)   (b)   (c)

Fig. 1. The challenges of this work are stated. A vision-based thermography system is used to control the quality of a tape production process. A poor tape image captured by the thermography system is shown in (image (a)). The position of the tape in the image is changing due to the vibrations occurring in the production line (image (b)). Markers are made on the tapes to track the faults in them. These markers, as well as probable faults appearing in diverse shapes and sizes, must be detected.

Finding the location of the tape and improving the visual quality in the input image are two main primary concerns for preprocessing of the UD-Tape images. Also, understanding the foreground and the background plays a key role in the identification of the visual area.

Figure 1 shows the challenges we want to tackle in this paper. The quality of input images is most likely poor and they should be enhanced to improve the overall accuracy of the system. Since the location of the tape within the image is changing Fig. 1.b, an automatic module is required to find the location of tape and remove the background. Figure 1.c shows some spots over a tape after production. They have been highlighted by red and blue colors showing tape faults and markers respectively. Note that to trace the position of the faults over the tapes, some markers are intentionally added to the tapes after the production process. An automatic system is required to detect these spots, markers, and accordingly, find faults among them. In this paper, vision-based methods on thermographic images are deployed to detect and classify the tape faults and markers during the tape manufacturing process. A comparative study among famous feature descriptors are undertaken to generate fault candidates, then tested using machine learning techniques based on some features to recognize the faults and markers.

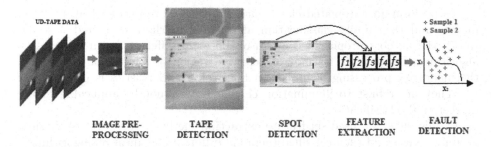

IMAGE PRE-   TAPE   SPOT   FEATURE   FAULT
PROCESSING   DETECTION   DETECTION   EXTRACTION   DETECTION

Fig. 2. The proposed framework for automatic fault detection in UD-Tape production process is shown. First, the input frames from the optic sensor are enhanced, and then within the enhanced image, the tape is located using a machine vision technique. Afterward, the spots on the tape image are detected and from each point, features are extracted and forwarded to a machine learning module to classify these spots into tape faults and intentionally created markers.

The rest of the paper is organized as follows: A general overview of related work is given in Sect. 2. Section 3 proposes our algorithms for the tape fault detection problem. Experimental results of this research are reported in Sect. 4 and finally, Sect. 5 concludes this paper.

2   Related Work

This section gives a literature review including state-of-the-art for similar works to this research. Prognosis enhances the structural monitoring and qualitative planning on composite materials from a production perspective. The main motivation for our work can be seen in [2], as it presents the development of a process integrated quality control of continuous fiber-reinforced plastics(CFRP) using eddy current inspection.

Non-destructive evaluation techniques help in determining the properties of UD-tapes showing different defects such as fiber misalignments, porosities and delaminations [3]. These methods used for inspection of composites are largely based on the detection and analysis of delamination in composite materials [4].

Thus leveraging learning approaches will help analyze relevant component characteristics. Furthermore, with the extraction of features for efficient production of composites, the vast space of composite structure data can be explained [5]. As, much research is being done on prognosis of composites laminates based on machine learning techniques, the area of unidirectional composite is untapped yet. One such research [6] shows the prediction of delamination size using machine learning models, such as Linear model, Support Vector Machines, and Random Forests. Here the authors present damage quantification from the delamination area in X-ray images.

To perform an automated classification one needs to extract features from images and then apply classification algorithms to those extracted features. In [7], the authors propose a comparison between different feature extraction algorithms. Local HOG features of defective region of interest obtained from thermography processing of in carbon fiber-reinforced plastics is presented in [8]. They are robust to illumination change but cannot be applicable to the marker/fault classification.

In this paper we used six feature extraction methods over a set of thermographic images and then classified them via different classification algorithms.

3   Proposed Method

In this section, we explore the characteristics of UD-Tape sensor data. Figure 2 shows the proposed framework for automatic inspection of the tape manufacturing process based on thermographic sensors. We start with enhancing the visual quality of the input sequence, and then prioritize a region of interest, aiming to find the location of the tape within the image which is then followed by exploiting feature extraction techniques to detect the intentionally made markers as well as the tape faults. The specific characteristic of the tape quality control is that the relative position of the tape and camera changes frame to frame. Therefore, first, the tape within the image must be localized, then the voids on the tapes must be detected and then classified. The results of tape location and void classification are presented in Sect. 3.1, 3.2 and 3.3 respectively.

The captured images suffer from a poor illumination condition. To elevate the overall detection of the proposed system, it is necessary to enhance the quality of the images before further steps. Thus we look into Histogram equalization [9], where in we equalize luminance, brightness and contrast values, of a poor quality image to that of the good quality image. Finally, the pixels' value of all input images fall between 0–255.

As part of image enhancement, we take into account different image filtering techniques for reduction of noise or any irregularities. Here, we have studied different image enhancement techniques and chosen the one that gave relevant features corresponding to our requirement. We begin with performing morphological operations such as erosion, dilation, and finally with the help of a linear filter, in convolution with a 1x1 structuring element, we were able to reduce its trivialness and tweak an image into yielding us the right results.

3.1   Tape Location

Two algorithms for prioritizing the region of interest have been exploited. Grab-Cut algorithm and Faster-RCNN. The Grab-Cut algorithm implements a Gaussian Mixture Model (GMM) [10] to create labels and cluster pixels according to their intensity. If a greater dissimilarity is found, then those pixels are segmented as either foreground or background. Figure 3 shows the results achieved from the interactive grab cut algorithm.

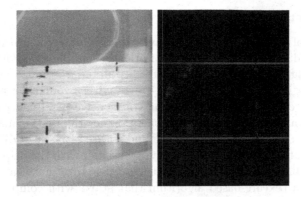

Fig. 3. Tape detection using the Grab-cut method is presented. The output mask image (right image) shows the upper and lower margins of the tape (left image) by two lines. These lines are normally thicker than one pixel. For each margin, the middle line is selected as the tape border.

Alternatively, in the context of tape detection, a famous deep neural network [11] has been employed to achieve a generalized model for tape detection. Faster-RCNN is employed, because of its precision and fast running in the test phase. This method uses convolutional neural networks for object detection and classifications to improve the performance of recognition. The manually labeled tapes in diverse images are fed to the classifier for training. The result of Faster-RCNN classifier is shown in Fig. 4.

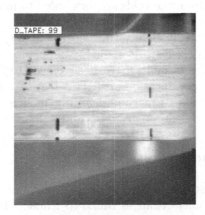

Fig. 4. Tape detection using Faster-RCNN technique with a confidence score of 99 is shown. This classifier has been trained with various tapes and finds their location in the test phase.

3.2 Void Detection

For detection of voids over tapes, we have explored different feature descriptors which are listed as follows: Histogram of oriented Gradients (HOG) [12], determines the orientation and magnitude of key points in an image. Features from Accelerated Segment Test (FAST) [13], starts with a round mask over a pixel, which is then compared to the following consecutive pixel using a comparison function. This algorithm focuses on corners rather than edges and is based on SUSAN(corner criteria). Oriented FAST and rotated BRIEF (ORB), makes use of the FAST feature descriptor to get key points and with the help of Harris corner measure [14], outputs multi-scale features. Scale Invariant Feature Transform (SIFT) [12] is a texture-based algorithm that starts with comprehending local features and comparing the neighboring pixels. Respectively, the key points are eliminated with key points examination. Speed-up robust features (SURF) [13], is a three-step feature extraction method that comprises of detection, matching, and description. SIFT and SURF employ detection in similar ways but with slight difference in image pyramids. Canny edge detection (CANNY) [15] makes use of multi-stage algorithms to detect edges present in images. Figure 5 shows the application of these features for void detection. According to this figure, HOG shows better results than FAST, ORB, and SIFT but has a large number of false positives. The SURF algorithm was not much successful as it detected many non-void regions, whereas the Canny edge detector detects the voids very successfully.

3.3 Void Classification

The extracted features from the voids are forwarded to machine learning models for classification and recognition. We have used supervised machine learning techniques to simultaneously detect the faults and markers over the located tapes. A set of geometrical features namely shape, size, relative location, and aspect-ratio of the markers/faults have been calculated and used to make a feature set for the fault classification task. This feature set is then, in a random cross-validated method fed to machine learning models for training and test. The details of this experiment are given in the Evaluation and Results Sect. 4.3.

4 Evaluation and Results

The specific characteristic of the tape quality control is that the relative position of the tape and camera changes frame to frame. Therefore, first the tape within the image must be localized, then the voids on the tapes must be detected and then classified. The results of tape location and void classification including discussion of each part are presented in Sections A, B and C respectively. To evaluate the proposed system, we have used 450 thermographic images of size 768x1024 pixels. For the evaluation of tape location algorithms, IoU measure (Intersection over union) [16] is used to decide if the predicted output corresponds to the ground truth. In the tape location scenario 3.1, if the predicted

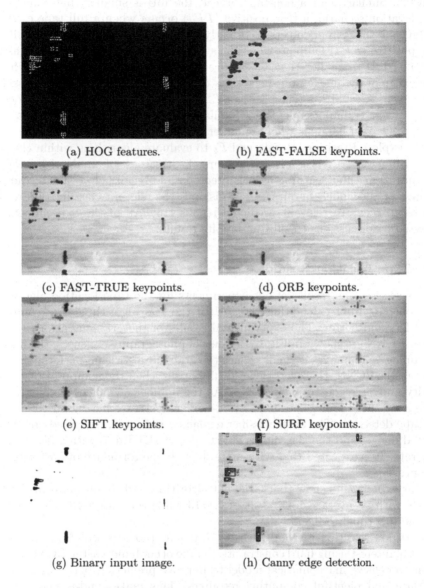

(a) HOG features.

(b) FAST-FALSE keypoints.

(c) FAST-TRUE keypoints.

(d) ORB keypoints.

(e) SIFT keypoints.

(f) SURF keypoints.

(g) Binary input image.

(h) Canny edge detection.

Fig. 5. Visual results of a comparison among Histogram of oriented Gradients, Features from Accelerated Segment Test, Oriented FAST and rotated BRIEF, Scale Invariant Feature Transform, Speed-up robust features, and Canny edge detection for marker and void detection. Canny edge shows the best results among them. This is because it reflects markers and faults very well and it doesn't produce redundant false alarm information.

tape location exactly matches with the actual tape location i.e., IoU is greater than 0.5. Similarly, for a non-tape object, the IoU is smaller than 0.5. In the tape location scenario, a true positive (TP_tl) occurs when a full tape exists in the image if the system detects it correctly i.e., IoU is greater than 0.5. Similarly, false positive (FP_tl) happens when the system detects a non-tape object as the tape i.e., IoU is smaller than 0.5. and a false negative (FN_tl) occurs when a tape exists in the image and the system cannot detect it. In the void classification scenario Sect. 3.2, a true positive (TP_vc) and a false positive (FP_vc) occurs when a (marker/fault) is detected. Similarly, a true negative (TN_vc) or a false negative (FN_vc) happens when a marker/fault is not detected.

We explore Precision, Recall, and F_1 to evaluate algorithms within the proposed framework. Precision is the measure of correctness of an algorithm. When it comes to machine learning models precision and recall help evaluate the correctness of our model. In our case, precision is the relevant classification results and recall refers to relevant predicted results [16]. All of the experiments are analysed by discussions to substantiate its results.

4.1 Tape Location Results

In this section, the results of our experiments for the tape location algorithms with outputs from calculating the IoU of the two algorithms, are compared to understand better. The experimental results of the grab-cut algorithm and Faster-RCNN are shown in Fig. 6. Because of photography expenses, the dataset containing valid images is small , hence the grab-cut algorithm performed better than Faster-RCNN. However, the difficulty arises when the computational complexity and time constraint increases. As the number of images is constrained, there is no such requirement for a black-box analysis. We have used 100 images for Tape detection. The deep classifier explained in Sect. 3, is trained on 60 augmented images and the classifier is able to detect UD Tapes with a 95.7% accuracy rate. It is expected that with more data the performance of Faster-RCNN is improved.

Our experiments show that Grab-cut algorithm slightly outperforms Faster-RCNN with an average IoU result of 0.8773 while the Faster-RCNN shows an average IoU equal to 0.808.

A region selection method such as Grab-cut, provides isolation around an object of interest using hard constraints. On the other hand Faster-RCNN defines a comprehensive mathematical model to improve prediction performance using big data and plentiful computing resources. This makes end-to-end learning overkill as traditional computer vision techniques can solve a problem more efficiently.

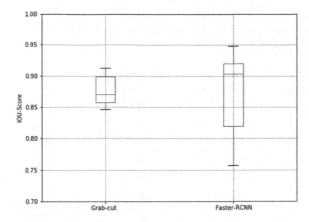

Fig. 6. Quantitative IoU results for predicted images for the both tape location algorithms i.e., Grab-cut and Faster-RCNN are shown. The Grab-cut shows averagely better results than Faster-RCNN.

4.2   Void Detection Results

The main purpose of this process is to localize voids on the tapes. Table 1 shows the performance of well-known low-level feature extraction algorithms. There is no single method sufficient enough to detect all cases, hence the comparison is based on the number of correctly identified voids/dents present in the UD Tape data set in regard to the total number of voids/dents present. The HOG and SIFT features are able to detect the main voids present but due to some local vicinities, there are too many false positives as compared to the others. In contrast, Canny edge detector outperformed all the other descriptors, although due to intensity variations in-between close by pixels, it barely detects wrong regions as positive. As a result, canny edge detector is chosen for void detection.

In the UD-Tape dataset, the horizontal smearing effect caused by fibers inline, show different gray-value distribution, with different level and type of noises. Canny edge detection provides the best edge completeness and noise suppression, where as SIFT, SURF and HOG produces more number of false positives and false negatives. One reason for this can be as SURF and SIFT recognize key points based on corner detection and HOG counts occurrences of gradient orientation in localized portions. FAST on the other hand puts more than one detection on the same marker/fault.

4.3   Void Classification Results

We have divided the voids into two classes as *'marker'* and *'non-marker/faults'*. The non-marker class represents the production fault which is assigned to 0. Whereas the marker class is set to 1. In our initial experiments, we obtained poor overall results because of the highly unbalanced distribution of data over the target classes. Figure 8 shows initial data distribution, wherein the number of

Table 1. Quantitative results of the experimented feature descriptors to detect the markers and the faults on a UD-tape are shown. Canny edge detector shows the best performance in terms of T_P and F_P.

Feature descriptor	Total number of detections $(TP_vc + FP_vc)$	Correctly detected (TP_vc)	Incorrectly detected (FP_vc)
HOG	14	9	5
FAST-TRUE	254	20	234
FAST-FALSE	1650	20	1630
SIFT	87	17	67
SURF	295	20	275
ORB	318	20	298
CANNY	24	20	4

samples present are 2192. However, there are more number of non-marker class samples than that of a marker class. Hence, to have a stratified distribution Fig. 9, we consider a similar number of marker class to that of non-marker class samples (Fig. 7).

As accuracy would not be sufficient, hence we take the F_1-score as well into consideration. Table 2 shows the outcome of machine learning models using a uniform distribution of data set Fig. 9. Based on these results we observe that

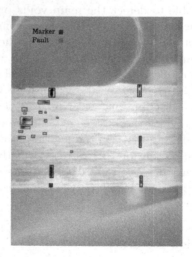

Fig. 7. Classifying the detected spots into markers and faults using a supervised machine learning method. For this task 6 classifiers namely Support Vector Machine, Logistic-Regression, Bayesian Network, Random Forest, Decision Tree and K-Nearest Neighbours have been used.

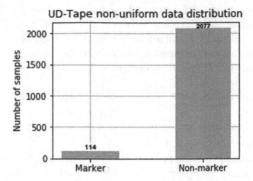

Fig. 8. Initial non-uniform distribution of data over the two classes of markers and tape faults is shown. For one class, very poor classification results were obtained.

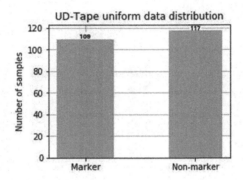

Fig. 9. Uniform distribution of data over two classes of markers and tape faults is used. Thus, the overall performance of classification has been improved.

all the models performed comparable well to each other. In conclusion, logistic-regression and random forest classifier stood out and this can be seen in Fig. 10.

The data points are categorical and discrete in the UD-tape dataset. With this, logistic regression and random forests work well with simple explanatory variables. Furthermore, the first important aspect of our analysis concerns the similarity of performance for logistic regression and random forest. In Fig. 10 we see the prediction accuracy of both models tend to perform similar.

For non-uniform distributed data set Fig. 8, we have considered micro averaged accuracy for each class outcome because it is expected to perform better for an imbalanced classification task [17]. Figure 11 shows the corresponding performance.

Table 2. Comparison among different machine learning classifiers i.e., Support Vector Machine, Logistic-Regression, Bayesian Network, Random Forest, Decision Tree and K-Nearest Neighbours for void classification into marker/fault. Logistic-Regression and Random Forest show better performance among them in terms of Accuracy (A_vc), Precision (P_vc), Recall (R_vc) and F1-score (F_vc).

Classifier models	A_vc	P_vc	R_vc	F_vc
SVM	91.11%	91%	91%	91%
Logistic Regression(L-R)	97.05%	97%	97%	97%
Bayesian Network (N-B)	91.11%	91%	91%	91%
Random Forest(R-F)	97.05%	97%	97%	97%
Decision Tree (D-T)	91.17%	91%	91%	91%
K-Nearest Neighbour (KNN)	91.11%	91%	91%	91%

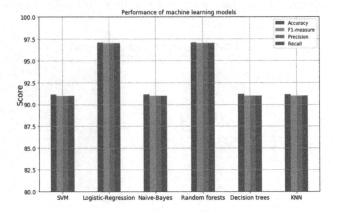

Fig. 10. A comparison among machine learning models with the uniform distributed data set is shown. Random forest and logistic regression outperform the other classifiers.

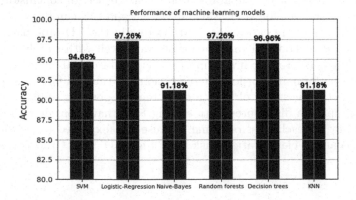

Fig. 11. The performance of various machine learning models with non-uniform data set is shown. Although the overall performance is high but, the results for one class are poor for all the classifiers.

5 Conclusion

This paper has proposed an automatic machine vision-based system to monitor the quality of UD-Tapes in the manufacturing procedure. With prioritizing our region of interest using detection algorithms, the tape is located and prepared for feature extraction. We have implemented some of the traditional descriptive feature descriptors for fault candidate generation and found Canny edge detector as the best one. Accordingly, relevant geometrical features of faults are extracted. The proposed framework can work with different machine learning strategies for classification over markers and tape faults. It is concluded that Logistic-Regression and Random-Forests performed better, in terms of micro averaged accuracy and F_1. For future research, one should focus on feature extraction algorithms to understand fiber alignment and delamination. This area of UD-Tape assessment could help to analyze the material grade. The distribution of data should be prepared such that machine learning models could perform multi-classification approaches showing real like scenarios and clustering patterns from UD-Tape features, thus enabling an automated feedback system to improve the production quality of the Tapes.

References

1. Kropka, M., Muehlbacher, M., Neumeyer, T., Altstaedt, V.: From UD-tape to final part-a comprehensive approach towards thermoplastic composites. Procedia CIRP **66**, 96–100 (2017)
2. Berger, D., Egloff, A., Summa, J., Schwarz, M., Lanza, G., Herrmann, H.-G.: Conception of an eddy current in-process quality control for the production of carbon fibre reinforced components in the rtm process chain. Procedia CIRP **62**, 39–44 (2017)
3. Grosse, C.U., et al.: Comparison of NDT techniques to evaluate CFRP-results obtained in a MAIzfp round robin test. In 19th World Conference on Non-Destructive Testing (WCNDT), Munich/Germany (2016)
4. Aymerich, F., Meili, S.: Ultrasonic evaluation of matrix damage in impacted composite laminates. Compos. Part B: Eng. **31**(1), 1–6 (2000)
5. Vikram Gopal, C.S.L. Continuous fiber thermoplastic composites (2015)
6. Liu, H., Liu, S., Liu, Z., Mrad, N., Dong, H.: Prognostics of damage growth in composite materials using machine learning techniques. In: 2017 IEEE International Conference on Industrial Technology (ICIT), pp. 1042–1047. IEEE (2017)
7. Kitanovski, I., Jankulovski, B., Dimitrovski, I., and Loskovska, S. Comparison of feature extraction algorithms for mammography images. In: 2011 4th International Congress on Image and Signal Processing, vol. 2, pp. 888–892. IEEE (2011)
8. Erazo-Aux, J., Loaiza-Correa, H., Restrepo-Giron, A.: Histograms of oriented gradients for automatic detection of defective regions in thermograms. Appl. Opt. **58**(13), 3620–3629 (2019)
9. Han, J., Yang, S., Lee, B.: A novel 3-d color hisogram equalization method with uniform 1-d gray scale histogram. IEEE Trans. Image Process. **20**(2), 506–512 (2011). ISSN 1941-0042

10. Wu, X.Y., Yang, L., Li, S.B., Xu, P.: An interactive video foreground segmentation system based on modeling and dynamic graph cut algorithm. In: Advanced Materials Research, vol. 532, pp. 1770–1774. Trans Tech Publications (2012)
11. Ren, S., He, K., Girshick, R., Sun, J.: Faster R-CNN: towards real-time object detection with region proposal networks. In Advances in Neural Information Processing Systems, pp. 91–99 (2015)
12. Ozturk, S., Bayram, A.: Comparison of HOG, MSER, SIFT, FAST, LBP and CANNY features for cell detection in histopathological images. HELIX 8(3), 3321–3325 (2018)
13. El-Gayar, M., Soliman, H., et al.: A comparative study of image low level feature extraction algorithms. Egypt. Inf. J. 14(2), 175–181 (2013)
14. Amaricai, A., Gavriliu, C.-E., Boncalo, O.: An FPGA sliding window-based architecture harris corner detector. In 2014 24th International Conference on Field Programmable Logic and Applications (FPL), pp. 1–4. IEEE (2014)
15. Jain, R., Rangachar Kasturi, B.S.: Machine Vision. McGraw-Hill Inc., New York (1995). ISBN 0-07-032018-7
16. Gan, K., et al.: Artificial intelligence detection of distal radius fractures: a comparison between the convolutional neural network and professional assessments. Acta Orthop. 90, 1–12 (2019)
17. García, V., Mollineda, R.A., Sánchez, J.S., Alejo, R., Sotoca, J.M.: When overlapping unexpectedly alters the class imbalance effects. In: Martí, J., Benedí, J.M., Mendonça, A.M., Serrat, J. (eds.) IbPRIA 2007, Part II. LNCS, vol. 4478, pp. 499–506. Springer, Heidelberg (2007). https://doi.org/10.1007/978-3-540-72849-8_63
18. Wålinder, A.: Evaluation of logistic regression and random forest classification based on prediction accuracy and metadata analysis (2014)

Author Index